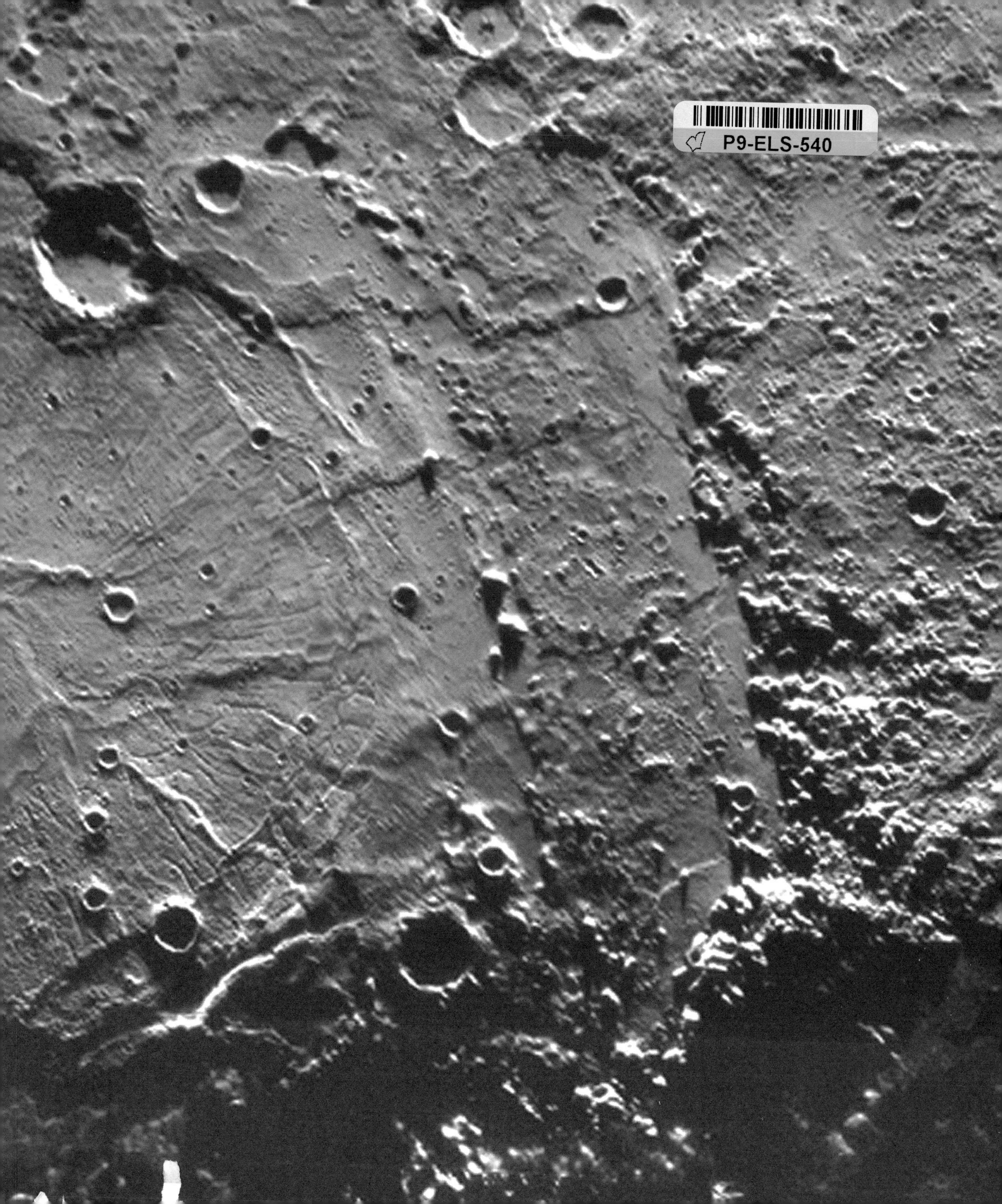

SPACE
S P A C E

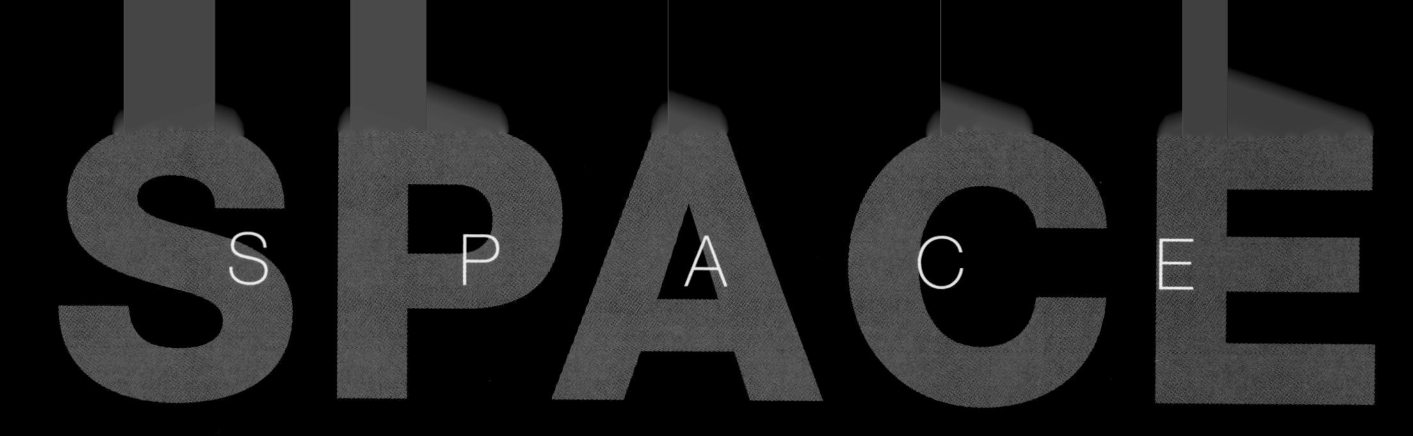

SPACE

from earth to the edge of the universe

CAROLE STOTT
ROBERT DINWIDDIE
DAVID HUGHES
GILES SPARROW

LONDON, NEW YORK, MELBOURNE, MUNICH, AND DELHI

Editorial consultant Iain Nicolson

Senior editor Peter Frances
Senior art editors Mark Lloyd, Vicky Short

Editors Jemima Dunne, Megan Hill, Nicola Hodgson, Patrick Newman, Steve Setford, Andrew Szudek
Proof-reader Sean O'Connor
Index Indexing Specialists (UK) Ltd
US editor Jenny Siklós

Designers Vivienne Brar, Mandy Earey, Alison Gardner, Richard Horsford, Matthew Robbins, Liz Sephton, Alison Shackleton
Design assistant Anushka Mody
Jacket designers Silke Spingies, Duncan Turner

Picture researchers Ria Jones, Myriam Megharbi, Roland Smithies, Sarah Smithies
Illustrators The Art Agency, SJC Illustration

Creative technical support Adam Brackenbury
Production editor Phil Sergeant
Senior production controller Inderjit Bhullar

Managing editors Camilla Hallinan, Julie Oughton
Managing art editors Karen Self, Louise Dick

Associate publisher Liz Wheeler
Art director Phil Ormerod
Publisher Jonathan Metcalf

First American edition, October 2010
Published in the United States by DK Publishing, 375 Hudson Street, New York, New York 10014

10 11 12 13 14 10 9 8 7 6 5 4 3 2 1
176040—10/10

Published in Great Britain by Dorling Kindersely Limited

A catalog record for this book is available from the Library of Congress.

ISBN 978-0-7566-6738-2

DK books are available at special discounts when purchased in bulk for sales promotions, premiums, fund-raising, or educational use. For details, contact: DK Publishing Special Markets, 375 Hudson Street, New York, New York 10014 or SpecialSales@dk.com.

Printed and bound in China by Hung Hing Offset Printing Company

Discover more at **www.dk.com**

Endpapers: Rembrandt Crater, Mercury
Half-title page: Launch of the Space Shuttle *Endeavour*
Title pages: The surface of Mars
Above: The center of the Milky Way

CONTENTS

1 LAUNCHPAD EARTH

2 NEIGHBORING WORLDS

3 BEYOND THE BELT

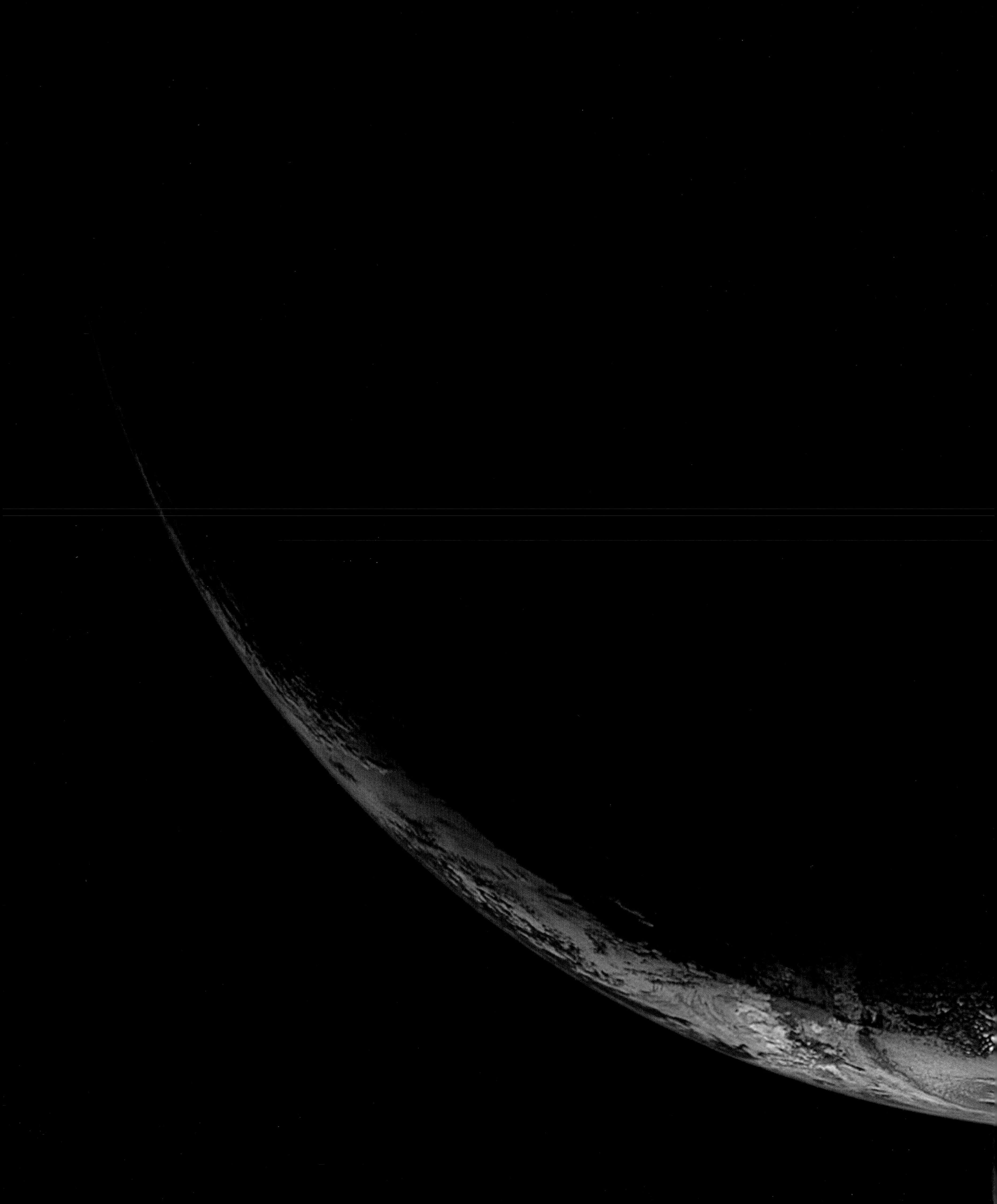

1

LAUNCHPAD EARTH

« Earth
Viewed from a distance of 220,000 miles (350,000km)

LOOKING OUT

Our understanding of the Universe is shaped by the view from our own small planet, Earth. Despite having only a snapshot of the heavens from a single viewpoint at a single moment in cosmic history, astronomers have pieced together a remarkable understanding of the Universe—and our place within it.

EARTH-BASED OBSERVATION

Though we have yet to travel far from Earth, we are constantly bombarded by information that can help reveal the secrets of the Universe. Most of this information comes in the form of visible light and other radiations that may have spent hundreds, millions, or even billions of years traveling across space to reach us. Collected by telescopes, these radiations reveal not just the appearance of distant objects, but also their motions through space, their physical properties, and even their chemistry. And radiation is not the only form of information reaching planet Earth—meteorites and high-energy particles, misleadingly called cosmic rays, also offer intriguing glimpses of our planetary neighborhood and the wider Universe.

NAKED-EYE VIEW
On a clear night, the human eye can see around 3,000 stars, several planets, the pale band of the Milky Way, a number of star clusters and nebulae, and even some of the galaxies closest to our own.

ENHANCED TELESCOPIC VIEW
This Hubble Space Telescope view, of an apparently empty patch of sky in the constellation Tucana, reveals thousands of entire galaxies crowding distant space.

THE SCALE OF THE UNIVERSE

The sheer size of the Universe is beyond our everyday experience. Earth itself is a mere speck in the vast space of the Solar System, while the Sun and its planets are one of 200 billion or more such systems in our galaxy, most of which are separated by gulfs of largely empty space. Likewise, our galaxy is one of more than 100 billion galaxies scattered across the Universe.

If measured in miles or kilometers, cosmic distances are incomprehensible. In order to make sense of them, astronomers often use larger units of measurement. Within the solar system, a basic unit of measurement is the astronomical unit (AU), which is equivalent to the average Earth–Sun distance of 93.0 million miles (149.6 million km). To describe interstellar and intergalactic distances, astronomers use the light-year—the distance that light, the fastest phenomenon in the Universe, travels in a year. Since light travels at an incredible 670 million mph (1,079 million kph), a light-year is equivalent to a distance of almost 5.9 trillion miles (9.5 trillion km).

NEAREST NEIGHBOR
The Moon is our closest neighbor in space. It orbits at an average distance of 238,800 miles (384,400km) from Earth—roughly 30 times the diameter of Earth itself.

DISTANT COSMOS
The galaxy cluster Abell 1689 lies some 2.2 billion light-years away. Its light takes half the age of the Earth to reach us.

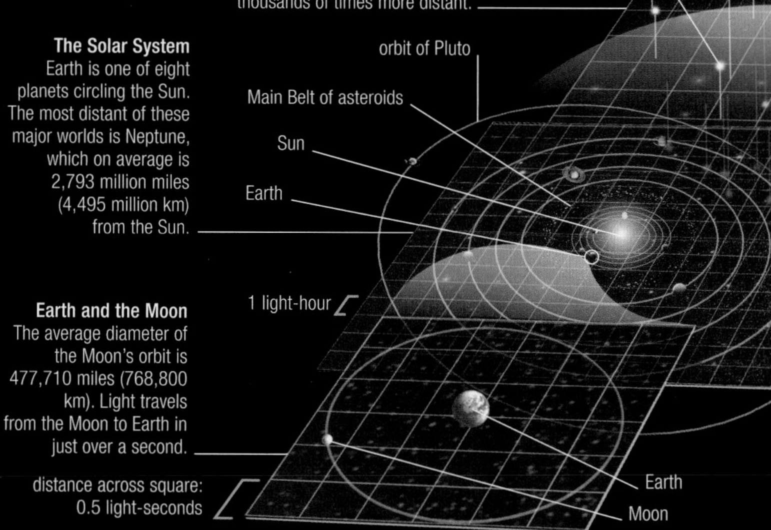

The Milky Way
The Sun and all its neighbors lie within the disc of an enormous spiral galaxy called the Milky Way, roughly 100,000 light-years across.

the stellar neighborhood lies in the Orion Arm of the Milky Way, some 26,000 light-years from its center

Our stellar neighborhood
The nearest star to the Sun is Proxima Centauri, some 4.24 light-years away. Most stars are tens, hundreds, or even thousands of times more distant.

The Solar System
Earth is one of eight planets circling the Sun. The most distant of these major worlds is Neptune, which on average is 2,793 million miles (4,495 million km) from the Sun.

Earth and the Moon
The average diameter of the Moon's orbit is 477,710 miles (768,800 km). Light travels from the Moon to Earth in just over a second.

GOING THERE

Most of the Universe is so far away that we only have the information about it that travels to us. Fortunately, we are able to investigate a few nearby objects in detail, through manned and unmanned exploration. Since the dawn of the Space Age in 1957, satellites have shown us our planet from a new perspective, human astronauts have visited the Moon, and automated spaceprobes have traveled to all the major worlds of the Solar System, and many of the smaller ones. Although not as versatile as human beings, these robot vehicles are far more durable and can be designed for a range of different tasks, including brief but revealing flybys, extended periods in orbit around other planets, and surface exploration from either a static platform or a wheeled rover. They can reveal not just the appearance of other worlds, but also their surface chemistries, internal structures, and geological histories.

> "I DON'T THINK THE HUMAN RACE WILL SURVIVE THE NEXT THOUSAND YEARS, UNLESS WE SPREAD INTO SPACE."
>
> **STEPHEN HAWKING**, BRITISH NEWSPAPER, THE DAILY TELEGRAPH, 2001

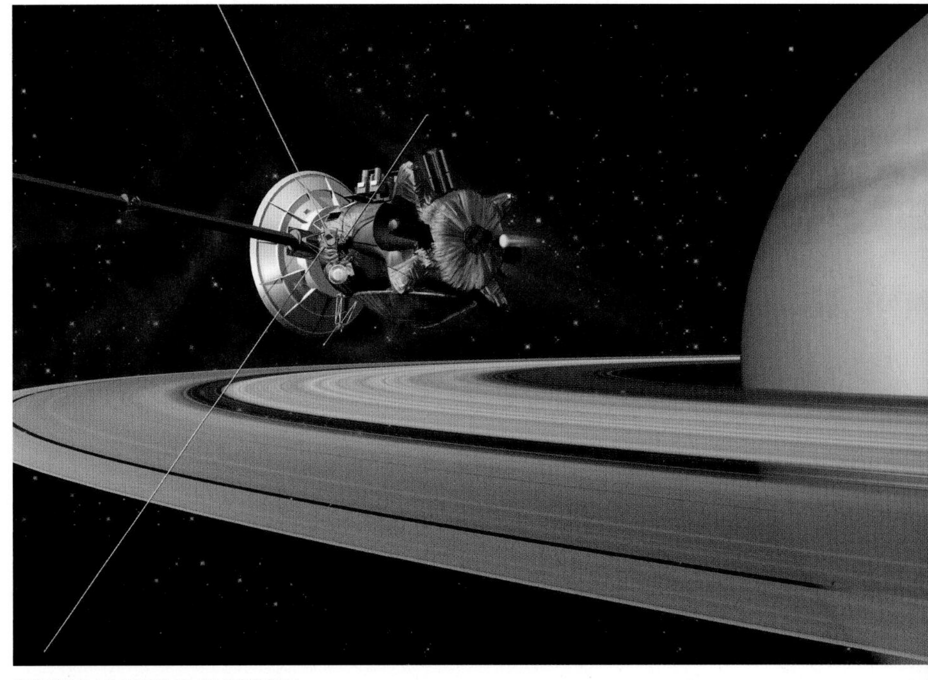

CASSINI-HUYGENS AT SATURN
This artist's impression depicts Cassini-Huygens orbiting Saturn. The spacecraft consisted of a probe, Huygens, that descended to the surface of Titan, one of Saturn's moons, and an orbiter, Cassini, that continues to circle Saturn today (see pp.156–157).

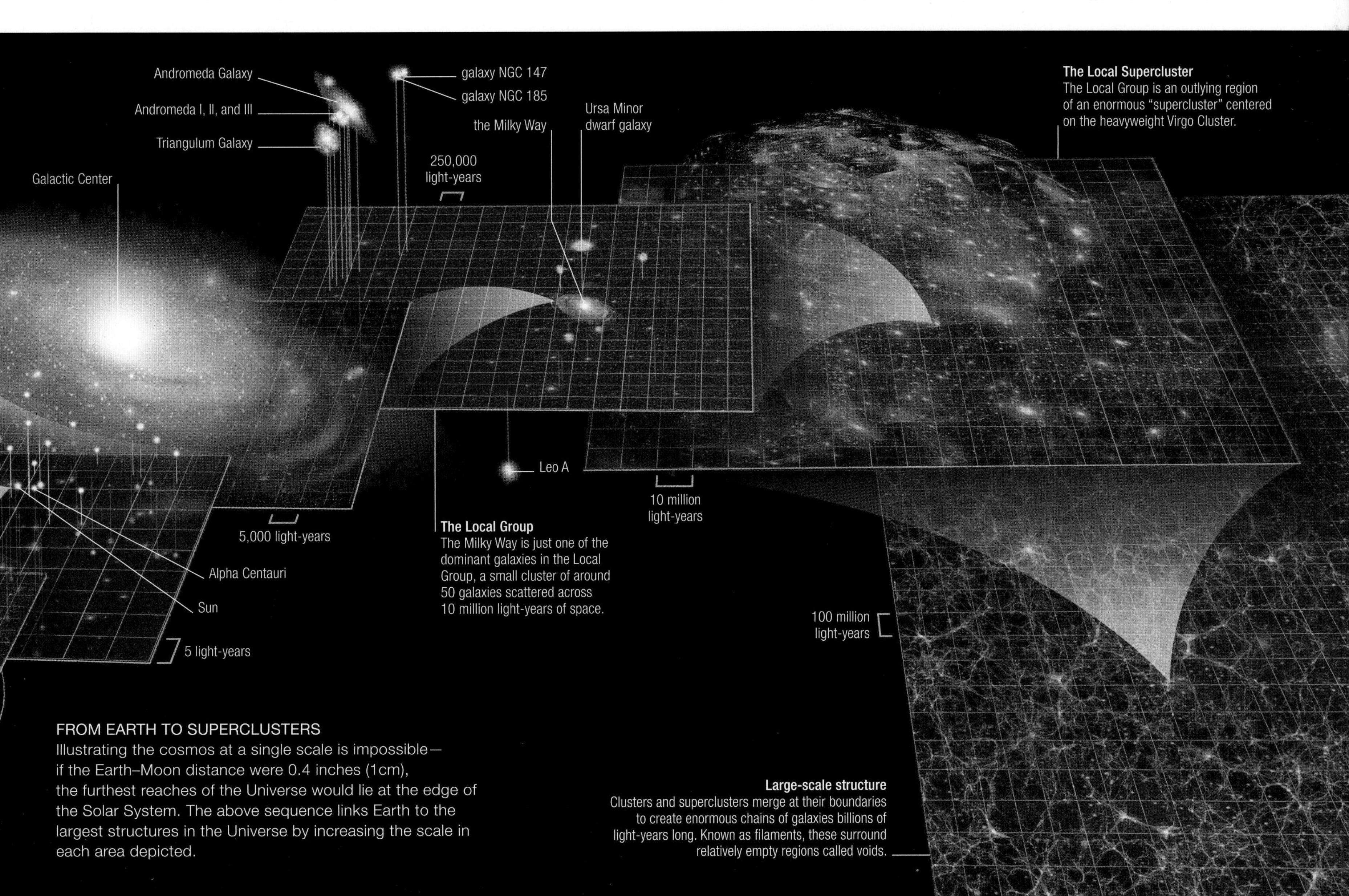

Andromeda Galaxy
Andromeda I, II, and III
Triangulum Galaxy
Galactic Center
galaxy NGC 147
galaxy NGC 185
the Milky Way
Ursa Minor dwarf galaxy
250,000 light-years
The Local Supercluster
The Local Group is an outlying region of an enormous "supercluster" centered on the heavyweight Virgo Cluster.
Leo A
10 million light-years
5,000 light-years
The Local Group
The Milky Way is just one of the dominant galaxies in the Local Group, a small cluster of around 50 galaxies scattered across 10 million light-years of space.
Alpha Centauri
Sun
5 light-years
100 million light-years
Large-scale structure
Clusters and superclusters merge at their boundaries to create enormous chains of galaxies billions of light-years long. Known as filaments, these surround relatively empty regions called voids.

FROM EARTH TO SUPERCLUSTERS
Illustrating the cosmos at a single scale is impossible—if the Earth–Moon distance were 0.4 inches (1cm), the furthest reaches of the Universe would lie at the edge of the Solar System. The above sequence links Earth to the largest structures in the Universe by increasing the scale in each area depicted.

OBSERVING THE SKIES

For most of history, astronomers were limited to recording the properties of stars, planets, and other objects with their unaided eyes and simple measuring instruments. Today, in contrast, they are able to study every aspect of the heavens with a range of high-tech machinery on Earth and in orbit.

OPTICAL TELESCOPES

The most important tool of modern astronomy is undoubtedly the telescope. Invented in the early 1600s and constantly modified and improved ever since, the telescope is essentially a collecting device capable of capturing much more light than the human eye can. It can produce magnified images of objects and areas of the sky, or pass the light it gathers on to other instruments that analyze this light in different ways. While earlier telescopes (and some modern amateur instruments) used lenses to collect incoming light and bend it to a focus, modern telescopes use enormous mirrors, many yards across, to perform the same task. Today, professional astronomers rarely look directly through an eyepiece to see the image formed by the telescope—instead they rely on electronic detectors and other instruments to gather light for them, convert it into a stream of digital data, and pass it to computers for further analysis.

> “A MOST SERENE AND QUIET AIR ... MAY PERHAPS BE FOUND ON THE TOPS OF THE HIGHEST MOUNTAINS ABOVE THE GROSSER CLOUDS.”
>
> **SIR ISAAC NEWTON**, ENGLISH ASTRONOMER

SELECTED MAJOR OPTICAL OBSERVATORIES

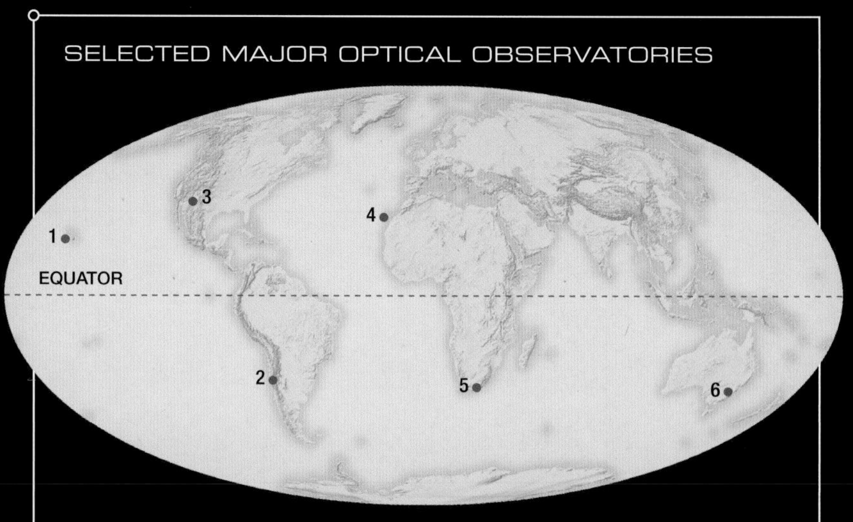

Earth's turbulent atmosphere and variable weather are a challenge to ground-based astronomers. Clouds can block the view of the heavens, and even clear skies absorb and distort starlight. Many of the world's most important telescopes are therefore grouped in a few locations—on high island peaks, such as Mauna Kea in Hawaii, and in high deserts, such as the Atacama in Chile. Telescopes here, placed above much of the atmosphere and most of its weather, benefit from the clearest night skies on the planet.

1. **Keck I and II** (10m), Mauna Kea, Hawaii
2. **Very Large Telescope** (8.2m), Paranal, Chile
3. **Large Binocular Telescope** (11.8m), Mount Graham, Arizona
4. **Gran Telescopio Canarias** (10.4m), La Palma, Canary Islands
5. **South African Large Telescope** (9.2m), Karoo, South Africa
6. **Anglo-Australian Telescope** (3.89m), Siding Spring Mountain, Australia

THE THIRTY METER TELESCOPE
This artist's impression shows the Thirty Meter optical telescope planned for completion atop Mauna Kea in 2018. When finished, it will dwarf even the largest telescopes currently on the Hawaiian mountain top.

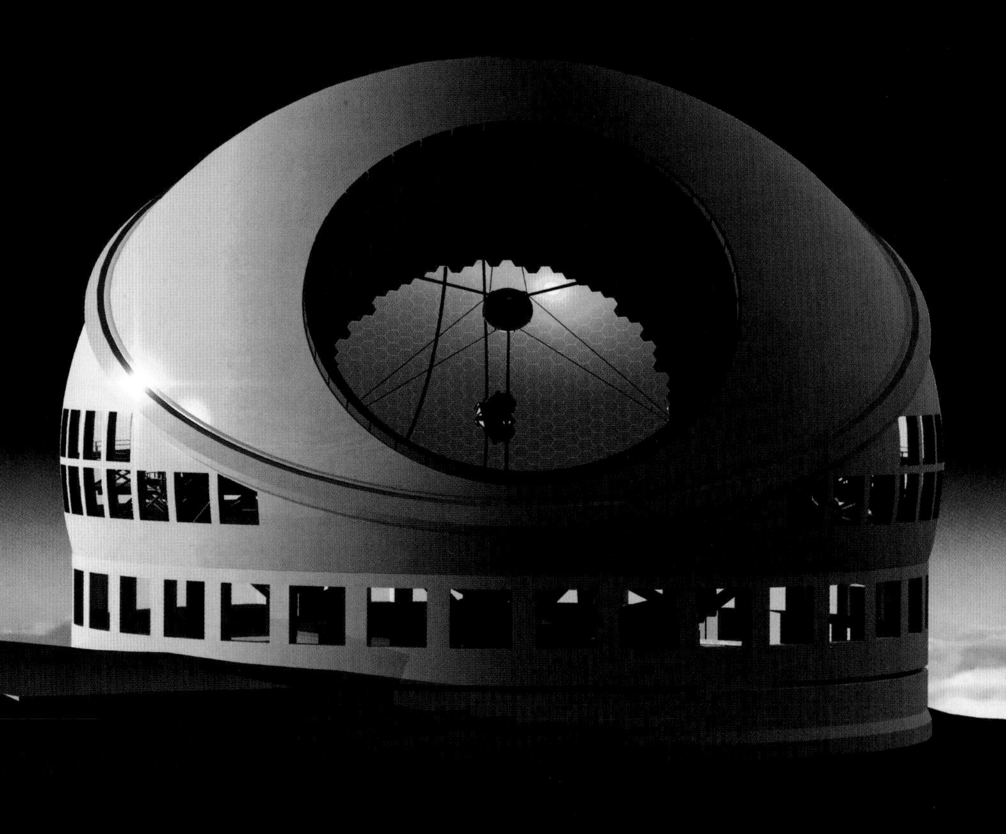

THE FULL SPECTRUM OF TELESCOPES

Since the 19th century, scientists have seen light as an electromagnetic wave—a pattern of electrical and magnetic disturbances that stimulates the nerve cells at the back of our eyes to produce images. Different colors of light are caused by waves with different wavelengths and energies. Visible light is part of a wider electromagnetic spectrum, and invisible forms of radiation can also reveal much about the Universe.

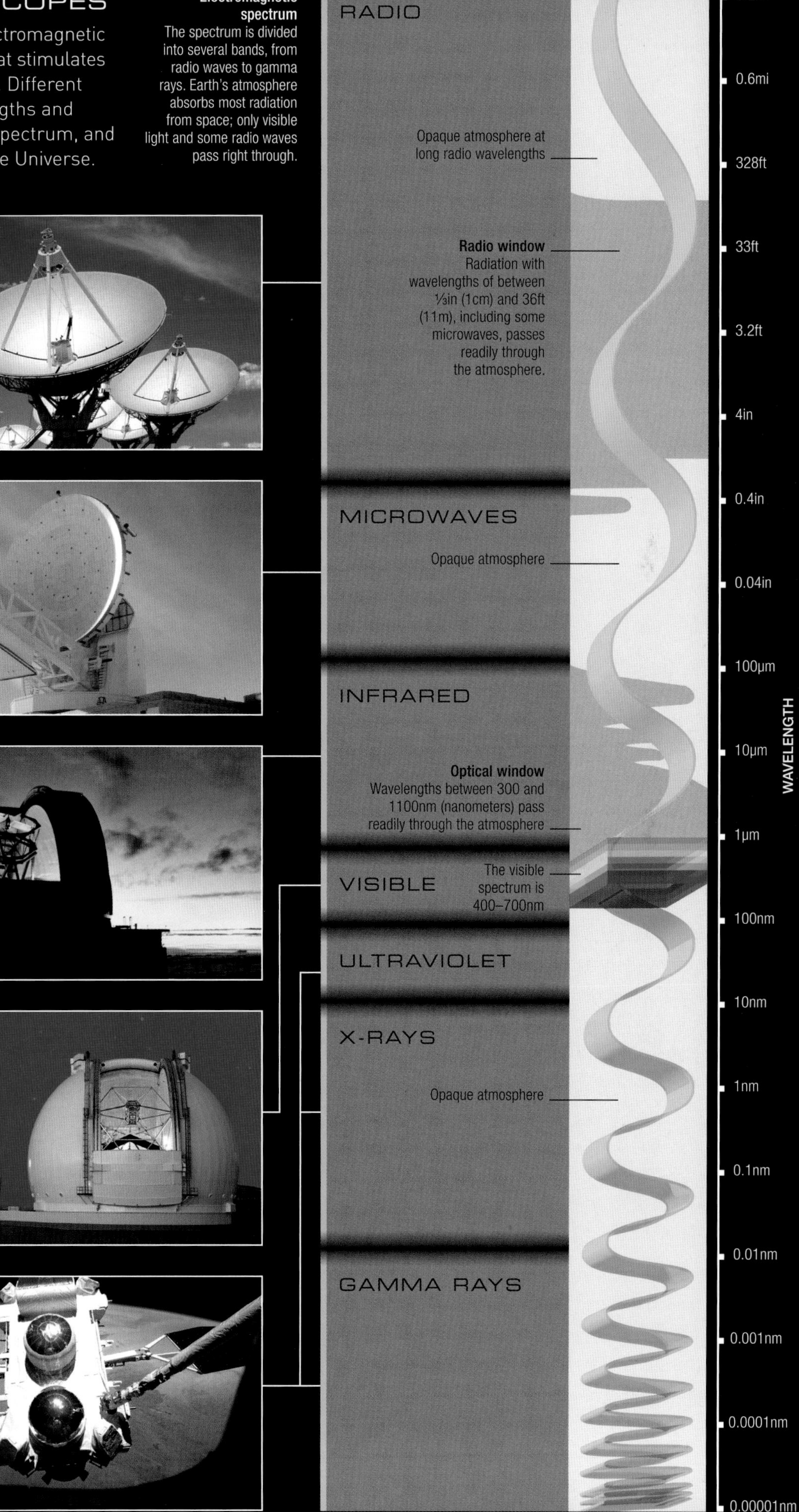

RADIO TELESCOPES

Radio waves are generated by astronomical objects and interstellar gas clouds. Most penetrate Earth's atmosphere and reach the surface intact, but their long wavelengths make their sources difficult to pin down. To create sharper images of celestial radio sources, astronomers use dish-shaped metal antennae like these at the Very Large Array in New Mexico (right), and combine signals from multiple telescopes.

MICROWAVE TELESCOPES

Microwaves, the shortest radio waves of all, are emitted by objects such as stars in the process of being formed. Other microwaves come from the cosmic background radiation—an echo of the Big Bang (see pp.316–317). Microwaves can be detected using Earth-based antennae, such as the South Pole Telescope (right), but only reveal all their secrets to satellite observatories such as WMAP (see p.316).

INFRARED TELESCOPES

Infrared waves are those closest to the red end of the visible-light spectrum. Infrared, effectively heat radiation, is emitted by objects that do not have the energy or temperature to glow in visible light. Some infrared can be detected by telescopes on high mountains (such as the United Kingdom Infrared Telescope on Mauna Kea, right). Lower-energy waves can only be studied with orbital telescopes at low temperatures.

OPTICAL TELESCOPES

Turbulence in Earth's atmosphere causes telescopic images to become blurred and distorted. The size of a telescope's mirror also places a limit on its light-collecting area and the amount of detail that it can resolve. Optical telescopes are constantly evolving to produce sharper images—the twin-telescope Keck Observatory (opposite) on Hawaii, right, uses 33ft (10m) mirrors and computerized adjustments.

HIGH-ENERGY TELESCOPES

Radiations with shorter wavelengths and higher energies than visible light are produced by the most violent events in the Universe. Ultraviolet, X-ray, and gamma-ray observatories (successors to the Compton Gamma Ray Telescope, right) orbit the Earth. Most X-rays and gamma rays cannot be reflected and bent to a focus by normal mirrors, so other approaches are used.

ON THE LAUNCHPAD

Every spacecraft must begin its journey at a launch site somewhere on Earth. The location of these sites is governed by factors including geography, politics, and logistics, and the largest have grown into towns in their own right, equipped for the preparation, servicing, launch, and ground control of space vehicles.

LAUNCH SITE LOCATIONS

In order to reach space from Earth, spacecraft need all the help they can get, and for this reason, most launch sites are located as close to the equator as possible within each space-faring nation's borders. These sites offer the advantage of a speed boost imparted by our spinning planet, but they inevitably put spacecraft into orbits over Earth's lower latitudes. Spacecraft destined for orbits that fly over the polar regions are often launched from sites at higher latitudes. Finding potential launch sites is complicated by the need for good transportation links and a safely uninhabited area under the flight path in case of accidents during launch.

CAPE CANAVERAL

The most famous launch complex of all is located at Cape Canaveral, on Florida's Atlantic coast. This site, with open ocean to the east, is home to NASA's Kennedy Space Center (KSC) and the Cape Canaveral Air Force Station, both of which see frequent launches. The Air Force Station was the departure point for manned US spaceflights up until 1967 and still hosts many unmanned launches. The KSC, opened in 1962, was used for the famous Apollo Moon missions and well over a hundred Space Shuttle flights. Its facilities include numerous hangars for preparing rockets and spacecraft that are either flown to its airstrip or brought in by barge. The KSC is also home to a powerful crawler transporter, which moves the largest vehicles, fully assembled, from the enormous Vehicle Assembly Building (VAB) to the launchpads.

MAJOR SPACEFLIGHT LAUNCH SITES

EQUATOR

1. Vandenberg, California
2. Edwards, California
3. Cape Canaveral, Florida
4. Wallops Island, Virginia
5. Kourou, French Guiana
6. Alcântara, Brazil
7. Hammaguir, Algeria
8. Torrejón, Spain
9. Andøya, Norway
10. Palmachim, Israel
11. Plesetsk, Russia
12. San Marco, Kenya
13. Kapustin Yar, Russia
14. Baikonur, Kazakhstan
15. Sriharikota Island, India
16. Jiuquan, Inner Mongolia
17. Xichang, China
18. Taiyuan, China
19. Svobodny, Russia
20. Tanegashima, Japan
21. Kagoshima, Japan
22. Woomera, Australia

A MOMENT IN HISTORY
A Saturn V rocket bearing the Apollo 11 lunar spacecraft leaves Pad A at Cape Canaveral's Launch Complex 39 on July 16, 1969, watched by part of a huge crowd.

THE VEHICLE ASSEMBLY BUILDING
The Space Shuttle *Endeavour*, attached to its external tank and rocket boosters, undergoes final preparations for launch in February 2010 in the VAB at NASA's KSC.

SPACEPORT JAPAN
An H-IIA rocket blasts off from Tanegashima Space Center, Japan's largest launch site. Located on an island to the south of Kyushu, Tanegashima lies as close to the equator as possible within Japanese territory.

BAIKONUR COSMODROME

Now the largest launch site in the world, Baikonur, the main Russian launch complex, is located at Tyuratam in the deserts of Kazakhstan. It has operated continuously since 1957 and was used for historic launches such as Vostok 1, the first manned spaceflight, in 1961. Several hundred miles of road and railroad links connect it to rocket and spacecraft factories in Russia. Since the collapse of the Soviet Union, Baikonur has been operated by the Russian Federal Space Agency. It also plays host to commercial spaceshots and manned Soyuz launches to the International Space Station (see pp.22–23).

TRAIN TRANSPORTATION
Unlike NASA's rockets, those at Baikonur are assembled horizontally then pulled at walking pace to their launchpads by powerful trains.

GUIANA SPACE CENTRE

Europe's main launch complex is located at Kourou, French Guiana, on the Atlantic coast of South America. Of all the world's major launch sites, it is the closest to the equator, and Russian Soyuz rockets launch unmanned cargoes from here, too. The Guiana Space Centre was founded by the French Space Agency in 1968, but today it is largely funded by the European Space Agency (ESA).

PREPARING FOR COUNTDOWN
An ESA Ariane 5 rocket carrying the Herschel and Planck satellites (see p.316) is readied for launch in the Final Assembly Building at the Guiana Space Centre in May 2009.

JOURNEY INTO SPACE

In order to reach Earth orbit, a spacecraft needs to overcome Earth's tremendous gravity, accelerating with enough force and for long enough to reach speeds of several miles per second.

ROCKET POWER

The only vehicles currently capable of reaching orbit and beyond are rockets. A rocket works on the principle of reaction. In most rockets, two chemicals—a fuel and an oxidant—are mixed in a combustion chamber and ignited. When the stream of exhaust gases generated by the resulting explosive burning escapes at high speed from the back of the rocket motor, reaction causes the rocket to move in the opposite direction. The most important manned spacecraft put into orbit by rockets have been the American Space Shuttle and the Russian Soyuz.

THE SPACE SHUTTLE

NASA's Space Shuttle, which first flew in 1981, was designed to reduce the costs of manned spaceflight to orbit. The system consists of a large, planelike orbiter, a huge external fuel tank, and a pair of reusable booster rockets. It has acted as a satellite delivery system, a recovery and repair vehicle, and an orbiting laboratory. But expensive running costs and two tragic accidents have crippled its operation, and the Shuttle is scheduled for retirement in late 2010.

ROCKET STAGES

First suggested by Russian rocket enthusiast Konstantin Tsiolkovskii as early as 1903, multi-stage rocket designs keep the amount of rocket propellant required to reach space to a minimum by reducing the amount of "dead weight" carried as the rocket speeds up. The rocket consists of a series of stacked units, each with its own propellant tanks and engines. Each stage is jettisoned before the next stage ignites, diminishing the vehicle's overall weight. Booster rockets that fire alongside the first stage can give a similar benefit.

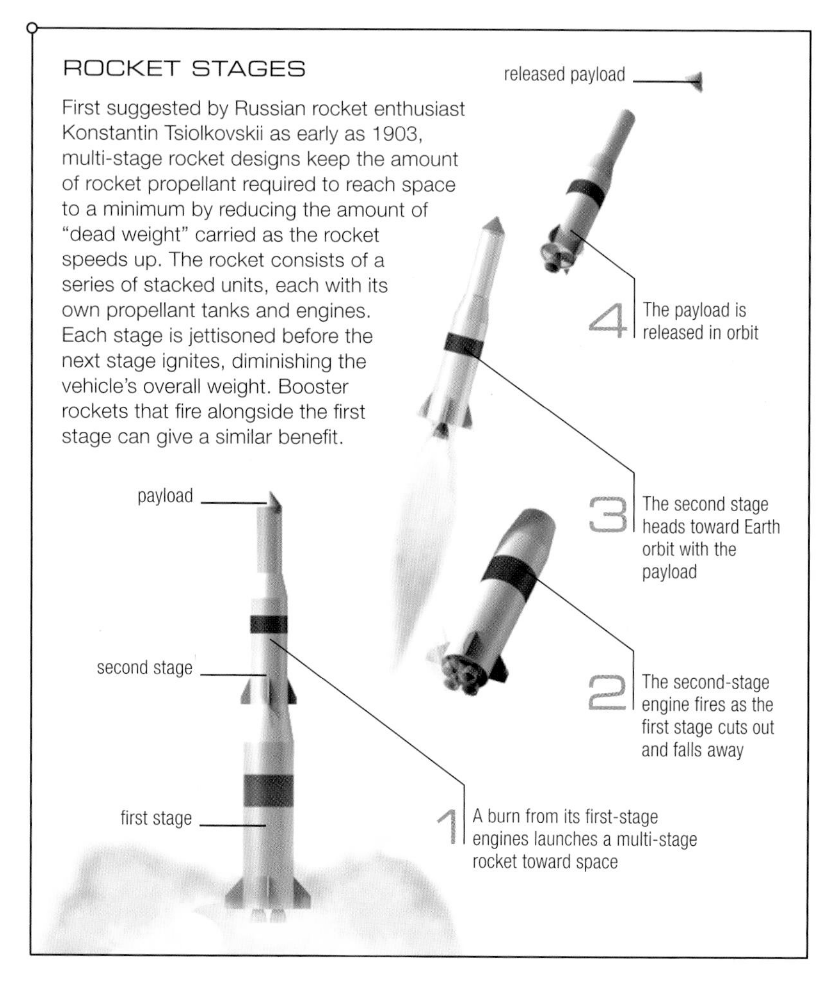
released payload
4 The payload is released in orbit
3 The second stage heads toward Earth orbit with the payload
2 The second-stage engine fires as the first stage cuts out and falls away
1 A burn from its first-stage engines launches a multi-stage rocket toward space
payload
second stage
first stage

SHUTTLE LAUNCH
The Space Shuttle *Atlantis* lifts off from the Kennedy Space Center in Florida in 2006. Exhaust from the booster rockets is more visible than the flames from the engines.

THE STORY OF SOYUZ

Russia's staple manned spacecraft, the Soyuz, has been in use, with modifications, since the mid-1960s. It is a modular spacecraft similar to the US Gemini and Apollo craft of the 1960s. Launched using the Soyuz rocket, it consists of an orbital module in which the crew spend most of their time, a re-entry module, and a cylindrical service module that carries maneuvering engines, power systems, solar panels, and other equipment. Unlike the Space Shuttle, Soyuz is not reusable—the service and orbital modules are left in orbit to break up in Earth's atmosphere, and only the re-entry module returns to Earth.

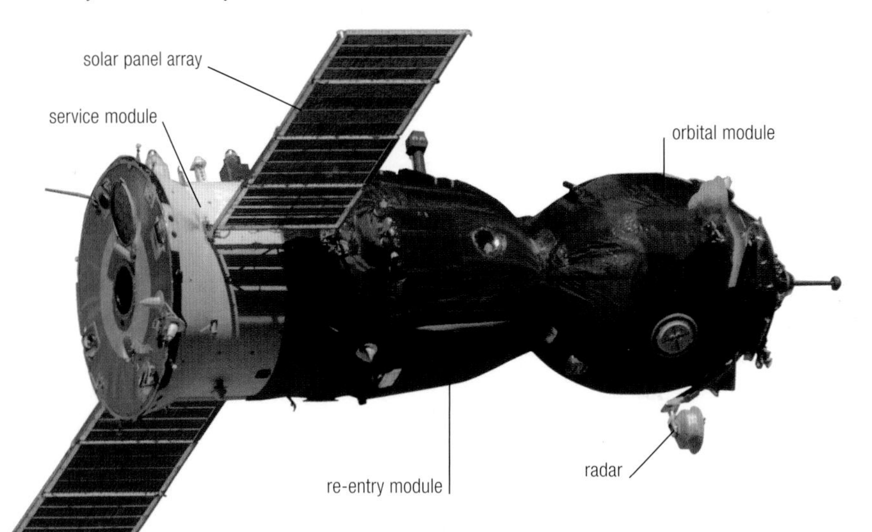

SOYUZ TMA
The latest version of the Russian Soyuz spacecraft, the Soyuz TMA acts as a ferry to and from the International Space Station.

SOYUZ LANDING
The Soyuz re-entry module returns to Earth over land. Parachutes slow its descent, and retro-rockets cushion the final impact.

THE FUTURE OF RE-USABLE SPACECRAFT

While the Space Shuttle has ultimately proved too expensive and unreliable to transform access to space, commercial spaceflight companies have taken up the quest for a fully reusable launch vehicle. SpaceShipOne (which made three suborbital flights in 2004) and its successor Virgin Galactic vehicles use a carrier aircraft to take a smaller, rocket-powered second stage to the edge of space. Other companies are working on single-stage-to-orbit (SSTO) rockets.

INTO ORBIT

Once a spacecraft has successfully reached the edge of space, 60 miles (100km) above Earth's surface, it needs to maintain its position. Achieving a stable orbit around Earth requires a vehicle to reach high speeds that allow it to counteract Earth's gravity, and this means that the final stages of a satellite's deployment can vary depending on its purpose and destination.

SHUTTLE SATELLITE RELEASE

Eight-and-a-half minutes after launch, the Space Shuttle orbiter's main engines cut out, leaving it in a stable low Earth orbit (LEO) at an altitude of around 210 miles (340km). Moving at speeds of about 17,400mph (28,000kph) parallel to Earth's surface, the orbiter's tendency to keep moving in a straight line out into space is perfectly balanced against the pull of Earth's gravity. From here, the shuttle can deploy payloads directly into LEO using its robot arm—an ideal way of releasing satellites that are intended for a short time in orbit or for later retrieval (such as the Hubble Space Telescope). Once deposited in orbit, the payload stays more or less where it is put, except for a slow decline due to drag from the thin upper atmosphere. Hubble in particular is put into a higher-than-usual orbit and boosted by the Shuttle during each servicing visit to slow this decline. Satellites intended for higher orbits are usually fitted with a small independent rocket motor known as a Payload Assist Module (PAM). They are ejected from the orbiter's cargo bay from a tilting platform that also sets them spinning for added stability. The rocket only fires once the two spacecraft have separated by a safe distance in orbit.

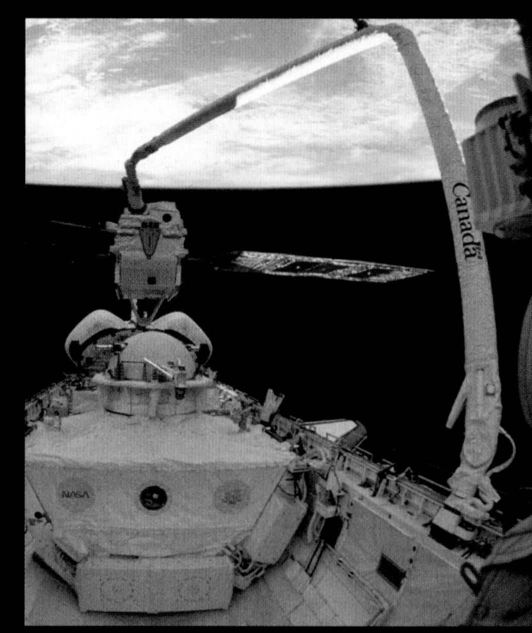

EURECA MOMENT
The Shuttle's robot arm lifts the EURECA (European Retrievable Carrier) satellite out of *Atlantis*'s cargo bay in July 1992. Placed at an altitude of 316 miles (508km), EURECA carried 15 experiments. It orbited Earth for 11 months before being retrieved by *Endeavour*.

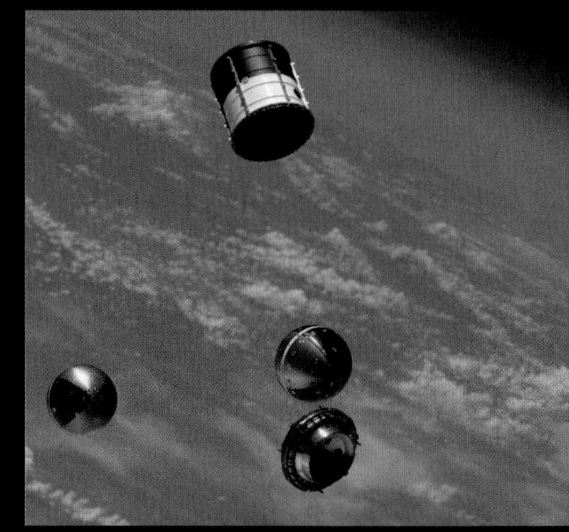

FLOATING FREE
At a safe distance from *Discovery*, the ANDE capsule opens to release its twin microsatellites. The two ball-shaped spacecraft have different densities and masses—changes to their orbits over months and years will reveal properties of Earth's tenuous upper atmosphere.

TWO FOR ONE
A launch canister containing the US Department of Defense's Atmospheric Neutral Density Experiment (ANDE) drifts away from the Space Shuttle *Discovery* following deployment in December 2006. The canister contains a pair of ball-shaped microsatelllites, each 17in (43.5cm) in diameter.

ROCKET RELEASE

The upper stage of any rocket-launch vehicle is usually fitted with an aerodynamic shell called a fairing. Satellites themselves do not need to be aerodynamic, since they suffer from little drag even in the tenuous upper atmosphere. However, their ungainly shapes would make them vulnerable to being torn apart by high-speed air flows if they were not protected during the launch. The lower stages of a rocket are usually designed to launch the payload into LEO, where the fairing separates and exposes its cargo. The satellite is typically fitted with a PAM-like single-stage or two-stage rocket, which fires to boost it out of LEO and into its correct orbit. A single burn of this engine at the right time transforms the satellite's orbit into an ellipse with its closest point to Earth (perigee) in LEO and its furthest point (apogee) much higher up. A second engine burn can then put the satellite onto a higher circular orbit.

PROTECTED FOR LAUNCH
Mounted on the upper-stage of a Delta II rocket, NASA's Fermi gamma-ray telescope is placed into an aerodynamic fairing before launch at Cape Canaveral.

SATELLITE REVEALED
This artist's impression depicts a Russian Proton rocket jettisoning the payload fairing to reveal ESA's Integral gamma-ray astronomy satellite.

SPACE STATION SUPPLY

At present, all manned missions in orbit involve a rendezvous with the huge International Space Station (ISS), 210 miles (340km) above the Earth. While the Space Shuttle's visits are focused on supplying new components, the crew usually travels to and from the ISS aboard Soyuz "ferries" that remain docked to the station for a few days while the astronauts and cosmonauts change over. Two Soyuz capsules are docked to the station at all times, providing emergency "lifeboats" for the crew. Unmanned vehicles—including the Russian Progress, the European Automated Transfer Vehicle, and the Japanese H-II Transfer Vehicle (HTV)—carry fresh equipment and supplies to the station. These may be joined by commercially developed spacecraft in the future.

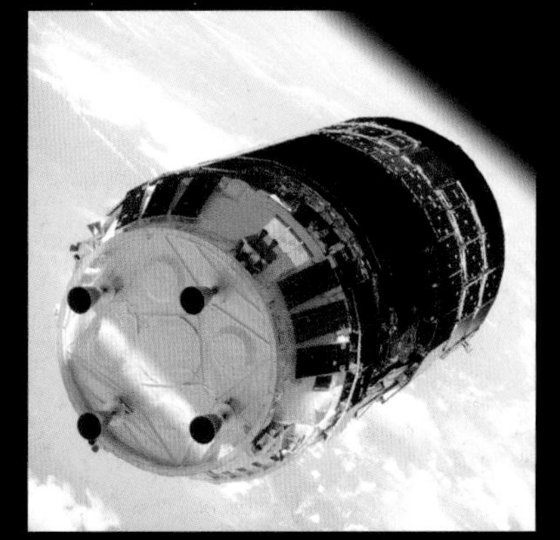

DELIVERY FROM EARTH
A Japanese H-II Transfer Vehicle approaches the ISS in September 2009. Once within reach, the HTV is grasped by the station's robot arm and moved to a docking port.

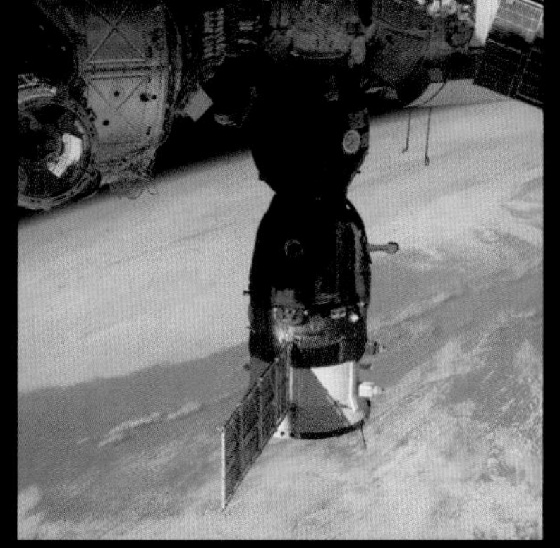

SOYUZ LIFEBOAT
A Russian Soyuz spacecraft hangs on the underside of the ISS in November 2009, on standby in case an emergency evacuation of the station is required.

ORBITING EARTH

In order to carry out their tasks, satellites can be placed into a wide variety of orbits, ranging from near-circular, low Earth orbits (LEOs) to highly elliptical, tilted orbits. Some craft follow paths that are not Earth orbits at all but nevertheless keep them in a stable relationship with the planet.

TYPES OF ORBIT

Satellite orbits range in height from a few hundred to thousands of miles. The laws of gravity mean that the further out a satellite orbits, the slower it moves through space. For example, vehicles such as the Space Shuttle in LEO around Earth orbit in about 90 minutes, while satellites in intermediate orbits, around 12,400 miles (20,000km) above Earth, orbit in about 12 hours. At a height of precisely 22,236 miles (35,786 km), satellites orbit exactly once in every 23 hours 56 minutes (the same period in which Earth rotates), or one day, and maintain the same position relative to Earth's surface. This type of orbit is known as a geostationary or geosynchronous orbit. Spacecraft on elliptical orbits that are much further from Earth at one extreme than at the other vary their speed during each orbit. Inclination (the tilt of a satellite's orbit relative to Earth's equator) is another important factor—most satellites orbit above the equator or in orbits with shallow tilts that carry them over Earth's low latitudes. Highly inclined polar orbits allow satellites to pass close to the poles and to overfly most of Earth's surface as the planet spins. Sun-synchronous polar orbits slowly change their inclination so that they keep the angle between satellite, Earth, and Sun constant.

VARIETY OF ORBITS
A range of typical orbits used by Earth satellites is shown here, though not to scale. Spacecraft usually move between these different types via elliptical transfer orbits.

Molniya-type orbit Satellites in this orbit spend most of their time close to the most distant point.

Polar orbit This orbit carries a satellite over most of the Earth's surface as the planet rotates.

N

Low Earth orbit This high-speed orbit is relatively easy to reach from Earth's equatorial regions.

S

Geostationary orbit A satellite here retains a constant position relative to Earth's surface.

SATELLITES AND THEIR ORBITS

Different satellite applications require different orbits. LEO, for example, is ideal for satellites that only need to get above Earth's atmosphere, and low-altitude polar orbits are suitable for satellites surveying the entire Earth's surface. Photographic-imaging missions use Sun-synchronous orbits. Meanwhile, geostationary orbits allow satellites to hang permanently over the same point on Earth's equator and keep an entire hemisphere in view—ideal for monitoring communications satellites. Highly inclined, extremely elliptical Molniya orbits, which place a satellite high above a single point on Earth so that it moves only slowly in the sky, are used for high-latitude communications.

GLONASS SATELLITE
This spacecraft is part of a Russian-built equivalent to the widely used American Navstar Global Positioning System (GPS). GPS satellites are placed in intermediate orbits. At present, they are on six planes with at least four satellites on each plane to give coverage around the Earth.

EURECA IN ORBIT
The European Retrievable Carrier satellite EURECA (see p.16) floats in LEO over the coast of Florida after its release from the Shuttle *Atlantis*.

LAGRANGIAN POINTS

At certain points within the Earth-Moon-Sun system, known as Lagrangian Points, the gravitational pull of two or more bodies combines to produce useful effects. The most important points, L1 and L2, are either side of Earth on the Earth-Sun line. Spacecraft here experience forces that cause them to circle the Sun in the same time as the Earth. The L1 point is a popular location for solar observatories, while the antisolar point L2 is used by observatories such as Gaia (see p.199), Herschel (see p.218), and WMAP (see p.316). At L3, a spacecraft will remain permanently opposite the Earth. L4 and L5 are Trojan points, where other craft can share Earth's orbit without being perturbed.

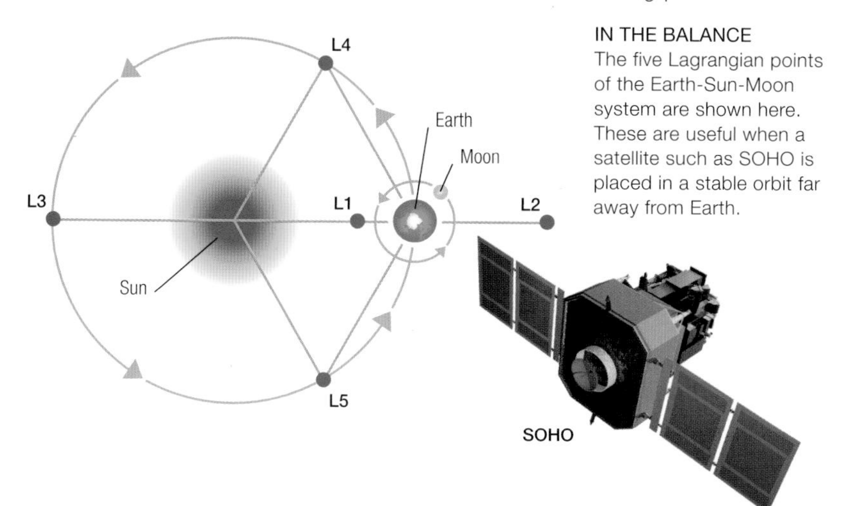

IN THE BALANCE
The five Lagrangian points of the Earth-Sun-Moon system are shown here. These are useful when a satellite such as SOHO is placed in a stable orbit far away from Earth.

SPACE DEBRIS

The several thousand satellites that have been launched to date have created a vast amount of orbital junk, ranging from flecks of paint to entire rocket stages. Much of the debris in LEO eventually re-enters the atmosphere of its own accord, and some satellites are deliberately brought down to Earth at the end of their lives, but the threat of a collision between working spacecraft and fast-moving fragments of orbital debris continues to grow.

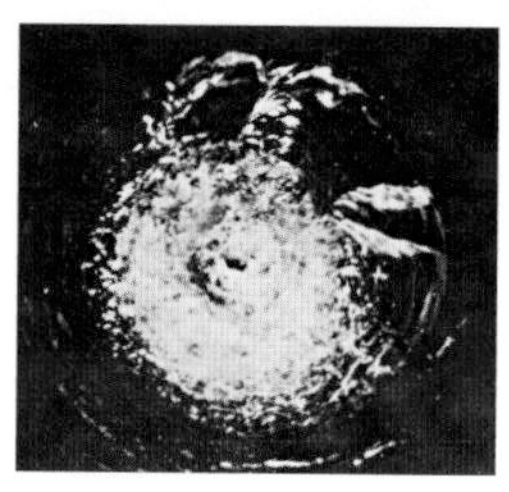

IMPACT DAMAGE
In 1983, a chip of paint hit a window on the *Challenger*, creating this microcrater ⅙in (4mm) wide.

DOWN TO EARTH
A group of Saudi Arabians inspect a large piece of space junk that crashed back to Earth in January 2001 after its orbit decayed with unexpected speed. The wreckage is all that remains of a Payload Assist Module used to launch a Navstar Global Positioning System satellite eight years earlier.

EARLY SPACE STATIONS

A succession of space stations has been in orbit around Earth for most of the past four decades. A few hundred miles above the planet, space stations provide a home and workplace where astronauts can live and carry out research.

SALYUT AND SKYLAB

A total of nine space stations have orbited Earth. The first was the Soviet Salyut 1, which was put into orbit in 1971. Small enough to be transported into space in one piece, it was also big enough to house a crew of three. Just one crew spent 28 days onboard. The second manned space station was the American Skylab, which orbited Earth from 1973 to 1979. Its first crew stayed onboard for 23 days, while two later crews stayed for 59 and 84 days. A further five Salyut stations followed between 1975 and 1991. The numbers of crews visiting these stations progressively increased, and the crews stayed in space for increasingly longer periods. The seventh and final Salyut, Salyut 7, was in orbit from 1982 to 1991 and was occupied for 816 days. Salyut 7 had two docking ports, one at either end of the station. The successful docking of modulelike craft at these ports paved the way for a new type of space station that would be constructed in space from modules.

SKYLAB
This was the view of Skylab as the last crew left the station in 1974. It remained in orbit until 1979, when it was guided into Earth's atmosphere. Skylab broke up on re-entry.

SALYUT 7
About 52ft (16m) long and averaging about 13ft (4m) wide, Salyut 7 completed 51,917 orbits of Earth in its 3,216 days in space. Six resident crews and four visiting crews stayed onboard. In addition, 15 unmanned missions transported supplies to the station.

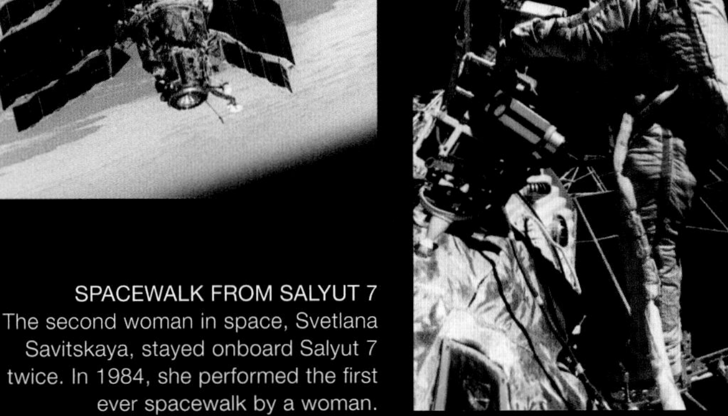

SPACEWALK FROM SALYUT 7
The second woman in space, Svetlana Savitskaya, stayed onboard Salyut 7 twice. In 1984, she performed the first ever spacewalk by a woman.

MIR OVER THE PACIFIC
This June 1998 view from the Space Shuttle *Discovery* shows the Mir space station with the Pacific Ocean in the background. In the foreground is a docked Soyuz transport craft.

ONBOARD MIR
The interior of Mir was similar in size and shape to a train carriage, but cramped with all the equipment onboard. Mir 18 crew members Vladimir Dezhurov (left) and Gennady Strekalov are seen at work here.

MIR

The first truly successful space station was Mir, which was occupied by astronauts almost continuously from February 1987 to June 2000. This Soviet station was constructed in space by adding new parts to the original living module between 1986 and 1996. In its final configuration, it consisted of seven modules providing living quarters and working areas. A crew of three were permanently onboard, but Mir could also accommodate three visiting crew. Between January 1994 and March 1995, it was home to Valeri Poliakov, who holds the record for the longest single stay in space. The primary transport vehicles serving the station were Soyuz craft, which delivered and returned crew from Mir, and the unmanned Progress, which delivered cargo. Eleven Space Shuttle missions also docked with Mir between 1994 and 1998. Mir was brought out of orbit and broke up in Earth's atmosphere in March 2001.

“YOU DON'T HAVE A LOT OF SUPPORT FROM THE GROUND. YOU REALLY ARE ON YOUR OWN.”

SHANNON LUCID, MIR ASTRONAUT, 1996

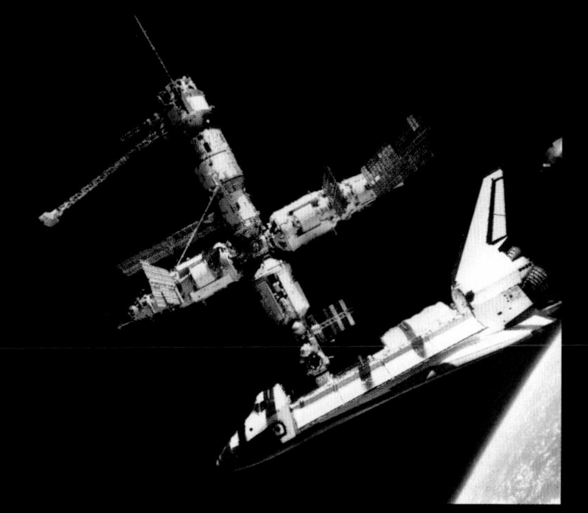

SHUTTLE DOCKS WITH MIR
The *Atlantis* orbiter docked with Mir in July 1995. This image was taken by a Russian astronaut, Nikolai Budarin, from a Soyuz craft that had been temporarily undocked for a short flight around the space station.

THE INTERNATIONAL SPACE STATION

The largest and most recent space station to orbit Earth is the International Space Station (ISS). Launched in parts, it has been assembled in space by astronauts. Crew from at least 15 different nations have lived onboard since November 2000.

BUILDING THE STATION

The ISS orbits Earth 15 times a day at an altitude of about 240 miles (390km). The project is a collaboration between 16 countries: the US, Russia, Canada, Japan, 11 European countries, and Brazil. Construction started in 1998 and is due for completion in 2011. So far, more than 100 parts have been delivered in more than 50 launches, and astronauts have completed more than 140 spacewalks to install these parts and carry out repairs.

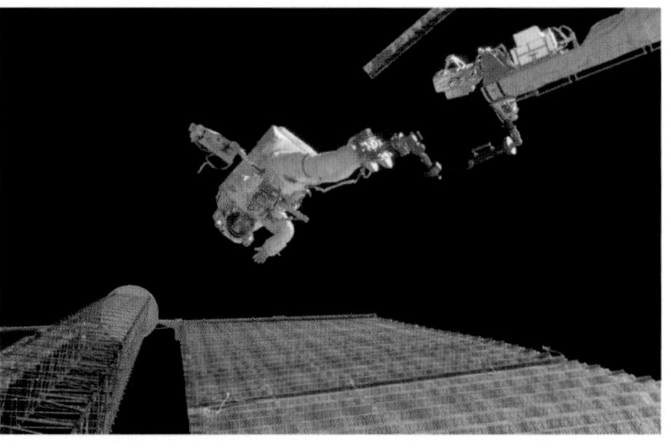

RUNNING REPAIRS
Astronauts regularly carry out maintenance work on the ISS. Here, an astronaut assesses a solar array that was damaged during its deployment a few days earlier.

THE SHUTTLE AND THE ISS
The Shuttle orbiter *Endeavour* moves away from the ISS after being docked with the station in February 2010. It had just delivered the Tranquility node and the Cupola.

LIFE ONBOARD

The inside of the ISS has the space of a five-bedroom house loosely divided into habitation and work areas, with the largest part being dedicated to work. Astronauts spend weeks to months at a time on the ISS. They do not regulate their stay by the Sun, which they see rising and setting every 90 minutes as they complete an orbit of Earth, but by their watches. The astronauts are in space to work and follow a five-day-a-week schedule made up of nine-hour working days. Other periods of time are set aside for activities such as taking care of personal hygiene, eating, relaxation, household chores, and getting a good night's sleep.

Exercise
Astronauts perform a daily exercise routine that includes sessions on a treadmill or on cycling apparatus, as shown here.

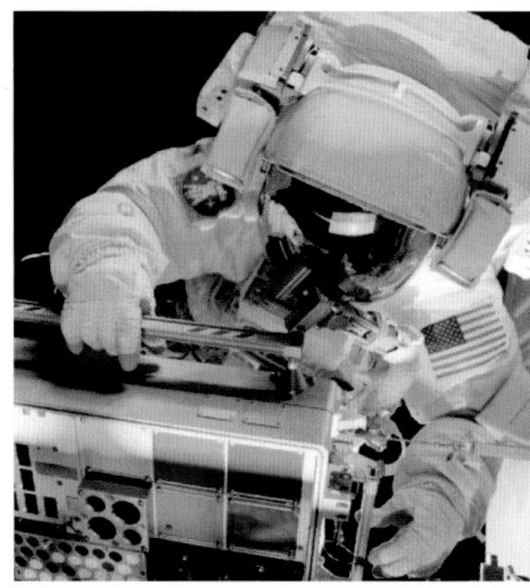

Outside experiments
Construction, maintenance, and experiments are done outside the ISS. Materials are being tested here to see how they react in space.

Inside experiments
Experiments with plant growth in the Lada greenhouse of the Zvezda module may lead to food being grown in space in the future.

THE MAIN COMPONENTS OF THE ISS

The station consists of 18 major parts, including the living module Zvezda and the three experiment laboratories—Destiny, Columbus, and Kibo, which comprises experiment modules and a facility exposed to space. Smaller modules include an airlock, where astronauts begin and end spacewalks, and docking ports. A central truss supports two sets of huge solar arrays that power the station. Canadarm2, a robotic arm with seven motorized joints, is used to move equipment around the station and support astronauts as they work.

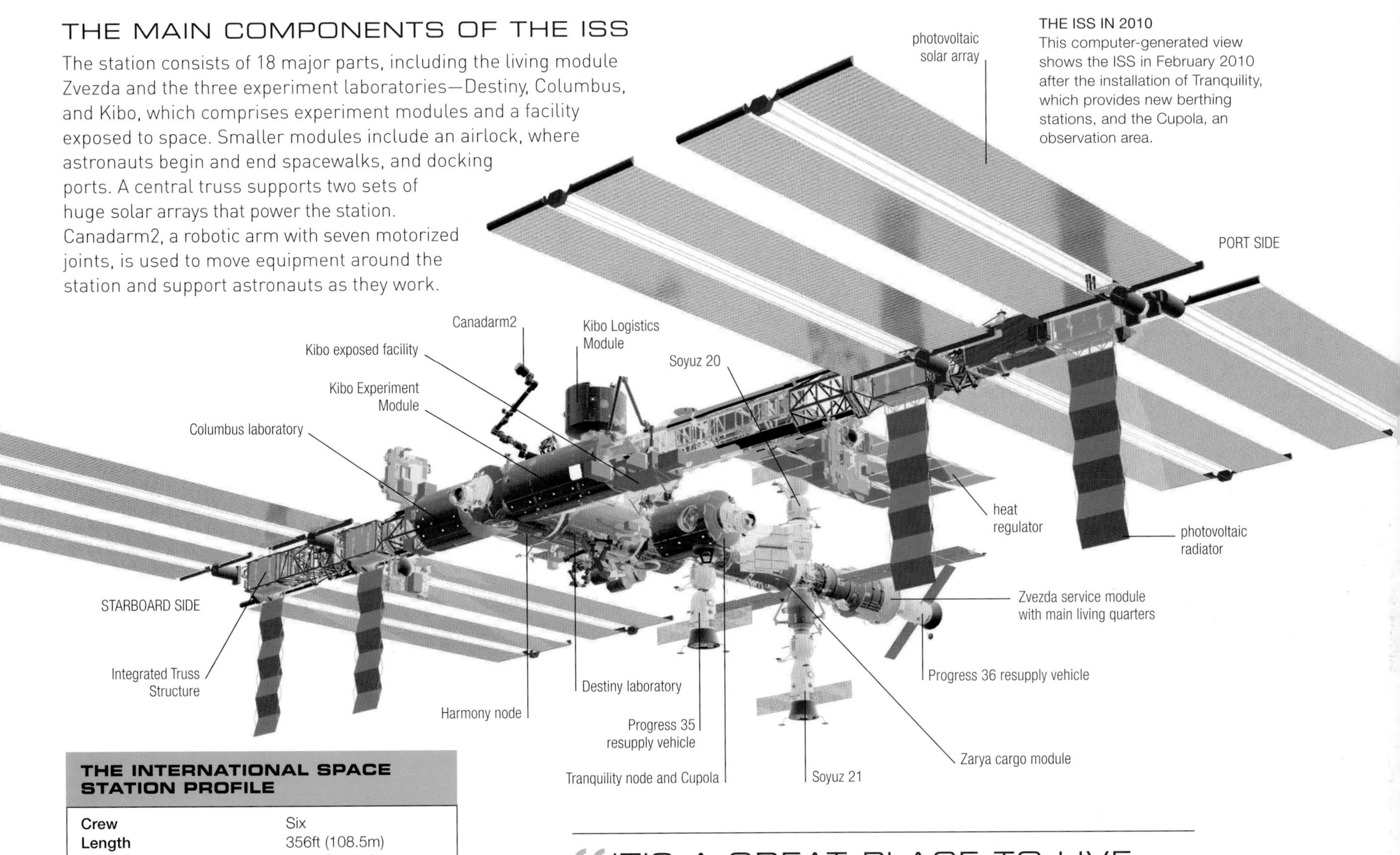

THE ISS IN 2010
This computer-generated view shows the ISS in February 2010 after the installation of Tranquility, which provides new berthing stations, and the Cupola, an observation area.

THE INTERNATIONAL SPACE STATION PROFILE

Crew	Six
Length	356ft (108.5m)
Width	239ft (72.8m)
Mass	927,316lb (420,623kg)
Habitable volume	14,400 cu ft (408 cu m)
Solar array area	32,528 square ft (3,023 square m)
Power	80 kilowatts
First assembly launch	November 1998
Final construction launch	December 2011

> “IT'S A GREAT PLACE TO LIVE AND WORK, AND TO EVEN HAVE THE SMALLEST ROLE IN THAT IS A WONDERFUL THING ...”
>
> **NICHOLAS PATRICK**, NASA ASTRONAUT, 2010

Ham radio operator
Astronauts keep in touch with home by email and occasional radio calls. Here an astronaut talks to school students on Earth.

Leisure time
When not working, time is spent reading, listening to music, watching films, looking out of the window, or playing long-distance chess.

Meal times
By the start of 2010, about 19,000 meals had been consumed on the ISS. Astronauts have snacks, drinks, and three main meals a day.

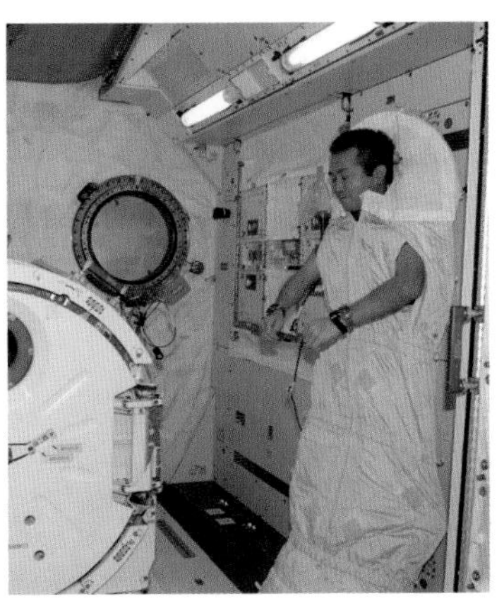

Sleeping
The schedule allows for eight hours of sleep each day. Sleeping bags are fixed in place, so that they do not float around.

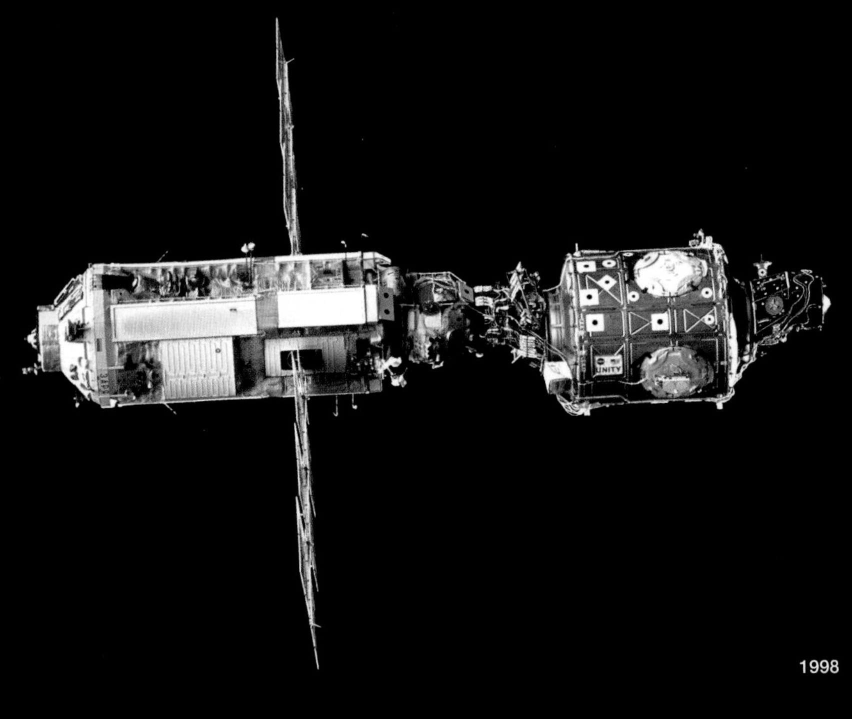
1998

2000

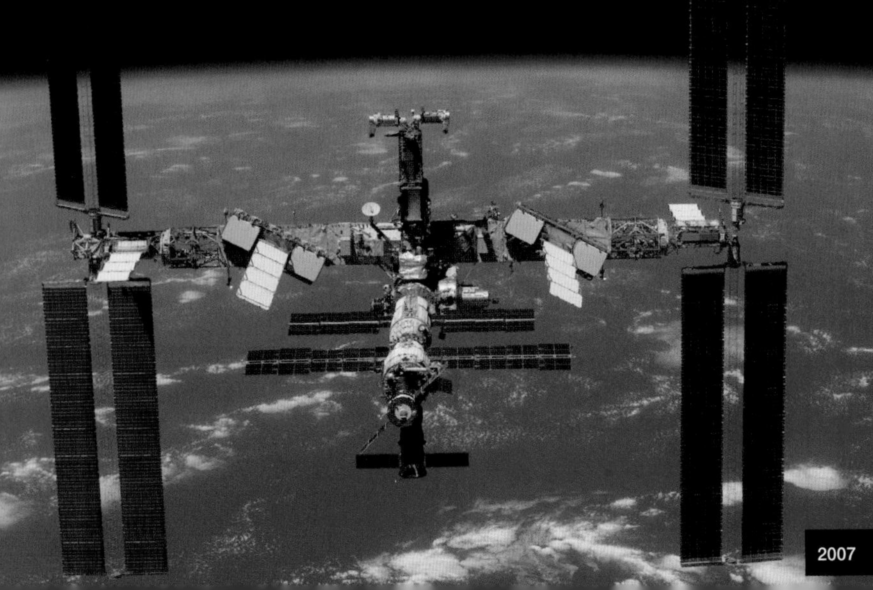
2007

BUILDING THE ISS
Assembly of the ISS began in 1998 with the launch of the Zarya and Unity modules, which together (top left) formed the station's core. Adding the Zvezda module in 2000 allowed a crew of three to live on the ISS. By 2007, the Pirs docking compartment, Destiny laboratory, Quest airlock, and much of the Integrated Truss Structure had been added (bottom left). By 2010 (main image), a new set of solar arrays had been installed, and 13 modules completed. The station is scheduled for completion in 2011.

SPACEWALKING

Known as extra-vehicular activity, or EVA, spacewalking means moving outside your craft—something 200 or so of the 500-plus astronauts who have been into space have experienced. Astronauts go outside to work; they carry out repairs, monitor experiments, or work on the International Space Station (ISS).

EARLY SPACEWALKS

The first spacewalk was performed on March 18, 1965. It did not last long, only about 10 minutes, just long enough for Soviet astronaut Alexei Leonov to go outside his spacecraft, wave to the camera recording the historic moment, and then move back inside. Nearly all walks have been like this one, made from a craft orbiting Earth. The exceptions were those made by Apollo astronauts traveling back from the Moon. In December 1972, for example, Apollo 17's Ron Evans made a 1-hour 6-minute walk to retrieve film from a camera positioned outside his craft. About two-thirds of all walks have been performed by American astronauts, just over a third by those from Russia, and fewer than 20 from other countries.

SOVIET PIONEER
Alexei Leonov makes the world's first spacewalk. Minutes later he struggled to return to his craft, Voskhod 2. His suit had inflated and required air to be released before he could re-enter.

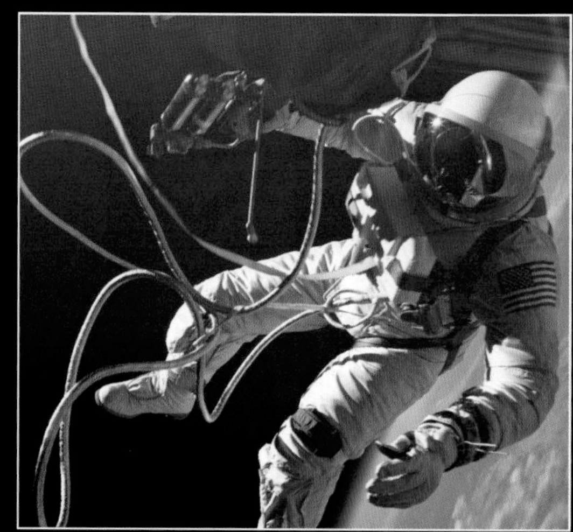

AMERICAN PIONEER
On June 3, 1965, Ed White becomes the first American and the second person to spacewalk. He uses the gas-gun he holds in his hand to move around outside his Gemini 4 spacecraft.

SPACESUITS

Spacewalking astronauts wear a pressurized spacesuit when outside their craft. It completely covers the body, providing an astronaut with his own Earth-like environment and oxygen to breathe. The outer layer is a blend of waterproof, bulletproof, and fire-resistant fabrics. Further layers are tear-resistant and are strong enough to hold in the astronaut's oxygen and maintain the correct pressure. Such suits are kept onboard the ISS (see pp.22–23), where each one is used about 25 times before being replaced. They were originally tailored for individual astronauts but are now of a standard size with adjustable leg length. The ISS suit is known as an Extravehicular Mobility Unit, or EMU suit. It incorporates a Primary Life Support System (PLSS) backpack, which controls the temperature and pressure and supplies oxygen.

Lights and camera Illuminate and record the astronaut's environment
Helmet and visor Include the astronaut's communications cap
Controls For operating the astronaut's EMU
Checklist Provides the astronaut with a spacewalking schedule
Gloves Molded from silicone rubber for sensitivity
Tether Mountaineering clips secure walker to craft
Boots Soft-soled, as feet are not used on spacewalk
Tube Supplies air during spacesuit test—not used in space

ISS SPACESUIT
Sergei Krikalev, the world's most experienced astronaut, tests an EMU spacesuit. Krikalev has spent 803.4 days on six missions and performed eight spacewalks.

LIFELINES

An astronaut remains secured to his or her craft during a spacewalk. Early walkers were attached by a cordlike tether, while a second cord supplied oxygen. Today, astronauts use shorter, mountaineering-style metal-and-rope tethers and carry their oxygen supply on their backs. They can also be fixed by their feet and back to the ISS's robotic arm. Fewer than 10 astronauts have been totally free of their craft. Astronauts on three Space Shuttle missions in 1984 used the Manned Maneuvering Unit (MMU)—a jet-propelled backpack—to fly free. On a later Shuttle mission, two astronauts tested a smaller, thruster-powered backpack called SAFER (Simplified Aid for EVA Rescue).

TESTING SAFER
American astronaut Mark Lee tests SAFER, a backpack propulsion unit with hand-controlled nitrogen thrusters. The system flies the astronaut back to safety if he or she mistakenly drifts away. ISS astronauts routinely wear the device.

FIRST FREE-FLIGHT
American astronaut Bruce McCandless flies 320ft (98m) from a Space Shuttle in 1984—the furthest any astronaut had spacewalked, and the first time any had flown freely.

WORKING IN SPACE

Spacewalks regularly last up to seven hours, and astronauts are kept busy the whole time. But even before they leave their craft, they undergo several hours of preparation. In the days prior to the scheduled walk, the batteries powering the spacesuit are fully charged and other equipment and tools are made ready. On spacewalk day, the astronaut exercises while breathing pure oxygen to reduce the amount of nitrogen in his or her bloodstream, before moving into the airlock. Here, the pressure is gradually lowered to about one-quarter of Earth's atmosphere at sea level. If this is done too quickly, the astronaut will suffer decompression sickness (the "bends"), caused by bubbles of nitrogen forming in the bloodstream. On the ISS (see pp.22–23), astronauts even sleep overnight in the airlock to avoid getting the bends. After a minimum of six hours in the airlock, and now wearing a spacesuit, the astronaut is ready to open the hatch and go to work.

> "... THE HANDRAIL ABOVE THE [SHUTTLE] HATCH IS DENTED FROM PEOPLE HOLDING ON TO IT SO HARD."
>
> **CINDY BEGLEY**, NASA'S OFFICER FOR SPACEWALKS, 2005

SATELLITE REPAIR AND SERVICING

From 1983 until 2010, astronauts regularly performed spacewalks from orbiting Space Shuttles. They worked on experiment palettes in the payload bay and on satellites already in space. In May 1992, over the course of three spacewalks, three astronauts retrieved the Intelsat VI satellite that had failed to reach its intended orbit. They fitted it with a new upper stage and re-launched it onto the correct orbit. The following year saw the first of five repair and servicing missions to the Hubble Space Telescope (see pp.276–77). Spacewalking astronauts corrected the telescope's mirror problems, upgraded its computers, and replaced cameras, spectrometers, gyroscopes, and solar arrays.

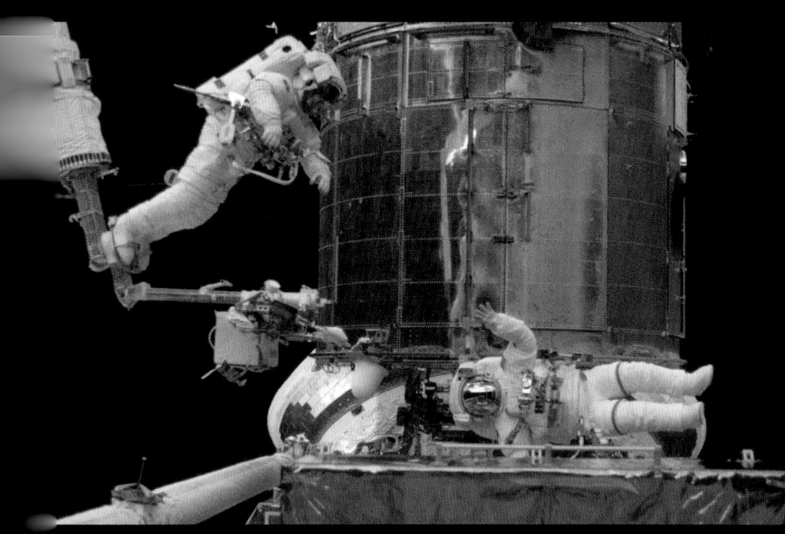

SERVICING THE HUBBLE SPACE TELESCOPE
The Hubble Space Telescope is locked into the payload bay of *Atlantis* in May 2009. American astronauts Andrew Feustal (left) and John Grunsfeld carry out repairs as part of the fifth and final servicing mission.

HEAD FOR HEIGHTS
American astronaut Nicholas Patrick works outside the ISS in early 2010. In a 5-hour 48-minute spacewalk, he removed insulation blankets and restraining bolts from the windows of the station's newly installed Cupola.

WORKING OUTSIDE THE ISS

Astronauts have spent a total of about 900 hours spacewalking outsid the ISS. They carry out two main types of work—adding new parts to the station, and maintaining and repairing the existing structure. The longest ISS spacewalk to date was undertaken by American astronau Susan Helms and Jim Voss on March 11, 2001, as they installed hardware to the external body of the Destiny module. Their walk laste 8 hours and 56 minutes. Spacewalking astronauts take everything the need with them. Tools are attached by strings that retract when not ir use, and astronauts stay hydrated by sucking on a drink straw in their helmet. Although they have a checklist on their wrist to remind them of what they have to do, the astronauts repeatedly practice beforehanc on life-size ISS parts in a huge water tank back on Earth, known as th Neutral Buoyancy Laboratory (NBL). The underwater conditions do nc give true weightlessness, but they offer the next best thing.

UNDERWATER TRAINING
In January 2003, American and Swedish astronauts Robert Curbeam and Christer Fuglesang practice on a full-scale model of the truss segment they will be attaching to the ISS. They are in the NBL and wearing white training versions of the Extravehicular Mobility Unit (EMU) spacesuit.

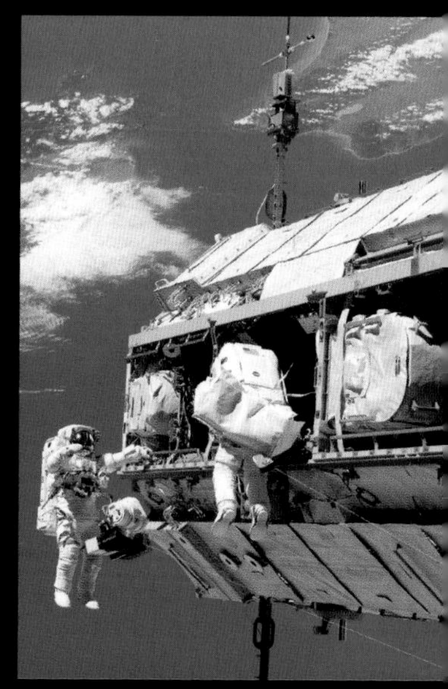

WORKING ON THE REAL THING
In December 2006, Curbeam (left) and Fuglesang are now outside the ISS and completing the jobs they spent years preparing for. On their six-and-a-half-hou walk, they attach the truss, complete power and data connections, and replace a faulty video camera.

BACK INSIDE THE ISS
Back in the Quest airlock after a 6-hour spacewalk in May 2009, American astronauts Richard Arnold (left) and Steve Swanson are helped out of their spacesuits by fellow Americans Michael Fincke (back) and Tony Antonelli.

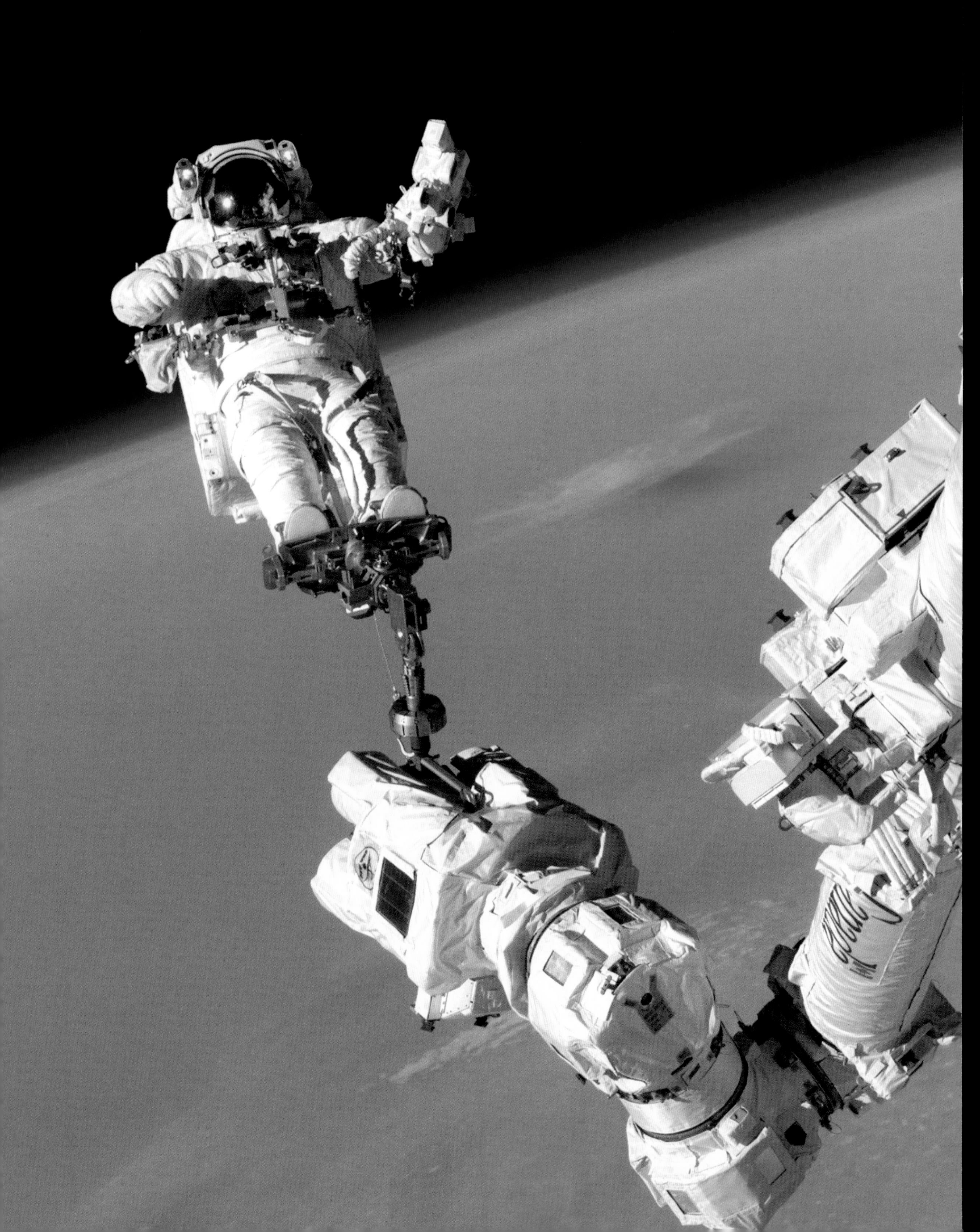

SPACEWALK AT THE ISS
American astronaut Stephen Robinson, of the STS-114 Space Shuttle mission, stands in a foot restraint at the end of the International Space Station's Canadarm2. On the same mission, he became the first astronaut to perform an in-flight repair to a spacecraft's exterior, removing two pieces of protruding material from *Discovery*'s heat shield.

AN ASTRONAUT'S VIEW OF EARTH

For almost 50 years, astronauts have been taking cameras into space. They have used them to record historic space moments, as well as the routine of the astronaut's life. Today, they take pictures not only as part of a program of continuously monitoring Earth, but also for pleasure.

ASTRONAUT PHOTOGRAPHY

The first astronaut to take a camera into space was the American John Glenn, in February 1962. Ever since, astronauts have all been trained in photography and have recorded their missions. Early flights carried cameras and film specially designed for space. Today, the astronauts use commercially available digital cameras and recorders, housed onboard the International Space Station (ISS). Astronauts spend some 30 minutes each day taking photographs requested by scientists.

A UNIQUE VIEW

Astronauts have some advantages over the satellites that record Earth's surface as they orbit. Most satellites look straight down, but astronauts can take pictures in any direction, as well as swap lenses and track curious features. They can see phenomena such as thin dust clouds that are only visible when viewed at an angle. Astronauts can also look out for things such as Sun glint—sunlight reflected off ocean water—which reveals unexpected currents. Photographs can also reflect individual taste. American astronaut Don Pettit, for example, has taken many spectaular images of Earth's cities at night.

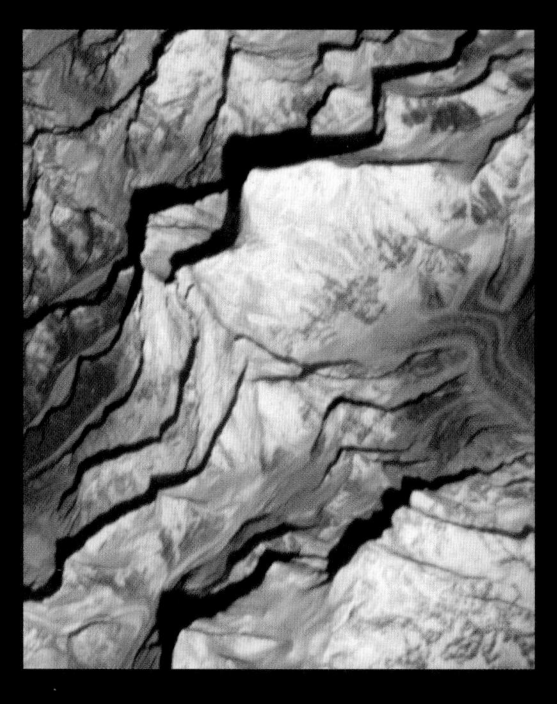

MOUNT EVEREST FROM THE ISS
The Sun was rising over Tibet when American astronaut Dan Bursch spotted the Himalayas 220 miles (350km) below the ISS, on March 20, 2002. He captured the view on a digital camera fitted with an 800mm lens and e-mailed it back to Earth. Everest is the bright peak above center, casting a shadow on the mountains beyond.

> "MOUNT EVEREST SEEMED TO JUMP OUT AT ME!"
>
> **DAN BURSCH**, AMERICAN ASTRONAUT, MARCH 2002

SHUTTLE OVER THE PACIFIC
This view from a Space Shuttle is of the cloud-speckled Pacific Ocean. The water glints in the sunshine, and a dark plume of ash rising from Mount Pagan, one of a pair of volcanoes that constitute Pagan Island, stretches over the water (center left).

PHOTOGRAPHY FROM THE ISS

Ten windows on the ISS allow astronauts to look down on Earth; another six or so point elsewhere. Astronauts take pictures from all the windows, but two are used in particular. The 20in- (50cm-) wide Destiny Laboratory window was specifically installed for photography. It is four panes thick, transmits 98.5 percent of the light hitting it, and creates no distortions when a telescopic lens is pointed through it. Earth observation photographs are routinely taken through it. The second vantage point for photography is the Cupola. Installed on the ISS in February 2010, it is a European Space Agency-built module with six side windows and a seventh on top. Primarily used as a control area for directing operations outside the ISS, it is also used for photography

CUPOLA ON THE ISS
Japanese astronaut Soichi Noguchi takes a photograph through a Cupola window on February 18, 2010. Features photographed that day included the Arabian Desert and Lake Pukaki, New Zealand. Noguchi regularly took photographs from here and put them on his web page for anyone to see.

EARTHKAM

One digital camera currently onboard the ISS takes pictures as part of an educational program. Called EarthKAM, it is used to take photographs of locations requested by students from around the world. The scheme has been in operation since 1995, first as KidSat, and then as EarthKAM since 1998. It has been used on the ISS since the first crew went onboard in 2001.

Installation of EarthKAM
The first EarthKAM flew onboard the Space Shuttle *Endeavour* in January 1998. Here, American astronaut Bonnie Dunbar fits it into position while in Earth orbit.

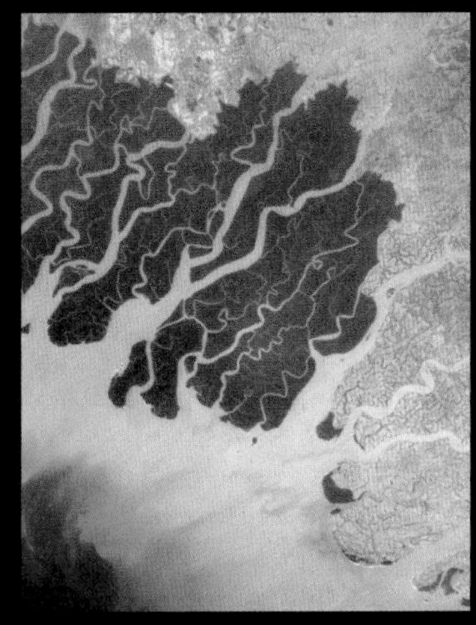

EarthKAM view of river delta
This view of the Ganges River delta was taken from the Space Shuttle *Atlantis*. The dark part is Sundarbans, a vast wildlife preserve.

EXPLORING EARTH FROM SPACE

Since the early 1960s, hundreds of satellites have been launched into orbit around Earth specifically to image, measure, and monitor aspects of the planet's land, sea, ice, snow, and atmosphere.

EARTH-OBSERVATION SATELLITES

The main purpose of many of these satellites has been to monitor and investigate weather phenomena, including warm and cold fronts, tropical storms, dust storms, snow storms, and the state of the sea. Other satellites are used to detect and measure longer-term changes in such things as forest cover, air and sea pollution, and the extent of Arctic ice. Still others are designed to measure changes in the shape of the oceans and land, or to produce improved maps.

Satellite instruments range from passive sensors, which simply detect light or heat, to active sensors, including radar and laser devices, that emit energy and record the reflected or back-scattered response. Over the decades, these instruments have improved greatly in both accuracy and spatial resolution (the minimum size of objects they can image). More than 150 observation satellites now orbit Earth (see pp.18–19), many carrying several highly sophisticated sensors.

HIGH-TECH WEATHER-MONITORING
One of the earliest applications of satellite technology was monitoring changes in the weather. Modern weather-satellites use radar to make 3D images of rainstorms, like this one recorded over Texas in 2004.

THE WHOLE EARTH IMAGED IN ONE DAY
Modern satellites gather data about Earth's land, oceans, atmosphere, snow, and ice continuously. This composite image is based largely on data collected in the course of just one day—July 11, 2005—by NASA's Terra and Aqua satellites.

EARLY SATELLITE IMAGING

The first satellite to transmit pictures of Earth from space was NASA's Explorer 6 in 1959, and the first television footage of weather patterns from space came from the same agency's TIROS-1 (Television Infrared Observation Satellite 1) in 1960. These early satellites used relatively simple cameras, but the 1960s and 1970s saw the key development of satellites carrying multispectral sensors. These detect radiation from Earth's surface over several wavelength bands, including infrared radiation and microwaves as well as visible light. One such satellite was Landsat 1. Launched in 1972, this was the first in a series of Landsats that have imaged continents and coasts at resolutions useful for many different practical purposes, such as studying land cover, assessing the effects of human activity, and mapmaking.

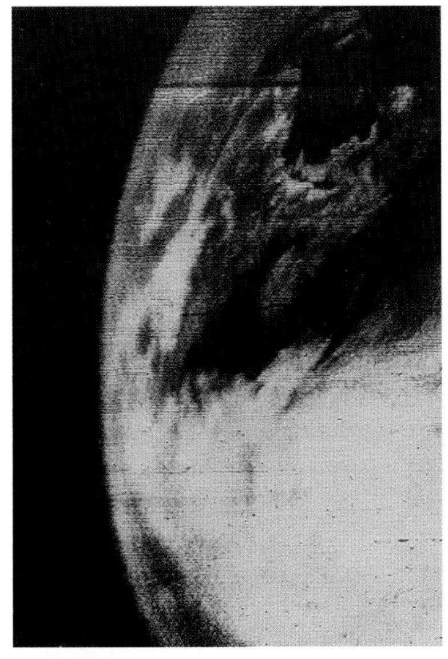

FIRST TELEVISION SPACE IMAGE
Taken by TIROS-1 on April 1, 1960, this was the first ever television picture from space. It shows clouds over the Atlantic Ocean (bottom) and part of Canada (top).

EARLY LANDSAT 1 THERMAL IMAGE
Landsat 1 took this infrared image of part of Utah, on August 7, 1972, 15 days after its launch. Vegetation shows red, desert gray, and the Great Salt Lake black.

SATELLITE IMAGING TODAY

NASA's flagship Earth-observation satellite, Terra, has covered the whole globe every one to two days since 2000. Its instruments include MODIS (Moderate Resolution Imaging Spectroradiometer), which measures such things as cloud properties; and ASTER (Advanced Spaceborne Thermal Emission and Reflection Radiometer), which maps land temperature, elevation, and reflectance. Sister to Terra is Aqua, which studies Earth's water cycle. The European Space Agency's Envisat, launched in 2002, also has multiple sensors.

MODIS AT WORK
MODIS, on NASA's polar-orbiting Terra and Aqua satellites, scans Earth in swathes, each one about 1,400 miles (2,300km) wide.

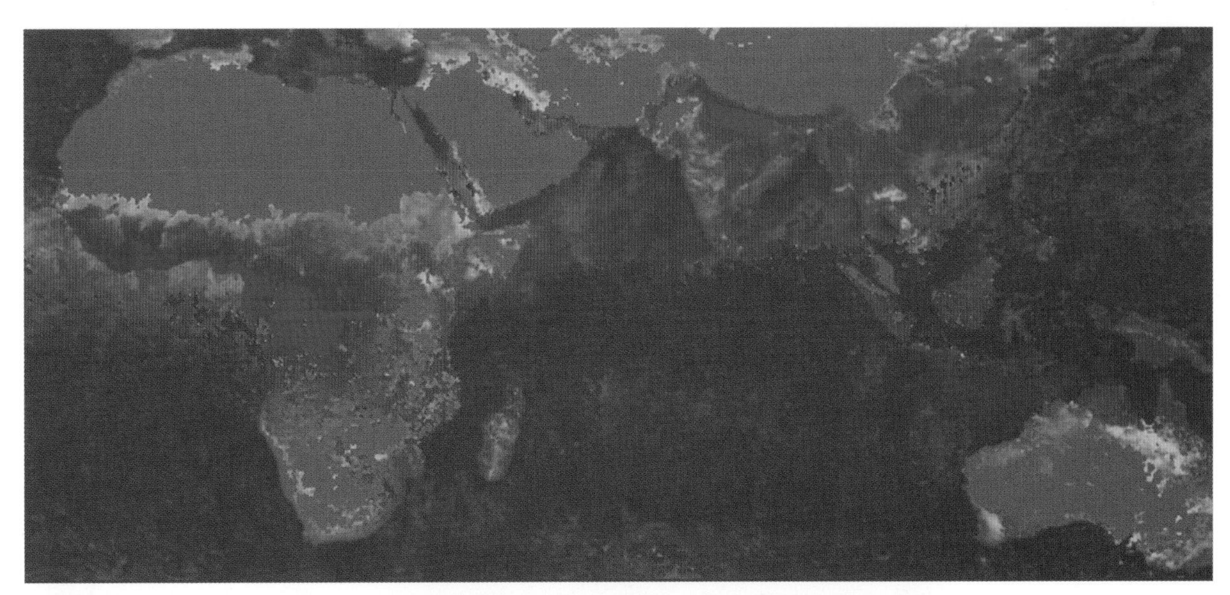

PARTICLE MAPPING
In December 2004, Terra's MODIS imaged airborne particles over five days. Red dots are fine smoke; gold dots are coarser desert dust and sea salt.

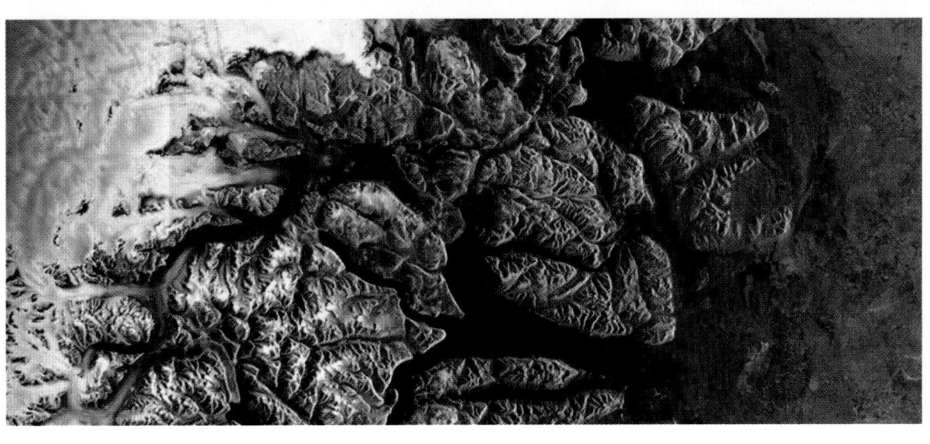

COMBINING IMAGES
Modern uses of satellite technology often involve combining images. Here, when three Envisat images of Greenland's coast are combined, differences show up as colors.

EARTH'S OCEANS

Oceans cover 71 percent of Earth's surface. As well as being a major source of food and of great importance to the transportation of goods, they have a huge influence on climate and weather patterns. This influence comes from the effects of warm and cold ocean currents, from the oceans' ability to absorb carbon dioxide (a greenhouse gas) from the atmosphere, and from the effects of the periodic oceanic/climatic disturbance in the Pacific known as El Niño. Other ocean-related phenomena, such as tides and changes in sea level, are of profound importance to the millions of people who live on or near coasts.

SATELLITE MONITORING

A great amount of ocean monitoring is now done by satellites. In a joint NASA and French Space Agency venture, Jason 1 and Jason 2 use radar-based instruments to monitor the exact shape of the ocean surface, as well as average wave heights. Scientists use the data to study tides, sea-level change, and changes associated with winds (important in providing shipping warnings). Warm water expands, so measuring small variations in ocean shape is also a way of indirectly measuring temperature variations at particular spots. Other satellites directly detect surface temperatures from the amount of infrared radiation (heat) given off. Either way, satellites monitor how warm and cool masses of sea water move in the oceans, which allows the study of large-scale movements of heat energy. This type of data is of great value in climate modeling. Satellites also monitor concentrations of phytoplankton. These microscopic plants form the base of the oceanic food web, influence the amount of carbon dioxide in the atmosphere, and are sensitive to environmental changes.

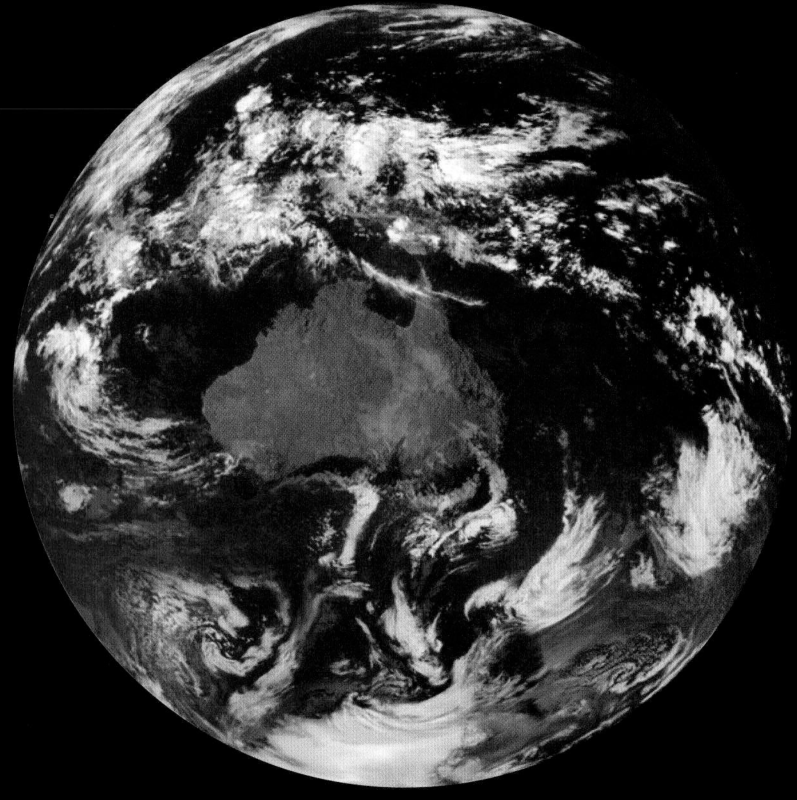

THE EXTENT OF THE WORLD'S OCEANS
The predominance of oceans on Earth's surface is clear in this composite image, which combines land and sea data from an instrument called SeaWIFS on GeoEye's OrbView-2 satellite with cloud textures from an instrument on NASA's Terra satellite.

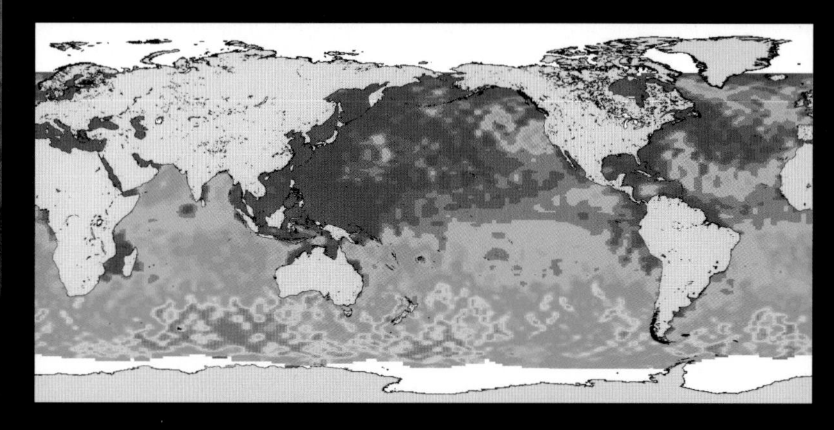

AVERAGE WAVE HEIGHTS
The data for this global map of average wave heights in July 2008 are from Jason 2. Strong winds whipped up particularly big waves in the Southern Ocean that month.

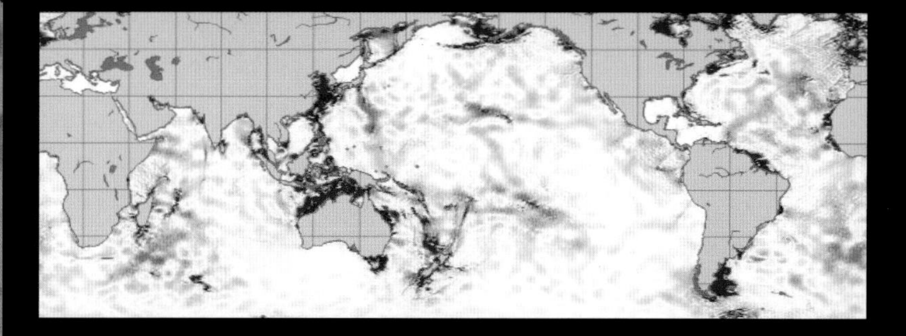

TIDAL-ENERGY LOSS
Tides raise and lower sea levels. Satellite sea-level data show that 30 percent of tidal energy is lost on deep-ocean seamounts and ridges, the rest on shallow sea floors.

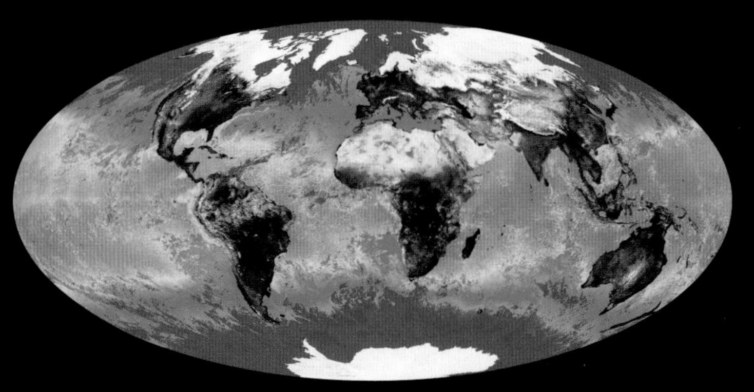

SEA-SURFACE TEMPERATURE
Terra's MODIS (Moderate Resolution Imaging Spectroradiometer) provided the data for this map of sea-surface temperatures in March 2000. Red areas are warmest, violet coldest.

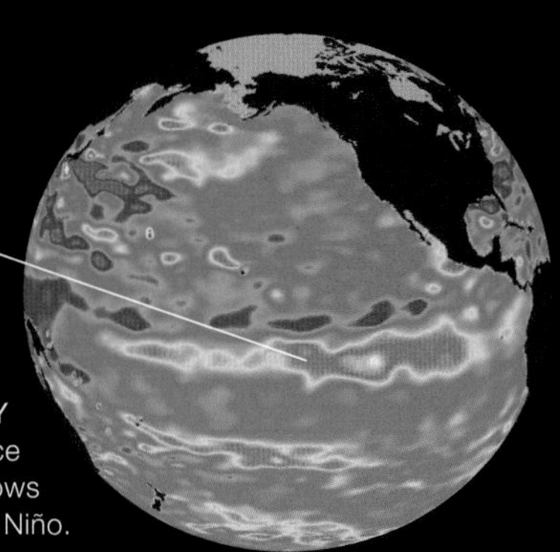

EL NIÑO ANOMALY
This Jason 1 map of sea surface temperatures in November 2006 shows a warm patch characteristic of El Niño.

PHYTOPLANKTON BLOOM
Envisat's MERIS (Medium Resolution Imaging Spectrometer) detects phytoplankton from the color they tint the sea. The European Space Agency satellite imaged this bloom in the Barents Sea in August 2009.

SHIP TRACKS SEEN FROM SPACE
Taken by NASA's MODIS satellite, this image shows streaky clouds above the Pacific Ocean. The clouds form as water molecules condense around exhaust particles emitted by ships. All such particles contribute to cloud-formation, but they are only clearly visible in relatively uniform air, such as that above the oceans.

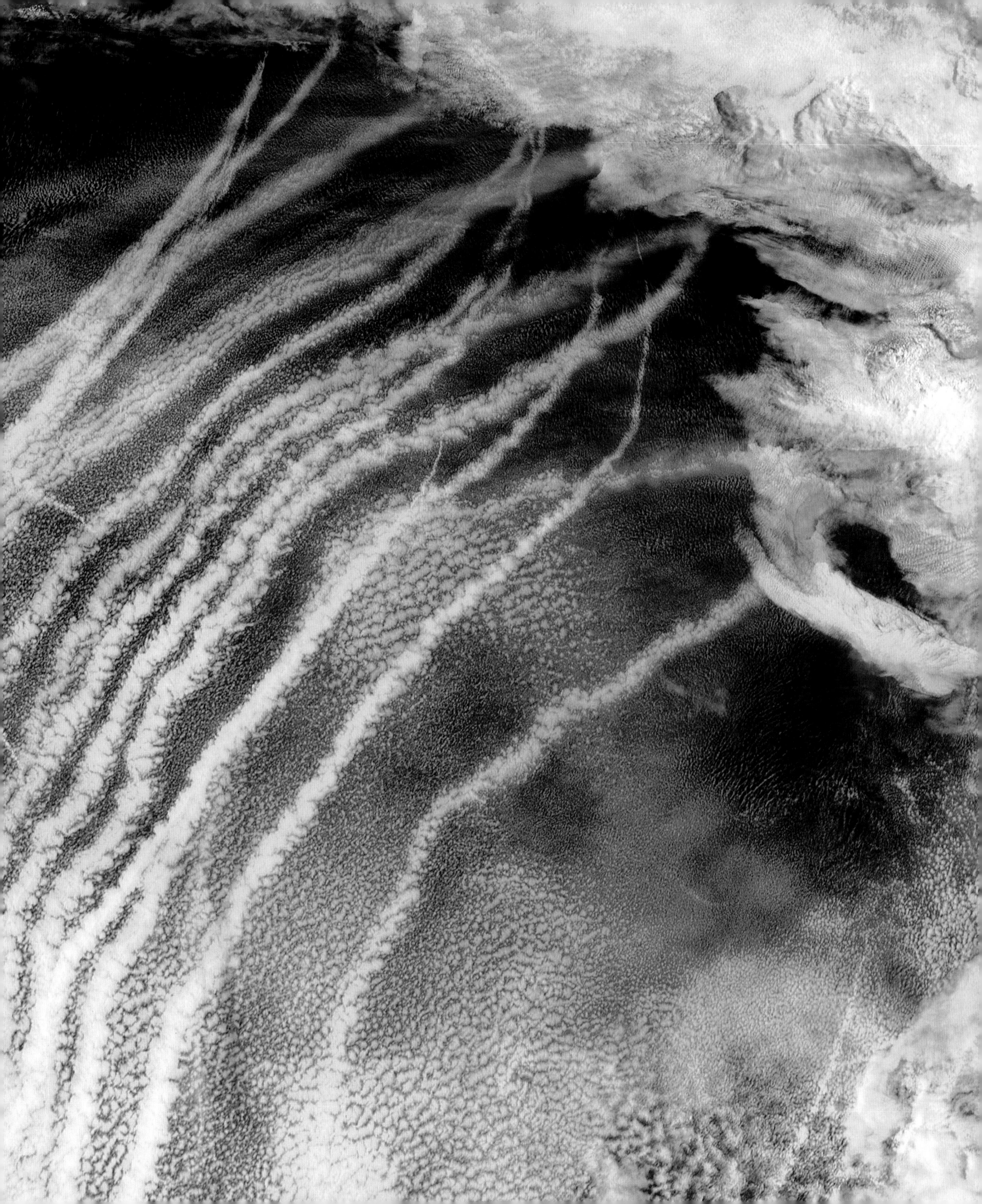

EARTH'S LAND

Almost 30 percent of Earth's surface is land. This land is far from smooth and includes a wide variety of physical forms and habitats, ranging from vast mountainous regions to extensive grasslands, deserts, forests, and tundra. Most mountain ranges are the result of continental plates pushing together to cause uplift. The erosional power of streams and rivers, and of slowly moving ice in the form of glaciers, has cut through and shaped these uplifted areas, producing many different landscapes. In lower-lying areas, deposition has produced features such as river floodplains and coastal deltas. Among other distinctive features on Earth's land surface are vast, thick deposits of ash or lava from volcanic eruptions. Increasingly, there are also many marks left by human activity, such as sprawling cities, road networks, and patchworks of highly cultivated fields.

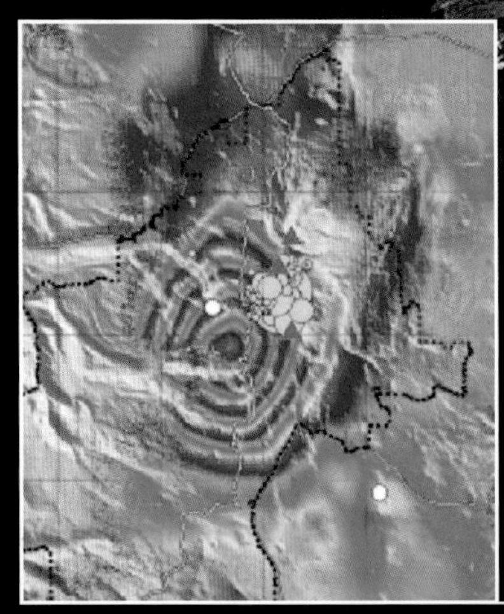

BULGING GROUND IN OREGON
A bull's-eye pattern near the Three Sisters volcanoes (shown as red triangles) is where intruding magma in 1996–2000 raised the land by around an inch, as revealed by satellite radar monitoring.

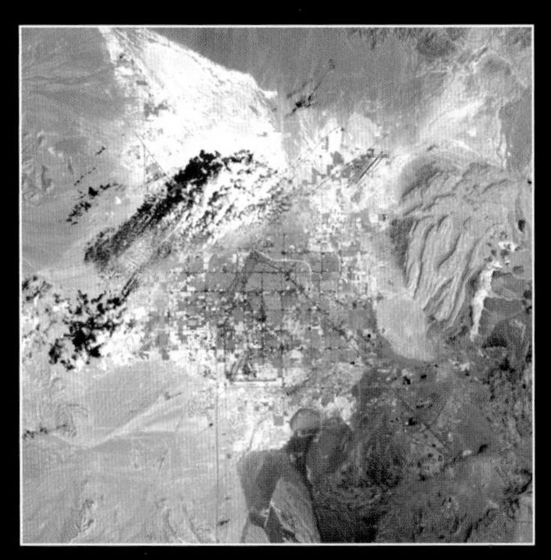

1984

2009

DESERT CITY SPRAWL
With its flourishing gambling and tourism industries, Las Vegas, Nevada is one of the fastest-growing metropolitan areas in the US. These two images, obtained by NASA's Landsat 5 satellite, show its growth between 1984 (above) and 2009 (above right). In both, roads and buildings appear blue-gray or blue-green, parks and golf courses are bright green, and the surrounding desert is brown or beige.

THERMAL FIRE IMAGING
This false-color infrared image, taken by the ASTER (Advanced Spaceborne Thermal Emission and Reflection Radiometer) on NASA's Terra satellite, shows a huge forest fire raging in California in October 2003. The fire front is the boundary between the green area (intact forest) and the crimson area (burned forest). Smoke shows blue.

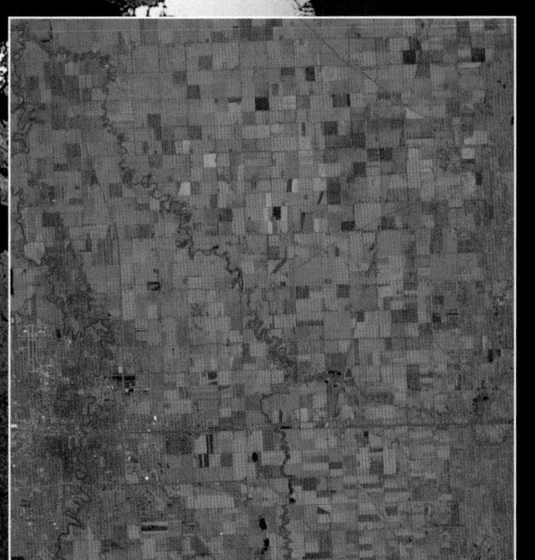

SATELLITE FARMING
Farmland in Minnesota is the subject of this false-color infrared image taken by Landsat 5 in 2009. Arable farms here have hundreds of huge fields covering thousands of acres, so farmers use such images to monitor healthy crops (red/pink), harvested fields (brown) and flooded fields (black).

SATELLITE OBSERVATION

All of Earth's land features, large and small, are easily observed and studied from space. Using satellites, we can monitor unfolding natural disasters such as volcanic eruptions, floods, or major forest fires. From space, we can also spot ancient faults in Earth's crust and craters made millions of years ago by asteroid impacts. Another use of satellites is to monitor how a combination of human activity and natural processes are affecting, for example, the extent of forests and deserts. It is even possible to monitor cloud-covered areas, since radar can penetrate cloud to measure the exact heights of features. By repeatedly bouncing radar signals off a land surface, scientists can detect small changes in ground shape over time.

NORTH AMERICA
Geographically diverse, North America has a range of features that can be imaged by satellites. NASA used several images from different satellites to form this overall picture of the continent.

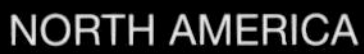

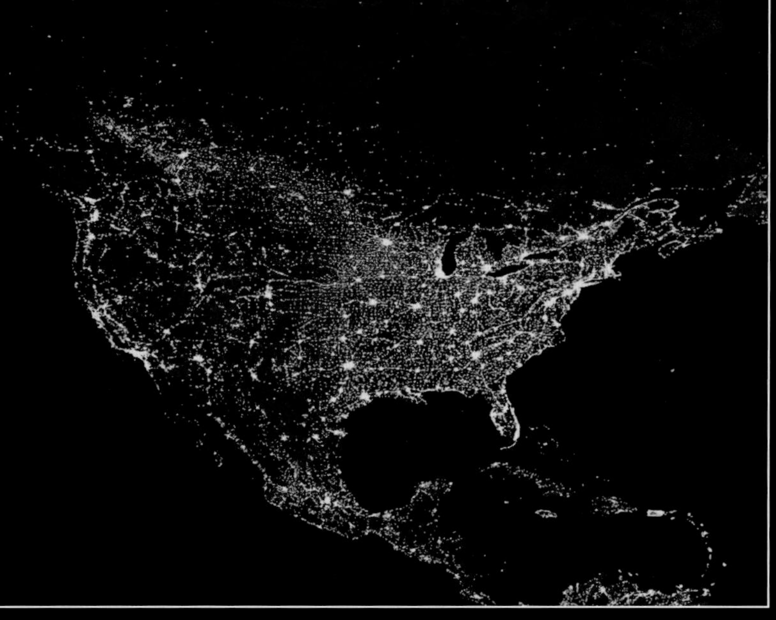

NIGHT LIGHTS
Hundreds of weather-satellite images were combined to create this picture of North America at night in 2001. The lights show the distribution of major population centers.

RIVER SEDIMENT
This image, acquired by the MODIS (Moderate Resolution Imaging Spectroradiometer) on NASA's Terra satellite in 2001, shows the brown, silt-laden water of the Mississippi River mixing with the clear, dark blue waters of the Gulf of Mexico. The Mississippi deposits about 500 million tons (550 million tonnes) of silt into the gulf each year.

CHANGES IN THE GREENLAND ICE SHEET
This ICEsat image shows changes in the Greenland ice sheet from 2003 to 2006. Pink indicates thickening, blue and purple thinning. ICEsat operated from 2003 until 2009. Its replacement, ICEsat-2, is due for launch in 2015.

EARTH'S ICE AND SNOW

Satellite observation enables scientists to monitor how global ice and snow coverage varies both seasonally and year on year. A persistent trend of year-on-year changes may be an indicator of climate change.

Sea ice—sea water that has frozen at the surface—covers a large but seasonally fluctuating area of the Arctic Ocean. In the winter, existing Arctic sea ice thickens and new sea ice forms; in the summer, it thins and melts. Extensive monitoring by instruments on satellites such as the European Space Agency's Envisat and NASA's ICEsat (Ice, Cloud, and land Elevation Satellite) shows that Arctic sea ice has been both thinning and decreasing in area for several years.

ICEsat and such instruments as ASTER (Advanced Spaceborne Thermal Emission and Reflection Radiometer) on NASA's Terra satellite and ASAR (Advanced Synthetic Aperture Radar) on Envisat have also been used to monitor the ice sheets covering the landmasses of Greenland and Antarctica, glaciers, and ice shelves—glaciers that extend over the sea. Again, the general picture has been one of overall loss of ice mass and retreat, even though some parts of the Greenland ice sheet, for example, are actually thickening.

Other ice-related satellite applications include studying the disintegration of Antarctic and other ice shelves, and tracking the movement of large icebergs—chunks of ice that break off the shelves—in areas where they might be a danger to shipping.

Finally, the thickness of snow falling over large areas around the world can now be accurately measured from space, as can its rate of melting. By combining such satellite measurements, scientists can predict river flows.

SNOW REFLECTIVITY

Because it reflects up to 80 percent of the sunlight and other radiation falling on it, snow cover affects the overall amount of heat that Earth absorbs from the Sun. One way that satellite-borne instruments can help study this is by measuring the different properties of snow that affect its reflectivity. For example, large grains and impurities both reduce reflectivity. Scientists used data from Terra's MODIS (Moderate Resolution Imaging Spectroradiometer) to produce these color-coded images, which show the grain size of Arctic snow and its impurity content.

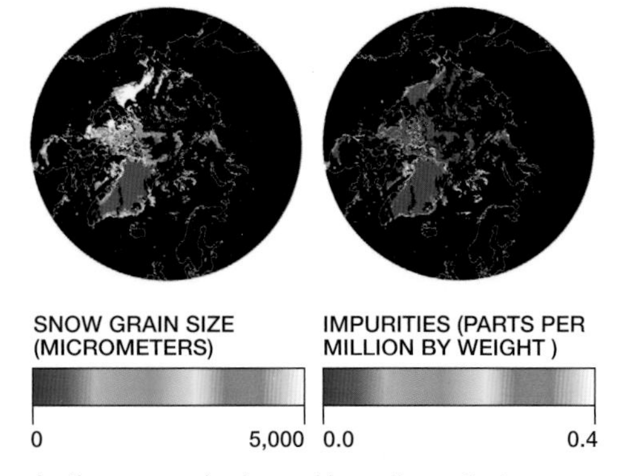

SNOW GRAIN SIZE (MICROMETERS)
0 5,000

IMPURITIES (PARTS PER MILLION BY WEIGHT)
0.0 0.4

Arctic snow grain size and impurity content
Preliminary studies of data from Terra's MODIS show that in the Arctic, average snow grain size (left) is smallest at high altitudes, such as at the top of the Greenland ice sheet. The impurity content of snow (right) is highest on sea ice near continental coasts.

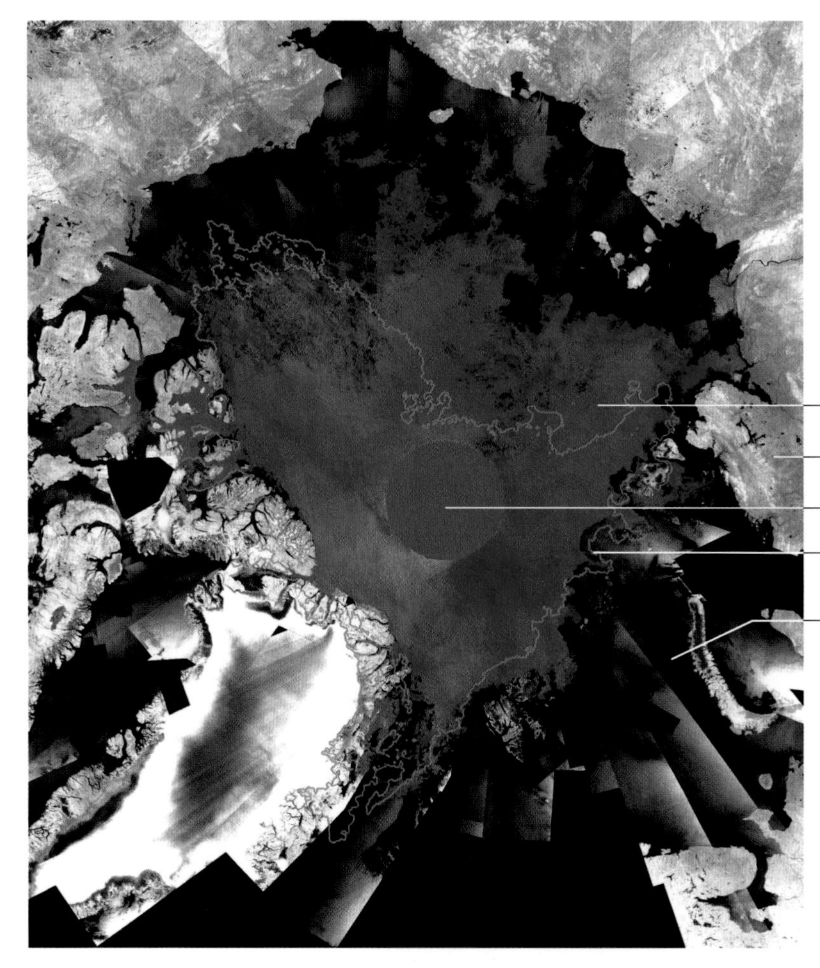

ARCTIC SEA ICE
This Envisat ASAR image shows the extent of Arctic sea ice (blue) in August 2008 compared with its record measured minimum (red outline) in September 2007. The Arctic is one of the least accessible areas on Earth, so obtaining such measurements was difficult before satellites.

HIMALAYAN GLACIER RETREAT
This June 2002 Terra ASTER image shows that the ends of several glaciers in the Bhutan section of the Himalayas have thinned or melted to form glacial lakes (blue), with retreat of the glaciers back up their valleys.

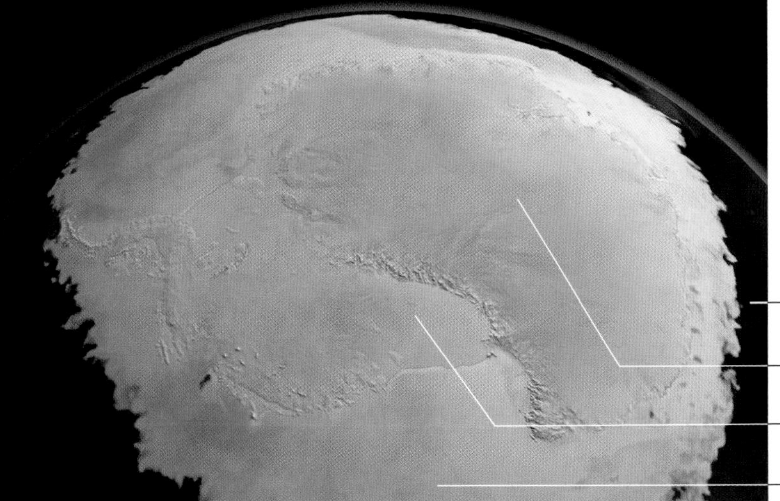

ANTARCTIC ICE
Scientists used data collected over four years by instruments on several satellites to produce this composite image of Antarctica. By animating series of such images, they can easily see seasonal and year-on-year changes.

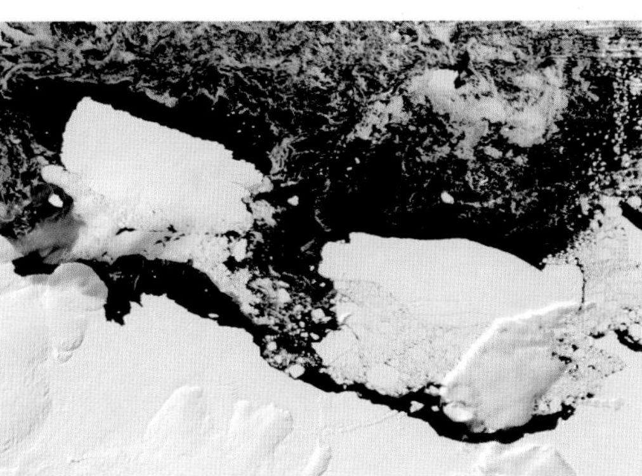

ANTARCTIC ICEBERGS
NASA's Aqua satellite imaged these two icebergs, each the size of a small country, with its MODIS (Moderate Resolution Imaging Spectroradiometer) in March 2010. The iceberg on the right, which formed in 1987, broke the other off when it hit the end of the Mertz glacier (bottom).

EARTH'S ATMOSPHERE

Earth's atmosphere is 78.1 percent nitrogen and 20.9 percent oxygen, with traces of other gases such as argon, carbon dioxide, water vapor, methane, and ozone. The last four act as greenhouse gases, increasing the temperature at ground level to much higher than it would otherwise be. All weather occurs in the atmosphere's lowest layer, the troposphere, which extends to 10 miles (16km) or so above the equator and some 5 miles (8km) above the poles. Above that, up to about 30 miles (50km), is the more stable stratosphere, or upper atmosphere. Beyond that are layers of very thin (low density) air.

SATELLITE MONITORING

The only visible atmospheric gas is water vapor, when it forms clouds of tiny droplets or ice crystals. Satellites monitor cloud cover and weather events that produce distinctive cloud patterns, such as hurricanes. From the movement of clouds in the troposphere, satellite data can also be used to calculate the speeds and directions of winds. Instruments such as MODIS (Moderate Resolution Imaging Spectroradiometer) on NASA's Terra and Aqua satellites also monitor sand and dust storms, smoke pollution, and volcanic ash plumes. Others study invisible gases. For example, Aqua's AIRS (Atmospheric Infrared Sounder) monitors carbon dioxide in the mid-troposphere, while OMI (Ozone Monitoring Instrument) on NASA's Aura satellite looks at ozone levels in the stratosphere. On the European Space Agency's Envisat, SCIAMACHY (Scanning Imaging Absorption Spectrometer for Atmospheric Chartography) measures stratospheric methane.

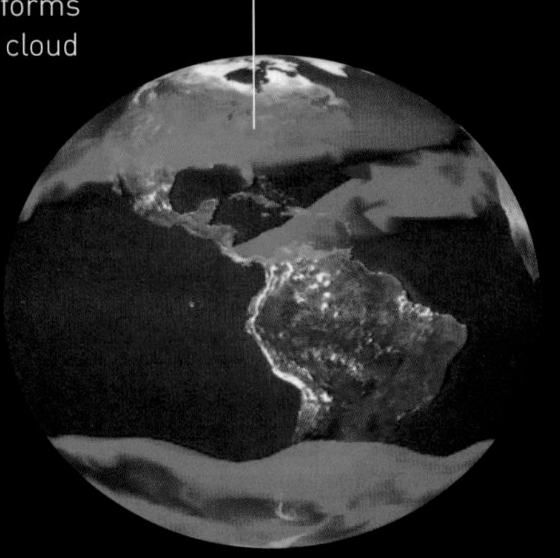

Mid-latitude methane
Methane concentrations in the northern hemisphere are high at mid-latitudes.

STRATOSPHERIC METHANE
This map of methane over the Americas is based on 1999 data from NASA's now-decommissioned UARS (Upper Atmosphere Research Satellite).

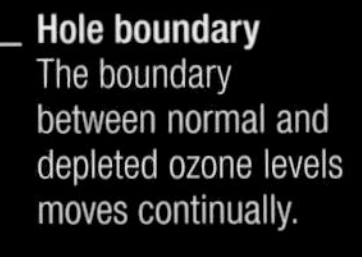

Hole boundary
The boundary between normal and depleted ozone levels moves continually.

Normal levels
Green areas denote normal or average ozone levels, yellow areas slightly higher.

Depleted ozone
Purple areas are where there is least ozone.

HOLE IN THE OZONE LAYER
In September 2006, OMI on NASA's Aura satellite observed the hole in the ozone layer over Antarctica at its largest ever recorded extent: 10.6 million square miles (27.5 million square km). The hole is caused by pollutants, especially chlorofluorocarbons (CFCs). Now CFCs are banned, ozone levels should recover.

STRATOSPHERIC OZONE LEVELS
HIGHEST LOWEST

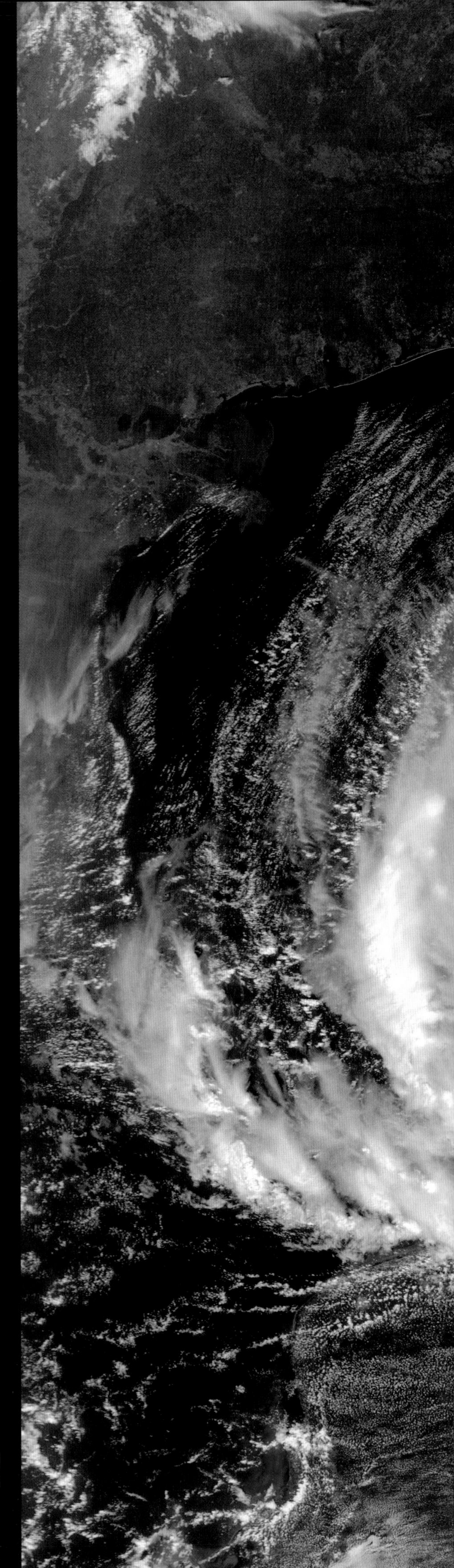

HURRICANE RITA
Hurricane Rita in September 2005 was the strongest tropical cyclone ever observed in the Gulf of Mexico. MODIS on NASA's Aqua satellite captured this image of the storm as it plowed northwestward into the Gulf.

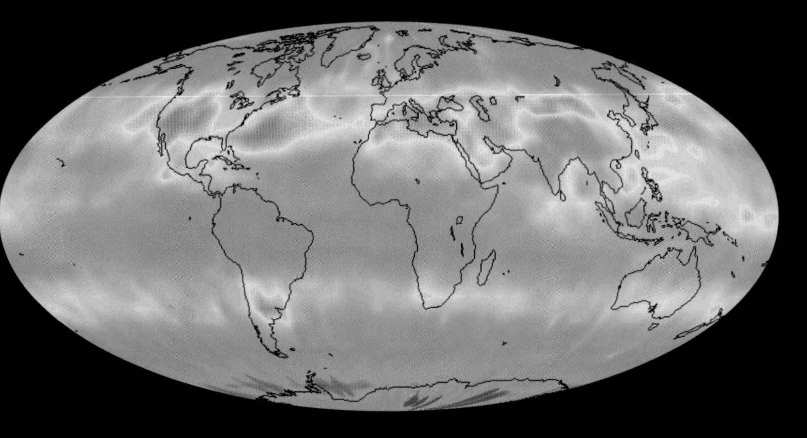

CARBON DIOXIDE
This Aqua AIRS map of the July 2003 global distribution of carbon dioxide at 5 miles (8km) shows the highest concentrations at northern-hemisphere mid-latitudes.

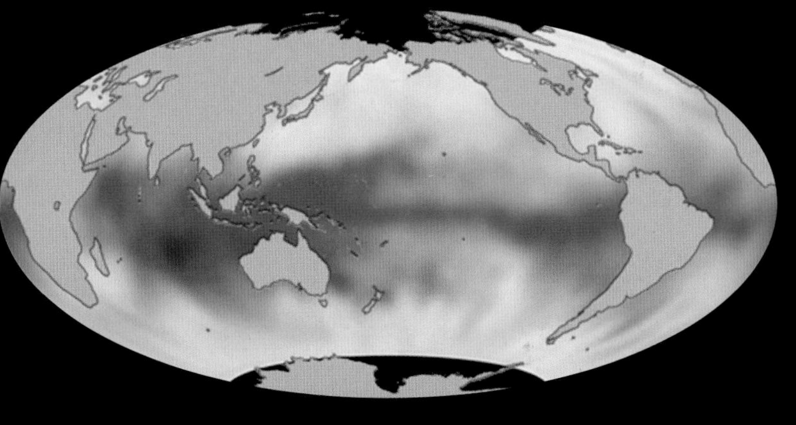

ATMOSPHERIC WATER VAPOR
This map of the global atmospheric water-vapor content in June 2008 is based on data from a sensor on the joint NASA/French Space Agency satellite Jason 2.

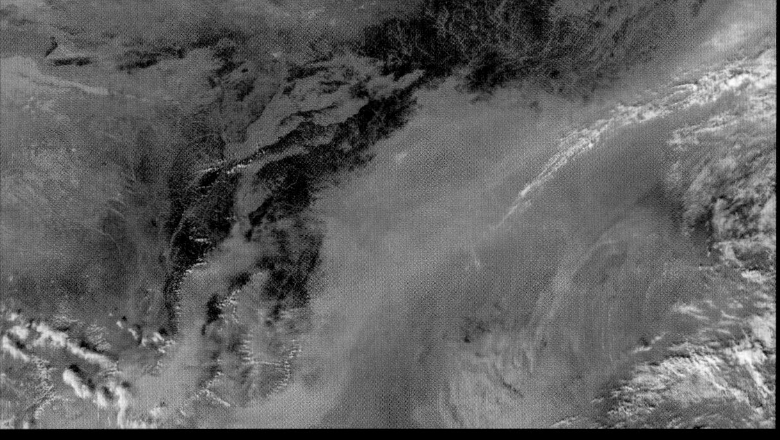

SMOG POLLUTION OVER CHINA
MODIS on NASA's Terra satellite imaged this band of thick, gray smog over part of eastern China in October 2009. To the right of center, 90 miles (150km) southeast of Beijing, is the Bohai Sea.

SURROUNDING EARTH

The boundary between the atmosphere and space lies at an altitude of approximately 60 miles (100km). Here, conditions are determined by the interaction of three physical systems: remnants of Earth's upper atmosphere, charged particles streaming from the Sun, and Earth's powerful magnetic field.

ABOVE EARTH

Scientists divide Earth's atmosphere into a series of layers. The lowest is the troposphere, which contains 90 percent of all atmospheric gases in a layer up to 10 miles (16km) deep. This region gives rise to nearly all of Earth's weather and produces a wide variety of climates as it churns and transports heat and water vapor around the planet.

Ascending through the troposphere, air pressure and temperature both reduce rapidly, until, at the boundary with the stratosphere, the atmospheric temperature is around -58°F (-50°C). Beyond this point, conditions are roughly uniform all around the planet, and vertical changes in the atmosphere become much more significant. The density of gases falls by a factor of 10 for each 10 miles (16km), but the temperature varies in a more complex way, rising through the stratosphere, then dropping again in the mesosphere, before rising once again in the thermosphere. The atmospheric composition also changes, since atoms and molecules of heavier gases remain close to Earth, while lighter ones, such as hydrogen and helium, rise higher.

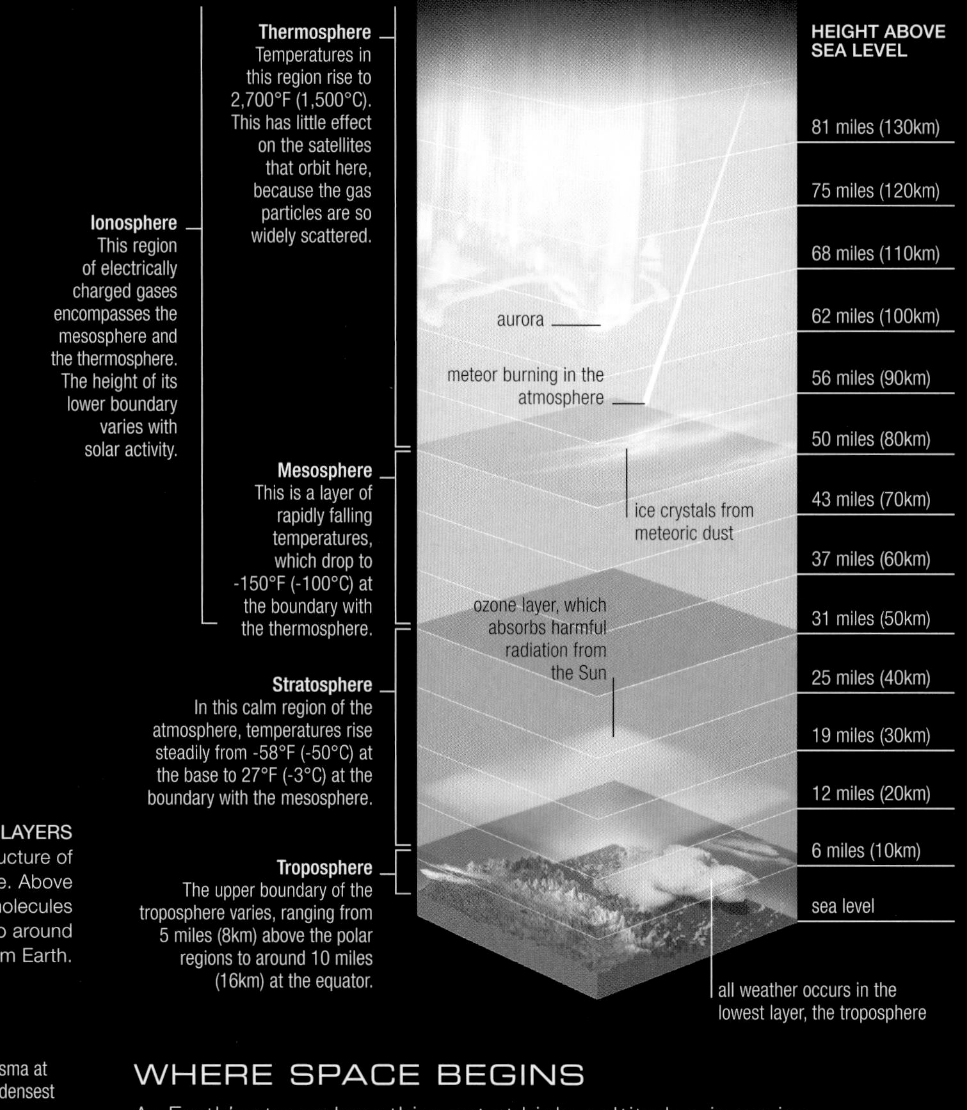
Thermosphere
Temperatures in this region rise to 2,700°F (1,500°C). This has little effect on the satellites that orbit here, because the gas particles are so widely scattered.
Ionosphere
This region of electrically charged gases encompasses the mesosphere and the thermosphere. The height of its lower boundary varies with solar activity.
Mesosphere
This is a layer of rapidly falling temperatures, which drop to -150°F (-100°C) at the boundary with the thermosphere.
Stratosphere
In this calm region of the atmosphere, temperatures rise steadily from -58°F (-50°C) at the base to 27°F (-3°C) at the boundary with the mesosphere.
Troposphere
The upper boundary of the troposphere varies, ranging from 5 miles (8km) above the polar regions to around 10 miles (16km) at the equator.
aurora
meteor burning in the atmosphere
ice crystals from meteoric dust
ozone layer, which absorbs harmful radiation from the Sun
all weather occurs in the lowest layer, the troposphere
HEIGHT ABOVE SEA LEVEL
81 miles (130km)
75 miles (120km)
68 miles (110km)
62 miles (100km)
56 miles (90km)
50 miles (80km)
43 miles (70km)
37 miles (60km)
31 miles (50km)
25 miles (40km)
19 miles (30km)
12 miles (20km)
6 miles (10km)
sea level

ATMOSPHERIC LAYERS
This illustration shows the detailed structure of Earth's atmosphere up to the thermosphere. Above this is a layer of scattered gas atoms and molecules called the exosphere, which stretches out to around 60,000 miles (100,000km) from Earth.

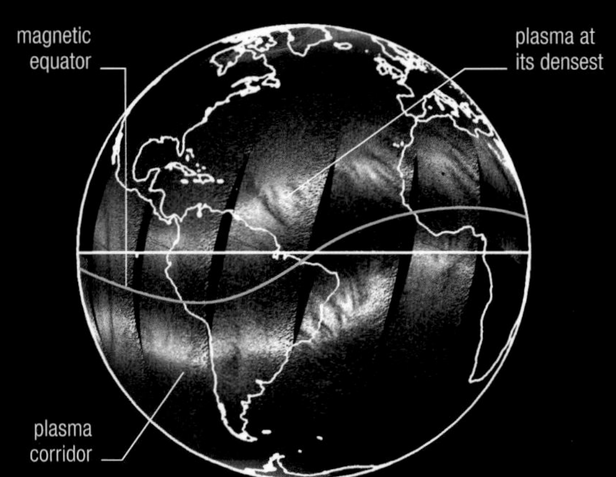
magnetic equator
plasma at its densest
plasma corridor

IONOSPHERE FROM SPACE
NASA's TIMED satellite uses ultraviolet and infrared instruments to observe the changing ionosphere, revealing the influences of both Earth's climate and the Sun. The plasma is densest at the tropics.

WHERE SPACE BEGINS

As Earth's atmosphere thins out at higher altitudes, incoming ultraviolet and X-ray radiation from the Sun splits gas atoms apart to create a mix of negatively charged electrons and positively charged ions. This gaslike state of matter is called plasma, and the atmospheric layer it dominates, ranging from about 30 to 600 miles (50 to 1,000km) in altitude, is the ionosphere. According to aeronautical authorities, it contains the official "edge of space," known as the Kármán line.

Charged particles in the plasma are influenced by Earth's magnetic field (opposite), creating windlike flows around the planet. The intensity of charge in the ionosphere also varies depending on the intensity of incoming radiation and particles from the Sun, so the structure of the upper atmosphere changes throughout each day, across the seasons, and according to the longer cycles of solar activity.

AURORA BOREALIS
Charged particles pouring into Earth's upper atmosphere above the poles collide with the rarefied gases, energizing them and causing them to glow in different colors. Around the North Pole, the result is a stunning display known as the aurora borealis.

EARTH'S MAGNETOSPHERE

Earth's molten core gives it the strongest magnetic field of any of the rocky inner planets—swirling currents of molten iron create a magnetic field that influences any magnetic or electrically charged object in its vicinity. This field extends to dozens of times the diameter of Earth in all directions, forming a magnetosphere that encounters and deflects the stream of charged particles from the Sun known as the solar wind (see p.89). Many of these particles are trapped in doughnut-shaped regions of the magnetosphere called the Van Allen belts. Prolonged exposure to these regions can be a threat to both spacecraft and astronauts. Particles passing through the field close to the poles interact with the atmosphere to create glowing lights called aurorae.

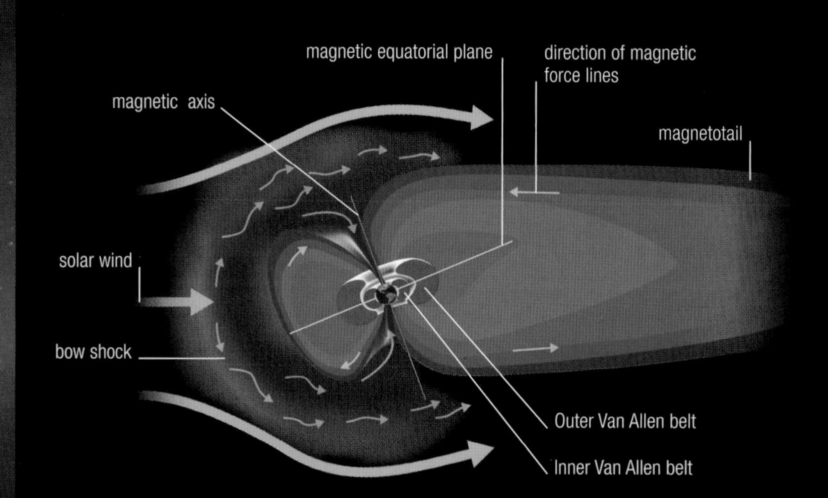

THE SOLAR WIND AND MAGNETOSPHERE
The illustration shows the teardrop-shaped structure of Earth's magnetosphere, distorted by its interactions with the solar wind. Most solar-wind particles flow around the magnetosphere, but those that run straight into it are slowed down at a region called the bow shock.

STUDYING THE MAGNETOSPHERE

The existence of a complex magnetic environment around Earth was largely unsuspected until the launch of Explorer I, the first US satellite, in January 1958. This small spacecraft, which carried instruments designed by physicist James van Allen, detected a torrent of electrically charged particles surrounding Earth in the region now known as the Van Allen belts. Later satellites investigated the shape of Earth's magnetic field further and studied its interactions with the solar wind. However, because the magnetosphere cannot be imaged directly, it can only be investigated by satellites reporting on conditions in a particular location—this makes it difficult to map changes in the magnetosphere's structure and to link them to outside influences.

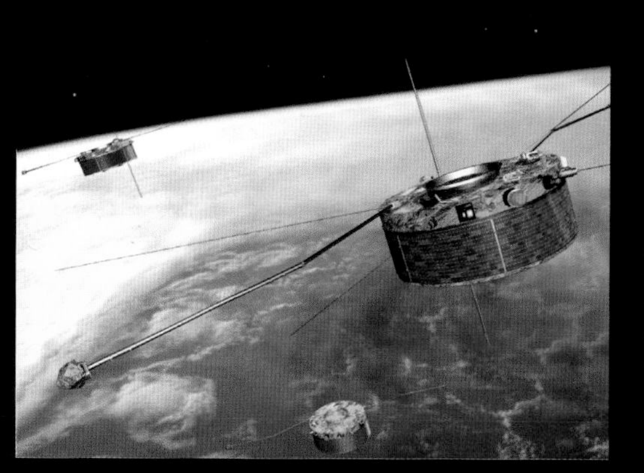

CLUSTER II
Launched in 2000 by the European Space Agency, Cluster II is a group of four identical probes that orbit Earth in formation. By measuring charged particles and magnetic conditions at several locations simultaneously, they have revealed unexpected features in the magnetosphere.

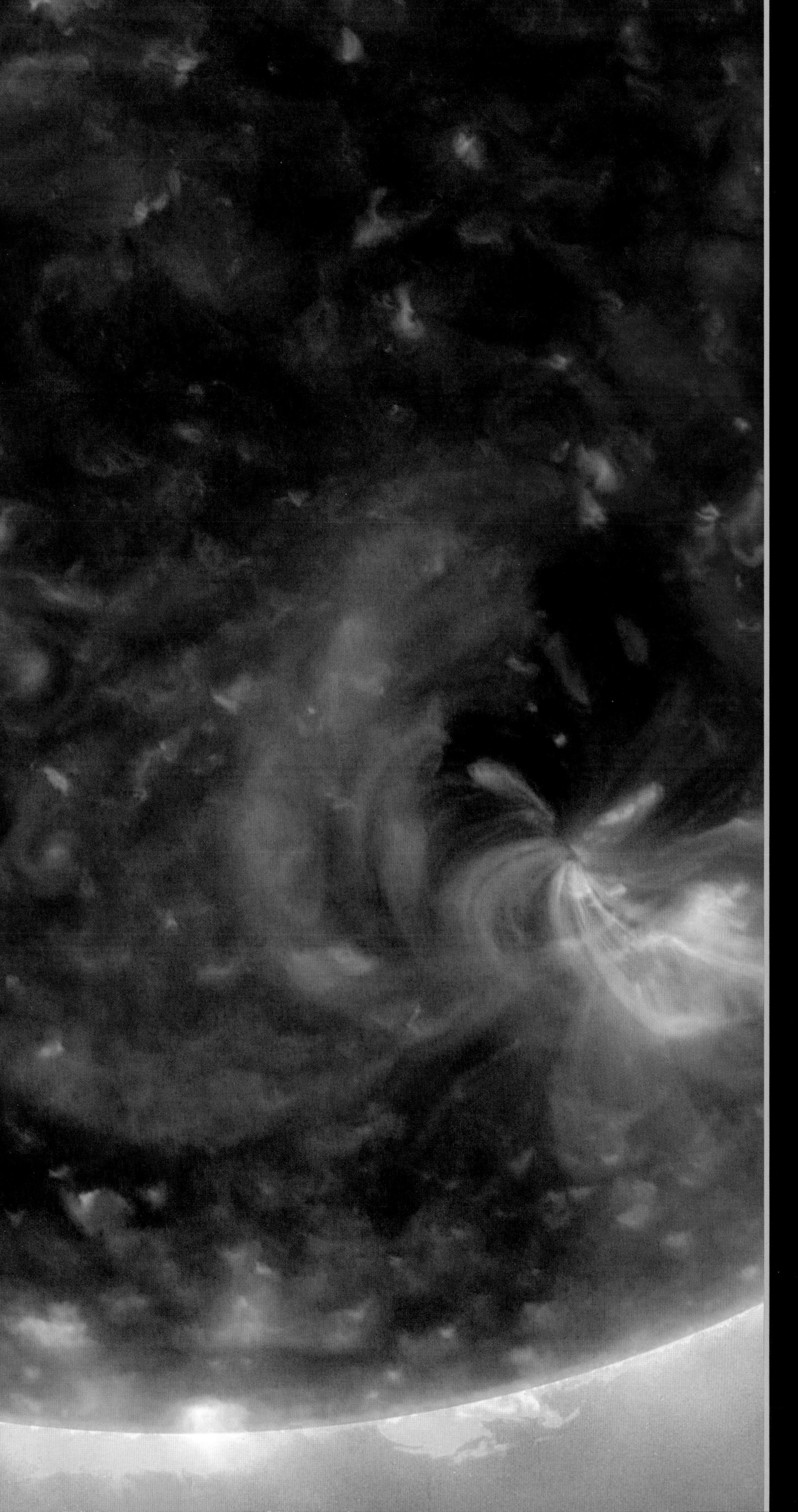

2

NEIGHBORING WORLDS

« The Sun
93.0 million miles (149.6 million km) from Earth

THE SOLAR SYSTEM

The Solar System consists of our local star, the Sun, along with a large number of objects that orbit around it. Just the part containing the planetary orbits is about 6 billion miles (9 billion km) across. Our planet, Earth, orbits within a relatively small central area of this colossal region of space.

THE SUN'S FAMILY

The Sun is by far the largest body within the Solar System, containing 99.85 percent of its mass. Other constituents include: eight planets and their moons; a handful of dwarf planets; and vast numbers of smaller objects that fall into three main groups—asteroids (composed largely of rock and metal), comets (made of ice, rock, and dust), and Kuiper Belt objects (largely ice bodies). Also present are appreciable quantities of rock fragments and particles of dust and ice left in the wake of comets. The planets fall into two distinct groups. These are the relatively small, rocky inner planets (Mercury, Venus, Earth, and Mars—see pp.52–53) and the much larger outer planets (Jupiter, Saturn, Uranus, and Neptune—see pp.124–25).

THE FORMATION OF THE SOLAR SYSTEM

The series of events shown here, called the nebular hypothesis, is the most widely accepted theory for how the Solar System originated. It neatly explains all of the system's most striking features, such as why it is flat and why all of the planets orbit the Sun in the same direction.

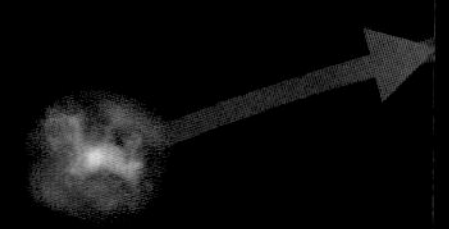

1 Solar nebula forms The starting point for the formation of the Solar System was the solar nebula—a spinning cloud of gas and dust.

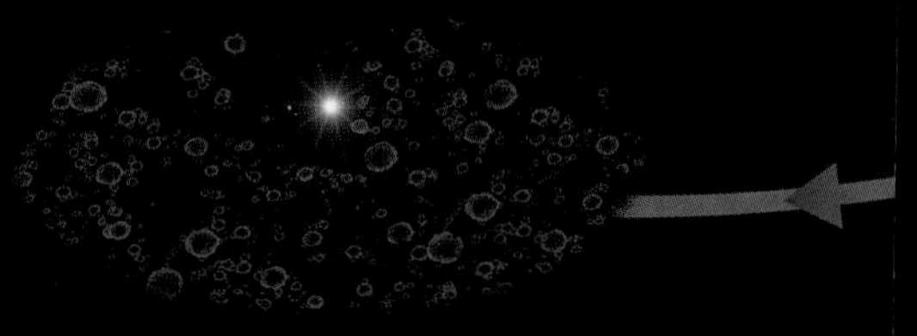

6 Unused debris Some of the leftover planetesimals in the outer part of the disc are thought to have formed a cloud of comets, called the Oort Cloud.

THE ORBITS OF THE PLANETS

The planets and all of the asteroids orbit roughly in the same flat plane (the ecliptic plane) and in a counterclockwise direction, if viewed from above the Sun's north pole. The time it takes a planet to orbit increases with its distance from the Sun. The planetary orbits are not shown to scale.

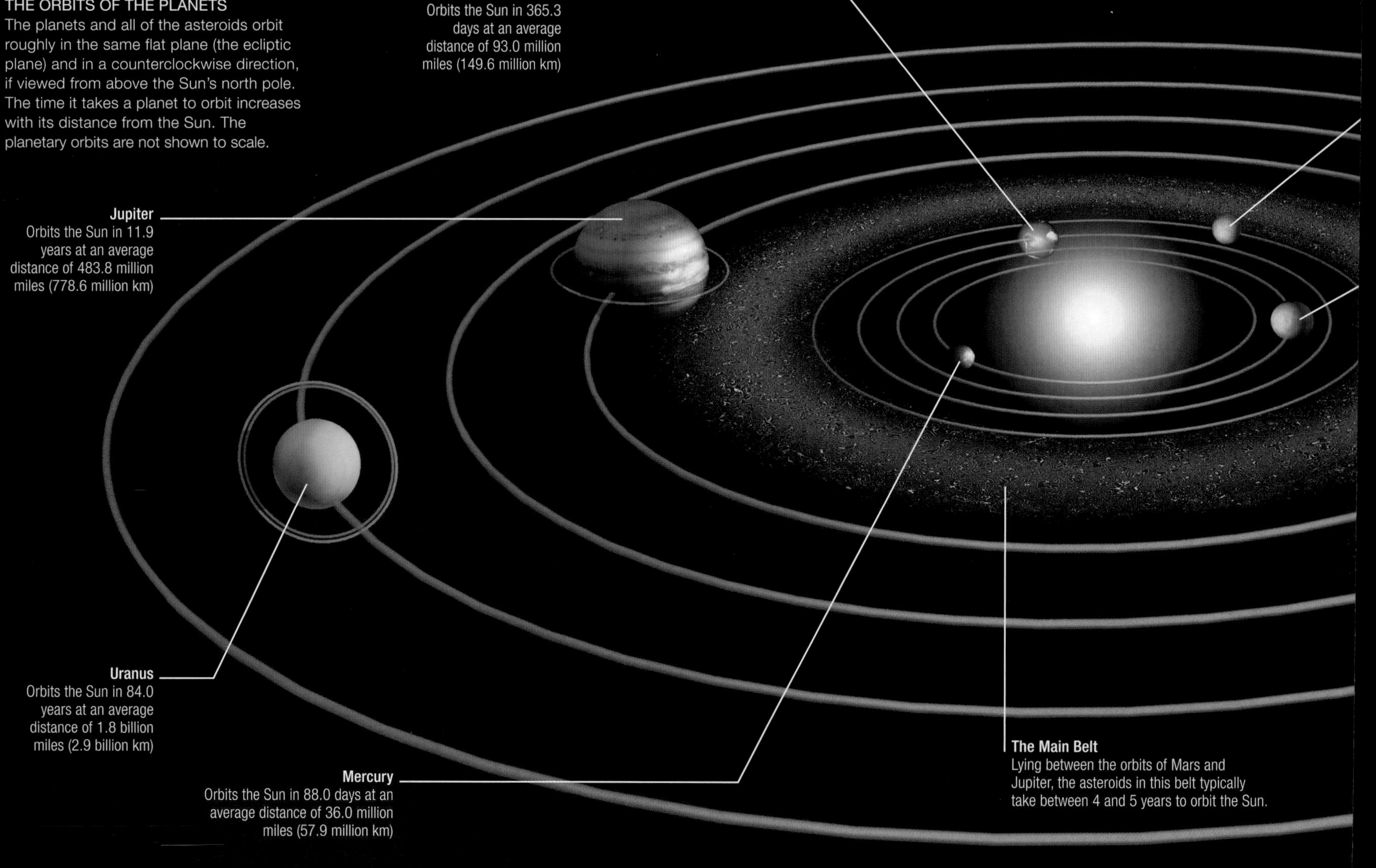

Earth Orbits the Sun in 365.3 days at an average distance of 93.0 million miles (149.6 million km)

Jupiter Orbits the Sun in 11.9 years at an average distance of 483.8 million miles (778.6 million km)

Uranus Orbits the Sun in 84.0 years at an average distance of 1.8 billion miles (2.9 billion km)

Mercury Orbits the Sun in 88.0 days at an average distance of 36.0 million miles (57.9 million km)

The Main Belt Lying between the orbits of Mars and Jupiter, the asteroids in this belt typically take between 4 and 5 years to orbit the Sun.

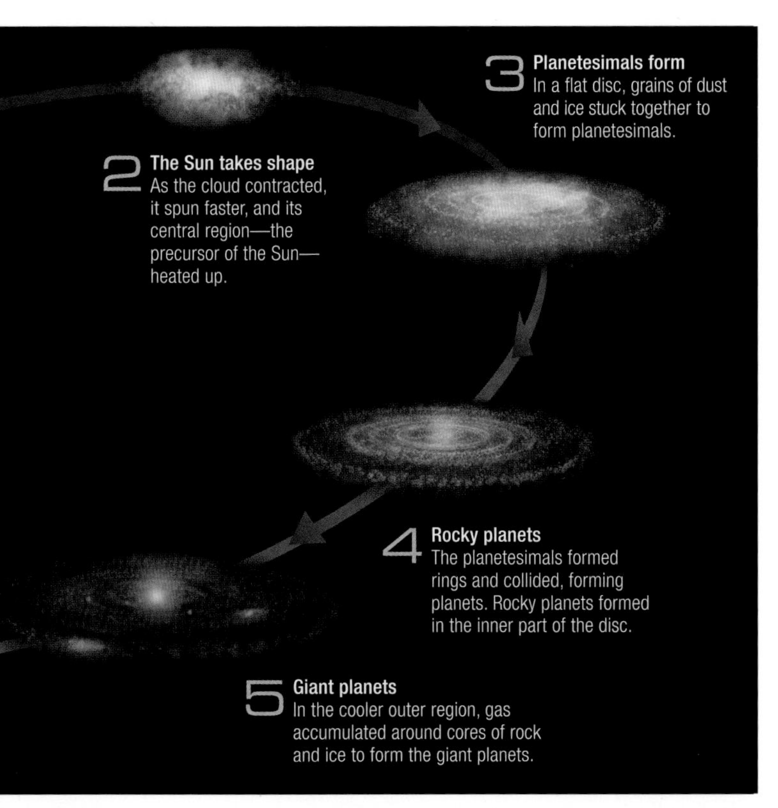

3 Planetesimals form
In a flat disc, grains of dust and ice stuck together to form planetesimals.
2 The Sun takes shape
As the cloud contracted, it spun faster, and its central region—the precursor of the Sun—heated up.
4 Rocky planets
The planetesimals formed rings and collided, forming planets. Rocky planets formed in the inner part of the disc.
5 Giant planets
In the cooler outer region, gas accumulated around cores of rock and ice to form the giant planets.

THE BIRTH OF THE SOLAR SYSTEM

The Solar System formed about 4.6 billion years ago out of an immense, slowly spinning cloud of gas and dust within the Milky Way. Gradually, gravity caused the cloud to contract and spin faster. As its central region grew denser, it also became hotter. This region eventually evolved into the Sun. Surrounding the central region was a spinning disc of gas, dust, and ice. Within this disc, grains of dust and ice stuck together to form particles called planetesimals, and these came together to make up larger objects called protoplanets. The protoplanets underwent further collisions and produced the four inner planets and the cores of the four outer planets. Finally, large amounts of gas were attracted to the cores of the outer planets, creating the vast gaseous atmospheres that surround those planets.

THE SOLAR SYSTEM IN PERSPECTIVE

Scientific understanding of the Solar System has changed immensely over the past 500 years. Until the mid-16th century, it was generally believed that Earth, rather than the Sun, was at its center, and as late as 1780, the Solar System was thought to contain just the Sun, six planets, and a few comets. Through the use of telescopes and spacecraft, the known size of the system and the number of objects it contains have grown by leaps and bounds, especially over the last 50 years. At the same time, our knowledge of the rest of the observable Universe has increased even more dramatically—and we also now know that there are many other systems of planets orbiting stars elsewhere in our galaxy. Consequently, the perceived significance of the Solar System has actually decreased a little over this time.

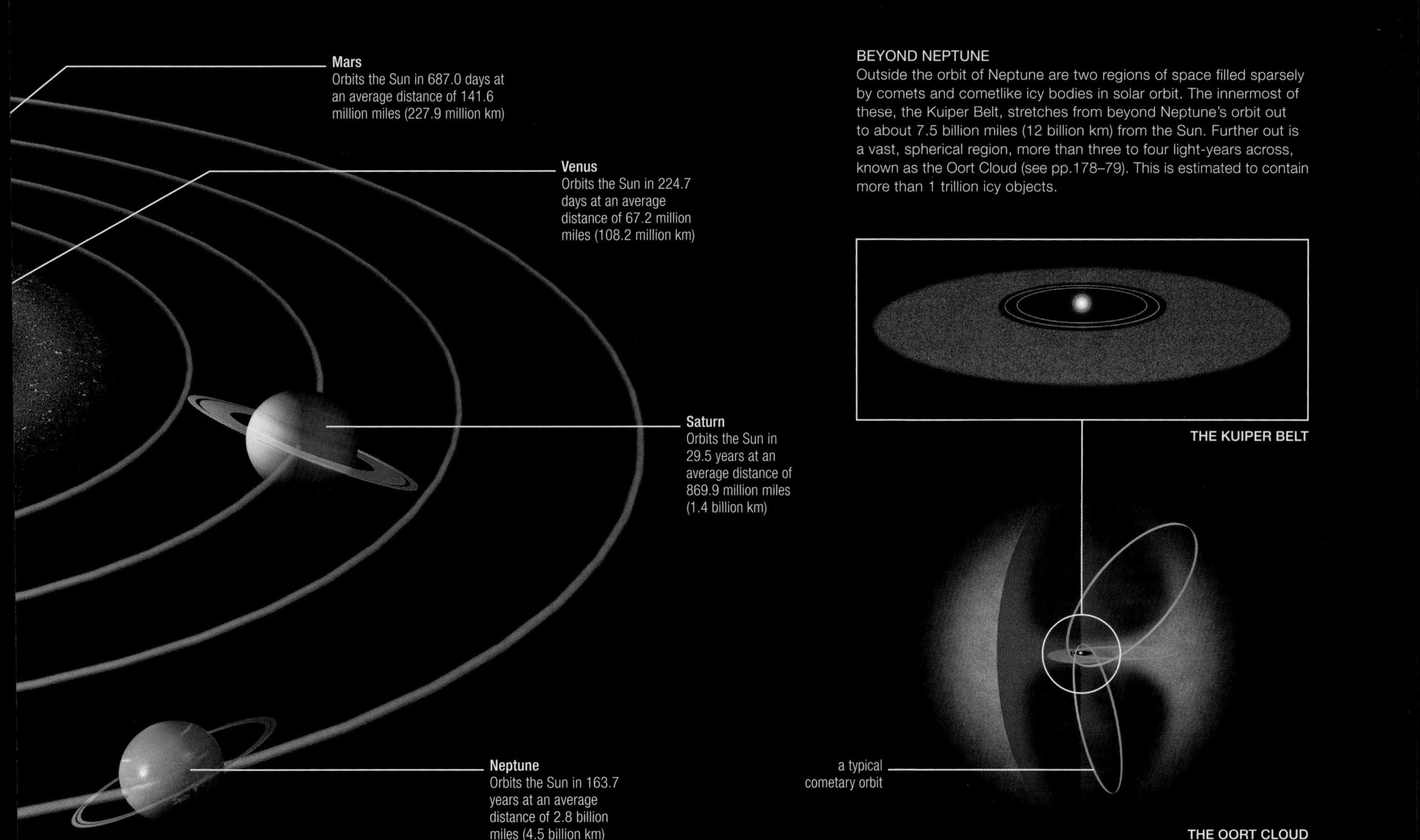

THE FOUR INNER PLANETS
Although they are similar in their internal structure, the inner planets differ greatly in their outward appearance. Mars is notable for its canyons and dormant volcanoes; Earth for its oceans; Venus for its hot, volcanically ravaged surface; and Mercury for its heavy cratering.

MARS
Diameter: 4,220 miles (6,792km)

EARTH
Diameter: 7,926 miles (12,756km)

THE INNER PLANETS

In decreasing order of distance from the Sun, the four inner planets, or rocky planets, are Mars, Earth, Venus, and Mercury. Each of these relatively small worlds is a compact, mostly solid body made principally of rock and metal. These planets have either one, two, or no moons and no rings.

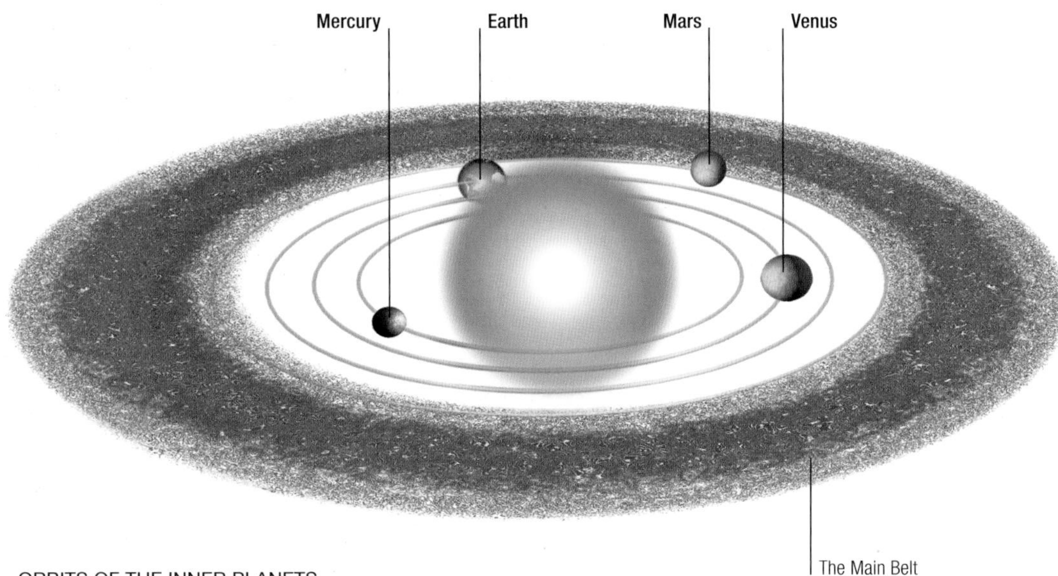

ORBITS OF THE INNER PLANETS
The rocky planets occupy orbits at roughly equally spaced intervals from the Sun. These orbits are very nearly circular, except for Mercury's, which is markedly elliptical (a stretched circle). The sizes of the planets and their orbits are not shown here to scale.

ORIGINS

All four of the rocky planets formed at roughly the same time, about 4.6 billion years ago, from the material in the spinning disc that surrounded the newly forming Sun (see pp.50–51). They formed from the inner part of the disc, where it was hot and only materials with high melting and boiling points—mineral grains and metals—could persist in solid or liquid form. Other materials were vaporized. In due course, the rocky and metallic particles coalesced, through the effects of gravity and random collisions, to form solid chunks called planetesimals. These were anything from a several feet to hundreds of miles in size, but gradually they came together to form moon-sized objects, thousands of miles wide, called protoplanets. Finally, through a series of increasingly violent collisions, the protoplanets accumulated into four large lumps, which became the inner planets.

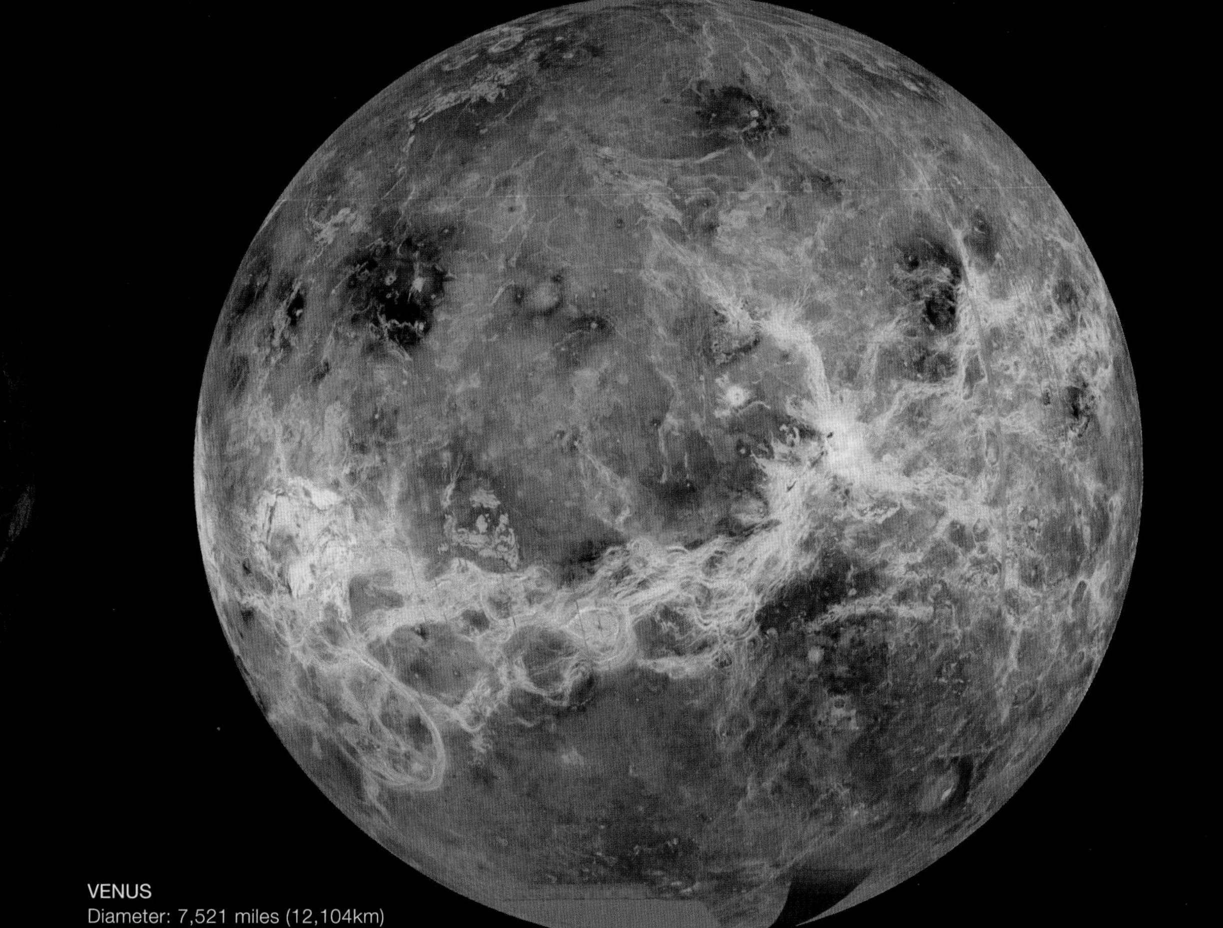

VENUS
Diameter: 7,521 miles (12,104km)

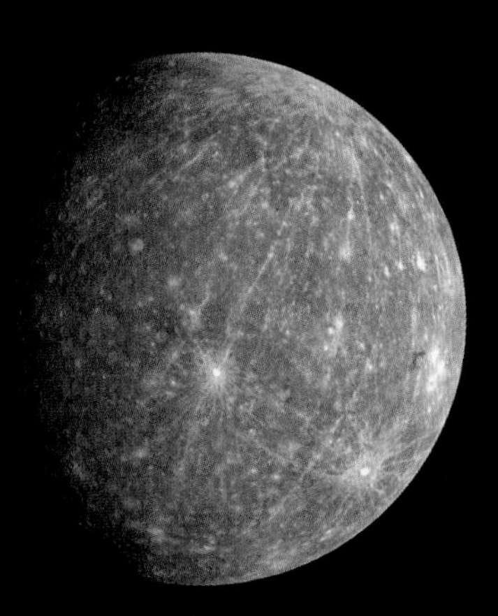

MERCURY
Diameter: 3,032 miles (4,879km)

The series of collisions that led to the formation of the inner planets generated a tremendous amount of heat, so each planet started off in an entirely molten (liquid) state. While molten, the material within the planets separated into two main parts, in a process called differentiation. The denser material, mainly iron and other metals such as nickel, sank toward the planets' centers, while lighter materials floated up toward their surfaces. Over time, the planets gradually cooled and began to solidify from their surfaces inward. The lighter materials closer to the planets' surfaces solidified into two layers of rocks—a thin outer layer, called the crust, and a deeper, much thicker layer, termed the mantle. Deep below these rocky layers, the materials at the centers formed dense cores, composed predominantly of iron. Part of Earth's core is still liquid today, and the cores of Mercury, Mars, and Venus may also still be partly molten.

IRON METEORITE
Objects like this are thought to originate from asteroids that were once molten—just like the inner planets—which allowed their iron content to settle into central cores.

ROCKY SURFACES

The inner planets were heavily bombarded during their early history by material left over from the planet-forming process, and they have continued to be impacted from time to time. Mercury most obviously bears the scars of these impacts in the form of a large number of visible impact craters. The other rocky planets were equally affected, but relatively few obvious craters remain on them today, particularly on Earth. This is due to their surfaces being heavily reworked over time by processes such as plate tectonics and erosion.

Volcanic activity has affected all of the inner planets. On Mercury, much of this activity occurred billions of years ago. Huge volcanoes exist on Mars, although they have been inactive for tens of millions of years, while both Earth and probably Venus are still volcanically active. Venus's surface in particular has been heavily shaped by volcanism. Mars, Earth, and Venus have atmospheres made up of gases released from their interiors by volcanic activity. On Earth and Mars, there has been substantial reshaping of the surface by winds. Only Earth has liquid water at its surface, although Mars's surface was also altered in the past by moving water.

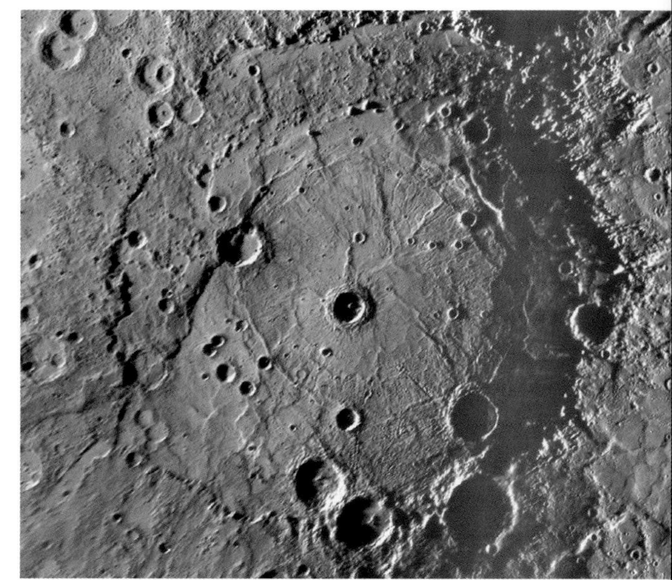

REMBRANDT CRATER ON MERCURY
With a diameter of 447 miles (720km), Rembrandt (the large background crater here) is the second largest impact crater on Mercury. It is thought to have formed 3.9 billion years ago, during the planet's relatively early history.

EARTH

The third major planet from the Sun, Earth has many unique features. It is the only planet known to have an active plate-tectonics system, water flowing freely on its surface, and abundant life.

ORBIT

Earth travels around the Sun in a near-circular orbit at an average distance of 93.0 million miles (149.6 million km), completing one orbit every 365.3 days. Its axis tilts at 23.5° to its orbital plane, which creates distinct seasons as each hemisphere takes its turn to point toward or away from the Sun, and therefore receive more or less sunlight.

STRUCTURE AND ATMOSPHERE

As the Solar System's largest solid body, Earth generated internal heat as it formed, and it has retained that heat during the 4.6 billion years since. This heat is intense enough to create a partly molten core of nickel and iron, surrounded by a mantle of hot rocks that grind past each other in vast convection cells. Earth's crust and upper mantle, down to a depth of about 60–200 miles (100–300km), are broken into huge segments called tectonic plates that are slowly pushed and pulled around the surface by the movement of the mantle beneath. Earth's atmosphere insulates it from the worst temperature changes and allows water to exist on its surface. The atmosphere has been greatly transformed by the existence of life throughout much of Earth's history.

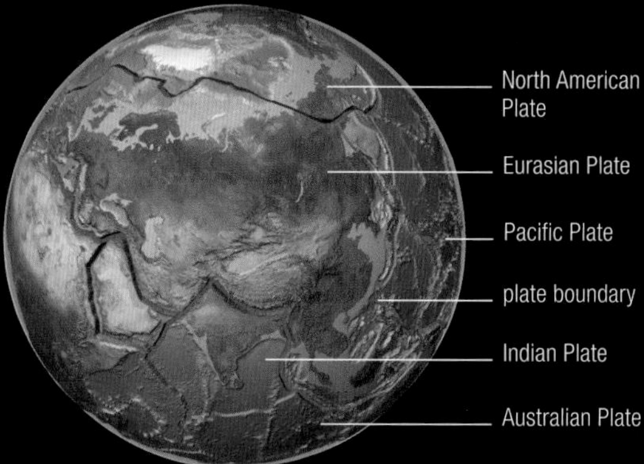

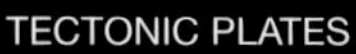

TECTONIC PLATES
Earth's crust and upper mantle are broken into seven large plates and several dozen smaller ones. As the plates move around on the surface, they create mountain ranges, volcanoes, and new seabeds.

VOLCANIC ERUPTION FROM SPACE
An ash plume rises over the Sarychev Volcano on the Pacific coast. The Pacific Ocean is surrounded by volcanic activity where the Pacific Plate collides with others.

THE LIVING PLANET
Earth appears unique even from space, with two-thirds of the globe covered by water, ice at its two poles, and much of the land colored green by plant life.

ORBIT AND ROTATION

Earth's orbit varies only slightly between perihelion and aphelion, so the planet benefits from a fairly even temperature. Perihelion occurs in early January, close to the southern summer solstice, while aphelion occurs near the northern summer solstice.

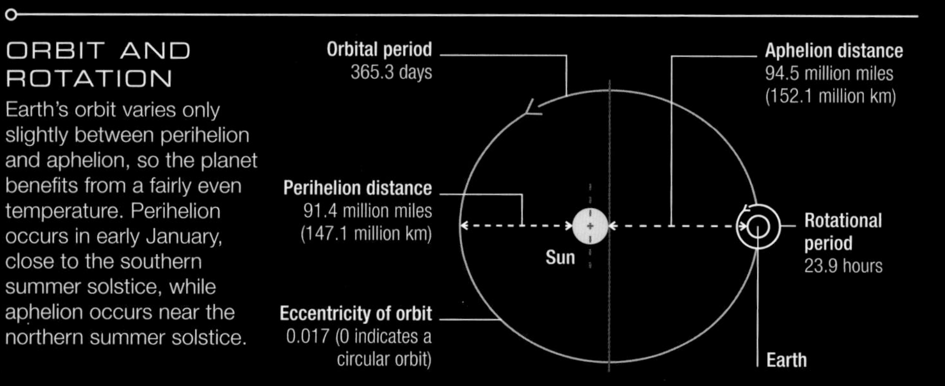

STRUCTURE

Earth's core is kept mostly molten thanks to internal temperatures of around 9,900°F (5,500°C), yet it is slowly cooling, resulting in the gradual growth of a solid inner core.

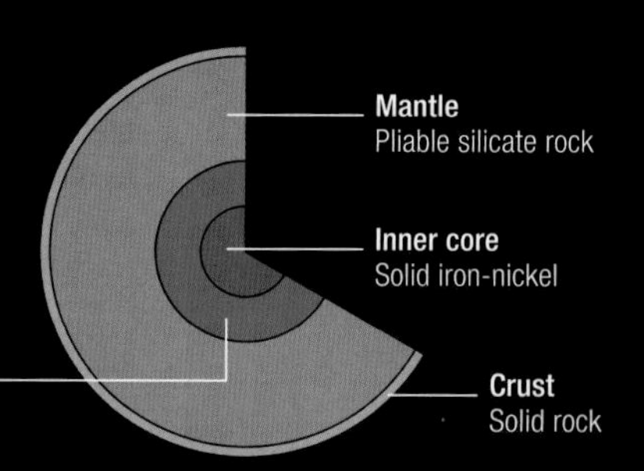

WATER AND LIFE

The abundant life on Earth's surface is inextricably linked to the presence of liquid water. Earth's distance from the Sun, its gravity, and an insulating atmosphere combine to create ideal conditions for water to exist in all three of its physical states: as liquid, ice, and vapor. The constant transfer of water between these different states helps to shape Earth's landscape and drive its complex weather systems. Just as importantly, liquid water provides a vital medium for the chemical reactions needed for life as we know it. Life on Earth ranges from simple, single-celled life forms to complex multicellular organisms. The dominant groups of multicellular organisms are plants, algae, and animals. Plants and algae generate energy from sunlight and atmospheric carbon dioxide, releasing oxygen as a by-product, in a process called photosynthesis. Animals produce energy by eating plants or other animals.

THE GLOBAL WATER CYCLE
Water on Earth can follow a variety of different pathways from ice to liquid to vapor and back, but the dominant water cycle involves evaporation from the oceans and precipitation back onto land and sea.

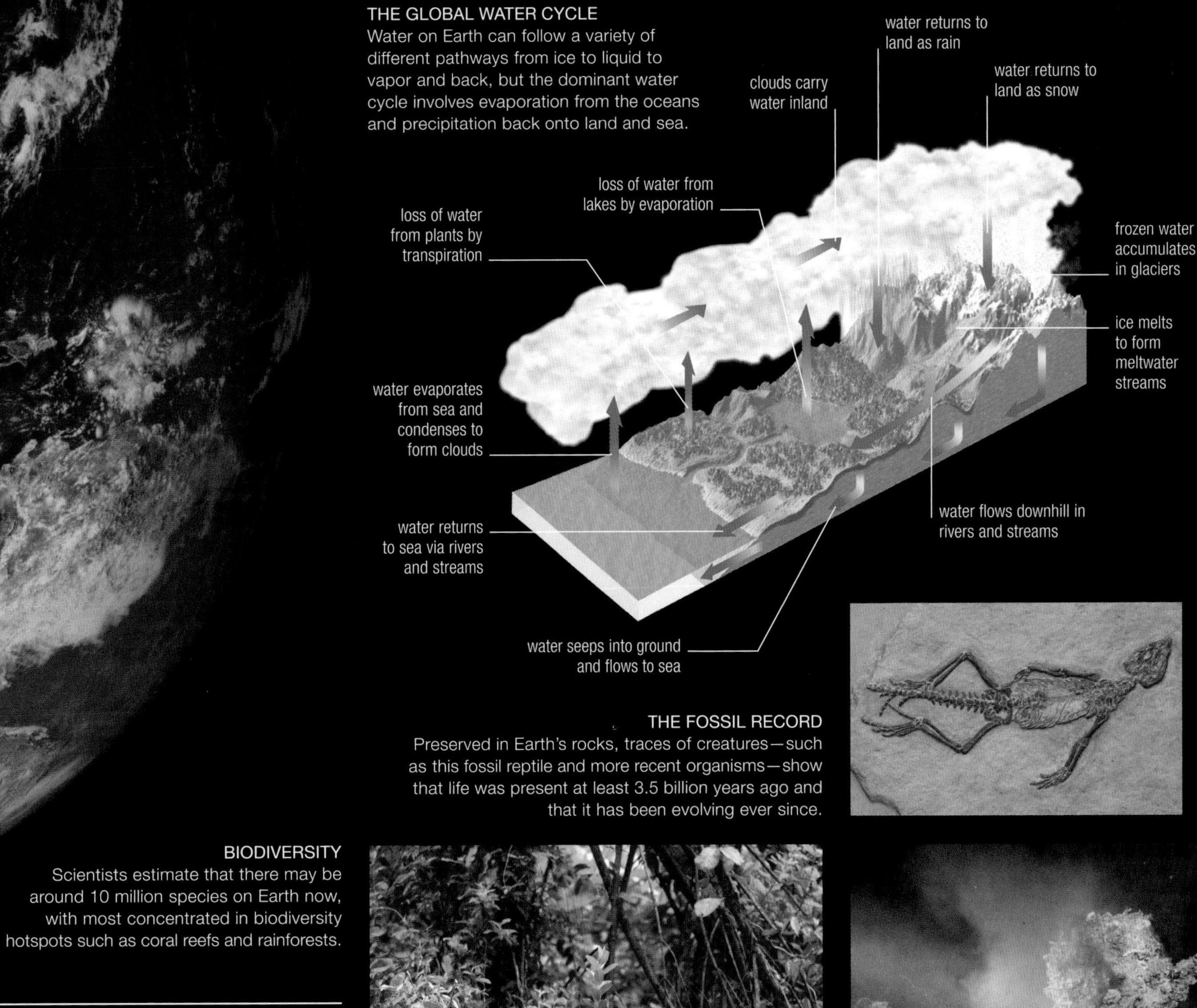

water returns to land as rain
clouds carry water inland
water returns to land as snow
loss of water from lakes by evaporation
loss of water from plants by transpiration
frozen water accumulates in glaciers
ice melts to form meltwater streams
water evaporates from sea and condenses to form clouds
water returns to sea via rivers and streams
water flows downhill in rivers and streams
water seeps into ground and flows to sea

THE FOSSIL RECORD
Preserved in Earth's rocks, traces of creatures—such as this fossil reptile and more recent organisms—show that life was present at least 3.5 billion years ago and that it has been evolving ever since.

BIODIVERSITY
Scientists estimate that there may be around 10 million species on Earth now, with most concentrated in biodiversity hotspots such as coral reefs and rainforests.

DEEP-SEA VENT
Some species inhabit extremely hostile environments. The oases of life around deep-sea volcanic vents, such as giant tube worms anchored around the vents, must endure extreme temperatures to survive.

ATMOSPHERE

Earth's mostly transparent atmosphere is predominantly nitrogen. It also contains a substantial amount of oxygen and low levels of trace gases. Some of these trace gases help to trap heat from the Sun and maintain a temperate climate.

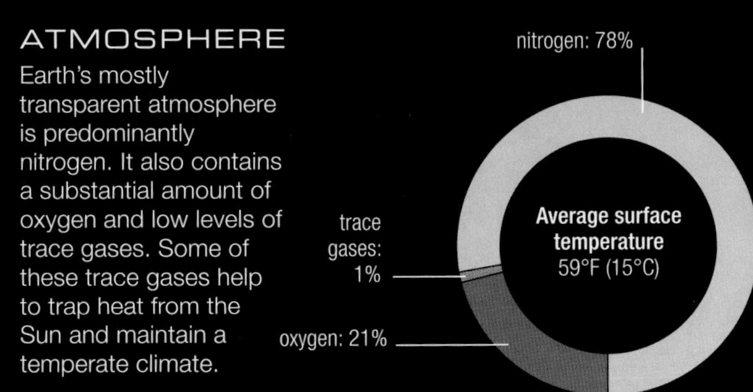

nitrogen: 78%
trace gases: 1%
oxygen: 21%
Average surface temperature
59°F (15°C)

ABOVE EARTH'S NORTH POLE
This image of Earth was taken by NASA's SeaWiFS sensor on the OrbView-2 satellite, which produces color images of Earth's surface. This is a view from directly above the North Pole, looking due south across the frozen Beaufort Sea. The Greenland ice sheet lies to the left, while the Canadian Shield streches from bottom right to the top of the picture.

THE MOON

Earth is accompanied by one natural satellite as it makes its yearly orbit around the Sun. This is the Moon, whose phases mark our months and whose tidal pull has slowed Earth's spin period from an original six hours to its present twenty-four.

EARTH'S NEAREST NEIGHBOR

The Moon is a dry, dead ball of rock about one quarter the size of Earth. As our nearest neighbor, it has been a ready target for exploration. It is the only other world that humans have walked upon. The Moon is locked to Earth in a synchronous spin mode, meaning it completes one spin on its axis in one orbit of Earth. As a result, the same lunar hemisphere always points toward Earth. The Moon's weak gravity means that it has a negligible atmosphere, and the vast majority of the water that has been released from its interior has evaporated into space. With no blanketing atmosphere, the surface temperature varies at the equator by some 536°F (280°C) between day and night.

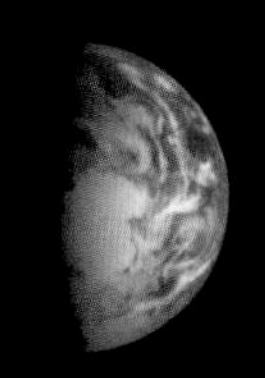

EARTH AND THE MOON
Earth's blue color is caused by the scattering of the shorter (bluer) wavelengths of light as they encounter molecules in the planet's atmosphere. By contrast, the lifeless Moon reveals only its dry, rocky, gas-free surface.

THE ORIGIN OF THE MOON

Astronomers think that a Mars-sized asteroid hit Earth just after our planet had differentiated into an iron core and a rocky surface mantle. The collision blasted hot rock out of the surface, and this formed into a ring of material around Earth. This material then slowly came together to form a world of its own in orbit around Earth—the Moon. This hypothesis explains why the Moon is iron-poor and why it started life much closer to Earth than it is now. It also explains why it fell quickly into a synchronous spin mode and why its outer shell became a magma ocean in its early days and lost many of its volatile elements.

LANDMARK MISSIONS TO THE MOON

SPACECRAFT	DATE	ACHIEVEMENT
Luna 3 (USSR)	October 7, 1959	First images of far side of Moon
Luna 9 (USSR)	February 3, 1966	First soft landing of robot craft
Apollo 8 (USA)	December 24, 1968	First humans to orbit Moon
Apollo 11 (USA)	July 20,1969	First humans to land on Moon
	July 24, 1969	First moonrocks returned to Earth
Luna 16 (USSR)	September 24, 1970	First automated sample return
Luna 17 (USSR)	November 17, 1970	First robotic lunar rover, Lunokhod 1
Apollo 15 (USA)	July 30, 1971	First manned lunar rover
Apollo 17 (USA)	December 11, 1972	Last humans to land on Moon

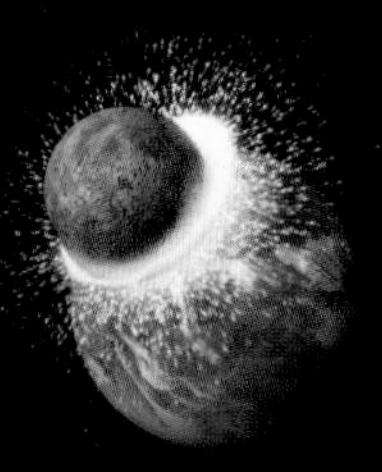

1 ASTEROID HITS EARTH
A Mars-sized asteroid gives Earth a glancing blow. An impact of such a large asteroid with an inner planet was rare, explaining why Mars, Venus, and Mercury do not have large satellites.

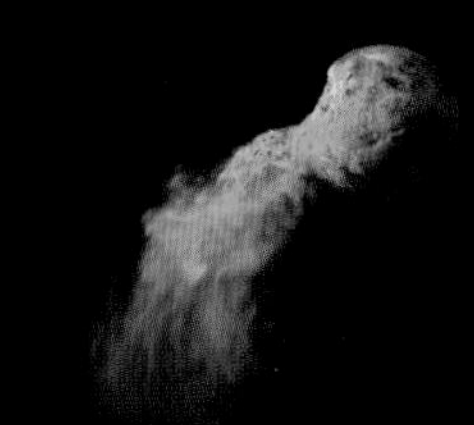

2 MATERIAL BREAKS AWAY
A few percent of Earth's mantle is ejected into space, but much returns to the surface. Some escapes and moves away from the planet to form a "cloud" of rock in orbit around Earth.

3 MATERIAL ORBITS EARTH
The particles in the cloud form a ring around the planet. Unlike Saturn's rings, this ring is very dense. It is also far enough away from the pull of Earth's gravity for particles to coalesce rather than break up.

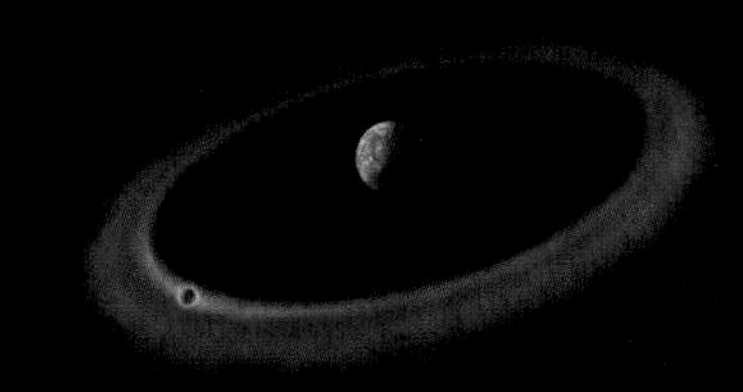

4 MATERIAL FORMS THE MOON
The size of the biggest body in the ring increases, and collisions become less frequent. And all the time, the material in the ring is cooling. After a few tens of millions of years, only one large body remains in orbit.

ORBIT AND ROTATION

The Moon follows an elliptical orbit around Earth; at its most distant, it is 10 percent further away than at its nearest point. Its orbit is inclined to the plane of Earth's orbit by 5.15°. The Moon orbits in the same time it takes to spin once

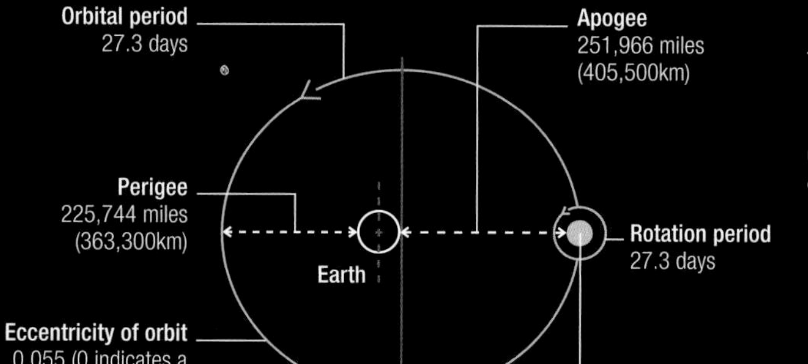

SIZE

The Moon is spherical. It is 3.7 times smaller than Earth and 81 times less massive. Its gravity is one-sixth of Earth's.

Earth
Diameter at equator:
7,926 miles (12,756km)

Moon
Diameter at equator:

THE SURFACE OF THE MOON

At the Moon's surface, there is no wind, water, or veathering. There is just rock. Over time, this has oeen cratered by the impact of asteroids and neteoroids. The impact rate was at its highest about 4 billion years ago and has been steadily decreasing ever since. At that time, the Moon vas hotter, and just below its thin crust here was a region of molten magma. Large craters penetrated this crust, and he magma flowed over the surface, illing low-lying areas with basaltic lava.

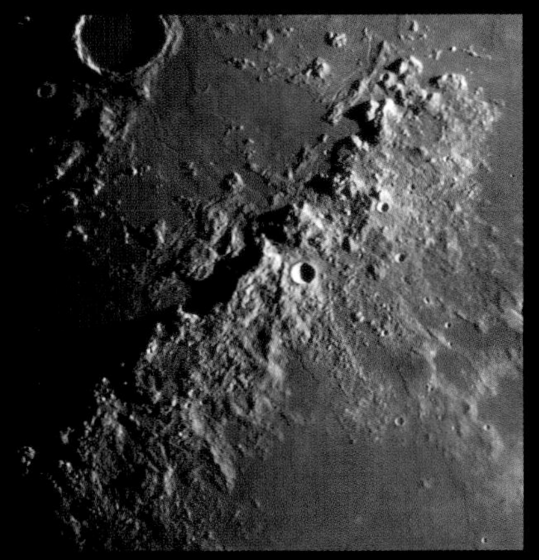

MONTES APENNINUS
This mountain range, 373 miles (600km) long, orms part of the rim of Mare Imbrium (the Sea of Showers). The highest peak is 3 miles (5km) above the floor of the mare.

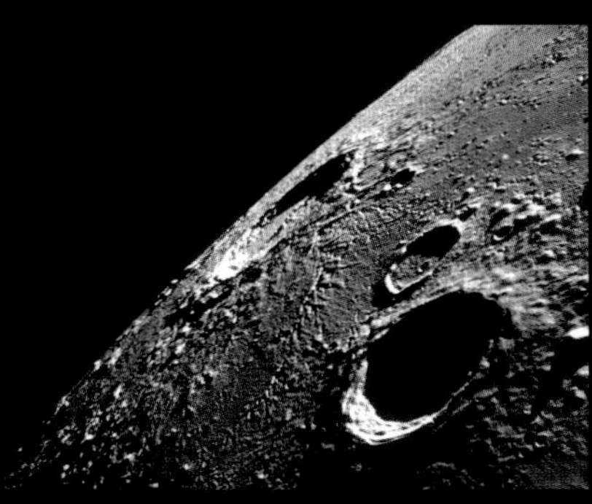

COPERNICUS AND REINHOLD CRATERS
The low angle of the Sun emphasizes the beauty of the craters Copernicus (upper horizon) and Reinhold (foreground). This view vas captured by Apollo 12's Richard Gordon.

APOLLO 16'S VIEW
The Moon's surface is covered by craters. Much of this Apollo 16 view is of the heavily cratered terrain of the lunar far side, a region that cannot be seen from Earth.

STRUCTURE

The Moon's crust is made of calcium-rich, granitelike rock, and the mantle is rich in silicates. Moonquakes indicate hat the solid core might be surrounded by a molten region.

ATMOSPHERE

The 25 tons of lunar atmosphere comes from radioactive decay, meteoroid impacts, and captured solar wind.

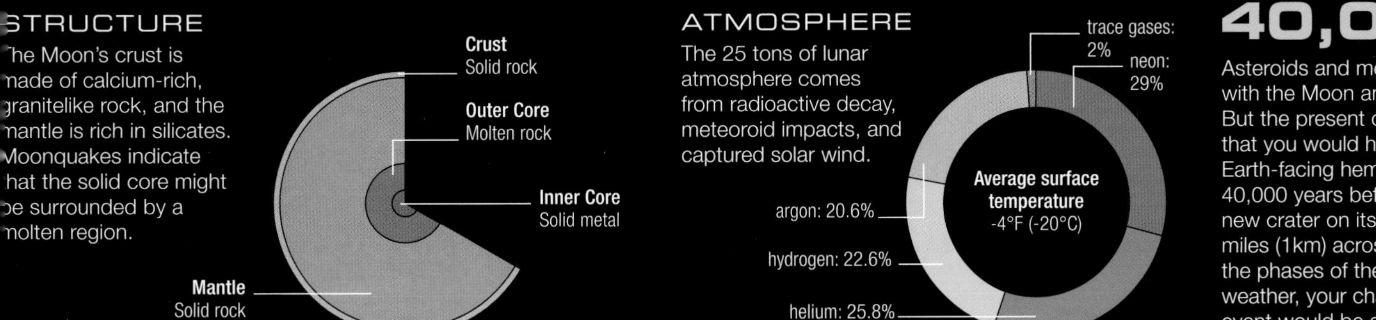

40,000

Asteroids and meteoroids still collide with the Moon and produce craters. But the present cratering rate is such that you would have to observe the Earth-facing hemisphere for around 40,000 years before you could see a new crater on its surface larger than 0.6 miles (1km) across. Taking into account the phases of the Moon and Earth's weather, your chance of seeing the event would be about 1 in 8.

THE MOON IN OUR SKIES

The Moon is our closest space neighbor, and appears to be the largest object in the night sky. It keeps the same face toward us at all times, and through reflected sunlight, we see varying amounts of this face over a 29.5-day cycle. Periodically, the Moon also covers the Sun in the sky, creating a spectacular solar eclipse.

THE NEAR SIDE OF THE MOON

Early in the Moon's history, when it was much closer to Earth, the Moon became synchronously locked with our planet (see p.58). Since then, the Moon has followed an orbit that takes exactly the same time as one rotation on its axis. As a result of this synchronicity, the same side of the Moon, known as the near side, faces Earth at all times. But rather than seeing 50 percent of the Moon's surface, we see 59 percent. This phenomenon, known as libration, occurs because the Moon follows an elliptical orbit, and because when closest to Earth (at perigee), the Moon moves faster than when at its furthest distance (at apogee). This enables us to see around one limb and then the other. Additionally, because the Moon's spin axis is tilted by 6.7° to the plane of its orbit, the poles alternately point toward Earth, allowing us to see a little of the Moon's far side near the poles.

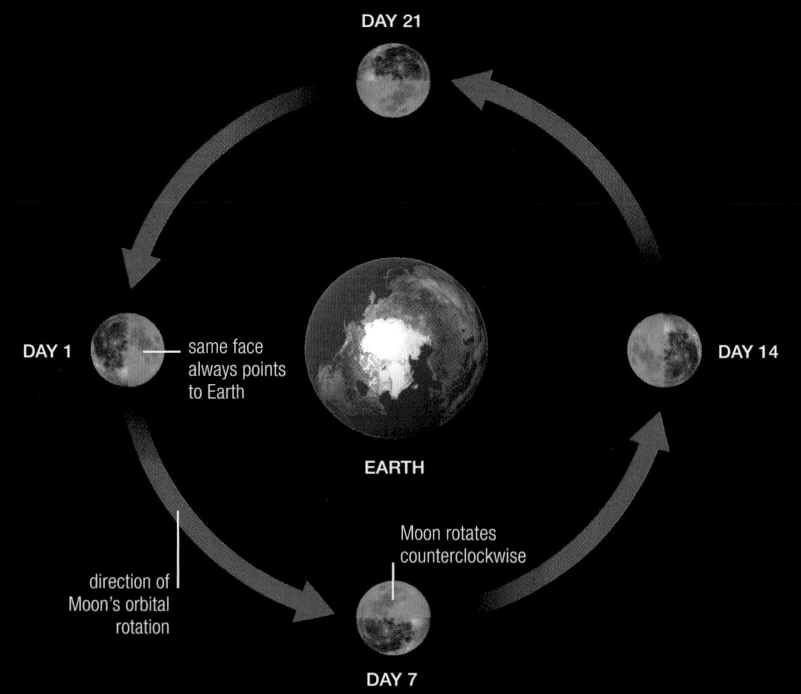

SYNCHRONOUS ROTATION
The Moon takes 27.3 days to orbit Earth. As it orbits, the Moon spins on its axis, also in 27.3 days, with the result that the same side of the Moon always faces Earth.

PHASES OF THE MOON

The Moon has no light of its own, but it is illuminated by the Sun. When one half of the Moon is in sunlight, the other half is in darkness. Viewed from Earth, the Moon's disc seems to assume a different shape from day to day. This is because as the Moon orbits Earth, a changing amount of the side we see is bathed in sunlight. The differing shapes of the Moon are what we call the Moon's phases, and a complete phase cycle takes 29.5 days to complete. During the first half of its cycle, the Moon appears to grow (wax) until it is fully lit at full Moon—that is, when the Moon is on the opposite side of Earth from the Sun. During the second half of its cycle, the Moon appears to shrink (wane) until it is completely unlit at new Moon—that is, when the Moon is positioned between Earth and the Sun.

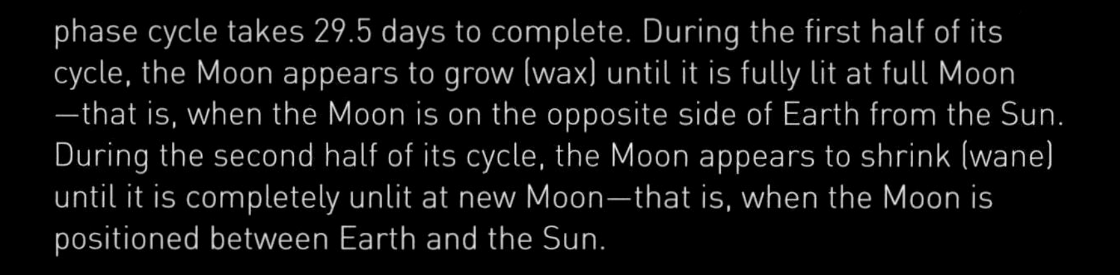

PHASE CYCLE
The Moon grows from new Moon to full Moon, and then returns to new Moon. At first quarter, it has completed a quarter of its cycle; at last quarter, a quarter remains. It is called gibbous when between half- and fully lit, and cresent when between new Moon and half-lit.

WAXING CRESCENT

FIRST QUARTER

WAXING GIBBOUS

FULL MOON

WANING GIBBOUS

LAST QUARTER

WANING CRESCENT

NEW MOON

SIZE IN THE SKY

The distance between Earth and the Moon varies as the Moon follows its elliptical orbit (see p.58). When the Moon is at perigee (opposite), it is 10 percent closer to Earth than when it is at apogee. This change in distance has an effect on the apparent size of the Moon in Earth's sky. During each phase cycle, the Moon moves through perigee and is then at its largest; about two weeks later, it is at apogee and at its smallest. This change goes largely unnoticed, although the difference becomes significant during solar eclipses (below), when more or less of the Sun is obscured depending on the relative size of the Moon. A more commonly reported phenomon is that the Moon seems larger when seen near the horizon rather than high in the sky. This is an optical illusion; our brain is tricked into thinking the horizon Moon is large because it is seen against a close foreground, whereas the higher Moon is seen against a void.

CHANGING SIZE
These two views show the comparative sizes of the Moon in Earth's sky at apogee and perigee. When seen together, the difference is obvious, but the gradual change from one to the other is difficult to notice in the sky.

ECLIPSES

The Moon and Sun appear the same size in the sky, although they are vastly different in size. This is because, by extraordinary coincidence, though the Moon is 400 times smaller than the Sun, it is also 400 times closer to Earth. For this reason, when the Sun, Moon, and Earth are directly aligned, the Moon's disc covers the Sun's and blocks it from view—perfectly so when the Moon is at perigee, fractionally less so when it is at apogee. The Sun is eclipsed and the Moon casts a shadow on Earth. Anyone in the darkest part of the shadow sees a total eclipse of the Sun; those in the lighter part see a partial eclipse. The period of time during which the Sun is completely obscured is called totality. A lunar eclipse occurs when the Sun, Earth, and Moon are lined up and the Moon moves into Earth's shadow.

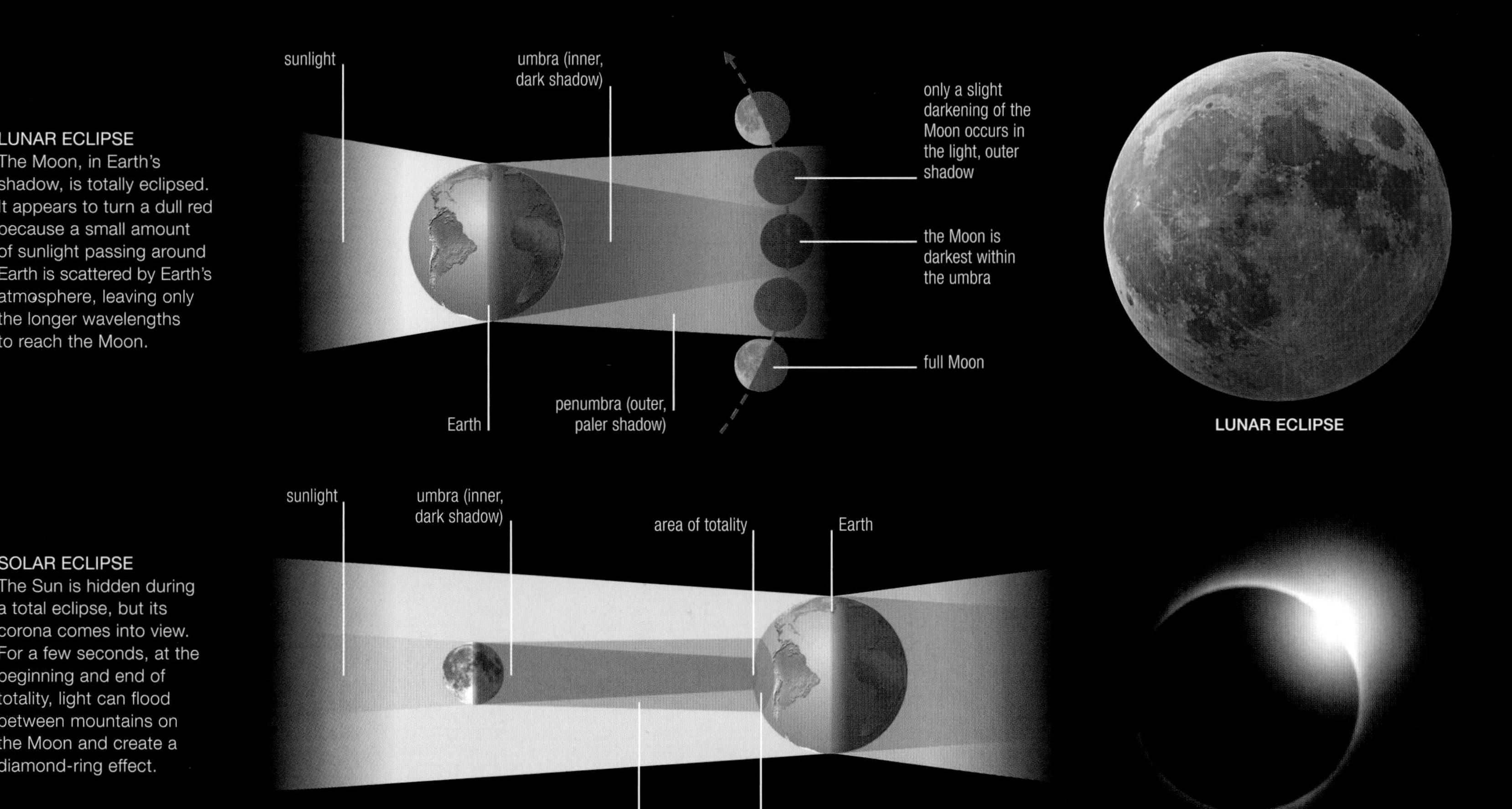

LUNAR ECLIPSE
The Moon, in Earth's shadow, is totally eclipsed. It appears to turn a dull red because a small amount of sunlight passing around Earth is scattered by Earth's atmosphere, leaving only the longer wavelengths to reach the Moon.

SOLAR ECLIPSE
The Sun is hidden during a total eclipse, but its corona comes into view. For a few seconds, at the beginning and end of totality, light can flood between mountains on the Moon and create a diamond-ring effect.

MAPPING THE MOON

The first maps of the Moon were little more than sketches recording the dark and light regions on the side of the Moon facing Earth. Detail was added once telescopes were turned toward the Moon. In recent years, spacecraft have been used to make detailed maps of the entire lunar surface.

EARLY MAPPING

Early printed maps recorded the view seen through telescopes and introduced Latin names for surface features. The dark areas were thought to be water and so were called mare (sea) or oceanus (ocean). These had associated lakes, bays, and marshes, respectively called lacus, sinus, and palus. Lighter regions included mountain ranges, called montes. This naming system is still in use today.

HEVELIUS AND RICCIOLI MAPS
Two influential maps were produced in the mid-17th century by astronomers Johannes Hevelius and Giovanni Riccioli. Riccioli's (right) introduced the Latin naming system.

MAPPING BY SPACECRAFT

When the US embarked on the Apollo project, the mapping of the Moon became a serious endeavor. In preparation for a manned landing, five Lunar Orbiter missions photographed 99 percent of the lunar surface, resolving features as small as 197ft (60m) across. Twenty potential landing sites were studied in even more detail. The films were developed onboard the craft and the data transmitted back to Earth.

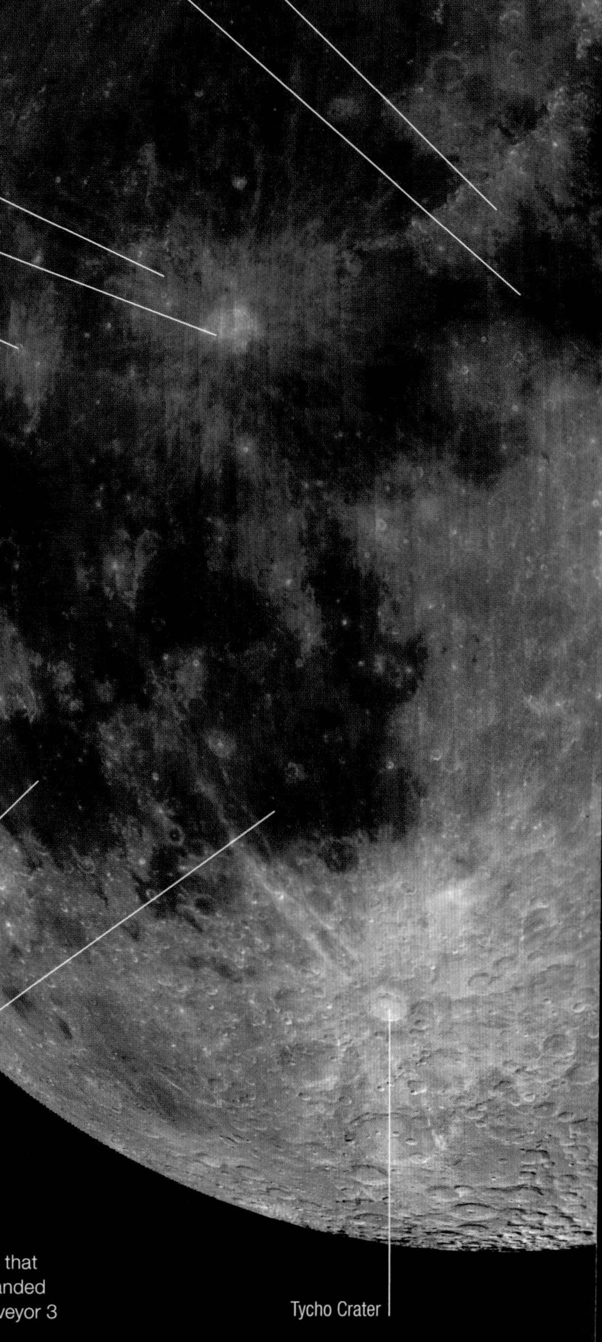

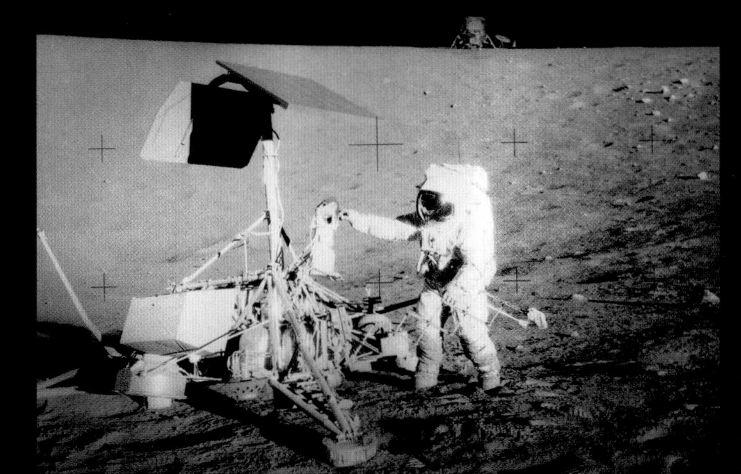

APOLLO 12 MEETS SURVEYOR 3
Lunar mapping in the 1960s was so accurate that in 1969 the Apollo 12 module (background) landed within easy walking distance of the robotic Surveyor 3 craft. The latter had landed two years earlier.

TOPOGRAPHIC MAPPING

The surface of the Moon is rough, like Earth's. The height range between its highest mountain and lowest valley is about 12 miles (20km). Laser altimeters working on orbiting spacecraft, such as Clementine in 1994 and the Lunar Reconnaissance Orbiter since 2009, are used to plot lunar surface irregularities. These are then converted into height contour lines on maps. Colors can also be used to distinguish between high and low regions. The Moon's highest regions occur on its far side, in the center-north, as does its deepest depression, the South Pole-Aitken Basin. When comparing maps of the near and far sides, it is immediately apparent that the two sides are very different.

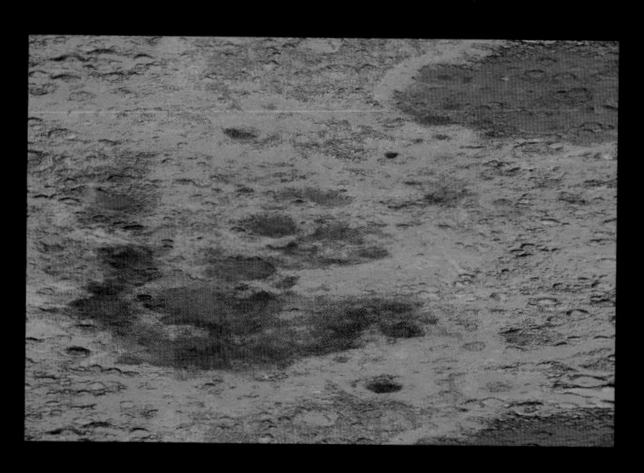

CLEMENTINE MAPPING
The relief in this oblique view of the lunar surface has been exaggerated ten times to show better the relative heights of the visible surface features. False color gives further emphasis: purple denotes the lowest land; blue, then green are higher land; and red is highest of all.

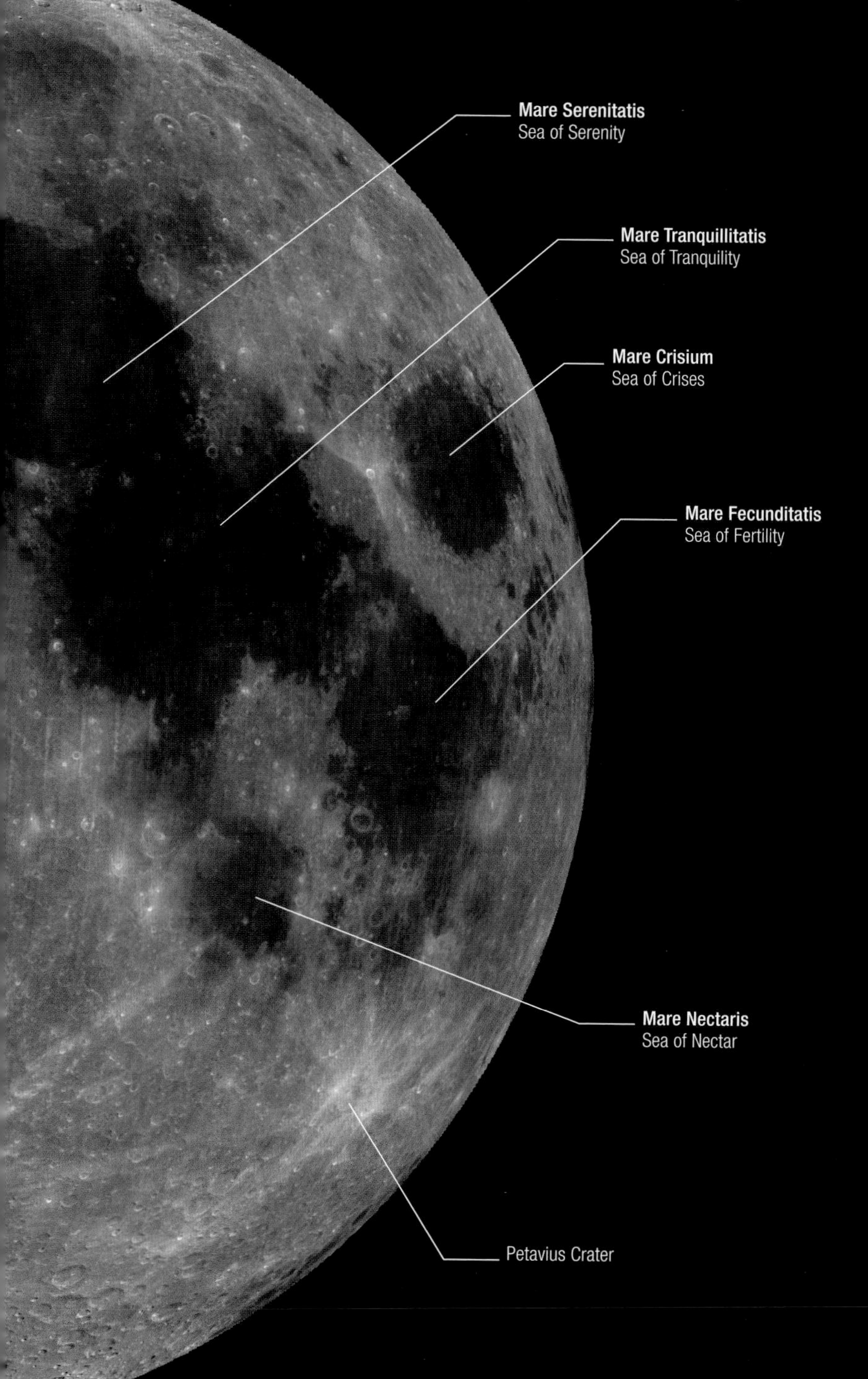

MAPPING INVISIBLE FEATURES

Lunar maps can show invisible features of the Moon, such as surface temperature changes and the Moon's gravitational field. The Moon's gravity tugs unevenly on orbiting spacecraft, with the result that they do not follow perfect elliptical paths. By measuring the way orbits deviate, astronomers can model how the density of lunar material varies. Magnetometers have also found regions where the lunar magnetic field is enhanced, called magnons. These too have been mapped.

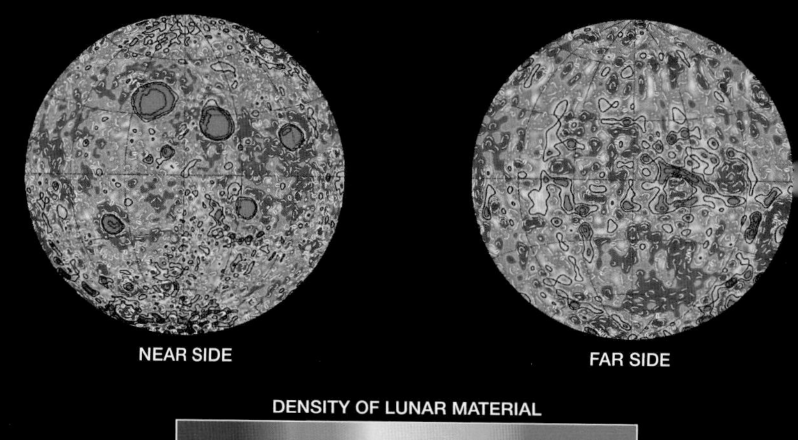

LUNAR PROSPECTOR GRAVITY MAPS
The colors in these maps by Lunar Prospector represent the varied density of the lunar material and the unevenness of the Moon's gravitational pull. The densest regions, called mascons, are colored red.

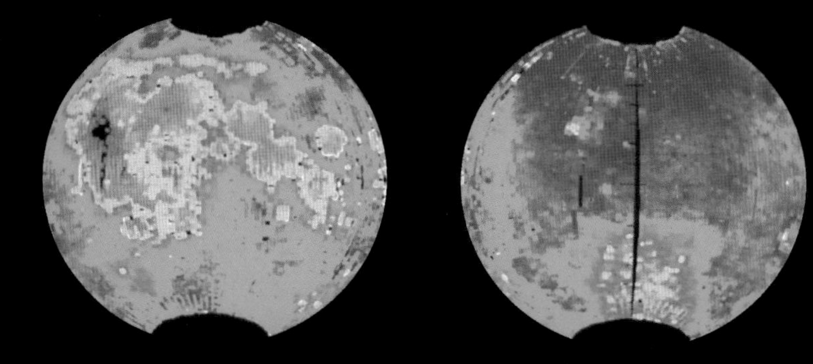

IRON CONTENT OF SOIL (PERCENT BY WEIGHT)

TO THE MOON AND BACK

On May 25, 1960, President John F. Kennedy committed the US to landing a man on the Moon by the end of the decade. The aim was achieved with five months to spare, and by 1972 a further five manned journeys had been made to the lunar surface.

THE APOLLO MISSIONS

The missions that took 12 men to the surface of the Moon between 1969 and 1972 were all part of NASA's Apollo program. The enterprise was governed by three revolutionary ideas. First, the Apollo spacecraft were designed so that sections of each craft could be jettisoned en route to reduce its overall mass. Second, instead of heading directly from Earth to the Moon, two intermediate "parking orbits" around Earth and the Moon were utilized to give the astronauts time to check their equipment. Third, the crew divided when they reached lunar orbit; two descended to the Moon, while the third remained in orbit in the connected Command and Service modules.

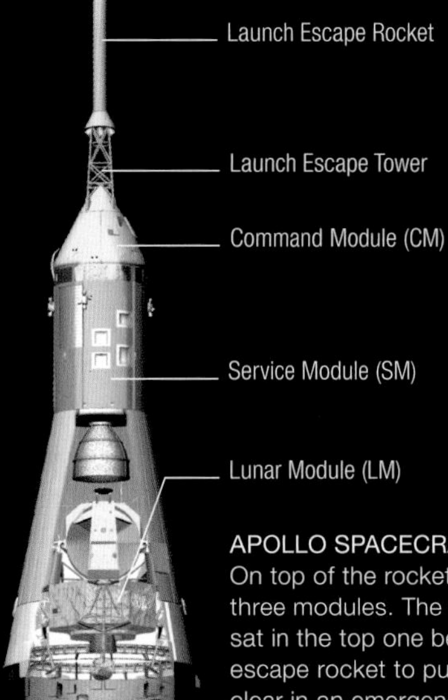
Launch Escape Rocket
Launch Escape Tower
Command Module (CM)
Service Module (SM)
Lunar Module (LM)

APOLLO SPACECRAFT
On top of the rocket were three modules. The crew sat in the top one below an escape rocket to pull them clear in an emergency.

THE LAUNCH AND JOURNEY

A three-stage Saturn V rocket lifted the Apollo craft away from Earth and into low Earth orbit. Its upper stage then blasted the craft on its long journey to the Moon. When close to the Moon, Apollo moved into lunar orbit. Two astronauts traveled to the Moon's surface in the Lunar Module and used the craft's ascent stage to return to the orbiting Command and Service Module. After a three-day return journey, the Service Module was jettisoned, and the Command Module parachuted to Earth.

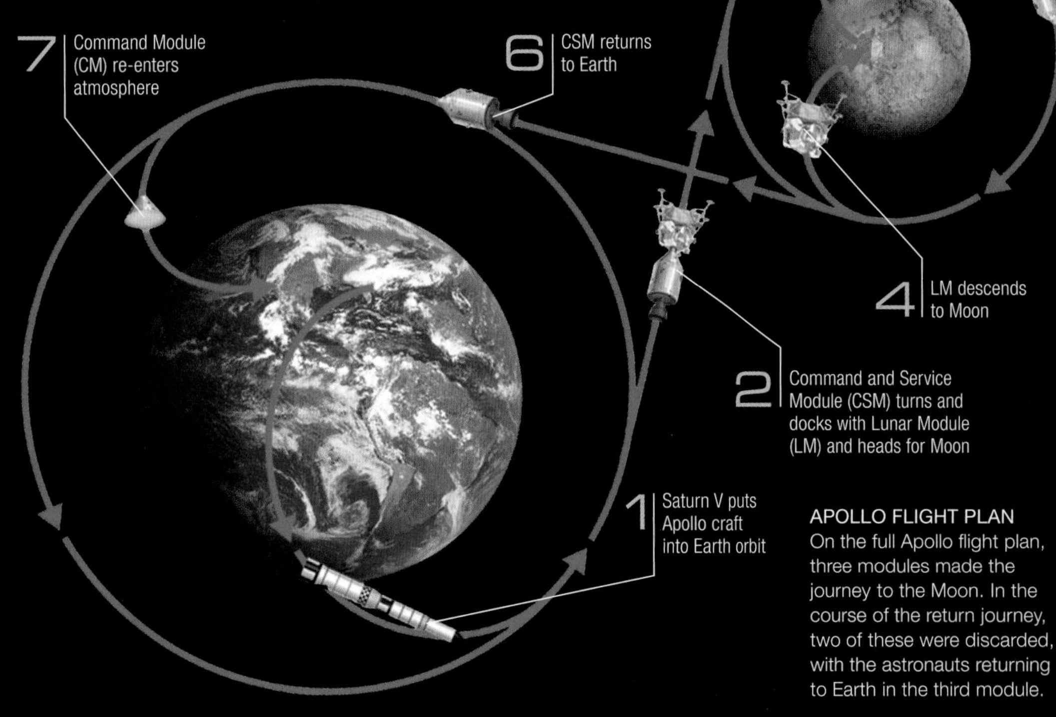
5 Ascent stage of LM returns to lunar orbit for re-docking and astronaut transfer
3 CSM separates from LM and stays in lunar orbit
7 Command Module (CM) re-enters atmosphere
6 CSM returns to Earth
4 LM descends to Moon
2 Command and Service Module (CSM) turns and docks with Lunar Module (LM) and heads for Moon
1 Saturn V puts Apollo craft into Earth orbit

APOLLO FLIGHT PLAN
On the full Apollo flight plan, three modules made the journey to the Moon. In the course of the return journey, two of these were discarded, with the astronauts returning to Earth in the third module.

APOLLO 11 LIFT-OFF
Blasting off from its launchpad at Cape Canaveral on July 16, 1969, Apollo 11's huge Saturn V rocket clears the tower at the beginning of its crew's historic 900,000-mile (1.5-million-km), week-long journey to the Moon and back.

AT THE MOON

Six missions (Apollos 11, 12, 14, 15, 16, and 17) took crews of three o the Moon and delivered two astronauts of each crew to the lunar surface. After a three-day trip from Earth, the Apollo craft fired its etro rockets and dropped into an elliptical orbit around the Moon. hen, once uncoupled from the Command and Service Module, the unar Module slowed and moved into an orbit that took it to within 8 miles (13km) of the surface. After a final check from Mission Control n Houston, the crew was given the go-ahead to land. Apollo 13 did not make it this far. An explosion in an oxygen tank prior to entering lunar orbit forced it to fly around the Moon and to head straight home.

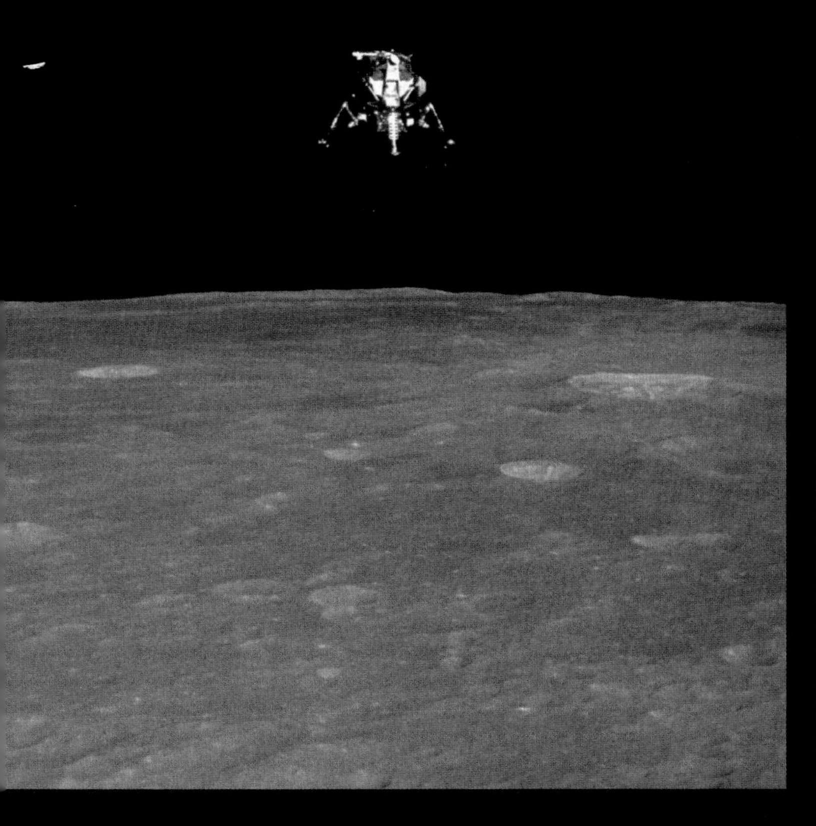

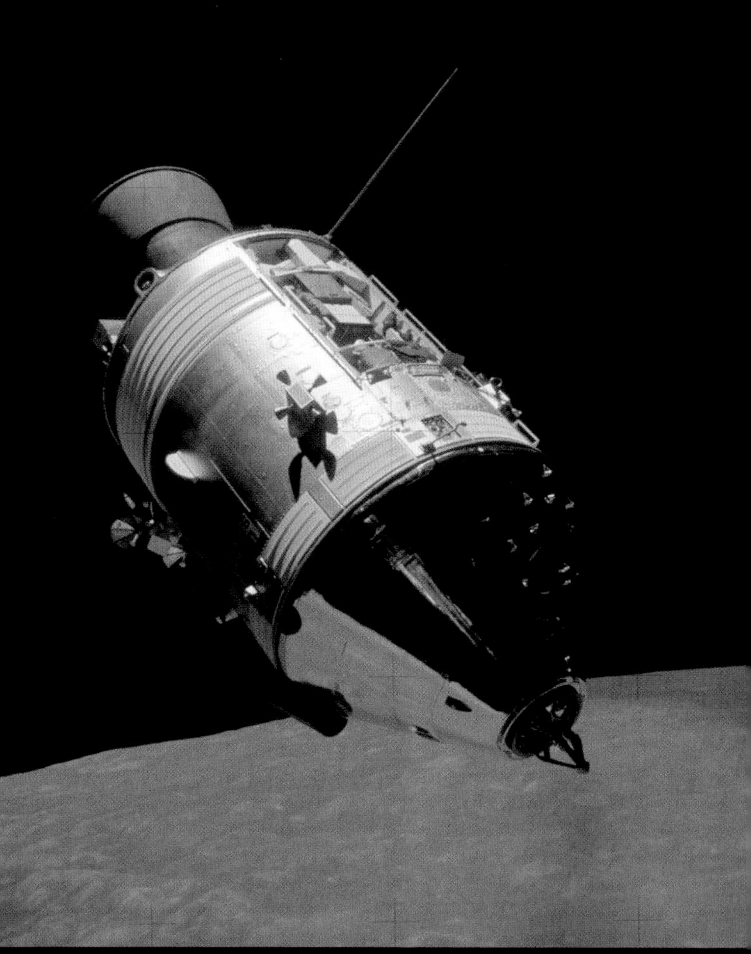

LUNAR MODULE DESCENT
The Apollo 12 Lunar Module, called *Intrepid*, flies across a desolate landscape pockmarked with craters. Moments later it touched down in Oceanus Procellarum. The scene was viewed from the Command and Service Module as it orbited the Moon.

COMMAND AND SERVICE MODULE
The crew of the Apollo 17 Lunar Module took this picture of their Command and Service Module. The Command Module and its docking port are in the foreground. A rectangular scientific instrument bay can be seen on top of the Service Module.

RETURNING HOME

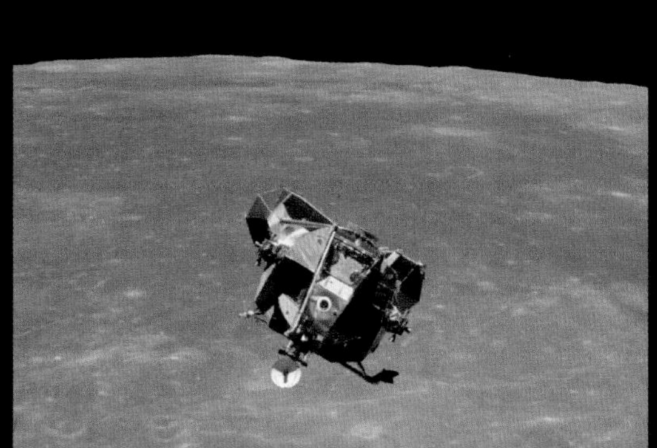

LANDING ON THE MOON

Careful planning went into choosing the sites for the Apollo Moon landings. The safety of the Lunar Module and its two-man crew was paramount. But no amount of planning can prepare for every eventuality, and the success of the historic first landing, by Apollo 11's *Eagle*, owed much to cool piloting.

CHOOSING THE FIRST LANDING SITE

Eagle's landing site was Mare Tranquillitatis, a huge impact basin filled with basaltic lava. The site had been chosen long before the mission had left Earth. It was relatively flat, which was important not only for landing, but also for taking off. The landing time had also been carefully judged—it had to take into account the temperature of the lunar surface, which varies drastically through a lunar day (29.5 Earth days). To stay within a safe temperature range, the Apollo missions needed to land in lunar early morning or late evening, and to stay for about three days.

Inside *Eagle* were Neil Armstrong and Buzz Aldrin. Apollo 11's third crew member, Michael Collins, was circling the Moon every two hours in the combined Command and Service Module, *Columbia*. Collins listened in on the radio as *Eagle* fired its descent engines and lowered its altitude. Closing in on the landing site, Armstrong saw it was scattered with unforeseen boulders, took manual control, and headed for a safer spot. With only 20 seconds of fuel left, *Eagle* touched down.

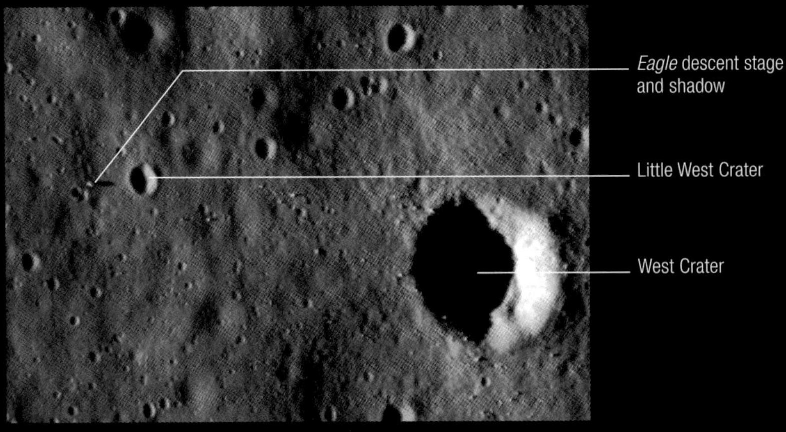

THE APOLLO 11 LANDING SITE
Forty years after Apollo 11, the Lunar Reconnaissance Orbiter (LRO) took this image of Apollo 11's landing site. The lower part of *Eagle* that was left behind when the astronauts lifted off on their return journey is visible as a white dot. The LRO was about 30 miles (50km) above the surface when it recorded this view.

> "THAT'S ONE SMALL STEP FOR [A] MAN, ONE GIANT LEAP FOR MANKIND."
>
> **NEIL ARMSTRONG**, APOLLO 11, 1969

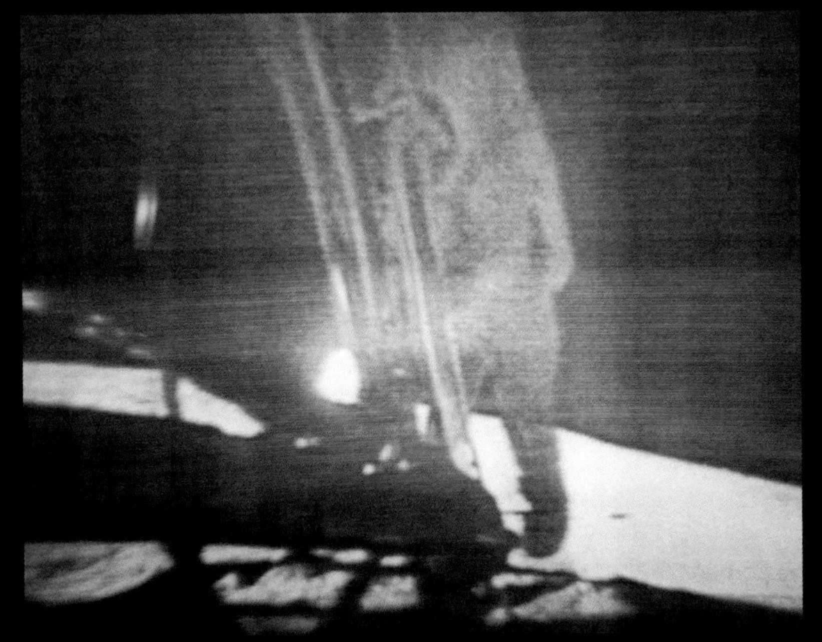

ARMSTRONG STEPS ONTO THE MOON
Armstrong's descent onto the lunar surface on July 21, 1969 was recorded by a monochrome television camera fixed on the side of *Eagle*. Some 19 minutes later, Buzz Aldrin became the second man to stand on the Moon.

HISTORIC FOOTSTEPS

Eagle touched down about 4 miles (6km) away from its target spot, but within the proposed landing area. Although scheduled to sleep, the astronauts were then given the go-ahead to prepare for their first surface excursion. Six hours and 21 minutes after landing, with life-support systems strapped to their backs, the astronauts de-pressurized the cabin and opened the module's hatch.

Watched by millions of television viewers on Earth, Neil Armstrong backed through the hatch and moved down the ladder on the outside of the Lunar Module to become the first human to step onto a world other than Earth. After speaking his first words on the Moon, he scooped up a sample of soil into a pant leg pocket.

In total, Armstrong and Aldrin logged 21.5 hours on the Moon, two-and-a-half hours of them outside their craft. After making a successful ascent from the surface and rendezvous with Collins and *Columbia*, they returned safely to Earth three days later.

THE APOLLO LANDING SITES

After Apollo 11, five more landings were made. All six landing sites were on the near side of the moon—direct communication between the lunar surface and Earth is not possible from the far side. The landing sites were also relatively close to the equator; the regions around the poles are far too cold.

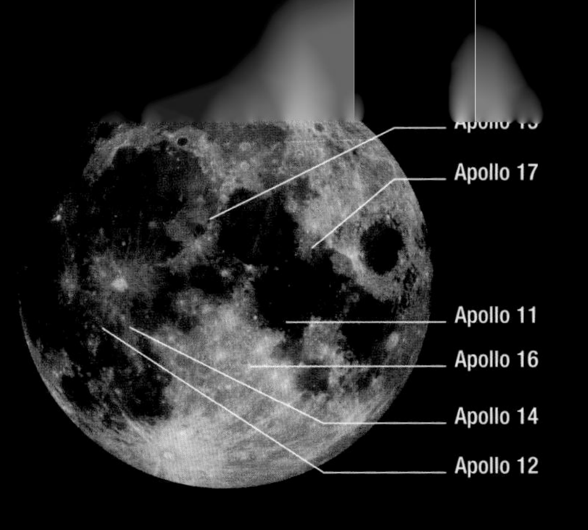

LUNAR DESTINATIONS
All the landing sites, except for Apollo 16's, were in low-lying, relatively flat maria. Apollo 16's highland site was deemed to be safe on the basis of images returned by Apollo 14.

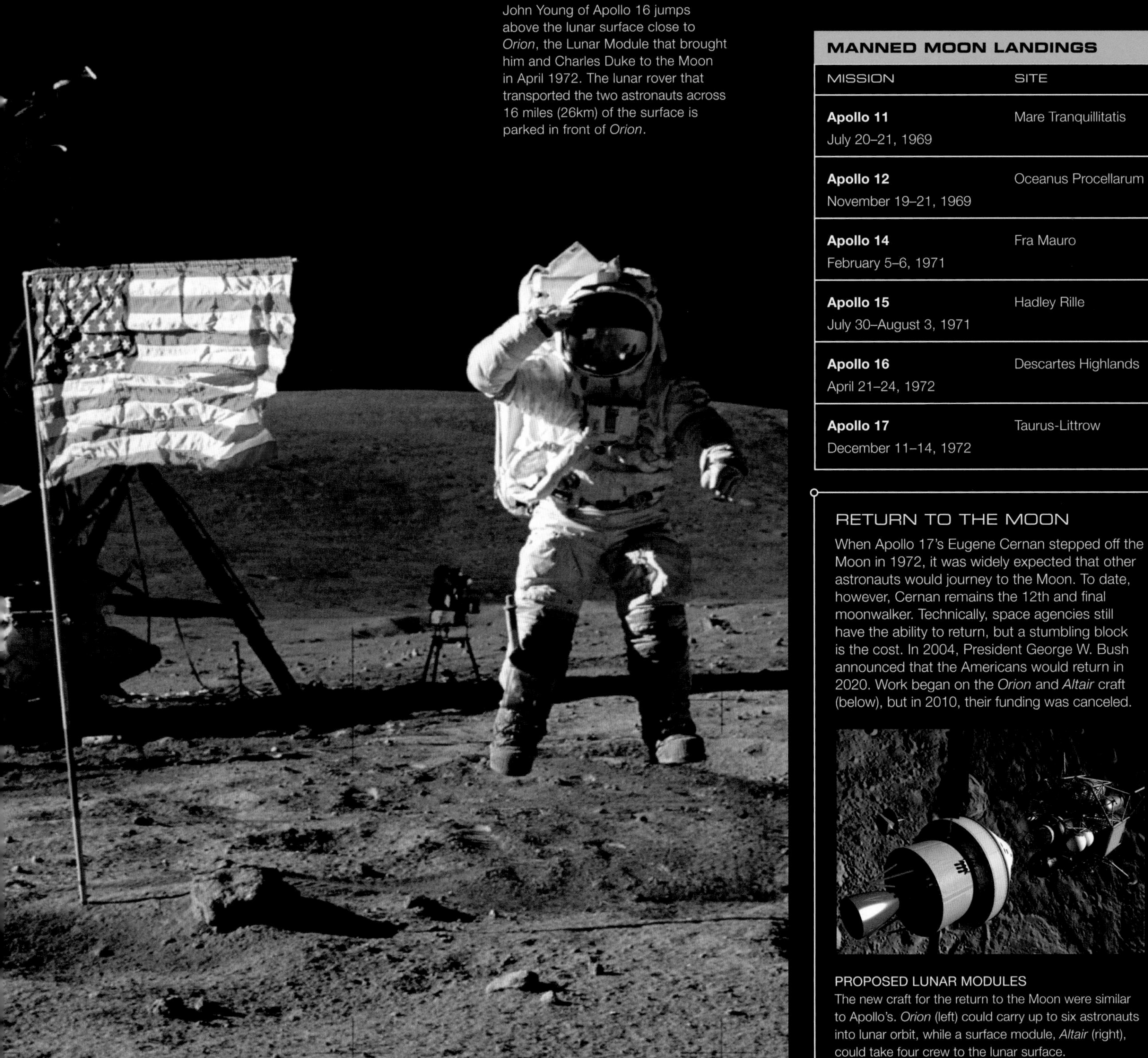

ON SAFE GROUND
John Young of Apollo 16 jumps above the lunar surface close to *Orion*, the Lunar Module that brought him and Charles Duke to the Moon in April 1972. The lunar rover that transported the two astronauts across 16 miles (26km) of the surface is parked in front of *Orion*.

MANNED MOON LANDINGS

MISSION	SITE
Apollo 11 July 20–21, 1969	Mare Tranquillitatis
Apollo 12 November 19–21, 1969	Oceanus Procellarum
Apollo 14 February 5–6, 1971	Fra Mauro
Apollo 15 July 30–August 3, 1971	Hadley Rille
Apollo 16 April 21–24, 1972	Descartes Highlands
Apollo 17 December 11–14, 1972	Taurus-Littrow

RETURN TO THE MOON

When Apollo 17's Eugene Cernan stepped off the Moon in 1972, it was widely expected that other astronauts would journey to the Moon. To date, however, Cernan remains the 12th and final moonwalker. Technically, space agencies still have the ability to return, but a stumbling block is the cost. In 2004, President George W. Bush announced that the Americans would return in 2020. Work began on the *Orion* and *Altair* craft (below), but in 2010, their funding was canceled.

PROPOSED LUNAR MODULES
The new craft for the return to the Moon were similar to Apollo's. *Orion* (left) could carry up to six astronauts into lunar orbit, while a surface module, *Altair* (right), could take four crew to the lunar surface.

EXPLORING ON FOOT

Once on the Moon, Apollo astronauts had only a limited time in which to explore. The Apollo 11 moonwalkers spent only two-and-a-half hours outside their module. By Apollo 14, this had been extended to nearly nine-and-a-half hours, while on Apollo 17, the final mission, there were 22 hours of exploration time. Astronauts on the first three successful missions (Apollos 11, 12, and 14) were also restricted to exploring on foot, staying within 200ft (60m), 1,500ft (450m), and 6,000ft (1,800m) of their landing sites respectively. Later missions had the advantage of the Lunar Roving Vehicle (see pp.70–71).

> **“TAKE A BREAK, GET THE MAP, AND SEE IF WE CAN FIND OUT EXACTLY WHERE WE ARE.”**
>
> **EDGAR MITCHELL**, APOLLO 14 ASTRONAUT, 1971

Having left the safety of the Lunar Module, an astronaut’s life depended entirely on his spacesuit. Though unwieldy, this protected him from the vacuum of space, from X-ray and ultraviolet waves from the Sun, and from harmful cosmic radiation. It was also strong enough not to tear or fray if an astronaut brushed against rocks made jagged by the Moon’s extreme temperature fluctuations. The Portable Life Support System strapped to his back was also bulky and made getting in and out of the Lunar Module a struggle, but this system kept the astronaut cool and ventilated, and provided enough oxygen for nine hours of activity if needed. Flexibility was

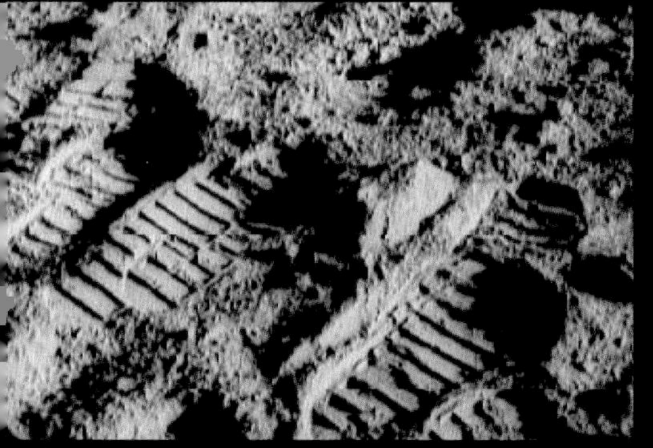

BOOTPRINTS ON THE MOON
The powdery, gritty nature of the lunar topsoil is emphasized by these bootprints, some 0.6in (16mm) deep, made by Apollo 14’s Alan Shepard and Edgar Mitchell in February 1971

WALKING ON THE MOON
Apollo 14’s Edgar Mitchell consults a map while out on the lunar surface. Apollo’s maps were created from photograps taken by the unmanned Lunar Orbiter spacecraft (1966–67) and were essential for estimating distances.

SURFACE EXPERIMENTS PACKAGE
On Apollo 12 and subsequent missions, instrumentation packages were set up to continue working long after the astronauts had left. These used a plutonium-238 thermoelectric generator (foreground) to provide enough energy to run the equipment and to send its data to Earth.

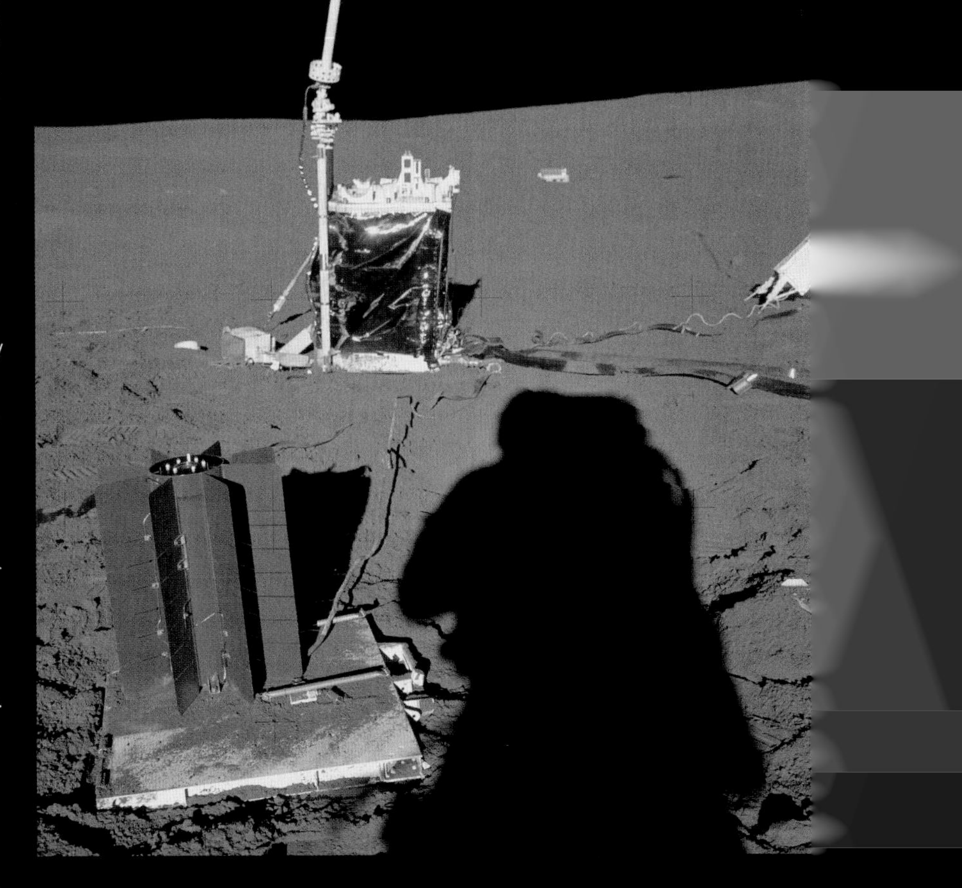

also important; the suit's rubber joints stretched like bellows, and its gloves were molded individually to fit the astronaut's hands. Thumb and fingertips were also thin enough to allow for a maximum sense of touch. As for walking, this was made easier by the Moon itself; with only one-sixth of Earth's gravity, its hills were six times easier to climb.

There were two main reasons for extra-vehicular activity, as leaving the spacecraft was called. One was to collect rock, soil, and core samples. The other was to set up instrumentation packages to record data after the missions had ended. These had to be placed a reasonable distance from the Lunar Module so that they were not disturbed when the rockets fired on the astronauts' departure.

The Apollo 14 moonwalkers, Alan Shepard and Edgar Mitchell, spent their first four hours of surface activity setting up their instrumentation package. Their second excursion was an arduous climb to the rim of Cone Crater, some 4,600ft (1,400m) from their Lunar Module. With a wheeled equipment-trolley in tow, which they pulled and loaded up with lunar samples, theirs was the Apollo program's longest moonwalk.

CARRYING EQUIPMENT
Apollo 12's Alan Bean sets up equipment to record moonquakes, the strength of the solar wind, and the Moon's magnetic field. As he walks, he disturbs the fine surface dust, some of which sticks to his spacesuit.

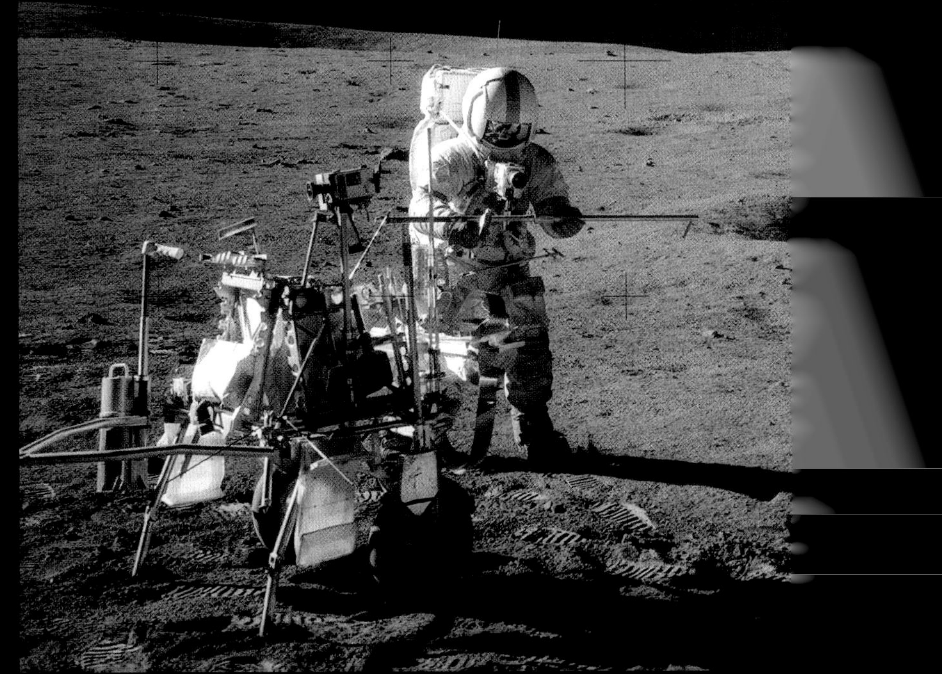

EQUIPMENT TROLLEY
Apollo 14 was the first mission to be equipped with the Modular Equipment Transporter, nicknamed the "rickshaw." This two-wheeled cart, here used by Alan Shepard, was used for carrying experimental equipment and tools, and for returning samples to the Lunar Module.

ROVING ACROSS THE MOON

The last three trips to the Moon, Apollos 15, 16, and 17, each took a Lunar Roving Vehicle. Commonly called the rover, the vehicle was about 10ft (3m) long, 6ft (2m) wide, and 3ft (1m) high—roughly the size of a Volkswagen Beetle car. It traveled to the Moon folded up and stowed on the side of the Lunar Module, from where it was lowered onto the lunar surface on a pulley. Using the rover, the astronauts could comfortably investigate an area 10 times larger than they could on foot.

Made of aluminum tubing, the rover was light, strong, and robust. It had four shock-absorbing wire-mesh wheels with titanium treads, and it was powered by two large 36-volt batteries. In total, it weighed 458lb (208kg), and could carry two astronauts plus their life-support systems, communication equipment, scientific equipment, and camera gear—a load three times its own weight. There was still room left for 60lb (27kg) of lunar samples. When loaded, the rover could travel over 12-in- (30cm-) high rocks, climb and descend 25° slopes, and pitch and roll through 45°. Its turning radius was only 10ft (3m), making it highly maneuverable. It was designed to travel a total distance of just over 56 miles (90km), at speeds of up to 11.5mph (18.5kph).

The astronauts were told not to travel more than 6 miles (10km) from their Lunar Module. If the rover then failed, they would be able to walk back before their oxygen supply ran out. The furthest they actually traveled on one excursion, during the Apollo 17 mission in December 1972, was 4.7 miles (7.6km).

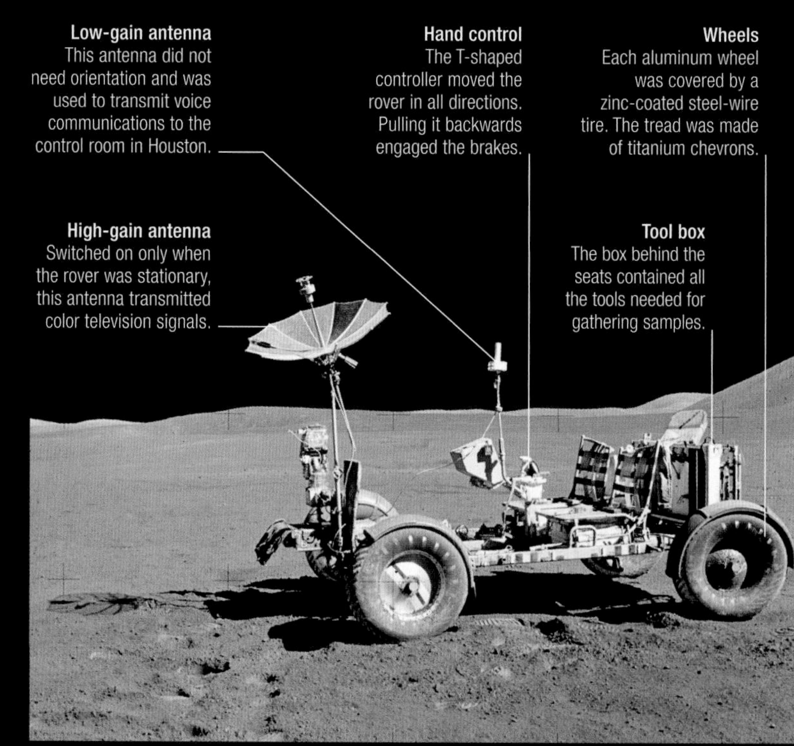

Low-gain antenna
This antenna did not need orientation and was used to transmit voice communications to the control room in Houston.
Hand control
The T-shaped controller moved the rover in all directions. Pulling it backwards engaged the brakes.
Wheels
Each aluminum wheel was covered by a zinc-coated steel-wire tire. The tread was made of titanium chevrons.
High-gain antenna
Switched on only when the rover was stationary, this antenna transmitted color television signals.
Tool box
The box behind the seats contained all the tools needed for gathering samples.

LOW-GRAVITY LUNAR TRANSPORTATION
The Lunar Roving Vehicle, sometimes called the Moon Buggy, was lightweight and reliable. The astronauts were strapped in tightly, but even then they felt nervous when on a slope. Only four flight-ready models were manufactured; three of these went to the Moon, where they remain today.

EXPLORING SHORTY CRATER
After landing on the southeastern rim of Mare Serenitatis, the Apollo 17 astronauts drove over to Shorty Crater. In the picture, Harrison Schmitt is seen collecting samples from the deep lunar soil. Most of the rock is breccia, made of fragments cemented together by the pressure of a meteorite impact.

The rover's T-shaped steering controller was located between the two seated astronauts, so that either could drive, and each of its wheels was powered independently, so that if one of the wheels failed, it could simply be put into neutral and turned off. The navigation system on the dashboard indicated speed, pitch, power, temperature, total distance traveled, and distance from the Lunar Module, as well as the direction of travel with respect to lunar north. When the astronauts brought the rover to a halt, the brakes had to be put on full, and the communication antennae pointed toward Earth, which was always visible in the sky. They then disembarked to photograph and collect interesting rocks, before moving on to the next site.

The astronauts had two main problems while roving across the Moon. The first was that the glare from the Sun made it difficult to choose the best driving route. When, in April 1972, John Young and Charles Duke of Apollo 16 drove directly away from the Sun, everything in front of them was directly lit by bright sunlight and there were no shadows to help them spot obstacles such as boulders. The second problem was the dust, which, being electrostatically charged, stuck to all surfaces. The rover's wheel guards were vital for keeping the astronauts and their equipment clean. Eugene Cernan of Apollo 17 accidentally broke one of the guards, and although he repaired it with gaffer tape and part of a laminated map, still the dust got everywhere.

> “MAN, THIS IS THE ONLY WAY TO GO, RIDING THIS ROVER.”
>
> **CHARLES DUKE**, APOLLO 16 ASTRONAUT, 1972

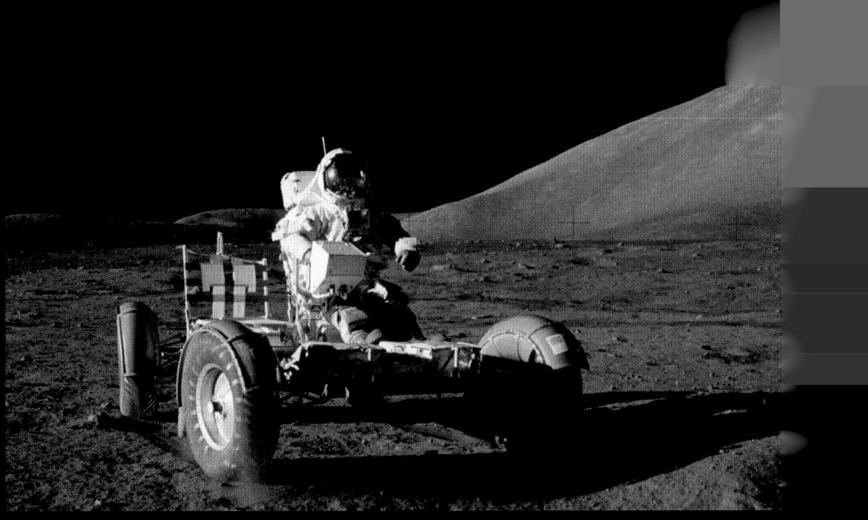

GETTING READY TO ROVE
Apollo 17's Eugene Cernan checks the rover before he and Harrison Schmitt use it to go exploring. It is yet to be loaded with the ground-controlled television system, the communication relay, the high- and low-gain antennae, the bags of tools, and all the scientific gear.

ROVER ROUTE MAP
This map records where the Apollo 15 astronauts David Scott and James Irwin drove in August 1971. They made three trips close to Hadley Rille, for a total of 17 miles (28km). Samples were collected at 12 different locations, some of which were chosen on arrival.

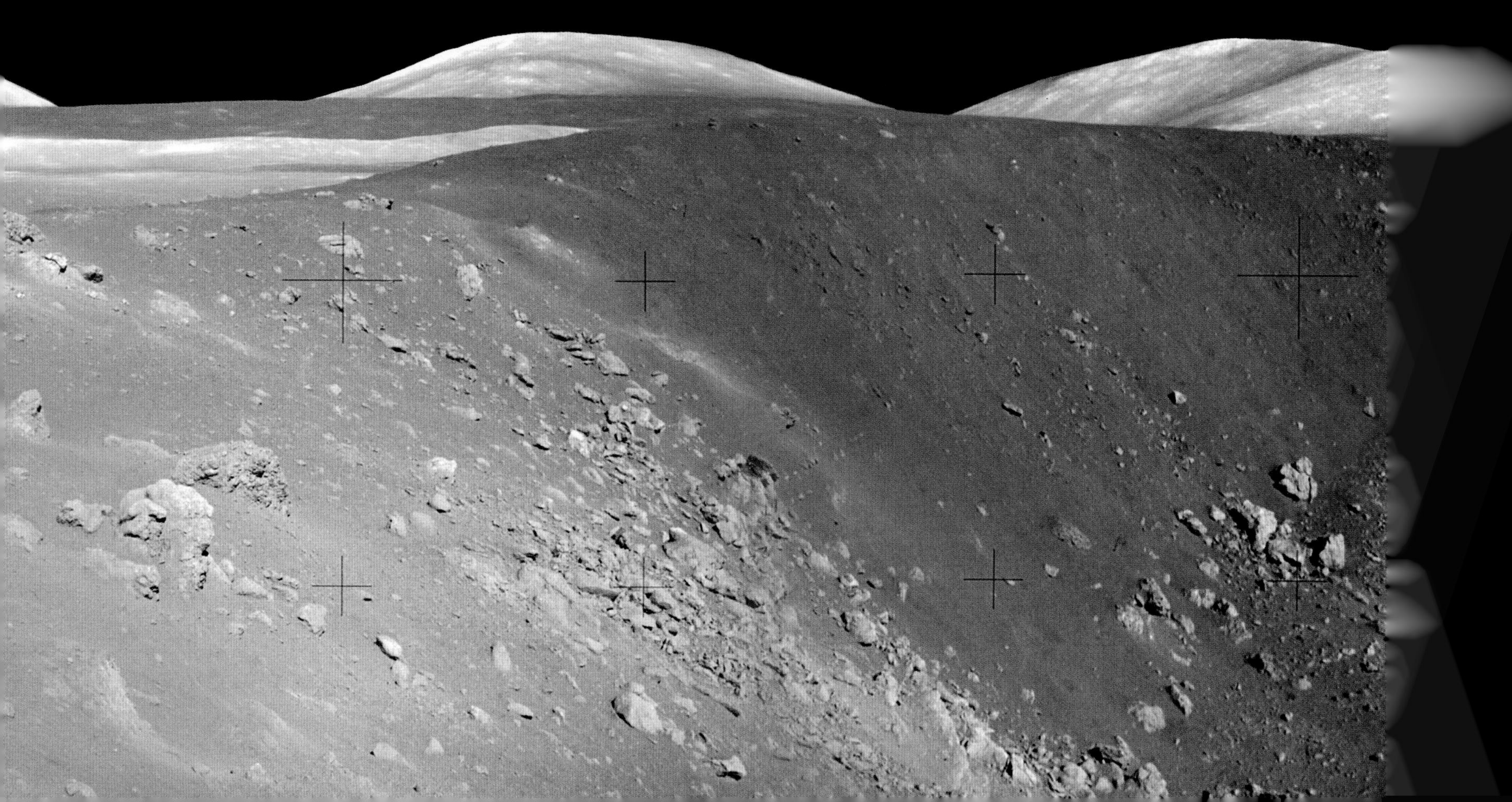

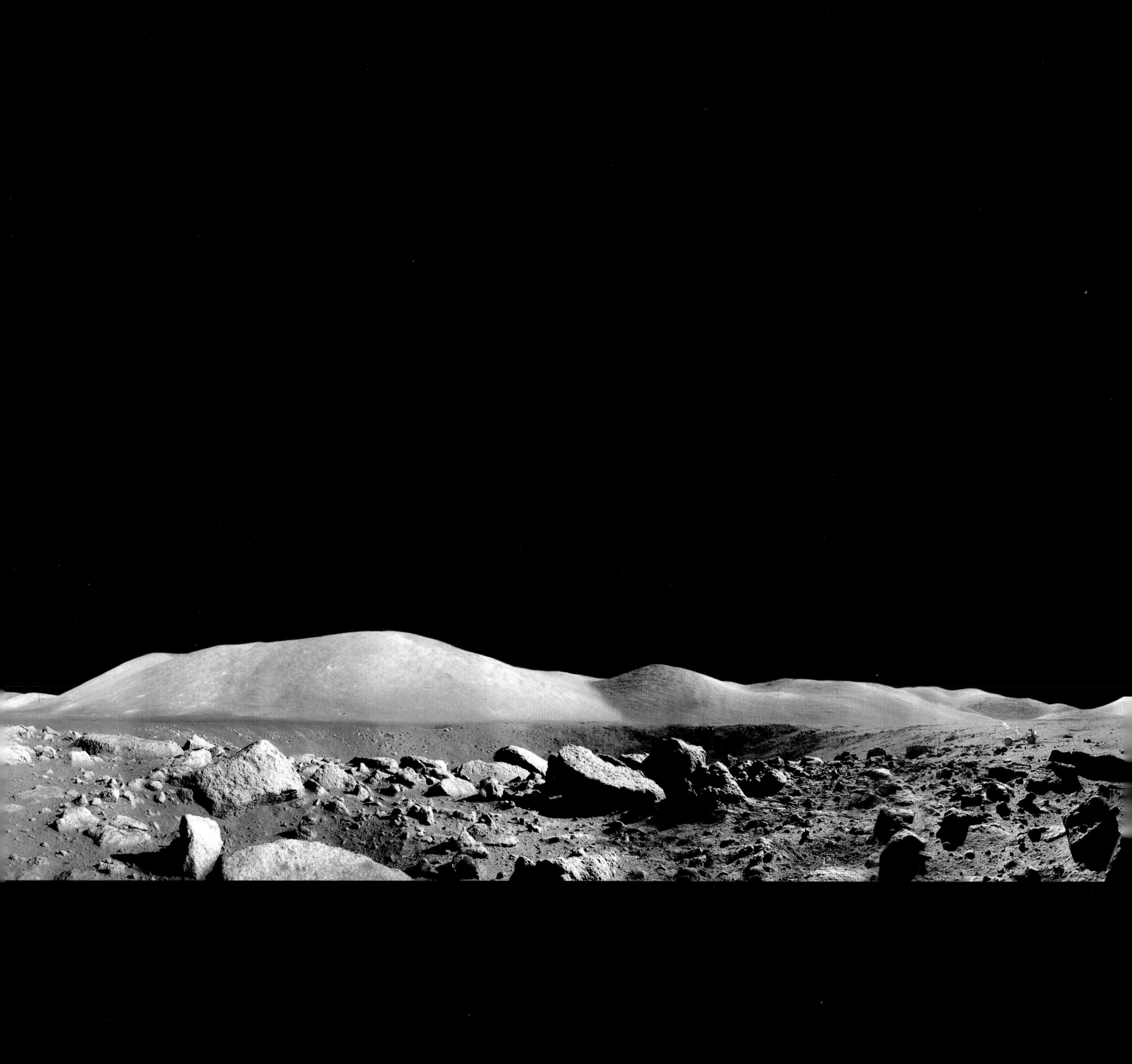

APOLLO 17 AT CAMELOT CRATER
This view of the Moon, taken by Apollo 17 commander Gene Cernan, shows Lunar Module pilot Harrison Schmitt running in the direction of the lunar rover (center), which is parked at the edge of Camelot Crater (left). The astronauts spent roughly 20 minutes working in the boulder field in the foreground, on the crater's southwest rim, taking samples.

THE MOON'S UNSEEN FACE

For the vast majority of its life, the Moon has spun around once for every circuit around Earth, so that one hemisphere has always pointed away from Earth. In October 1959, the lunar far side was imaged for the first time. Since then, astronauts have seen it for themselves, and orbiting craft have explored it.

THE FIRST VIEWS

The USSR's Luna 3 craft was the first to image the Moon's far side. It flew by, taking photographs with its twin-lens camera over a 40-minute period. The Moon's full disc was captured using one lens, while more restricted views were seen through the other. After exposure, the film was developed, fixed, dried, and slowly scanned before being transmitted back to Earth using a system similar to that of a fax machine. Astronomers were surprised by the lack of dark volcanic maria—much less volcanic activity had occurred on the far side than on the near side. The crew of Apollo 8 were the first to see the far side in person when they flew around the Moon in December 1968.

Mare Moscoviense
Sea of Muscovy

Tsiolkovsky Crater

HISTORIC FIRST PHOTO
The far side of the Moon is seen for the first time in this Luna 3 image taken on October 7, 1959. Only the right-hand three-quarters of the photo shows the far side. The most obvious features are the basaltic Mare Moscoviense at upper right and the dark Tsiolkovsky Crater, with its bright central peak, at lower right.

COMMEMORATIVE STAMP
Obtaining the first image of the far side of the Moon was a huge propaganda coup for the Soviets. In 1960 they issued this stamp to commemorate the event. The Luna 3 spacecraft is shown with its cameras, concealed in the lower end of the craft, pointing toward the Moon.

CRATERED FAR SIDE
The far side of the Moon appears more rugged than the near side. Volcanic lava covers a smaller area than on the near side. The far side also features one of the largest craters in the Solar System, the South Pole-Aitken Basin.

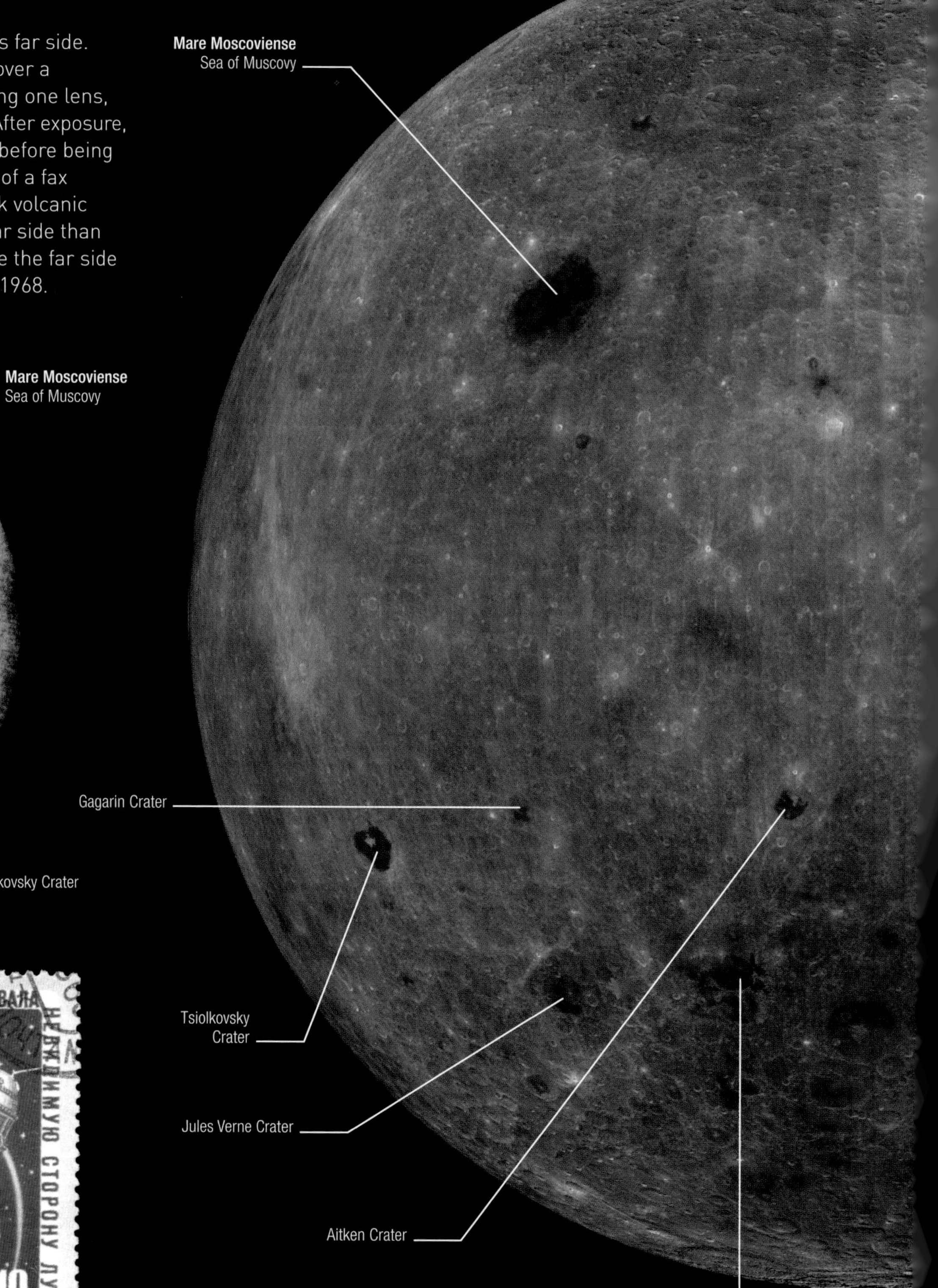

THE NATURE OF THE FAR SIDE

Today the Moon's crust is on average 31 miles (50km) thick, but the crust is about 9 miles (15km) thicker on the far side than on the near side. This difference is explained by events in the Moon's history. Some 4.5 billion years ago, the Moon was closer to Earth and became locked in a synchronous orbit. Over the next 200 million years, it cooled and formed a rocky crust, but the gravitational and heating influence of Earth made the Earth-facing crust form slightly thinner and more slowly than the far-side crust. Until about 3.9 billion years ago, the Moon was bombarded by asteroids and both sides looked similar. Then, from 3.9 to 3.2 billion years ago, the balance between surface cooling, interior radioactive-decay heating, and crustal growth meant that the Moon went through an era of volcanic activity. Due to the difference in crustal thickness, more lava flowed and filled the deep crater basins on the Earth-facing side.

Cockcroft Crater

Hertzsprung Crater

Doppler Crater

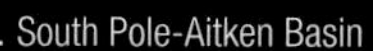
South Pole-Aitken Basin

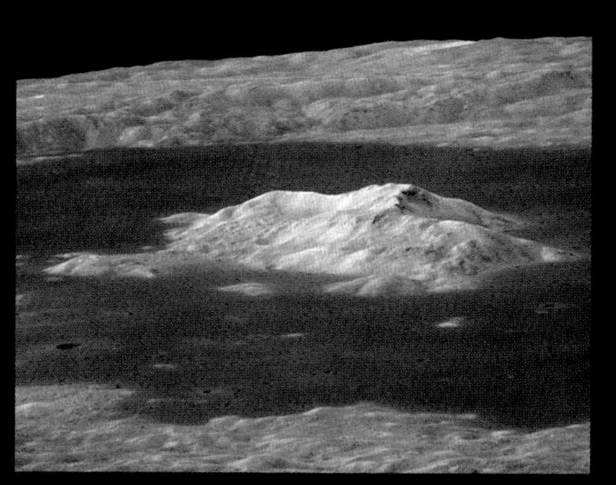

TSIOLKOVSKY CRATER
First identified in the Luna 3 pictures, the 115-mile- (185km-) wide Tsiolkovsky Crater was also imaged during the Apollo missions. This Apollo 15 image from 1971 emphasizes the height of the crater's central peak. It also shows that lava has flooded part of its floor, creating a smooth surface.

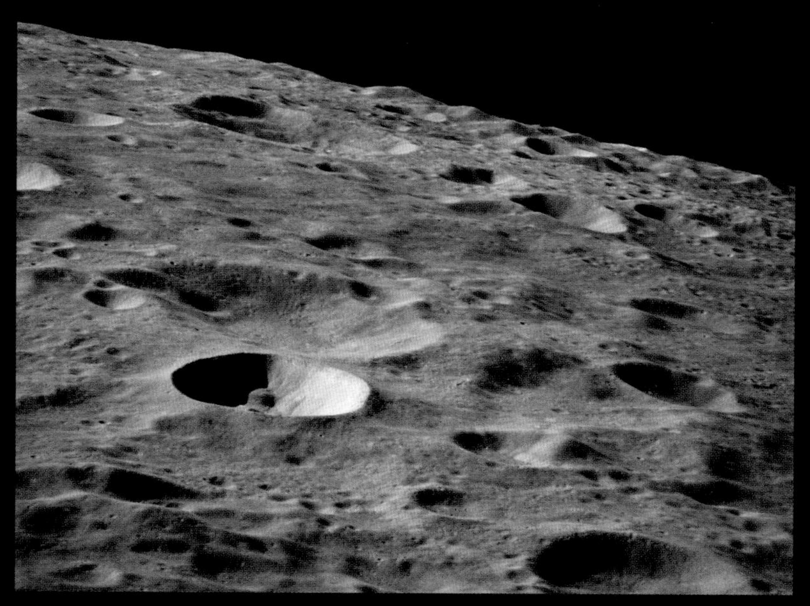

CRATER LANDSCAPE ON THE FAR SIDE
The camera on the outside of the Apollo 16 Service Module took this view of the Moon's far side in 1972. It reveals a landscape of impact craters that were then eroded by further meteoroid bombardment.

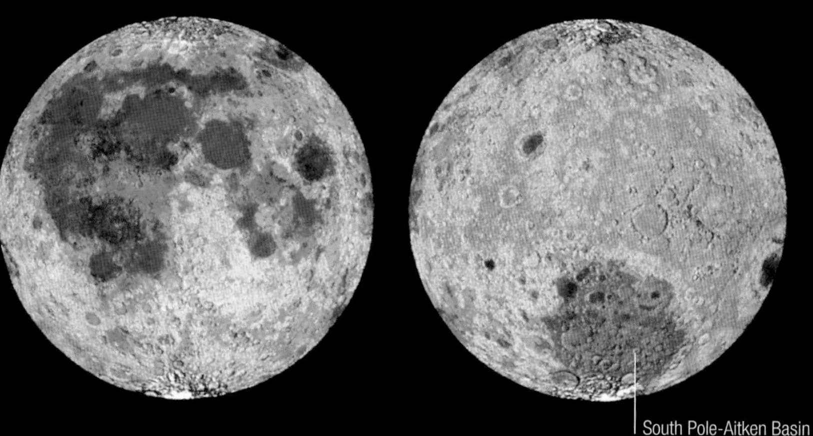

MAPPING THE LUNAR SURFACE
The Clementine orbiter mission to the Moon in 1994 had a laser onboard that measured the height profile of the lunar surface. The resulting lunar maps emphasize the most prominent craters and basins.

VENUS

Our planet's inner neighbor, Venus is sometimes described as Earth's twin, because the pair are similar in size, structure, and composition. But beneath Venus's dense clouds lies a scorching surface dominated by volcanism.

ORBIT AND STRUCTURE

Second in distance from the Sun after Mercury, Venus completes an orbit around our star every 224.7 days. Venus is tilted to its orbital plane by 177.4°, which means that this planet rotates on its axis in the opposite direction from the other planets—clockwise when seen from above its north pole. Its rotation speed is very slow: Venus takes longer to complete one spin than it does one orbit. Venus and Earth formed from the same material, and Venus's Earth-like size and density indicate that its material has differentiated into a layered structure like that of our own planet.

SURFACE TEMPERATURE

Venus is an oppressively hot world. Although not the closest planet to the Sun, Venus has the highest surface temperature of all the planets, averaging 867°F (464°C). This high temperature is relatively constant across the entire planet, both day and night. The small amount of sunlight that gets through the clouds warms the atmosphere and the rock. Any heat subsequently released from the ground is trapped by the clouds in a giant greenhouse effect, which adds to the warming process.

> "IT IS PROBABLY JUST A MATTER OF TIME BEFORE WE "SEE" A VOLCANO ERUPTING."
>
> **PROFESSOR FRED TAYLOR**, VENUS EXPRESS MISSION, APRIL 2008

ORBIT AND ROTATION

Of all the planets, Venus has the most circular path, and there is little difference between its aphelion and perihelion distances. Its rotation period is longer than that of any other planet.

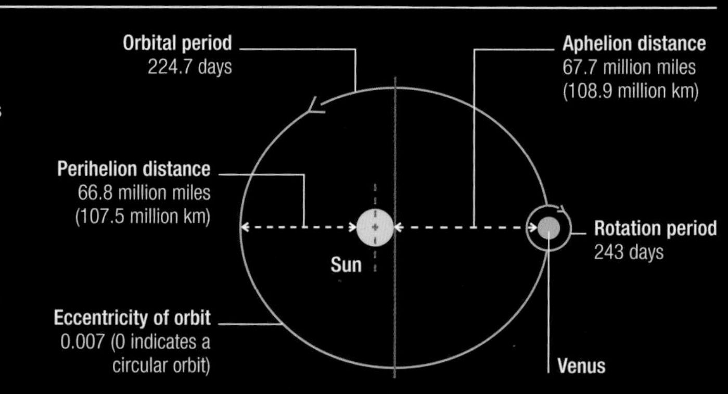

NATURAL VIEW
This natural-color view of Venus was captured by the Messenger spacecraft in 2007. The apparently featureless yet highly reflective clouds give the planet a serene beauty.

ATMOSPHERE

A thick, unbroken, carbon-dioxide-rich atmosphere envelops Venus. Extending from the surface to an altitude of about 50 miles (80km), it contains three distinct cloud layers between roughly 28 and 43 miles (45 and 70km). The clouds are mainly minute droplets of sulfuric acid and other liquids. The upper cloud layer reflects around 80 percent of the sunlight hitting Venus, making this an overcast, gloomy world. Although featureless to the eye, the clouds' structure and dynamic nature are revealed at ultraviolet wavelengths. Infrared observations also provide information on cloud temperature and height. The lower atmosphere moves slowly with the planet's spin. Higher up, the clouds speed around Venus in a matter of days. A vortex forms over each pole, but it is not certain why. It may be that gas heated at the equator by the Sun rises and moves toward the poles. Here it converges, sinks, and is deflected sideways by the planet's rotation.

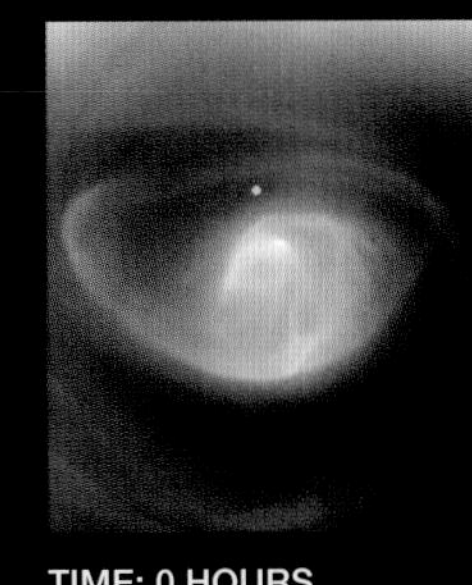

TIME: 0 HOURS

TIME: + 4 HOURS

TIME: + 24 HOURS

TIME: + 48 HOURS

POLAR VORTEX
Some 1,200 miles (2,000km) wide, the vortex above Venus's south pole looks like a huge eye. It can change shape rapidly, appearing oval, round, hourglass-shaped, or anything in between. These Venus Express images of February 2007 show the vortex over a 48-hour period. The view is about 37 miles (60km) above the planet's surface. The yellow dot shows the location of the south pole.

EXPLORING VENUS

Venus's proximity to Earth made it the obvious target for spacecraft exploration, and it was the first planet to have a craft fly by it, in 1962. Mariner 2 found that Venus had a hot surface and cooler clouds. In 1974, Mariner 10 took close-up images and recorded atmospheric circulation patterns. The Venera series, launched over a 20-year period up to 1983, made flybys of Venus, descended into its atmosphere, and even touched down on its surface. Pioneer Venus Orbiter, which arrived in December 1978, produced the first global map of Venus's surface. A second Pioneer Venus craft, the Multiprobe, released three identical probes into the atmosphere, revealing the three-layered cloud structure. The first comprehensive survey was achieved by Magellan in the 1990s.

Venus Express has been investigating the atmosphere since 2006. Its infrared spectrometer studies the southern hemisphere's clouds and winds. Sulfur-dioxide gas has been detected, possibly from recent volcanic eruptions. Temperature differences in the southern hemisphere may also be due to volcanism or simply to differences in surface material.

GROUND TEMPERATURE
Centered on Venus's south pole, this map is a composite of more than 1,000 images from Venus Express's spectrometer. It shows ground temperatures of 792–828°F (422–442°C), ranging from red (hottest), to yellow, green, and blue (coolest). Higher temperatures correspond to lower altitudes, and lower temperatures to higher altitudes.

VENUS EXPRESS
The European Space Agency's craft Venus Express follows an orbit around Venus that brings it to within 155 miles (250km) at its closest and 41,000 miles (66,000km) when furthest away. Its seven instruments are mounted on the central body, which is about 5ft (1.5m) across.

SIZE

Venus is slightly smaller than Earth. Its polar and equatorial diameters are the same, while Earth has a slight bulge at its equator.

Earth
Diameter at equator: 7,926 miles (12,756km)

Venus
Diameter at equator: 7,521 miles (12,104km)

STRUCTURE

How much of the core is solid or molten is unclear. Internal radioactive heat melts the upper mantle, which erupts onto the surface as lava.

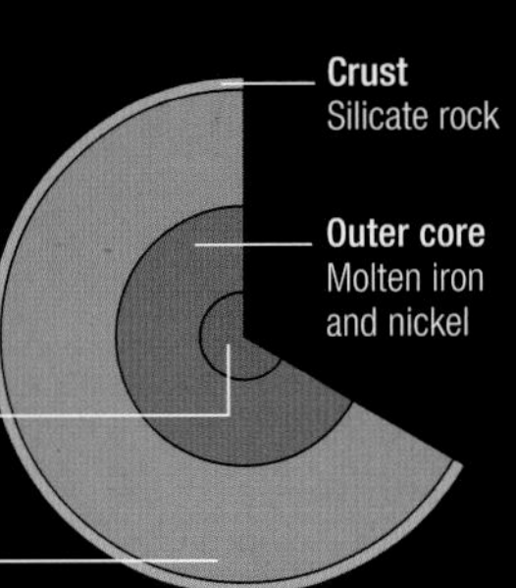

ATMOSPHERE

The Venusian atmosphere is mostly carbon dioxide, with some nitrogen and traces of water vapor, argon, and sulfur dioxide.

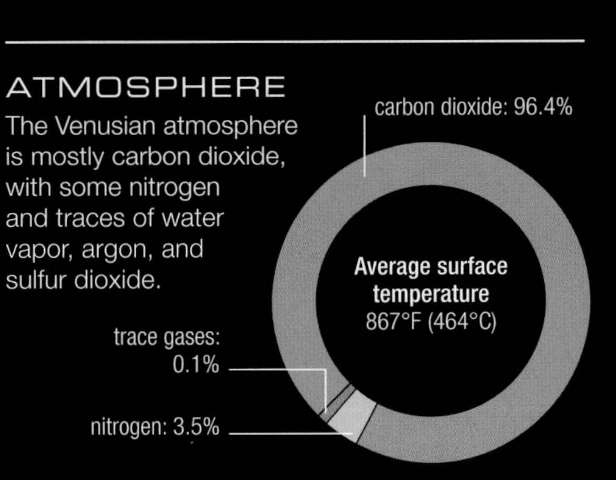

BENEATH VENUS'S CLOUDS

From above its clouds, Venus is a beguiling world—an appearance that is fitting for a planet named after the Roman goddess of love and beauty. However, probes sent into the Venusian clouds and other craft carrying atmosphere-penetrating radar have revealed a world that no human would want to visit.

FIRST SURFACE VIEW

Venus is a hostile environment for spacecraft as well as for humans. A craft attempting to land on Venus must survive not only the planet's scorching heat, but also the intense pressure encountered at its surface. The atmospheric pressure here is 90 times greater than that at Earth's surface. The Russian Venera craft were the first to penetrate the atmosphere. In 1967, Venera 4 was within 15 miles (24km) of the surface when it was crushed by the pressure. Four more Venera craft followed—of which two successfully made it to the surface—before Venera 9 returned the first black-and-white image to Earth in 1975. In the early 1980s, the landers Veneras 13 and 14 sent back the first color images, while the orbiters Veneras 15 and 16 used radar equipment to map the northern part of the planet.

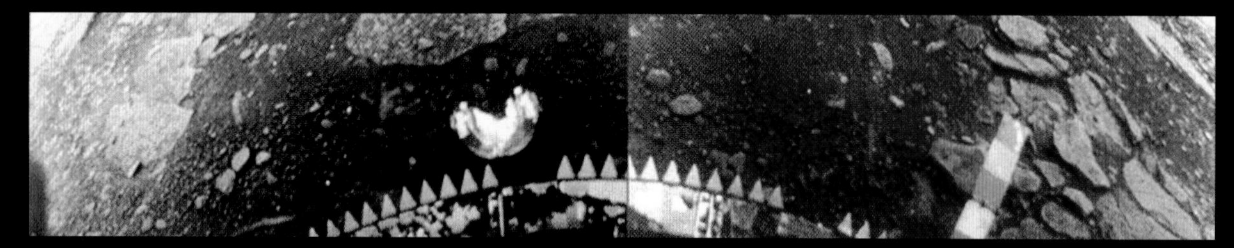

FIRST COLOR SURFACE IMAGE
This 170° panoramic view shows slabs of basaltlike rock and soil. It was taken on March 1, 1982 by Venera 13, which survived on the surface for just 2 hours 7 minutes. Part of the craft and its camera-lens cover are visible in the foreground.

LOOKING THROUGH THE CLOUDS

The Magellan spacecraft continued the radar mapping started by Veneras 15 and 16, and by Pioneer Venus 1. In May 1989, it set out on its mission to make the first comprehensive survey of the planet and at a hugely improved resolution. Constructed in part from material and equipment left over from the Voyager and Galileo missions, Magellan was highly successful. The craft followed a polar orbit around Venus, making one complete circuit in less than 3.5 hours. Magellan's path was elliptical, taking it to within 183 miles (294km) of the Venusian surface and as far out as 1,300 miles (2,100km). The craft completed four eight-month mapping cycles between 1990 and 1994. During the first three cycles, its radar recorded surface details. As Venus made its 243-day rotation under the orbiting spacecraft, the planet was imaged as a series of long, narrow strips. Each strip was 12 miles (20km) wide and 10,600 miles (17,000km) long, and neighboring strips overlapped. All the strips were then combined to produce a global view. Magellan mapped 98.3 percent of the planet's surface, recording details as small as 300ft (90m) in diameter and surface heights to within 160ft (50m). On its fourth cycle, Magellan collected data on Venus's gravity. Then it was deliberately directed into Venus's atmosphere, from where it sent back measurements before burning up.

MAGELLAN SPACECRAFT
In the picture above, Magellan is about to be released from the cargo bay of the Shuttle orbiter *Atlantis* while the two craft are in low Earth orbit. Once free of *Atlantis*, Magellan ignited its booster-rocket stage and moved onto a course for Venus. Its job complete, the rocket stage separated from Magellan.

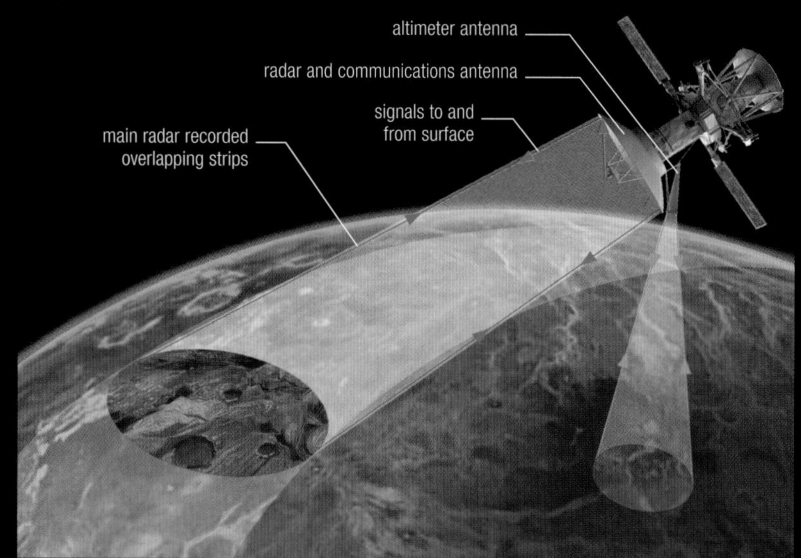

MAPPING THE SURFACE
The 12ft 2 in (3.7m) radar antenna looked down and to the side of Magellan's orbital path, emitting several thousand pulses of energy every second. As the signals bounced back to the craft, their strength and frequency were measured. The altimeter pointed vertically down and measured the craft's height above the ground.

NORTHERN HEMISPHERE

VOLCANIC WORLD

Data returned by Magellan were processed by computer to produce maps and images of the Venusian surface, revealing a dry, lifeless landscape dominated by volcanism. About 85 percent of the surface is volcanic plain covered by extensive lava flows. There are hundreds of volcanoes dotted around the planet. Many are broad-based, shallow-sloped shield volcanoes formed through successive eruptions. Some of the smaller volcanic features, such as arachnoids and pancake domes (below), are unique to Venus. There are also highland regions, mountain belts, and impact craters. Venus Express has shown that the highland rock is old compared with that of the rest of Venus, and it is reminiscent of the granite that forms Earth's continents. Since water is involved in the formation of granite, proving the presence of granite on Venus would indicate that water and plate tectonics played a role in Venus's past.

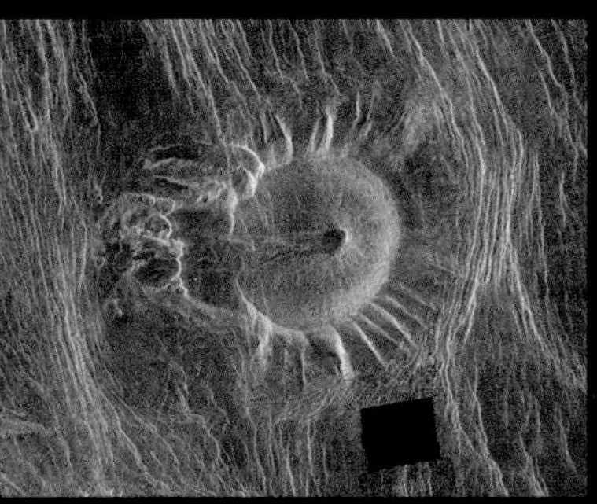

ARACHNOID FEATURE
This spiderlike volcano's slightly concave summit is 22 miles (35km) across. A lava flow extends to the west (left), while ridges and valleys radiate from the summit.

PANCAKE DOMES
These features are flat-topped domes that formed when viscous lava erupted onto the surface. The two above are 40 miles (65km) across but less than 0.6 miles (1km) high.

SOUTHERN HEMISPHERE

MAGELLAN MAPS
The surface of Venus is shown on these two "flower-petal" maps, which fit together to form a globe. The result of more than a decade of radar investigations, the maps are made from a mosaic of Magellan images—gaps in its coverage are filled by data from earlier craft. High land is shown in red and low-lying land in blue.

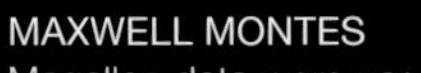

MAXWELL MONTES
Magellan data were used to produce this computer-generated image of Maxwell Montes. It is the view that would be seen from an airplane flying over Venus. The mountains measure 7 miles (11km) from the plain to the highest peak. The clouds and haze were added to suggest Venus's high temperature.

VENUS'S SURFACE

Venus's vast plains of volcanic lava are known as planitiae. Channel-like features called valles (singular vallis) are often seen on these plains. Typically 0.6–1.9 miles (1–3km) wide, they formed when lava melted or eroded a path across the landscape. The volcanoes are mainly in the upland areas called regiones (singular regio). Higher still are three highland areas named terrae (singular terra). Ishtar Terra is home to the Maxwell Montes range—the highest point on Venus. Other features such as troughs, rifts, and chasms are tectonic, meaning that they formed as the crust was pulled apart or compressed. Radar images do not show Venus's features as they appear to the naked eye. In such images, bright areas appear the roughest; dark areas are relatively smooth.

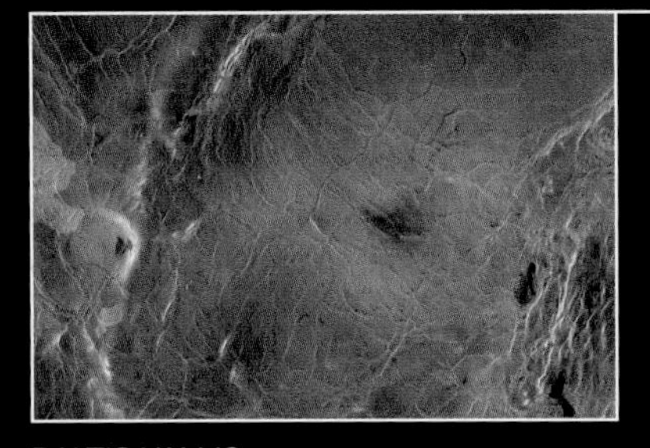

BALTIS VALLIS
At 3,700 miles (6,000km) long, Baltis Vallis is the longest of Venus's lava channels. It is named after the Syrian word for the planet. A 370-mile (600-km) section of the channel runs from top to bottom in this image.

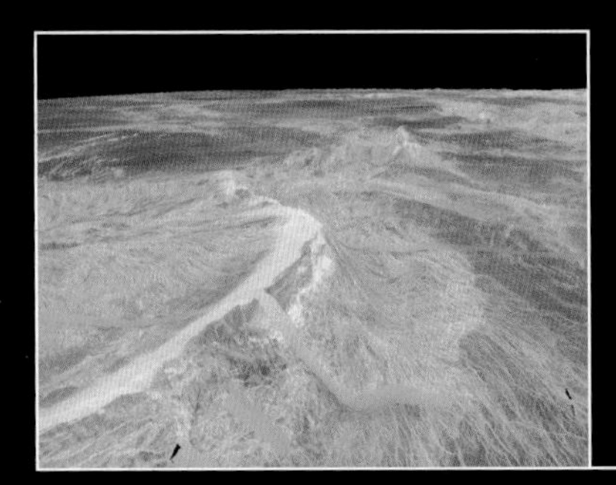

DALI CHASMA
The Dali Chasma is a system of canyons and troughs that runs for 1,291 miles (2,077km). A 1.9-mile- (3km-) deep portion is shown here to the right of the bright, curving rim of the Latona Corona.

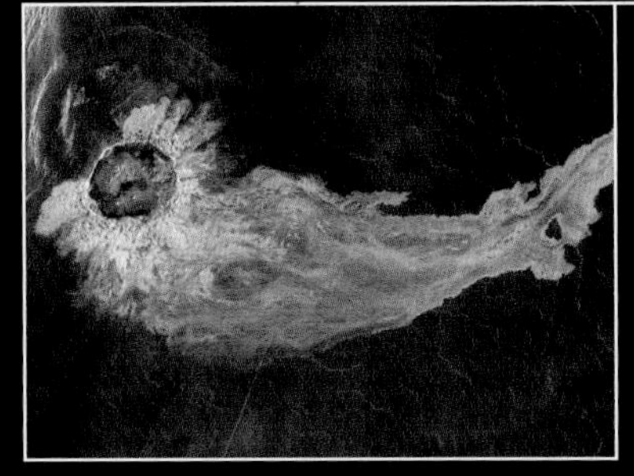

ADDAMS CRATER
A mermaidlike tail of material flows away from the 54-mile- (87km-) wide Addams Crater. This was once molten material ejected from the crater. Its brightness indicates that it has a rugged surface.

VOLCANIC VENUS
This global view of Venus's surface was made using data collected almost exclusively by Magellan. The coloring is based on that recorded in the images from the Venera 13 and 14 spacecraft.

GANIS CHASMA
This huge rift valley, which crosses the Atla Regio volcanic rise, was formed by a fracture in the planet's crust. It is 2,300 miles (3,700km) long—the third longest chasma on Venus.

VOLCANIC MOUNTAINS

The largest volcanoes on Venus are its shield volcanoes. Five huge one linked by a complex system of crustal fractures, occur in the Atla Reg a broad volcanic rise north of the equator. They include Maat Mons a Sapas Mons. Venus also has hundreds of coronae—roughly circular volcanic structures formed when magma lifted crustal rock above the surrounding land. It has long been thought that Venus experienced volcanic activity as recently as 500 million years ago, but some exper believe that its volcanoes could still be active. Venus Express scientis are looking for hotspots that could be current lava flows and for localize increases of sulfur dioxide that could be coming from a volcano.

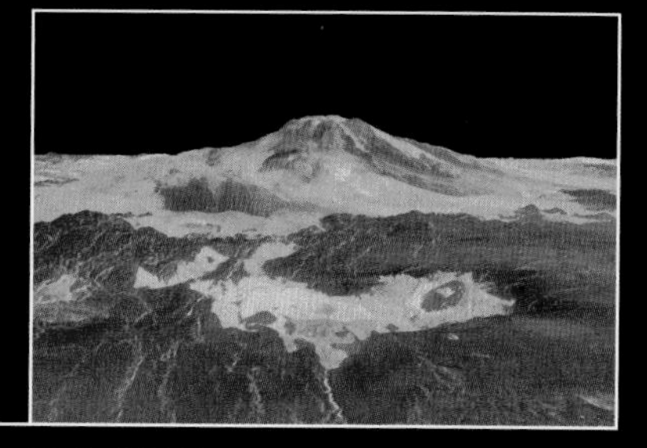

MAAT MONS
Lava flows extend hundreds of miles beyond the base of Maat Mons—Venus's largest volcano, which takes its name from an Egyptian goddess. This computer-generated view shows it rising to almost 3 miles (5km) above the plains.

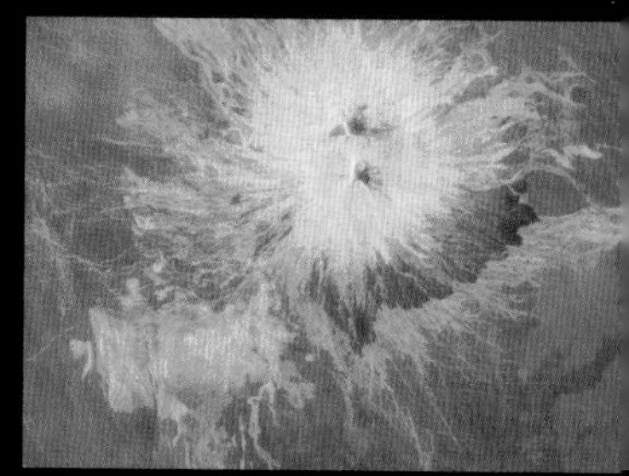

SAPAS MONS
The summit of Sapas Mons is seen at to center in this bird's-eye view. The two da areas on the volcano's summit are mesa The brightness of the surrounding lava indicates rougher material. There are lava flows on the flanks and on the summit.

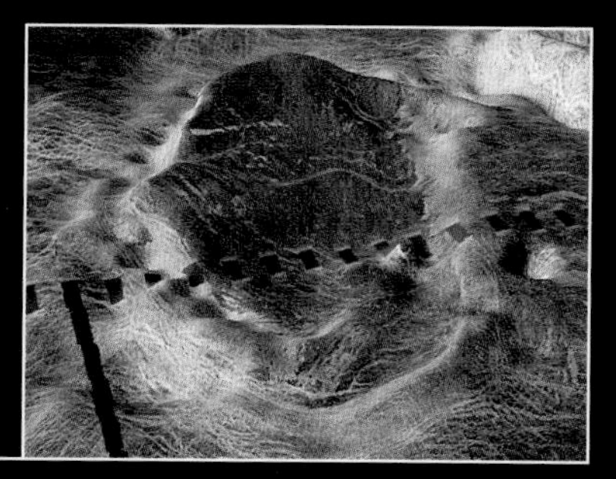

ATETE CORONA
Named after an Ethiopian fertility goddes Atete Corona is almost oval in shape and up to 370 miles (600km) across. This blisterlike structure formed when subsurface magma pushed up partially molten crust. The dark blocks shown are produced by the imaging process.

IMPACT CRATERS

Like the other inner planets, the young Venus was bombarded by asteroids. But while Mercury still bears the scars, Venus's surface has been renewed by volcanic activity. The hundreds of craters identified o Venus formed in the past 500 million years or so. They range from a few miles wide to Mead Crater, at 168 miles (270km) across. There are few small, bowl-shaped craters, since the small asteroids that form them rarely make it through the atmosphere. Ejected material can be blown by the winds to form tails stretching away from the craters.

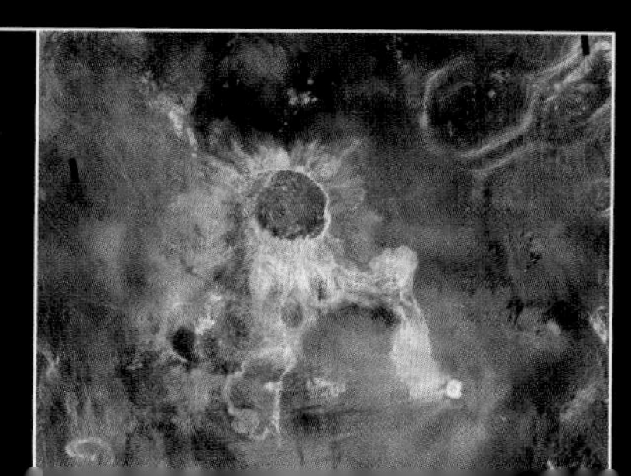

ISABELLA CRATER
Named after the 15th-century Spanish queen, Isabella is the second largest crater on Venus. It is 108 miles (175km) across, with flows of material extending al around it. The material may be rock melte by the heat of impact, or a mix of hot gas, rock, and melted rock pushed across the landscape during impact.

The smallest planet in the Solar System, Mercury is also the closest to the Sun. With little atmosphere and a slow rate of spin, on its long days and nights, it goes from being hot enough to melt lead to cold enough to liquefy air.

PIN, ORBIT, AND STRUCTURE

rcury has the most eccentric and highly inclined orbit of all the nets in the Solar System. Its distance from the Sun varies from 6 million miles (46.0 million km) to 43.4 million miles (69.8 million). As with other planets, its closest position changes very slightly h each orbit. But in Mercury's case, a tiny part of this is caused by curvature of space near the n. Mercury spins once in 58.6 ys. The axis of its spin is almost pendicular to the plane of its oit, so it has no seasons. For its e, it has a relatively large iron e, which generates a magnetic d about 100 times weaker than rth's. Astronomers are not sure ether Mercury is iron-rich, due its proximity to the Sun, or k-poor, due to its extensive mbardment by asteroids nen it was young.

TRANSIT OF MERCURY
Mercury passes between Earth and the Sun 13 to 14 times each century. This will next happen on May 9, 2016, November 11, 2019, and November 13, 2032.

EMPERATURE AND ATMOSPHERE

e day on Mercury, from sunrise to sunrise, is 176 Earth days long. ring this time, the planet makes two orbits of the Sun. When the n is directly overhead, the temperature at the surface is a scorching 6°F (430°C). In the middle of a Mercury night, though, it is a chilling 92°F (-180°C)—only a few degrees above the temperature of liquid rogen. Mercury's atmosphere is very thin, but it is much denser on e night side of the planet than on the day side. Its overall density is out a trillion times less than that of Earth's. Mercury's atmosphere also highly unstable. The planet's gravity is not strong enough to ld onto it for very long, and the rate at which it leaks away into space ries according to solar activity as well as whether it is on the night or y side of the planet. Mercury's atmosphere is constantly replenished th gases captured from the solar wind, emitted by radioactive decay the planet's crust, baked out of the surface rocks by the Sun, or ected by meteoroid impacts.

CRATERED WORLD
Relatively recent impact craters and ejecta rays appear brightest in this photograph of Mercury taken by Messenger (see pp.84–85) on its second flyby (October 2008).

IERCURY

ercury's highly eccentric bit means that there is a ge difference between perihelion and aphelion stances from the Sun. In ery two orbits, the planet tates exactly three times.

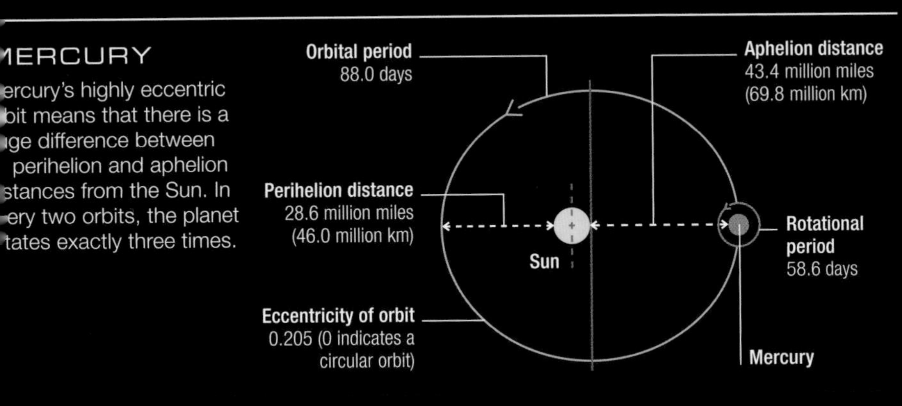

SIZE

Mercury is about a third bigger than the Moon, with a diameter 38 percent that of Earth and a mass 5.5 percent that of our planet.

Earth
Diameter at equator:
7,926 miles (12,756 km)

Mercury
Diameter at equator:
3,032 miles (4,879 km)

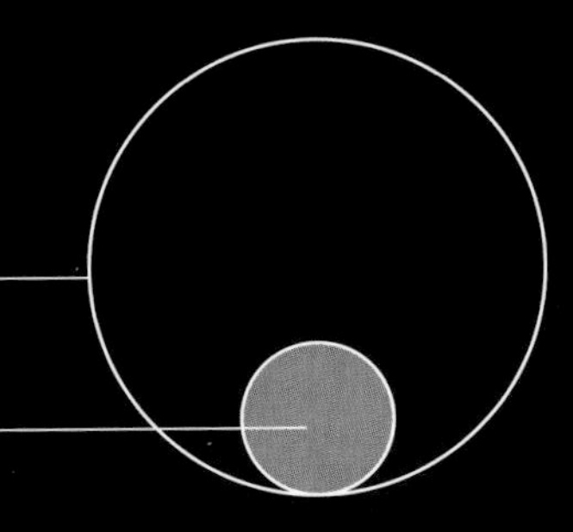

Like Earth's moon (see pp.56–57), Mercury is heavily cratered as a result of a long period of asteroid and meteorite bombardment about 4 billion years ago. Mercury craters range from small, bowl-shaped ones to the enormous Caloris Basin, whic at some 960 miles (1,550km) across is nearly one-third of the width of the whole planet. And like the Moon, in certain regions of Mercury, there are smooth plains between the craters, the result of ancient periods of volcanism and lava flow. The surface of Mercury also has ridges and cliffs that are hundreds of miles long. These are the result of two processes: uneven shrinkage of the crust as the plane gradually cooled after formation; and a change in the planet's shape from oblate (looking like a squashed ball) to its present almost spherical form—a change tha took place when the planet's rate of spin slowed (below).

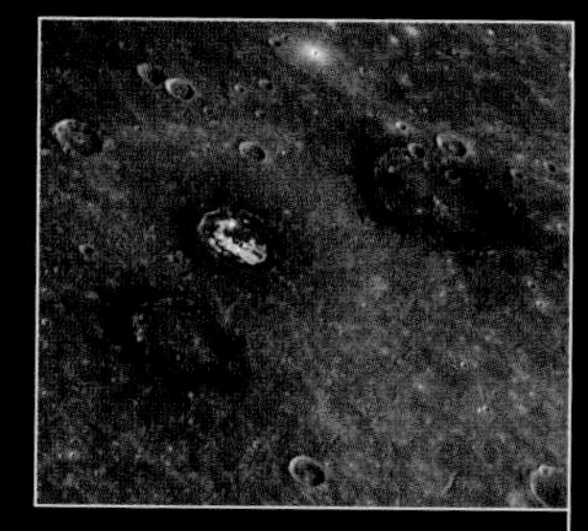

THREE IMPACT CRATERS
Munch (left), Sander (middle), and Poe (right) are three large impact craters that formed about 3.9 billion years ago—well after the impact that formed the Caloris Basin.

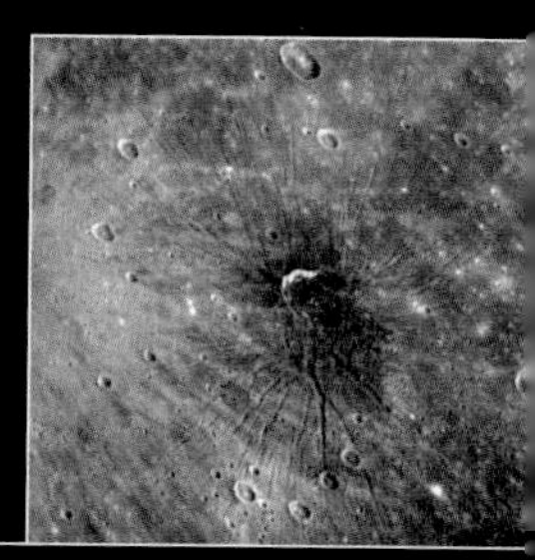

PANTHEON FOSSAE
Pantheon Fossae is a series of long, shallow troughs that appear to radiate from a crat called Apollodorus that is some 25 miles (41km) wide. Scientists are not sure if the impact that formed the crate also formed the troughs.

THE CALORIS BASIN
The Caloris Basin appears as a large orange circle in this false-color image taken by Messenger on its first flyby (January 2008).

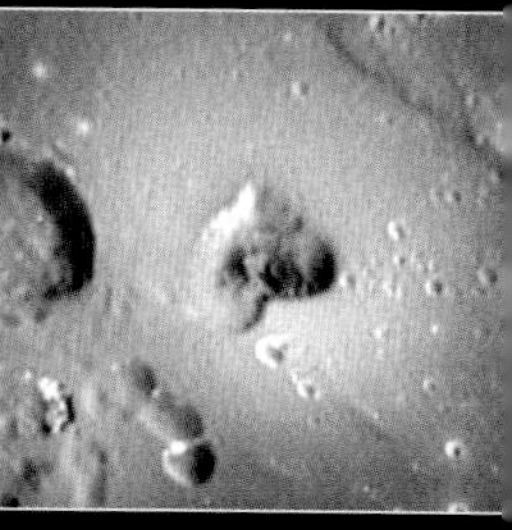

CALORIS VOLCANO
This is the largest volcano so far discovered on Mercury. This image shows vents inside the central caldera, which is surrounded by ejected volcanic material.

STRUCTURE

Mercury has a relatively large iron core that makes up some 65–70 percent of the planet's mass. Earth's core accounts for only 32.5 percent of its mass.

ATMOSPHERE

Mercury's entire atmosphere only weighs around 2,205lb (1,000kg).

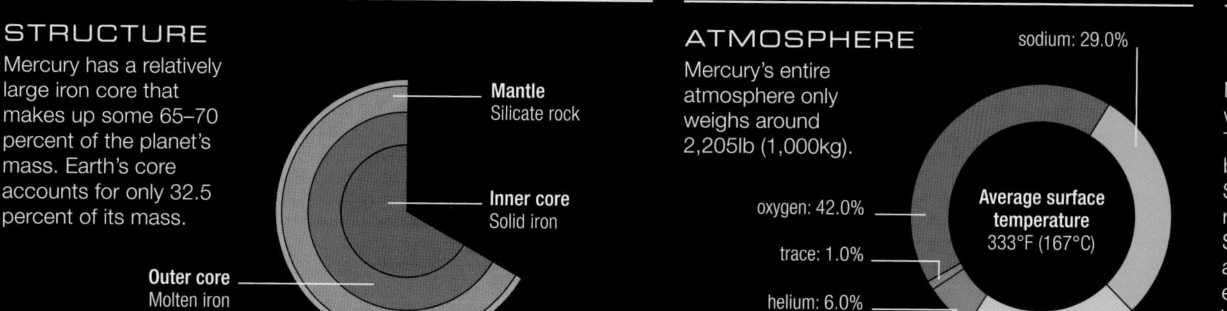

58.6

Mercury rotates once in 58.6 days, which is two-thirds of its orbital period. This is unusually long for one spin because most planets in the Solar System, with the exception of Venus, rotate in approximately one day or less Something has drastically slowed Venus and Mercury down. One possible explanation is that Mercury once orbitec

ИESSENGER AT MERCURY

ıly two spacecraft have ever visited Mercury. Mariner 10 flew past the anet in 1974–75 and revealed it to be much like Earth's moon: heavily atered and sculpted by lava flows. Now Messenger has flown past it ree times, ready to move into orbit for a full year's investigation.

HE MISSION

ssenger's mission is to investigate Mercury's face, atmosphere, and magnetosphere. Once in elliptical orbit—at heights ranging from 125 les (200km) to 9,400 miles (15,200km)—it will nplete one circuit every 12 hours on a path lined at 80° to the equator. In the craft's year orbit, a battery of instruments will be at work: de- and narrow-angle cameras will photograph e entire surface; neutron, gamma-ray, and X-ray tectors will reveal the composition of the exposed ust; a laser altimeter will produce an accurate ntour map; and an ultraviolet spectrometer will easure the composition of the atmosphere.

HE SPACECRAFT

Mercury, the Sun is around 11 times brighter and tter than at Earth, so Messenger has a large nshade, but relatively small solar panels. Except its magnetometer, all its scientific instruments int straight downward. Most are fixed in place, t the dual imaging system can pivot to take otographs of different areas of the surface.

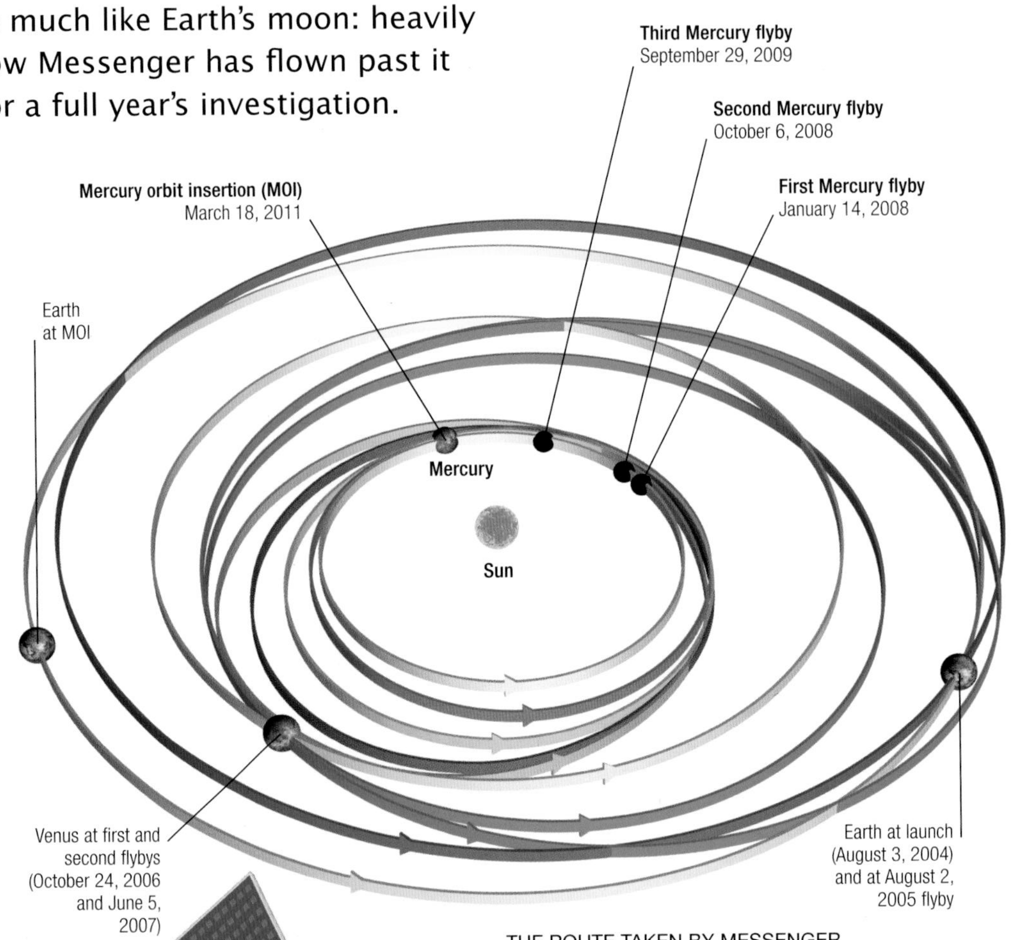

Third Mercury flyby
September 29, 2009

Second Mercury flyby
October 6, 2008

First Mercury flyby
January 14, 2008

Mercury orbit insertion (MOI)
March 18, 2011

Earth at MOI

Mercury

Sun

Venus at first and second flybys (October 24, 2006 and June 5, 2007)

Earth at launch (August 3, 2004) and at August 2, 2005 flyby

THE ROUTE TAKEN BY MESSENGER
Since its launch in August 2004, Messenger has flown past Earth once, Venus twice, and Mercury itself three times. These planetary flybys are essential braking maneuvers to counter the gravitational pull of the Sun, slowing the spacecraft down so that it can finally go into an elliptical orbit around Mercury in March 2011.

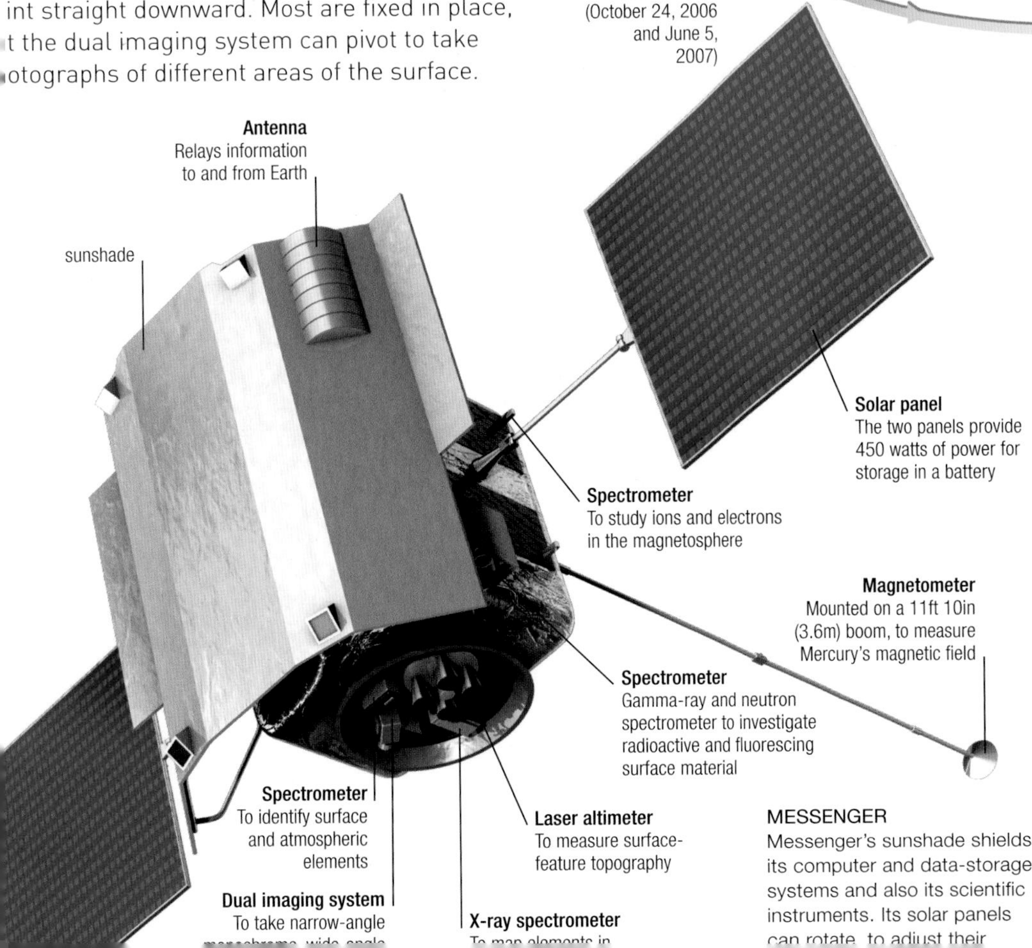

Antenna
Relays information to and from Earth

sunshade

Solar panel
The two panels provide 450 watts of power for storage in a battery

Spectrometer
To study ions and electrons in the magnetosphere

Magnetometer
Mounted on a 11ft 10in (3.6m) boom, to measure Mercury's magnetic field

Spectrometer
Gamma-ray and neutron spectrometer to investigate radioactive and fluorescing surface material

Spectrometer
To identify surface and atmospheric elements

Laser altimeter
To measure surface-feature topography

Dual imaging system
To take narrow-angle

X-ray spectrometer

MESSENGER
Messenger's sunshade shields its computer and data-storage systems and also its scientific instruments. Its solar panels

MESSENGER PROFILE

MISSION	
Launch date	August 3, 2004
Arrival at Mercury	March 18, 2011
Mission ends	2012
Launch vehicle	Delta II 7925-H
MESSENGER	
Agency	NASA
Main body	6ft X 4ft 8in X 4ft 2in (1.8m X 1.42m X 1.27m)
Weight	2,410lb (1,093kg)
Power source	Bipropellant (hydrazine and nitrogen tetroxide) thruster

SCALE

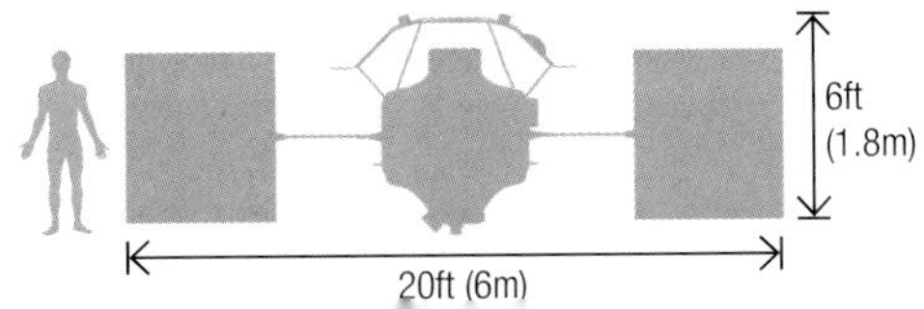

6ft (1.8m)

20ft (6m)

MESSENGER'S VIEW

Mariner 10 imaged about 45 percent of the surface of Mercury, but once in orbit, Messenger will look at the whole globe, including the polar regions. Messenger will photograph the surface when the terrain is lit by the Sun from various angles. Its wide-angle camera has 11 color filters—nine more than Mariner 10's. Along with data from its altimeter, Messenger's cameras will reveal Mercury's surface in unprecedented detail, identifying features only 59ft (18m) across—90 times the resolution of Mariner 10 images.

Messenger has already returned images of previously unseen parts of Mercury, taken from around 125 miles (200km) away on its flybys. These revealed the Caloris Basin in full for the first time, and showed clear evidence of volcanic activity, as well as many impact craters.

TRUE-COLOR VIEW
On the second flyby, the wide-angle camera imaged Thakur, an impact crater 73 miles (118km) wide, through its red, blue, and green filters. The result is how the scene would look to the human eye.

FALSE-COLOR VIEW
Combining images taken through all 11 filters of Messenger's wide-angle camera, from visible to near infrared, creates a single image that enhances subtle color differences in the surface rock.

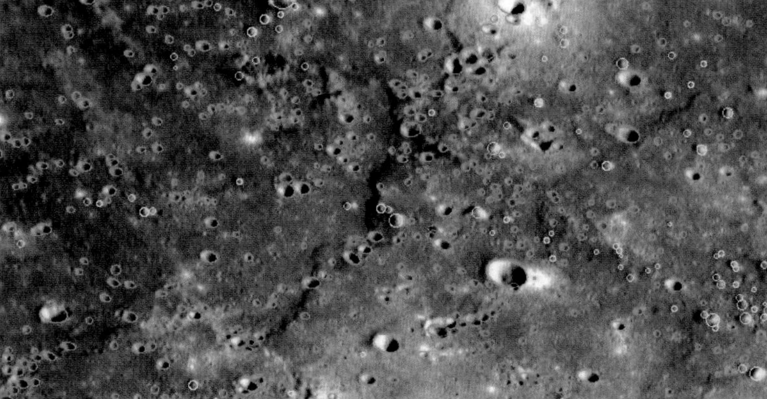

COUNTING CRATERS AND HILLS
Messenger's narrow-angle camera took this photograph on the first flyby. It shows 763 impact and ejecta craters (marked green) and 189 hills (marked yellow) in an area some 137 miles (220km) wide.

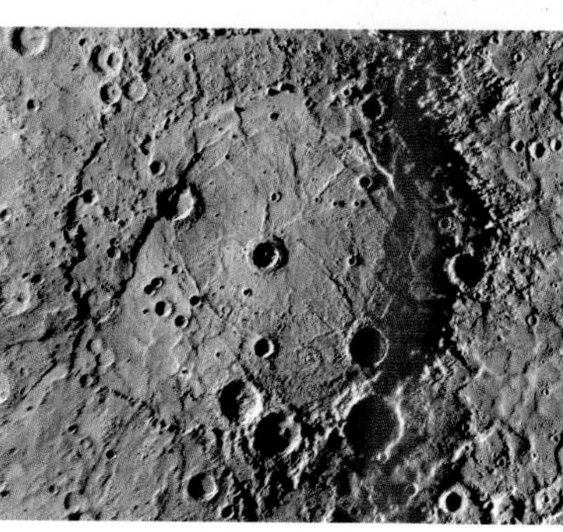

NEWLY DISCOVERED IMPACT BASIN
On the second flyby, the narrow-angle camera photographed a previously unknown impact basin in Mercury's southern hemisphere. Called Rembrandt, it is about 447 miles (720km) wide.

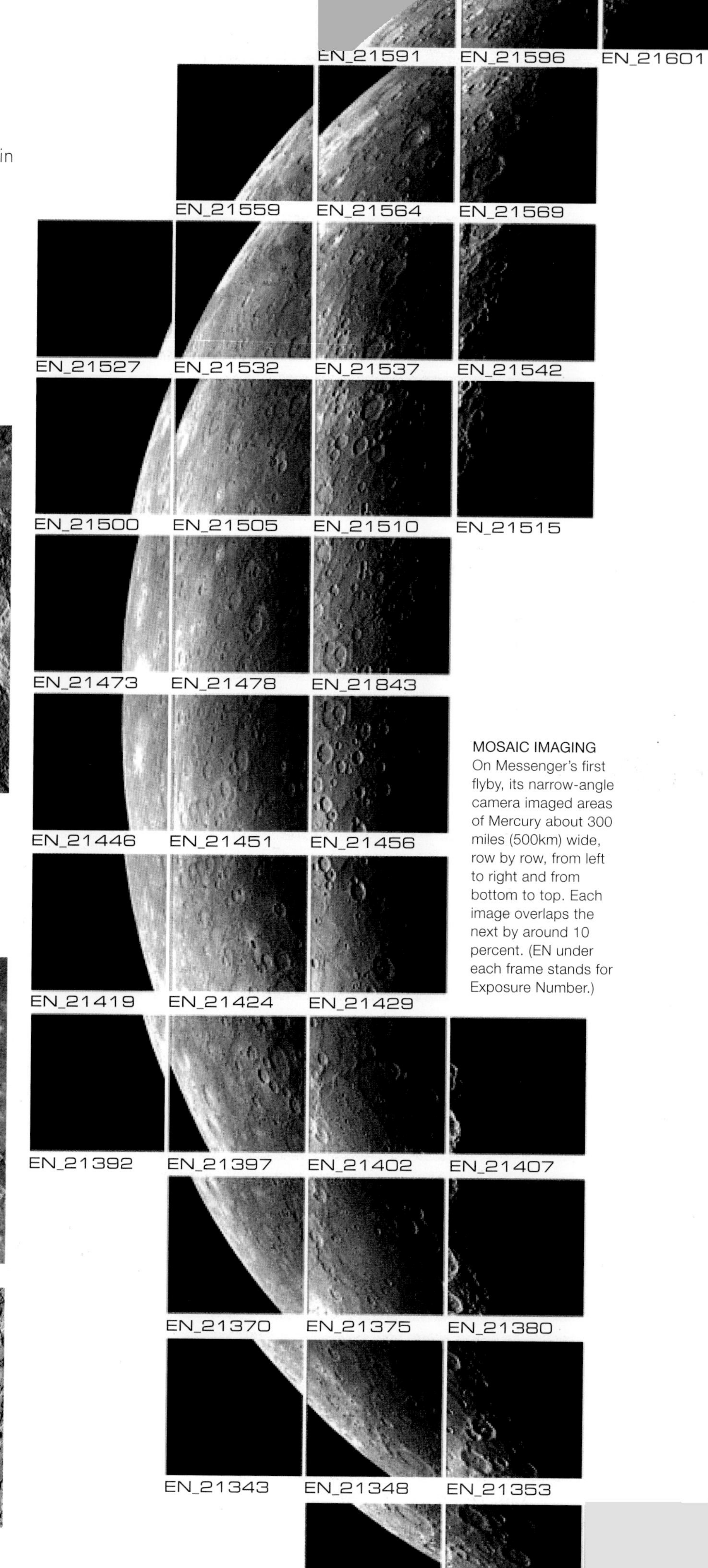

MOSAIC IMAGING
On Messenger's first flyby, its narrow-angle camera imaged areas of Mercury about 300 miles (500km) wide, row by row, from left to right and from bottom to top. Each image overlaps the next by around 10 percent. (EN under each frame stands for Exposure Number.)

THE SUN

The hot, incandescent ball of gas that we call the Sun provides Earth with life-sustaining heat and light, and dictates its seasons and climate. But, as awesome as the Sun seems to us, it is a fairly typical star—one of at least 200 billion in the Milky Way alone.

A TYPICAL STAR

The Sun lies in the disc of the Milky Way (see pp.192–93), in a region where the average space between stars is about 8 light-years. Like more than 90 percent of all stars, the Sun is in a stable stage of its life. In its hot, dense core, nuclear-fusion reactions convert hydrogen into helium. This releases energy, which gradually makes its way to the surface, where it escapes into space, mostly as infrared radiation (heat) and light. The Sun's rate of energy production and its surface temperature have both stayed relatively constant for about 4.6 billion years—and will stay so for as long again, until the hydrogen runs out. The Sun will then swell into a red giant, increasing its energy output as it burns helium and destroying the inner planets. After ejecting its outer layers, it will end its days as a white dwarf (see pp.238–39).

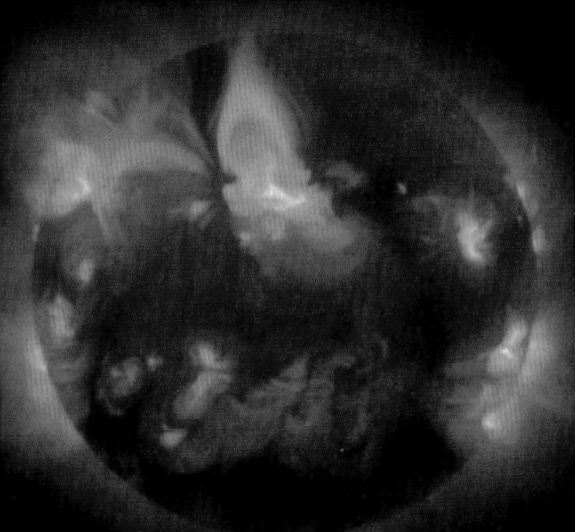

SOLAR X-RAY
The bright areas in this X-ray image show concentrations of hot plasma (ionized gas) in the corona, the Sun's outer atmospheric layer. The dark patches are coronal holes—regions from which particles stream out and form the solar wind (see p.89).

THE SUN IN VISIBLE LIGHT
This is the Sun's surface, or photosphere, as we see it. Although the Sun is gaseous throughout, its density increases so quickly with depth that the disc's edge looks sharp, albeit darker. The spots are cooler regions produced by the Sun's magnetic field.

SOLAR DATA	
Diameter	864,000 miles (1.39 million km)
Mass (Earth = 1)	330,000
Composition	73.5% hydrogen, 24.9% helium, 1.6% trace elements
Energy output	385 million billion gigawatts
Surface temperature	9,941°F (5,505°C)
Core temperature	28.3 million °F (15.7 million °C)
Polar rotation period	34.3 Earth days
Equatorial rotation period	25.05 Earth days
Surface gravity (Earth = 1)	28
Escape velocity (Earth = 1)	55
Axis tilt	7.25°
Distance from Earth	93.0 million miles (149.6 million km)
Age	About 4.6 billion years
Life expectancy	About 10 billion years

ATMOSPHERE OF THE SUN
This image from the SOHO spacecraft (see p.89) shows ultraviolet radiation emitted by the Sun. The ultraviolet is mainly produced by helium plasma in the chromosphere, the lower layer of the atmosphere, at about 930 miles (1,500km) above the surface.

VISIBLE SURFACE AND INTERIOR

The Sun's surface layer, known as the photoshpere, emits the visible light we see from Earth. It is about 300 miles (500km) thick, around 6,000 times less dense than the air that we breathe, and has an averag temperature of 9,941°F (5,505°C). Density, pressure, and temperature gradually rise as you descend into the Sun's interior. Beneath the photosphere is the convective zone, where circulating gas currents car heat from the interior to the surface. Hot gas bubbles up to 600 miles (1,000km) across rise to the surface, cool for about 8 minutes, and ther sink back again, making the photosphere appear to boil. The next lay down is the radiative zone, a calm region in which energy is gently transported between the core and the convective zone by radiation. The core itself—the Sun's nuclear reactor—makes up just 2 percent of the Sun's volume, but contains about 60 percent of its mass.

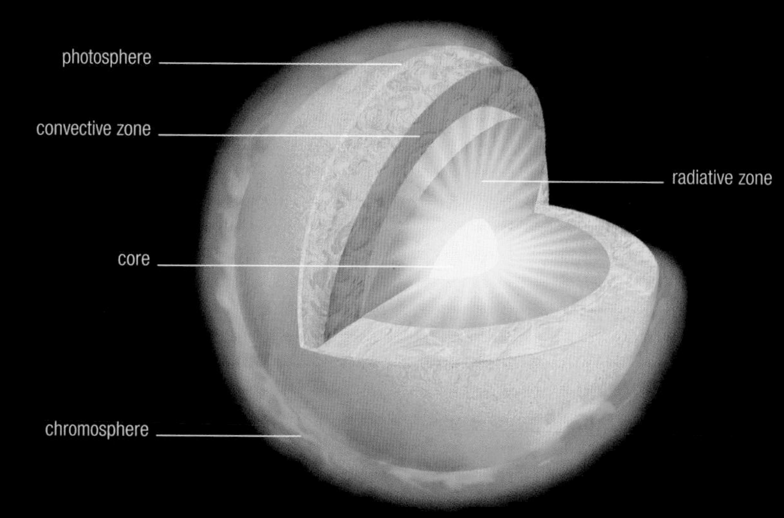

THE SUN'S LAYERS
The core has a temperature of 28.3 million °F (15.7 million °C) and a density of more than 9,375lb per cubic foot (150,000kg per cubic meter), but by the top of the radiative zone, these values have dropped to 3.6 million °F (2 million °C) and 12.5lb per cubic foot (200kg per cubic meter). Energy can take about 100,000 years to travel from the core to the photosphere.

ATMOSPHERE

The thin solar atmosphere consists of a chromosphere and a corona, which the photosphere's brilliance normally prevents us from seeing. Beginning just above the photosphere, the chromosphere stretches fo 1,200 miles (2,000km) to the corona. The temperature gradually rises through the chromosphere, reaching about 18,000°F (10,000°C) at the top, but the density drops by a factor of 5 million. The Sun's outer atmospheric layer is the corona, which is a million million times less dense than the photosphere and extends for millions of miles into space. Believed to be heated by a magnetic process, the corona has a temperature in excess of 1.8 million °F (1 million °C). Coronal matter is flowing out into space at a rate of millions of tons every second.

THE CORONA
In a solar eclipse, the Moon covers the Sun's photosphere and reveals the corona as a surroundin halo. The highly ionized material of the corona is sculpted by the Sun's magnetic field, giving it a "spiky" structure. The pink, blotchy area just visible

ie Sun may appear to be a serene golden globe, ut up close, it is a violent world. The surface eems to boil, flares erupt from around magnetically ctive regions called sunspots, and tremendous xplosions hurl atmospheric material far into space.

UNSPOTS

e Sun radiates heat reasonably steadily over time, but its surface, or otosphere, changes. Dark depressions called sunspots appear either ngly or in groups and last from a few hours to about two months. The ots are typically 930–31,000 miles (1,500–50,000km) across, and their mber varies over a cycle of roughly 11 years. They were once thought be inner planets crossing the Sun's disc, but in 1612, Italian scientist alileo correctly identified them as surface features. Richard Carrington, English astronomer, noted in around 1863 that spots near the equator oved faster than those close to the poles. This is because different rts of the Sun rotate at different speeds. Sunspots are caused by the ay this differential rotation affects the Sun's magnetic field (below).

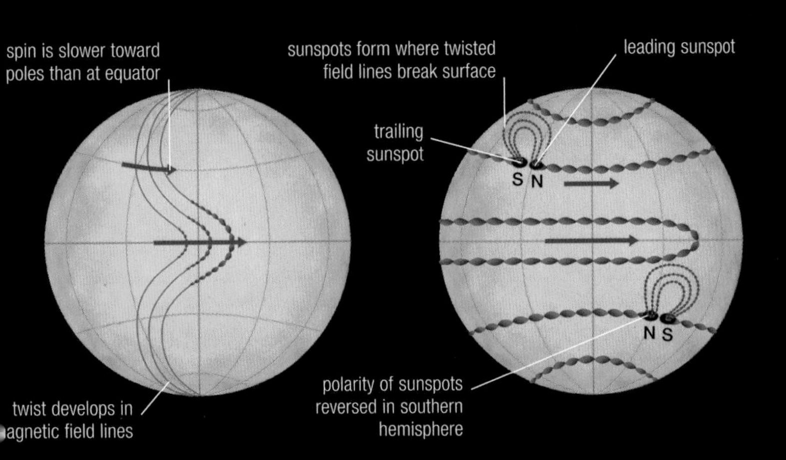
spin is slower toward poles than at equator
sunspots form where twisted field lines break surface
leading sunspot
trailing sunspot
S N
N S
twist develops in agnetic field lines
polarity of sunspots reversed in southern hemisphere

OW SUNSPOTS DEVELOP
agnetic field lines run from pole to pole just beneath the solar surface. ne Sun spins faster at the equator than at the poles, winding up the ld lines and making them more closely packed. Eventually they rise nd burst through the surface, producing spots. Each sunspot cycle eaks down the magnetic field, then rebuilds it with a reversed polarity.

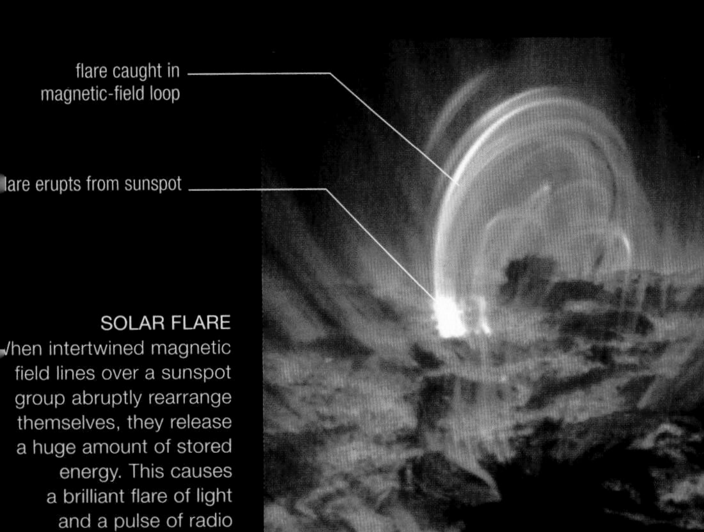
flare caught in magnetic-field loop
lare erupts from sunspot

SOLAR FLARE
/hen intertwined magnetic field lines over a sunspot group abruptly rearrange themselves, they release a huge amount of stored energy. This causes a brilliant flare of light and a pulse of radio

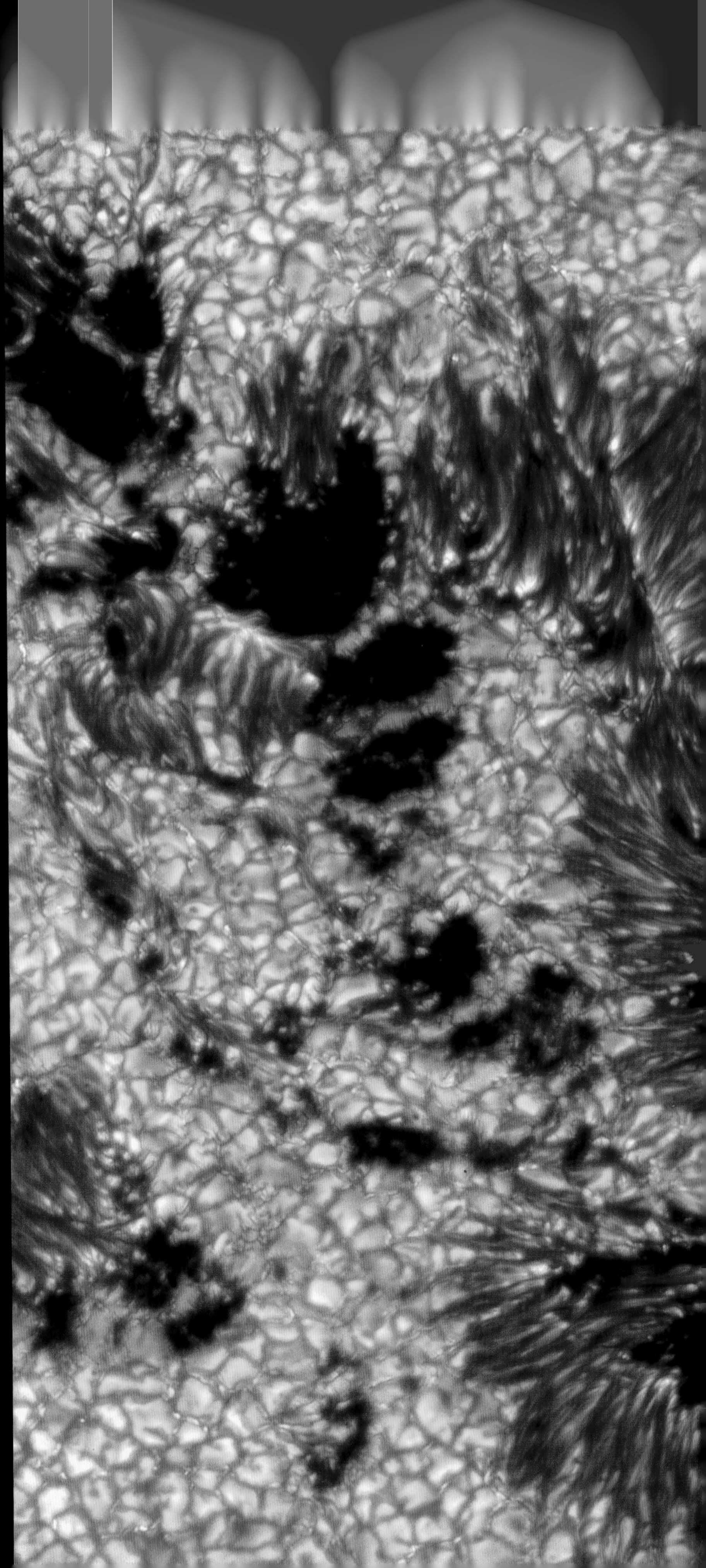

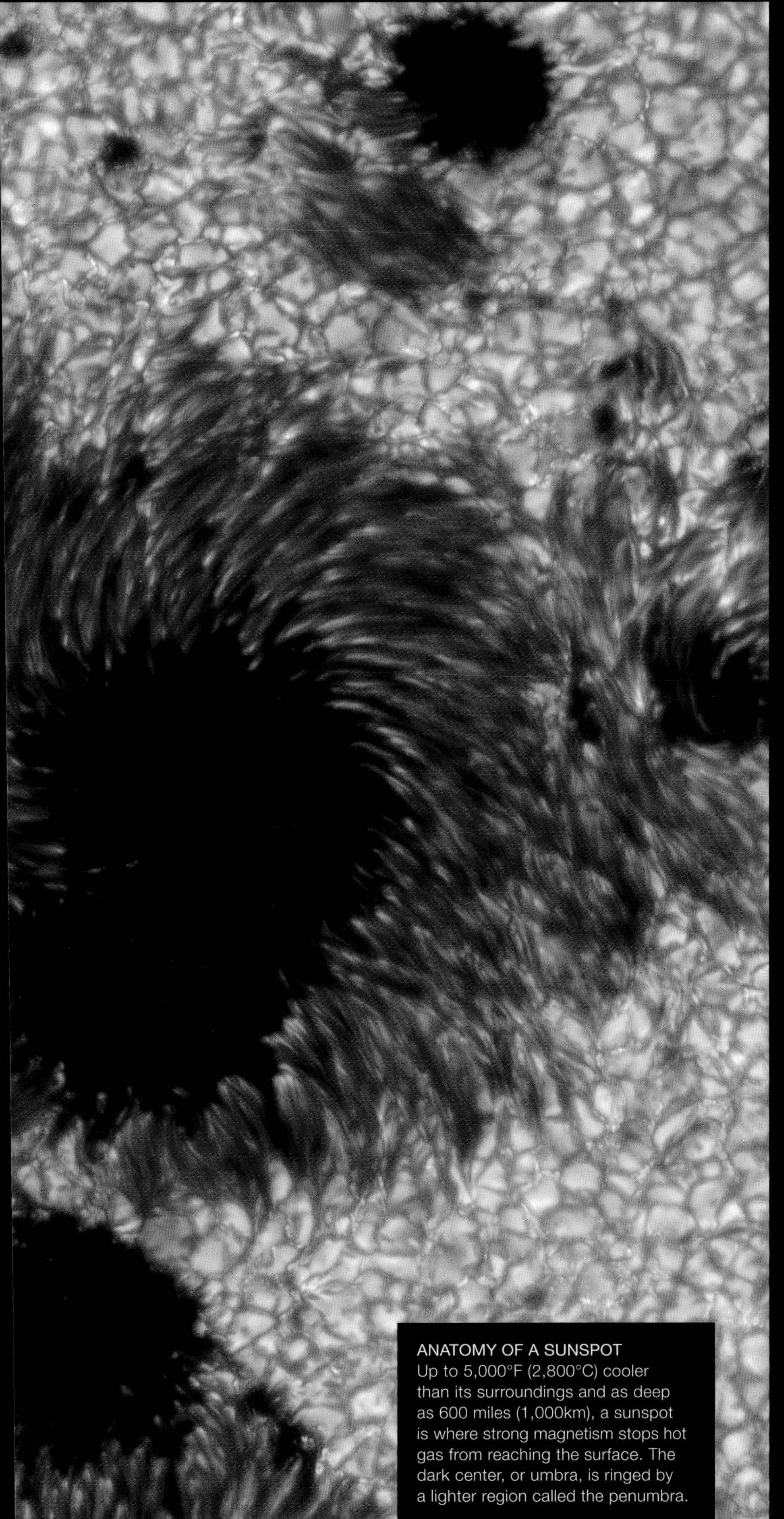

ANATOMY OF A SUNSPOT
Up to 5,000°F (2,800°C) cooler than its surroundings and as deep as 600 miles (1,000km), a sunspot is where strong magnetism stops hot gas from reaching the surface. The dark center, or umbra, is ringed by a lighter region called the penumbra.

SOLAR ACTIVITY AND EARTH

Sunspot activity varies in cycles. The spots break out at latitudes of around 40° north and south, but as the activity declines, they occur progressively closer to the equator. Between 1645 and 1715, few spots were recorded, and this low point coincided with a series of very cold winters on Earth known as the Little Ice Age. This has led to speculation that sunspots affect Earth's climate. Although the Sun's total energy output changes little throughout each cycle, its X-ray and ultraviolet emissions vary. These energetic radiations interact with the ionosphere, Earth's upper atmospheric layer. They may also affect atmospheric ozone levels and influence cloud formation. In 2008, solar activity was at a minimum and few sunspots were seen. However, in 2001, there were many spots, and astronomers expect a similar number around 2013.

SOLAR WIND AND AURORAE
Charged particles such as electrons and protons stream out from the corona, forming the solar wind. Some particles become trapped over Earth's magnetic poles. Their collisions with atoms in the ionosphere generate colorful, flickering light displays known as aurorae.

STUDYING THE SUN

Telescopes have been used to observe the Sun for 400 years, but they are hindered by Earth's weather, an atmosphere that absorbs X-rays and ultraviolet radiation, and the fact that the Sun can only be studied by day. However, earth-based telescopes can be fitted with spectrometers that break sunlight into its myriad colors, revealing much about the Sun's chemical composition and the physical properties of its spots and photosphere. Today, much solar investigation is carried out by orbiting spacecraft that monitor solar activity continuously. They can image the Sun in a range of different wavelengths and study the chromosphere and corona too.

SOHO SPACECRAFT
The Solar and Heliospheric Observatory (SOHO) spacecraft has been observing the Sun from 9.3 million miles (15 million km) above Earth since May 1996. SOHO is jointly operated by ESA and NASA.

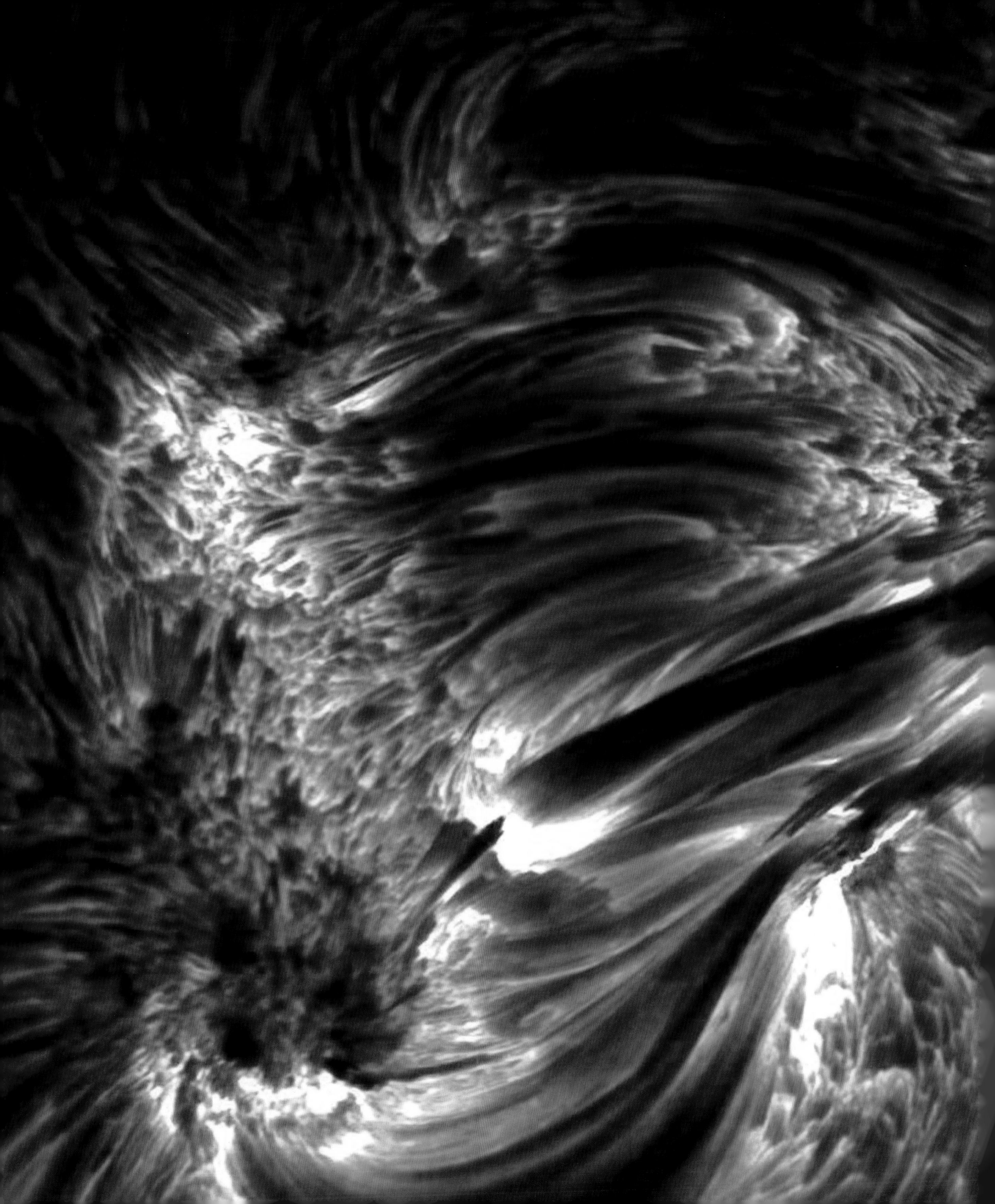

THE SURFACE OF THE SUN
This close-up of the Sun was taken by the Swedish Solar Telescope in Spain. The dark area above right of center is a sunspot—a cooler region caused by a tube of magnetized gas welling up from the core of the Sun and breaking the surface. Monitoring sunspots allows scientists to predict conditions in Earth's ionosphere, which in turn affect satellite communications.

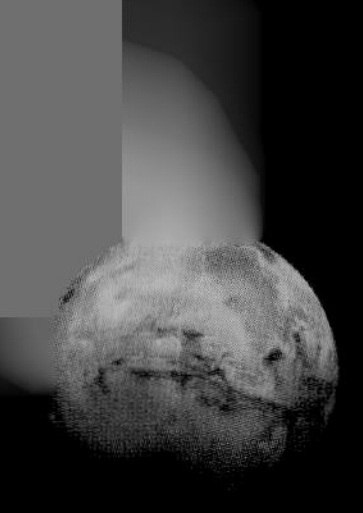

water flowed across its surface in the long-distant past, but today the Red Planet—as Mars is known—is a cold, dry world, its surface sliced by a huge system of deep canyons and studded with giant, extinct volcanoes.

SPIN, ORBIT, AND STRUCTURE

The outermost of the Solar System's four rocky planets, Mars is about half the size of Earth and, like our planet, has a distinct crust, mantle, and core. Being smaller and more distant from the Sun, it cooled more rapidly than Earth, but its outer core is thought to still be molten. Mars has an elliptical orbit that takes it some 26.4 million miles (42.6 million km) closer to the Sun at its closest than at its most distant. It spins on its axis in about the same time as Earth. As with Earth, the axis of its spin is tilted, so its northern and southern hemispheres alternately point to the poles, producing seasons. This tilt has varied in the past, causing significant changes in the planet's climate.

ATMOSPHERE AND WEATHER

Mars has a thin atmosphere made up almost entirely of suffocating carbon dioxide. The atmospheric pressure on Mars is low, at about 0.6 percent of that on Earth. From the surface, the atmosphere looks pink because of a haze of iron-oxide (rust) particles suspended within it. Occasional thin, white clouds of frozen carbon dioxide and water ice form at high altitudes. The atmosphere thins in the martian winter, when the temperature plummets to an icy -128°F (-89°C) and much of the carbon dioxide forms dry ice at the planet's poles (see pp.100–101). Dust storms also occur in a regular seasonal pattern, sometimes on a global scale (see pp.106–107). Constant observation by spacecraft has also revealed that Mars experiences daily weather changes, including a rise in temperature on a typical summer's day to 68°F (20°C).

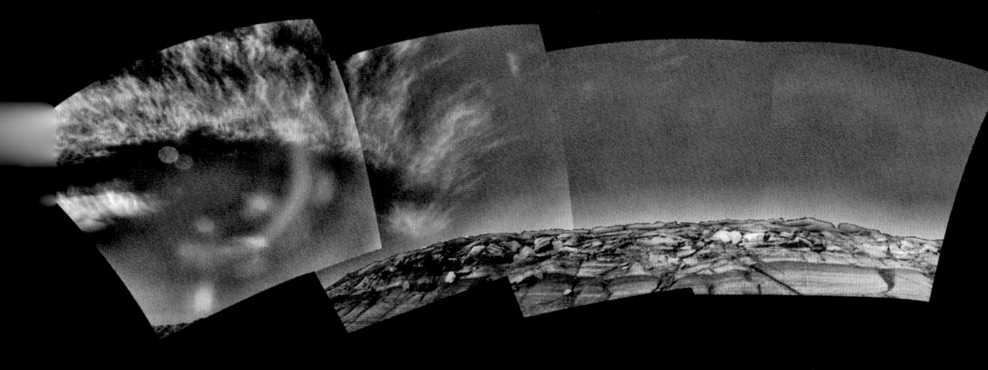

CLOUDS ON MARS
NASA's robot rover *Opportunity* took this mosaic of wispy clouds in the martian sky from inside Endurance Crater (see pp.112–113).

ORBIT AND ROTATION

Mars takes about the same length of time as Earth to turn on its axis, so its days are about the same length as Earth's. Its eccentric orbit means that it receives 45 percent more solar radiation at its closest point to the Sun than at its furthest.

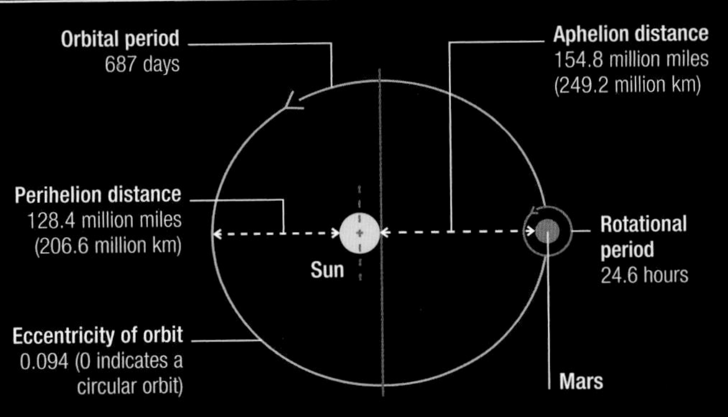

SIZE

The second smallest planet in the Solar System, Mars is half as big as Earth, but with no oceans, it has about the same total area of land.

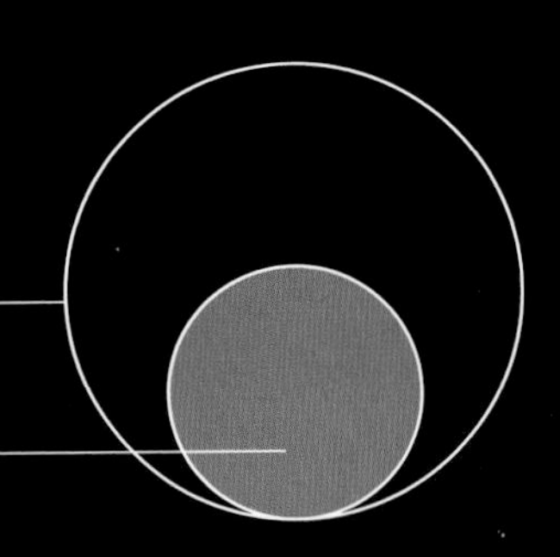

THE RED PLANET
This mosaic of 102 Viking orbiter images (see p.94) centers on Valles Marineris, which runs west to east just south of the equator and spans more than 2,500 miles (4,000km).

SURFACE FEATURES

The northern hemisphere of Mars is mainly low volcanic plains. To the south, the ground is higher and older, with more impact craters, which date from a period of intense asteroid bombardment some 3.9 billion years ago. Tectonic processes formed the major surface features of the planet. Valles Marineris, an enormous system of canyons, formed as the surface split when Mars was young (see pp.96–97). The vast volcanoes of the Tharsis region grew through volcanic activity during much of Mars's history (see pp.98–99). Nearly half of the surface of Mars is covered by lava. Its dusty soil contains iron oxide, which gives the planet its distinctive color.

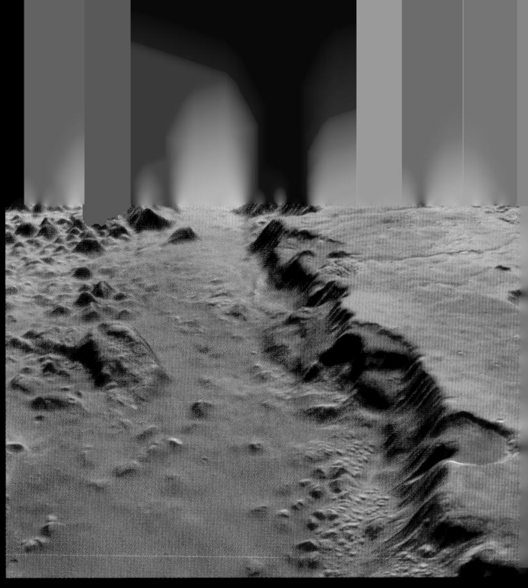

EOS CHASMA
Images taken by the orbiting Mars Express spacecraft (see p.94) form this perspective view of a canyon that is part of the eastern end of the sprawling Valles Marineris.

WATER ON MARS

Ancient floodplains and networks of dry riverbeds are clear evidence that large amounts of water flowed across the surface of Mars billions of years ago, when the planet was warmer than it is today. Water still exists on Mars, but only in the form of vapor and ice. Vapor forms low-lying mists and freezes out of the atmosphere to make early morning frost. Ice is also obvious in the planet's polar ice caps, and recent imaging by orbiting spacecraft has revealed that there are thick masses of ice at various locations across the planet, but just below its surface.

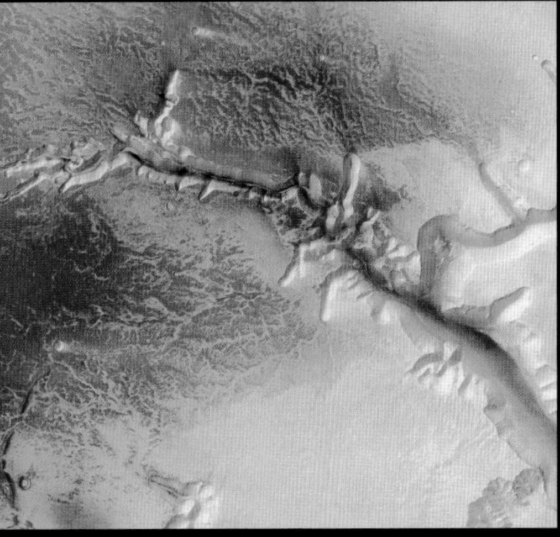

ANCIENT WATER CHANNELS
This Mars Express image is of the ancient water channels of Echus Chasma, north of Valles Marineris. At 60 miles (100km) long and 6 miles (10km) wide, Echus Chasma was one of the largest source regions for water on Mars.

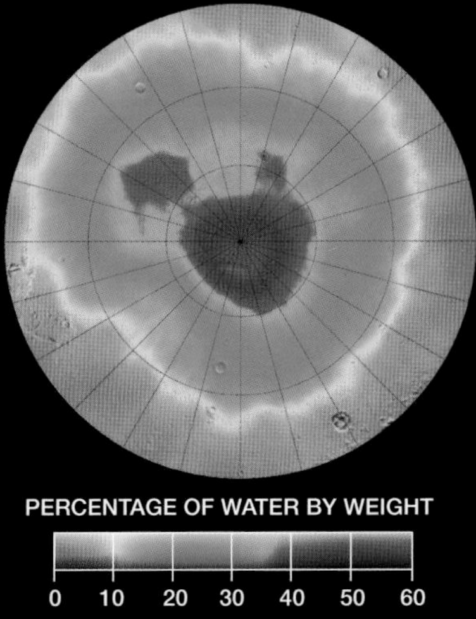

NEAR-SURFACE WATER MAP
Centered on the north pole, this map of the water content of near-surface material is based on gamma-ray spectrometry by NASA's 2001 Mars Odyssey (see p.94).

STRUCTURE

Mars is made up of a metal core surrounded by a silicate-rock mantle wrapped in a silicate-rock crust. The mantle was the source of volcanic lava millions of years ago.

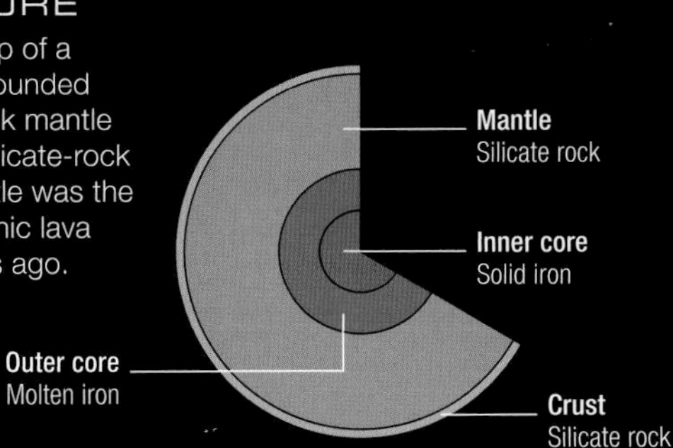

ATMOSPHERE

Trace gases in Mars's mainly carbon dioxide atmosphere include oxygen, water vapor, and carbon monoxide.

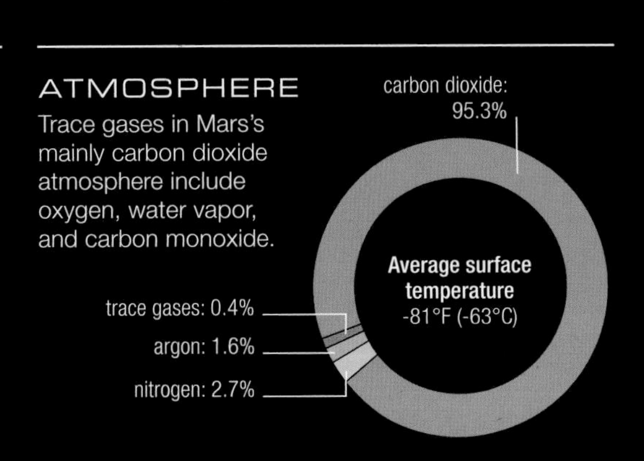

2

Mars has two small, potato-shaped moons—Phobos and Deimos (see pp.120–121). Phobos is the larger of the two at just 16.6 miles (26.8km) long, is closest to Mars, and is the better known. In a series of eight flybys in 2008 and 12 in 2010, Mars Express investigated Phobos's surface as well as what lies beneath. Armed with new information, astronomers are piecing together the history of both moons.

MARS FROM ABOVE

Spacecraft have been orbiting Mars on and off since 1971—for a few months in 2006 there were four circling the red planet at the same time. Between them, the various orbiters have imaged the whole globe—and in ever sharper detail.

EARLY ORBITERS

Mars was the first planet to be orbited by a spacecraft. This was NASA's Mariner 9, and when it arrived at Mars in mid-November 1971, the whole planet was in the grip of a dust storm. After delaying its mapping program, Mariner 9 returned images that produced the first global map of Mars. In the late 1970s, NASA's Viking 1 and 2 orbiters took thousands more images. Twenty years later, NASA's Mars Global Surveyor became the first of a second phase of missions. In 2006, it completed nine successful years of orbits. By then, its mineral-mapping instrument had found hematite, a mineral that often forms in wet conditions, and the craft had tracked changes through many martian seasons.

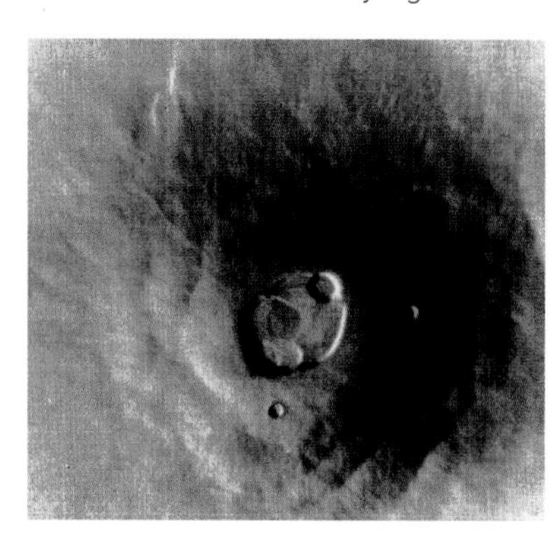

MARINER 9 IMAGE OF OLYMPUS MONS
Mariner 9 yielded the first close-up images of features such as the enormous Tharsis volcanoes (see pp.98–99), including this January 1972 view of Olympus Mons.

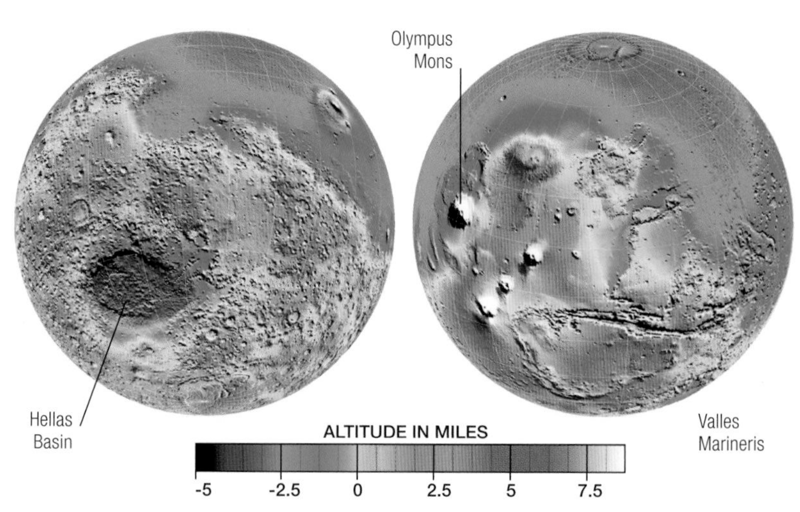

MARS GLOBAL SURVEYOR TOPOGRAPHY MAPS
Mars Global Surveyor's laser altimeter measured heights all over the planet. The giant Tharsis volcanoes (right) are the highest land and show as white peaks. The lowest terrain is mainly in the northern hemisphere, with the clear exception of the huge Hellas Basin (left).

2001 MARS ODYSSEY

NASA's 2001 Mars Odyssey has been in polar orbit around Mars since October 2001. The mission was first scheduled to last two years, but it was extended until 2010 to allow changes to be monitored over several years. The craft has also supported the two Mars Exploration Rovers, *Spirit* and *Opportunity*, which have been on the planet's surface since 2004 (see pp.110–111), relaying more than 95 percent of their data to Earth. Its three primary instruments have mapped the amount and distribution of chemical elements and minerals in the surface of Mars. Hydrogen mapping by the orbiter led to the discovery of huge amounts of water ice beneath the surface at the planet's poles.

MARS EXPRESS

Mars Express is the European Space Agency's (ESA's) first mission to a planet. It has been in polar orbit around Mars since 2004 and will continue until 2012. Its instruments map and study the planet's surface and measure Mars's atmosphere. Its High Resolution Stereo Camera (HRSC) images features as small as 6ft 7in (2m) across, its radar altimeter assesses the composition of the sub-surface, and its three spectrometers determine the composition of the surface and the composition, temperature, and pressure of the atmosphere. The craft's elliptical orbit means it has also made regular flybys of Phobos, the largest of Mars's two moons (see pp.120–121).

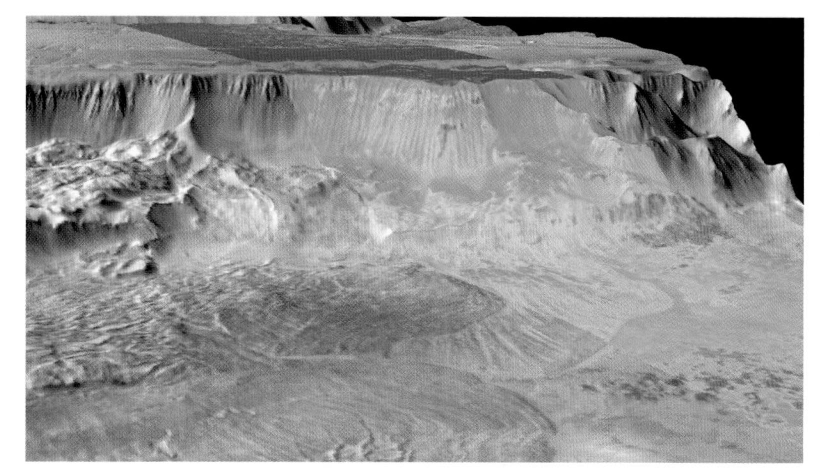

2001 MARS ODYSSEY THERMAL IMAGING OF MELAS CHASMA
Two images of part of Valles Marineris (see pp.96–97) are combined here. In the monochrome one, taken in daylight, slopes in shadow appear dark at -31°F (-35°C), while sunlit slopes appear bright at 23°F (-5°C). In the color nighttime image, rock retains daytime heat (red/yellow), unlike dust (blue).

MARS EXPRESS IMAGE OF PART OF IANI CHAOS
Mars Express's HRSC took this image of one of several aptly named regions of disrupted terrain. It shows light-toned deposits that were probably formed by ancient groundwater springs. Some scientists consider such places may have provided a habitat for microbial life.

MARS RECONNAISSANCE ORBITER

NASA's Mars Reconnaissance Orbiter started orbiting the red planet in March 2006. Its arrival meant that for a while, three NASA orbiters, one ESA orbiter, and two NASA rovers were all working at Mars at the same time. Mars Reconnaissance Orbiter's mission is twofold. With its imaging spectrometer, it is searching for minerals that form through long-term interaction with water. The presence of such minerals would prove water persisted on Mars over a long period. It is also looking for future landing sites—globally mapping, regionally surveying, and spot-imaging with its HiRISE camera (High Resolution Imaging Science Experiment), which homes in on targets and resolves objects 3ft (1m) across.

MARS RECONNAISSANCE ORBITER
NASA's Mars Reconnaissance Orbiter is the biggest and most sophisticated orbiter so far sent to Mars. One instrument not visible here is its Mars Climate Sounder, a spectrometer for examining atmospheric water-vapor content, dust content, and temperature.

MARS RECONNAISSANCE ORBITER PROFILE

MISSION	
Launch date	August 12, 2005
Mars orbit insertion	March 10, 2006
Mission ends	Ongoing
Launch vehicle	Atlas V-401
MARS RECONNAISSANCE ORBITER	
Agency	NASA
Dimensions	21ft (6.5m) X 45ft (13.6m)
Weight	4,800lb (2,180kg)
Main propulsion	Monopropellant (hydrazine) thrusters

SCALE

21ft (6.5m)

45ft (13.6m)

high-gain antenna

Shallow Subsurface Radar
Probes the planet's ice caps (see pp.100–101) and crust.

Mars Color Imager
A wide-angle camera.

Electra
A radio for relay contact with Earth.

HiRISE
Has a telescopic lens with a 20in (50cm) aperture.

Context Camera
Works in tandem with HiRISE.

Compact Reconnaissance Imaging Spectrometer for Mars
Identifies water-related surface minerals.

solar panel

solar panel

SAND DUNES IMAGED BY HIRISE
The dark streaks in this HiRISE image, which has been colored to highlight contrasts, are bands of dark sand dislodged from the crests of frosted dunes when carbon-dioxide ice in the dunes was warmed by spring sunshine and turned to gas. The streaks are up to about 164ft (50m) long here.

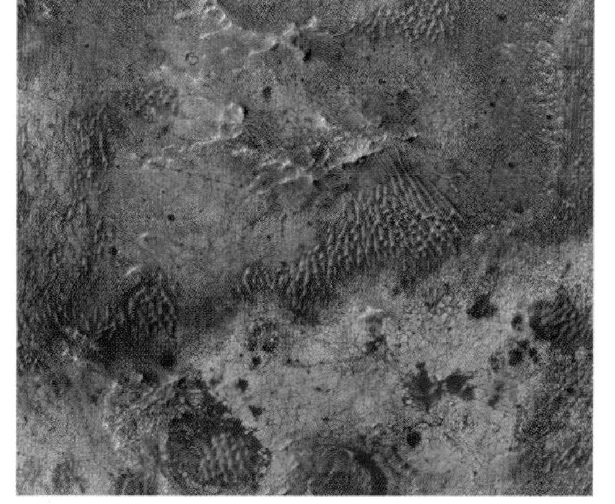

COMBINATION IMAGING
This view of a region called Nili Fossae, made by combining HiRISE and imaging-spectrometer images, shows areas of exposed, partly fractured clay in green and the olivine of sand dunes blowing across it in red.

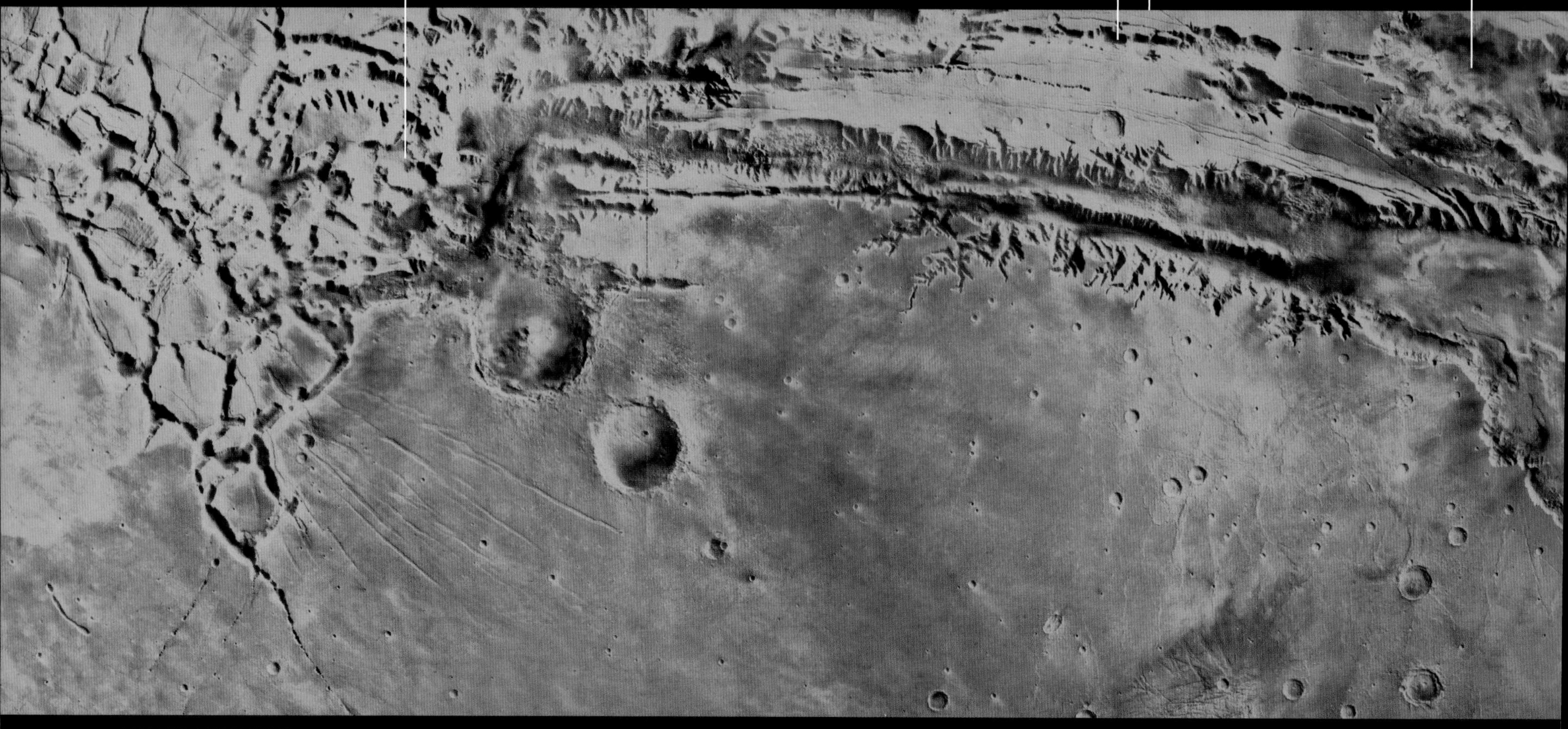

VALLES MARINERIS
The entire Valles Marineris canyon system is seen in this composite image made from views taken by the two Viking orbiters in 1976. It extends from Noctis Labyrinthus in the west to Eos Chasma in the east. The top of the image is close to the martian equator; the bottom is 20° south.

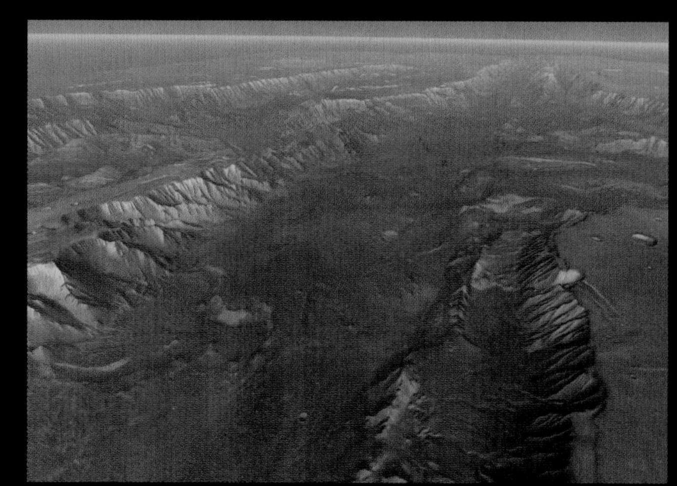

MELAS CHASMA
This view of Melas Chasma was produced using images taken by 2001 Mars Odyssey's infrared camera and a computerized topographic model.

tectonic and volcanic history of the planet and, through analysis of its rocks and erosion, have gained insights into the martian climate.

At the western end of the system is Noctis Labyrinthus, a triangular region of canyons, troughs, and pits. Its origins are uncertain, but it may have developed when there was activity in the neighboring volcanic Tharsis region (see pp98–99)—activity that stretched and fractured the crust, so that sub-surface water ice melted and escaped, leaving cavities and causing the area to collapse.

The central part of Valles Marineris consists of many individual canyons each measuring up to 60 miles (100km) wide. The surface here

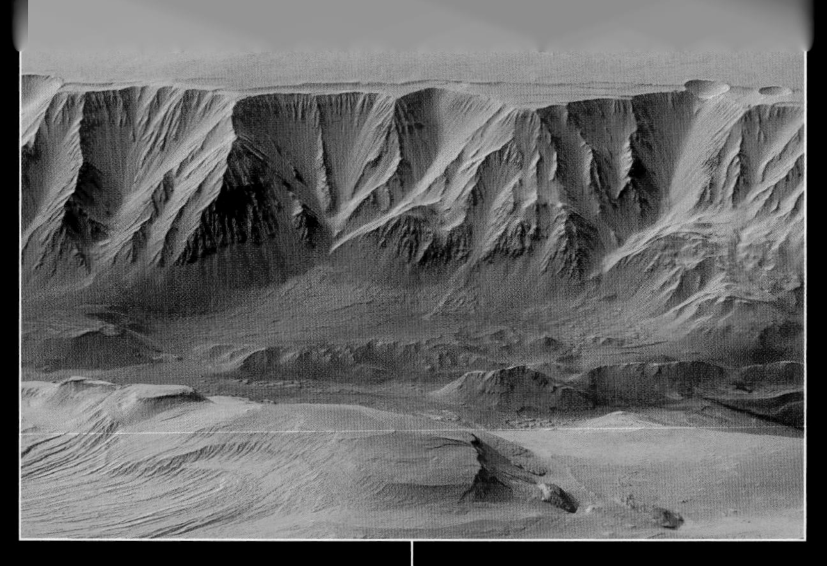

CANYON VIEW
Mars Express's High Resolution Stereo Camera imaged Candor Chasma as the spacecraft orbited Mars on July 6, 2006. Material dumped by huge landslides is evident on the lower part of the cliffs. Sharp-edged rocks shaped by the wind can be seen on the lower right.

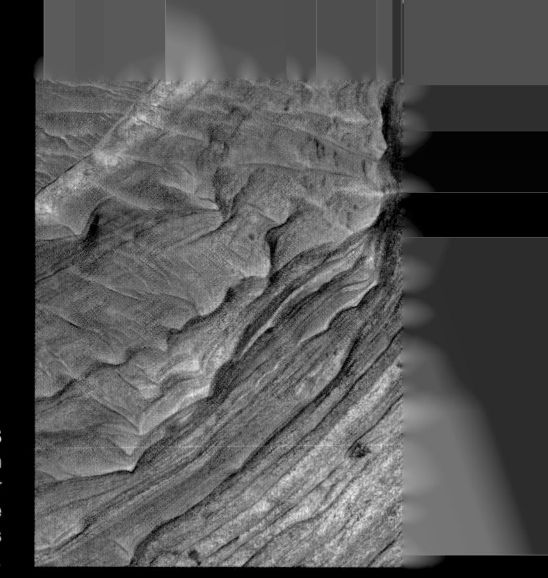

ROCK LAYERS
Studies of crustal rock exposed on the walls and floor of eastern Candor Chasma show that the rock is made of layers of material. The wavy edge is partly due to surface erosion.

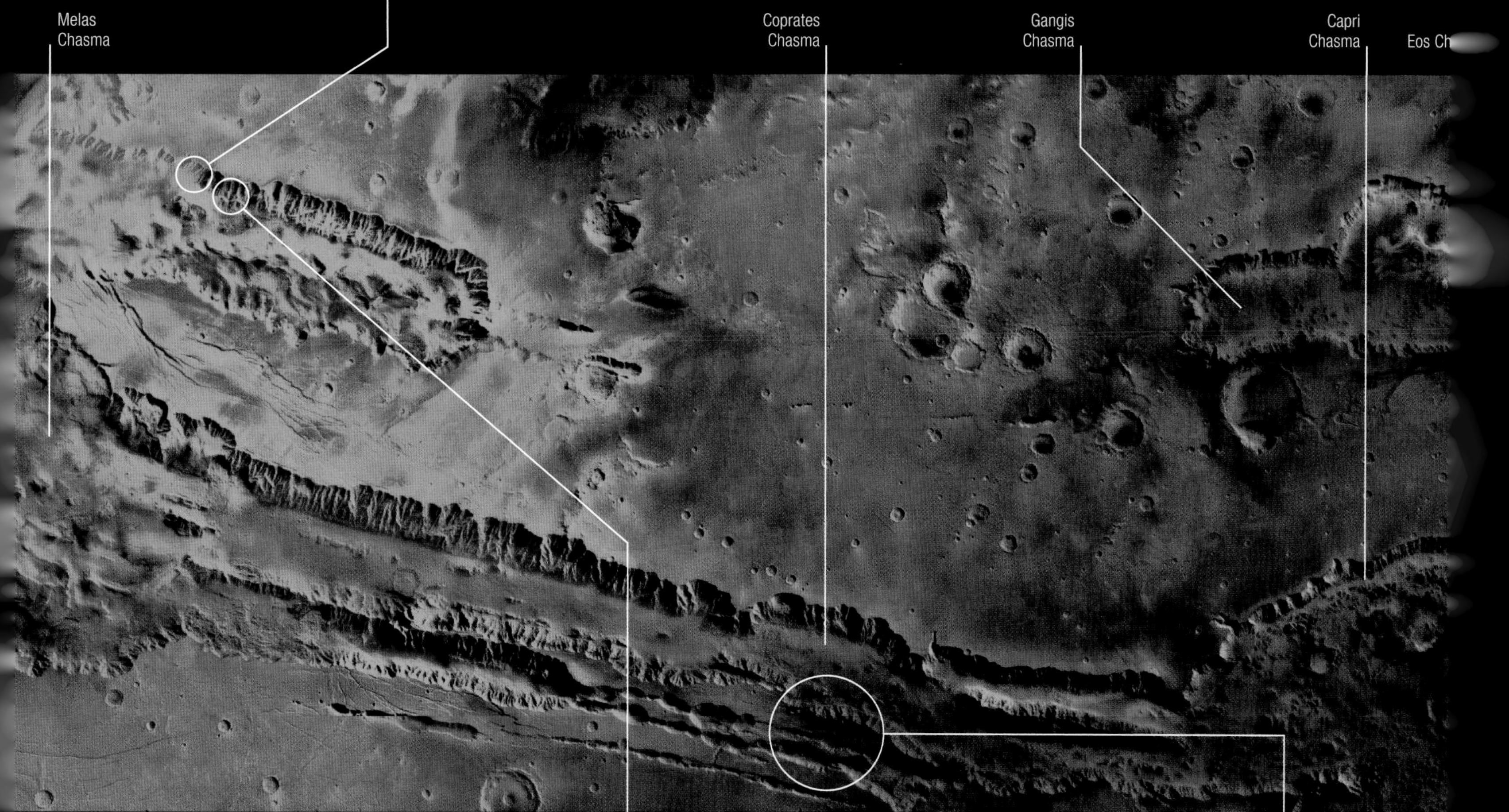

collapsed and filled with layered material. Grabens—depressed blocks of land—formed as the land faulted and landslides occurred. Wind-carried material still moves along the canyon floor today. The floor of eastern Candor Chasma, for example, is filled with material of volcanic or sedimentary origin. Bounded by walls over 4 miles (6km) high, this canyon's floor also has landslide deposits and structures created by the wind.

At the eastern end of Valles Marineris is chaotic terrain where canyons give way to outflow channels that once carried water away from the area. Whole regions here have been stripped of material by water action

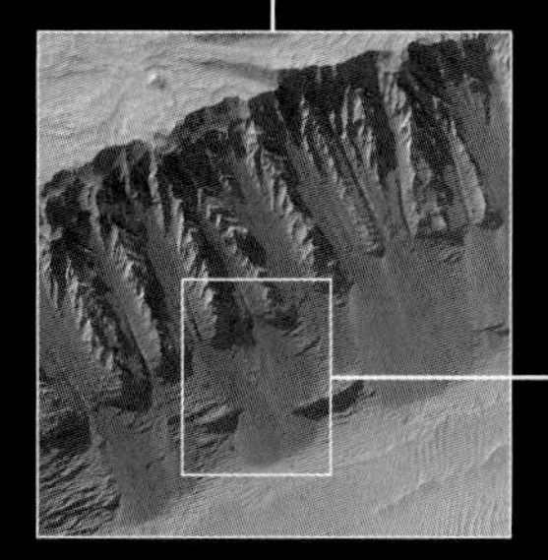

LANDSLIDES
This Mars Global Surveyor view from 2004 shows layered sedimentary rocks in eastern Candor Chasma. The edge of the rock has broken up and slid down the steep slope.

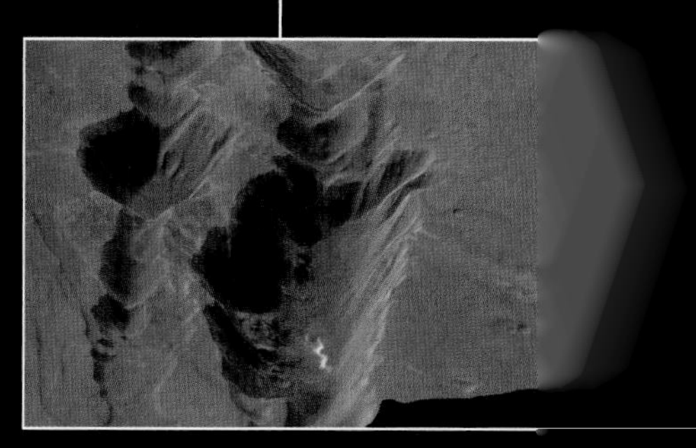

CHAIN OF STRUCTURES
This perspective view by Mars Express shows Coprates Catena, a chain of collapsed structures that formed as underlying material was removed. Up to 2 miles (3km) deep, the chain is shallower than the main canyon system

THE THARSIS REGION

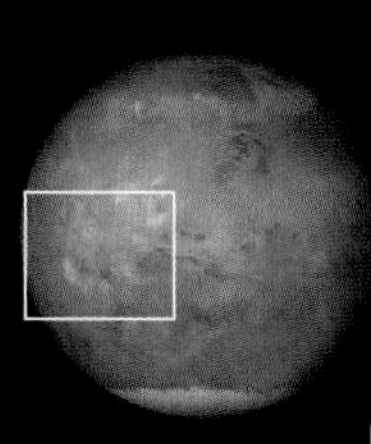

Centered on the planet's equator to the west of Valles Marineris (see pp.96–97), the Tharsis region of Mars is a huge, domed plateau— 2,500 miles (4,000km) wide and, at it highest, 5 miles (8km) higher than the northern-hemisphere plains—topped by giant but long-extinct volcanoes. Like other volcanic areas of Mars, it formed over 3 billion years ago, when the planet was relatively young, through a combination of crustal uplift and persistent but sporadic volcanism over hundreds of millions of years.

As they formed, the volcanic regions of Mars remained fixed in place in the same part of the planet's surface, so volcanoes at specific spots continued to grow with each successive lava flow. This is in contrast to Earth, where tectonic plates move over rising magma and new volcanoes form in different spots as the plates shift position. Mars's weaker gravity also allowed its volcanoes to grow higher than those on Earth, and its crust is better able to support the weight of massive volcanoes.

The huge volcanoes that we see today in Mars's Tharsis region include the mighty Olympus Mons, the largest known volcano in the Solar System. Towering about 14 miles (22km) into the martian sky, Olympus Mons is more than three times the height of the highest volcanoes on Earth. With a broad base some 370 miles (600km) wide and gently sloping sides, it is similar in appearance to the shield volcanoes of Hawaii, but many times wider. One of the youngest of Mars's large volcanoes, Olympus Mons built up from thousands of lava flows.

The next three largest volcanoes in the region—Ascraeus Mons, Pavonis Mons, and Arsia Mons—are in a line on the top of the plateau. Tharsis also has smaller volcanoes called paterae, with gentler slopes, and even smaller ones, called tholi.

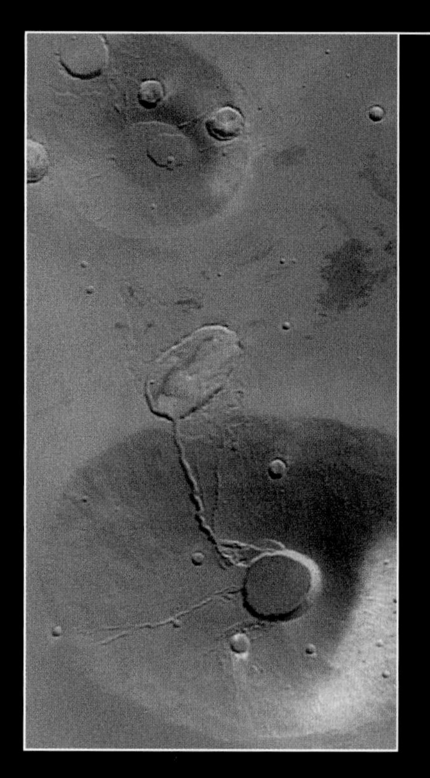

URANIUS THOLUS AND CERAUNIUS THOLUS
Old lava flows and impact craters can be seen on the flanks of Uranius Tholus (top) and Ceraunius Tholus (bottom) in this 2002 Mars Global Surveyor image (see p.94). Ceraunius is as big as Hawaii—its caldera (crater) is 16 miles (25km) wide. The light area (bottom right) is dust from the 2001 global dust storm (see p.106).

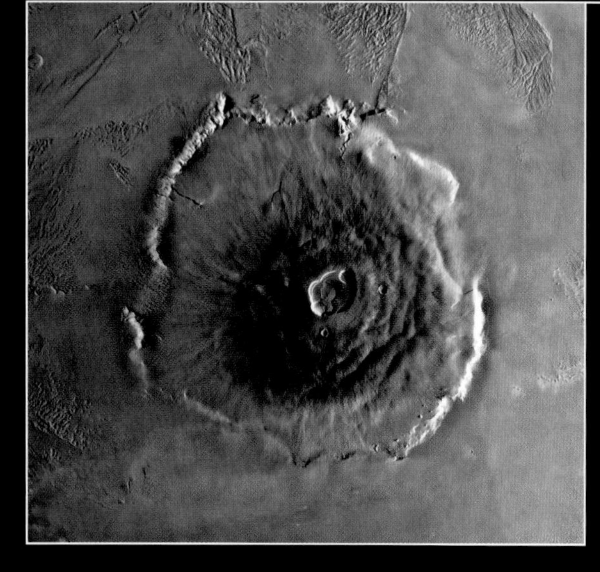

OLYMPUS MONS
This Viking 1 orbiter color mosaic looks straight down onto Olympus Mons. The caldera is about 50 miles (80km) wide and is encircled by wide terraces of solidified lava. The exposed edges of a steep slope around the volcano reveal a stack of lava layers. Further out, lava covers the vast Tharsis plateau.

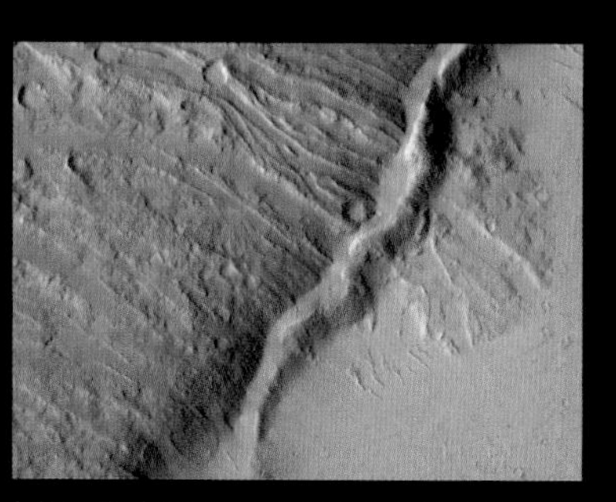

LAVA FLOWS ON OLYMPUS MONS
A fault bisects the base of Olympus Mons in this Mars Reconnaissance Orbiter image (see p.95). At left are old lava flows; at right, a later flood of lava.

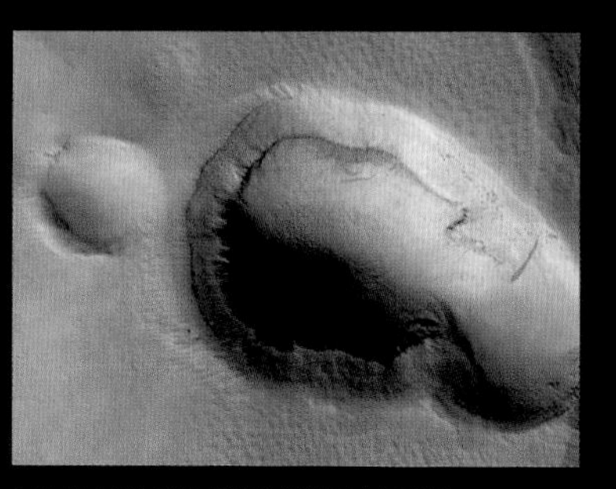

LAVA PIT ON OLYMPUS MONS
This Mars Reconnaissance Orbiter image reveals the lava layers of Olympus Mons and shows that the volcano is covered in thick dust or volcanic ash.

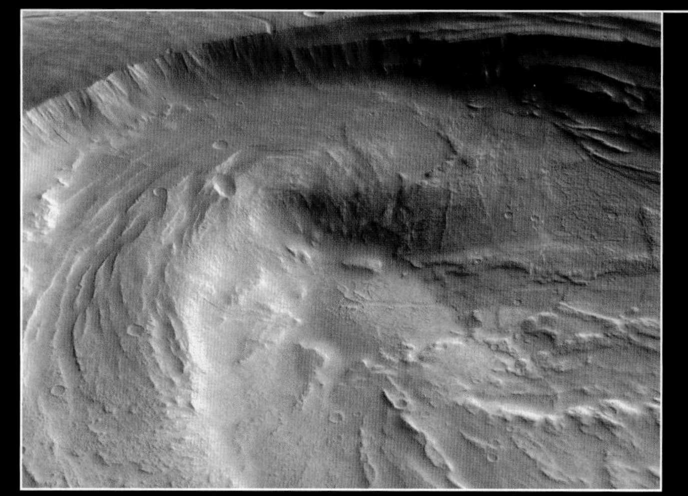

BIBLIS PATERA
This Mars Express image (see p.94) shows the caldera of this Tharsis volcano. The crater is 33 miles (53km) wide and up to 2.8 miles (4.5km) deep. It formed when the volcano's magma chamber collapsed.

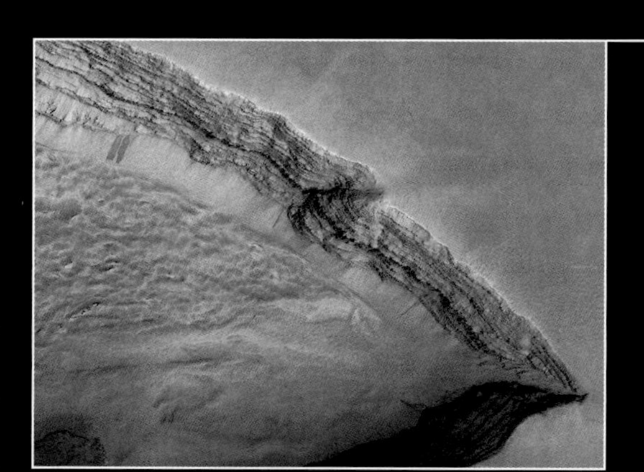

LAVA LAYERS OF ARSIA MONS
This image of part of a pit on the flank of Arsia Mons covers an area about 2.5 miles (4km) wide. Taken by Mars Global Surveyor, it shows layer after layer of solidified lava, indicating repeated ancient eruptions.

CLOUDS OVER THARSIS
Numerous smaller images taken by Mars Global Surveyor were combined to produce this global view of Mars showing Tharsis on a northern summer day. White clouds of water ice hang over the region's larger volcanoes.

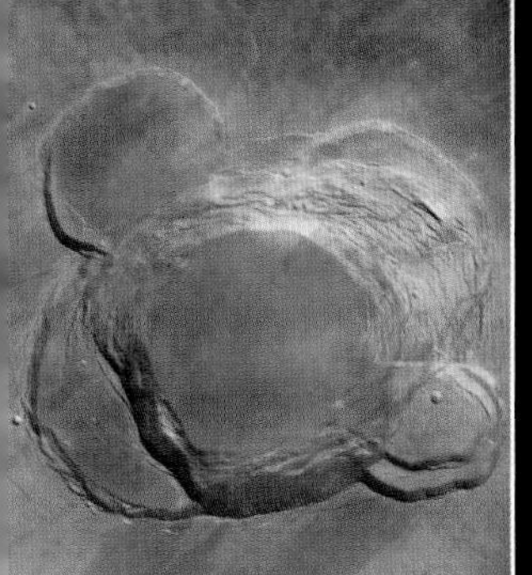

ASCRAEUS MONS CALDERA
The nested appearance of the caldera of Ascraeus Mons in this Mars Express image shows that the crater collapsed repeatedly. The lowest floor is lava

THE ICE CAPS OF MARS

Much of the water on Mars is frozen into two polar ice caps. The northern cap is called Planum Boreum (Northern Plain) and consists of a permanent mound mainly of water ice that stands about 1.2 miles (2.0km) above the surrounding land. The bright ice forms a distinctive swirling pattern when viewed from above. During the northern winter months, when the polar latitudes are in permanent darkness, the cap is covered and also extended in size by carbon-dioxide ice. As winter approaches, the temperature lowers a few degrees to about -193°F (-125°C), and atmospheric carbon dioxide turns to frost and snow, which covers the polar regions to about 65° north. Six months later, when the Sun is permanently in the summer sky, the carbon dioxide turns back to gas, and the cap shrinks. The permanent part of the cap is about 90 percent ice; the rest is sand and dust. At its edge is a steep slope where the cap's

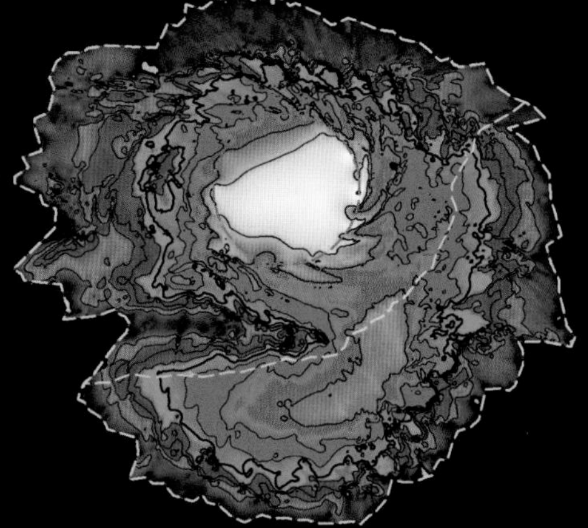

THICKNESS IN MILES

0 1.2

NORTH CAP STRUCTURE
This radar-generated Mars Reconnaissance Orbiter map shows the layered deposits in the north cap. The whole region is about 560 miles (900km) across and contains the equivalent of 30 percent of Earth's Greenland ice sheet. The colors indicate the thickness of the ice layers.

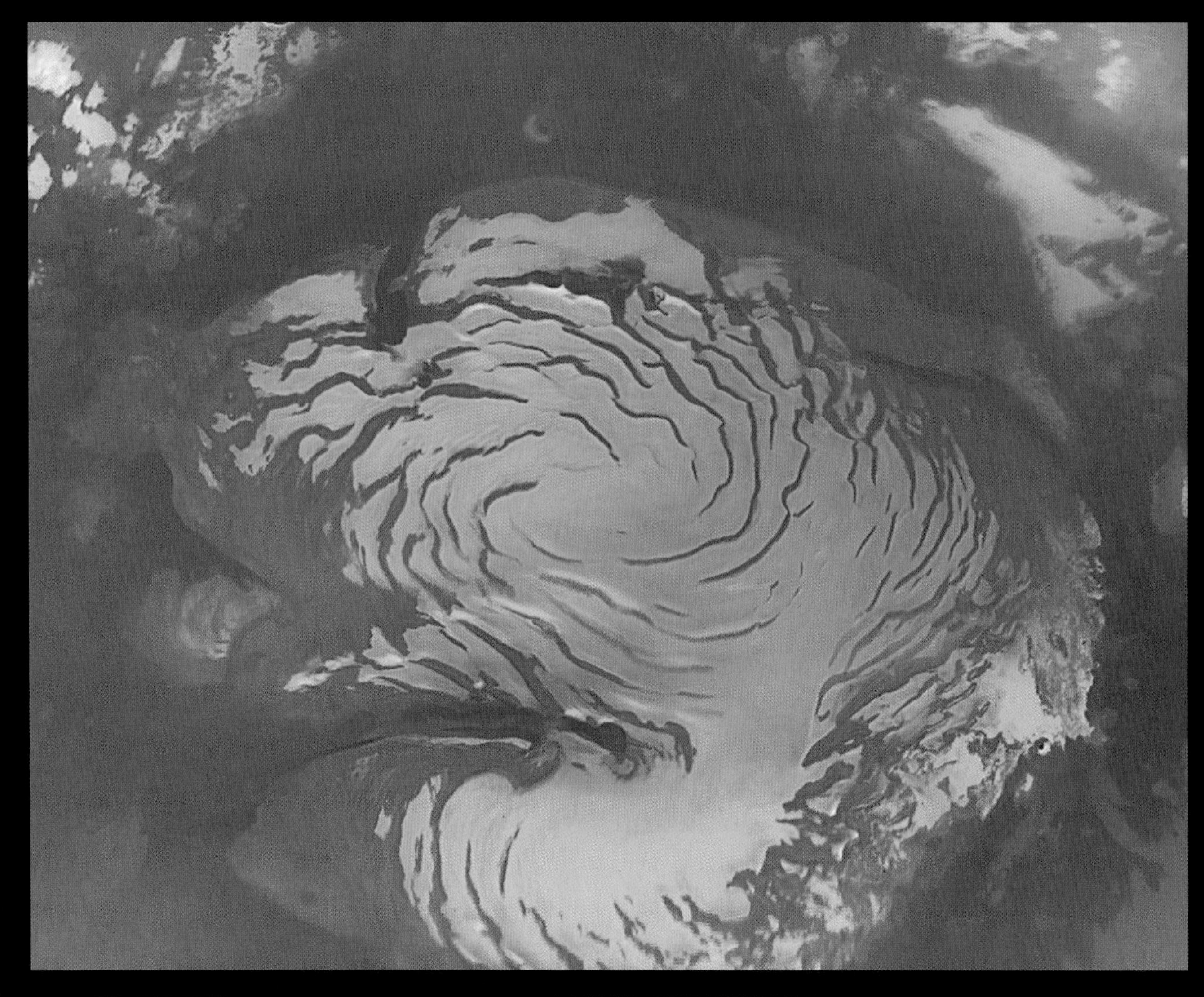

MARTIAN NORTH POLE
The white, permanent water ice cap rises above the more extensive layered rock and sand of Mars in this Mars Reconnaissance Orbiter image of the north polar region. This picture is a composite of four images taken at different times of the day. They show the cap during the martian summer, when the north pole is in permanent sunlight.

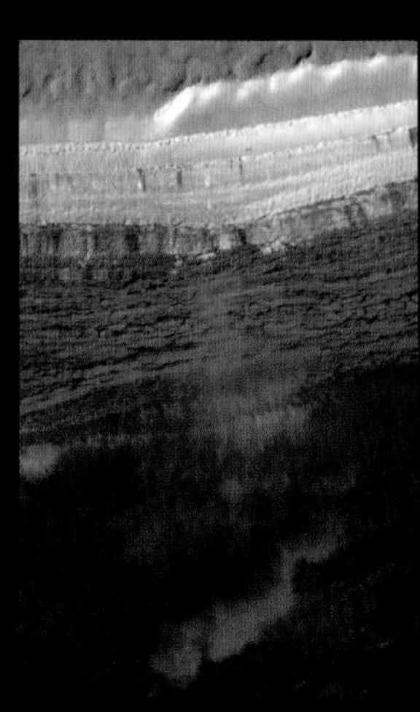

AVALANCHE ON SCARP
This false-color image, by Mars Reconnaissance Orbiter, highlights material as it falls down a 2,300ft- (700m-) tall slope on the edge of the north polar ice cap. The material is fine-grained ice and dust with some larger blocks. The accompanying cloud of finer material traces the path of the debris as it hits the lower slopes and continues downhill.

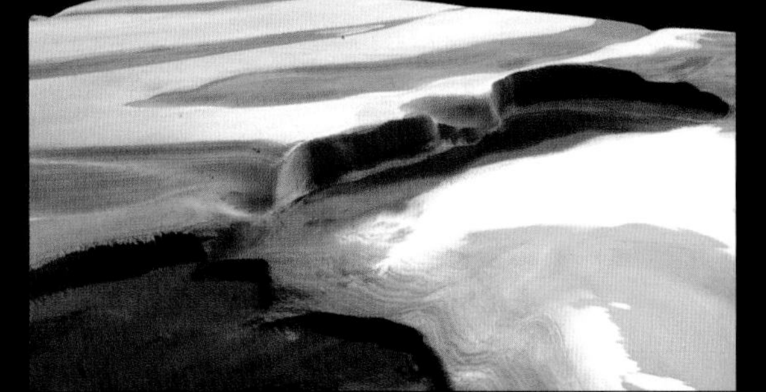

LAYERING AT THE NORTH POLE
Data from the high-resolution camera onboard Mars Express (see p.94) has been used to create this perspective view of the north polar cap. The white areas are water ice; the red are rock and sand. The cliffs, which are about 1.2 miles (2km) high, are thought to be the sides of a volcanic caldera. For this reason, the darkest material is thought to be volcanic ash.

internal structure is exposed. The cap consists mostly of parallel layers. The radar on Mars Reconnaissance Orbiter (see p.95) has looked at these (see opposite)—their number suggests the cap is about 4 million years old.

Like its northern counterpart, the southern cap is a permanent mound of ice, in this case a thick base of water ice with an upper 26ft (8m) layer of carbon-dioxide ice. Additional carbon-dioxide frost and snow cover this on a seasonal basis. Known as Planum Australe (Southern Plain), it measures at its minimum about 260 miles (420km) across. At its edge, steep slopes of ice lead to the surrounding plain, and beyond them are hundreds of square miles of permafrost—water ice mixed with the martian soil and frozen to the hardness of solid rock.

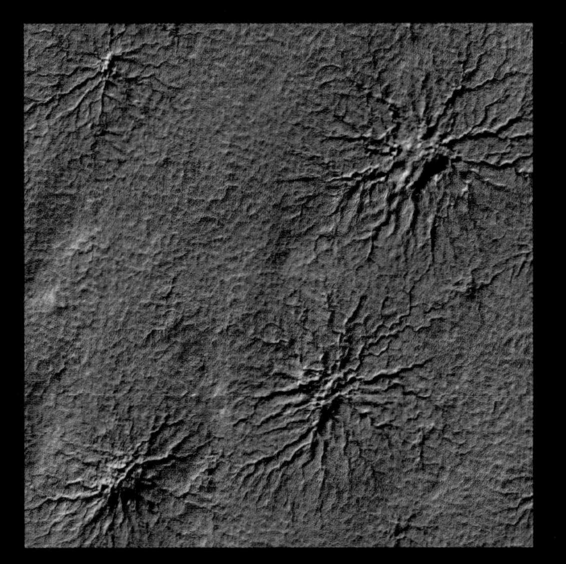

CHANNELS CARVED BY ICE
Mars Reconnaissance Orbiter imaged these spiderlike channels near Mars's south pole in August 2009. They form as summer sunshine warms seasonal carbon-dioxide ice and turns it to gas. Gas builds up under the ice and carves channels in the ground as it seeks a way to the surface. Once the seasonal frost has gone, the channels remain. Each channel is about 5ft (1.5m) deep, and a "spider" is some 1,300ft (400m) across.

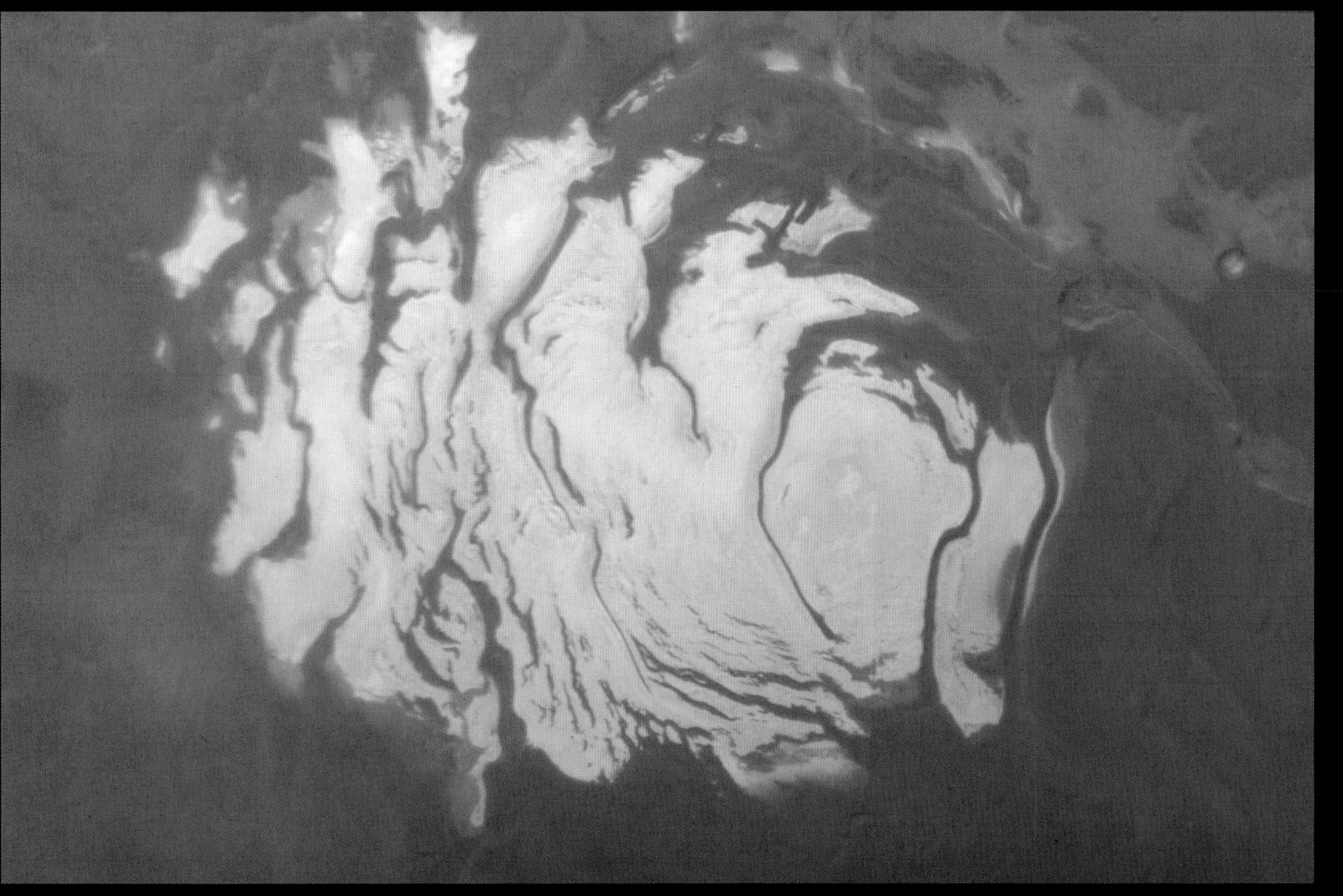

SOUTH POLE CAP
Midsummer sunshine illuminates the south cap in this Mars Global Surveyor image (see p.94). It is April, and the cap is at its minimum size. As the seasons progress, the area is increasingly covered by frost. By December, it will be winter, and the entire region will be covered in frost. It will stay that way until the Sun's warmth returns in the spring, when the frost recedes, and the cycle starts again.

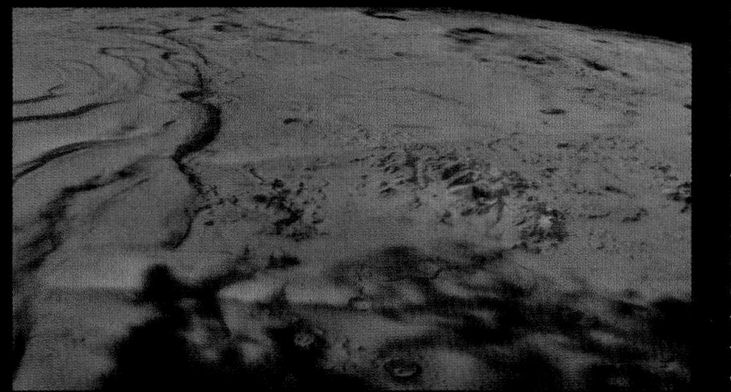

SOUTH POLE VIEW
Data from Mars Global Surveyor have been used to create this natural-color perspective view looking over the south polar cap. As it orbited the planet every 117 minutes, at an average altitude of 235 miles (378km), Mars Global Surveyor used its laser altimeter to determine the height of the polar terrain. The spacecraft mapped the planet for four-and-a-half years of its nine years in orbit.

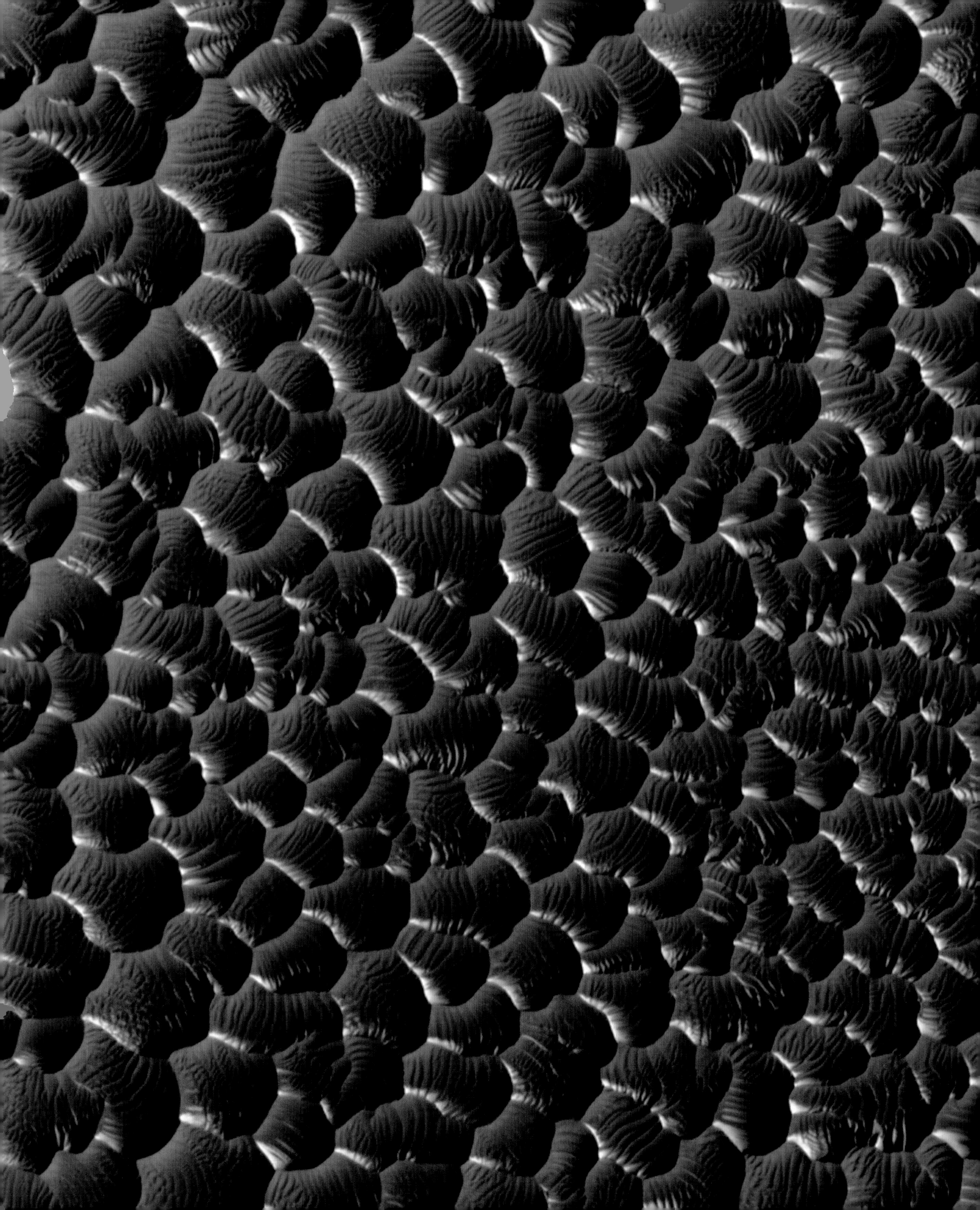

MARTIAN SAND DUNES
Taken by Mars Reconaissance Orbiter's HiRISE camera from an altitude of 185 miles (300km), this false-color image shows frosted sand dunes in Mars's southern hemisphere. The savage martian winds, coupled with the planet's thin atmosphere and weak gravity, produce dunes that are up to ten times larger than any on Earth.

THE DUNES OF MARS

Sand dunes have been found in locations all around Mars. They were a surprise discovery when Mariner 9 explored the planet from orbit, in 1971–72. More dunes were discovered later in the 1970s by the Viking orbiters, but the image resolution of the cameras on these early craft (see p.94) was such that only two types of dune were identified—barchan (arrowhead shaped) and linear. Higher-resolution images from Mars Global Surveyor revealed additional types. Today's spacecraft, especially Mars Reconnaissance Orbiter (MRO), show gullies, ripples, and other features in dunes in stunning detail (see p.95).

Many martian dunes look like the wind-created sand dunes of Earth. These form as wind-blown sand accumulates, and take on different shapes according to wind action and

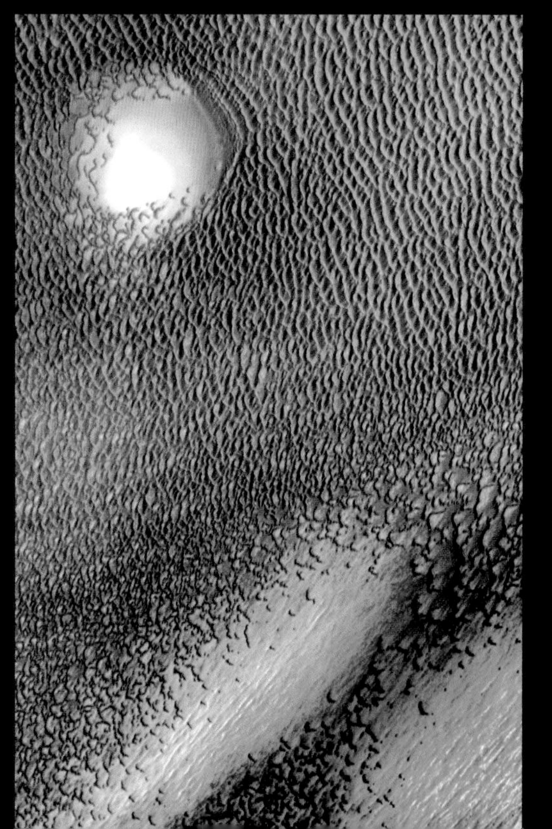

POLAR DUNE FIELD
Imaged in false color by 2001 Mars Odyssey (see p.94), this sea of dunes is about 200 miles (300km) from the north polar cap. The crests of the orange-colored dunes are about 1,600ft (500m) apart. The white and blue areas are frost-covered. The bright spot at top left is a hill

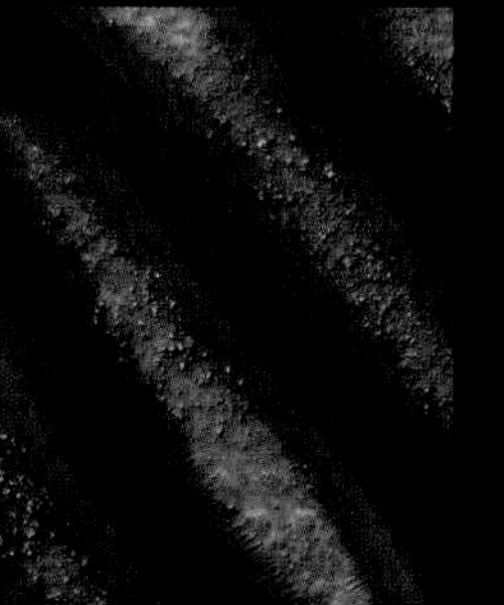

LINEAR DUNES
This MRO false-color image is of linear dunes within a crater in the Noachis Terra region. The dark dunes have a reddish dust band on their northeast-facing slopes. The crater floor between the dunes is littered with boulders.

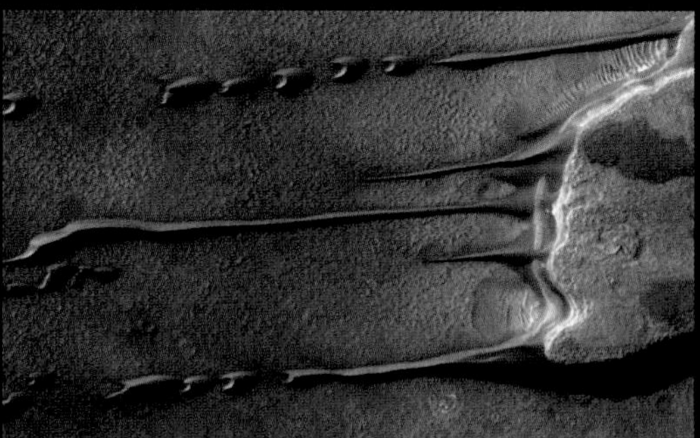

LINEAR AND BARCHAN DUNES
Sand is swept around the leeward side of a mesa (a flat-topped mound sculpted by wind) and blown into long, linear dunes parallel to the wind direction in this MRO image. These linear dunes break up into barchans away from the mesa

GUSEV CRATER DUNES
Spirit visited this dune field, named El Dorado, inside Gusev Crater (see pp.116–117) in early 2006. The dunes consist of fine, well-rounded sand grains a few hundred microns across.

DUNE GULLIES

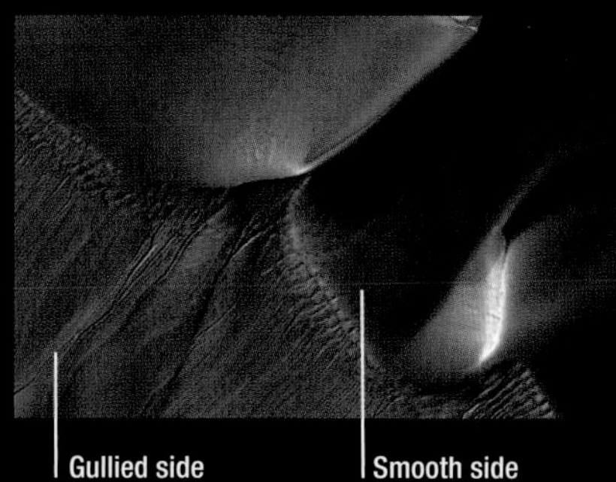

Gullied side
Gullies stop abruptly with no sign of deposition.

Smooth side
This side of the dune is gully-free.

This MRO image of an area some 2.5 miles (4km) across shows channels running from the top of a dune in Russell Crater. It is not certain how they formed. One theory is that frost settles on the dunes in shadow or at night and turns to gas in sunlight, triggering avalanches that form the gullies in the dune.

direction. Obstructions such as cliffs further influence the collection of the sand and the shape it forms. Earth's dunes change continually, therefore. Mars's dunes, by contrast, appear to be static. It could be that Mars's dunes formed in the distant past when the martian atmosphere was denser—or it could just be that they form extremely slowly in the planet's present extremely thin atmosphere.

Small fields of sand dunes are often found in the center of impact craters. Mariner 9 found the first, in Proctor Crater in 1972. Images from MRO, in 2009, of this and other crater fields have shown that the dunes have much smaller ridges of sand, called ripples, associated with them. Data from orbiters, and close examination by the rovers *Spirit* and *Opportunity* (see pp.110–111), suggest that the dunes are composed of basaltic sand derived from volcanic rock.

One type of dune seen often on Mars, but uncommon on Earth, is the linear dune—a long, straight line of sand formed as the wind consistently blows in one direction. These dunes can form on the leeward (sheltered) side of an obstruction. Sand funneled around the obstruction then forms long lines parallel to the flow of the wind. With distance from the obstruction, they can break up and form barchans as they are influenced by wind from another direction.

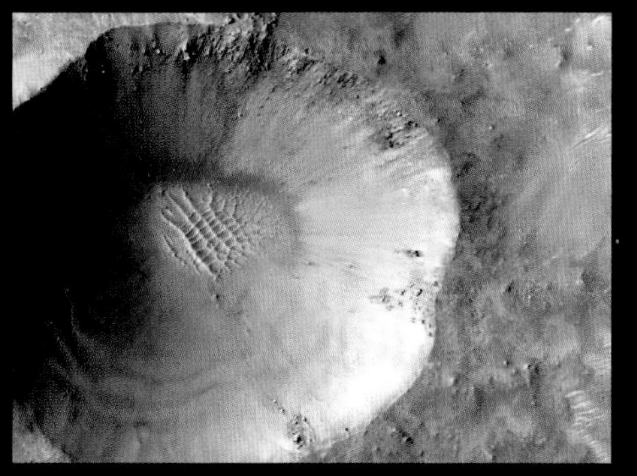

DUNES ON CRATER FLOOR
A false-color image by MRO of an unnamed crater shows a dune field on the crater floor. Sand trapped within the crater has formed a criss-cross pattern of dunes about 820ft (250m) wide.

BEAGLE CRATER
Sand found in even the smallest of craters is blown into patterns. This view into the 115ft- (35m-) wide Beagle Crater shows rippled sand at its center. It was imaged by the rover *Opportunity* as it drove to the much larger Victoria Crater (see pp.114–115).

DUST STORMS ON MARS

Dust storms are a common feature on Mars. They can occur all year round, but are also seasonal, and they can occur in specific areas or engulf the whole planet. Storms originating in the south polar region, during southern spring and summer, move across the planet, triggering new storms as they travel, eventually covering the whole globe in dust. Mars Global Surveyor (see p.94) monitored the planet in June 2001, hoping to witness such a storm. As it orbited, the craft recorded a local storm turn global over the course of a few weeks.

The south polar storms start when carbon dioxide ice is warmed by the summer sun and turns to vapor, creating fast-blowing winds. These sweep up dust and transport it around the planet. Localized storms can also be predictable. As winter turns to summer at the north pole, the increasing amount of sunlight creates storms lasting a day or so on the edge of the northern ice cap. Spacecraft have also witnessed shorter events—such as dust devils which are like those seen on Earth but 10 times larger—skim across the surface of Mars.

MARS UNDER CLEAR SKIES (JUNE 10, 2001)

South polar ice cap
The onset of the spring melt sees high winds and dust storms.

Into obscurity
As the 2001 dust storm raged, it engulfed even the highest peaks, including Olympus Mons (see p.98).

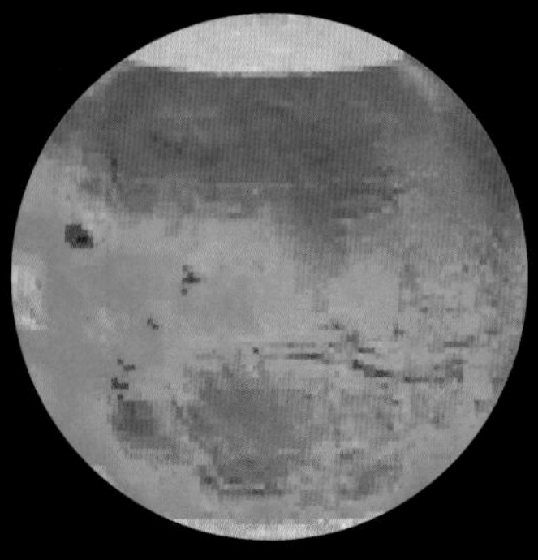

JULY 1, 2001

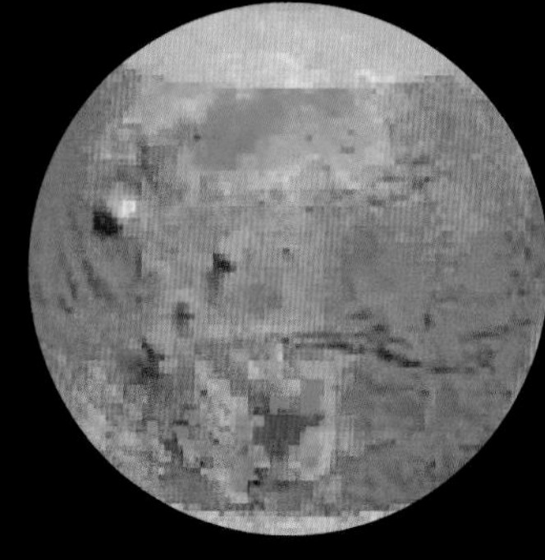

JULY 3, 2001

JULY 8, 2001

STORMY PLANET
The Thermal Emission Spectrometer on Mars Global Surveyor measured the temperature and dust content of the atmosphere during the 2001 global storm. Maps made using the collected data show the storm's evolution. At its height, the daily average temperature dropped by 5°F (3°C).

VISIBILITY

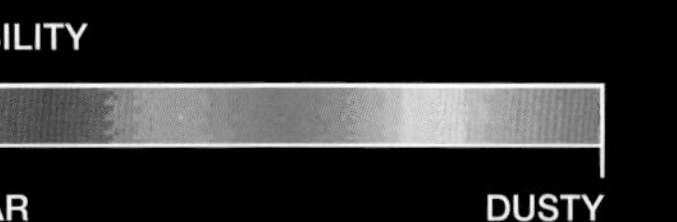

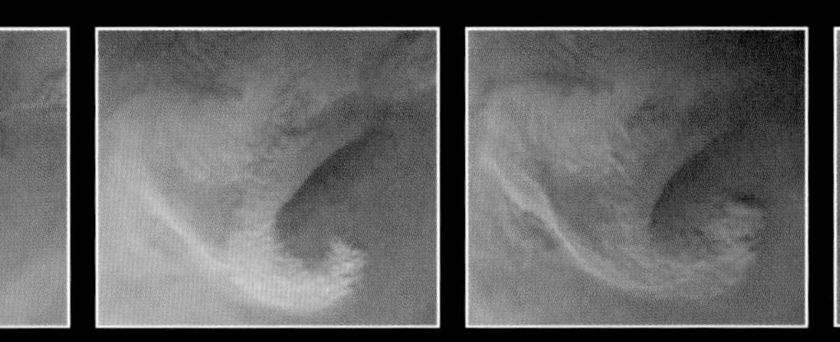

06:51:59 UTC

08:49:34 UTC

10:47:11 UTC

12:44:52 UTC

GATHERING STORM
These images, taken at two-hourly intervals—given here in Coordinated Universal Time (UTC)—show a short-lived summer storm developing near the north polar ice cap. The cap is at the upper left, and the storm is moving to the upper right—a curling cloud forms behind the storm front. White clouds are mainly water ice; pale brown ones contain dust

Start of 2001 storm
The storm's origin was a dust cloud inside the Hellas Planitia, an impact basin at this latitude on the opposite side of Mars.

A GLOBAL DUST STORM
A veil of dust high above Mars completely covers the planet on July 31, 2001. Beneath it are regional dust storms. Only a few weeks earlier, Mars Global Surveyor had imaged a relatively dust-free Mars, albeit with local dust storms, occurring mainly around the south polar cap. On June 21, one of these storms spread northward, then eastward. Five days later, it crossed the equator. During the following week, a high veil of dust moved around the planet; beneath it, a wind front created local storms.

ROVING THROUGH THE DUST

Dust storms on Mars affect all spacecraft working at the planet. Those in orbit are unable to image the ground, while the rovers on the surface (see pp.110–11), being solar powered, have their energy source obscured. Then, as the dust settles, it covers the rovers' solar panels, reducing the amount of electrical power that they can generate and so affecting their overall performance.

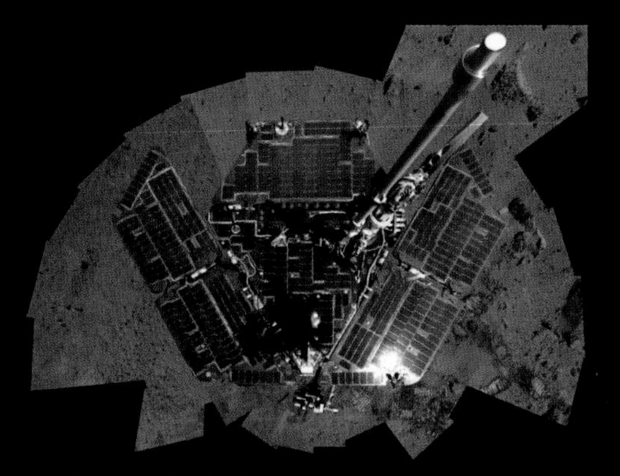

CLEAN SOLAR PANELS
A mosaic of images, taken by its own panoramic camera, shows the rover *Spirit* on August 27, 2005, 18 months after landing on Mars. Its solar panels gleam in the sunlight.

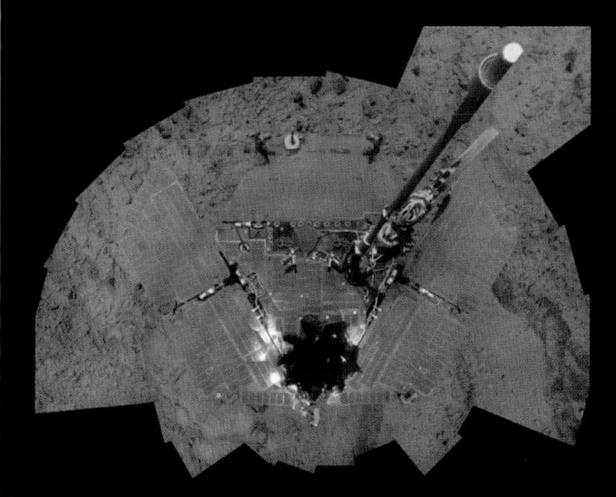

DIRTY SOLAR PANELS
Spirit is almost lost from view in this self-portrait, taken on October 29, 2007. There is so much dust on the panels that it appears the same color as Mars's surface.

MARS UNDER GLOBAL DUST STORM (JULY 31, 2001)

South polar ice cap
The 2001 global storm spread as the carbon-dioxide ice on the cap shrank.

DUST DEVILS
These whirling columns of dust, which are up to 6 miles (10km) tall, form as atmospheric gas warmed by daytime ground heat rises and cooler gas falls to take its place. A horizontal gust of wind makes the column rotate and as it moves, it leaves a trail known as the devil's tracks.

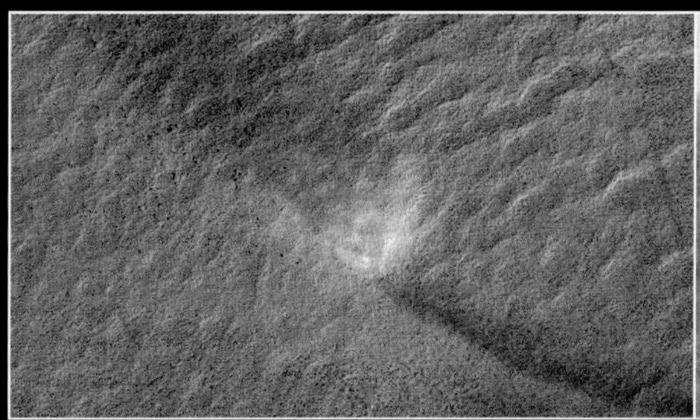

DUST DEVIL IMAGED BY AN ORBITING SPACECRAFT

DUST DEVIL IMAGED FROM THE GROUND BY *SPIRIT*

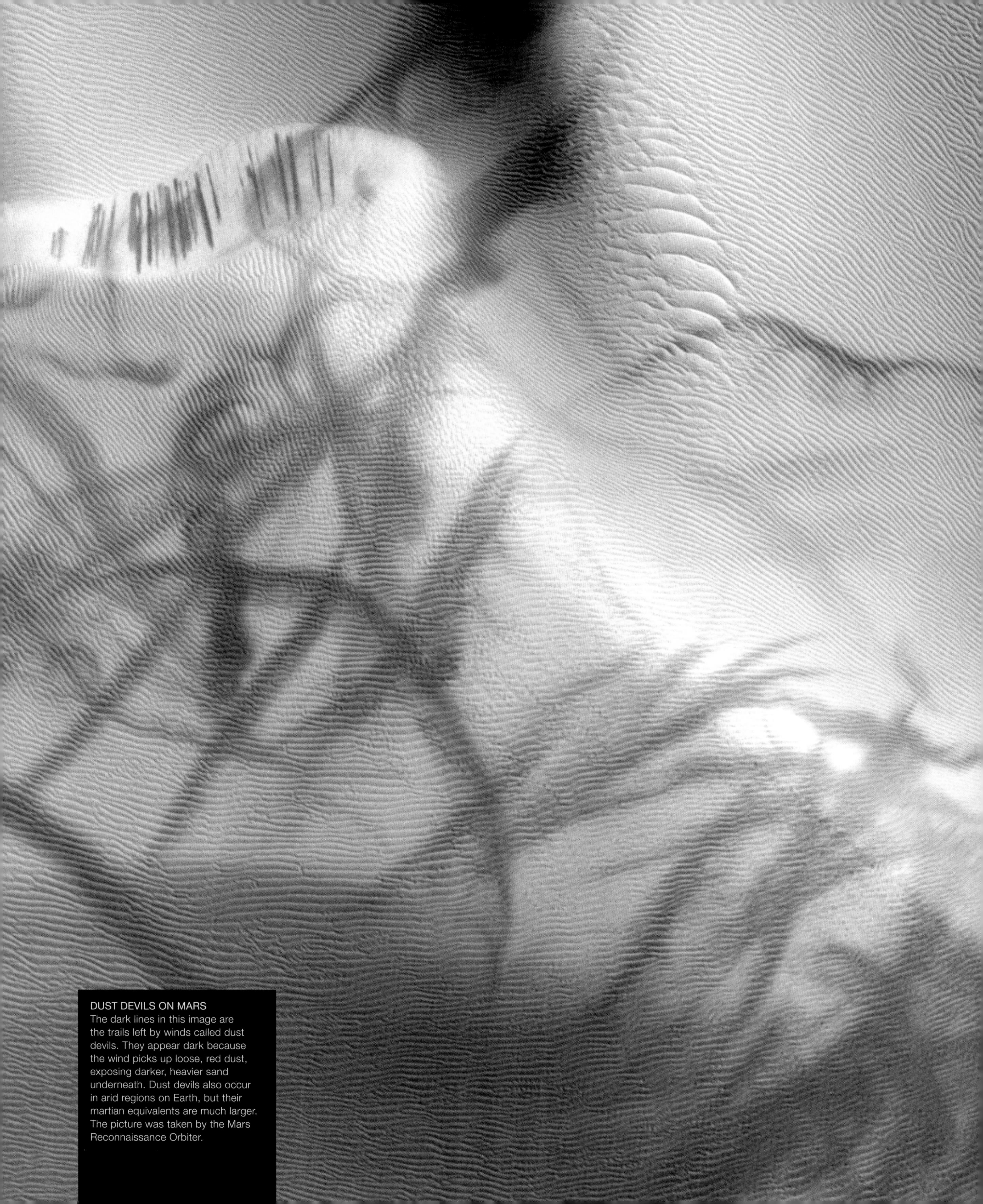

DUST DEVILS ON MARS
The dark lines in this image are the trails left by winds called dust devils. They appear dark because the wind picks up loose, red dust, exposing darker, heavier sand underneath. Dust devils also occur in arid regions on Earth, but their martian equivalents are much larger. The picture was taken by the Mars Reconnaissance Orbiter.

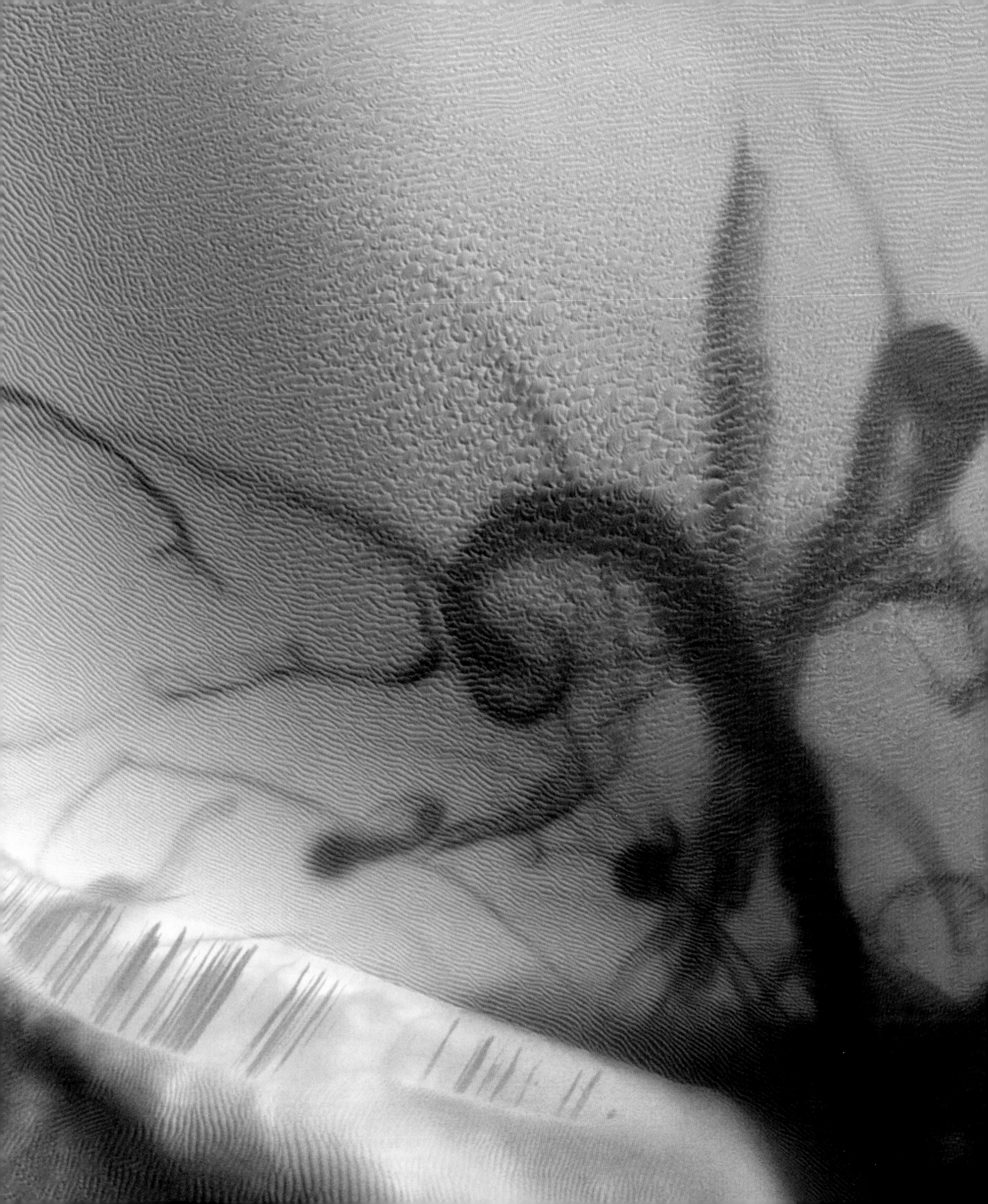

ROVING OVER MARS

Six craft—all of them NASA's—have made successful landings on Mars. Three were designed to stay put where they landed; three were built to rove across the planet. The most recent rovers, *Spirit* and *Opportunity*, arrived on opposite sides of Mars in January 2004 and remain operational to this day.

MARS EXPLORATION ROVERS

The first Mars rover was Sojourner, which arrived on the planet with the Pathfinder lander in July 1997. About the size of a household microwave oven, it investigated 16 areas of the surface close to its landing position. The larger twin Mars Exploration Rovers (MERs) *Spirit* and *Opportunity* are far more advanced. Like roving robot geologists, they have each moved across several miles of the planet, stopping frequently to make scientific investigations.

Sojourner relied on Pathfinder for functions such as communications, but an MER carries everything it needs with it. Its body is covered by an equipment deck supporting various antennae. At either side of the deck are hinged solar panels. At the front is a mast topped by panoramic and navigation cameras and a thermal emission spectrometer. Pairs of Hazard Cameras at the front and rear warn of anything in the rover's path when it drives forward or backward. At the end of a jointed arm, which tucks under the deck when the rover is traveling, are instruments for examining rock and soil, including a tool for studying rocks.

SUCCESSFUL MARS LANDERS AND ROVERS

NAME	OPERATIONAL DURATION	TYPE
Viking 1 (USA)	July 20, 1976–November 13, 1982	Lander
Viking 2 (USA)	September 3, 1976–April 11, 1980	Lander
Sojourner/Pathfinder (USA)	July 4, 1997–September 27, 1997	Rover/lander
***Spirit* (USA)**	Ongoing since January 4, 2004	Rover
***Opportunity* (USA)**	Ongoing since January 25, 2004	Rover
Phoenix (USA)	May 25, 2008–November 10, 2008	Lander

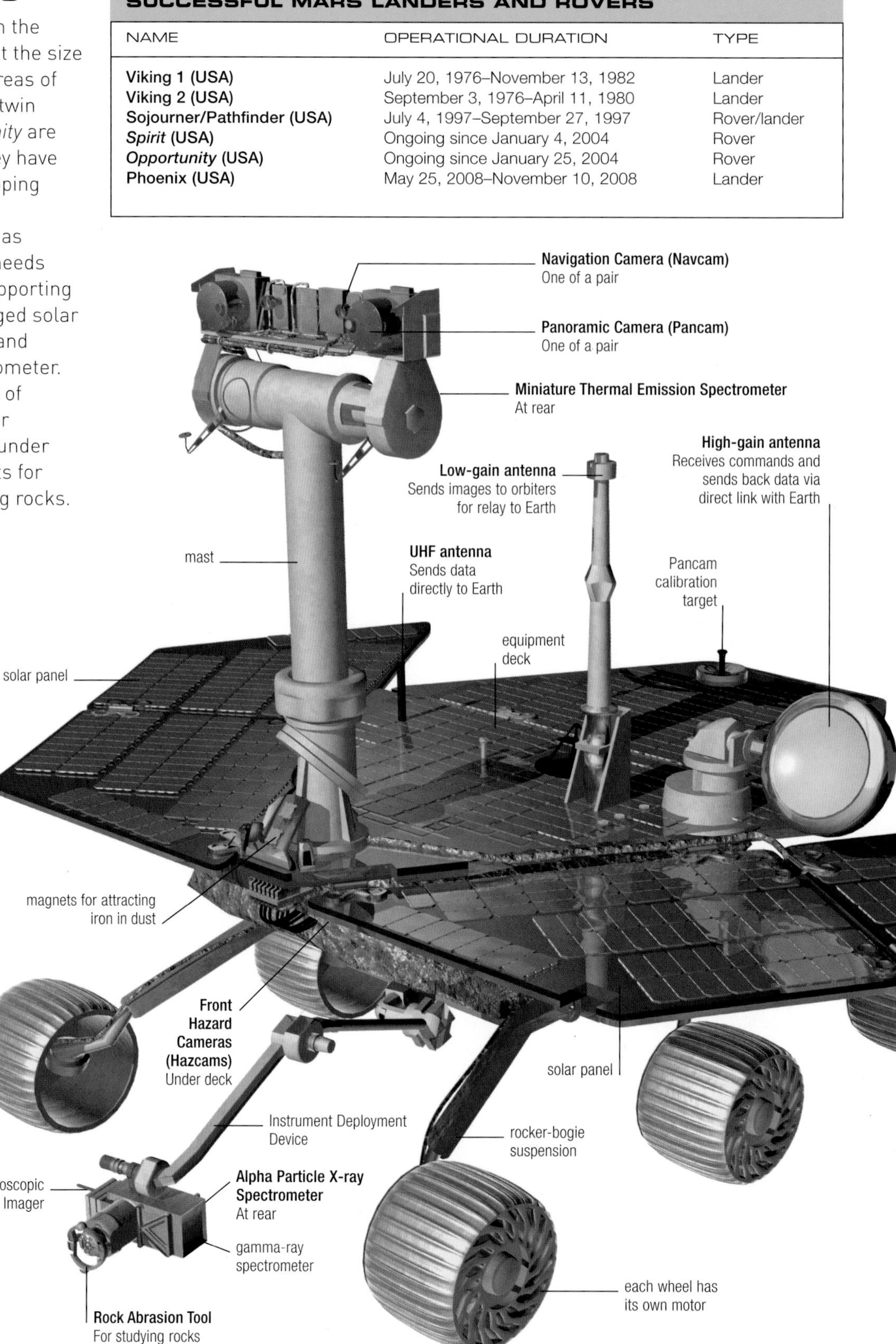

MARS EXPLORATION ROVERS PROFILE

MISSION	
Launch date	*Spirit*: June 10, 2003 *Opportunity*: July 7, 2003
Landed on Mars	*Spirit*: January 4, 2004 *Opportunity*: January 25, 2004
Intended duration	90 sols (90 martian days)
Launch vehicles	Delta II 7925 rockets
MARS EXPLORATION ROVERS	
Agency	NASA
Height	4ft 11in (1.5m)
Length	5ft 3in (1.6m)
Width	7ft 6in (2.3m)
Weight	384lb (174kg)
Power source	Solar panels and rechargeable lithium ion batteries

SCALE

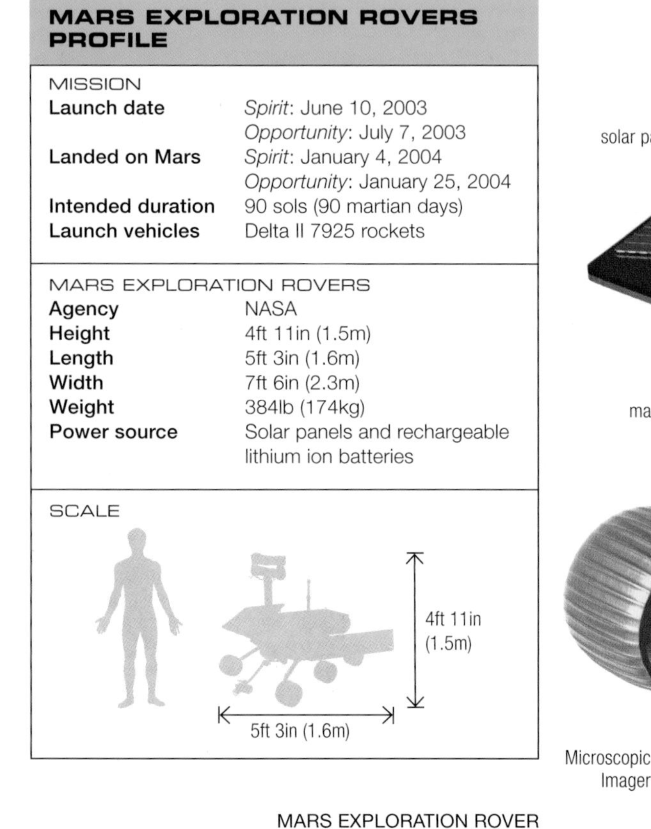

MARS EXPLORATION ROVER

The Mars Exploration Rovers were designed so that their panoramic and navigation cameras are almost at the height an astronaut's eyes would be. Their rocker-bogie suspension ensures that all six wheels always stay in contact with the ground, even on rocky terrain.

ARRIVING ON MARS

Each Mars Exploration Rover was launched separately and traveled to Mars in a hinged casing wrapped in deflated airbags and packed in an aeroshell, which was attached to a cruise rocket stage for the seven-month journey. Just before entering Mars's atmosphere, the cruise stage was jettisoned. As the aeroshell fell to Mars, atmospheric friction slowed it down. A parachute slowed it further, its heat shield was jettisoned, and the airbag-wrapped casing was released from the aeroshell on a tether.

1 Bouncy landing
Close to the ground, the airbags inflate and the tether is cut. The package hits the ground and bounces high.

2 Safely down
After bouncing several times, the package rolls across the surface and finally comes to a halt. The airbags then deflate.

3 Airbags retract
The deflated airbags are retracted to leave the way clear for the three outer "petals" of the rover's hinged casing to unfold.

4 Casing unfolds
If the casing does not land on the "base petal," the rover's computer orders that petal to open first, to right the rover.

CHOOSING THE LANDING SITES

Scientists spent two years choosing the landing sites for *Opportunity* and *Spirit*. The sites had to be of scientific interest, not so rugged as to endanger landing or restrict roving, and on low-lying land so that the aeroshells carrying the rovers passed through enough atmosphere to slow their fall. Data from Mars Global Surveyor and 2001 Mars Odyssey (see p.94) yielded a shortlist of 155 sites. Those finally chosen fulfilled all the criteria: the low, smooth Gusev Crater, which may once have held a lake, for *Spirit* (see pp.116–117); and the low, even Meridiani Planum, with exposed mineral deposits that may have formed in watery conditions, for *Opportunity* (see p.112).

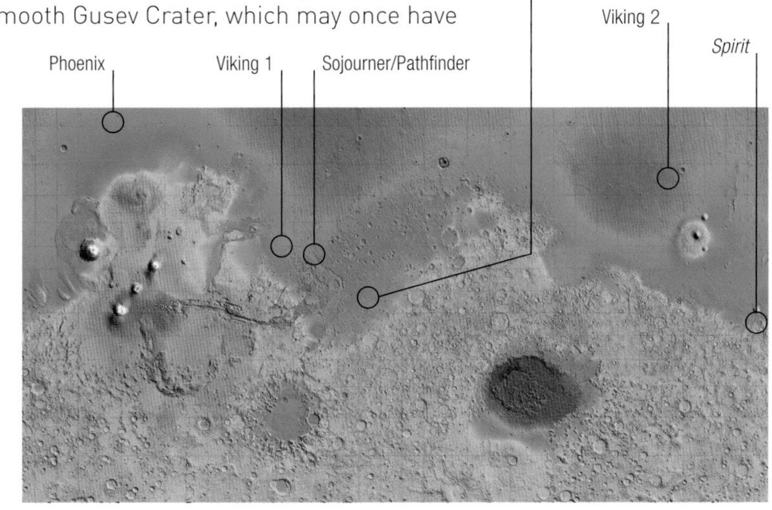

MARS LANDING SITES
Like Viking 1, Viking 2, and Sojourner/Pathfinder, *Spirit* and *Opportunity* were sent to sites relatively near the equator, where conditions are least extreme. Phoenix was sent close to the north pole.

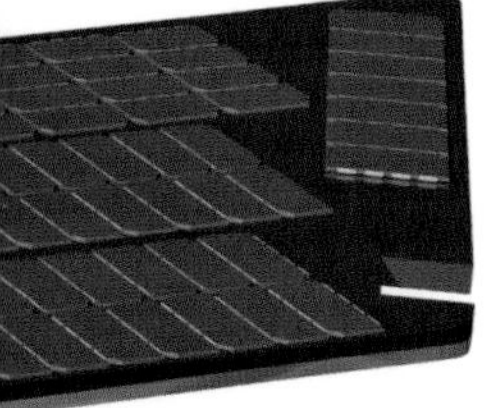

IN OPERATION

Once mission scientists had pinpointed where the MERs had landed, they identified target features within roving distance for the rovers to investigate. The instruments on top of each rover's mast—the cameras and thermal spectrometer, which classifies rock type from a distance—picked out likely places to stop on the way. Guided by controllers on Earth, the rovers moved forward cautiously, at an average speed of ½in (1cm) per second.

The rovers were intended to function for 90 sols (martian days), or 92 Earth days, but have far exceeded that. Until it got stuck in soft sand in May 2009, *Spirit* had covered 4.8 miles (7.7km). It is now a research platform. As of April 28, 2010, *Opportunity* had covered 12.8 miles (20.6km) and was still traveling.

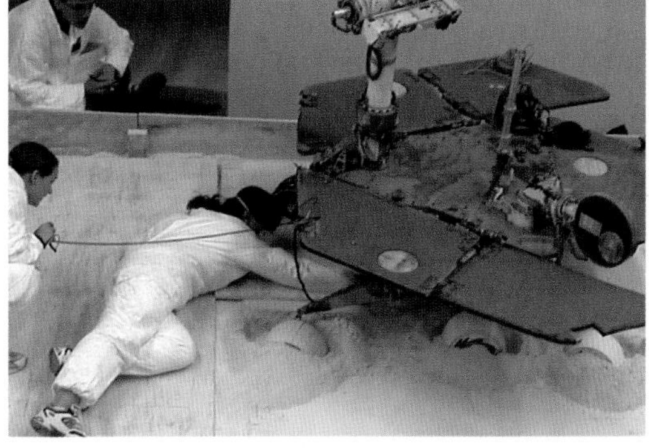

MOCK-UP OF ROVER STUCK IN MARTIAN SAND
For several months in 2009, mission scientists tried to get *Spirit* moving again, using a simulated surface and a test craft to try to work out how to free it from the soft sand in which it is stuck. In January 2010, they accepted defeat.

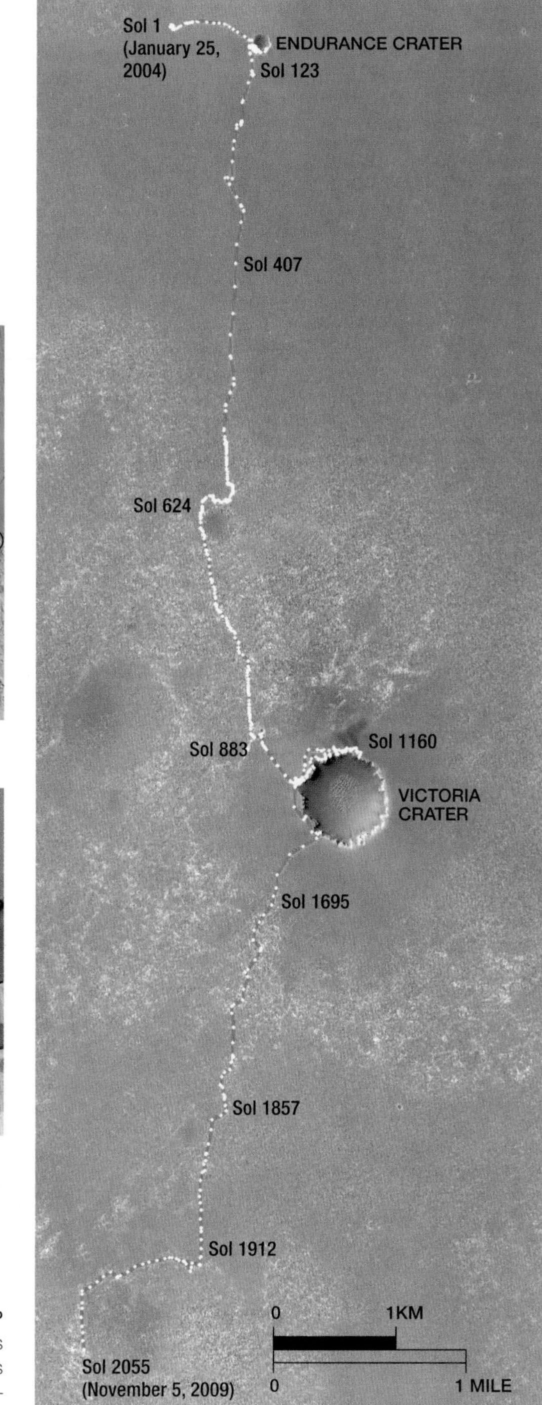

EXPLORATION ROVER ROUTE MAP
This NASA map charts *Opportunity*'s progress up until November 5, 2009. Its major stops have been at Endurance Crater (see pp.112–113) and Victoria Crater (see pp.114–115).

ENDURANCE CRATER

Looking east across Meridiani Planum from its January 25, 2004 landing site, on the horizon about 0.5 miles (0.8km) away, the Mars Exploration Rover *Opportunity* could see the rim of a football-field-sized impact crater. Mission scientists studied Mars Global Surveyor images of the crater, which they named Endurance, to decide the scientific merits of *Opportunity* exploring it and to see if the rover could drive safely in and out of it. Once *Opportunity* had arrived at the crater, on April 30, back on Earth the scientists built a test surface that

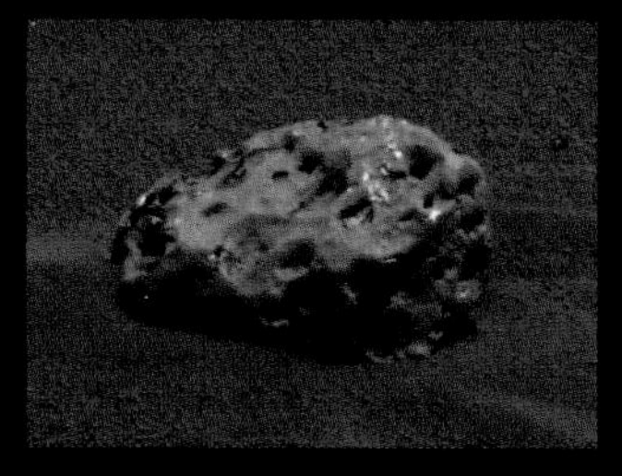

METEORITE ON MARS
On January 6, 2005, *Opportunity* found a basketball-sized pitted rock near its heat shield. Analysis confirmed that it is mostly made of iron and nickel and is a meteorite. Dubbed Heat Shield Rock, it was the first meteorite found on another planet, although both Mars rovers have since found others.

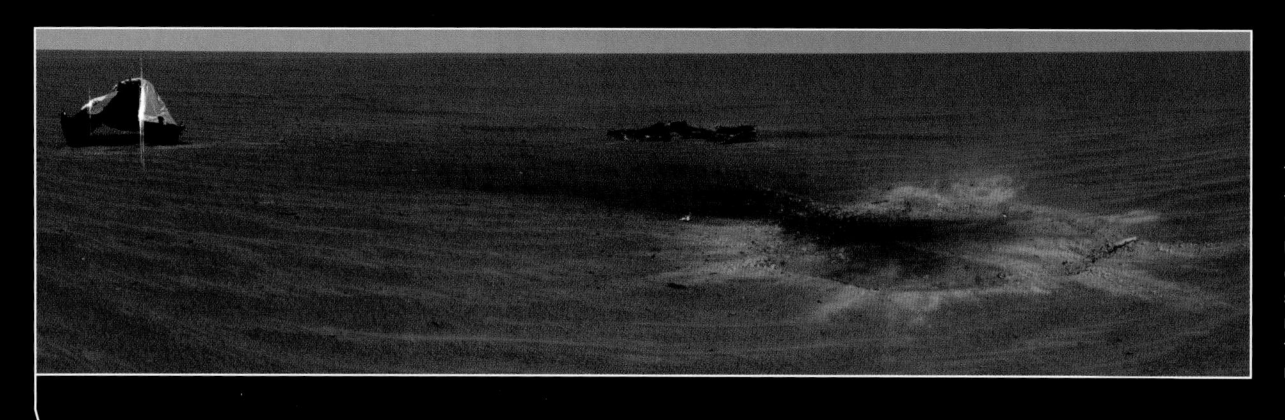

HEAT-SHIELD IMPACT SITE
This mosaic is made from images taken by *Opportunity* just south of Endurance Crater at the site where its heat shield—discarded on the rover's descent to Mars—landed. Part of the shield (left) is inverted, revealing its metallic insulation. Another part is at the center. At the right is the crater made when the shield hit the ground.

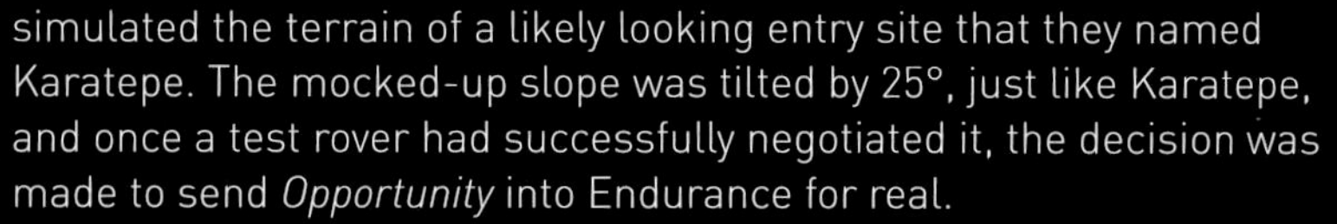

simulated the terrain of a likely looking entry site that they named Karatepe. The mocked-up slope was tilted by 25°, just like Karatepe, and once a test rover had successfully negotiated it, the decision was made to send *Opportunity* into Endurance for real.

Opportunity made its way down Karatepe's slope in June to begin 180 martian days of exploration of the crater. It stayed on the crater's sloping sides, to avoid getting stuck in the sand dunes on the central floor 65ft (20m) below the rim. *Opportunity*'s work inside Endurance showed that the layers of rock exposed when the crater was formed were ancient sedimentary rocks laid down by water more than 3.7 billion years ago, when Mars was warmer and wetter than it is today.

Its work at Endurance done, in December 2004 *Opportunity* set off to examine its discarded heat shield—and discover the first ever meteorite found on a planet other than Earth.

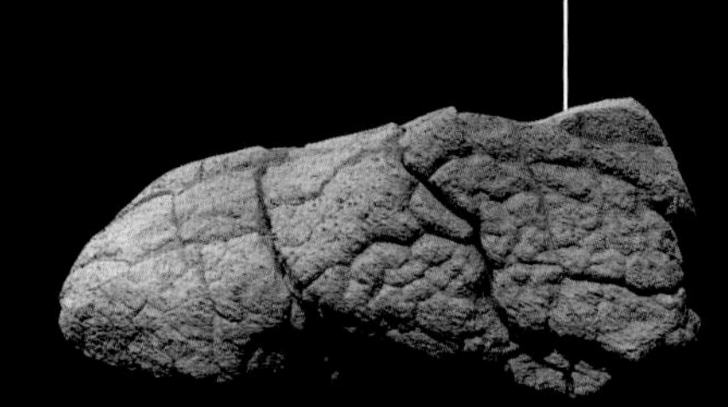

WOPMAY BOULDER
This lumpy 3ft (1m) rock inside the crater so intrigued mission scientists that they sent *Opportunity* to investigate. Its shape is probably due to exposure to water. This false-color image shows martian "blueberries"—iron-rich concretions, or hematites—embedded in the rock.

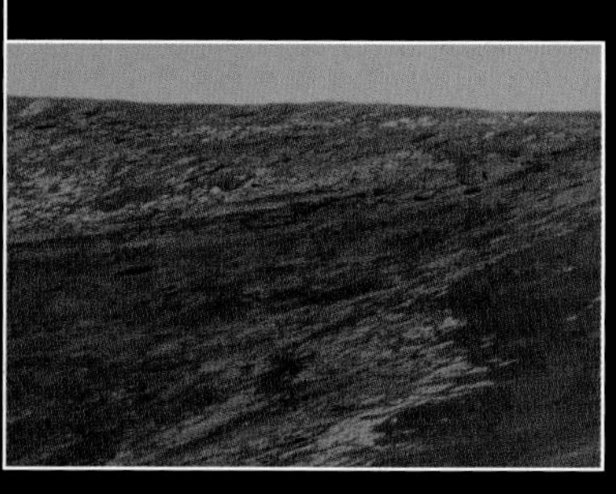

KARATEPE
Opportunity entered Endurance Crater across a band of rock that scientists named Karatepe. Here, Karatepe is the central region on the upper part of the crater's rim. Its lower edge is marked by a sloping line of light-colored rock just above center.

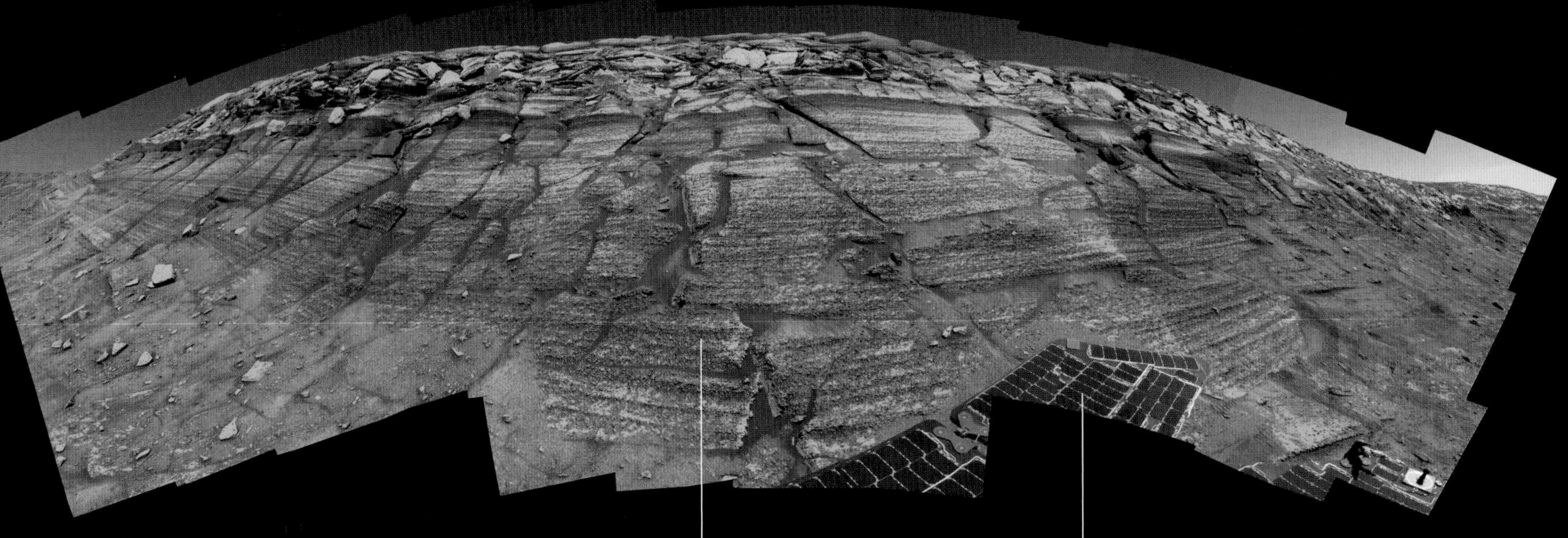

BURNS CLIFF
This wide-angle, true-color view is of part of the crater rim that scientists named Burns Cliff. It shows layer upon layer of ancient sedimentary rock, on top of which is a jumbled layer of material that was ejected by the impact of the meteorite that formed the crater.

Bedrock
The deeper the layer inside the crater, the older the rock

Partial self-portrait
Part of the rover's solar panels

IMPACT CRATER
Opportunity's Pancam took this true-color view of Endurance Crater, which is around 425ft (130m) wide. Parts of the rover's solar panels are in shot at the lower left and right corners.

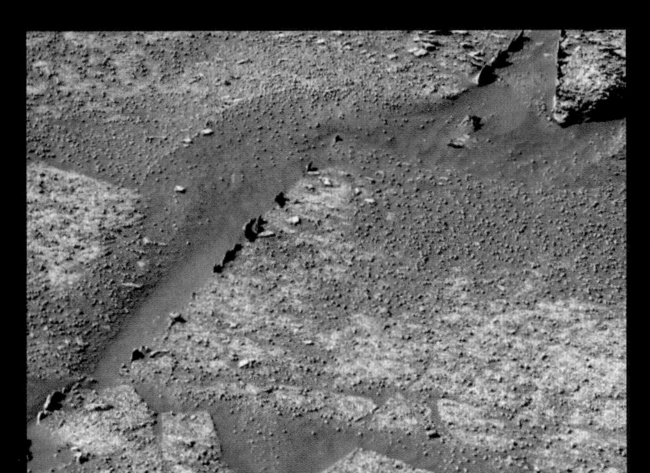

RAZORBACK
Dark, rocky points a few inches high stick up from the edge of a flat slab of "blueberry"–covered rock inside the crater in this false-color image. Dubbed Razorback, the points may be the remains of minerals that were deposited by water seeping through bedrock fractures and then survived later weathering.

DRILLING HOLES
Opportunity's Rock Abrasion Tool drilled these three holes into layered rock soon after entering Endurance Crater. The image has been processed to bring out color differences. The powdered rock from the top hole is redder than the other two because the tool ground through two marble-sized pebbles rich in hematite.

VICTORIA CRATER

After exploring Endurance Crater (see pp.112–113) in 2004, the Mars Exploration Rover *Opportunity* headed south. On September 26, 2006, after looking at various other craters and features on the way, it arrived at the much larger Victoria Crater—some 5 miles (9km) from the rover's landing site. About 0.5 miles (800m) wide and 230ft (70m) deep, Victoria is an impact crater with promontories and bays around its rim—features that give it an unusual scalloped shape, and many of which mission scientists named.

Opportunity spent almost two years exploring in and around the crater's outer edge. From Duck Bay, it drove clockwise around the rim, examining promontories and looking for a way into the crater. In a quarter circuit, it found that the promontories have almost sheer faces, exposing layers of bedrock, and are topped by ejecta rubble. But it found no suitable way in, so mission scientists decided it should return and enter at Duck Bay.

FIRST STOP
Opportunity's first stop inside the crater was to study an exposed layer of bright rock with its instrument arm. The rock was probably part of Mars's surface when the crater formed millions of years ago.

LAST LOOK
Opportunity took this parting shot of Duck Bay, Cape Verde, and its own tracks as it exited the crater in August 2008, retracing the way it went in almost a year earlier (September 2007).

EXPOSED BEDROCK
Mission scientists judged this band of smooth, exposed bedrock on the crater's inner slope too steep for *Opportunity* to drive onto. Sedimentary layers are evident in the rock in this false-color image.

VIEW FROM ABOVE
The central dunes and the scalloped rim are clearly defined in this Mars Reconnaissance Orbiter HiRISE image (see p.95). The bays formed when rim material crumbled and fell inward. The dark streaks beyond the rim are wind-exposed rock.

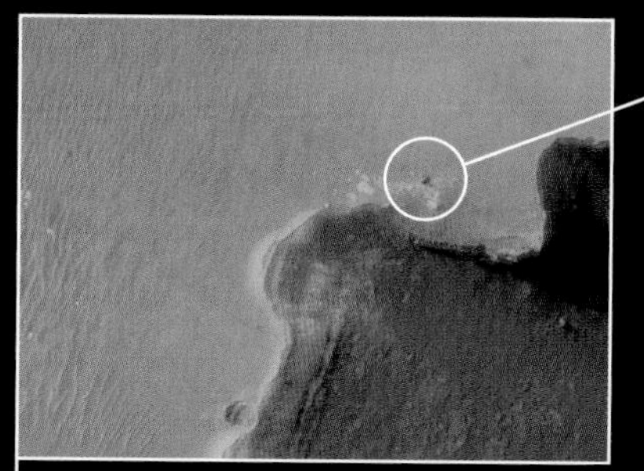

THE ROVER AT VICTORIA CRATER
Mars Reconnaissance Orbiter imaged the crater five days after *Opportunity* arrived. To the rover's lower left is Duck Bay. To its lower right is Cape Verde. *Opportunity* next moved to near the tip of Cape Verde to photograph the inside of the crater.

CAPE ST. VINCENT
A false-color view of this promontory emphasizes the structure of the exposed rocks. At the top is a crumbly layer of rubble ejected when the crater was formed. Below that are numerous layers of sedimentary rock.

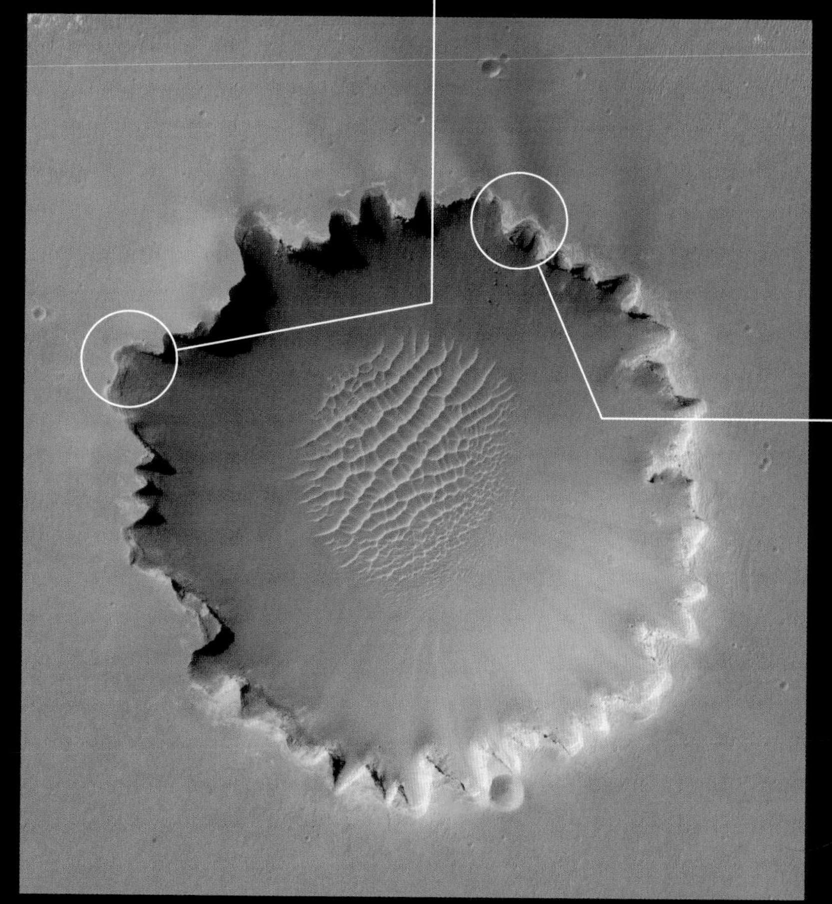

INTO THE CRATER

The scientists decided that Duck Bay's gentle 15–20° slopes and exposed bedrock made it a safe place for *Opportunity* to enter the crater, which it did on September 13, 2007. Two weeks later, after driving a few yards every few days, the rover arrived at its first target: a band of rock part of the way down the inner slope (see opposite).

Although *Opportunity*'s controllers were concerned that the rover had already operated for 12 times longer than its planned 90 martian days, they directed it even deeper into the crater to study ever older exposed rocks. It then drove to Cape Verde—a promontory 20ft (6m) high, next to Duck Bay—to examine its layers, before leaving the crater after almost a year inside. Its findings confirmed that the crater's rock layers were deposited by water more than 3.7 billion years ago.

VICTORIA CRATER FROM DUCK BAY
This is the first view that *Opportunity* had of the inside of the crater. Duck Bay is in the foreground. On the left is Cape Verde. On the right is Cabo Frio

GUSEV CRATER PANORAMA
This 240° panoramic view is a composite of images taken between February and October 2008 from the edge of Home Plate plateau. *Spirit* spent three southern-hemisphere martian winters here when solar energy to power it was limited.

GUSEV CRATER

Formed 3.9 billion years ago when an asteroid struck the surface of Mars, Gusev Crater is 103 miles (166km) wide. Scientists believe that floodwaters once poured into the crater through a large channel called Ma'adim Vallis, forming a lake. Seeking evidence to support or disprove this theory, mission specialists chose Gusev—which was named after a Russian astronomer and has smaller, younger craters inside it—as the landing site for the Mars Exploration Rover *Spirit*.

On receiving the first panoramic images sent back by the rover from its landing site, the rover mission scientists decided to send *Spirit* off in the direction of a range of seven low hills within Gusev—hills that they named both collectively (like the landing site) and individually in honor of the seven astronauts who died when the Space Shuttle *Columbia* disintegrated on re-entry to Earth's atmosphere in 2003.

On its way to the Columbia Hills, *Spirit* paused at the edge of Bonneville Crater, which is 690ft (210m) wide and is the resting place of the rover's heat shield, jettisoned on its descent to the surface of Mars. Moving on, in August 2005 *Spirit* reached the summit of Husband Hill, pausing periodically to investigate the rock. From there it drove to Home Plate, a low plateau some 295ft (80m) wide within the hills.

In March 2006, one of *Spirit*'s two front wheels stopped working, and even though the rover had traveled some way to a new target, McCool Hill, the trip there was abandoned. After that, *Spirit* had to drive backward, dragging its broken wheel. And since a rear wheel stopped working in 2009, it has been stuck in sand at the side of Home Plate.

As expected, *Spirit* found basaltic rocks on Gusev Crater's plain and, as hoped, it also identified bedrock—sedimentary material from a much earlier period—on the higher hills. Analysis of the bedrock (see right) revealed that its composition had changed, indicating that there were once substantial amounts of water inside Gusev Crater.

DRILLING FOR EVIDENCE

Spirit used its robotic deployment arm to investigate martian rocks and soil. Joints at the arm's shoulder, elbow, and wrist enabled it to extend, bend, and angle its instruments and tools precisely. Once in position, the tools and instruments got to work.

The Rock Abrasion Tool (RAT) ground away the surface to expose fresh material. This is because the surface may have weathered, have undergone chemical change through contact with the atmosphere, or be covered with dust. On Clovis outcrop (above) on a small hill in Gusev Crater, the RAT's grinding teeth rotated at speed to create a hole 1¾in (45mm) across and ½in (9mm) deep. (To the right of the drilled hole are circles made by the RAT's scrubbing brush.)

Spirit's two spectrometers then took readings over the next 10–12 hours to determine the composition of the exposed rock. Another tool, the Microscopic Imager, provided close-up images. These allowed analysis of the size and shape of the rock's grains. The results showed that liquid water once interacted with the rock, altering its composition.

THE ROVER'S LANDING SITE
The deflated airbags and the casing that carried *Spirit* to Mars remain at the landing site, named the Columbia Memorial Station. On the horizon, left of center, is the rim of Bonneville Crater, about 820ft (250m) away. *Spirit* stopped at Bonneville on its way to the Columbia Hills, which are on the horizon to the right, about 2 miles (3km) away.

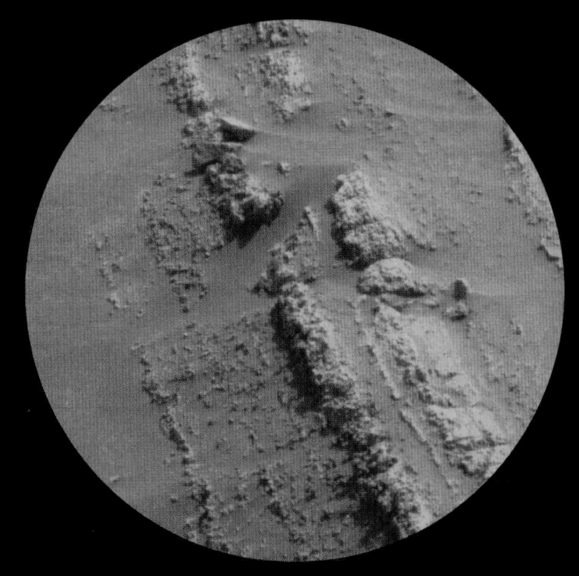

COARSE LAYERING
Spirit found layers of coarse-grained rock—possibly ejected material from a volcanic eruption or from a meteorite strike—around the edge of Home Plate.

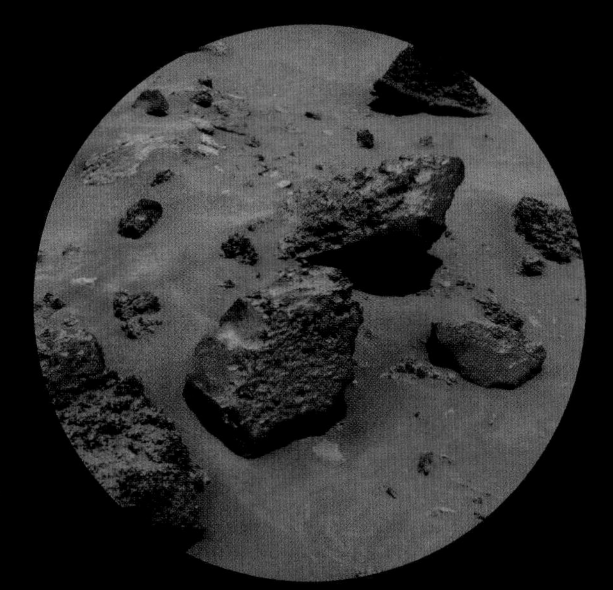

VOLCANIC ROCKS
Basaltic rocks are evidence of the volcanic lava that once flowed across Gusev Crater. *Spirit* found the rocks shown here, named FuYi, near Husband Hill. They have many small holes formed by lava solidifying around gas bubbles.

SUNSET ON MARS

The Sun sets behind the rim wall, 50 miles (80km) distant, of Gusev Crater on Mars. The halo of blue light is caused by sunlight, scattered by dust high up in the martian atmosphere. The image was taken by *Spirit*, one of two robotic rovers involved in NASA's ongoing Mars Exploration Rover mission.

THE MOONS OF MARS

Mars has two moons, called Phobos and Deimos. Many astronomers believe that these small, irregular bodies have not always been martian moons, but are captured asteroids – and Phobos's declining orbit indicates that it will only be a martian moon for another 11 million or so years.

SIZE, ORBITS, AND ORIGINS

Phobos and Deimos follow near-circular, synchronous orbits (that is, like Earth's moon, they keep the same face pointed toward Mars). Phobos is the larger of the two and the closest to the planet. It is 16.6 miles (26.8km) long, and orbits Mars every seven hours and 39 minutes at a distance of 5,827 miles (9,378km). Deimos, a little over half Phobos's size and more than twice as far away, takes four times as long to orbit the planet. Observations show that Phobos's orbit becomes smaller by around an inch each year, which means that it will eventually crash into Mars or—more likely—disintegrate. Both moons were discovered in 1877 by American astronomer Asaph Hall, but little was known about them until recently. Even now, astronomers are uncertain about their origins. Some think that Phobos formed from debris after Mars, not at the same time. Others think that both Phobos and Deimos are asteroids that were captured by Mars's gravity.

MOON OVER MARS
Phobos hovers above Mars in this view of the planet and its largest moon. The enhanced-color image is centered on Herschel Crater—at some 190 miles (300km) wide, it is Mars's seventh largest. The image was recorded by the Viking Orbiter 1 spacecraft in September 1977.

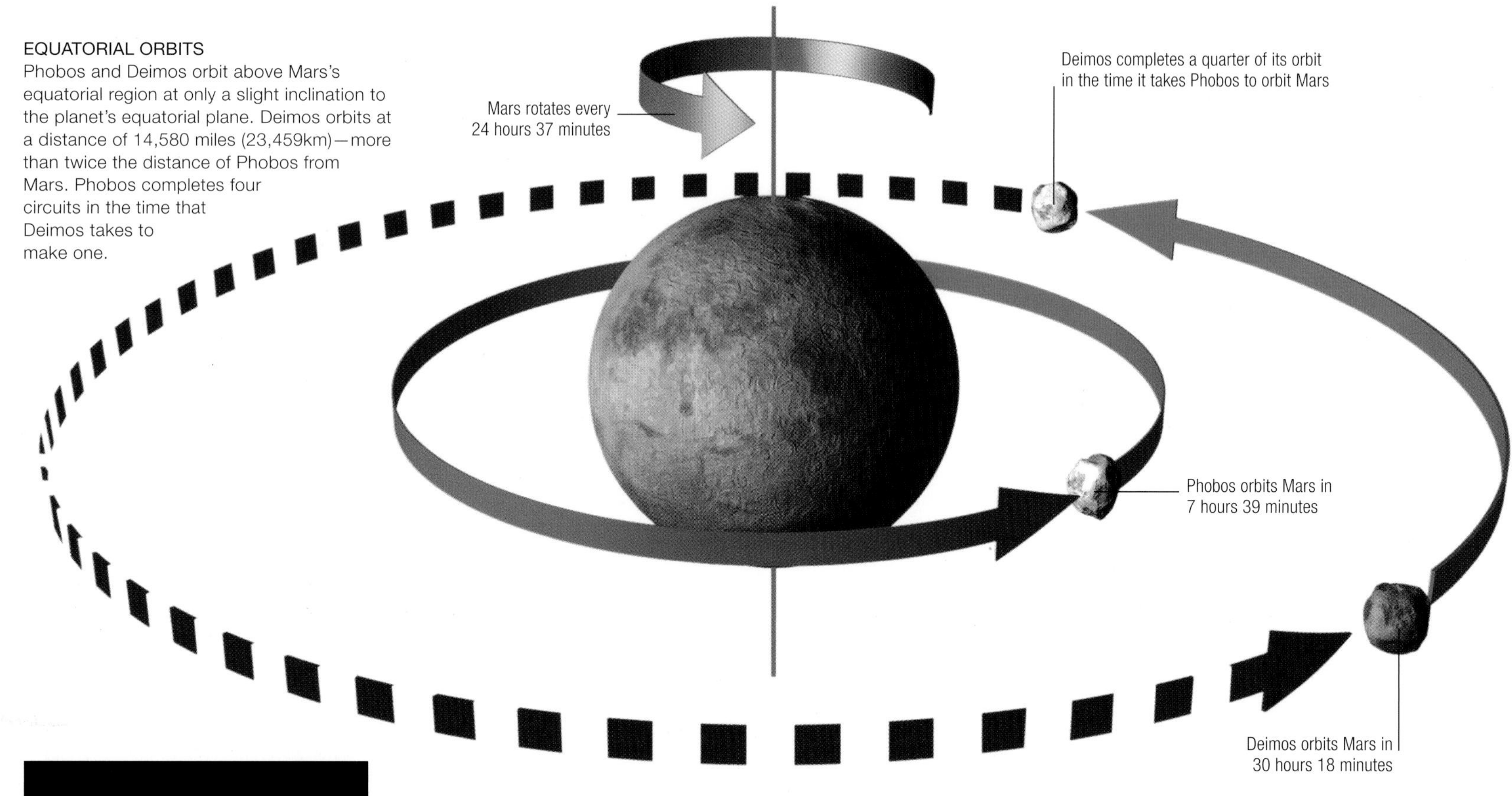
Mars rotates every 24 hours 37 minutes
Deimos completes a quarter of its orbit in the time it takes Phobos to orbit Mars
Phobos orbits Mars in 7 hours 39 minutes
Deimos orbits Mars in 30 hours 18 minutes

EQUATORIAL ORBITS
Phobos and Deimos orbit above Mars's equatorial region at only a slight inclination to the planet's equatorial plane. Deimos orbits at a distance of 14,580 miles (23,459km)—more than twice the distance of Phobos from Mars. Phobos completes four circuits in the time that Deimos takes to make one.

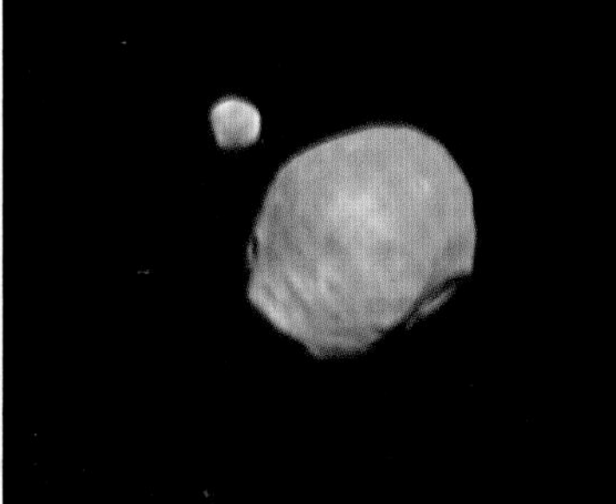

TWO MOONS TOGETHER
Here Phobos (right) and Deimos are seen photographed together for the first time in one of 130 images taken at one-second intervals by Mars Express (see p.94) in 2009. Phobos was 7,332 miles (11,800km) from the spacecraft; Deimos was more than twice as far. Such images yield new information about the moons' exact orbits.

> “PHOBOS IS PROBABLY A SECOND-GENERATION ... OBJECT. IT CAME FROM DEBRIS, IT WILL RETURN TO DEBRIS.”
>
> **MARTIN PÄTZOLD**, PRINCIPAL INVESTIGATOR, MARS EXPRESS, 2010

PHOBOS

Phobos is better understood than Deimos—especially since Mars Express made a series of flybys to within 42 miles (67 km), the closest ever, in 2010. Phobos's gravity tugged the spacecraft very slightly off course, but less so than astronomers had expected. From this they inferred that while the moon looks solid, it is not dense enough to be so and must be hollow in parts. It seems that rather than being a single lump of rock, Phobos is actually a cavity-ridden pile of rubble, which is held together by gravitational attraction.

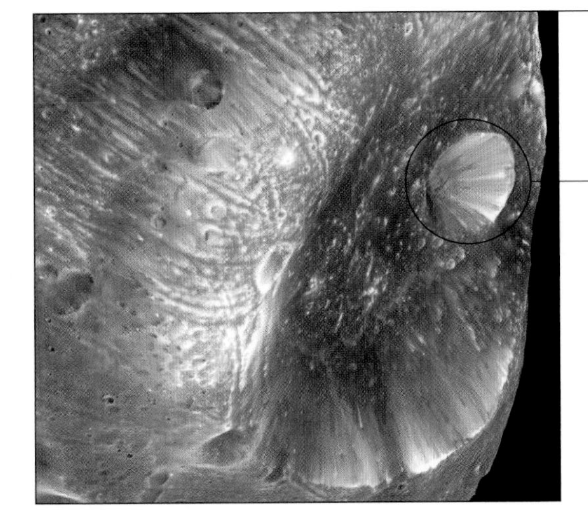

Limtoc Crater
A younger crater than Stickney, Limtoc is some 1.2 miles (2km) wide

STICKNEY CRATER
About 6 miles (10km) wide, Stickney Crater is Phobos's main feature. Lines on its inner walls—and on those of Limtoc Crater inside it—are landslides of rock and rock dust. Shiny areas are exposed bedrock.

PHOBOS
This enhanced-color image of Phobos was taken by the Mars Reconnaissance Orbiter (see p.95) in 2008. The linear grooves and crater chains were probably formed by ejecta from asteroid impacts on Mars.

DEIMOS

At some 9 miles (15km) long, Deimos is about the size of a large city. Like many asteroids, it is composed of carbon-rich rock, which suggests that it may itself have once been an asteroid. And, like Phobos, it is quite heavily cratered and covered in a layer of reddish regolith, or loose rock and dust. Swift and Voltaire, its two largest and only named craters, each measure less than 1.2 miles (2km) across. The surface of Deimos is generally much smoother than that of Phobos, because of the partial infilling of many of its craters by regolith.

Impact crater
This unnamed impact crater is about 0.6 miles (1km) wide

DEIMOS
The enhanced colors in this 2009 Mars Reconnaissance Orbiter image highlight variations in the moon's surface. Darker, redder areas are undisturbed rock and dust; lighter ones bedrock exposed by relatively recent impacts.

DEIMOS TRANSITING THE SUN
The Mars Exploration Rovers *Spirit* and *Opportunity* (see pp.110–11) have both photographed Deimos and Phobos racing across the Sun. *Opportunity* captured Deimos speeding across on March 4, 2004. The images are at 10-second intervals in Coordinated Universal Time (UTC).

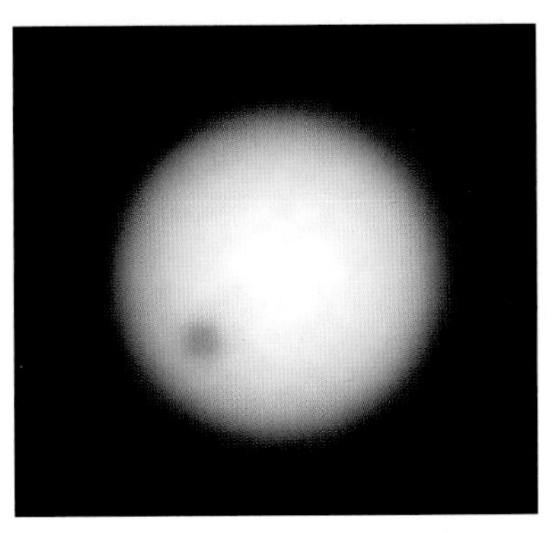

03:03:43 UTC

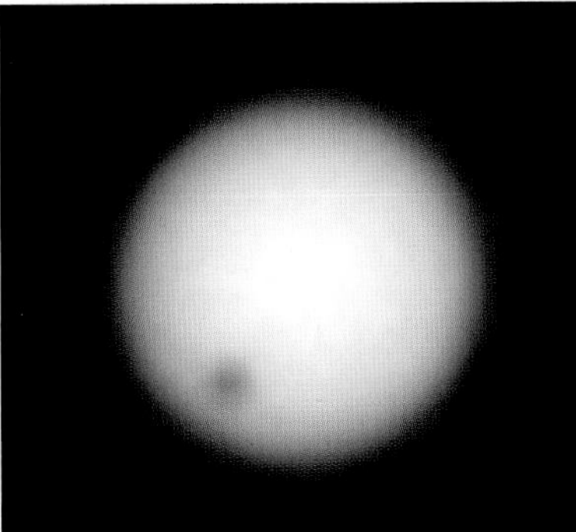

03:03:53 UTC

03:04:03 UTC

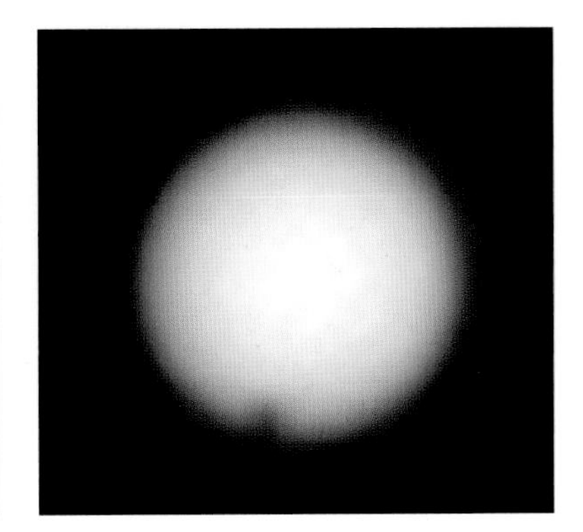

03:04:13 UTC

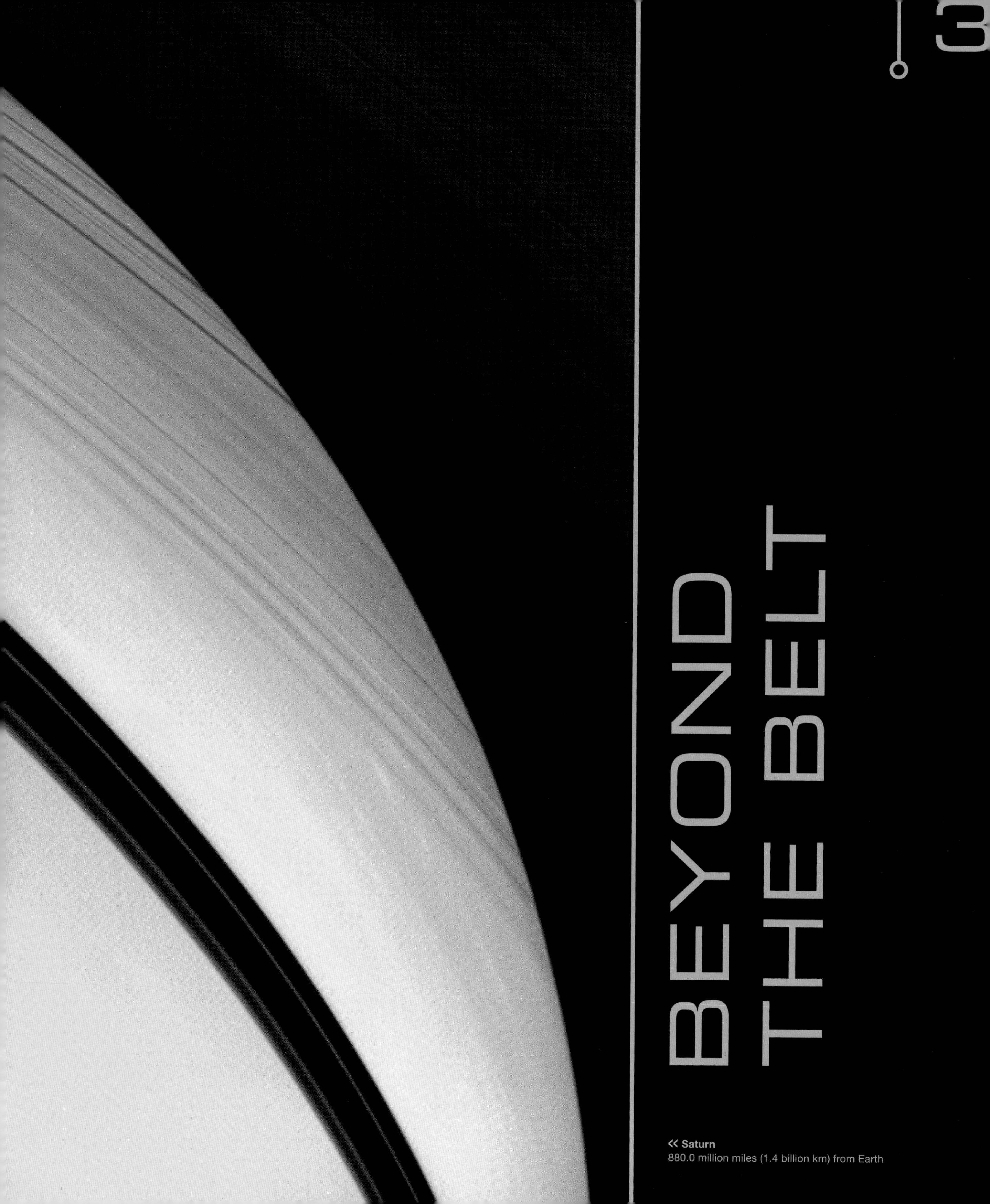

3

BEYOND THE BELT

« Saturn
880.0 million miles (1.4 billion km) from Earth

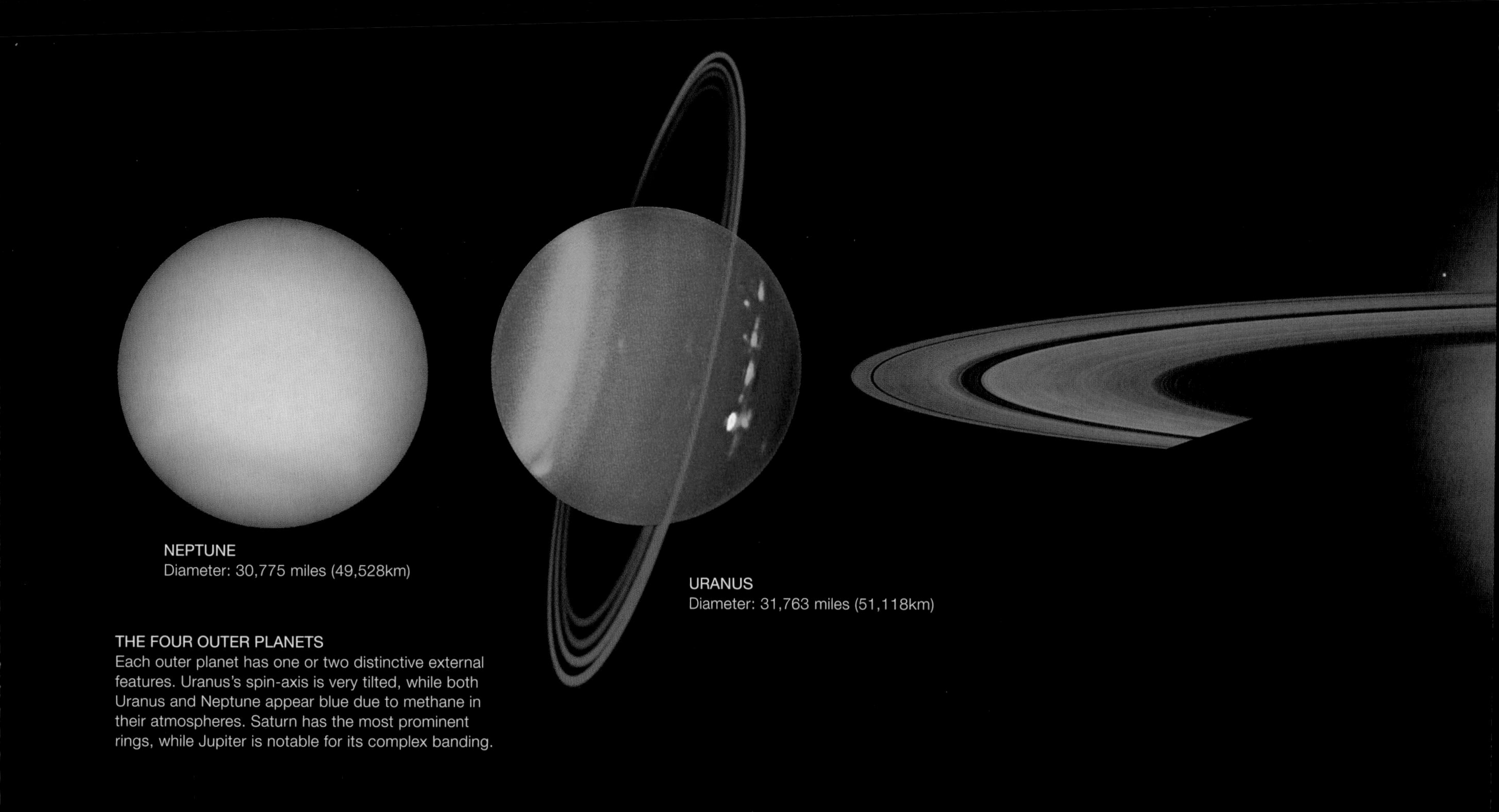

THE FOUR OUTER PLANETS
Each outer planet has one or two distinctive external features. Uranus's spin-axis is very tilted, while both Uranus and Neptune appear blue due to methane in their atmospheres. Saturn has the most prominent rings, while Jupiter is notable for its complex banding.

THE OUTER PLANETS

In decreasing order of distance from the Sun, the four giant outer planets are Neptune, Uranus, Saturn, and Jupiter. Each of these colossal bodies has an extensive atmosphere, a significant magnetic field, a large family of moons, and a system of rings.

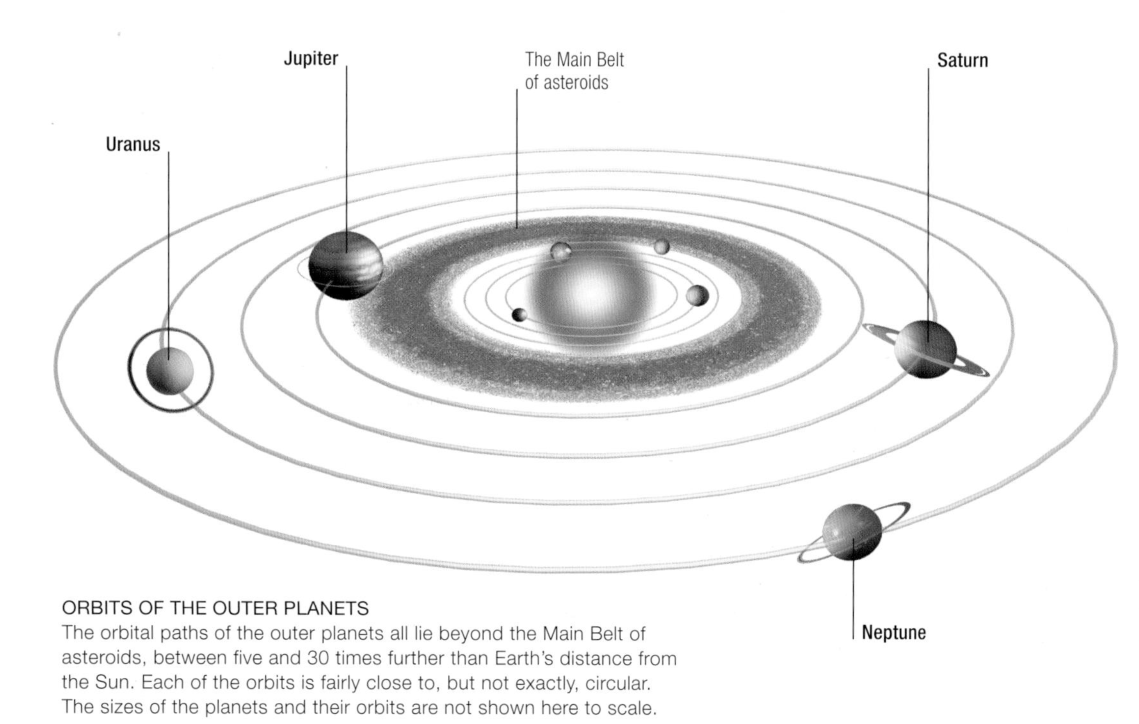

ORBITS OF THE OUTER PLANETS
The orbital paths of the outer planets all lie beyond the Main Belt of asteroids, between five and 30 times further than Earth's distance from the Sun. Each of the orbits is fairly close to, but not exactly, circular. The sizes of the planets and their orbits are not shown here to scale.

ORIGINS

The outer planets all formed at roughly the same time, in the outer part of the spinning disc of material that surrounded the newly forming Sun (see pp.50–51). This part of the disc, which was considerably cooler than the inner part, contained many tiny particles of ice, rock, and metal. Initially, these particles coalesced through collisions and gravitational attraction to form solid objects made of ice and rock. As they grew, these bodies attracted huge quantities of gases through their gravitational pull. These gases, mainly hydrogen and helium, enveloped the cores of ice and rock to form deep atmospheres.

The outer planets formed only approximately where they are today. Uranus and Neptune are thought to have moved some distance away from the Sun since their formation, and Jupiter may have moved slightly closer to the Sun. Astronomers think that an additional planet

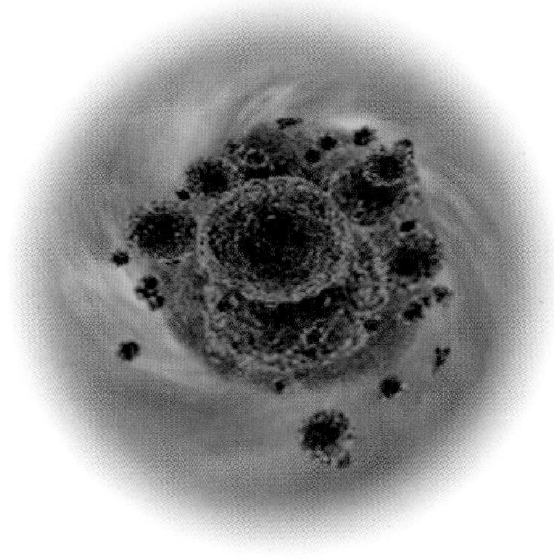
BIRTH OF AN OUTER PLANET
The process by which each of the four outer planets formed began with the accumulation of a core of rock and ice. The core then attracted, and became surrounded by, large quantities of gas.

could have formed between the orbits of Mars and Jupiter, had Jupiter's gravity not produced perturbations that kept the pre-planet material apart. The remains of this material is the present-day Main Belt of asteroids.

GIANT WORLDS

By volume, the outer planets are between 58 times (Neptune) and 1,320 times (Jupiter) the size of Earth. The outer planets each have a deep atmosphere of mainly hydrogen and helium, although the atmospheres of Uranus and Neptune also contain methane. Images of the four outer planets show the tops of their atmospheres, which (particularly in the case of Saturn and Jupiter) are constantly changing, being affected by strong winds and storms.

Between their gaseous atmospheres and solid cores of rock and ice, each outer planet also has an intermediate liquid or semi-solid layer. In Jupiter and Saturn, these layers are composed of hydrogen and helium, but in Uranus and Neptune, they consist mainly of water, methane, and ammonia ices. All four of the planets have significant magnetic fields, but Jupiter's is exceptional, being 14 times stronger than that of Earth.

MOONS AND RINGS

Each outer planet has numerous moons. The larger moons formed at the same time as their parent planets, from material surrounding the planets, in the same way that the planets formed from material around the Sun. Many of these large moons are complex, fascinating worlds in their own right. The many smaller moons are unused Solar System material that was captured into orbit around the outer planets. Most of these smaller moons are irregular bodies made of rock and ice.

The four outer planets each have a system of thin rings around them. The composition of the rings varies between the planets and from one ring to another; some rings are made of dust, others of boulder-sized particles of ice or rock and ice. Some rings may have been present since the planets formed; others are probably the remains of moons broken up by collisions or by the planets' gravitational fields.

SATURN'S RINGS
Although all of the outer planets have rings, Saturn's are the most extensive. This image was contructed from data obtained by passing radio signals through the rings between the Cassini spacecraft and Earth.

ASTEROIDS

Also known as minor planets, asteroids are small bodies made of rock and metal that follow independent orbits around the Sun. The vast majority lie in a concentration called the Main Belt. The largest of these, Ceres, was only discovered in 1801 and was reclassified as a dwarf planet in 2006.

SIZES OF ASTEROIDS

The combined mass of all the asteroids equates to around 4 percent of the Moon's mass. Ceres makes up about 30 percent of this total. The Main Belt has about 200 asteroids in excess of 60 miles (100km) across, 200,000 bigger than 6 miles (10km), and 200 million more than 0.6 mile (1km). Asteroids larger than 180 miles (290km) are nearly spherical; smaller ones are irregular.

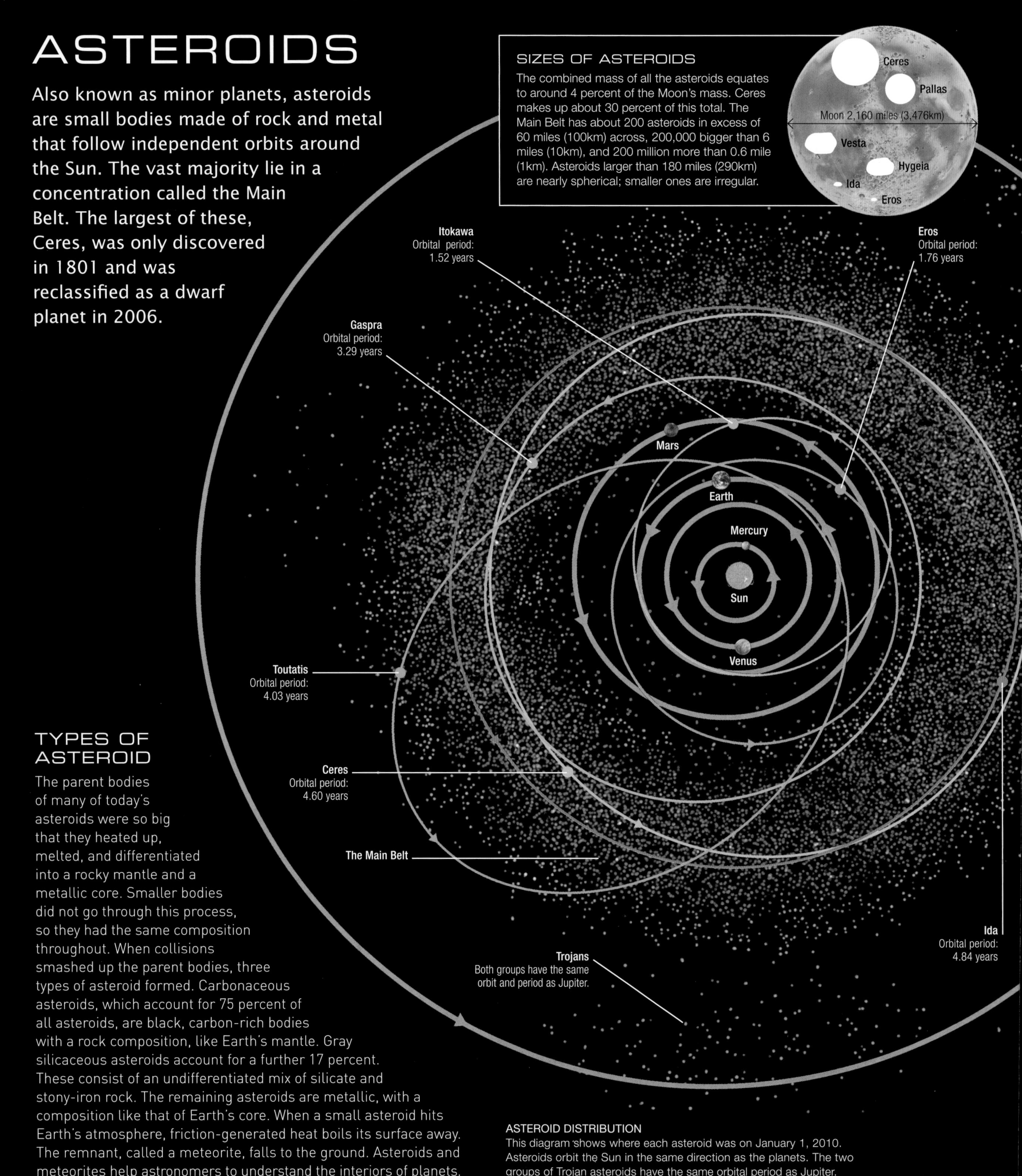

TYPES OF ASTEROID

The parent bodies of many of today's asteroids were so big that they heated up, melted, and differentiated into a rocky mantle and a metallic core. Smaller bodies did not go through this process, so they had the same composition throughout. When collisions smashed up the parent bodies, three types of asteroid formed. Carbonaceous asteroids, which account for 75 percent of all asteroids, are black, carbon-rich bodies with a rock composition, like Earth's mantle. Gray silicaceous asteroids account for a further 17 percent. These consist of an undifferentiated mix of silicate and stony-iron rock. The remaining asteroids are metallic, with a composition like that of Earth's core. When a small asteroid hits Earth's atmosphere, friction-generated heat boils its surface away. The remnant, called a meteorite, falls to the ground. Asteroids and meteorites help astronomers to understand the interiors of planets.

ASTEROID DISTRIBUTION
This diagram shows where each asteroid was on January 1, 2010. Asteroids orbit the Sun in the same direction as the planets. The two groups of Trojan asteroids have the same orbital period as Jupiter.

> **"ASTEROIDS ARE FRAGMENTS OF A PLANET WHICH HAS BEEN BURST ASUNDER."**
>
> **HEINRICH WILHELM OLBERS**, ASTRONOMER, 1803

THE MAIN BELT

The majority of asteroids occupy a doughnut-shaped region between Mars and Jupiter called the Main Belt, roughly 195–300 million miles (315–480 million km) from the Sun. Most asteroids that originally found themselves beyond this belt were destroyed long ago in collisions with planets and moons, causing the cratering of these bodies seen today. Some Main Belt asteroids experience such a strong gravitational interaction with Jupiter that they are pushed or pulled out of the Main Belt onto planet-crossing orbits. As a result, the Main Belt has several gaps, known as Kirkwood Gaps, which are swept free of asteroids by Jupiter's gravity.

Trojans

Jupiter

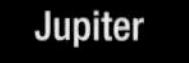

MAIN BELT FORMATION

The asteroids of the Main Belt were once thought to be the remnants of a planet that suffered an internal explosion or a devastating cometary impact. Astronomers now think that they are the result of an interrupted episode of planetary formation at the dawn of the Solar System. There was enough rocky and metallic material in the region now occupied by the Main Belt for a planet about four times more massive than Earth to form. Planetary formation initially progressed steadily, as material clumped together to form large bodies called protoplanets. However, the rapid growth of the young Jupiter disrupted this process: its gravity stirred up the protoplanets, changing their neat, near-circular orbital paths into ellipses, causing them to collide at high velocities and fragment into smaller bodies (see below).

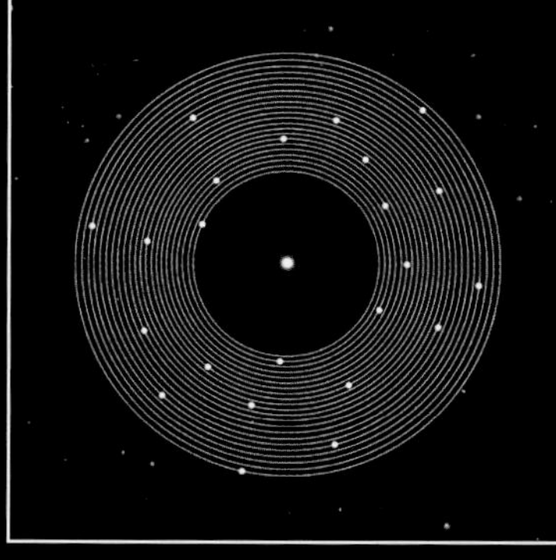

ASTEROID ORBITS BEFORE JUPITER
Before the formation of Jupiter, asteroidal objects had near-circular orbits. Collisions took place at low velocities, so particles stuck together and some became as big as Mars.

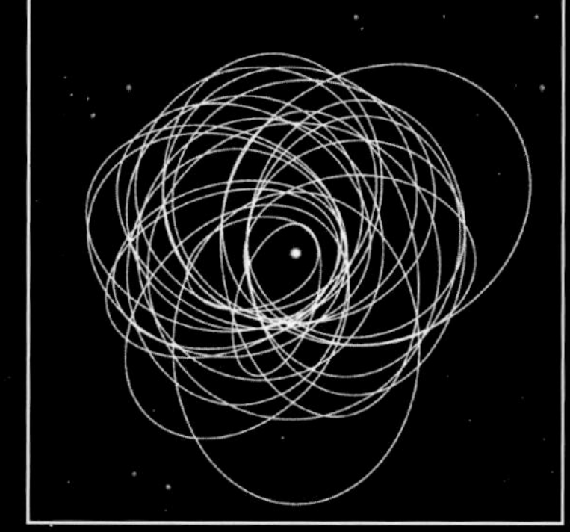

ORBITS AFTER JUPITER'S FORMATION
Jupiter's gravitational influence changed the circular orbits into ellipses. As the collision velocities grew to 11,200mph (18,000kph), impacting bodies smashed into pieces, rather than sticking together.

ASTEROID COLLISIONS

Asteroids have non-circular orbits and a range of orbital inclinations, so their orbits cross and collisions frequently occur. As a result, the total mass of the Main Belt has dropped by a factor of 1,000 since its origin. Over time, the size of the largest asteroid has fallen and the number of smaller asteroids has gone up. When a large parent asteroid breaks up, the pieces tend to have similar orbits to the parent, forming a "family" of asteroids. Gravity may also pull asteroids onto planet-hitting orbits.

IMPACT OUTCOMES
When one asteroid is struck by another, the outcome depends on the size of the impacting asteroid. Small impacting asteroids are much more common than large ones, and so cratering is the most common occurrence.

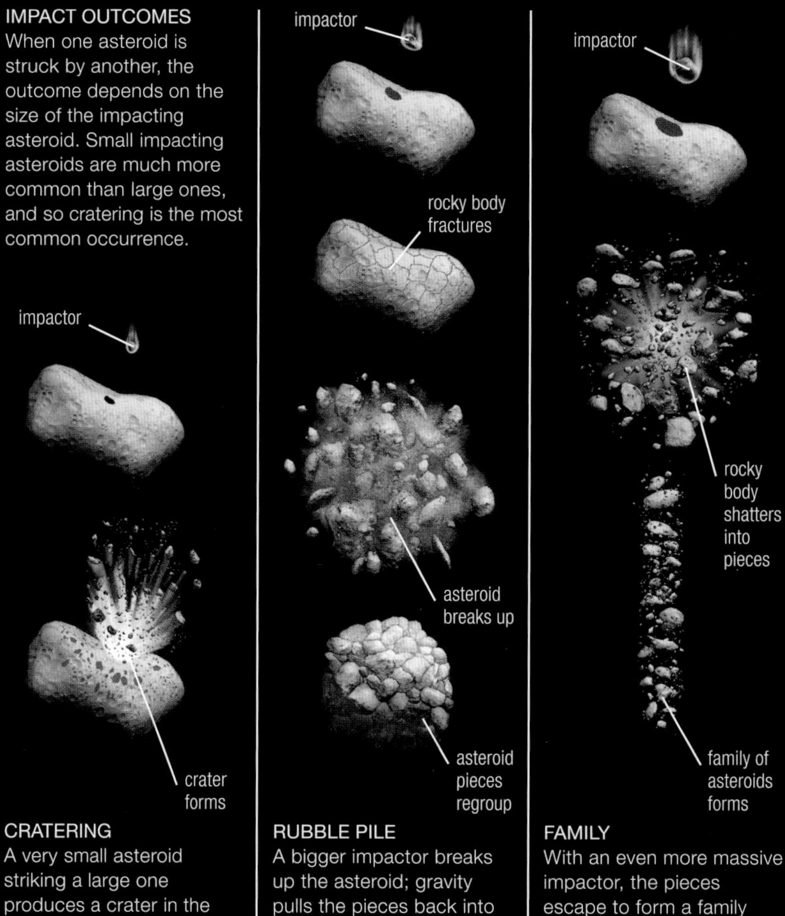

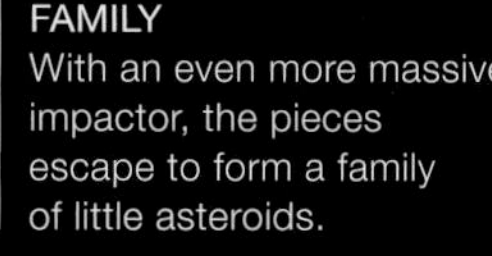

CRATERING
A very small asteroid striking a large one produces a crater in the large asteroid's surface.

RUBBLE PILE
A bigger impactor breaks up the asteroid; gravity pulls the pieces back into a rubble-pile asteroid.

FAMILY
With an even more massive impactor, the pieces escape to form a family of little asteroids.

NEAR-EARTH ASTEROIDS

Although most asteroids stay in the Main Belt, Jupiter's gravity can force some asteroids onto more elliptical orbits that bring them into the inner Solar System. Asteroids exceeding 490ft (150m) in diameter that approach Earth to within 20 times the Earth–Moon distance are termed Potentially Hazardous Objects (PHOs). Astronomers are constantly scanning the sky for these bodies in a program called Spaceguard. Some nations are even researching ways that rockets carrying explosive devices could be used to deflect PHOs from collision courses.

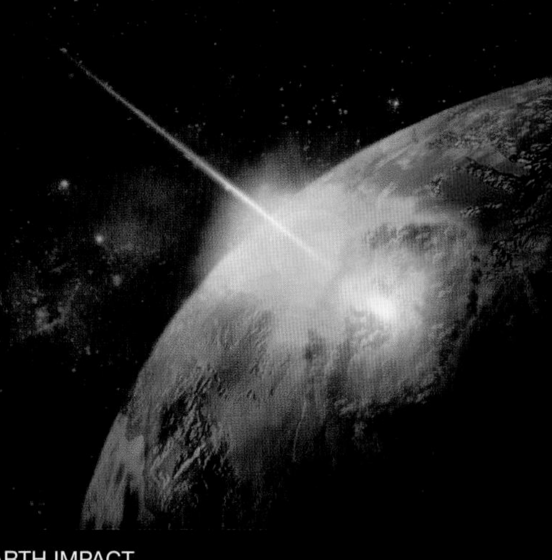

EARTH IMPACT
Asteroids over 0.6 miles (1.0km) across punch through Earth's atmosphere like a bullet through paper. They hit the surface at around 42,700mph (72,000kph), forming land craters 9–12 miles (15–20km) across.

MISSIONS TO ASTEROIDS

The Space Age has taught us much about asteroids. Spacecraft on their way to the outer giant planets took a few images of the asteroids they met as they crossed the Main Belt. More recently, dedicated craft have gone to specific asteroids and been placed into orbit around them. The first landed on an asteroid in 2001.

FLYBY OF GASPRA AND IDA

When crossing the Main Belt on its way to Jupiter, the Galileo spacecraft (see p.131) flew past two asteroids, Gaspra and Ida. Both are smooth and cratered, having been "sandblasted" by the impact of countless smaller bodies. Gaspra, which rotates in seven hours, has a density 2.7 times that of water, indicating that it is solid rock; Ida spins faster, in 4.6 hours, and has a much lower density. Galileo revealed that Ida has a small moon, Dactyl, which is just 0.9 miles (1.4km) in diameter. Ida was the first asteroid to be found with a satellite.

MISSIONS TO ASTEROIDS

ASTEROID	SPACECRAFT	ARRIVAL DATE	MISSION TYPE
Gaspra	Galileo	October 1991	Flyby
Eros	NEAR Shoemaker	February 2001	Orbit and land
Itokawa	Hayabusa	November 2005	Rendezvous and land
Vesta	Dawn	July 2011	Orbit; will encounter Ceres in February 2015

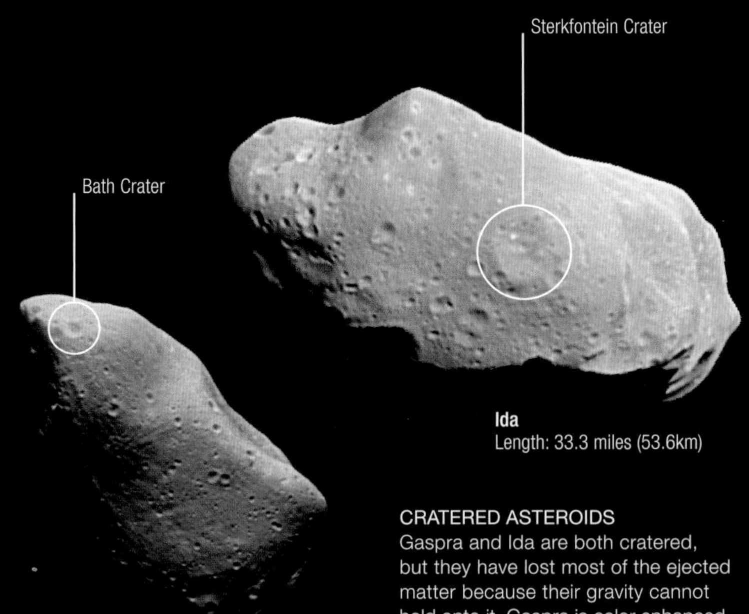

CRATERED ASTEROIDS
Gaspra and Ida are both cratered, but they have lost most of the ejected matter because their gravity cannot hold onto it. Gaspra is color enhanced in this image: to the eye, it looks gray, like Eros (below).

LANDING ON EROS

To date, the most successful mission to an asteroid has been NEAR Shoemaker (NEAR stands for Near Earth Asteroid Rendezvous). Launched in February 1996, the craft flew past the asteroid Mathilde and then, four years after leaving Earth, went into orbit around Eros at a height of about 210 miles (340km). After five months of reconnaissance, the orbit was lowered until the craft passed within 30 miles (50km) of Eros's surface, enabling its camera to take clearer images. Seven months later, the spacecraft descended to the surface, taking 59 highly detailed images during the last 3 miles (5km). Impact occurred on February 12, 2001 at a speed of 4mph (6kph). Originally intended just to orbit Eros, the craft's landing was a bonus in this mission. An onboard instrument used gamma rays to measure the composition of the surface rock for a further 16 days.

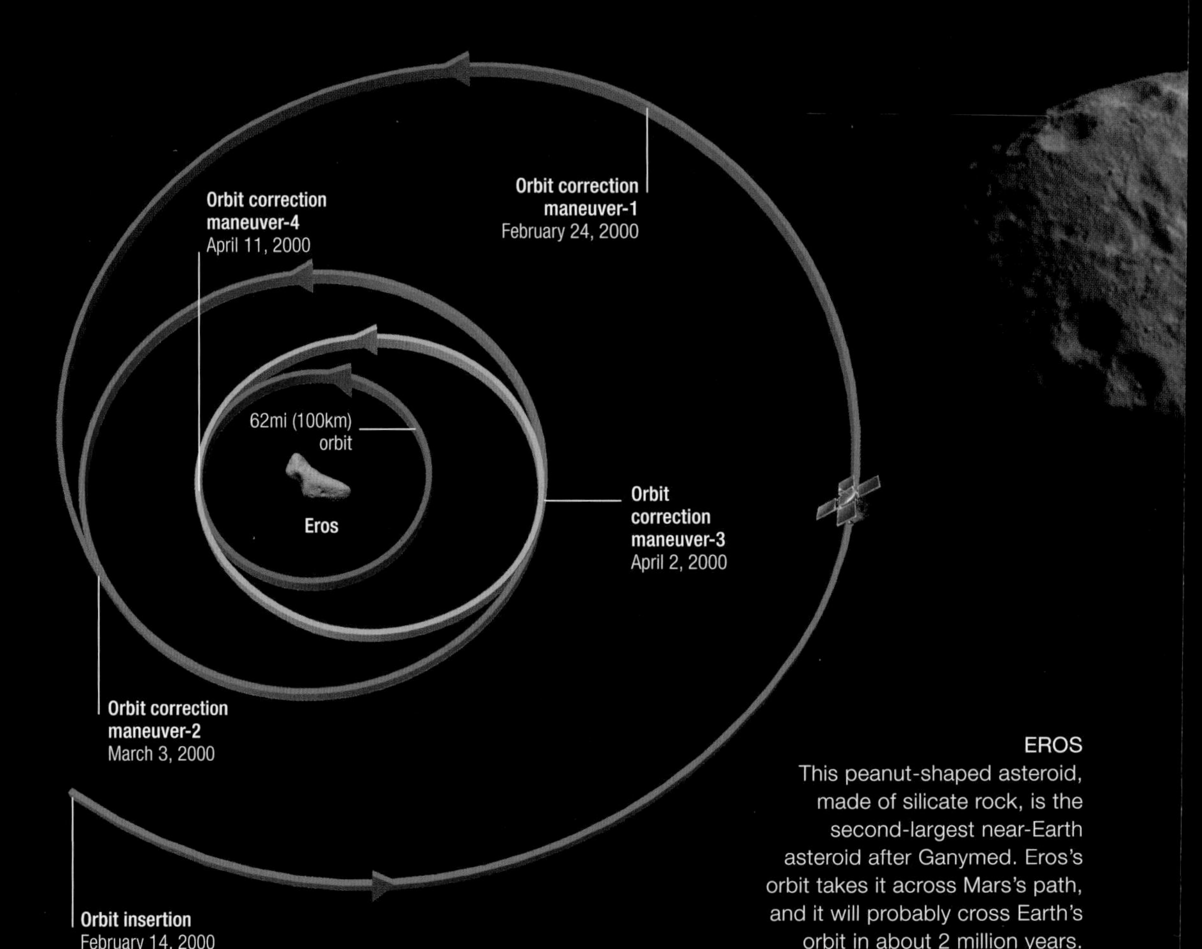

ORBIT AROUND EROS
Correction maneuvers in February and March 2000 moved NEAR Shoemaker from its initial orbital path onto a closer, 125-mile- (200km-) orbit. More maneuvers on April 2 and 11 lowered the orbit to 60 miles (100km); and maneuvers on April 22 and 30 (not shown) took it to 30 miles (50km).

EROS
This peanut-shaped asteroid, made of silicate rock, is the second-largest near-Earth asteroid after Ganymed. Eros's orbit takes it across Mars's path, and it will probably cross Earth's orbit in about 2 million years.

MISSION TO ITOKAWA

Itokawa is a recently discovered, irregularly shaped asteroid made of silicate rock. The Japan Aerospace Exploration Agency launched the Hayabusa craft in May 2003. In November 2005, the spacecraft rendezvoused with the tiny, 1,770ft- (540m-) long asteroid and kept station with it while carrying out a series of observations from a range of 12 miles (20km) above the surface. Hayabusa recorded the shape, spin, topography, color, composition, and density of Itokawa before descending and landing on the asteroid's surface. Hayabusa fired metallic particles onto Itokawa and captured samples of the surface material that these blasts stirred up. It then headed back to Earth. A re-entry capsule is due to land in Australia in 2010.

HAYABUSA COLLECTING SAMPLES
In this artist's impression, Hayabusa fires metallic particles at Itokawa's surface. This pioneering mission was the first attempt to bring asteroid samples back to Earth.

OBSERVING ASTEROIDS FROM EARTH

When viewed from Earth, most asteroids are little more than starlike dots in the sky, simply because they are too small and too far away to appear otherwise. But to date, nearly 400 have been detected by radar. Radar gives the range of the asteroid and, in the case of near-Earth asteroids, it can also be used to produce a crude image of the Earth-facing features. The asteroid Toutatis orbits the Sun every four years, and its low-inclination orbit crosses Earth's own orbital path. This means that Toutatis also makes a close Earth flyby every four years. In December 1992, it came within 2.5 million miles (4 million km) of Earth and was imaged once each day as it passed over the 230ft (70m) radar dish of the Goldstone Deep Space Communications Complex in the Mojave Desert in California. Radio pulses reflected off the asteroid gave the distance to the surface and the asteroid's relative velocity.

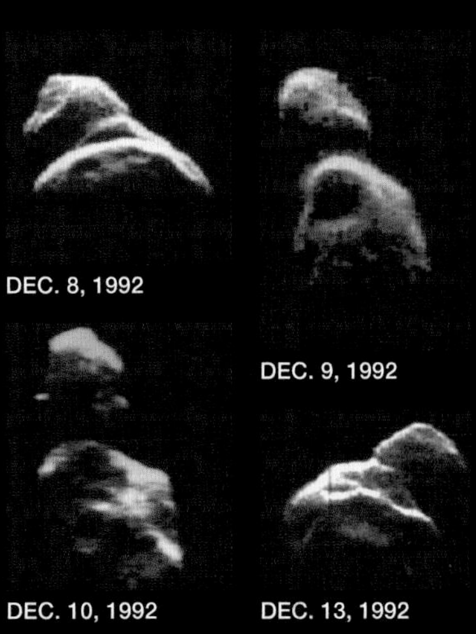

ASTEROID TOUTATIS
These images show the unusual shape of Toutatis. Some 2.8 miles (4.5km) long, it is probably made of two separate pieces that coalesced soon after another asteroid had fragmented due to a collision.

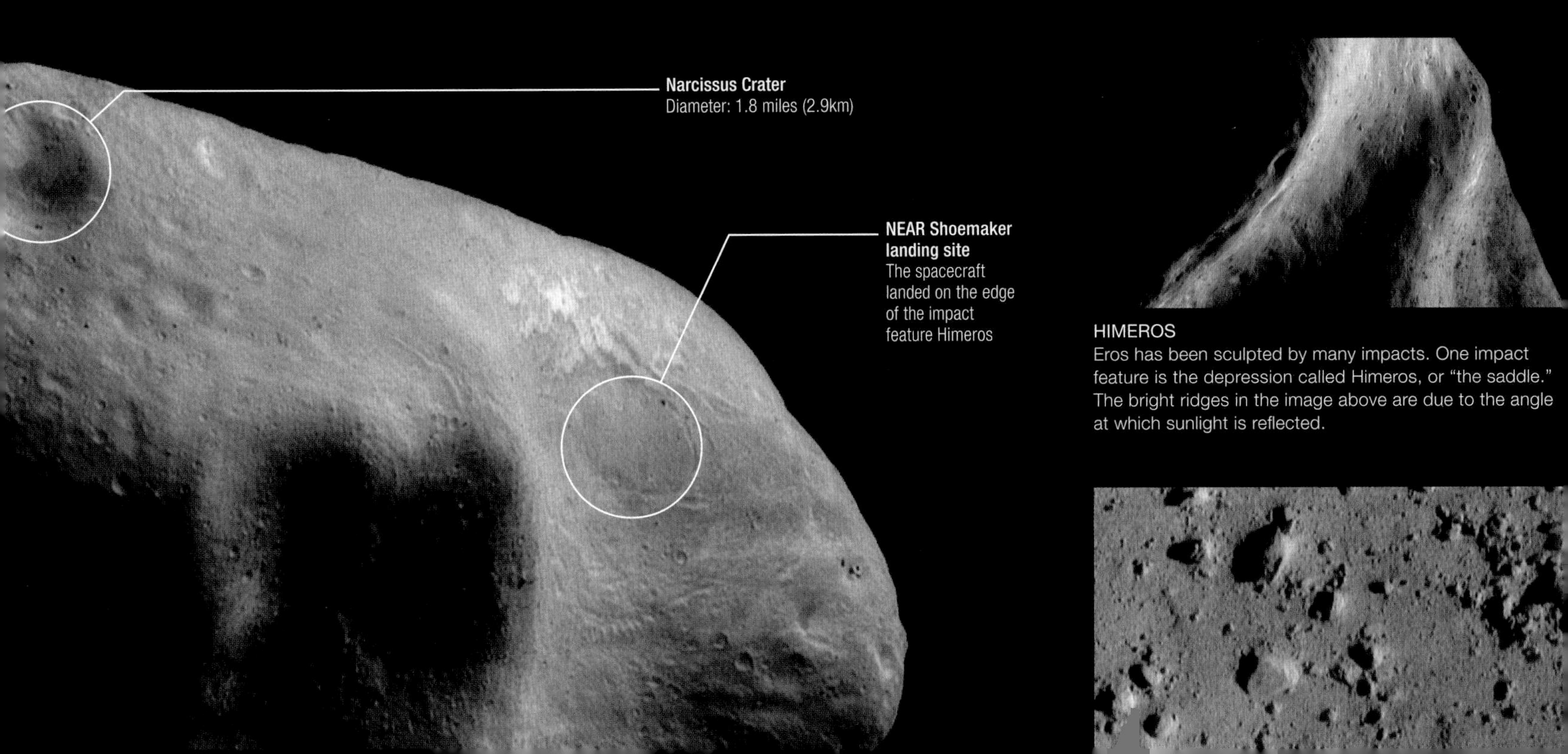

HIMEROS
Eros has been sculpted by many impacts. One impact feature is the depression called Himeros, or "the saddle." The bright ridges in the image above are due to the angle at which sunlight is reflected.

JUPITER

Within the Solar System, Jupiter is second only to the Sun in size and mass. Named after the most important Roman god, it is almost 2.5 times the mass of the other seven planets combined and has a large family of orbiting moons. The top of Jupiter's atmosphere forms distinctive colored bands.

ORBIT

Fifth out from the Sun, Jupiter is also about five times the distance of Earth from the Sun. The planet moves along an elliptical orbit, with its nearest and furthest distances from the Sun differing by 47.3 million miles (76.1 million km). Jupiter completes one orbit in just under 12 years. Like all other planets, it rotates as it travels, turning once in just less than 10 hours. This rapid spin pushes Jupiter's equatorial region out, making the planet 6.5 percent wider at its equator than at its poles. Jupiter's spin-axis is tilted from the vertical by only 3.1°. This means that Jupiter has no seasons, because neither hemisphere is tilted significantly toward or away from the Sun as Jupiter orbits.

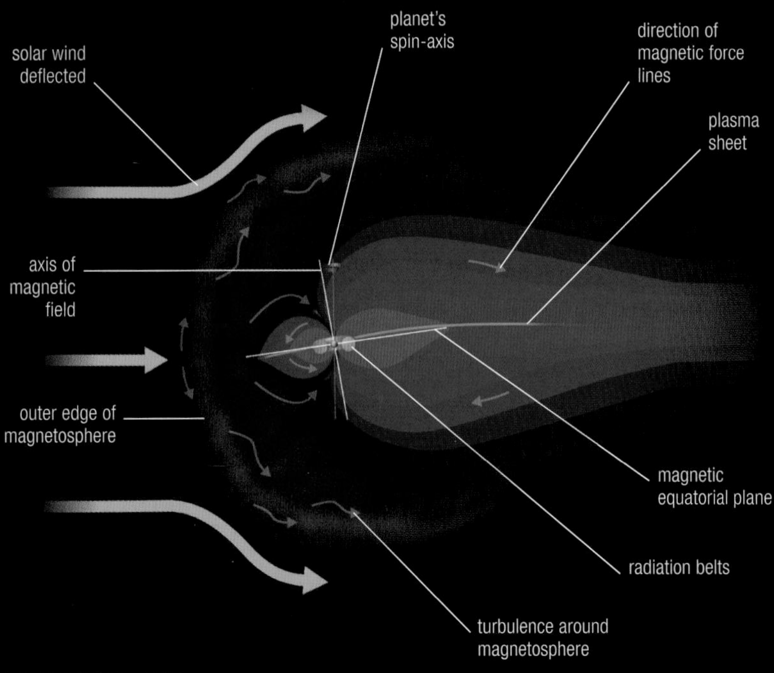

MAGNETOSPHERE
Electric currents in Jupiter's inner layer of metallic hydrogen generate a magnetic field. This field dominates a vast, bubblelike volume of space around Jupiter. Known as the magnetosphere, its tail stretches as far as Saturn's orbit. Solar wind particles plow into it and are directed along magnetic lines of force. Some form a plasma sheet, others radiation belts.

THE JOVIAN SYSTEM

Jupiter is surrounded by a thin, faint ring system of four distinct parts. The main, flat ring is about 4,350 miles (7,000km) wide and less than 19 miles (30km) deep. On its inner edge is the sparse, doughnut-shaped halo, which starts above Jupiter's atmosphere. On its outer edge is the broad, two-part gossamer ring. Orbiting within the main ring are the innermost of Jupiter's moons. Most distant of all the moons is S/2003 J2, at 17.7 million miles (28.5 million km) away.

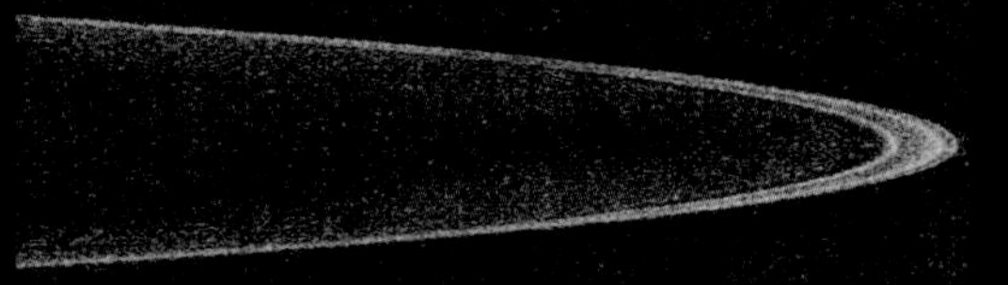

MAIN RING
In this New Horizons image of 2007, sunlight reflects off dust particles. The main ring also contains pieces up to the size of boulders.

STRUCTURE AND ATMOSPHERE

Jupiter is made of hydrogen, helium, and small amounts of other elements. The material is in gaseous form in the atmosphere, but it changes its physical state beneath this outer layer. The banding of the upper atmosphere is created by a combination of the large-scale movement of warm and cool gases, Jupiter's rapid rate of rotation, and hydrogen compounds condensing to form different-colored clouds.

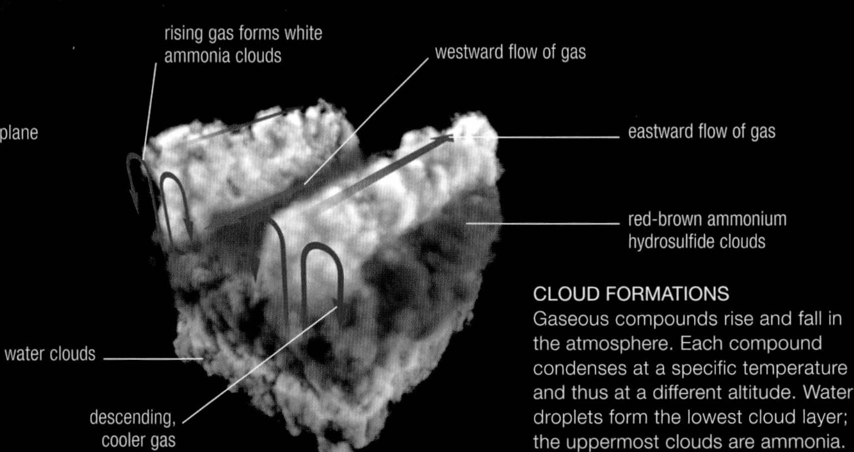

CLOUD FORMATIONS
Gaseous compounds rise and fall in the atmosphere. Each compound condenses at a specific temperature and thus at a different altitude. Water droplets form the lowest cloud layer; the uppermost clouds are ammonia.

ORBIT AND ROTATION

The eccentricity of Jupiter's orbit results in a large difference between its nearest and furthest distances from the Sun. Jupiter's rotation period is the shortest of all the Solar System planets.

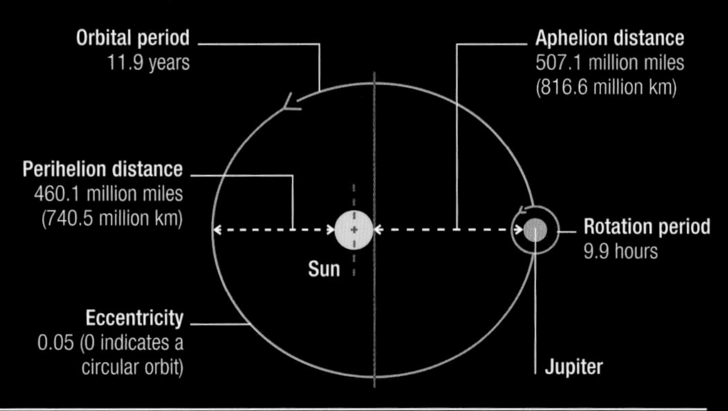

SIZE

Jupiter is a little more than 11 times the size of Earth. It has 318 times Earth's mass, but its density is low. About 1,300 Earths could fit inside Jupiter.

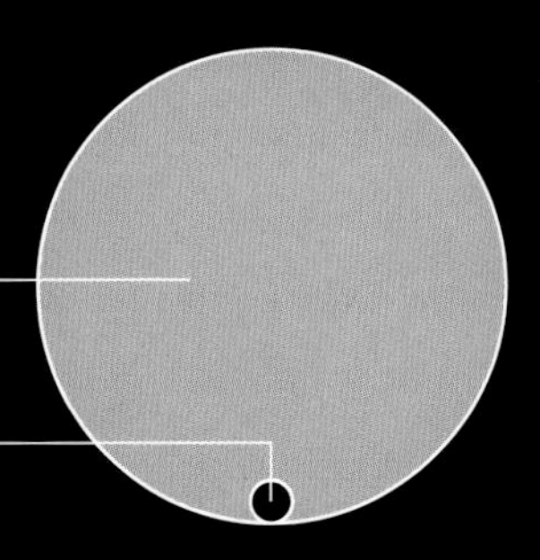

EXPLORATION

Five key spacecraft have investigated Jupiter. Four flew by the planet, while the fifth, and most recent, orbited around it. The first craft to arrive were Pioneer 10 and 11, in 1973 and 1974 respectively. These were followed by Voyager 1 and 2 in 1979. The Voyagers made a more in-depth study of the Jovian system than the Pioneer craft, discovering, among other things, active volcanoes on the moon Io.

The Galileo mission arrived in December 1995, after a six-year journey. As the first craft to orbit the planet, it provided the first long-term observations of the Jovian system. Galileo consisted of a main craft that orbited Jupiter and flew by some of its moons and a probe that was released into the atmosphere. A total of 16 instruments studied Jupiter, its magnetosphere, and its moons. The mission's most notable achievement was providing evidence that a liquid ocean exists under the surface of the moon Europa. About 14,000 images had been returned when the mission ended in 2003.

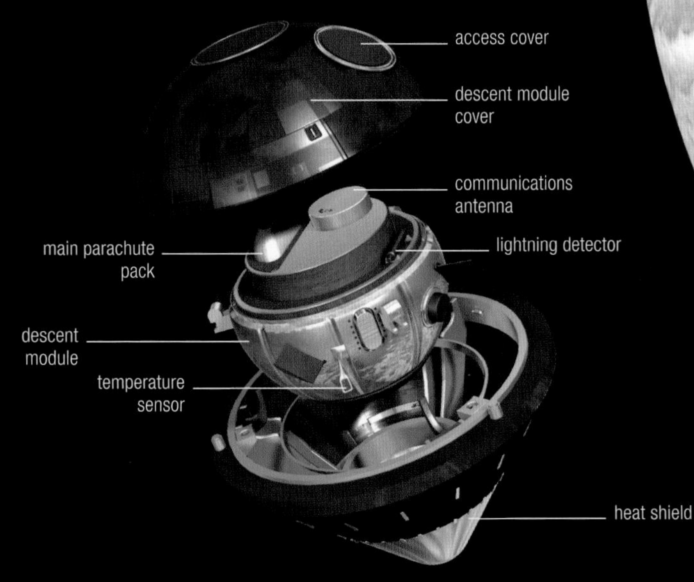

GALILEO'S ATMOSPHERIC PROBE
A cover with a heat shield protected the probe's descent module as it entered the atmosphere. The cover fell away after three minutes and a parachute opened out. The module transmitted data for 58 minutes to a depth of about 125 miles (200km).

TURBULENT WORLD
The top of Jupiter's colorful atmosphere forms its visible surface. The atmosphere is extremely turbulent, and although its bands and the Great Red Spot endure, the planet's surface is in a constant state of change.

STRUCTURE

As pressure, temperature, and density increase with depth, Jupiter's material gradually changes from gaseous to liquidlike. The inner layer resembles molten metal. Jupiter's core is about 15 times the mass of Earth.

ATMOSPHERE

Hydrogen and helium dominate the atmosphere. Methane and other hydrogen compounds give color to the upper atmosphere.

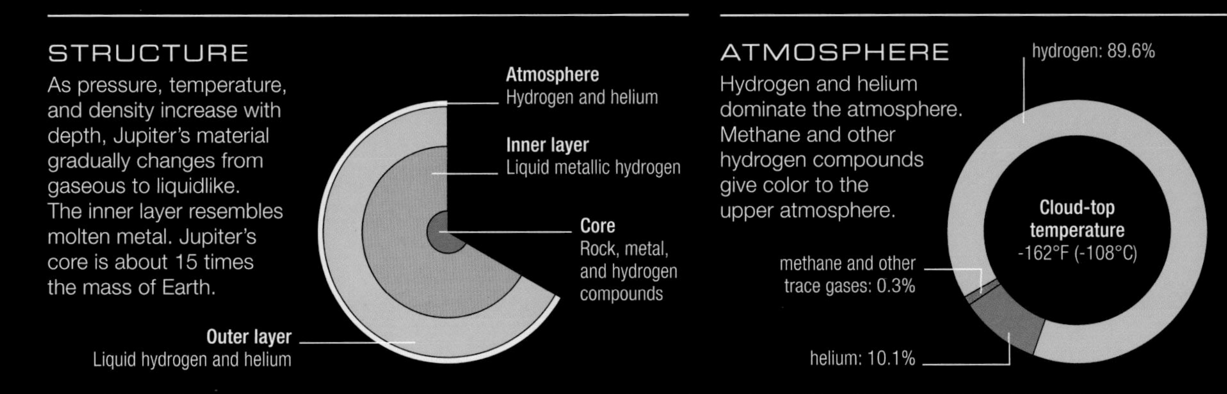

63

Jupiter has 63 moons, more than any other Solar System planet. At least two-thirds of these have been discovered since January 2000. Most are named after the descendants and lovers of the god Jupiter in Roman mythology (Zeus in Greek). The largest is Ganymede, the biggest moon in the Solar System. The closest moons orbit within Jupiter's ring system; the most distant is more than 74 times the Moon–Earth distance.

THE GREAT RED SPOT

Jupiter's rapid spin, its internal heat combined with heat from the Sun, and the planet's winds all combine to produce turbulent regions in Jupiter's upper atmosphere. These regions include giant storms, which we see as oval, cloudlike structures on Jupiter's visible face. Some are short-lived, but others last for decades. The largest and apparently longest-lived storm is the Great Red Spot. It is huge, about twice the size of Earth, and the largest known storm in the Solar System. It has been observed on and off for more than 300 years.

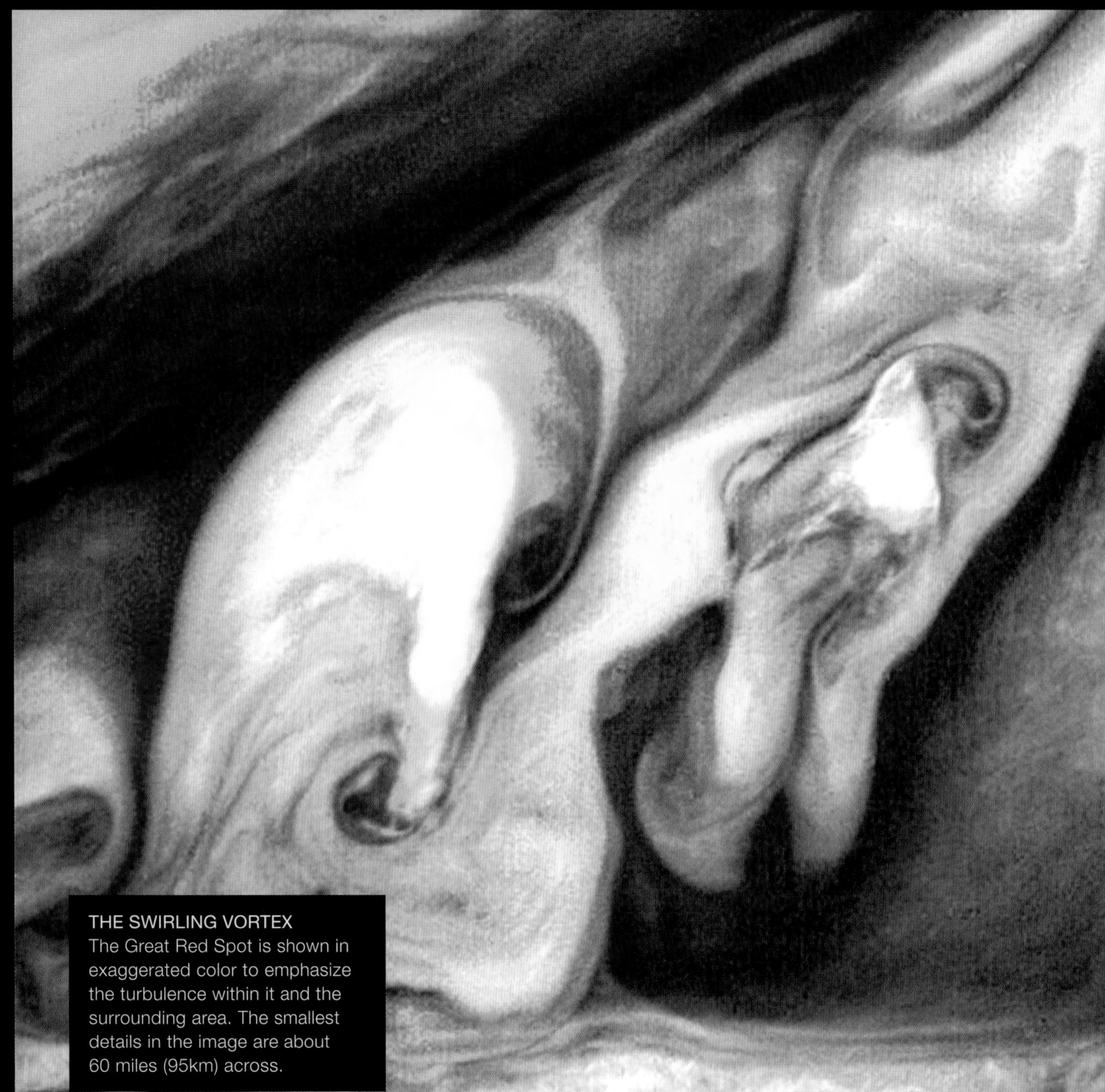

THE SWIRLING VORTEX
The Great Red Spot is shown in exaggerated color to emphasize the turbulence within it and the surrounding area. The smallest details in the image are about 60 miles (95km) across.

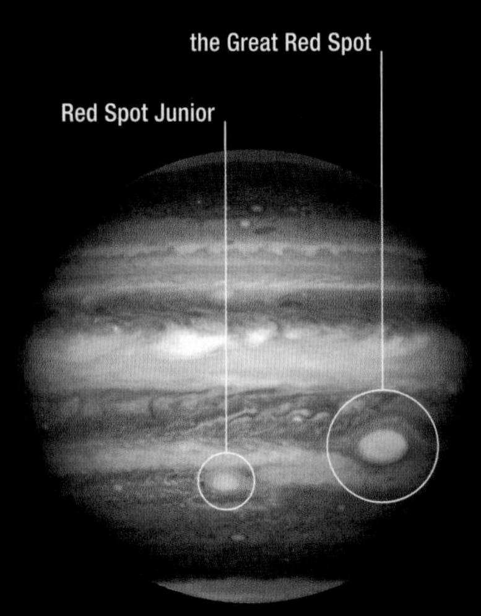

RED SPOT JUNIOR
When a second large storm turned red in late 2005, it was soon nicknamed Red Spot Junior. Half the size of its more famous neighbor, it formed when three white, oval-shaped storms merged during 1998 and 2000. The oldest of these can be traced back more than 90 years.

THE GREAT RED SPOT OVER TIME

The spot changes its shape, size, and color over time, as evident in the Hubble Space Telescope images on the right. The red coloring at the center is a result of material being brought up from the deeper atmosphere, which is then altered chemically by ultraviolet sunlight. In the northwest region of the spot, turbulence resulted from a westward jet stream pushing northward into an eastward jet.

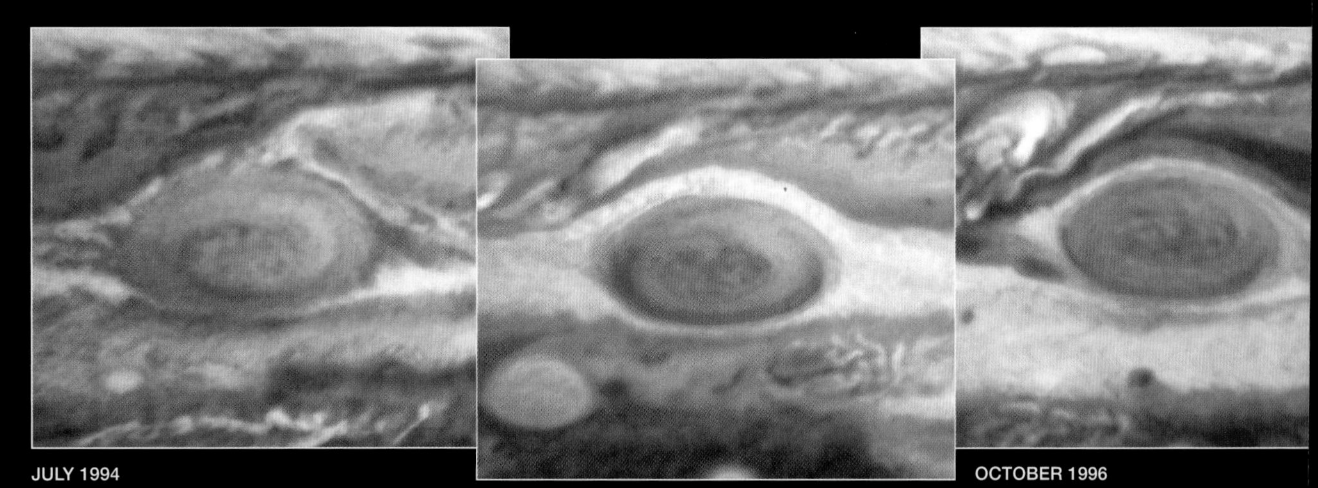

JULY 1994

FEBRUARY 1995

OCTOBER 1996

THE EYE OF THE STORM

The Great Red Spot is a high-pressure region that as a whole rotates counterclockwise once every six days. Winds in its outer regions reach speeds of about 270mph (434kph), but the center remains calm, with winds moving at 10mph (16kph). The central region is also relatively high compared to the surrounding area. Cloud heights in the spot and the area immediately around it vary by about 20 miles (30km). Earth-based observations from 2010 suggest that the center is a few degrees warmer than the rest of the spot and circulates in a clockwise direction.

HEIGHT OF THE GREAT RED SPOT
This false-color Galileo image reveals the varying cloud heights. The deepest clouds, which are dark blue, surround the spot. Higher clouds are light blue; high, thin hazes are pink; and high, thick clouds are white.

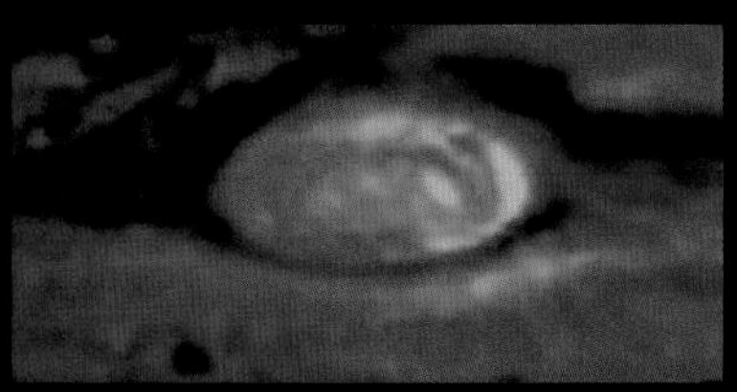

AMMONIA ICE

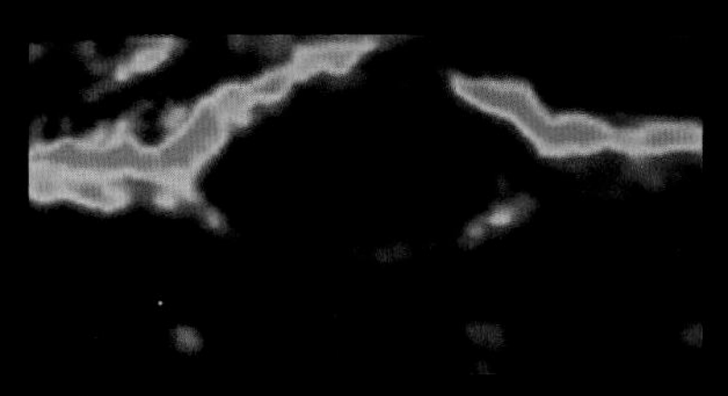

CLOUD THICKNESS

GALILEO'S VIEW
An instrument aboard Galileo recorded data in different infrared wavelengths. The top (false-color) image uses a wavelength that reveals the presence of ammonia ice (blue indicates none; yellow, the most). The bottom image is from a wavelength that shows Jupiter's heat coming through the clouds, indicating cloud thickness (red indicates the thinnest; blue, the thickest).

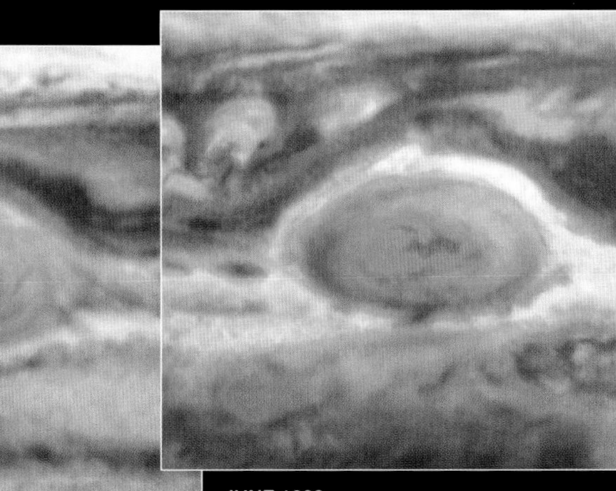

APRIL 1997

JUNE 1999

LIGHTNING
Jupiter's storms are often accompanied by lightning strikes hundreds of times brighter than those on Earth. In this natural-color image, multiple lightning strikes are coming from different parts of a storm. The lightning originates in Jupiter's water-cloud layer.

IO PASSES JUPITER
Io, the innermost of Jupiter's four Galilean moons, floats above the equatorial zone of its parent planet. Its surface is being constantly covered by volcanic eruptions. These are caused by tidal heating—the gravity-driven changes of Io's surface as it orbits Jupiter. The image was taken by Cassini-Huygens on its journey to Saturn.

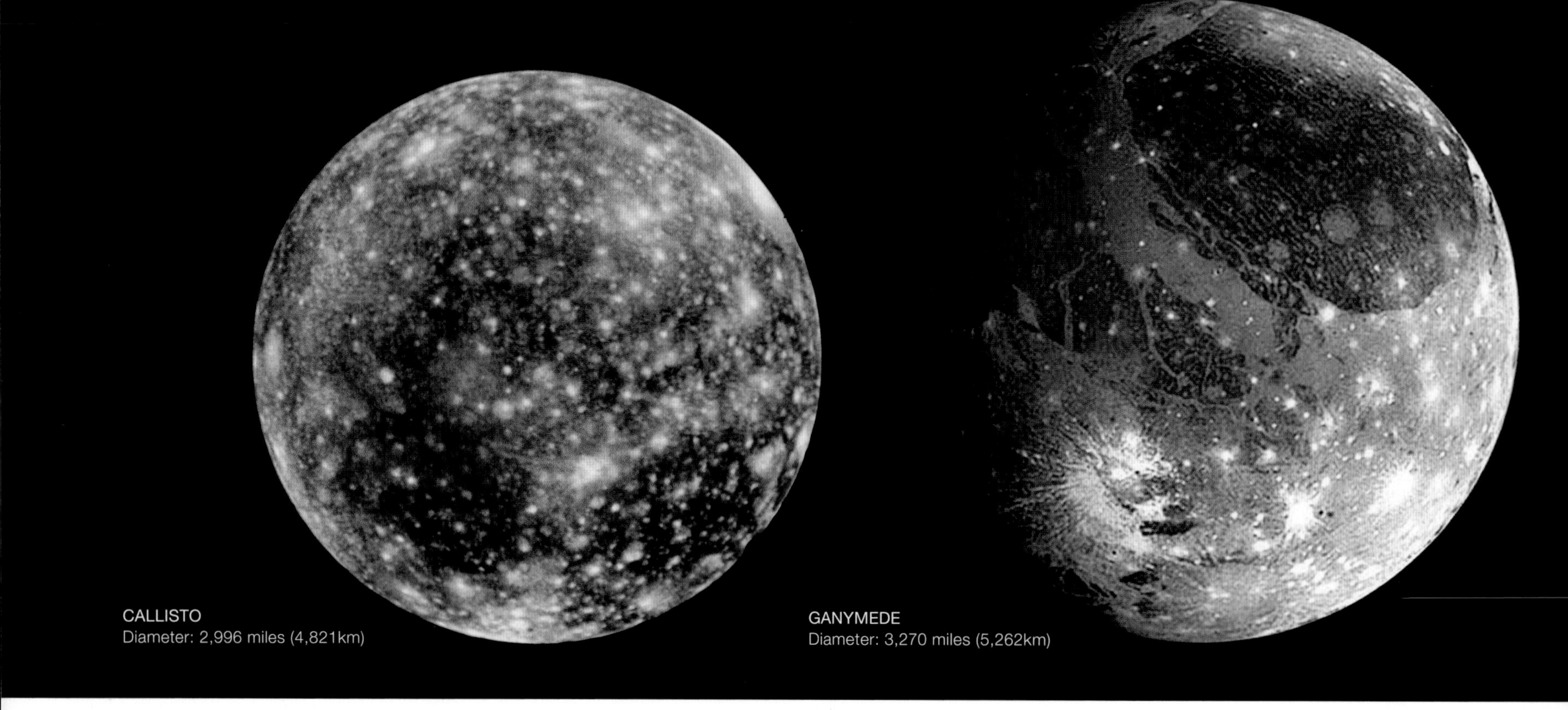

JUPITER'S MOONS

More than 60 moons orbit Jupiter. The majority are small and a great distance from the planet, but the four Galilean moons, which formed at the same time as Jupiter, orbit close by and are large worlds in their own right. They include Ganymede, the biggest moon in the Solar System, and Io, the most volcanically active.

THE FOUR GALILEANS

Jupiter's four largest moons, in order of increasing distance from the planet, are Io, Europa, Ganymede, and Callisto. Their collective name, the "Galileans," originates from the Italian scientist, Galileo Galilei, who observed them in January 1610 from Padua, Italy. Although possibly not the first to see them, he is widely acknowledged as their discoverer because he published his observations and brought the moons to the attention of the scientific and wider community. The four moons are more distant from Jupiter than the Moon is from Earth. The outermost of the four, Callisto, is five times further away.

INNER AND OUTER MOONS

Four small moons orbit within Jupiter's ring system, closer to Jupiter than the Galileans. Known as the inner moons, these are Metis, Adrastea, Amalthea, and Thebe.

Orbiting far beyond Callisto are the many outer moons. The closest, Themisto, is four times further from Jupiter than Callisto is, and the most distant, S/2003 J2, is 15 times further. All but the innermost seven of these orbit clockwise and complete an orbit in 1.5 to 2.5 years.

The outer moons are small—35 of them are less than 3 miles (5km) long—and irregular in shape. Their size, shape, and retrograde motion suggest that they were originally asteroids.

GALILEO MEETS THE GALILEANS

The four Galileans were the first Solar System moons to be discovered after Earth's Moon. We have observed them for 400 years. Our first close look came when the two Voyager craft flew by in 1979, but it was not

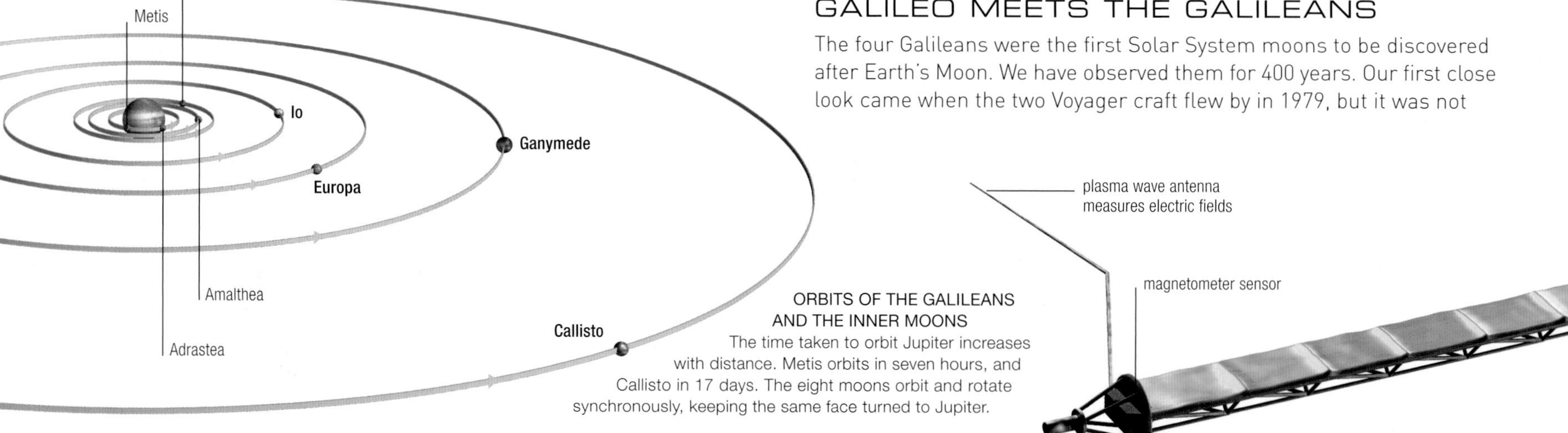

ORBITS OF THE GALILEANS AND THE INNER MOONS
The time taken to orbit Jupiter increases with distance. Metis orbits in seven hours, and Callisto in 17 days. The eight moons orbit and rotate synchronously, keeping the same face turned to Jupiter.

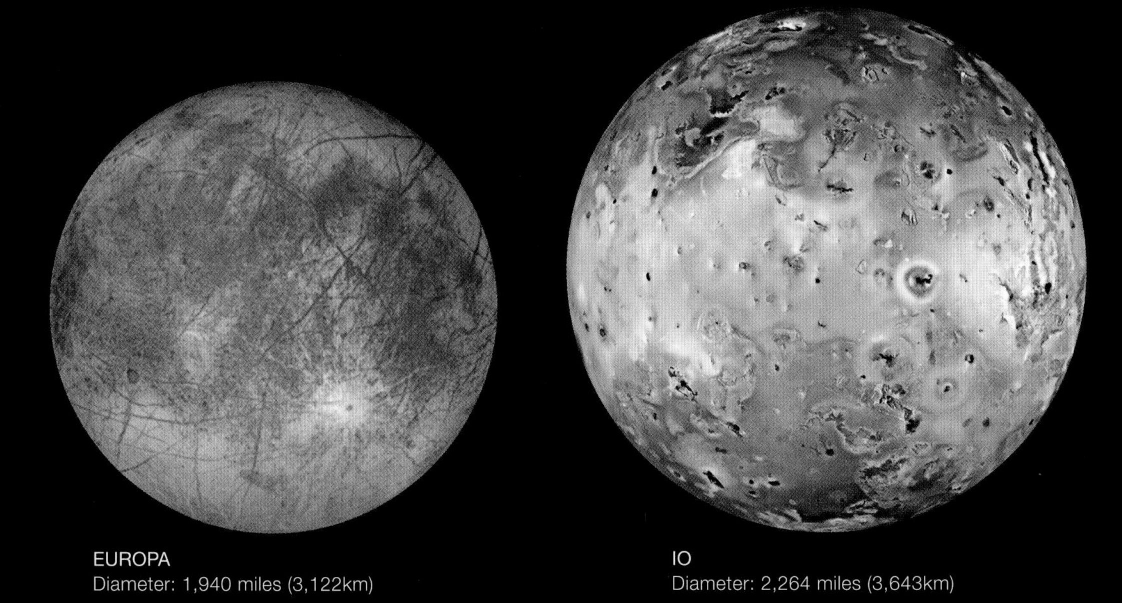

EUROPA
Diameter: 1,940 miles (3,122km)

IO
Diameter: 2,264 miles (3,643km)

SIZES OF THE GALILEAN MOONS
Ganymede, the largest Galilean, is larger than the planet Mercury and three-quarters the size of Mars. Of the four, only Europa is smaller than Earth's Moon. Ganymede, Callisto, and Io are the first-, third-, and fourth-largest Solar System moons.

SIZES OF THE INNER MOONS

Thebe
68 miles
(110km)

Amalthea
163 miles (262km)

Adrastea
16 miles
(26km)

Metis
25 miles
(40km)

Unlike the Galileans, which are all spherical bodies, the inner moons are irregular in shape. They are also tiny in comparison. Adrastea is about the size of a large city, and Amalthea is similar in size to the Mediterranean island of Mallorca.

until the Galileo mission of 1995–2003 that the moons were studied in detail. During Galileo's eight years at Jupiter, the main craft made 35 orbits of the planet and, as it did so, made flybys of the Galileans. Its four instruments, including a camera that provided stunning images, collected information on the type, texture, and size of the moons' surface materials and recorded surface temperatures. Data recorded as Galileo passed by a moon was then sent to Earth by its low-gain antenna.

Europa was of particular interest to the Galileo mission, and the spacecraft made 11 flybys of it. Galileo flew within 124 miles (200km) of its surface, which is the closest it flew by any of the moons. Io was the last moon to be studied by Galileo. When Galileo's work was complete, and with little fuel left, the craft was put on a collision course with Jupiter to avoid an unwanted impact into Europa and the potential contamination of its subsurface ocean.

GALILEO PROFILE

MISSION	
Launch date	October 18, 1989
Arrival at Jupiter	December 7, 1995
Mission ends	September 21, 2003
Launch vehicle	Space Shuttle *Atlantis*, on STS-34
GALILEO	
Agency	NASA
Length	50ft 10in (15.5m)
Height	17ft 5in (5.3m)
Weight	5,986lb (2,715kg)
Power source	Radioisotope thermoelectric generators
SCALE	

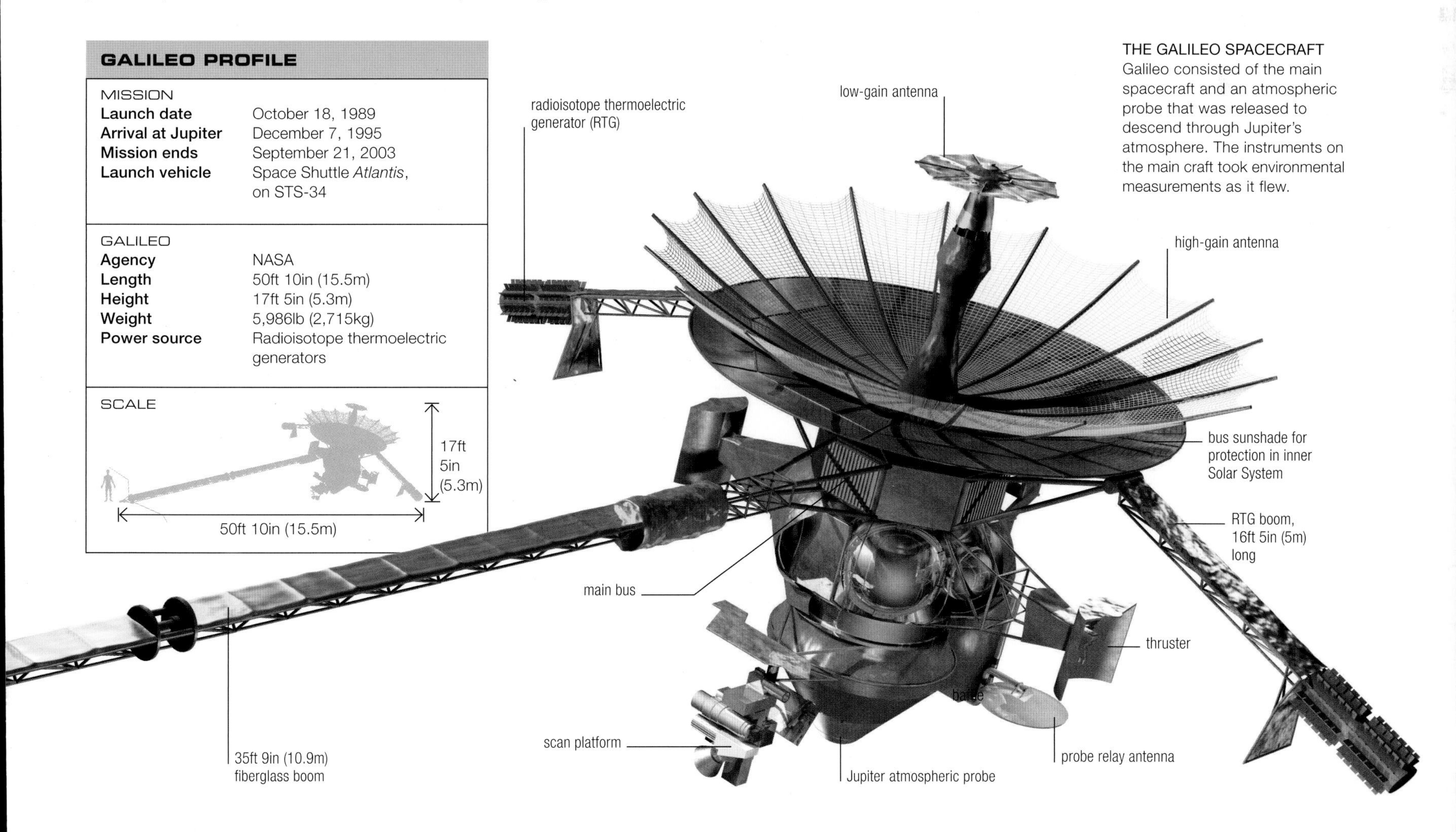

THE GALILEO SPACECRAFT
Galileo consisted of the main spacecraft and an atmospheric probe that was released to descend through Jupiter's atmosphere. The instruments on the main craft took environmental measurements as it flew.

IO

Jupiter's third largest moon, Io is the most volcanically active place in the Solar System. The moon is a colorful world of volcanic calderas and vents, lava flows, and high-reaching plumes. It is a little larger than Earth's moon, and orbits Jupiter in just 42 hours 30 minutes. The planet's powerful gravity pulls on Io and causes two tidal bulges in Io's surface, which rise and fall by about 330ft (100m) as it travels around its slightly elliptical orbit. Friction caused by the flexing of Io's globe heats the subsurface material. It is this molten material that erupts through the thin silicate crust and onto its surface. Over 100 mountainous masses and peaks, ridges, and elevated plains have been identified on Io. The names of features are associated with the story of Io, one of Jupiter's loves in Roman mythology, or with gods and goddesses of fire, sun, thunder, and volcanoes.

IO WITH JUPITER
This Cassini image shows much of Io's anti-Jupiter hemisphere—the side facing away from Jupiter throughout its orbit. Io orbits at a distance just slightly greater than the Moon's distance from Earth.

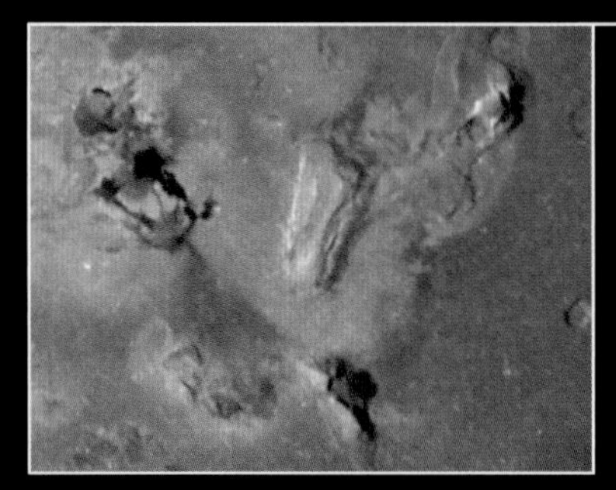

DORIAN MONTES
About half of the Ionian mountains are elevated plains. Dorian Montes is about 6 miles (9km) high, with a rugged surface, but no steep or prominent peaks. It measures 56 miles (90km) long by 35 miles (57km) wide.

IO PROFILE

DISTANCE FROM JUPITER
261,970 miles (421,600km)

ORBITAL PERIOD
1.77 Earth days

ROTATION PERIOD
1.77 Earth days

DIAMETER
2,264 miles (3,643km)

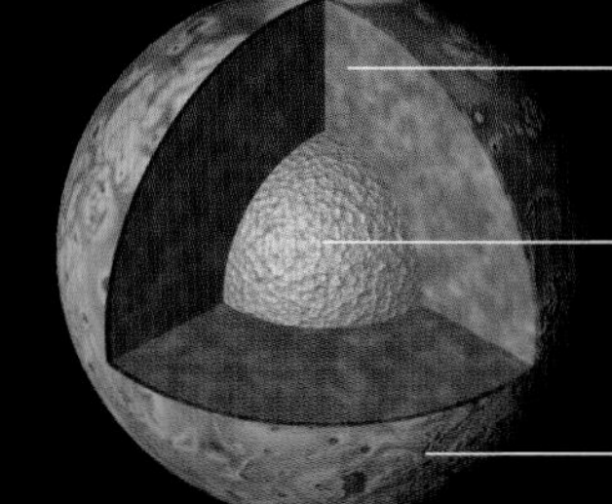

Pele
A huge red ring of deposited plume material surrounds this active volcanic site.

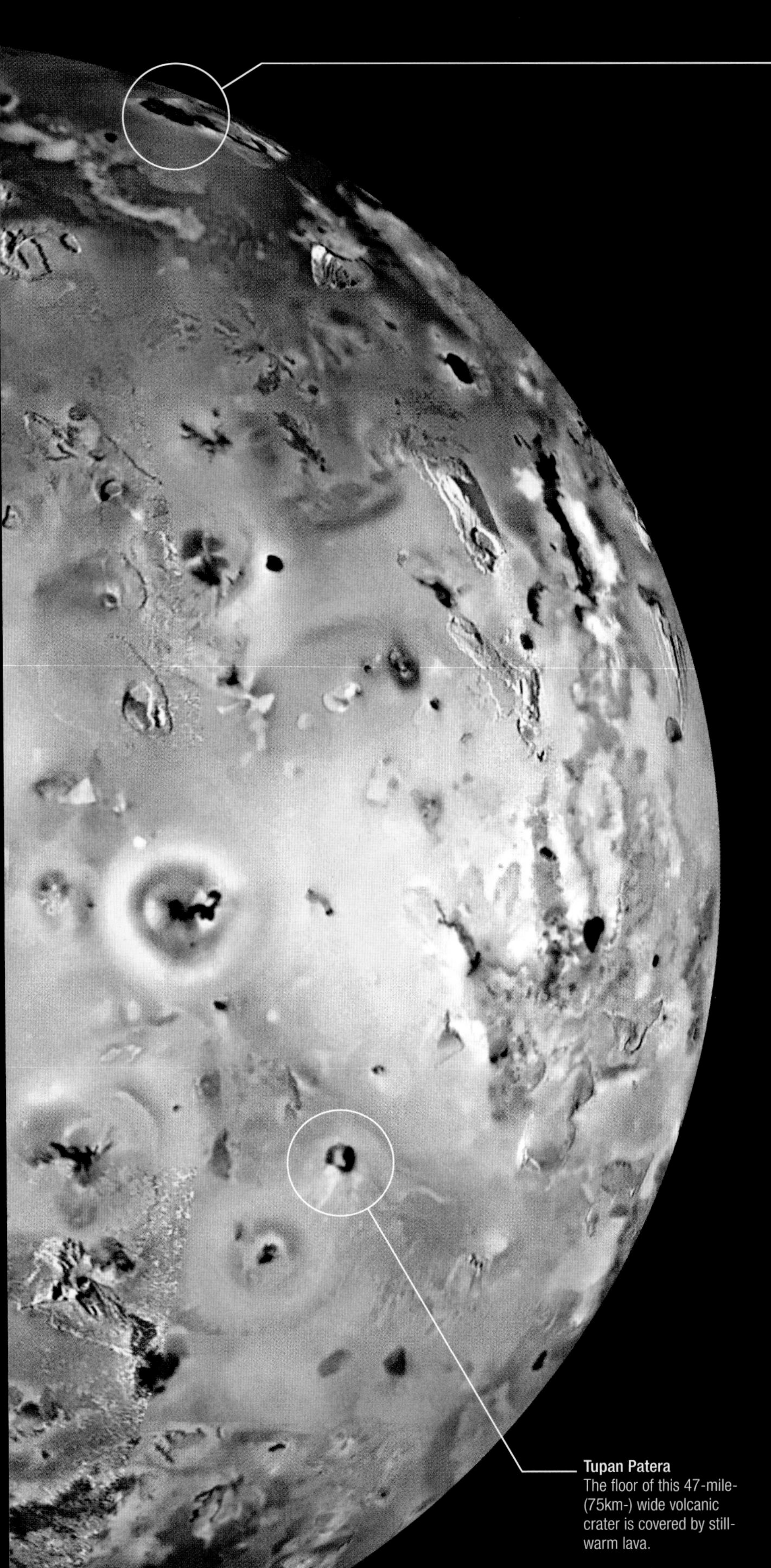

Tupan Patera
The floor of this 47-mile- (75km-) wide volcanic crater is covered by still-warm lava.

VIOLENT MOON
This enhanced-color image shows that Io has a predominantly yellow and light green surface. Black, gray, and red regions are due to recent volcanism.

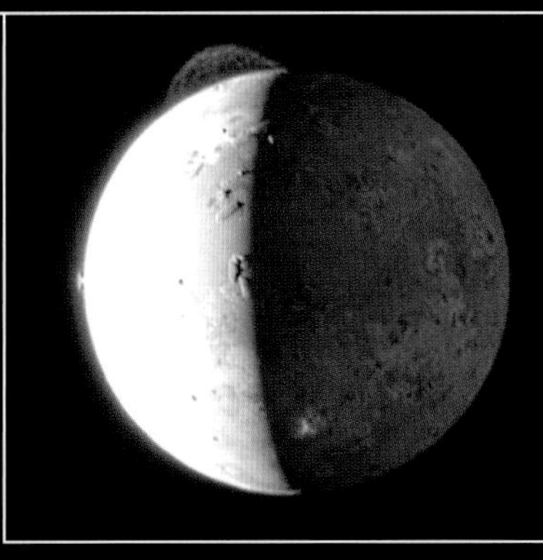

TVASHTAR CATENA
The Tvashtar Catena is a chain of active volcanic craters covering an area seven times that of the largest caldera on Earth. Above, a huge plume 180 miles (290km) high shoots out from the site in 2007.

LAVA FLOW
Newly erupted, orange-colored lava is visible at the left of this image taken by Galileo in February 2000. The dark gray region immediately to its right erupted in December 1999, but has now cooled.

VOLCANIC ACTIVITY

Io's volcanic nature was discovered in 1979 by Voyager 1. Almost two decades later, Galileo's more extensive study found huge areas that had been volcanically resurfaced since Voyager's visit. These two—and the more recent craft Cassini and New Horizons—have all witnessed volcanic eruptions, cooling surface lava, and volcanic plumes. More than 400 volcanic vents have now been identified.

Io's lava is molten silicate rock, mixed with sulfur and sulfur dioxide. Heated sulfur dioxide shoots through fractures in the moon's crust and cools instantly into volcanic plumes of cold gas and frost grains. The plume material falls back to Io's surface and leaves circular or oval frost deposits.

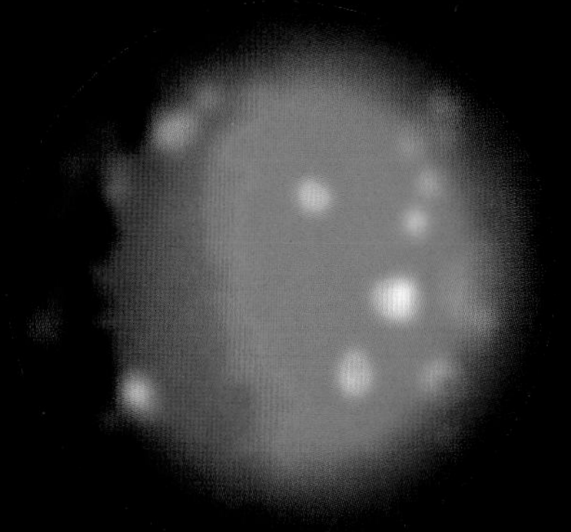

DETECTING VOLCANOES
An infrared image taken in 2001 by Galileo reveals hot eruption sites around Io. White, red, and yellow indicate hotter regions, and blue is cold. Four of the hot spots are previously unknown volcanoes. The right half of Io is in daylight and is warmer.

IO'S SODIUM CLOUD

Gas ejected by Io's volcanoes has formed a cloud around the moon. The yellow coloring comes from sodium within the cloud. Although there is only a trace amount, sodium is highly efficient at scattering sunlight. The bright light near Io's shadowed eastern limb is sunlight being scattered by the 60-mile- (100km-) high plume from the volcano Prometheus.

EUROPA

A little smaller than Earth's moon, Europa is Jupiter's fourth largest satellite. It orbits Jupiter every 3.5 days and in the same time rotates once on its axis. Europa's outermost layer is an icy crust, but beneath the ice, an ocean of liquid water extends over the entire moon. Deeper within, its material has differentiated into a rocky layer and an iron core. Europa's icy crust reflects light well and makes the moon's surface one of the brightest in the Solar System. It is also one of the smoothest, with none of its features extending more than a few hundred feet up or down. The surface is characterized by huge ice plains, chaotic terrain, and dark linear structures. Its surface features take their names from Celtic mythology or are associated with Europa, a girl seduced by the god Jupiter in Roman mythology.

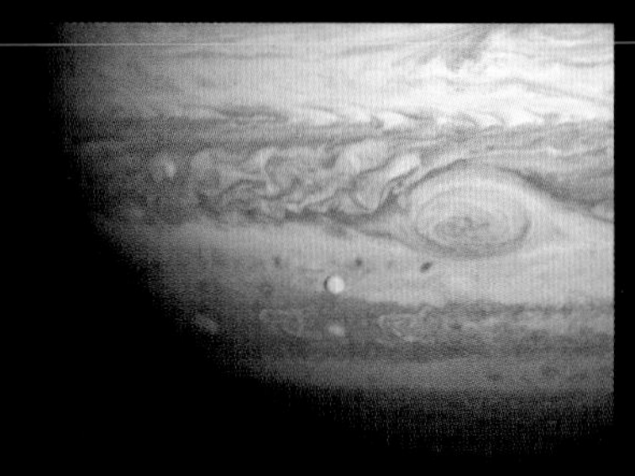

EUROPA WITH JUPITER
Europa is seen here as a small bright disc in front of Jupiter's cloud tops, to the lower left of the Great Red Spot. The image was taken by Cassini as it flew by en route to Saturn in December 2000.

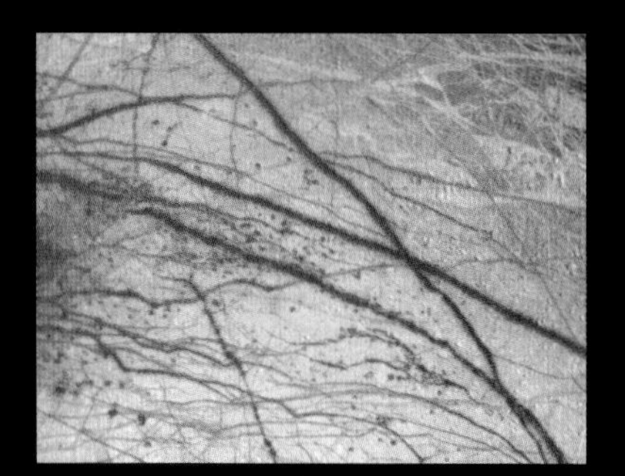

ICE PLAINS
Galileo's false-color image shows Europa's surface features. Deep red linear structures, called linea, cross the blue-colored icy plains. Red indicates the presence of icy material from inside the moon.

EUROPA PROFILE

DISTANCE FROM JUPITER
416,900 miles (670,900km)

ORBITAL PERIOD
3.55 Earth days

ROTATION PERIOD
3.55 Earth days

DIAMETER
1,940 miles (3,122km)

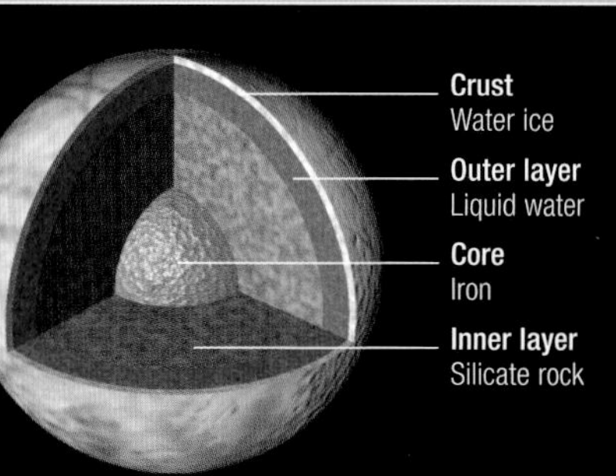

Callanish Macula
Concentric rings have formed around this large crater.

Minos Linea
This typical elongated marking formed when the crust cracked and warm ice erupted through. It is 1,348 miles (2,170km) long.

Pwyll Crater
This circular impact crater is 28 miles (45km) across and is surrounded by bright ejected material.

ICY MOON
This face of Europa points away from the direction of the moon's movement along its orbit and is known as the trailing hemisphere. This Galileo image shows it in natural color.

CHAOTIC SURFACE

Europa's surface is thought to be only about 50 million years old. It has very few impact craters, which indicates that it is relatively young. Craters that formed on the moon in its more distant past have been destroyed as the surface is continually renewed. Huge areas of Europa's icy crust have broken up, and large blocks have shifted and floated into new positions. Cracks have also opened up and filled with ice, and elsewhere ridges have taken shape as warm ice or water has pushed up the surface. These disrupted areas of the surface have formed distinctive regions of chaotic terrain, termed chaos, such as in the Conamara Chaos. Red-colored spots and shallow pits are found in the Conamara Chaos and elsewhere on Europa's surface. The spots are named lenticulae, the Latin word for freckles.

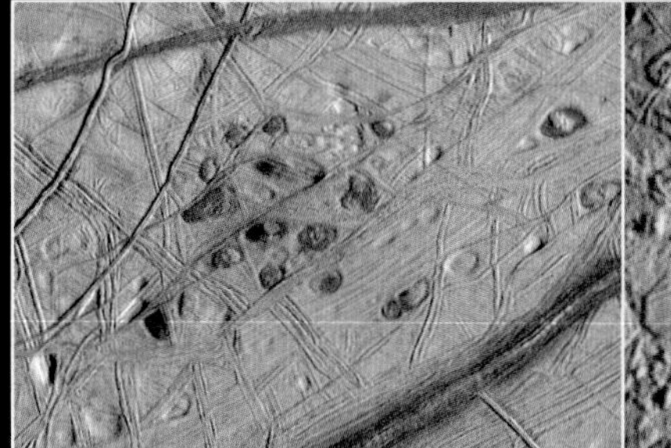

CONAMARA LENTICULAE
This view combines images of lenticulae on Europa's surface collected by Galileo on two separate orbits. Each spot measures about 6 miles (10km) across. They form when warm, ruddy ice within the moon's ice shell moves upward and erupts.

CONAMARA ICE CRUST
This area of the Conamara Chaos measures approximately 17 miles (27km) by 22 miles (35km). The colors in this Galileo picture have been enhanced and white and blue areas indicate where a fine dust of ice particles thrown up from the Pwyll Crater has settled.

LIQUID OCEAN

The average temperature at Europa's surface is -275°F (-170°C). Yet only a few tens of miles below it is warm enough for water to be liquid. Europa's water ocean contains about twice the amount of water found in all of Earth's oceans. The water is kept liquid by tidal heating. As Europa follows its orbit around Jupiter, the planet pulls the moon's interior in different directions. This flexing leads to frictional heating within Europa, which in turn keeps its ocean liquid. The ocean is of great interest to astronomers because it could support the sort of microbial life found in Earth's oceans. Recent studies suggest that it contains enough oxygen to do so.

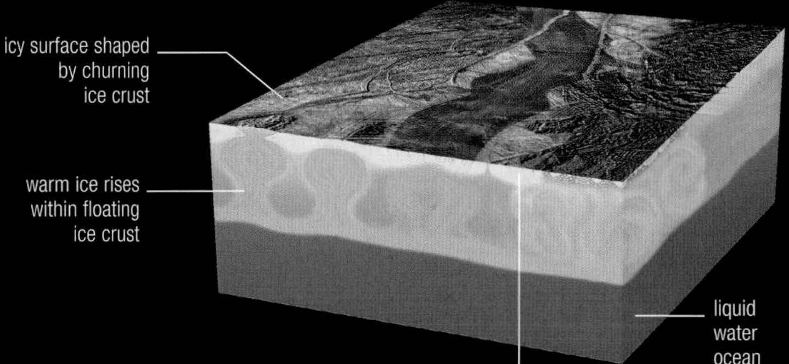
icy surface shaped by churning ice crust
warm ice rises within floating ice crust
liquid water ocean
cold near-surface ice can crack and crumble

OCEAN AND CRUST
Studies suggest that the ice crust could be 20 miles (30km) thick and the ocean 60 miles (100km) deep. Relatively warm ice at the base of the crust rises slowly to the surface and creates domes or breaks through.

FROZEN SURFACE OF EUROPA
Encompassing an area roughly 100 miles (160km) by 150 miles (240km), this false-colour Galileo image shows Europa's icy crust. Blocks of ice are thought to have broken apart and drifted into position. The long ridges are fractures in the crust, and the red spots (termed lenticulae) represent volcanic material forced up through the ice and onto its surface.

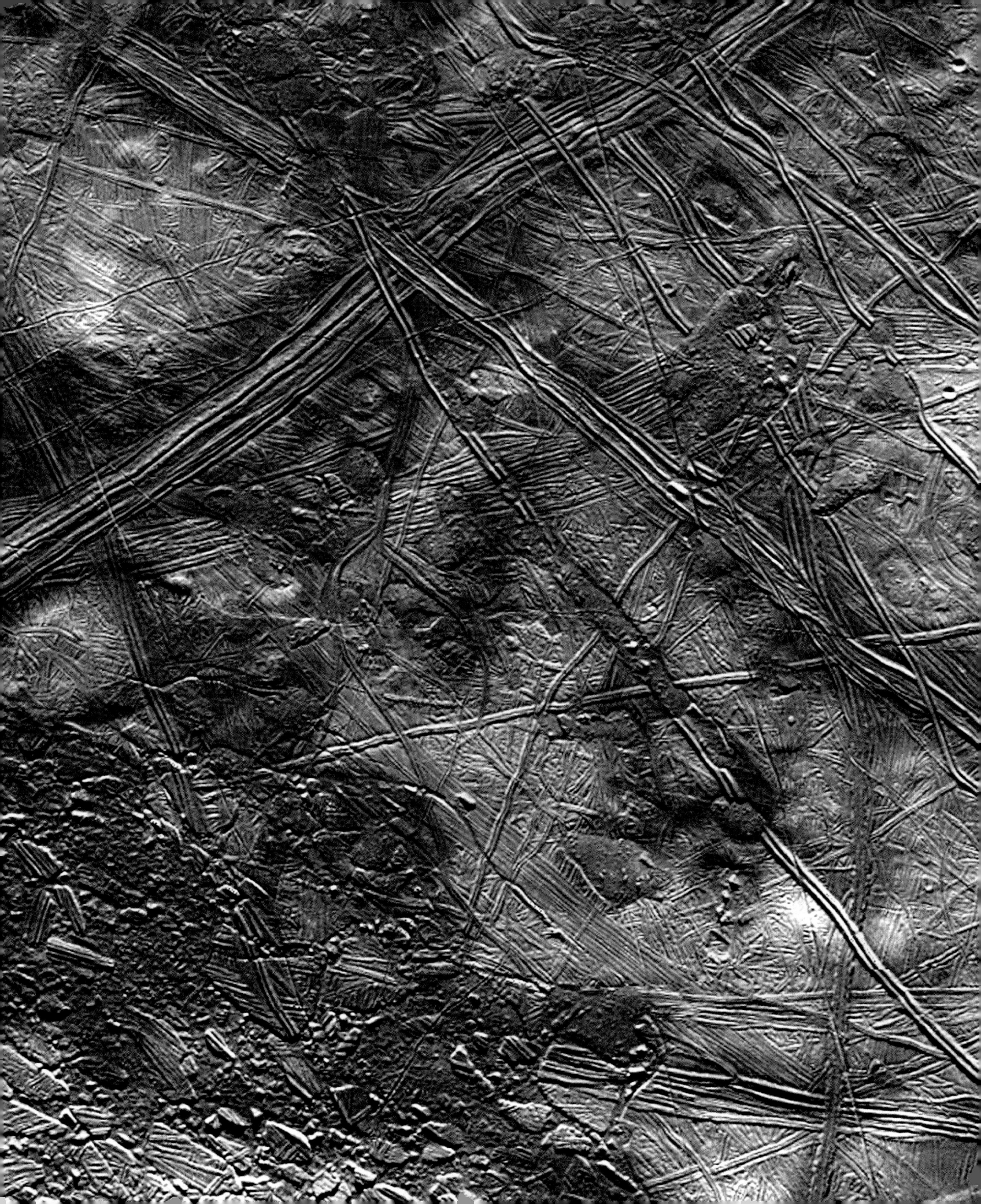

GANYMEDE

A huge ball of rock and water ice, Ganymede is the largest moon in the Solar System. It takes its name from a young boy in Greek mythology who was cup-bearer to the Gods. Deep inside, it is divided into an iron core surrounded by rock. Above the rock is warm, soft ice, which further out becomes an ocean of water, on top of which is a floating crust of ice. The heat required to maintain the layer of water some 90–120 miles (150–200km) below the surface comes from radioactivity in the rocky interior.

The moon's surface is composed of distinctive bright and dark terrain. The dark regions are slightly more ancient than the bright areas. Additionally, frost covers Ganymede's polar regions.

GANYMEDE WITH JUPITER
Ganymede is about 50 percent larger than our own moon and is dwarfed next to Jupiter in this Cassini image. Ganymede is the seventh moon in distance from the planet and orbits in just over seven days.

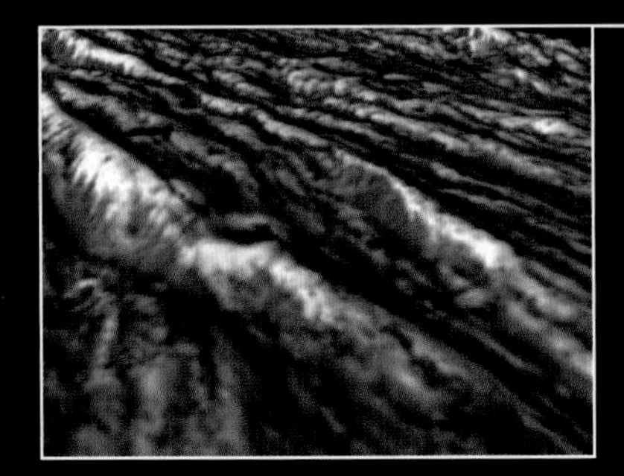

URUK SULCUS
The grooved regions of the bright terrain on Ganymede are termed sulci. At 1,400 miles (2,200km) long, Uruk Sulcus is one of the longest. In this perspective view, based on Galileo images, icy material is visible on the top of the ridges.

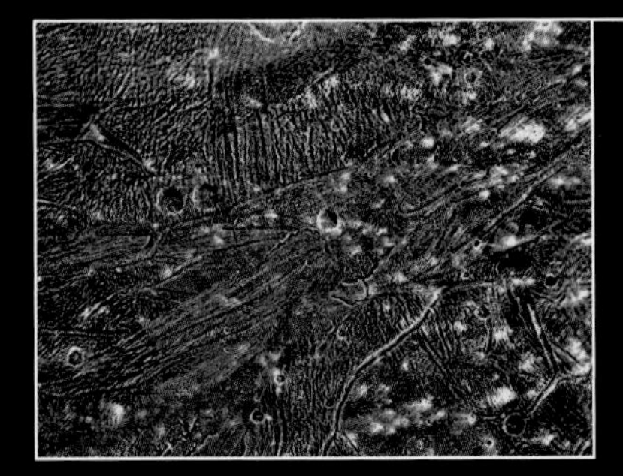

SIPPAR SULCUS
A stretch of fairly smooth terrain (lower left to upper right) cuts across the grooved terrain in this Galileo image. Bright patches are craters formed from material ejected when Osiris Crater (not in this view) was formed.

THE LARGEST MOON
Ganymede's contrasting dark and light terrain stands out in this natural-color image taken by Galileo. Impact craters dot the moon's surface, but are mainly found in the dark regions. The bright spots are relatively young craters.

Osiris Crater
Bright rays surrounding the 67-mile- (108km-) wide Osiris Crater are material thrown out as the crater formed.

Harakhtes Crater
At 67 miles (108km) wide, this is one of 27 named craters on the moon that are wider than 60 miles (100km).

DARK AND BRIGHT TERRAIN

The ancient dark terrains, consisting of furrows and craters, are termed regiones (plural of regio) and are named after astronomers who discovered moons of Jupiter. The furrows, which are long depressions, are thought to have formed as the result of geological stresses as the moon's crust solidified. These took shape before the large craters were formed when asteroids impacted the moon. Some older craters appear different from the others. Known as palimpsests, they have been smoothed out and filled with bright icy material.

The brighter, younger terrain often cuts across the dark regions. It consists of almost parallel ridges and grooves, which took shape as the moon's surface underwent tectonic stretching early in its history. Long swaths of such landscape are termed sulci (plural of sulcus) and are named after places associated with myths of ancient people. The longest is Mysia Sulcus at 3,148 miles (5,066km) long.

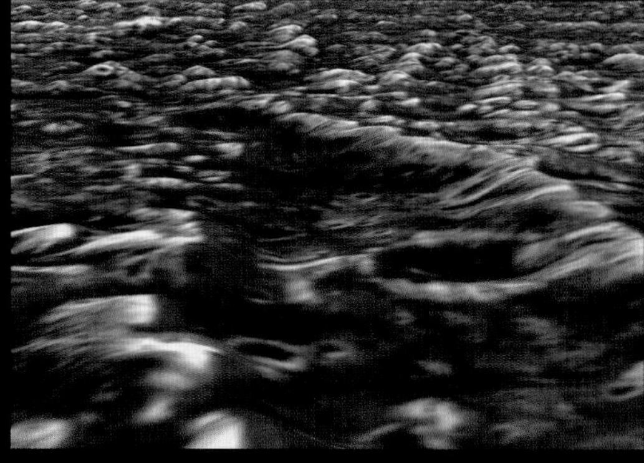

GALILEO REGIO
The largest area of dark terrain is the Galileo Regio, which measures more than 1,900 miles (3,000km) across and covers much of the northern hemisphere. Its furrows and impact craters are visible in this computer reconstruction, which uses images from the Galileo spacecraft.

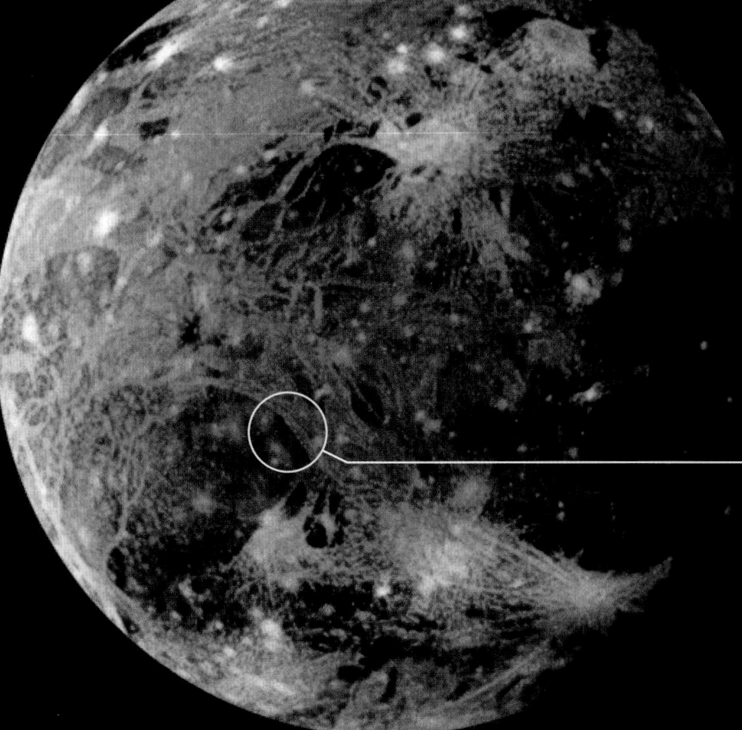

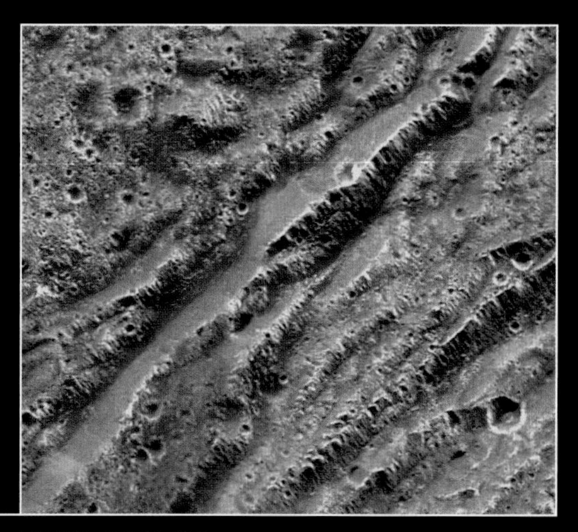

NICHOLSON REGIO
This Galileo image shows a detail of a series of scarps located within the border terrain of the dark Nicholson Regio and the lighter Harpagia Sulcus. Individual scarps resemble tilted stacks of books. Similar features on Earth formed when its crust was broken by faults and blocks between the faults were pulled apart and rotated.

THE TRAILING SIDE OF GANYMEDE
This color-enhanced image of Ganymede's trailing hemisphere, which is the one facing away from its forward orbital movement, reveals its frosty polar caps. The violet coloring may result from particles of frost scattering sunlight at the violet end of the spectrum.

Nah–Hunte Crater
This crater is 49km (30 miles) across and lies within bright terrain separating the Galileo and Nicholson Regiones.

GANYMEDE PROFILE

DISTANCE FROM JUPITER
665,000 miles (1.07 million km)

ORBITAL PERIOD
7.15 Earth days

ROTATION PERIOD
7.15 Earth days

DIAMETER
3,270 miles (5,262km)

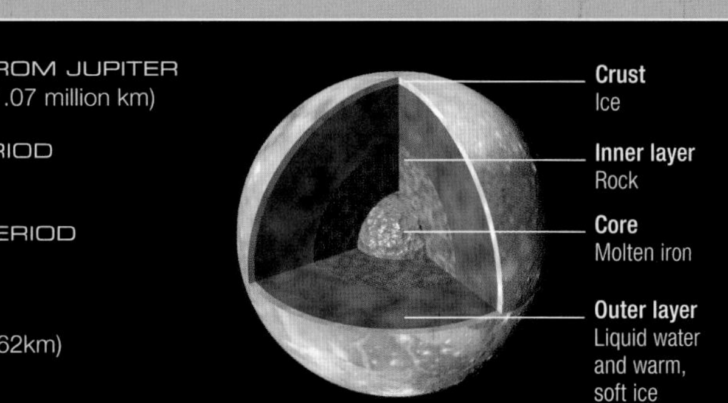

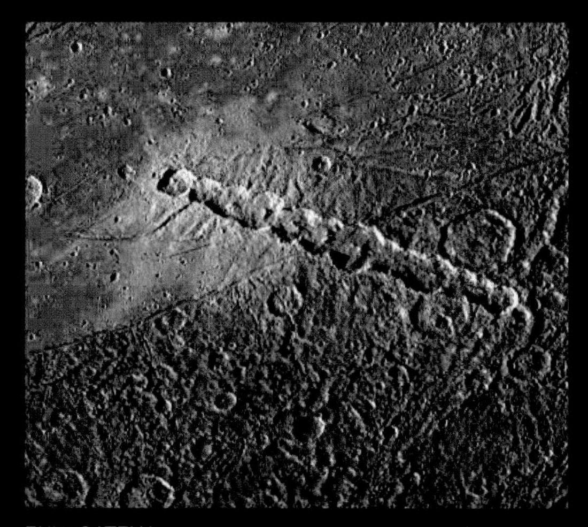

ENKI CATENA
The chain of 13 craters named Enki Catena formed when pieces of a comet slammed into Ganymede in quick succession. As can be seen in this Galileo image, the chain cuts across a boundary between light and dark terrain. The boundary is marked by a thin trough. Material ejected from the craters is seen on the bright terrain (top left).

CALLISTO

Named after a nymph who was seduced by Jupiter in Roman mythology, Callisto is the outermost and second largest of the Galilean moons. Made of rock and ice, it is similar in size to the planet Mercury. Callisto's surface has not been noticeably changed by tectonic movement or volcanic activity, unlike the surfaces of other Galileans, but it is one of the most heavily cratered. Impact craters, formed when asteroids smashed into the moon early in its history, pockmark its entire surface. The largest are multi-ringed craters, of which the biggest is Valhalla, measuring 1,900 miles (3,000km) across. Bright ray craters are also visible.

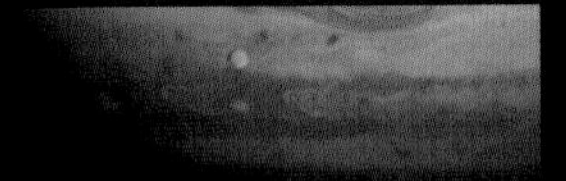

CALLISTO WITH JUPITER
Callisto completes one orbit of Jupiter in just less than 17 days. Cassini took this image of the two worlds together in 2000 as it flew by on its way to Saturn. Europa is also visible against Jupiter's banded atmosphere.

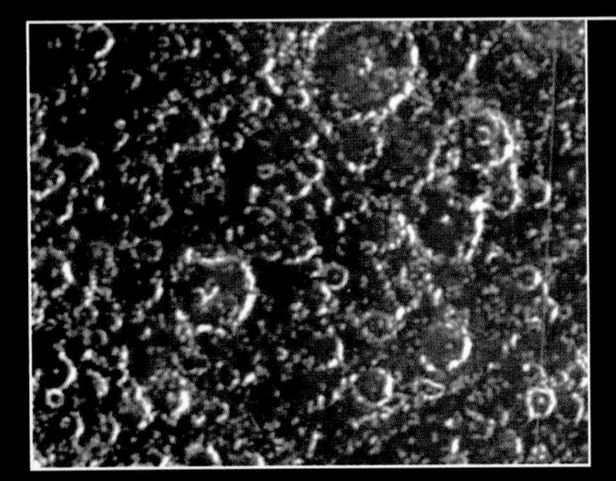

ANTIPODE OF THE VALHALLA REGION
This part of Callisto's surface is the opposite point, or antipode, of the Valhalla region. It shows Callisto's heavily cratered surface, but exhibits no effect from shock waves resulting from the impact that formed Valhalla.

CALLISTO PROFILE

DISTANCE FROM JUPITER
1.17 million miles (1.88 million km)

ORBITAL PERIOD
16.69 Earth days

ROTATION PERIOD
16.69 Earth days

DIAMETER
2,996 miles (4,821km)

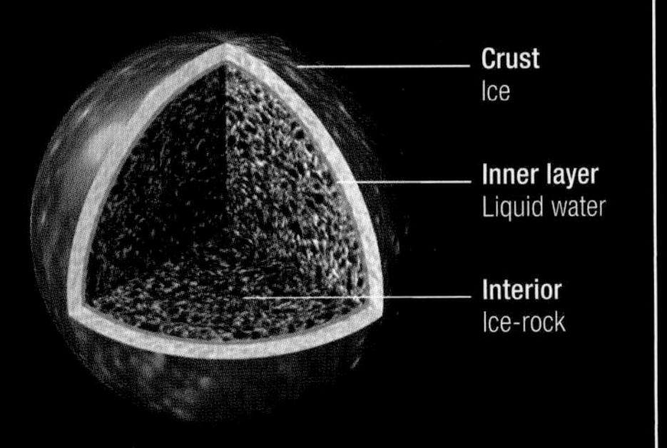

UNCHANGED MOON
In this Galileo image, taken in 2001, Callisto's surface appears uniformly cratered, but has dark and bright areas. Brighter land is rich with water ice, whereas darker areas have less.

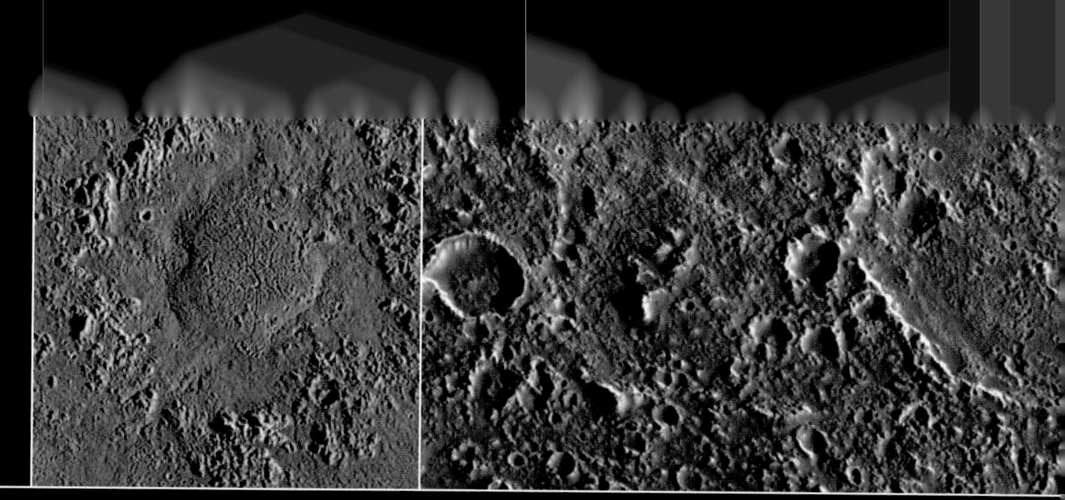

DOH CRATER
A saucer-shaped depression, Doh Crater is 37 miles (59.5km) wide. Its domed center probably formed because the impacting asteroid slammed into an area with slushy subsurface material.

THE ASGARD REGION
Doh is one of the many craters found inside the Asgard region. At 870 miles (1,400km) across, Asgard is Callisto's second largest multi-ringed impact crater. The small, icy bumps seen in Galileo's 50-mile- (80km-) wide view of Asgard's inner part make the region look bright.

THE VALHALLA REGION

Valhalla is one of the largest impact structures in the Solar System. Like Callisto's other surface features, it has a name taken from mythology—Valhalla was the enormous ha ruled by the Norse god Odin. The impact that formed this multi-ringed basin would have sent shock waves through Callisto. When Mercury and Earth's moon were similarly bombarded, shock waves formed grooved and hilly terrain on the opposite point from the impact, yet Valhalla's antipode shows no effect. Computer models indicate that a liquid-water layer would disperse the shock waves. This evidence, and measurements collected by Galileo, suggest that Callisto has a subsurface layer of liquid water.

CALLISTO'S VAST IMPACT STRUCTURE
The huge Valhalla region dominates this side of Callisto. The color has been enhanced to bring out the variation of the surface features.

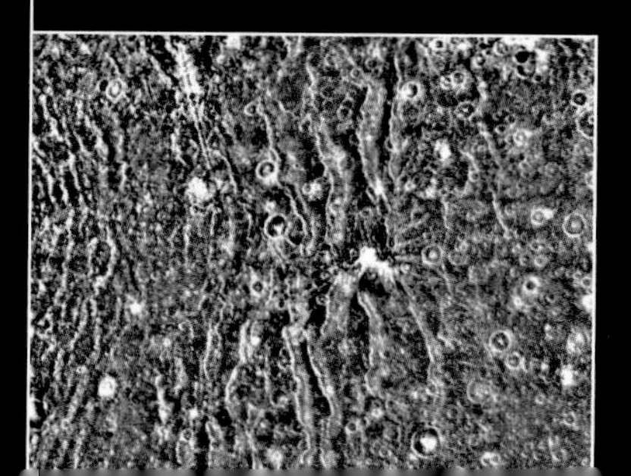

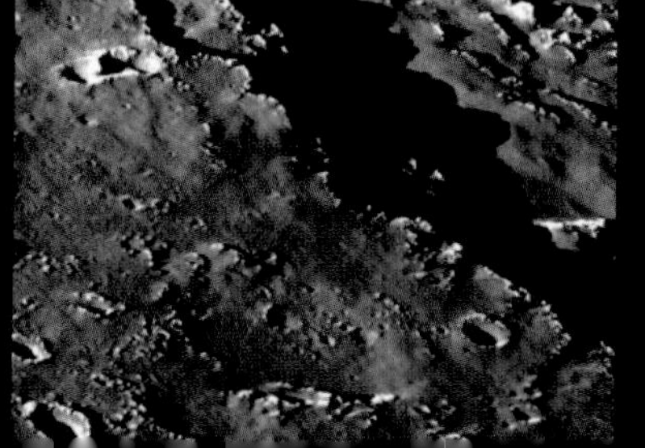

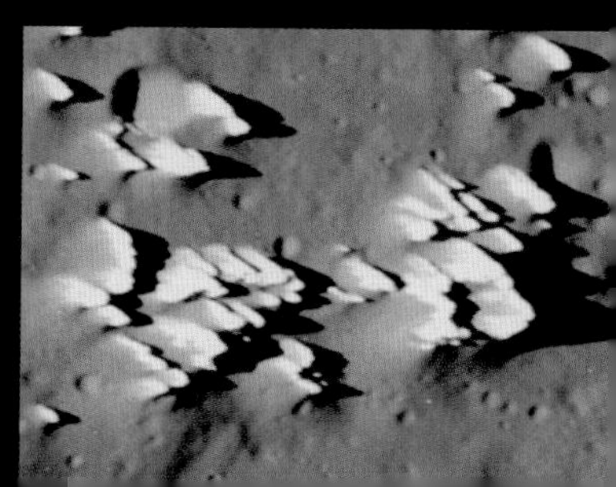

SATURN

Pale yellow Saturn is distinguished from all other Solar System planets by the magnificent rings that girdle its equator. Saturn is the second-largest planet, with a large family of moons. It has been under intense scrutiny ever since the Cassini spacecraft moved into orbit around it in 2004.

ORBIT

Named after the one-time ruler of the Roman gods and father of Jupiter, Saturn is the sixth planet from the Sun. It is nearly twice as far away as its larger inner neighbor, Jupiter, and it takes almost 29.5 years to complete one orbit around our star. At such a great distance, the intensity of the Sun's radiation is only about 1 percent of that experienced on Earth. Saturn spins as it orbits, with one rotation taking just over 10.5 hours. The planet's spin-axis is tilted by 26.7° to its orbital plane, which means that its poles alternately point toward and away from the Sun. This tilt, the weak sunshine, and an internal heat source combine to produce seasonal differences on the planet.

CHANGING VIEWS OF THE RINGS

As Saturn orbits the Sun, our view of its rings constantly changes—we see them from above, below, or edge-on, because the planet's northern and southern hemispheres point at the Sun once each per orbit. These Hubble images (right) show how the view of the rings changes as an increasing amount of the southern hemisphere faces the Sun, and spring in this part of Saturn changes to summer. The rings were seen edge-on again in 2009, and the northern hemisphere will fully face the Sun in 2017.

NOVEMBER 2000

NOVEMBER 1999

OCTOBER 1998

OCTOBER 1997

OCTOBER 1996

RINGED WORLD
Thirty images taken by Cassini's wide-angle camera are combined in this natural-color view of Saturn. Storms are visible in Saturn's upper atmosphere, and a moon is seen above the rings to the left.

ORBIT AND ROTATION

Saturn is approximately 9.5 times further from the Sun than Earth is. But its elliptical orbit means that its closest and greatest distances differ by 100.5 million miles (161.9 million km). When at perihelion, Saturn's south pole faces the Sun.

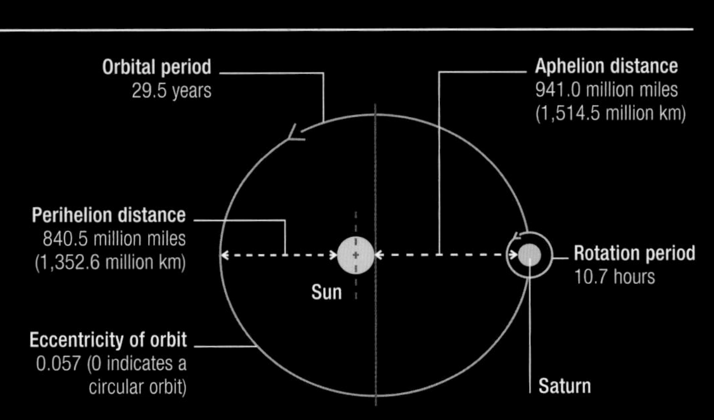

SIZE

Saturn is almost 9.5 times the width of Earth. The planet is also noticeably oblate in shape, being about 10 percent wider at its equator than at its poles.

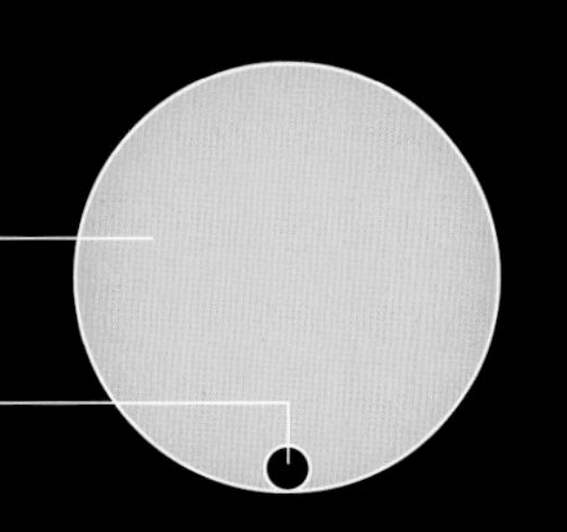

STRUCTURE

When we look at Saturn, we are peering at the top of its atmosphere. Made predominantly of hydrogen with some helium, this visible layer forms muted bands parallel to the equator. The physical state of Saturn's material changes with depth as the temperature, density, and pressure increase. The outer gaseous layer of hydrogen and helium slowly merges into another deeper layer, where the hydrogen and helium are liquidlike. Much deeper still, the atoms are stripped of their electrons and the hydrogen and helium now act like molten metals. It is within this metallic inner layer that electric currents generate Saturn's magnetic field. Here, too, the kinetic (motion) energy of helium raindrops is converted into heat. At the very center of the planet is a core of rock and ice.

NIGHT AND DAY IN INFRARED
Saturn's sunlit side appears green in this false-color view; the red unlit side radiates heat from the interior. The dark areas are where clouds stop heat from rising to the surface.

OBSERVATION AND EXPLORATION

Three flyby missions, Pioneer 11 in 1979 and Voyagers 1 and 2 in 1980–81, revealed Saturn, its rings, and some of its moons in detail for the first time. But our knowledge of Saturn was transformed by the Cassini–Huygens mission, which began its major study in 2004. In early 2009, Hubble took advantage of a rare opportunity to view Saturn with its rings edge-on, a view only seen every 14.7 years. This enabled the behavior of both poles and the aurorae surrounding them to be analyzed simultaneously. The Chandra X-ray Observatory, which orbits Earth, is also occasionally turned toward Saturn, and Earth-based optical instruments are used to discover new moons.

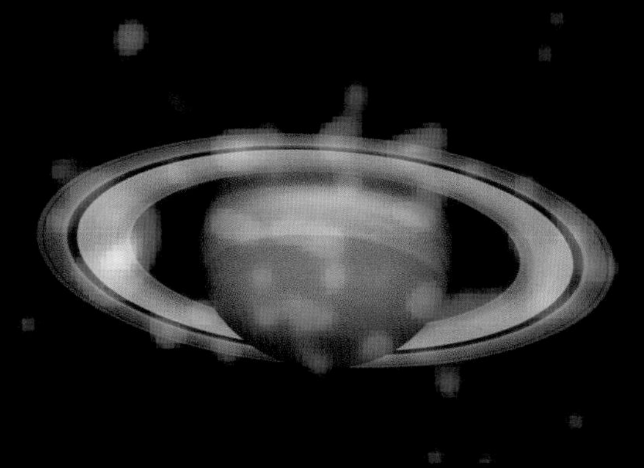

BLUE-FLUORESCING RINGS
Chandra revealed that Saturn's rings fluoresce in X-ray light. The blue fluorescence is probably caused by X-rays from the Sun striking oxygen in the water ice of the rings.

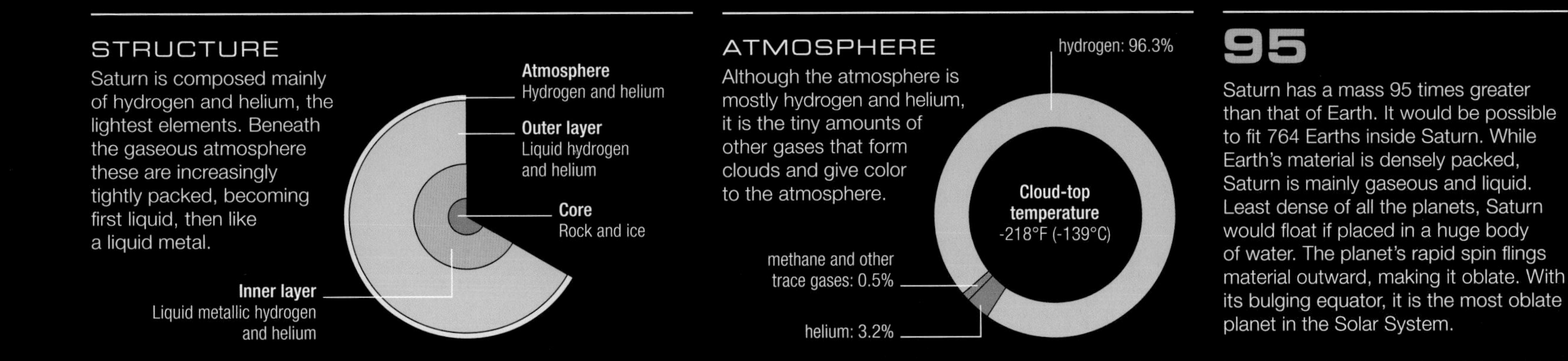

STRUCTURE
Saturn is composed mainly of hydrogen and helium, the lightest elements. Beneath the gaseous atmosphere these are increasingly tightly packed, becoming first liquid, then like a liquid metal.
Atmosphere
Hydrogen and helium
Outer layer
Liquid hydrogen and helium
Core
Rock and ice
Inner layer
Liquid metallic hydrogen and helium
ATMOSPHERE
Although the atmosphere is mostly hydrogen and helium, it is the tiny amounts of other gases that form clouds and give color to the atmosphere.
hydrogen: 96.3%
Cloud-top temperature
-218°F (-139°C)
methane and other trace gases: 0.5%
helium: 3.2%
95
Saturn has a mass 95 times greater than that of Earth. It would be possible to fit 764 Earths inside Saturn. While Earth's material is densely packed, Saturn is mainly gaseous and liquid. Least dense of all the planets, Saturn would float if placed in a huge body of water. The planet's rapid spin flings material outward, making it oblate. With its bulging equator, it is the most oblate planet in the Solar System.

SATURN ECLIPSES THE SUN
This image of Saturn blocking the Sun was taken by Cassini-Huygens. It is a mosaic of 165 pictures taken over a three-hour period. The image includes two faint and previously unknown rings. Earth is visible on the left-hand edge of the bright, main rings. The outermost ring is the E-ring, made from material jettisoned from the south polar region of the moon Enceladus.

SOUTHERN STORMS
In this infrared view of the southern hemisphere, the colors reveal different levels of methane gas. The bright blue rings (above right) lack this gas. One huge storm dominates the scene; others appear as dark patches nearby.

red indicates an abundance of methane above deep clouds

pale blue signifies little methane and high clouds

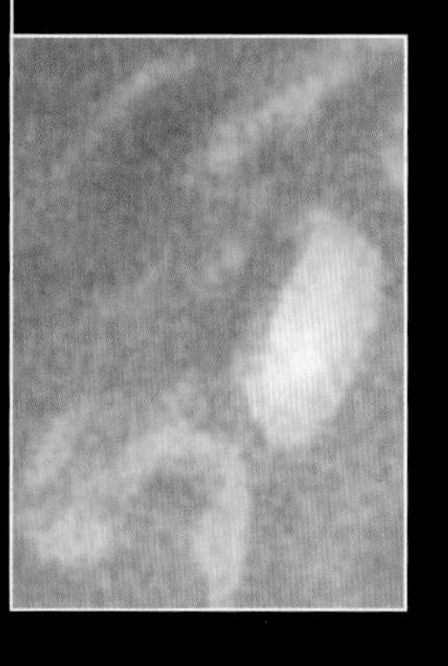

DRAGON STORM
This detail shows a vast thunderstorm in the region known as Storm Alley. Called the Dragon Storm due to its shape, it is thought to be a long-lived storm deep in the atmosphere that periodically flares up.

SATURN'S ATMOSPHERE

High-altitude haze in Saturn's atmosphere gives the planet a serene appearance, but fierce winds and huge electric storms rage here. The atmosphere moves around in bands parallel to the equator, generating vortices at both poles, while the solar wind interacts with the atmosphere to produce spectacular aurorae.

STORMY ATMOSPHERE

Saturn's speedy rotation, coupled with internal heat moving through its lower atmosphere, generates high-speed winds. These easterly winds blow in the same direction as the planet's rotation. Upper-atmosphere gas forms bands that encircle the planet, and within these bands clouds and storms take shape. Storms in the upper atmosphere may be seen on the visible surface as light-colored clouds, while those hidden from view are detected by their radio emissions. A region below latitude 30° south, nicknamed Storm Alley, has been the site of nine huge storms since 2004. One, Dragon Storm, emitted powerful on-off radio waves for several weeks. Scientists think the storm's precipitation generated electricity like it does in Earth's storms—but on a much larger scale. A storm lasting for 7.5 months in 2008 had lightning bolts with 10,000 times the power of those on Earth.

CLOUD HEIGHT
This infrared view shows clouds at three different atmospheric levels. Red indicates some of the deepest clouds found on Saturn; green denotes clouds normally seen in reflected sunlight; and blue is cloudy haze, 6 miles (10km) higher. The south pole is at the extreme left.

deepest clouds and high atmospheric pressure

highest clouds and low atmospheric pressure

upper-level clouds and medium atmospheric pressure

POLAR REGIONS

The atmospheric bands that encircle Saturn spiral into the planet's two poles. Each polar region is dominated by a hurricanelike vortex. The vortex at the north pole is a six-sided structure. Discovered by the Voyager craft, it was still there when Cassini arrived more than 20 years later. Consisting of a nest of hexagonal bands centered on the pole, the vortex is primarily a clearing in the clouds that extends for at least 47 miles (75km) into the atmosphere. Images taken over a 12-day period in 2006 showed that the feature is locked to the pole and does not drift.

The southern polar vortex resembles a giant eye. Its center, about 900 miles (1,500km) across, is mostly cloud-free but surrounded by a wall of clouds. Two spiral arms of clouds extend from this central ring. Winds blow around the ring at 342mph (550kph), and the extended area is littered with storms. Cassini continues to observe both vortices as the seasons change. It will monitor the effects of increasing amounts of sunshine on the north vortex and investigate what part internal heat plays in powering these huge polar storms.

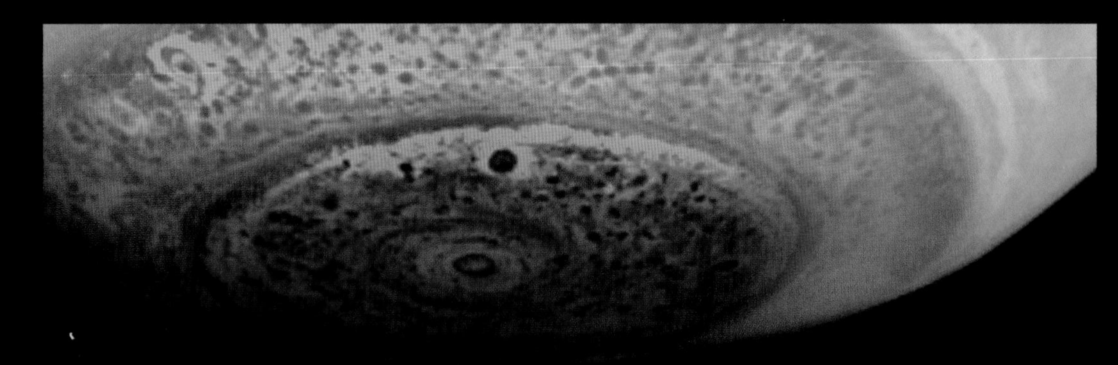

THE SOUTH POLE
In this infrared view from Cassini, the circle of red centered on the south pole shows the warm glow of Saturn's interior heat. In the very center is the eye of the polar vortex. The aquamarine color represents bright haze and clouds in the upper atmosphere. Dark spots are thick clouds at lower atmospheric levels.

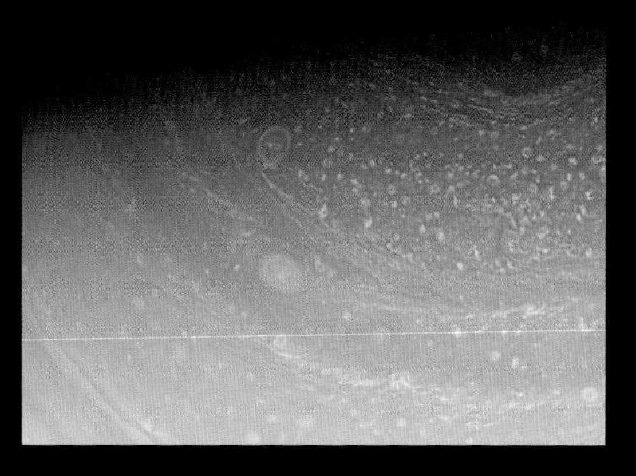

THE NORTH POLE
A small portion of the north polar hexagon is just visible in the upper right corner of this Cassini image (located at latitude 78° north). The north pole itself is out of view. The white circles and specks are some of the hundreds of storms within the region.

AURORAE

Spectacular displays of colored, dancing lights called aurorae are common in the skies above Earth's poles. Similar oval-shaped displays over both of Saturn's poles have been investigated by Cassini and the Hubble Space Telescope. Aurorae occur when solar-wind particles interact with the upper-atmosphere gases, producing flashes of energy in the form of light and radio waves. On our planet, the lights are mainly from oxygen and nitrogen gas; on Saturn, the lights, which rise more than 1,000 miles (1,600km) above the cloud tops, are from hydrogen. Saturn's aurorae last for hours or days and are revealed predominantly in ultraviolet or infrared, while Earth's are typically over in minutes but can be seen in visible light.

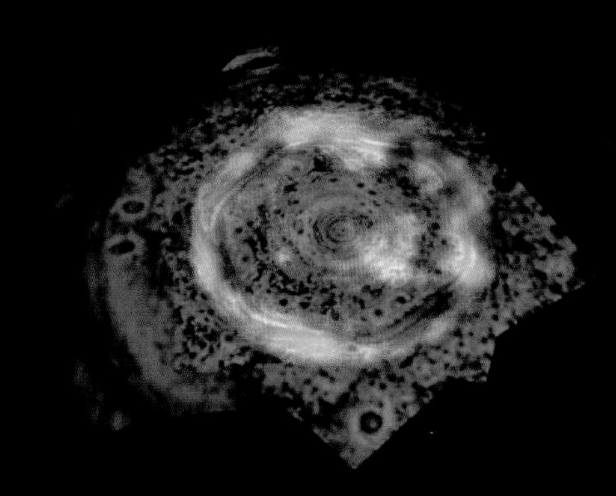

NORTH POLAR AURORA
Two images taken by Cassini's visual and infrared mapping spectrometer, were combined to produce this view of the north polar region. An aurora and the underlying atmosphere are shown at different wavelengths of infrared light. The aurora glows blue, and a bright ring and auroral emission within the cap are visible. The clouds below appear red.

JANUARY 28, 2004

JANUARY 26, 2004

JANUARY 24, 2004

DYNAMIC AURORA
The Hubble Space Telescope and Cassini recorded how the southern aurora changes over time. The increased activity on January 28, seen here in these ultraviolet Hubble images, corresponds with a large disturbance in the solar wind, measured by Cassini.

SATURN'S RINGS

Saturn's rings are the largest, most complex, and most impressive of any encircling the four giant planets. First observed 400 years ago, they were initially assumed to be a single solid ring. We now know that they are made up of millions of individual orbiting pieces and that the ring system is more extensive than was thought just a few years ago.

SIZE AND STRUCTURE

The ring system consists of many individual rings and a number of gaps between these rings. Seven rings are identified by a letter allocated in order of discovery, and the gaps take the names of astronomers. The breadth and the depth of the rings differ. The B ring is the broadest, and about 33ft (10m) deep.

Individual rings include pieces of dirty water ice ranging in size from tiny grains to three-foot-sized boulders. The ice reflects light, making the rings shine brightly. The D ring is closest to the planet, but those most easily seen are the three main rings, C, B, and A. Beyond these lie rings F and G and the more diffuse E ring. In 2009, a huge doughnut-shaped ring was discovered far beyond Saturn and its ring system. The Spitzer Space Telescope spotted the infrared glow of the ring's cool dust. This ring had not been detected because the sparse distribution of its particles makes it almost impossible to see.

Encke Gap
outer part of the A ring consists of ringlets made of icy particles
Keeler Gap
Encke Gap is filled with sparse ringlets of small, dirty particles

RING COMPOSITION
When seen in ultraviolet, the rings reveal their relative ice content. The bright red B ring and the pale red Cassini Division (both on the left) have less ice than the turquoise A ring (center and right).

THE D RING TO THE F RING
This natural-color view of Saturn's rings combines 45 separate Cassini images. The distance of the rings from Saturn increases toward the right of the image. The G and E rings lie beyond the right-hand edge, up to 300,000 miles (480,000km) beyond Saturn. The gaps are not complete voids, but contain ringlets of material.

Colombo Gap

Maxwell Gap

D ring | C ring | B ring

46,300 miles (74,700km) from Saturn's center

57,000 miles (92,000km)

DYNAMIC SYSTEM

The rings are an active, changing system. Moons orbiting within the system interact with it, shepherding particles into rings and maintaining gaps. The small moon Prometheus reshapes the narrow F ring, tugging at its inner edge, while Pandora, which orbits beyond the ring, molds the outer edge. Clumps within the A and B rings, called self-gravity wakes, are constantly colliding. At 130ft (40m) across, they are too small to be visible, but their distribution has been mapped. Collisions between large particles near the Keeler Gap cause fragmentation, while fingerlike shapes, stretching across the rings like spokes on a wheel, appear depending on the Sun's angle to the rings.

Cassini revealed two new aspects to the A ring. In 2006, it detected moonlets the size of football fields in the center of propeller-shaped features about 3 miles (5km) long. In 2009, the craft found that the gravity of the moon Daphnis, orbiting in the 26-mile- (42km-) wide Keeler Gap, pulls particles at the ring edges into wavelike structures.

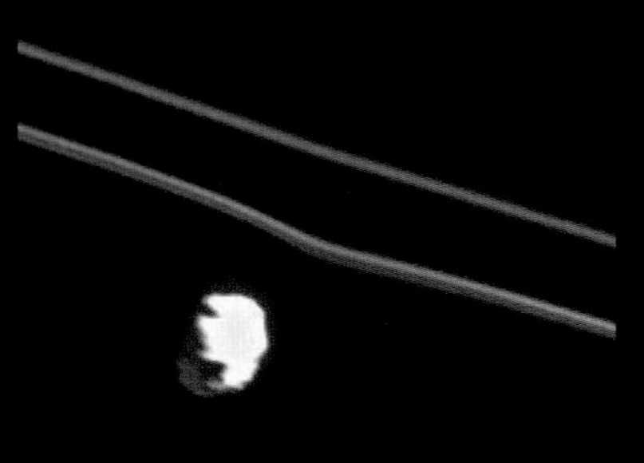

PROMETHEUS PULLS AT THE F RING
When more than half a million miles from Saturn, Cassini captured this view of Prometheus giving the F ring a gravitational tug. The close proximity of the moon has distorted the inner edge of the ring.

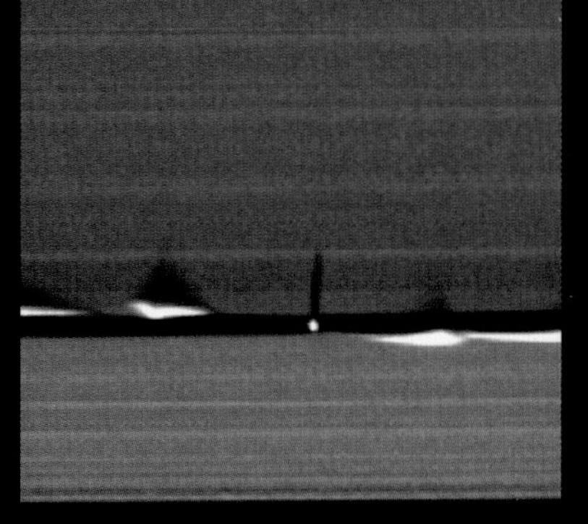

DAPHNIS CREATES WAVES
As Daphnis (center) moves through the Keeler Gap, it sculpts the edges of the A ring into waves (the bright blotches) about 0.6 mile (1km) tall. Both Daphnis and the waves cast shadows onto the A ring.

SIDE-ON DETAIL OF GIANT DUST RING
This Spitzer Space Telescope infrared image shows a slice of the giant ring. Its vertical height is 20 times Saturn's width and its orbit tilts by 27° to the other rings' planes.

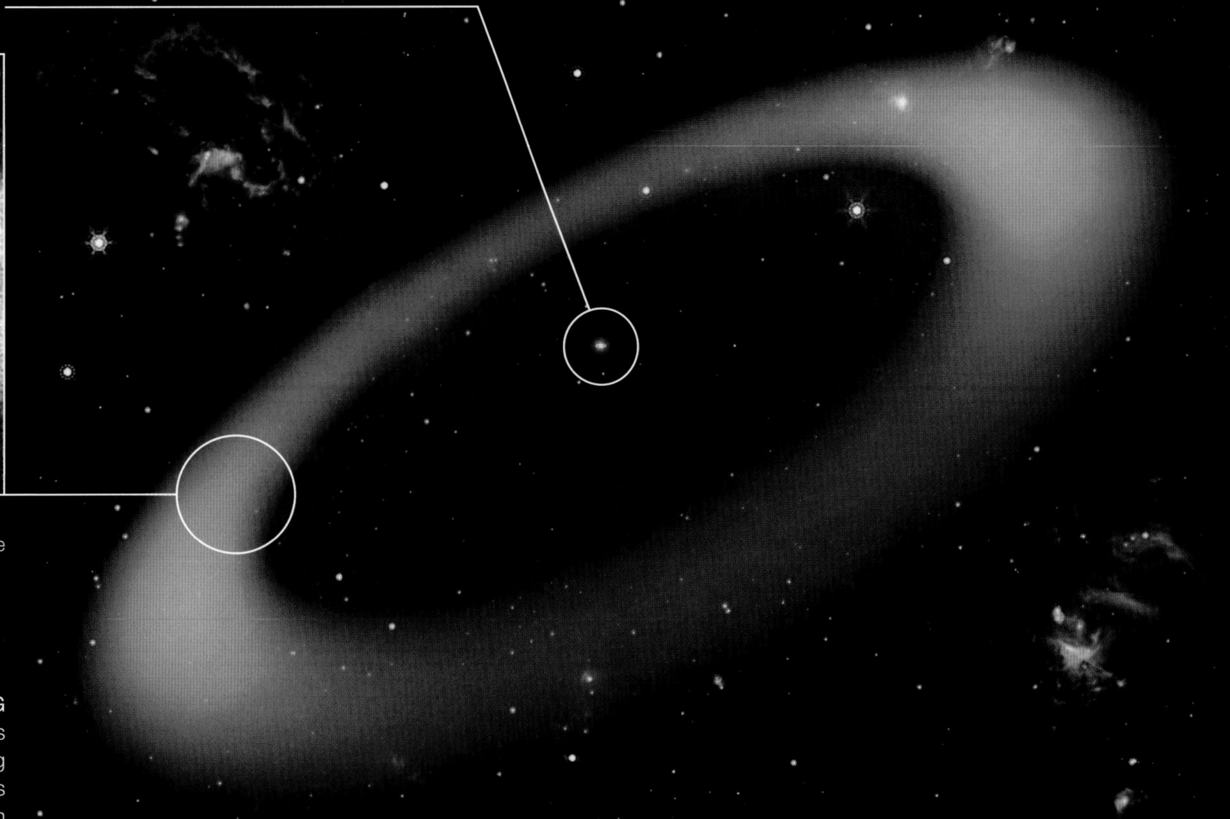

GIANT DUST RING
Here, an artist's impression simulates an infrared view of the huge dust ring discovered in 2009. The ring starts 6 million km (4 million miles) from Saturn and extends twice as far again.

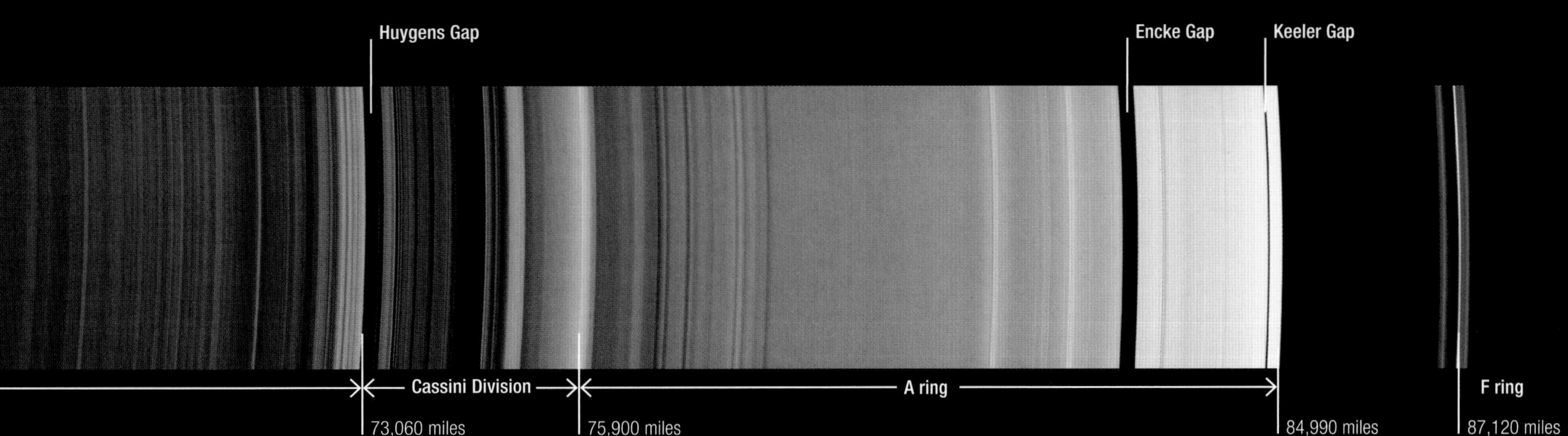

CASSINI AT SATURN

One of the largest and most complex craft to explore the Solar System moved into orbit around Saturn on July 1, 2004. Cassini-Huygens was at the start of its mission to explore all elements of the Saturnian system—Saturn itself, its rings, and its moons.

CASSINI-HUYGENS PROFILE

MISSION	
Launch date	October 15, 1997
Arrival at Saturn	July 1, 2004
Mission ends	May 2017
Launch vehicle	Titan IVB/Centaur
CASSINI	
Agency	NASA
Width	13ft (4m)
Height	22ft (6.7m)
Weight	12,593lb (5,712kg)
Power source	Radioisotope thermoelectric generators
HUYGENS	
Agency	ESA, ASI
Diameter	8.9ft (2.7m)
Weight	705lb (320kg)
Power source	Lithium sulfate batteries
SCALE	

22ft (6.7m)

13ft (4m)

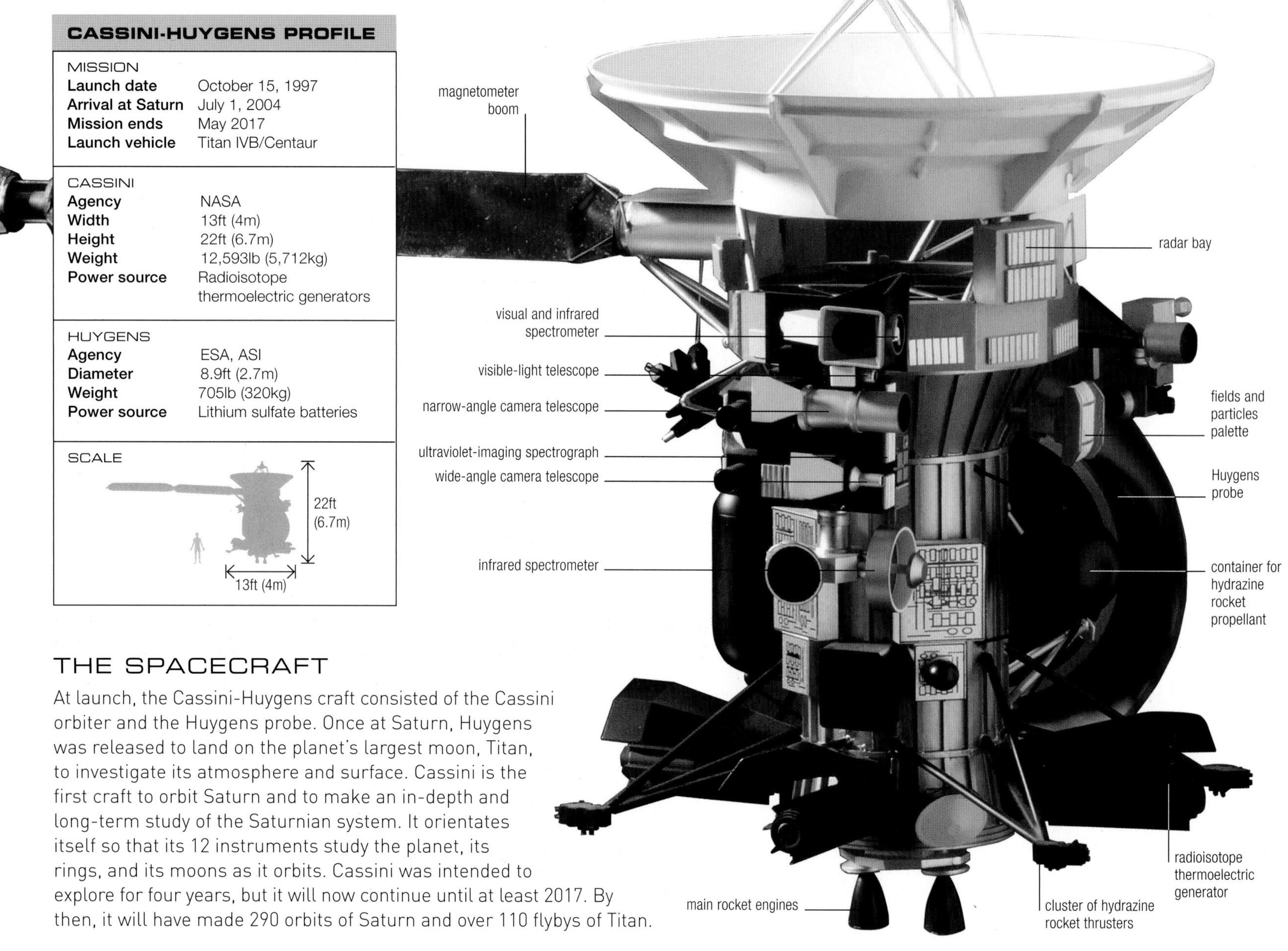

CASSINI-HUYGENS
The craft is roughly cylindrical, with its instruments mounted around the length of its body. On one side is the Huygens probe, and on the top is an antenna for communication with Earth. The signals take almost 1.5 hours to reach us.

THE SPACECRAFT

At launch, the Cassini-Huygens craft consisted of the Cassini orbiter and the Huygens probe. Once at Saturn, Huygens was released to land on the planet's largest moon, Titan, to investigate its atmosphere and surface. Cassini is the first craft to orbit Saturn and to make an in-depth and long-term study of the Saturnian system. It orientates itself so that its 12 instruments study the planet, its rings, and its moons as it orbits. Cassini was intended to explore for four years, but it will now continue until at least 2017. By then, it will have made 290 orbits of Saturn and over 110 flybys of Titan.

CASSINI'S INSTRUMENTS AT WORK

Cassini's instruments include telescopic cameras and spectrometers that work in a range of wavelengths to produce stunning images. Other instruments study the dust and magnetic field around Saturn, collect data on the sizes of particles in the rings, and see through Titan's atmosphere.

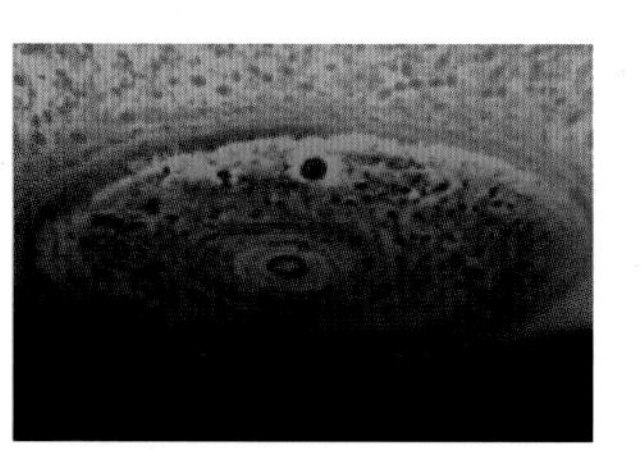

Near-infrared
The visual and infrared spectrometer here shows different cloud depths at Saturn's south pole.

Infrared
Data collected by the infrared spectrometer was used to make this temperature map of Saturn's rings.

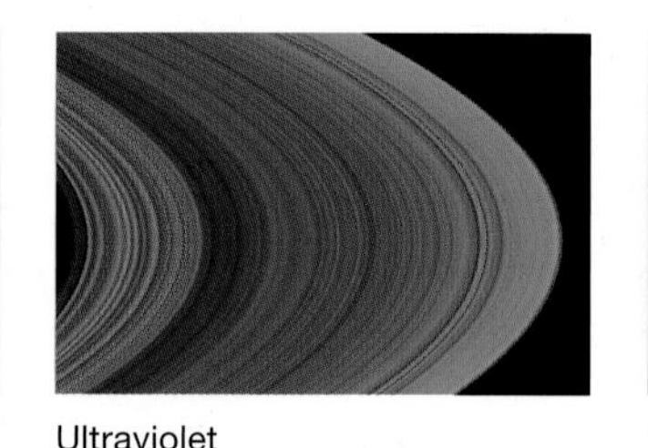

Ultraviolet
The ultraviolet imaging spectrograph collected light from Saturn's rings to reveal their ice content.

Visible light
Cassini's narrow-angle camera took this view of Mimas as it approached the moon in 2005.

THE JOURNEY TO SATURN

Cassini-Huygens's launch date was determined by the positions of Earth and Saturn. As it was too heavy to take a direct route to Saturn, it was launched on a gravity-assist flight path past Venus, Earth, and Jupiter.

As it approached Saturn in June 2004, Cassini-Huygens flew by Saturn's moon Phoebe. A path-correction maneuver followed, and one month later, the craft arrived at Saturn. Nearing Saturn from below its ring plane, the craft passed through the gap between the F and G rings. A main-engine burn slowed the craft enough for it to be captured by Saturn's gravity and move into orbit to make its tour of the system.

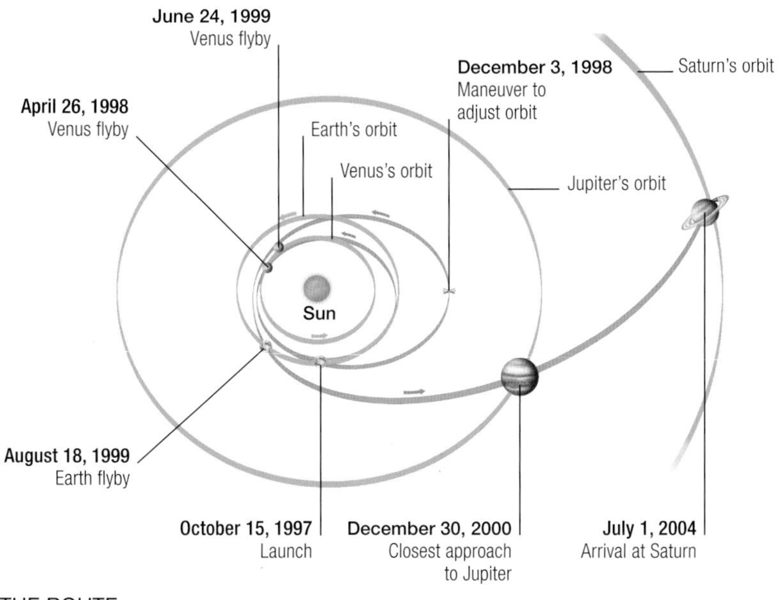

THE ROUTE
Cassini-Huygens flew past Venus twice, and past both Earth and Jupiter once. As it swung by the planets, some of the planets' orbital momentum transferred to the craft, boosting its speed.

FLYING BY TITAN

On its initial orbit around Saturn, Cassini-Huygens flew close to Titan. During its second flyby in December 2004, it maneuvered into position to release Huygens. Since then, Cassini has made numerous gravity-assist flybys of Titan, using the moon to change its direction and speed. Changes to the inclination of Cassini's orbit have meant its flight path has taken it out of the plane of the rings and given new perspectives on Saturn, its rings, and its moons.

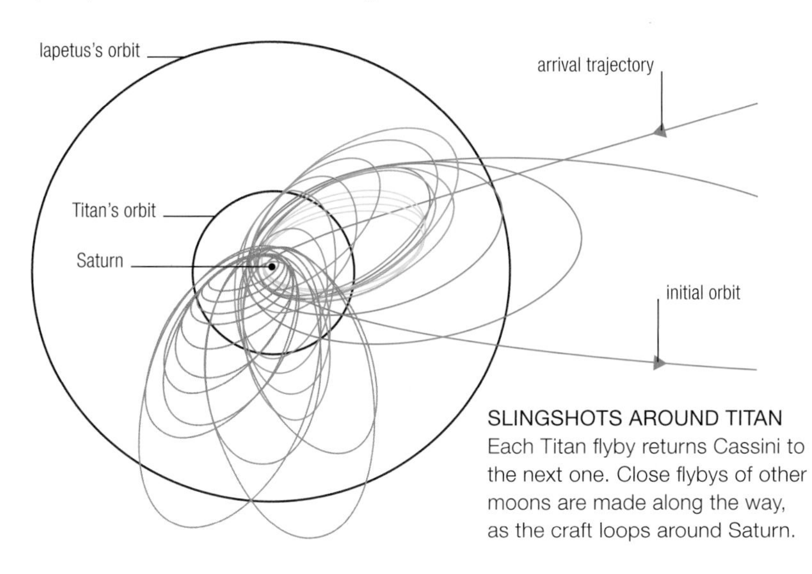

SLINGSHOTS AROUND TITAN
Each Titan flyby returns Cassini to the next one. Close flybys of other moons are made along the way, as the craft loops around Saturn.

HUYGENS'S DESCENT TO TITAN

After its release from Cassini on December 25, 2004, Huygens coasted toward Titan. Twenty days later and with its front shield pointing toward the moon's surface, it started its descent. The shield and the rest of the aeroshell casing slowed Huygens and protected the more delicate descent module and its instruments until about 100 miles (160km) above the ground. A pre-programed descent sequence then started. The two parts of the casing were ejected, parachutes slowed the craft further, and it touched down on January 14, 2005.

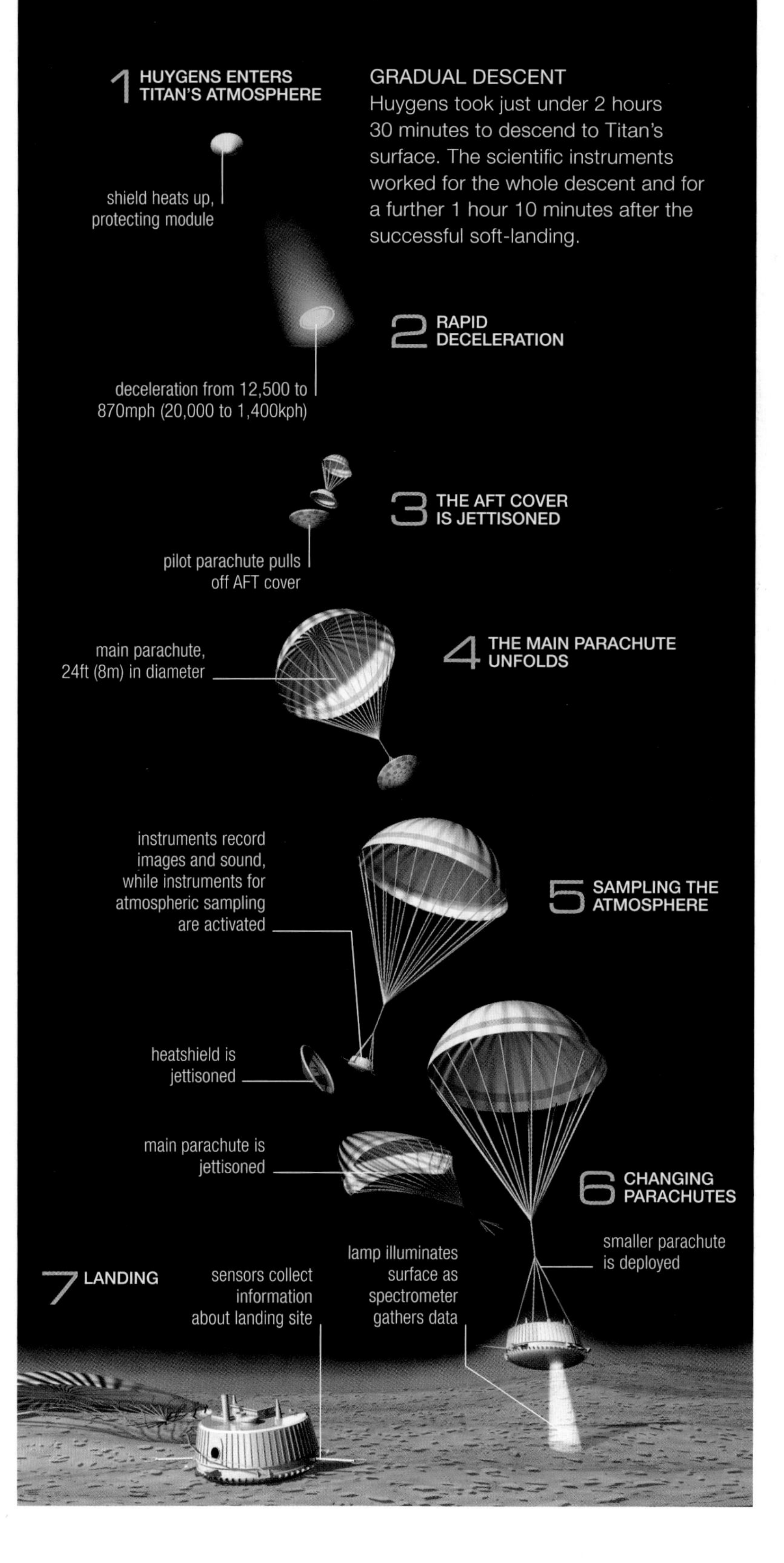

GRADUAL DESCENT
Huygens took just under 2 hours 30 minutes to descend to Titan's surface. The scientific instruments worked for the whole descent and for a further 1 hour 10 minutes after the successful soft-landing.

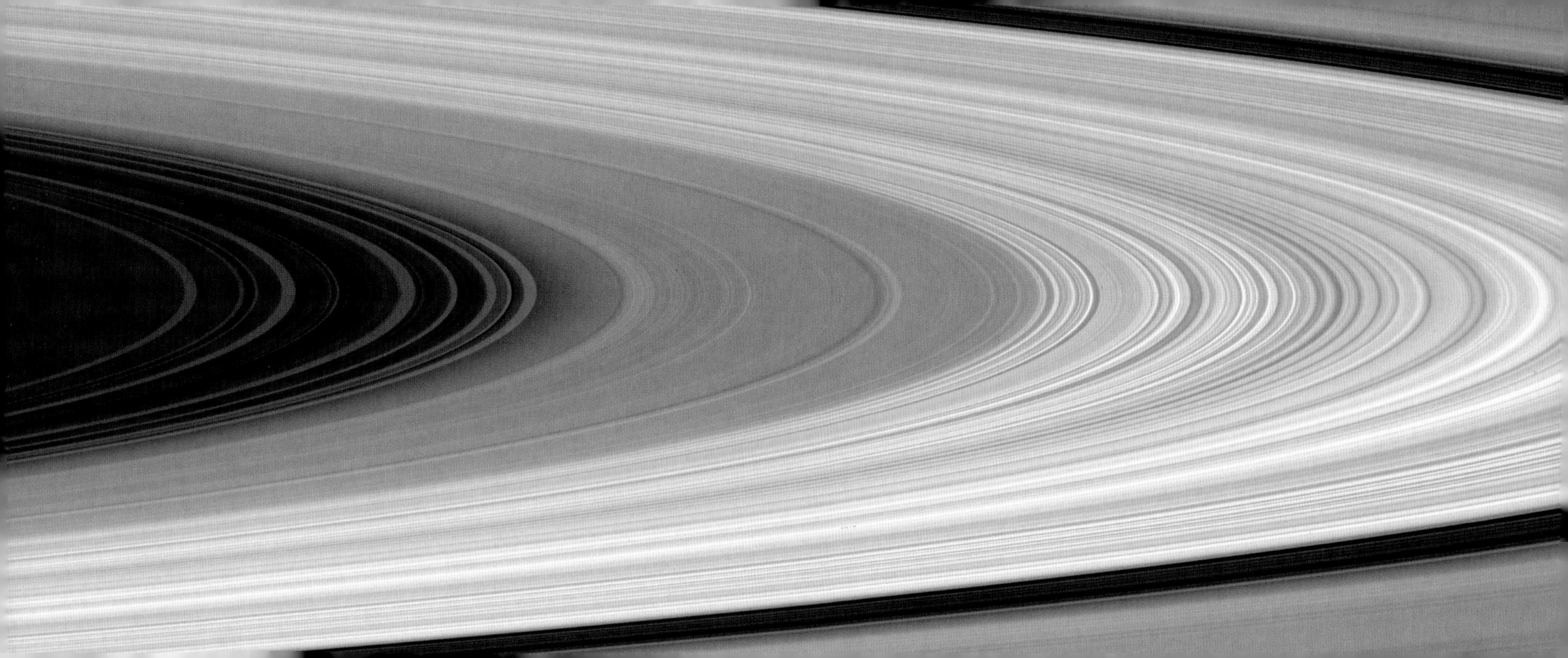

This natural-color mosaic of Saturn's rings was taken by the Cassini spacecraft from a distance of 1.1 million miles (1.8 million km). Made almost entirely of water ice, the rings are believed to be debris from a moon destroyed by meteorite bombardment some 4 billion years ago. The color variation among the rings is caused by dust and chemical contamination.

SATURN'S MOONS

Saturn has over 60 moons. They differ greatly in size and distance from the planet. Some orbit within Saturn's rings, others at a huge distance. Most are small and irregularly shaped. Large or small, the moons are named after mythological giants from different cultures.

DISCOVERY

Titan's size, which is similar to the planet Mercury's, meant that it was the first of Saturn's moons to be discovered, in 1655. Within 30 years of this, the four next-largest moons, each measuring between about 600 miles (1,000km) and 900 miles (1,500km) across, had been found.

Although four more smaller moons were identified before the 20th century began, as well as nine in its closing decades, the majority of Saturn's moons have been discovered since January 2000. A handful of these were found by the Cassini spacecraft, but astronomers have discovered more than 20 from Earth due to improved telescope technology and observing techniques. The majority of these moons are very small, measuring less than 6 miles (10km) across.

At the start of 2010, Saturn had 62 known moons, but this number is likely to rise as more small moons are found. Tiny moonlets that orbit within the ring system and affect nearby ring material will probably be given "moon" status—the first was seen directly in 2009.

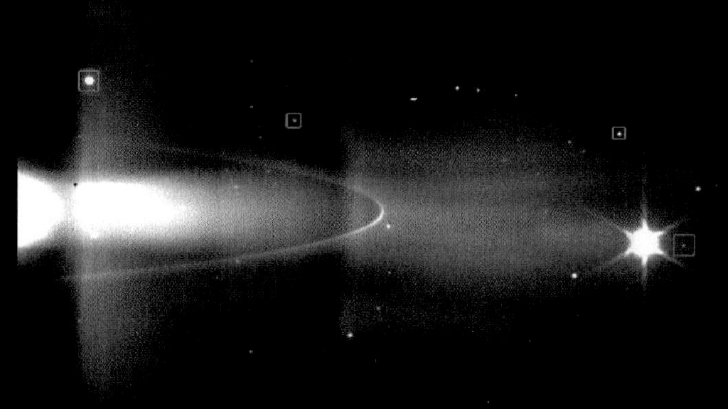

DISCOVERY OF SATURN'S SIXTIETH MOON
Cassini's wide-angle camera was pointed at Saturn's rings to capture images that could be searched for new moons. The moon in the red box, S/2007 S4, later named Anthe, was discovered in this 2007 image. To its left is Mimas. The green boxes indicate moons discovered previously.

INNER AND OUTER MOONS

Saturn has seven large major moons (see opposite). The rest are small, irregularly shaped bodies of rock and ice that fall into two groups: the inner and outer moons. The inner moons orbit relatively close to Saturn, and all but one, Hyperion, sit inside Titan's orbit and within the ring system. Innermost of all is Pan, just 16 miles (26km) across. It orbits Saturn every 13.7 hours, within the ring's Encke Gap. Prometheus and Pandora are shepherd moons, one at either side of the F ring. Just beyond the F ring are Epimetheus and Janus, which virtually share an orbit, about 30 miles (50km) apart and take it in turns to be the closest to Saturn. The outer moons follow orbits far beyond Titan. The innermost, Kiviuq, is nine times further from Saturn than Titan is, and the outermost, Fornjot, is 20 times the distance, at 15.6 million miles (25.1 million km) from Saturn. These are tiny worlds, and many are no more than 4 miles (6km) across. At 140 miles (230km) long, Phoebe is the largest. The nature of the moons and their mostly retrograde orbits suggest that they are bodies captured by Saturn's gravity.

MOONS WITHIN THE RINGS
Seven of Saturn's moons are visible in this image taken by Cassini's wide-angle camera. They are all small inner moons, except for Mimas, the smallest of the major moons.

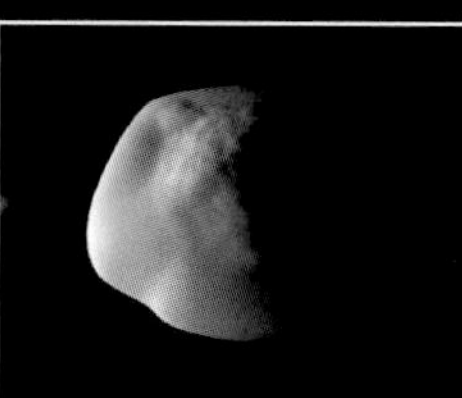

ATLAS

PAN

MAJOR MOONS

Saturn's major moons are noticeably larger than its other moons, and all are round in shape. In order of size from the largest to the smallest, they are Titan, Rhea, Iapetus, Dione, Tethys, Enceladus, and Mimas.

Five major moons orbit within the ring system. Mimas is closest to Saturn and orbits in 22.6 hours. Only Titan and the more distant Iapetus are beyond the rings. Titan is 750,000 miles (1.2 million km) from Saturn and orbits in 15.9 days. Iapetus is almost three times further from Saturn than Titan is and takes 79 days to circuit Saturn once. Iapetus is also the only major moon to follow an orbit that is highly inclined to Saturn's equatorial plane. All seven major moons are in synchronous rotation and so like Earth's moon, keep the same side facing their planet.

All of the major moons have been imaged by spacecraft: first by Voyagers 1 and 2 and more recently by Cassini. Composed of rock and ice, these cold worlds have cratered surfaces, some of which show signs of resurfacing by tectonics. Titan stands out from these and all other Solar System moons because it is the only one with a substantial atmosphere. Its surface is hidden from view, but the Huygens probe has shown us that beneath lie an intriguing, youthful surface and a methane cycle reminiscent of Earth's water cycle. Titan is also the only Solar System body other than Earth to have liquid lakes on its surface.

SATURN AND FOUR MOONS
Titan dwarfs three of Saturn's other moons in this Cassini image. The view is deceptive —in order of distance from Saturn, they are Prometheus, Janus, Mimas, and Titan. The dark stripes on Saturn are shadows cast by its rings.

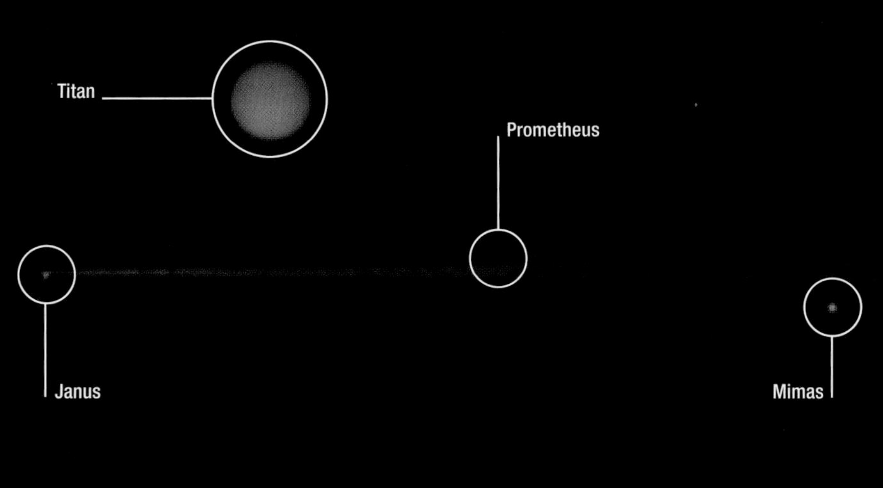

DIONE AND SATURN'S RINGS
Saturn's fourth-largest moon, Dione, is partly obscured by the planet's rings, here seen edge-on. The bright and dark sides of this icy moon are visible, as are the bright lines of cliffs.

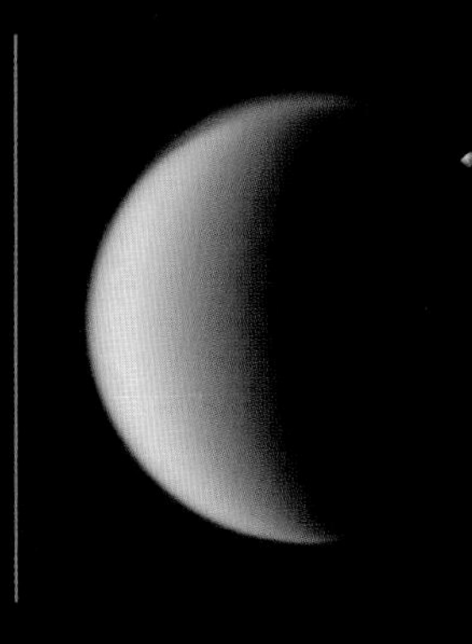

TETHYS MOVES BEHIND TITAN
Cassini's narrow-angle camera caught Tethys moving behind Titan in 2009. Two images, 18 minutes apart, show Tethys approaching the bigger moon (right image) and then emerging from behind it (left image).

ENCELADUS

Orbiting Saturn within its tenuous E-ring is Enceladus, measuring 318 miles (512km) across. This moon is one of the Solar System's brightest objects, because its water-ice surface reflects more than 90 percent of the sunlight that hits it. Enceladus is slightly larger than its near neighbor Mimas and was once thought to have an old, cratered surface similar to Mimas's. But recent Cassini flybys have shown that huge areas of Enceladus have been resurfaced and that it is a geologically active world. Near the south pole are four fissures —the tiger stripes—where ice particles, water vapor, and organic compounds spray through the surface. These eruptions create a temporary atmosphere and supply the E ring with material.

ENCELADUS
Long fractures called sulci (singular sulcus) and softened craters are visible on the side of Enceladus facing away from Saturn. The false colors in this image emphasize the coarse-grained ice exposed on fracture walls, compared to the powdery covering of Enceladus's flat regions.

Cashmere Sulci
At 162 miles (260km), Cashmere is the moon's third-longest sulcus.

Alexandria Sulcus
The shortest tiger stripe, Alexandria Sulcus is 69 miles (111km) long.

Cairo Sulcus
This tiger stripe stretches for 103 miles (165km).

Damascus Sulcus
Like the other tiger stripes, Damascus Sulcus consists of two parallel ridges separated by a deep, V-shaped trough. It is 78 miles (125km) long.

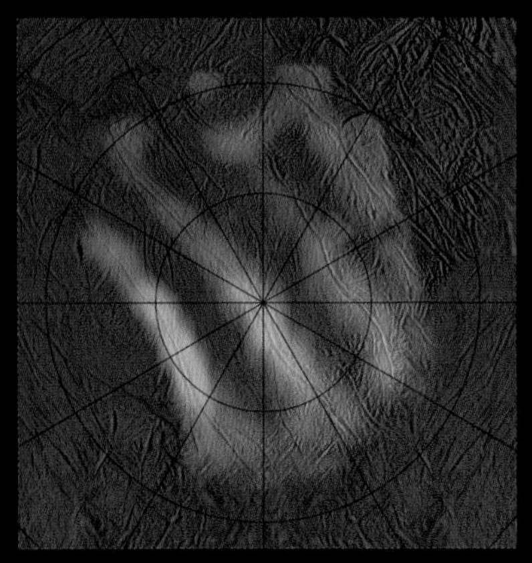

INFRARED MAP OF THE SOUTH POLE
Heat from the interior escapes along the tiger stripes. The temperature at Damascus Sulcus (the far left stripe) is -135°F (-93°C), compared to -330°F (-201°C) in the surrounding region. The warmest parts are where jets of water ice erupt (below).

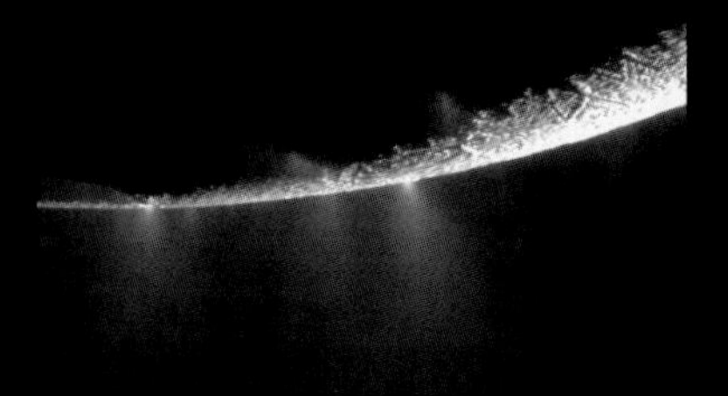

JETS AND PLUMES
Jets of water ice surrounded by plumes of water vapor spray out from a tiger stripe. This Cassini image was taken in November 2009, when the craft made a close flyby of the moon and passed through the jets.

BAGHDAD SULCUS
A 10-mile- (16km-) portion of the Baghdad Sulcus is seen bottom center in this 3D Cassini image. Running for 109 miles (176km), Baghdad Sulcus is the longest of the four tiger stripes. Each stripe is a few miles wide, about 1,640ft (500m) deep, and flanked by high ridges.

MIMAS

Mimas is the innermost and smallest of Saturn's major moons. It orbits within the ring system and cleared the region between the A and B rings to produce the Cassini Division. This ball of water ice and rock averages 247 miles (397km) in diameter. But Mimas is not quite spherical: it is about 19 miles (30km) longer than it is wide and deep. The moon's surface is an icy -344°F (-209°C) and is pitted by impact craters. The huge Herschel Crater dwarfs all the rest. It is Mimas's most conspicuous surface feature, at 86 miles (139km) wide and almost 6 miles (10km) deep. The crater takes its name from William Herschel, the British astronomer who discovered the moon in 1789.

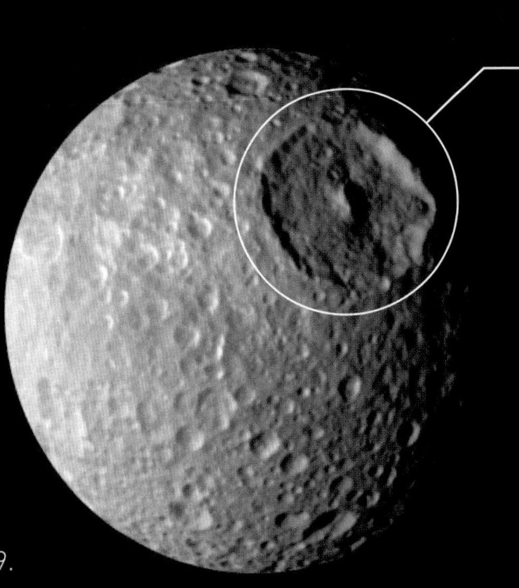

Herschel Crater
Mountain peaks, 4 miles (6km) high, are visible in the crater's center.

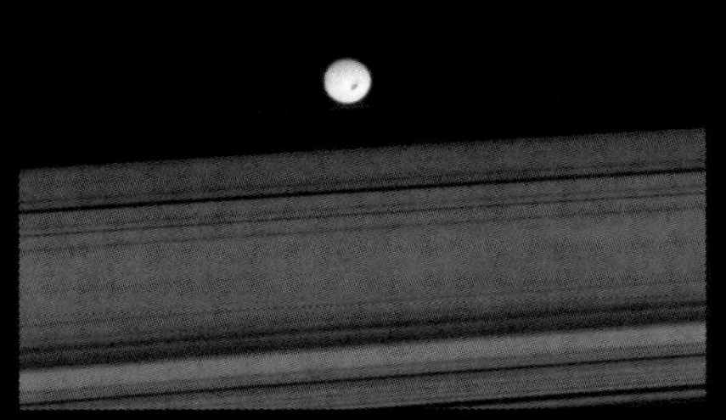

MIMAS FROM BEYOND SATURN'S RINGS
In this view of Mimas, Herschel Crater is on its leading side (left), pointing in the direction of the moon's orbital movement. It can be seen as a slight flattening of the moon's profile.

TETHYS

Measuring 666 miles (1,072km) across, Tethys is Saturn's fifth-largest moon. Orbiting within the ring system, it takes 45 hours to circle the planet and shares its path with two smaller moons, Telesto and Calypso —the Tethys Trojans. Odysseus, Tethys's largest impact crater, is 277 miles (445km) wide. Its rim and central peak have collapsed, which suggests that Tethys was partially molten when the crater formed. The Ithaca Chasma canyon may have been formed by the impact that shaped Odysseus, or by the interior of Tethys freezing and splitting the surface.

THE TETHYS TROJANS

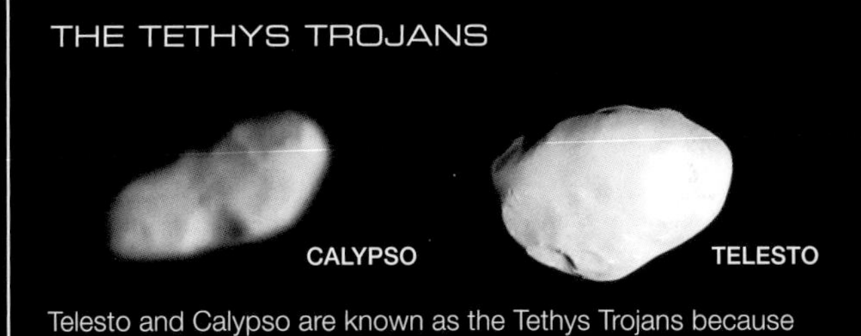

Telesto and Calypso are known as the Tethys Trojans because they lead and trail Tethys on its orbit in positions like those held by the Trojan asteroids that lead and trail Jupiter. Telesto is 60° ahead of Tethys; Calypso is 60° behind it. Telesto and Calypso are both irregular in shape and 19 miles (30km) long. They were discovered in 1980 by Earth-based observations.

Ithaca Chasma
This 757-mile- (1,219km-) long canyon system is about 60 miles (100km) wide.

TETHYS
This moon is a huge ball made up mainly of water ice, with some rock. Ithaca Chasma dominates the side of Tethys seen here, running roughly between the moon's north and south poles.

Antinous Crater
This crater is 86 miles (138km) wide.

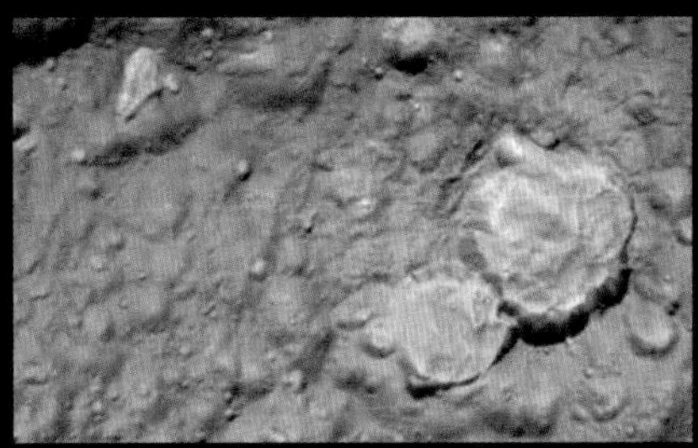

SURFACE DETAIL OF TETHYS
This false-color Cassini close-up reveals the variety of surface materials on Tethys. The outlines of the two craters on the right have been modified by landslides, during which material has moved to the crater floors.

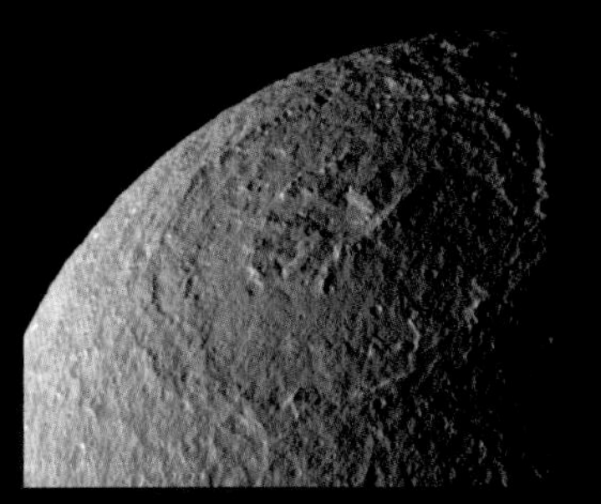

ODYSSEUS CRATER
The huge Odysseus Crater is on Tethys's leading side. Rings of mountainous cliffs form its edge, while a central mountain range and more recent craters lie within the crater basin.

DIONE

Fourth-largest of Saturn's major moons, Dione orbits within the planet's E ring every 2.7 days. It is accompanied on its orbital path by two much smaller moons—Helene, ahead of Dione by 60°, and Polydeuces, 60° behind.

Impact craters cover Dione, but the outstanding features on this 696-mile- (1,120km-) wide ball of rock and ice are the bright, wispy lines that streak across one side. Cassini showed them to be interwoven canyons with bright cliff walls.

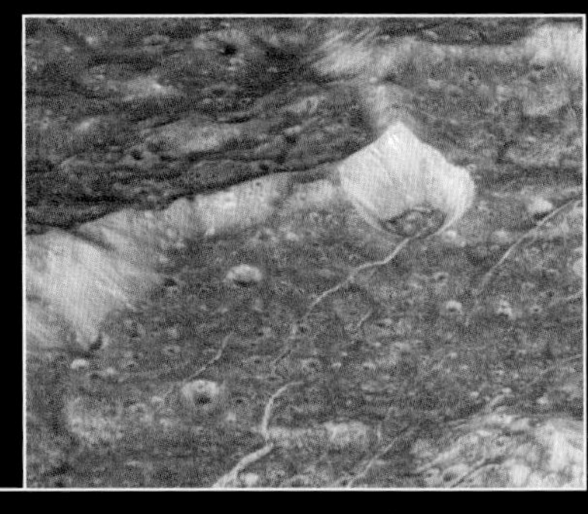

DETAIL OF PADUA CHASMATA
This detail shows terrain inside a 37-mile- (60km-) wide crater within the Padua Chasmata. The rim of the crater runs from lower left to top right. A mountainous peak in the crater's center is at lower right.

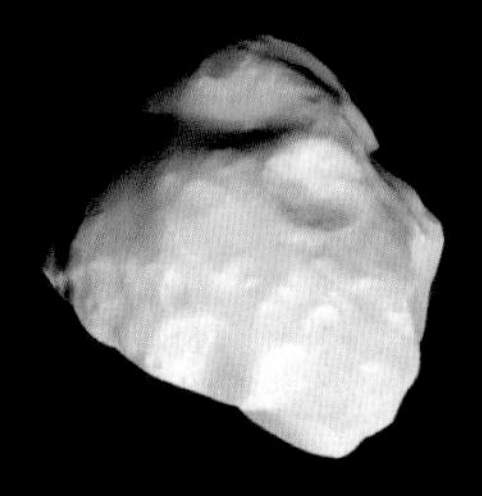

HELENE LEADS DIONE
Helene is a rock-and-ice moon measuring 22 by 19 miles (36 by 30km). Like Dione, it takes 2.7 days to orbit Saturn. Helene leads Dione along its orbit.

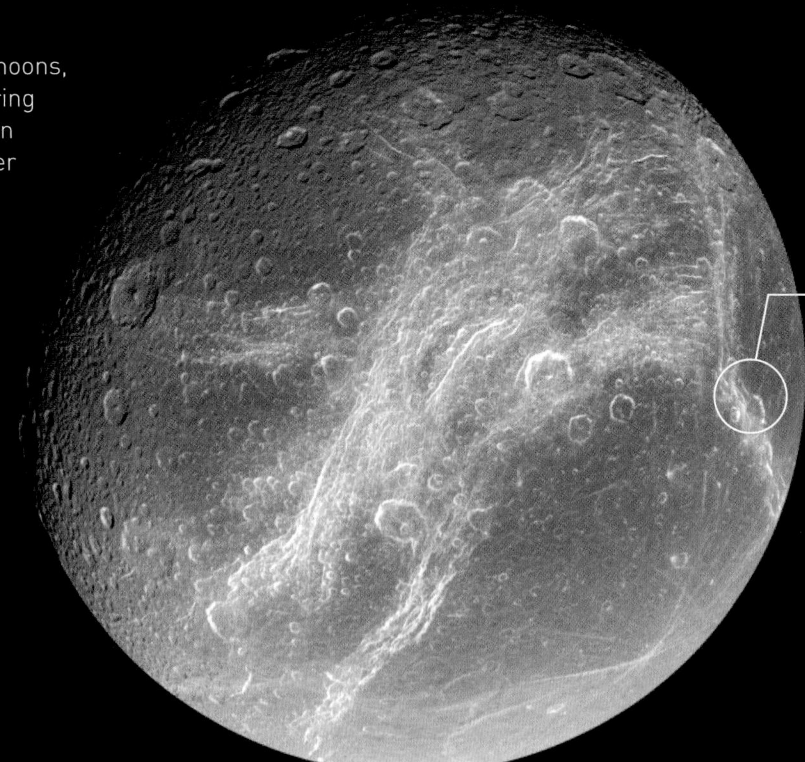

DIONE'S TRAILING SIDE
The long, white streaks that dominate Dione's trailing side (the side facing backward on the moon's orbital path) are termed chasmata and consist of many individual white lines. These are bright cliffs, which line fractures in Dione's surface.

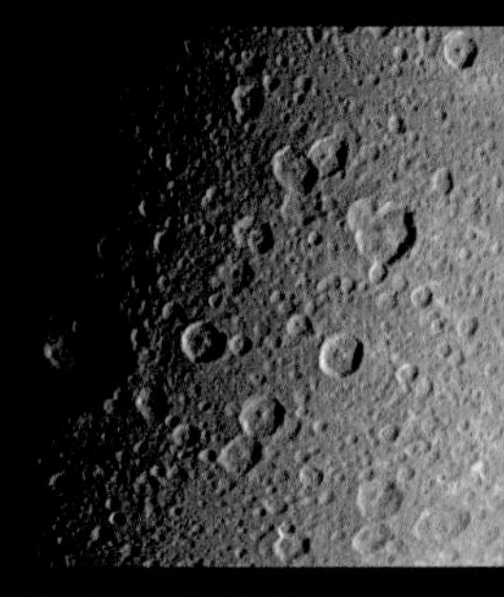

DIONE'S SATURN-FACING SIDE
These craters are in Dione's northern hemisphere, on the moon's face that is permanently opposite Saturn. They have bright walls, and the larger craters have central peaks.

RHEA

Saturn's second largest moon, Rhea is the first to lie beyond the ring system. It is a world of rock and ice, 949 miles (1,528km) across, and orbits Saturn in 4.5 days. Its surface is heavily cratered with no obvious smooth or resurfaced areas, suggesting that it is geologically old. Cassini's flyby in 2005 led to an investigation suggesting that Rhea has a tenuous ring system. If proved to be true, it will be the first moon known to have rings.

Yu-Ti Crater
This relatively young crater is 50 miles (80km) wide.

Ormazd Crater
The crater's degraded edge is visible here.

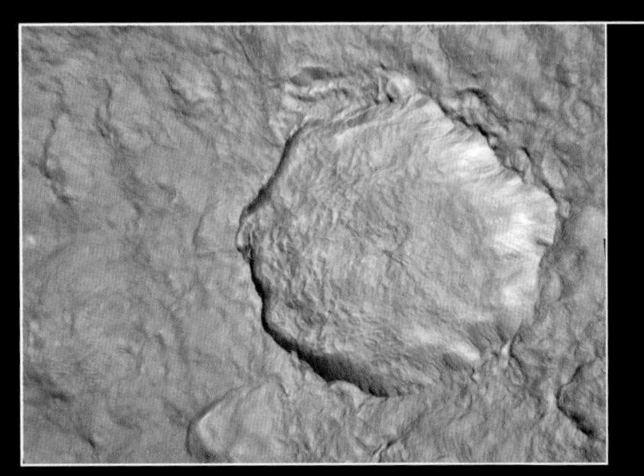

A YOUNG CRATER
This stereo view made from Cassini images gives depth to a young, 30-mile- (48km-) wide crater. Its sharp rim, brightness, and the low number of small craters within it indicate that it is geologically recent.

CRATERED SURFACE
This Cassini image is of Rhea's leading side—the hemisphere that faces forward on its orbital path. Bright ice ejecta thrown out by a young crater stands out.

HYPERION

This icy body is the most distant of Saturn's inner moons and orbits outside the rings in 21.3 days. Hyperion has a strange appearance, and, at 230 miles (370km) long, it is one of the largest irregularly shaped bodies in the Solar System.

Impact craters cover Hyperion but they differ from those on neighboring moons. Their odd appearance is due to the moon's unusually low density. Hyperion is highly porous and has weak surface gravity. Surface material has been blasted away and escaped, rather than compressed into the moon.

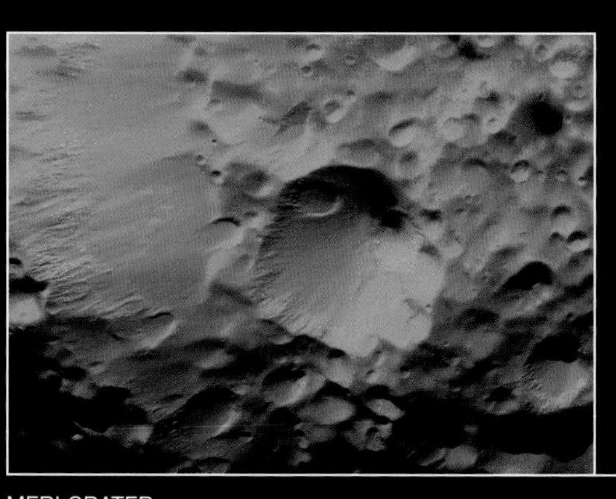

MERI CRATER
False color is used in this picture, centered on Meri Crater, to emphasize natural color differences on Hyperion. As can be seen here, many of Hyperion's craters have dark material in their floors.

SPONGELIKE MOON
Hyperion's shape, its porosity, and its surface craters combine to make the moon appear spongelike. This view was taken by Cassini's narrow-angle camera as the craft made its close flyby on September 26, 2005.

IAPETUS

At first glance, Iapetus does not seem exceptional. It is a cold, ice and rock world, 892 miles (1,436km) across, and the most distant of Saturn's major moons, orbiting every 79.3 days. But there is a surprising contrast between the moon's two sides—its leading hemisphere is as dark as charcoal, and its trailing side is predominantly bright. Since Cassini's exploration, one explanation is that dust from the outer moons deposited on the leading hemisphere speeds up the evaporation of ice there. Another is that dark material from Saturn's huge dust ring, orbiting in the opposite direction to Iapetus, is drifting inward and coating the moon.

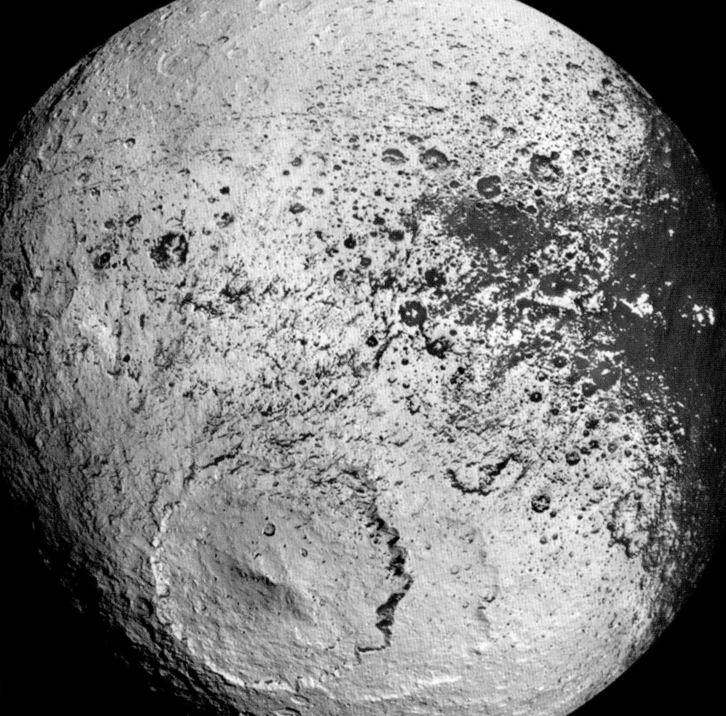

THE TWO-TONED MOON
The bright hemisphere of Iapetus appears heavily cratered in this Cassini image. The Engelier Crater, visible on the lower half, is 313 miles (504km) across.

THE LIGHT TRAILING SIDE OF IAPETUS

THE DARK LEADING SIDE OF IAPETUS

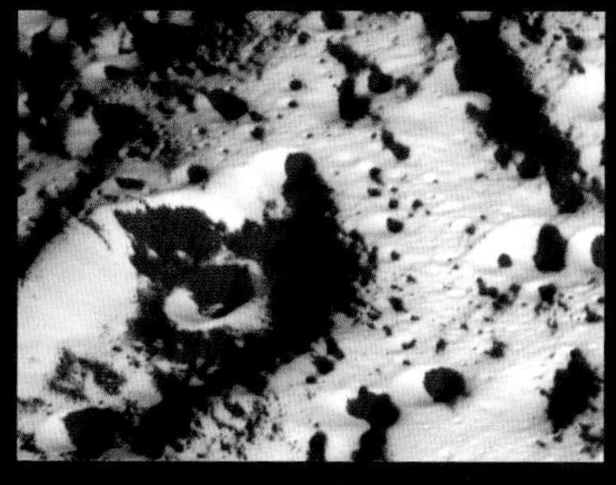

WHERE THE TWO SIDES MEET
This Cassini image of the terrain on Iapetus where its dark and bright sides meet shows a bright, icy landscape coated by dark material.

TITAN

Just bigger than the planet Mercury, Titan is Saturn's largest moon and the second largest of all Solar System satellites. It is a cold, rock-and-ice world surrounded by a thick, unbroken atmosphere. Discovered in 1655, the moon was first seen close-up by Voyager 1 in 1980. The world beneath the atmosphere was revealed after the Cassini–Huygens craft arrived at Saturn in 2004. Its mission included repeated flybys of the moon and the release of the Huygens probe to Titan's surface.

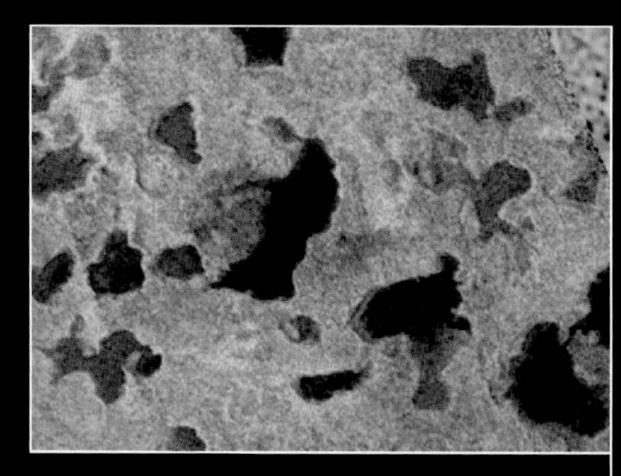

LAKES IN THE NORTH POLAR REGION
Huge bodies of liquid near Titan's north pole can be seen in this image produced by Cassini's radar mapper. The lakes are colored blue to make them stand out. Just like lakes on Earth, they have bays, inlets, and the occasional island.

TITAN BEYOND SATURN'S RINGS
Saturn's rings cut across Titan's face. This view, taken by Cassini, has been colored to show how it would appear to the human eye. The small moon Epimetheus is also visible above the rings (center).

Bazaruto Facula
Small, bright surface features are known as faculae (singular facula). Bazaruto Facula is the site of an impact crater.

Aztlan
Titan's prominent bright or dark surface regions such as Aztlan are named after enchanted, mythical places.

Tsegihi
This large, bright area takes its name from a sacred place of the Navajo people.

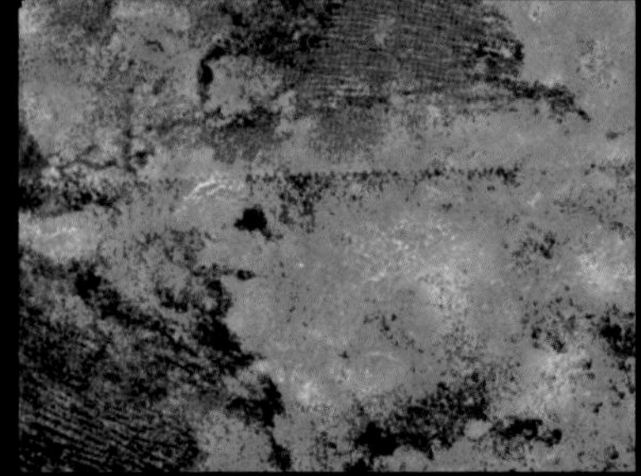

RADAR MAP OF TITAN'S SURFACE
The colors indicate height, with purple the lowest elevation and red the highest. The purple area is covered in sand dunes and is part of a dark equatorial region called Belet.

BENEATH TITAN'S CLOUDS
Images taken during two Cassini flybys are here combined to show Titan's surface. The lower right has many mountain ranges. The bright band of clouds is thought to be methane gas, cooled to form clouds as winds drove it over the mountains.

ATMOSPHERE

Titan's blanket of gas completely enshrouds the moon. It is 98 percent nitrogen, hundreds of miles deep, and has a density four times that of Earth's atmosphere. A yellow, smoglike haze in the upper atmosphere gives Titan its distinctive honey-color; further layers of haze lie beyond. Although Titan is a large moon, its atmosphere is still surprisingly thick and dense. It holds onto its gas because it is such a cold world. The surface temperature is about -289°F (-178°C), and the cold gas molecules do not move fast enough to escape Titan's gravity.

Titan's clouds form and move around much like those on Earth, but they travel more slowly. The atmosphere flows eastward, in the same direction as Titan's spin, and the clouds are distributed around the moon according to expected global circulation and seasonal weather patterns. Composed of methane and ethane, the clouds condense to produce strong but infrequent rainstorms. The rain and the wind create Earth-like surface features such as river channels and sand dunes. With its nitrogen atmosphere and liquids, Titan is like a cold version of an early Earth.

NORTH POLAR CLOUD
Cassini found a huge cloud system, produced by atmospheric circulation, covering Titan's north pole. The clouds rain liquid methane that pools into the lakes.

POLAR ATMOSPHERE
This mage shows the structure of Titan's atmosphere over the north polar region. It comprises multiple detached hazes or waves that move through stable layers.

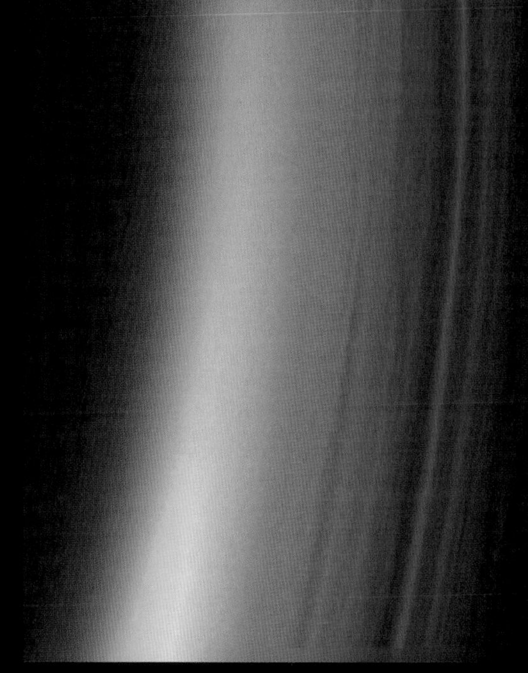

HAZY HEIGHTS
Light is scattered as it passes through Titan's atmosphere, revealing layers of haze at high altitude. More than 10 layers are visible. This is an ultraviolet image of part of Titan's night-side, but it shows the haze in true color.

Huygens' landing site

FULL-DISC SURFACE VIEW
Images from Cassini's first close flyby are here combined to give the first full-disc surface view. Features in the center are sharpest, since Cassini looks directly at this area; toward the edge they become increasingly fuzzy.

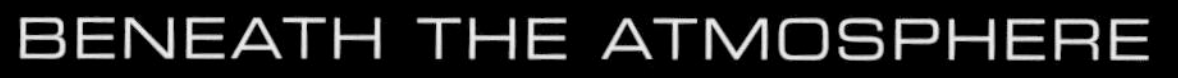

BENEATH THE ATMOSPHERE

The little sunlight that falls on Titan cannot penetrate the dense haze and clouds, making this a gloomy world. Observations made by Cassini's instruments as the craft flew by and data returned by Huygens (which landed on Titan's surface in January 2005, see p.157) show us that below the atmosphere, there are mountains, lakes, sand dunes, and features associated with cryovolcanism (low-temperature eruptions). The low number of impact craters suggests Titan's is a geologically young surface.

Observations made in the 1980s indicated that something was supplying Titan's atmosphere with methane, such as a global ocean or smaller lakes and seas. The debate ended when, in 2006, Cassini revealed that lakes exist near Titan's north pole. Larger than any of the Great Lakes, they consist of liquid hydrocarbons, including methane and ethane. Images also revealed channels that appear to have been formed by a liquid. Some of these channels drain into the lakes, while others fade out.

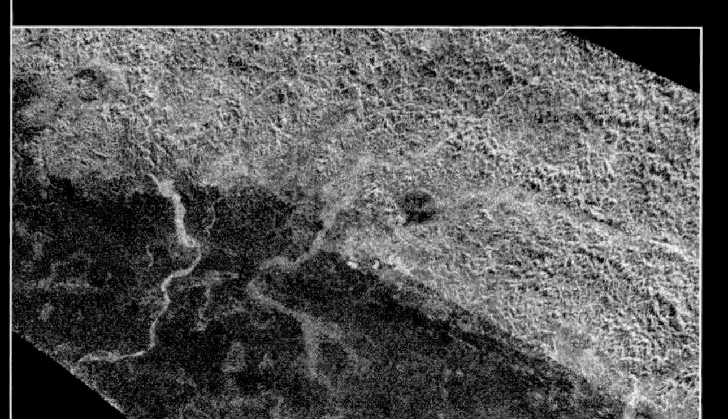

CHANNELS ON XANADU
A set of meandering channels crosses the Australia-sized bright region named Xanadu. Measuring up to 3 miles (5km) wide, the channels are more roughly textured than the surrounding area. They may be dry riverbeds, as some resemble dry lakes in northern Titan.

TITAN'S SURFACE
This was the first picture taken by Huygens of Titan's surface. It shows smooth, rounded rocks that resemble stones on Earth shaped by flowing water.

URANUS

Uranus is a blue, featureless world surrounded by a ring system and a family of moons. Its unremarkable appearance is deceptive, and its sideways stance and long seasons make it stand out from the other planets.

ORBIT

Twice as far from the Sun as its inner neighbor Saturn, Uranus takes 84 years to complete one orbit of the Sun. Uranus's spin-axis is tilted by 98°, and it seems to orbit on its side. This also has the effect of making its moons and rings, which are around its equator, appear to circle it from top to bottom. On each orbit, the north and south poles experience 42 years of continuous sunlight and 42 years of darkness, which has the effect of producing long-lasting seasonal contrasts.

STRUCTURE

Methane crystals in Uranus's hydrogen-rich atmosphere absorb the red wavelengths of incoming sunlight, and so the planet appears blue. A surrounding haze conceals a banded structure and signs of dynamism, such as bright methane clouds. As the density and temperature increase with depth, the physical state of the material changes. Underneath the surface clouds are thicker cloud layers, below these may be an ocean of liquid water, and at the center, a dense core.

THE TILTED PLANET
An infrared image taken by the Keck II Telescope reveals clouds in Uranus's northern hemisphere. The white clouds are the highest, while dark blue signifies the deepest clouds. A by-product of image processing is that the rings appear red.

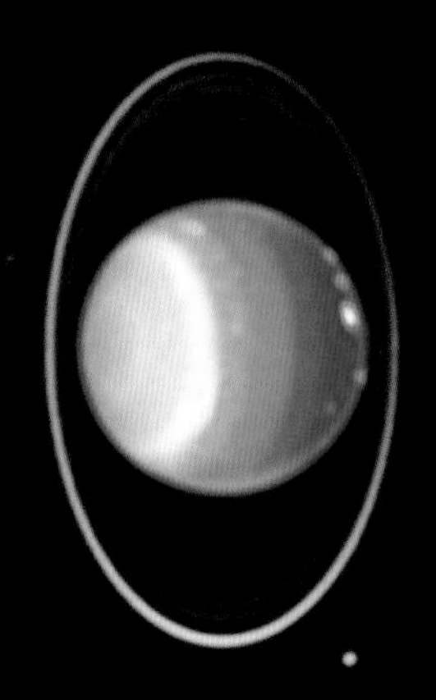

URANUS'S HAZE
This false-color image highlights atmospheric layers. Deep pink around the planet's edge represents a high-altitude thin haze; white and gray are deeper haze; and blue is clear atmosphere and deeper still.

ORBIT AND ROTATION

At its vast distance from the Sun, Uranus receives just 0.25 percent of the sunlight falling on Earth. Due to the extreme tilt of its spin-axis, Uranus appears to spin clockwise.

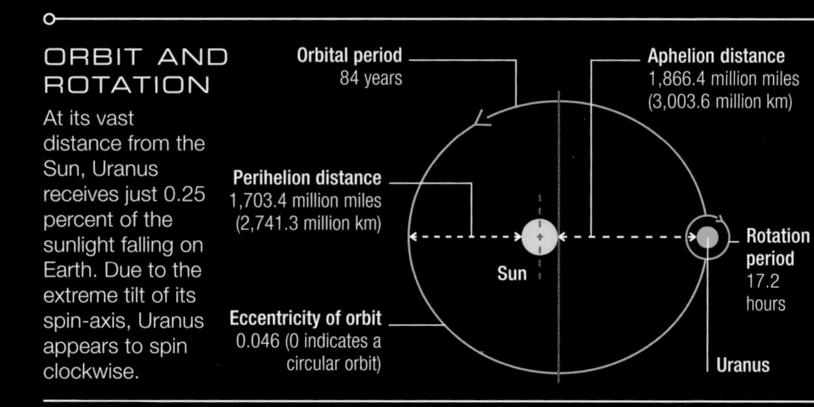

SIZE

Uranus is the third largest planet. It is four times the size of Earth and has 14.5 times Earth's mass.

Uranus
Diameter at equator:
31,763 miles (51,118km)

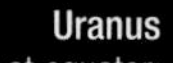

Earth
Diameter at equator:
7,926 miles (12,756km)

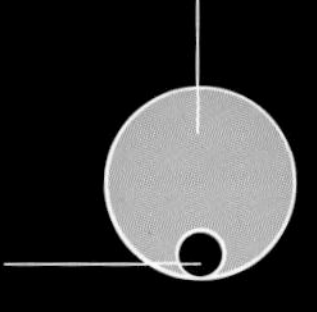

EXPLORATION

Uranus became the first planet to be discovered in modern times when the English astronomer William Herschel observed it in March 1781. Over the next 200 years, its five major moons were discovered, but the details of these and of Uranus itself were largely unknown until the Voyager 2 mission in 1986. Since then, observations made with the Hubble Space Telescope and from Earth-based telescopes using improved optical systems have added to our knowledge.

Voyager 2 is the only spacecraft to have explored Uranus. It flew by after first visiting Jupiter and Saturn and before continuing on to Neptune. Its journey to all four giant planets became known as the Grand Tour. Voyager 2 made its closest flyby of Uranus on January 24, 1986, when it came within 50,640 miles (81,500km) of the planet. The images it returned gave us our first close-up views of this distant world. The craft also took the temperature of the atmosphere, detected its magnetic field, investigated its rings, and discovered 11 moons.

A twin craft, Voyager 1, was launched 15 days after Voyager 2 and took a faster path to Jupiter and Saturn. The two craft transmit data daily back to Earth. Voyager 1 is now reaching the very edge of the Sun's influence, the heliosphere (see pp.188–89), and is about to enter interstellar space. Voyager 2, moving at a slower pace, is about 80 percent as far from the Sun.

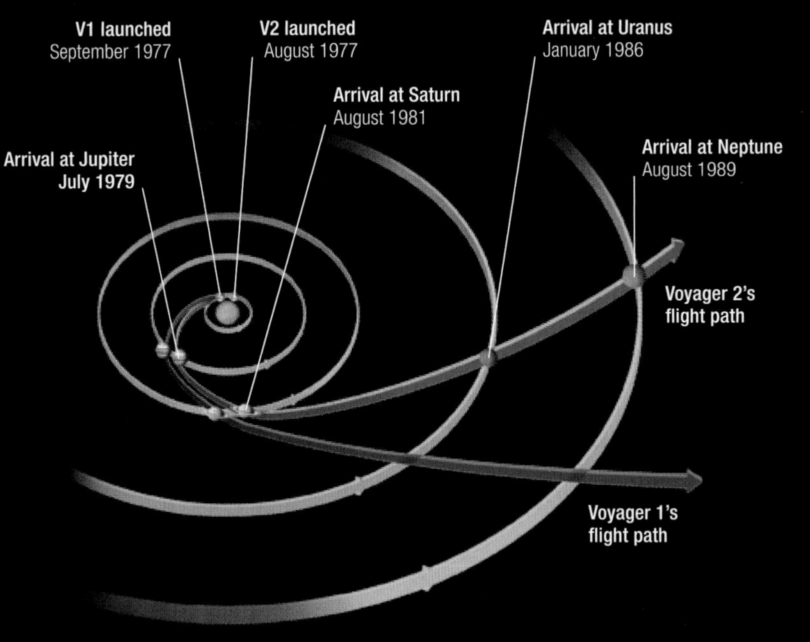

FLIGHT PATHS OF VOYAGER 1 AND 2
Conceived to go to just Jupiter and Saturn, Voyager 2's mission was extended to Uranus and Neptune. The planetary alignment gave it a gravity boost from one planet to the next.

VOYAGER 2 PROFILE

MISSION	
Launch date	August 20, 1977
Arrival at Uranus	January 24, 1986
Mission ends	2025
Launch vehicle	Titan III E-Centaur
VOYAGER 2	
Agency	NASA
Length	24ft 7in (7.5m)
Height	12ft 6in (3.8m)
Weight	1,797lb (815kg)
Power source	Radioisotope thermoelectric generator

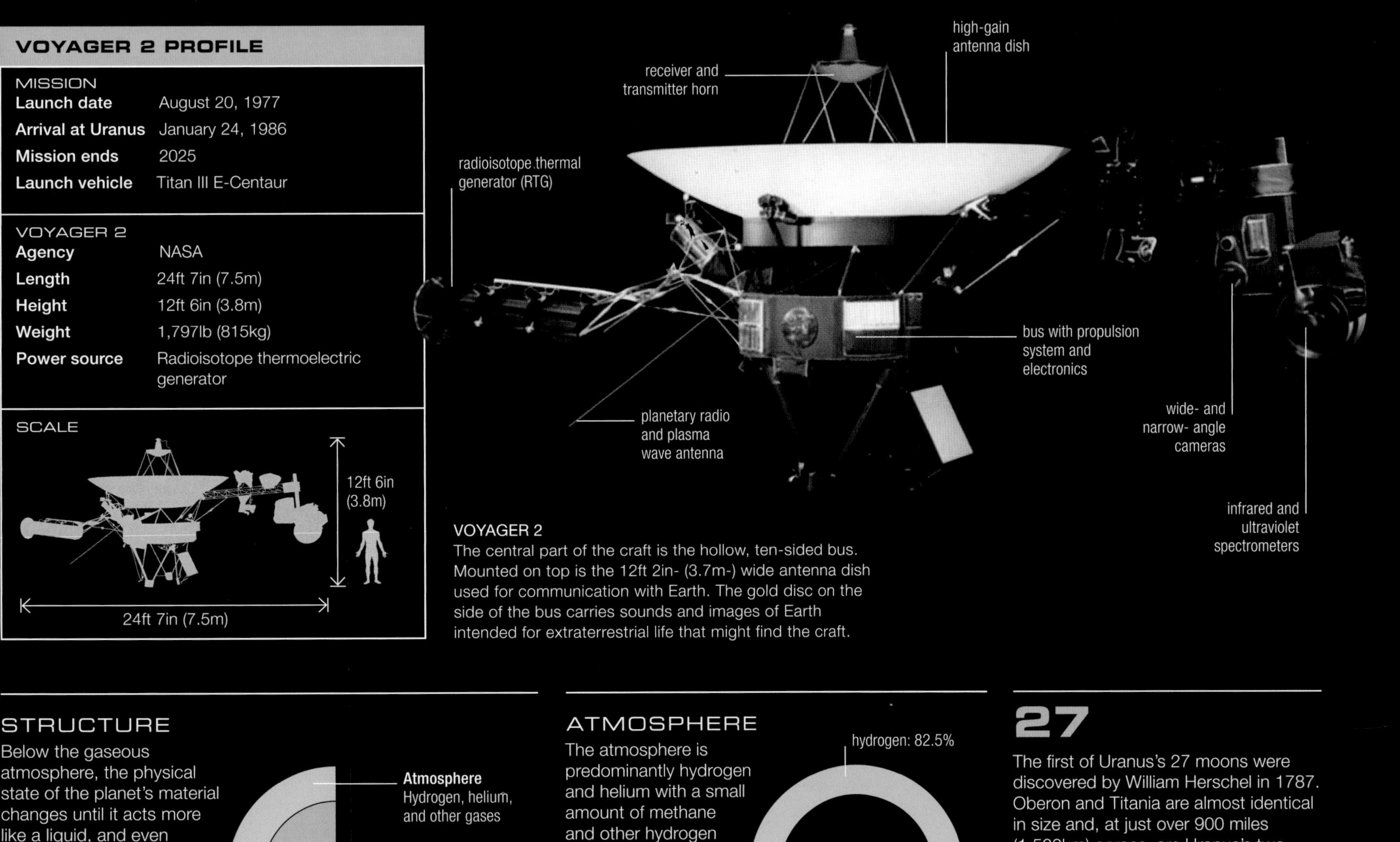

VOYAGER 2
The central part of the craft is the hollow, ten-sided bus. Mounted on top is the 12ft 2in- (3.7m-) wide antenna dish used for communication with Earth. The gold disc on the side of the bus carries sounds and images of Earth intended for extraterrestrial life that might find the craft.

STRUCTURE

Below the gaseous atmosphere, the physical state of the planet's material changes until it acts more like a liquid, and even deeper is dense core that is about half Earth's mass.

Atmosphere
Hydrogen, helium, and other gases

Core
Rock and possibly ice

Liquid inner layer
Water, methane, and ammonia

ATMOSPHERE

The atmosphere is predominantly hydrogen and helium with a small amount of methane and other hydrogen compounds, including ethane and acetylene.

hydrogen: 82.5%
Cloud-top temperature
-323°F (-197°C)
methane and trace gases: 2.3%
helium: 15.2%

27

The first of Uranus's 27 moons were discovered by William Herschel in 1787. Oberon and Titania are almost identical in size and, at just over 900 miles (1,500km) across, are Uranus's two largest moons. Twenty-two of the moons are small, and the most distant of these follow retrograde orbits, traveling in the opposite direction from the others.

URANUS'S RINGS AND MOONS

Twenty-seven moons orbit Uranus. Most are just tens of miles across, but a handful are much larger. Those closest to Uranus orbit within a system of 12 narrow and widely separated rings. The rings and some moons can be seen from Earth, but Voyager 2 and Hubble have given astronomers a much closer view.

THE RINGS

The first of Uranus's rings were identified in 1977, when astronomers were observing as Uranus occulted (passed in front of) a star. As the star neared the planet's disc, its light blinked on and off as individual rings blocked out the starlight. A second set of blinks on the planet's far side confirmed the presence of rings. Further rings were discovered by Voyager 2 and more recently by the Hubble Space Telescope. In 2003, Hubble photographed a pair of rings much further away from the planet than the rings that were then known. These two outer rings are made of dust that constantly spirals into Uranus. Two new moons, Mab and Cupid, were also discovered in images taken at about the same time. Mab shares an orbit with the outer ring, and its dust replenishes this ring.

Uranus's inner rings consist of dust and dark pieces of material that range from a few inches to a few yards across. The brightest is the Epsilon ring, which is straddled by two small moons. Cordelia, the innermost of all Uranus's moons, orbits near the ring's inner edge. Ophelia orbits outside the ring. Together, the two moons shepherd the ring particles, keeping them in place.

Our view of the rings changes as Uranus and Earth orbit the Sun. Every 42 years the rings are seen edge-on; our most recent chance to see them like this came in 2007. Astronomers were then able to look at the dimmest rings, which appear brighter when edge-on, as their material merges into a thin band.

RING DETAIL FROM VOYAGER 2
The inner set of rings contains lanes of fine dust particles. Also visible in this Voyager 2 image are short, bright streaks, which are star trails that have been caught in the camera's 96-second exposure.

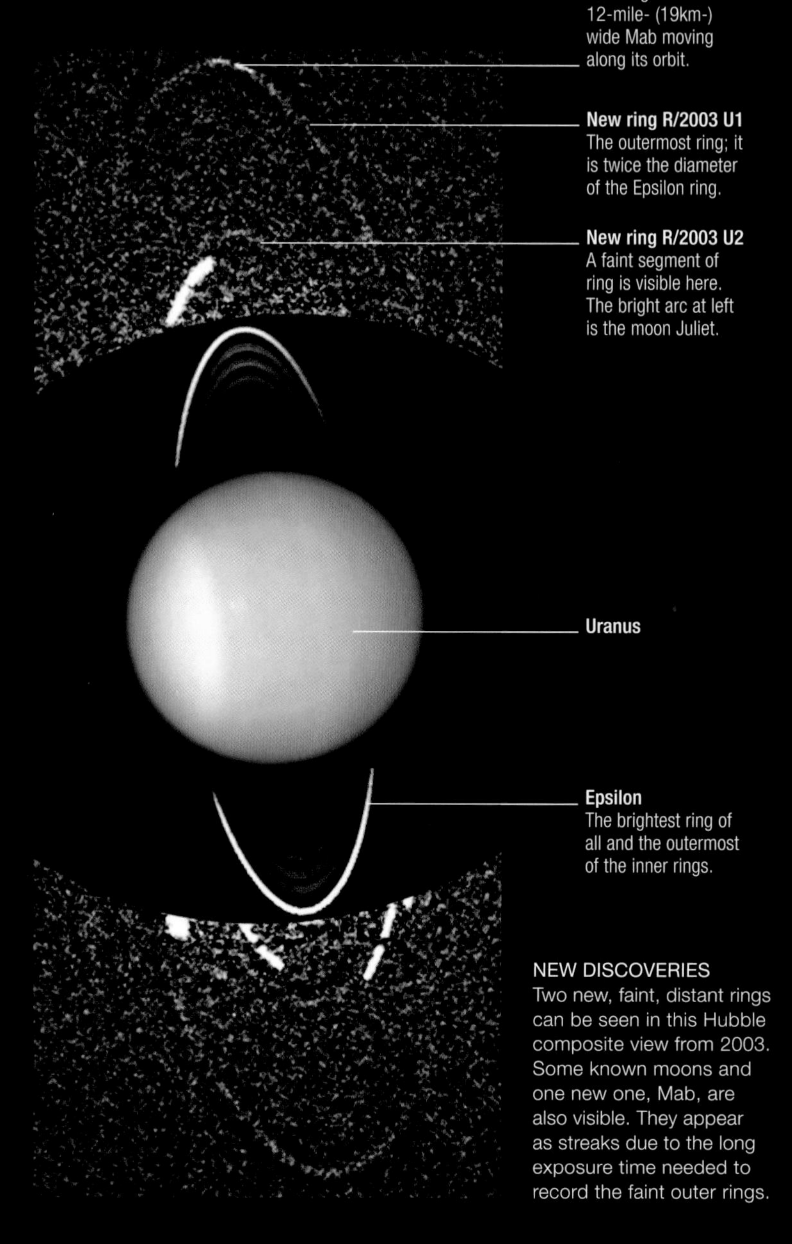

Mab
This bright arc is the 12-mile- (19km-) wide Mab moving along its orbit.

New ring R/2003 U1
The outermost ring; it is twice the diameter of the Epsilon ring.

New ring R/2003 U2
A faint segment of ring is visible here. The bright arc at left is the moon Juliet.

Uranus

Epsilon
The brightest ring of all and the outermost of the inner rings.

NEW DISCOVERIES
Two new, faint, distant rings can be seen in this Hubble composite view from 2003. Some known moons and one new one, Mab, are also visible. They appear as streaks due to the long exposure time needed to record the faint outer rings.

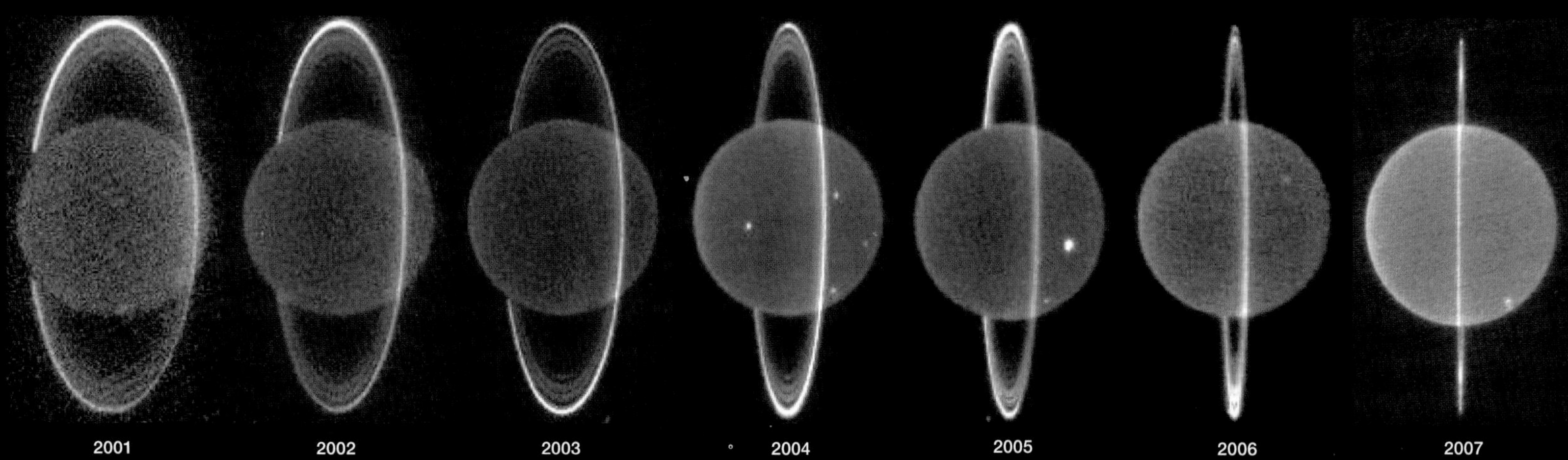

THE RINGS SEEN FROM EARTH
This series of infrared images from the Keck II Telescope shows how Uranus and its rings change appearance when viewed from Earth as the planet moves along its orbit. Uranus's south pole is at the left.

THE MOONS

Just five of Uranus's 27 moons are roughly spherical worlds. The remainder are small, irregular-shaped bodies. Most of these small bodies orbit closer to Uranus than the five major moons, and a further nine are much more distant. The innermost, Cordelia, orbits in just eight hours; the most distant, Ferdinand, takes 7.7 years. Eleven moons were discovered in Voyager 2 data, and two using the Hubble Space Telescope. Although the five majors—Miranda, Ariel, Umbriel, Titania, and Oberon—were all discovered using Earth-based telescopes, much of what we know of them came from the Voyager 2 mission. They are rocky bodies with icy surfaces marked by impact craters and surface fractures. All the moons are named after characters in the plays of the English dramatist William Shakespeare or the verse of the English poet Alexander Pope.

URANUS AND MOONS
This Earth-based, near-infrared view shows Uranus and its rings surrounded by seven moons. Small Puck and Portia are barely visible. Sunlight is absorbed by methane in Uranus's atmosphere, which is why the planet's disc appears unusually dark.

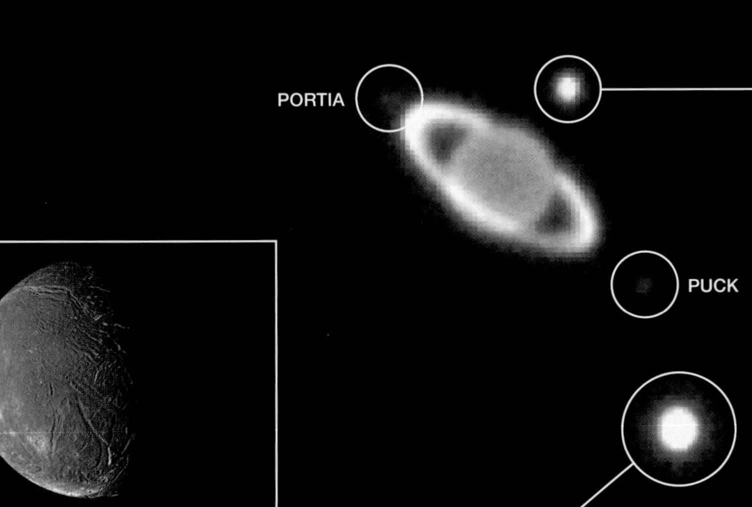

TITANIA
Uranus's largest moon, Titania is a little less than half the size of Earth's moon. Its icy surface is marked by cracks and impact craters. The Messina canyon system cuts across the moon at right.

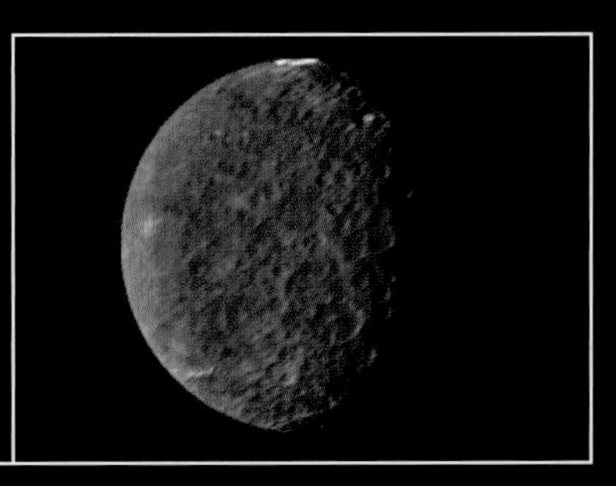

UMBRIEL
Dark Umbriel is covered in craters. Its one bright feature, named Wunda, is visible at the top in this image. Wunda lies near the moon's equator. Umbriel takes just over four days to make one orbit of Uranus.

MIRANDA
This is an odd-looking world. Different types of terrain dating from different periods butt up against one another on its surface. The bright, chevron-shaped Inverness Corona can be seen at lower right.

ARIEL
The surface of Ariel is sliced by faults, which formed when the moon's crust expanded, and pock-marked by relatively small impact craters. The large crater, at lower left, is 30 miles (50km) wide and surrounded by bright ejecta.

TRINCULO
Estimated to be 11 miles (18km) across, Trinculo (ringed at right) is one of the smallest of Uranus's moons. Discovered using an Earth-based telescope in August 2001, Trinculo was tracked for several months before its moon status was confirmed.

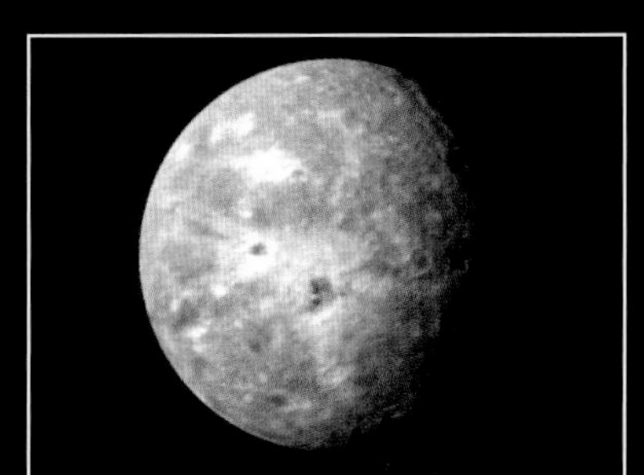

OBERON
The first Uranian moon discovered, Oberon is the most distant of the major moons. Its icy surface is pitted by impact craters. The dark floor of the Hamlet Crater is visible just below center.

NEPTUNE

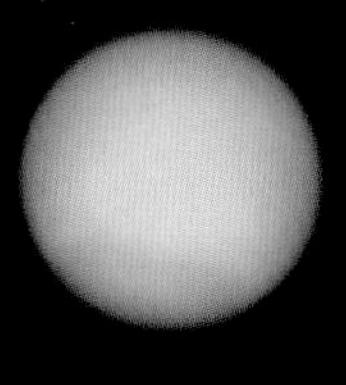

The outermost planet, Neptune is one of the coldest places in the Solar System. It is 30 times further from the Sun than Earth, and its cloud-top temperature is a freezing -330°F (-201°C). Just one spacecraft, Voyager 2, has journeyed out to Neptune and shown us the planet and its rings and moons.

ORBIT

Neptune has the longest planetary orbit around the Sun, taking 164.8 years to travel around it once. Because the planet was only discovered in 1846, it has so far only been observed during a single circuit. Neptune is tilted to its orbital plane by 28.3°, a little more than Earth's tilt. Like Earth's, Neptune's northern and southern hemispheres alternately point toward the Sun as it orbits. Even at its remote location, where the Sun appears 900 times dimmer in the sky than it does on Earth, Neptune undergoes seasonal changes. Its long orbit gives it long seasons, each lasting around 40 years.

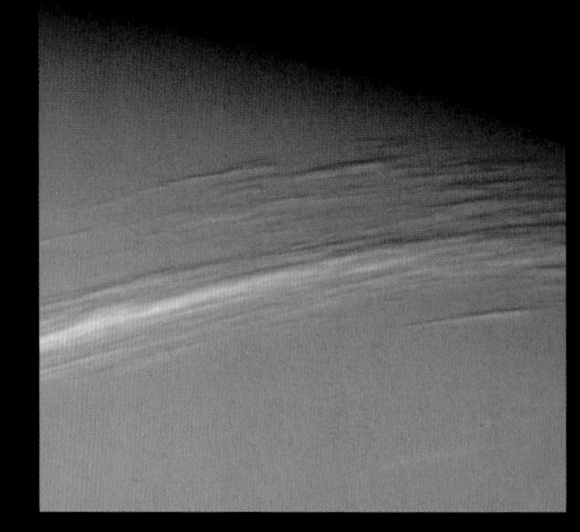

BRIGHT STREAK OF CLOUDS
Voyager 2 recorded these white clouds in Neptune's northern hemisphere. Made of methane ice, they form when gas rises and cools. Because the scene is sunlit from below, the clouds cast shadow.

STRUCTURE AND ATMOSPHERE

When we observe Neptune, we see its hydrogen-rich outer atmosphere. Below is a deep layer of liquid water, ammonia, and methane that surrounds a central rocky core. Neptune's bland appearance and the limited solar radiation that the planet receives suggest a benign world. Yet it has a dynamic atmosphere, with huge storms and some of the fastest winds in the Solar System. An unknown internal heat source, which makes the planet radiate more than twice the energy it receives from the Sun, seems to drive its weather.

Within the atmosphere, bands form parallel to the equator. The white clouds that stretch along lines of latitude form above its visible blue layer, and westward equatorial winds blow at 1,340mph (2,160kph). A huge storm seen by Voyager 2 in 1989 was named the Great Dark Spot. When the Hubble Space Telescope observed Neptune in 1994, the Spot had disappeared. The Great Dark Spot remains one of the most interesting features seen on Neptune to date.

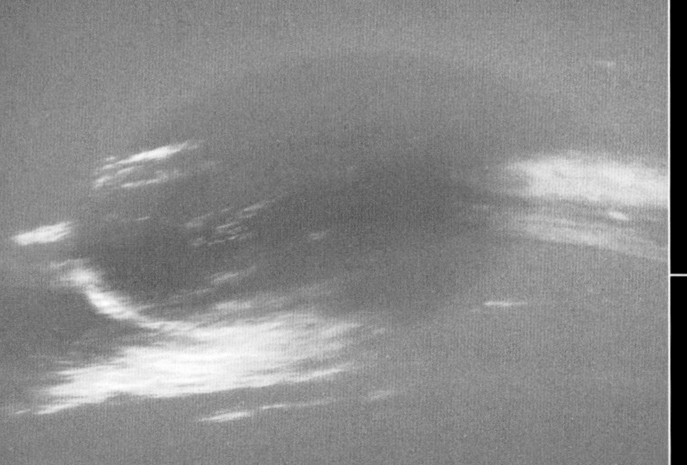

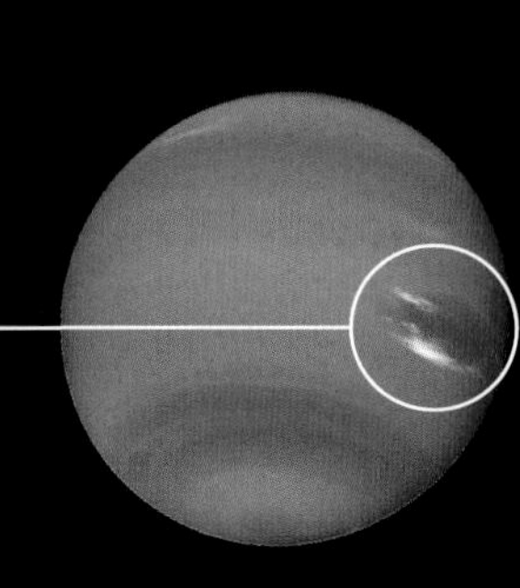

THE GREAT DARK SPOT
This huge anti-cyclonic storm seen by Voyager 2 was almost as big as Earth. At lower altitude than the blue methane layer, it changed shape as it was carried around Neptune by the planet's atmosphere. The white, higher-altitude clouds also regularly changed appearance.

ORBIT AND ROTATION

Neptune has the second least elliptical orbit of all the Solar System planets—only Venus's orbit is more circular. Consequently, there is little difference between its aphelion and perihelion distances.

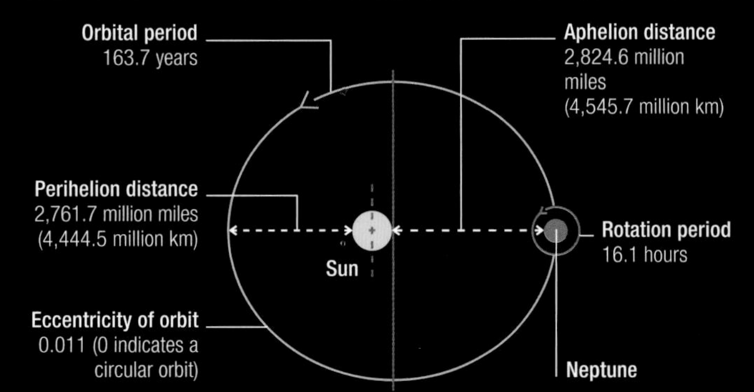

SIZE

Neptune is the smallest of the four giant planets. It is almost four times the size of Earth and just a little smaller than its inner neighbor, Uranus.

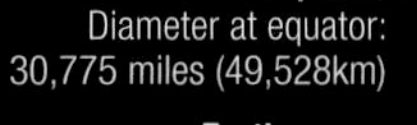

THE DISCOVERY OF NEPTUNE'S RINGS

Neptune is encircled by a system of thin and sparsely populated rings named after astronomers associated with the study of the planet. In order of distance from Neptune they are: Galle, Le Verrier, Lassell, Arago, and Adams. A sixth indistinct and unnamed ring lies within the Adams ring and shares its orbit with the moon Galatea. The rings are made of dust and small, dark particles of unknown composition.

The existence of Neptune's rings was anticipated once the first of Uranus's rings had been discovered in 1977—a discovery that confirmed that three of the four giant planets had ring systems.

Throughout the 1980s, astronomers observed Neptune as it passed in front of various stars. The sight of a star's light blinking on and off just before it was covered by Neptune's disc, or as it reappeared, could indicate the presence of a ring. But the observations produced inconsistent results and the situation only became clear when Voyager 2 visited Neptune. It discovered the five main rings and the 30-mile- (50km-) wide outer Adams ring, which contains five dense regions.

NEPTUNE'S RINGS FROM VOYAGER 2
The brightest rings in this image are Adams (left) and Le Verrier. Between them is the faint Lassell—the broadest ring of all. The Galle ring is closest to Neptune (right).

THE BLUE PLANET
This Hubble Space Telescope image shows Neptune, which is named after the God of the Sea in Roman mythology, in natural color. Methane gas within its upper atmosphere absorbs the red wavelengths in sunlight and gives the planet its blue coloring.

STRUCTURE

Neptune's gaseous atmosphere forms its outer layer. Beneath this, its material becomes increasingly liquid as pressure, density, and temperature increase with greater depth.

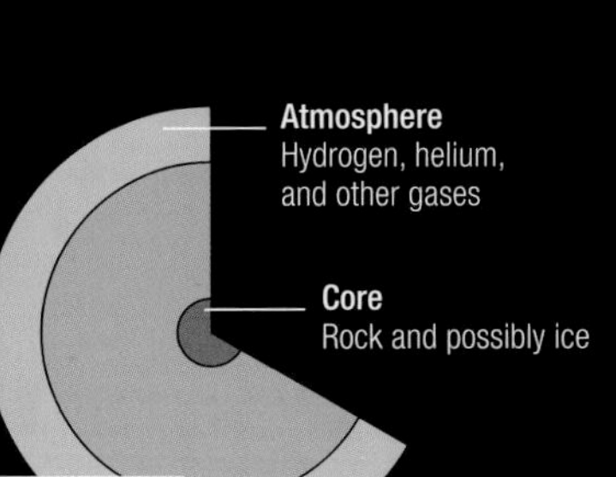

ATMOSPHERE

Neptune's atmosphere is mainly hydrogen and helium. It is 1.5 percent methane and contains small amounts of gases such as hydrogen deuteride and ethane.

hydrogen: 79.5%

Cloud-top temperature
-330°F (-201°C)

methane and trace gases: 2.0%

helium: 18.5%

164.8

Neptune completes one orbit around the Sun every 164.8 years. In July 2011, the planet will complete its first orbit since its discovery. As Neptune travels, its gravity can affect objects that orbit beyond it in the Kuiper Belt. As a result, these objects can become locked in orbital resonance with the planet. Neptune is in a 2:3 resonance with Pluto: Pluto makes 2 orbits of the Sun for every 3 orbits made by Neptune.

NEPTUNE'S MOONS

Thirteen moons orbit Neptune, at a great variety of distances from the planet. Most are small, irregularly shaped lumps, but one, Triton, is large and spherical and exhibits a range of surface features. The Voyager 2 spacecraft gave us our only view of these cold and distant worlds when it flew by in 1989.

FAMILY OF MOONS

The first of Neptune's moons to be discovered was Triton, the planet's largest moon, in October 1846. Over 100 years later, a second, Nereid, was found orbiting beyond Triton. These two remained the only moons known until six more, orbiting closer to Neptune, were discovered in Voyager 2 data in 1989. The four innermost—Naiad, Thalassa, Despina, and Galatea—all orbit within Neptune's ring system.

Since 2002, five moons more distant than Triton have been discovered through Earth-based observations. The most recent is Psamathe, which was detected in images taken in August 2003 with the Subaru telescope at Mauna Kea, Hawaii. Psamathe orbits Neptune in just less than 25 years. Also confirmed as a moon in 2003 is Neso, which is 30 million miles (48 million km) from Neptune and the most distant moon of any Solar System planet. It orbits Neptune in 25.7 years.

Except for Triton, Neptune's moons are small. Proteus is the second largest at 273 miles (440km) across, and Psamathe is one of three that is just 25 miles (40km) across. The moons are named after characters associated with Neptune, or Poseidon, in Greek and Roman mythology.

VOYAGER'S DISCOVERY

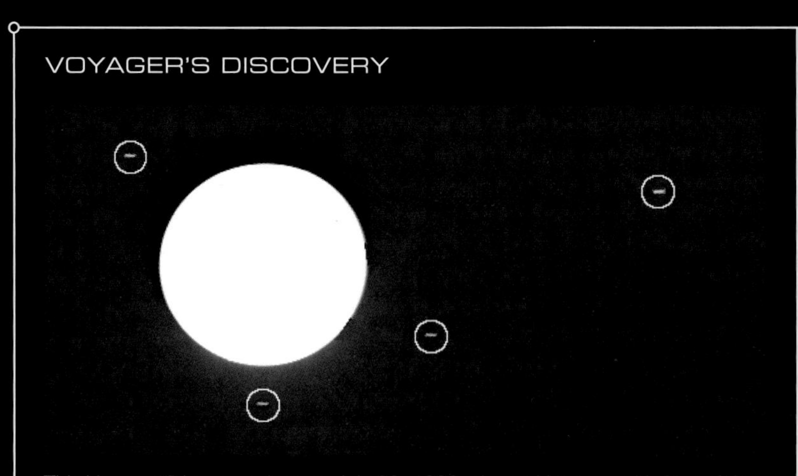

This Voyager 2 image, taken on July 30, 1989, shows Neptune with four of the moons the spacecraft discovered. Due to Voyager 2's movement during the long exposure time necessary to create this image, the moons appear as streaks. They were later named Proteus, Larissa, Despina, and Galatea.

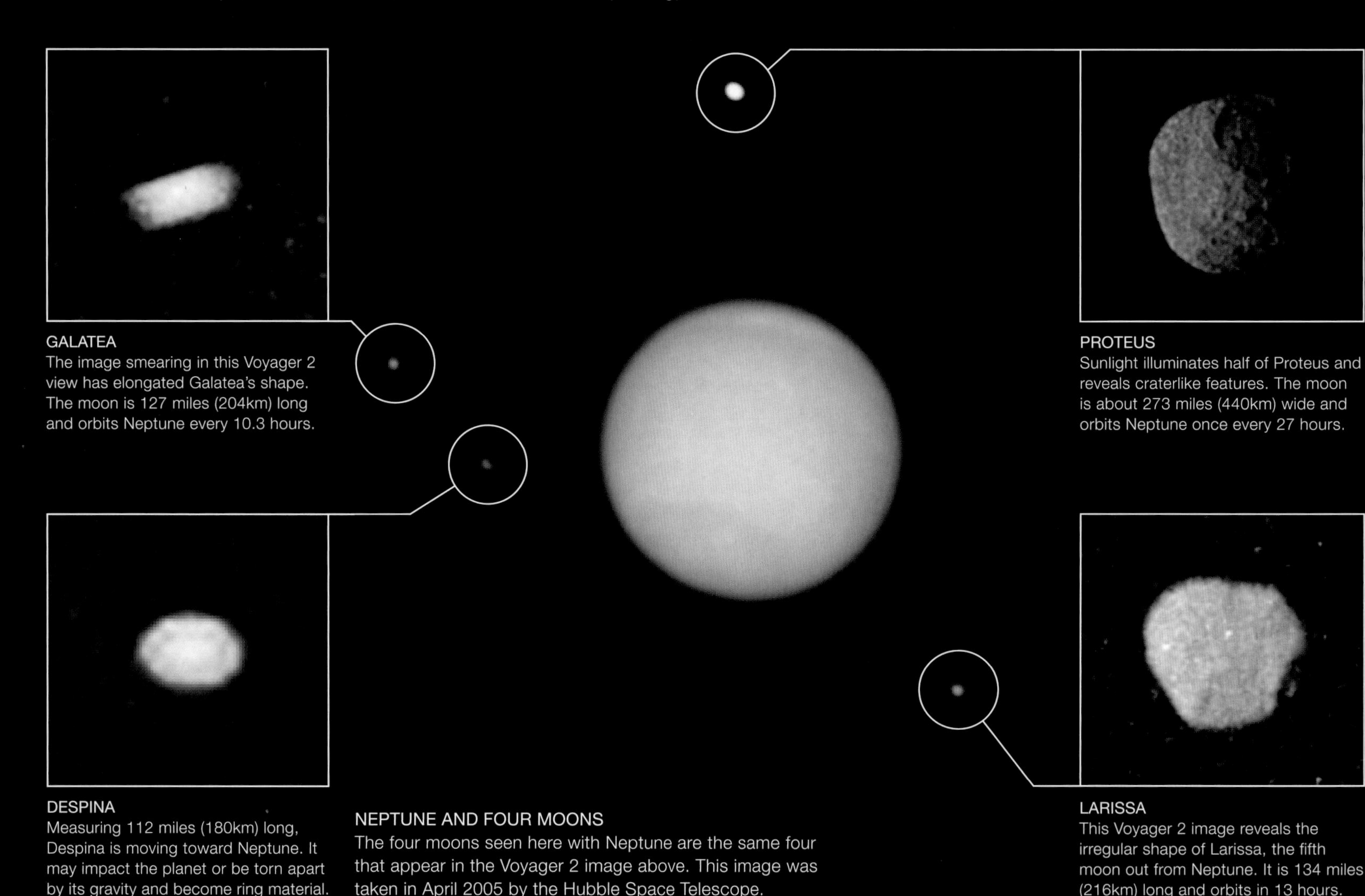

GALATEA
The image smearing in this Voyager 2 view has elongated Galatea's shape. The moon is 127 miles (204km) long and orbits Neptune every 10.3 hours.

PROTEUS
Sunlight illuminates half of Proteus and reveals craterlike features. The moon is about 273 miles (440km) wide and orbits Neptune once every 27 hours.

DESPINA
Measuring 112 miles (180km) long, Despina is moving toward Neptune. It may impact the planet or be torn apart by its gravity and become ring material.

NEPTUNE AND FOUR MOONS
The four moons seen here with Neptune are the same four that appear in the Voyager 2 image above. This image was taken in April 2005 by the Hubble Space Telescope.

LARISSA
This Voyager 2 image reveals the irregular shape of Larissa, the fifth moon out from Neptune. It is 134 miles (216km) long and orbits in 13 hours.

TRITON

Neptune's only major moon, Triton, is a ball of rock and ice, 1,682 miles (2,707km) across. It is about three-quarters the size of Earth's moon. In almost six days, it completes its circular path around Neptune and makes one rotation on its axis. This synchronous rotation, coupled with Triton's orbital inclination, means that the moon's polar regions take turns facing the Sun.

Triton is the only large moon in the Solar System to orbit in a clockwise direction, when viewed from above Neptune's north pole. This unusual feature suggests that Triton may not have always orbited Neptune but is in fact a Kuiper Belt object captured by the planet's gravity millions of years ago.

Triton is named after the sea-god son of Poseidon, and its surface features take aquatic names such as that of a mythical sea monster, water spirit, or river.

TRITON'S ICE SURFACE

Triton was the final world to be observed in detail by Voyager 2 on its tour of the outer planets. The craft recorded surface temperatures of -391°F (-235°C), making Triton one of the coldest places in the Solar System. Its surface and crust are made mainly of nitrogen ice, which overlies an icy mantle surrounding a core of rock and metal.

Voyager 2 imaged about 40 percent of Triton's surface. It revealed a young, sparsely cratered, relatively flat world with terrain sculpted by cryovolcanism, a volcanic process in which subsurface ice, heated by the Sun, erupts onto the surface. It found smooth plains of volcanic ice, mounds and pits formed by icy lava, and geyserlike plumes rising up to 5 miles (8km) high. The plumes form when warmed subsurface nitrogen ice turns to gas and erupts through surface cracks.

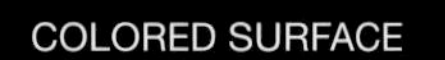

COLORED SURFACE
Triton's south polar cap appears pink; dark streaks are dust deposits from plumes. The bluish-green band, which extends right around Triton, is nitrogen frost or snow.

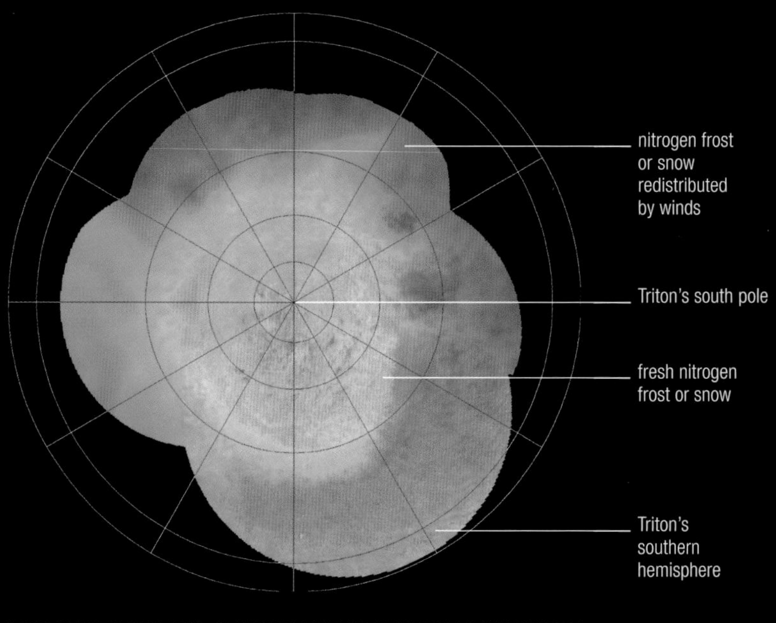

NITROGEN FROST AND SNOW IN THE SOUTHERN HEMISPHERE
Triton's south pole is covered by a cap of frozen nitrogen and methane. Extending out from the cap is a bright fringe of fresh nitrogen frost or snow. Some of this material has been blown by winds moving north, forming raylike features hundreds of miles long.

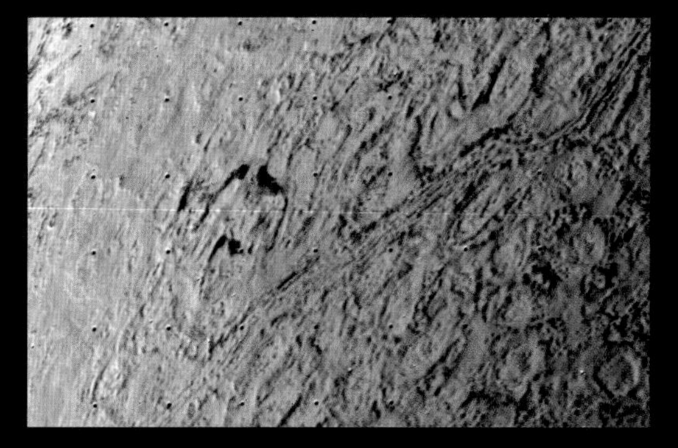

CANTALOUPE REGION
This region of Triton is nicknamed cantaloupe due to its resemblance to the skin of a cantaloupe melon. It has grooves, ridges, and smooth, circular depressions about 20 miles (35km) across.

ICY DIAPIRS
The vertical relief in this 100-mile- (160km-) wide scene of the cantaloupe terrain is exaggerated 25 times to show the diapirs better. These rising blobs of ice are hundreds of feet high and the same across.

NEPTUNE AND DESPINA

This composite of several Voyager 2 images shows the shadow of Despina, one of Neptune's small inner satellites, moving across the face of the planet. The moon, which is thought to have formed from the remains of several of Neptune's original satellites, is slowly spiraling inward. One day it may crash into the planet, or it may break up and form a planetary ring.

THE OUTER REGION

More than a trillion small, icy objects exist beyond Neptune in the dark and cold outer region of the Solar System. Kuiper Belt objects and dwarf planets populate the region extending out from Neptune's orbit. Even more distant are the huge number of comets that make up the vast, spherical Oort Cloud.

THE KUIPER BELT

A flattened belt of objects known as the Kuiper Belt begins about 3.5 billion miles (6 billion km) from the Sun. The outer edge of this broad belt is about the same distance again from the Sun. The first Kuiper Belt object, named 1992 QB1, was discovered in 1992. More than 1,000 are now known and many tens of thousands more are thought to exist. Most of these objects are a mix of ice and rock, and they are non-spherical in shape. They typically measure less than 600 miles (1,000km) across and take more than 250 years to orbit the Sun.

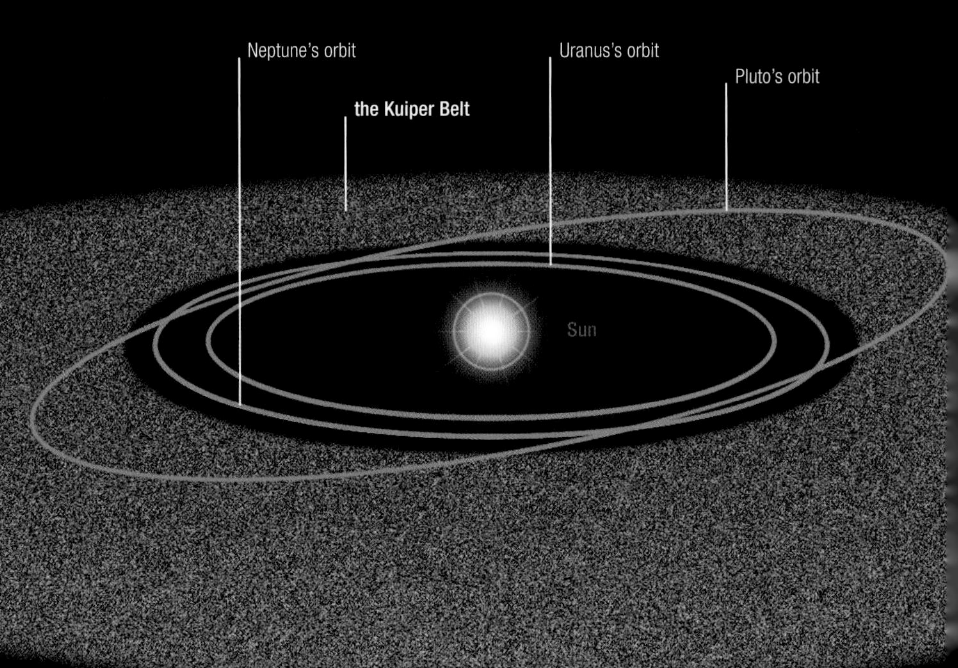

LOCATION OF THE KUIPER BELT
The Kuiper Belt lies just beyond Neptune's orbit. It consists of Kuiper Belt objects, including a small number of dwarf planets such as Eris, Pluto, Makemake, and Haumea. The orbits in this diagram are not drawn to scale.

THE OORT CLOUD

The disc-shaped planetary part of the Solar System is surrounded by the Oort Cloud. The outer edge of this vast sphere of comets is about 1.6 light-years away from Earth—nearly halfway to the nearest stars. The comets are the remains of material unused when the planets were formed at the dawn of the Solar System. The comets' combined mass amounts to a few times the mass of Earth.

The paths followed by the comets around the Sun differ from those of the planets. Their orbits are elongated and not in the planetary plane but at all angles. A passing star can disturb the comets and put them onto orbits that take them beyond the cloud or into the inner Solar System. Those comets that travel close to the Sun develop a huge head and tails.

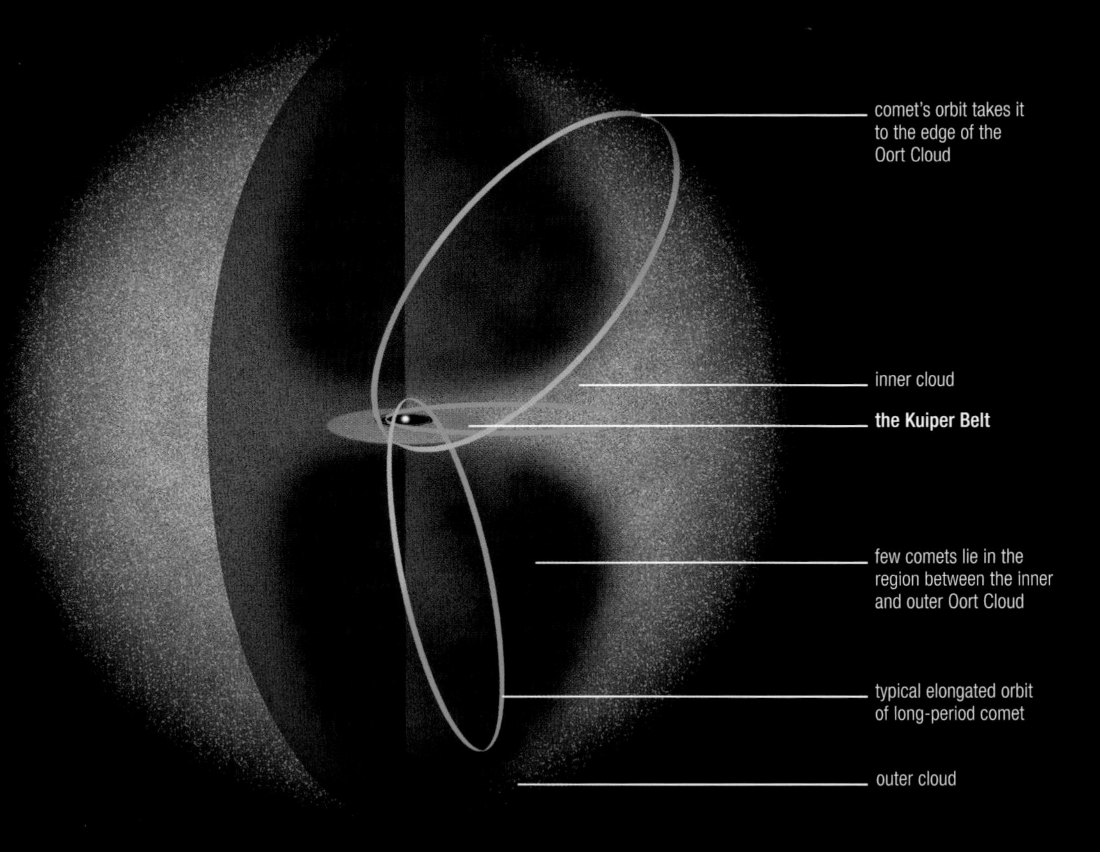

OORT CLOUD COMETS
The spherical Oort Cloud encircles the planetary part of the Solar System. It consists of more than 1 trillion comets following individual orbits around the central Sun. Comets with short orbits form the inner cloud; those with long orbits make up the outer cloud.

DWARF PLANETS

The largest objects in the Kuiper Belt are the dwarf planets. These are almost-round bodies that orbit the Sun in the neighborhood of other orbiting objects. The biggest of all is Eris, an ice-and-rock world discovered in 2005 whose solar orbit takes 550.9 years. Slightly smaller than Eris is Pluto, the first object discovered beyond Neptune. Found in 1930, it was considered the Solar System's smallest and most distant planet until 2006. Its change of status to dwarf planet followed the introduction of this new class of object in August of that year.

COMETS

Made of dust and ice, comets are small bodies that orbit the Sun on elongated circles. When they reach the inner Solar System and approach the Sun, they heat up and produce glowing comas and tails of gas and dust. Bright, naked-eye comets are seen in Earth's sky about once every decade.

COMET ANATOMY

The heart of a comet is an irregularly shaped, spinning nucleus of dirty, dusty snow, typically ranging from hundreds of feet to tens of miles across. The "dirt" is particles of silicate rock, like that which forms the crusts and mantles of the terrestrial planets. The "snow" is mainly water ice, but about 1 in 20 of its molecules are of substances such as carbon dioxide, carbon monoxide, methane, and ammonia. These cometary snowballs formed in the outer protoplanetary disc at the dawn of the Solar System. Many clumped together to form the cores of the giant planets. The remainder were flung onto a range of orbits by these new planets. Those we see today are among the survivors, still hurtling through the Solar System after 4.6 billion years.

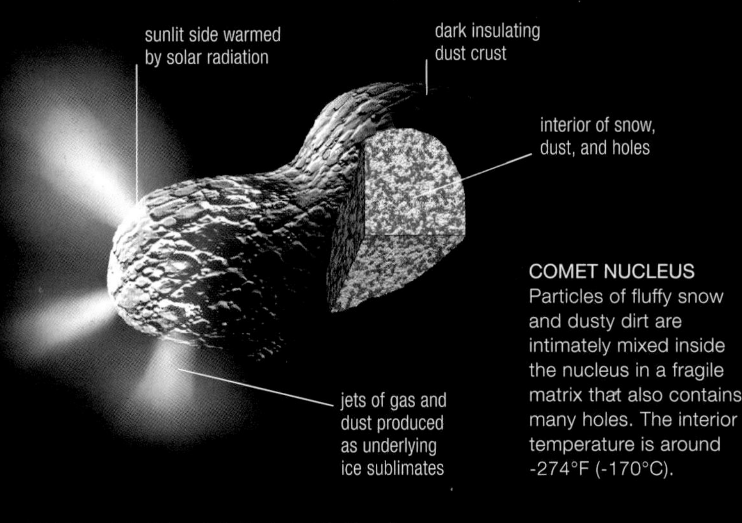

COMET NUCLEUS
Particles of fluffy snow and dusty dirt are intimately mixed inside the nucleus in a fragile matrix that also contains many holes. The interior temperature is around -274°F (-170°C).

COMAS AND TAILS

Beyond Jupiter, a comet is cold and dormant. As it approaches the Sun, solar radiation transmitted through the layer of dust around the nucleus causes the underlying ice to sublimate—that is, change from solid to gas. The escaping gas then exerts pressure on the loose surface dust, pushing it away to form a spherical cloud around the nucleus called a coma. Pressure exerted by the solar wind increases the force on both the gas and the small dust particles, creating two tails that extend away from the comet in the opposite direction from the Sun. The closer the comet gets to the Sun, the longer the tails become. When larger dust particles are released, they either slowly gain on the nucleus, or fall behind it, eventually forming a dust ring, or annulus, around its orbital path. When Earth's own orbit intersects this ring, the dust impacts with the upper atmosphere and produces shooting stars.

HALE–BOPP
With a nucleus 37 miles (60km) in diameter, this huge comet was visible to the naked eye for 18 months. It was at its most spectacular when passing the Sun in April 1997. Hale–Bopp's previous visit was 4,200 years ago, but Jupiter's influence means it will return in another 2,530 years.

COMETARY ORBITS

To date, astronomers have recorded about 200 comets with orbital periods of less than 20 years and a further 2,000 with longer orbital periods. The orbital planes of longer-period comets are randomly orientated to the Earth's orbit, whereas short-period comets usually orbit in the same plane as the planets. The Solar System is thought to contain at least a trillion comets, most of which do not travel close to the Sun and remain undetected. The far ends of their orbits can stretch halfway to nearby stars. These comets form a huge spherical reservoir known as the Oort Cloud (see pp.178–79).

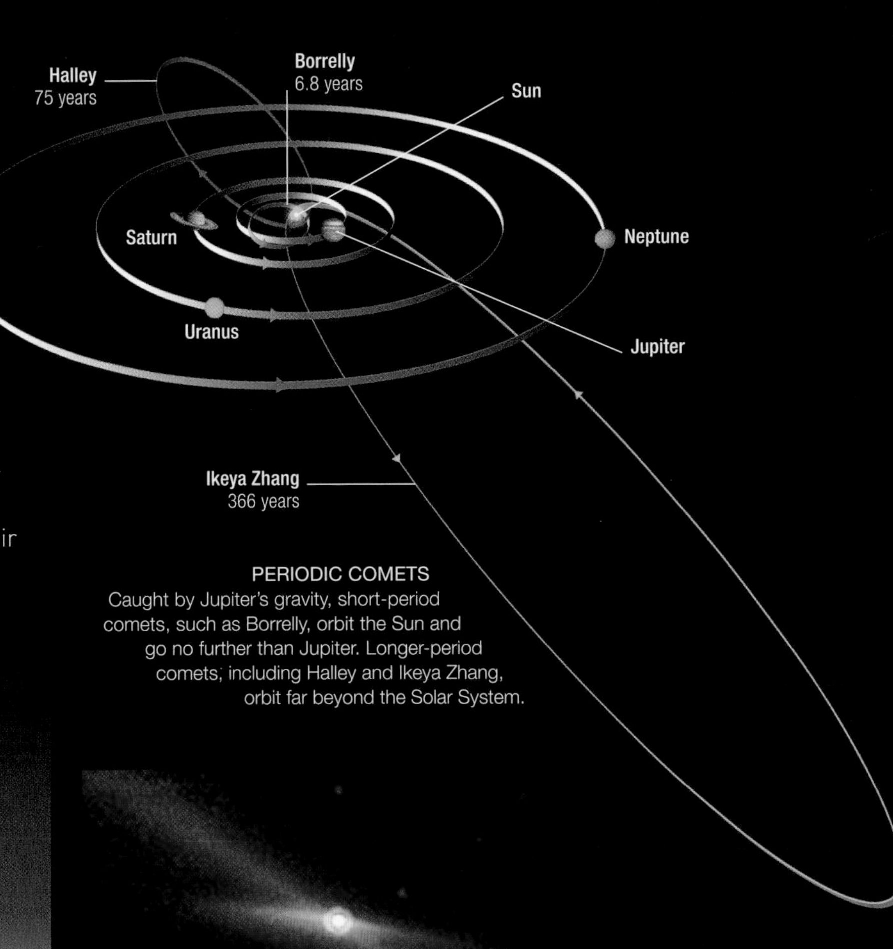

PERIODIC COMETS
Caught by Jupiter's gravity, short-period comets, such as Borrelly, orbit the Sun and go no further than Jupiter. Longer-period comets, including Halley and Ikeya Zhang, orbit far beyond the Solar System.

MCNAUGHT
In 2006, this great comet, with its stunning dust tail, was the brightest seen in the southern hemisphere for 40 years. The shape of its orbit means that it is heading off to the galactic disc and will never return to our skies.

ENCKE
Short-period Comet Encke rarely develops a visible tail, but this infrared Spitzer Space Telescope view, reveals the particles stripped from the nucleus by close encounters with the Sun. These form the Taurid meteoroid stream.

SHOEMAKER–LEVY 9 SHATTERED INTO 21 FRAGMENTS

NUCLEUS DECAY

The fact that the inner Solar System comets produce comas and tails means that their nuclei must be losing mass—they are decaying. The spacecraft GIOTTO found that the nucleus of Halley's Comet has an average radius of 3.4 miles (5.4km). Each time Halley passes the Sun on its 75-year orbit, the comet loses a layer 6ft 6in (2m) deep from all over its surface. At this rate, Comet Halley will cease to exist in 200,000 years' time. The debris from this cometary decay replenishes a large dust cloud surrounding the Sun.

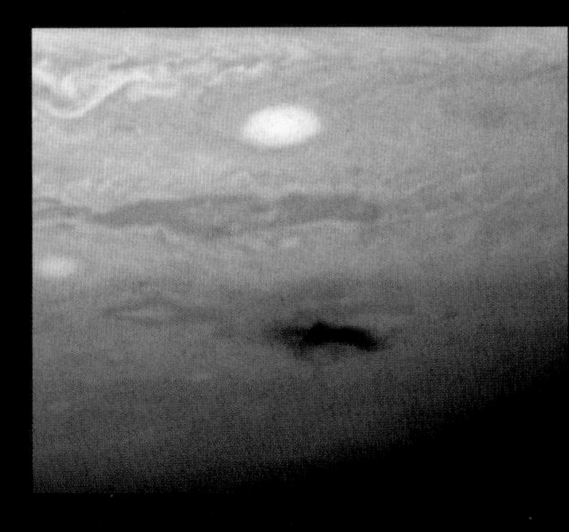

SHOEMAKER–LEVY 9
This comet was discovered orbiting Jupiter. In 1992, when it passed within 25,000 miles (40,000km) of the planet's cloud tops, the comet broke up. The pieces hit Jupiter in July 1994, staining the clouds with dust for many months.

RECENT IMPACT
The brown spot (below center) in this July 2009 Hubble image shows the impact of a comet or asteroid with Jupiter. The pieces of Shoemaker–Levy 9 created similar spots in 1994.

A COMET AND ITS GAS TAIL
Comet C/2006 M4 (SWAN) is seen here in 2006 as it made its closest approach to the Sun on its journey through the Solar System. Like most comets, it developed a tail of ionized gas as it entered the inner Solar System. The tail streams away in the direction of the solar wind. In Earth's night sky, the tail appeared about twice the width of the full Moon.

MISSIONS TO COMETS

Before 1986, astronomers had no idea what a cometary nucleus looked like. Then, in February of that year, the European Space Agency sent the GIOTTO spacecraft to take pictures of Halley's Comet as it neared the Sun. Since then, further, more complex missions to other cometary nuclei have been undertaken.

VISITING HALLEY

Cometary nuclei are so small that they cannot be picked out by Earth-based telescopes. To be seen, they must be visited by spacecraft. The European Space Agency's GIOTTO was the first spacecraft to visit a cometary nucleus. It flew across the Solar System and passed within 375 miles (600km) of the sunward side of the nucleus of Halley's Comet on March 13, 1986, five weeks after the comet had reached its closest point to the Sun. Halley was orbiting the Sun in the opposite direction from both Earth and the spacecraft, and so Halley and GIOTTO headed toward each other at a dangerous combined speed of 150,000mph (245,000kph). GIOTTO took measurements from the dust, gas, and plasma around the nucleus. It only just survived the dust battering, but went on to visit Comet Grigg-Skjellerup, in July 1992.

FIRST VIEWS

GIOTTO provided the first ever pictures of a comet's nucleus. The mapping was very limited as only half of Halley's nucleus was sunlit, and only half of this area was pointing toward the spacecraft. However, GIOTTO did record surface details, showing that the nucleus was generally smooth with only the occasional hill and valley. Also visible were bright jets of gas and dust. Only a small area of the nucleus was actively emitting gas and dust, and at the time of flyby, the comet was losing 6,600lbs (3 tonnes) of material per second. The dark, potato-shaped body, 9.5 miles (15.3km) in length, was spinning roughly once every three days.

MISSIONS TO COMETS

COMET	SPACECRAFT	ARRIVAL DATE	FLYBY DISTANCE
Halley	GIOTTO	March 1986	375 miles (600km)
Borrelly	Deep Space 1	September 2001	1,365 miles (2,200km)
Wild 2	Stardust	January 2004	150 miles (240km)
Tempel 1	Deep Impact	July 2005	Impact
67P/C–G	Rosetta	August 2014	Orbit and land

COMET IMPACT

In January 2004, NASA's Stardust mission flew through the inner coma of Comet Wild 2, capturing dust in a material called aerogel. These particles were then parachuted to Earth for analysis. The samples confirmed that the comet is an ancient body, but suprised scientists by proving to be made of a mixture of materials. As well as the expected abundance of ice, formed in regions beyond Neptune's orbit, the samples contained rocky material, formed at an extremely high temperature, closer to the Sun. This sheds light not only on the formation and nature of comets, but also on the origins of planets.

Just over a year later, NASA's Deep Impact mission flew by Comet Tempel 1 and imaged its surface. It then fired a self-guided impactor into the comet's surface, expecting a crater to form, which would then be imaged by the spacecraft. The images were meant to show the sub-surface regions of the nucleus, but this was unsuccessful (below).

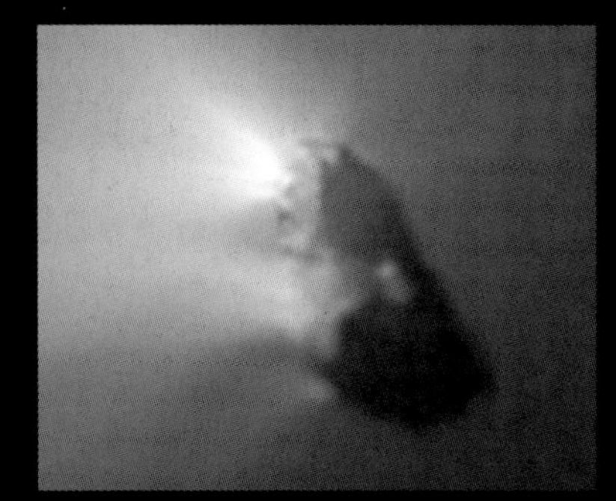

Halley's nucleus
9.5 miles (15.3km) long

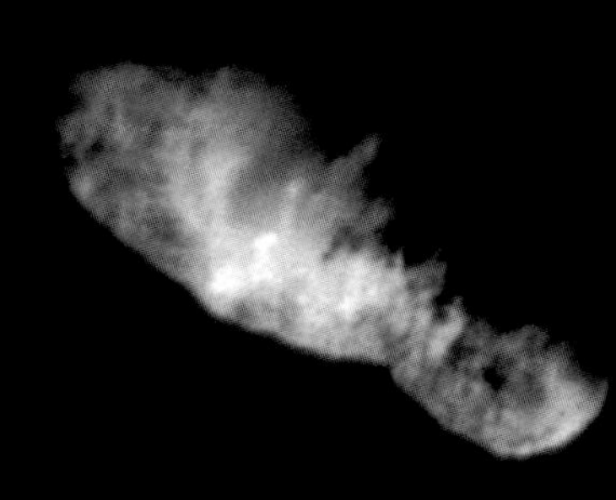

Borelly's nucleus
5.0 miles (8.0km) long

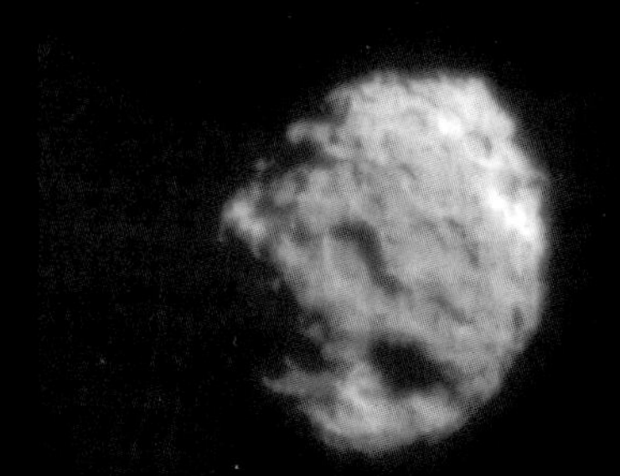

Wild 2's nucleus
3.4 miles (5.5km) long

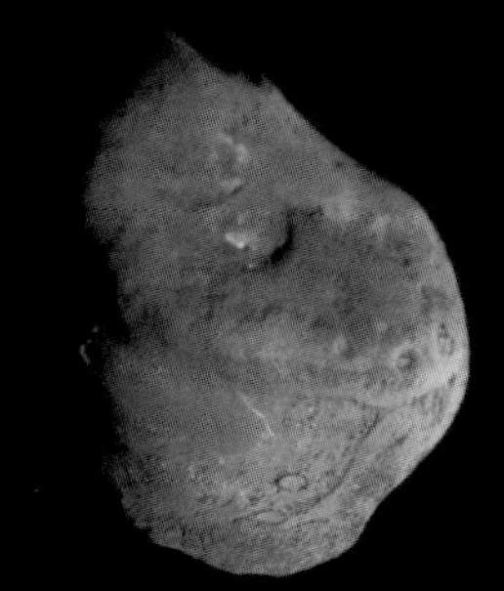

Tempel 1's nucleus
4.7 miles (7.6km) long

TEMPEL 1 AFTER IMPACT
This image shows Tempel 1 shortly after its collision with Deep Impact. Unfortunately, the damage inflicted by the impactor pulverized the fragile dust and snow, releasing a huge, obscuring cloud of talcum-powderlike particles. The retreating spacecraft could see nothing through this cloud.

LANDING ON A COMET

So far we have only taken snapshots of four cometary nuclei during brief flybys, but ESA's Rosetta mission to Comet 67P/Churyumov–Gerasimenko (67P/C–G) will place a spacecraft in orbit around a cometary nucleus for the first time. Rosetta will rendezvous with the comet in August 2014 at a distance of 370 million miles (600 million km) from the Sun. It will then remain as an orbiting satellite of the nucleus while the comet travels along its own orbit. All regions of the nucleus's surface will be monitored in detail, hour after hour, year after year.

In addition, a lander called Philae will land on the nucleus's fragile surface. It will carry ten instruments, including radar to probe the interior and chemical analyzers to investigate the composition.

ROSETTA AND PHILAE
When the spacecraft is near Comet 67P/C–G, the instruments on top of the main body will directly face the comet, and the antenna and the enormous solar panels, which can tilt 180°, will point toward the Sun and Earth.

67P/CHURYUMOV–GERASIMENKO
67P/C–G orbits the Sun every 6.6 years and is one of many comets whose orbit is influenced by Jupiter. The nucleus is about 2.5 miles (4km) across.

ROSETTA PROFILE

MISSION	
Launch date	March 2, 2004
Arrival at 67P/C–G	August 2014
Mission ends	December 2015
Launch vehicle	Ariane-5G
ROSETTA	
Agency	ESA
Length	105ft (32m)
Height	9ft 2in (2.8m)
Weight	6,600lb (3,000kg)
Power source	Solar panels
PHILAE	
Agency	ESA
Size	3ft x 32in (1m x 80cm)
Weight	220lb (100kg)
Power source	Solar panels

SCALE

ROSETTA'S PATH
It will take Rosetta 12 years to rendezvous with the comet, during which time it will receive three gravitational assists from Earth and one from Mars.

SAMPLE COLLECTION

Astronomers usually investigate Solar System bodies from afar, using telescopes and robotic craft. But they can also study samples of a small number of bodies that either fall to Earth or are returned by spacecraft.

METEORITES

Space rocks that land on Earth's surface are called meteorites. Most originate as asteroids, but some are pieces of the Moon or Mars that were blasted into space billions of years ago. Annually, around 3,000 meteorites each weighing over 2lb (1kg) land on Earth. Most, classed as stony meteorites (or stones), have a composition similar to Earth's rocky mantle. Less common are iron meteorites, with a make-up similar to Earth's core. Stony-iron meteorites are the rarest and originate in asteroids not big enough to melt and divide into mantles and cores.

LUNAR METEORITE
Found in Antarctica in 1982, this sample is called Allan Hills A81005. Lunar meteorites are similar to rocks collected on the Moon.

MARTIAN METEORITE
Allan Hills 84001, a 1984 Antarctic find, is from Mars. Martian meteorites contain gas like that of the Martian atmosphere.

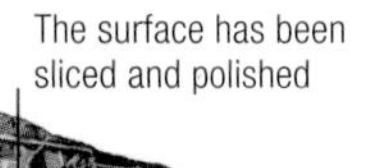
The surface has been sliced and polished

IRON GIBEON
This sample is one of many meteorites found near the town of Gibeon, Namibia, since the 1830s.

STONY BARWELL
In 1965, many stones were collected after meteorites were seen to fall over the village of Barwell, England.

STONY-IRON ESQUEL
Golden-colored crystals of olivine are embedded in this meteorite found in Esquel, Argentina, in 1951.

FALLS AND FINDS

Meteorites are divided into falls and finds, according to how they are collected. With a fall, the fireball produced by the incoming rock is seen, the landing location is calculated, and the meteorite is retrieved. Finds are meteorites discovered either by chance, or as part of an organized search. Cameras scan Earth's sky for meteorite fireballs, and scientists make ground searches in regions such as Antarctica where rocks are easily seen in the icy landscape. To date, across Earth, around 1,150 falls and at least 32,000 finds have been collected; more than 30 came from Mars and over 60 from the Moon.

METEORITES IN THE DESERT
Searches for meteorites are often made in desert regions, where rocks stand out amid the vast expanse of sand. Here, a 2-ton meteorite is examined in the Empty Quarter, Saudi Arabia.

MOON ROCK

Samples were returned to Earth from the Moon's surface by the Apollo astronauts and by Soviet robotic spacecraft. Beween 1969 and 1972, six Apollo missions brought back around 2,200 samples from six different sites, weighing 842lb (383kg) in total. The material consisted of rocks, pebbles, sand, and dust, plus some 54ft (16.5m) of core samples from below the lunar surface. The rocks included breccias produced when asteroids crashed into the Moon, melting and welding the surface rock and soil together, as well as younger basalts—volcanic rocks formed from lava that had seeped through the Moon's crust. In addition to the Apollo samples, the robotic craft Luna 16, 20, and 24 brought back 10.5oz (300g) from three other sites between 1970 and 1976. By studying all of these samples, scientists have learned much about the Moon's formation and early history.

BRECCIA SAMPLE
This 10in- (25cm-) long lump of breccia was returned to Earth by the Apollo 17 mission. It was collected from Taurus-Littrow, which the astronauts explored in the lunar rover.

COLLECTING LUNAR SAMPLES
Apollo astronauts had tongs to pick up rocks, hammers to chip off pieces, scoops to collect soil, and core tubes to take subsurface samples. Here, Harrison Schmitt uses a lunar rake at the Apollo 17 Taurus-Littrow landing site.

CAPTURING COMET DUST

The first mission to take samples of comet particles was named Stardust. The robotic spacecraft encountered Comet Wild-2 on January 2, 2004. As the craft made its 13,700mph (22,000kph) flyby, it collected dust from the comet's inner coma. The dust was captured in a sample collector standing proud of the craft. When dust particles hit the collector, they were slowed and caught by a low-density, porous silicon material called an aerogel. The aim was to collect at least 500 particles, but many more than this were obtained. Once the collector had been stowed away in the Sample Return Capsule, the craft headed for home. Stardust released the capsule to land safely back on Earth on January 15, 2006.

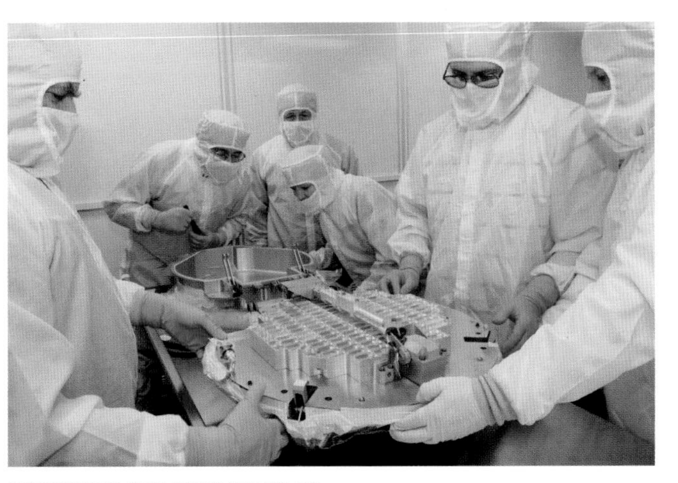

RETRIEVING THE DUST SAMPLES
Stardust's Sample Return Capsule was opened at NASA's Johnson Space Center in Houston, Texas. Inside was the sample collector, which is the size of a tennis racket and like an ice-cube tray in appearance. The collector's compartments contained the aerogel holding the comet dust.

SOLAR WIND PARTICLES

Particles of the solar-wind were collected in 2004 by Genesis, the first NASA sample-return mission since the Apollo Moon landings. Genesis completed a three-year mission that saw it travel toward the Sun, collect samples of its wind, and return these to Earth. The wind particles were caught in 250 hexagonal wafers, each 4in (10cm) wide and made from material such as silicon, diamond, and sapphire. They were stored on five trays, in a capsule that was released by the main craft to land on Earth. A parachute should have slowed the capsule's descent, and then in a daring maneuver, a long pole coming from underneath a helicopter was to hook the capsule and carry it to the ground. Unfortunately, the parachute did not deploy, so the capsule crash-landed.

Return to Earth
The capsule containing the solar-wind particles crashed into the desert in Utah, on September 8, 2004. In the picture above it seems to be buried in the sand, but the capsule has actually broken apart. The 250 wafers it contained were smashed into 15,000 fragments, but once cleaned, the fragments still yielded some solar-wind particles.

LOOKING BACK

After investigating the four outer planets, the Voyager 1 and Voyager 2 spacecraft continued traveling away from the Sun. Before heading off for interstellar space, Voyager 1 took one final look back.

BACKWARD GLANCE

Voyager 1 and Voyager 2 left Earth in 1977. Since then they have been getting increasingly distant from Earth. On February 14, 1990, when 4 billion miles (6.5 billion km) away, Voyager 1 directed its cameras to the planetary part of the Solar System it had left behind. The 60 images it recorded gave us our first view of the Sun and planets from the outer part of the Solar System (opposite). At this distance, the Sun looks about 40 times smaller than it does from Earth and the planets are only just detectable.

Voyager 1 is moving away from the Sun at 38,000 mph (61,200kph). Along with Voyager 2, it is traveling through the heliosphere, the part of space influenced by the Sun, and fast approaching interstellar space. The heliosphere itself is moving through space and is currently passing through the Local Interstellar Cloud, a wispy mix of hydrogen and helium about 30 light-years wide.

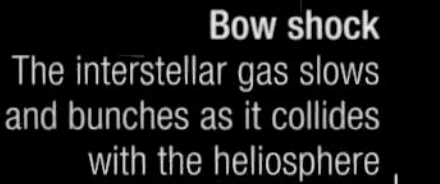
Bow shock
The interstellar gas slows and bunches as it collides with the heliosphere
Heliosheath
The outer heliosphere, where the interstellar gas slows the solar wind
Termination shock
The solar wind slows and bunches as it runs into interstellar gas
Voyager 1
Sun
Cassini
Voyager 2
Heliopause
The boundary between the heliosphere and interstellar space

HELIOSPHERE
The heliosphere is the bubble-shaped volume of space dominated by the solar wind and the Sun's magnetic field. Its radius is approximately 14 billion miles (22.5 billion km). As the heliosphere moves through interstellar space, it produces a bow shock similar to that produced by a boat traveling through water.

THE HELIOSPHERE MOVING THROUGH INTERSTELLAR SPACE
This artist's impression shows the Sun, its planets, and the heliosphere. The solar wind, shown as pale yellow lines, inflates the bubble from the inside while interstellar gas presses on it from the outside. The gray bow shock (left) consists of material from the Local Interstellar Cloud.

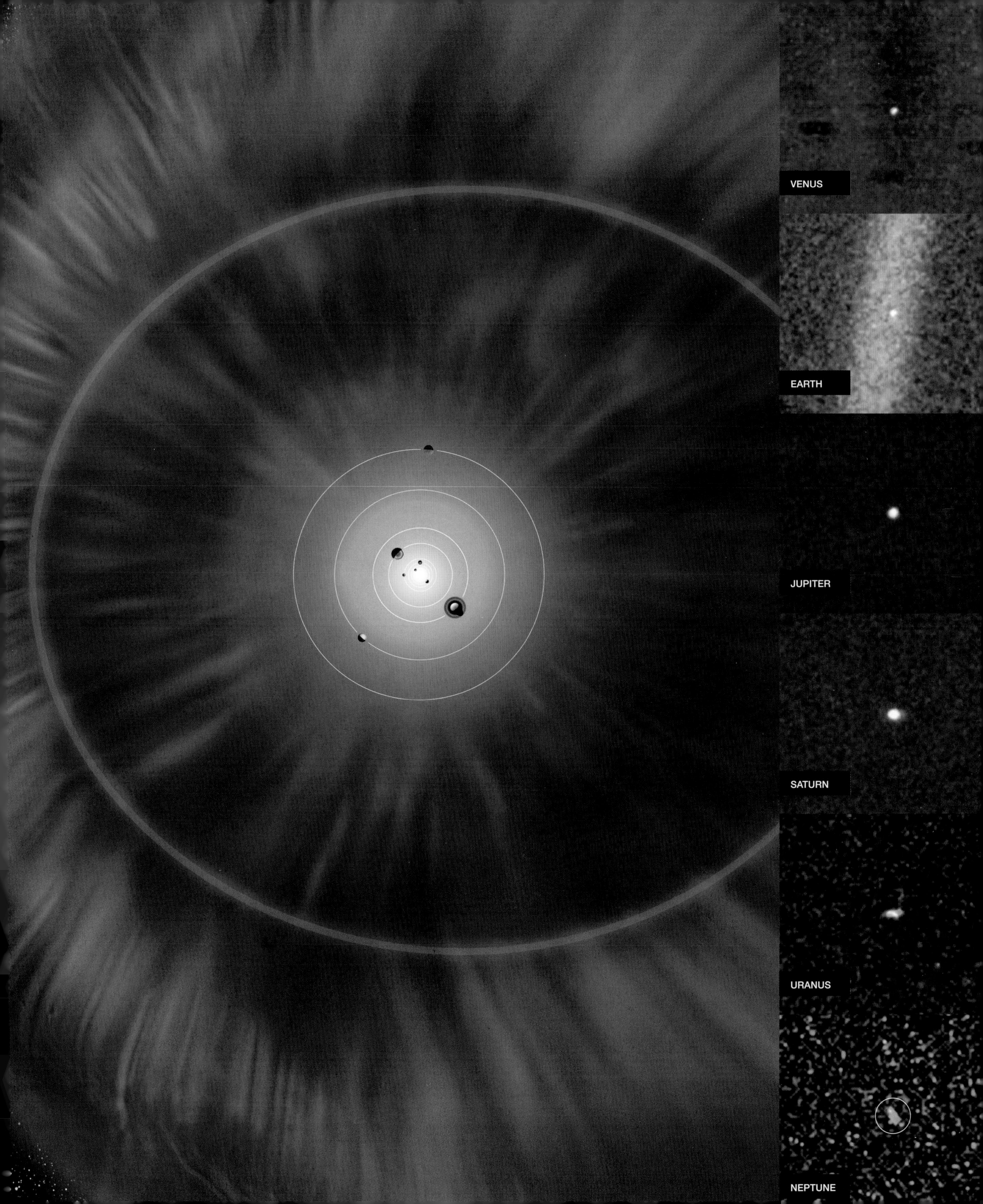
VENUS
EARTH
JUPITER
SATURN
URANUS
NEPTUNE

4 A GALAXY OF STARS

<< The center of the Milky Way
26,000 light-years from Earth

THE MILKY WAY

Vast though it seems, the Solar System is just a tiny speck within the enormous spiral galaxy called the Milky Way. This huge cosmic pinwheel is over 13 billion years old, is many thousands of light-years across, and is thought to contain as many as 200 billion individual stars.

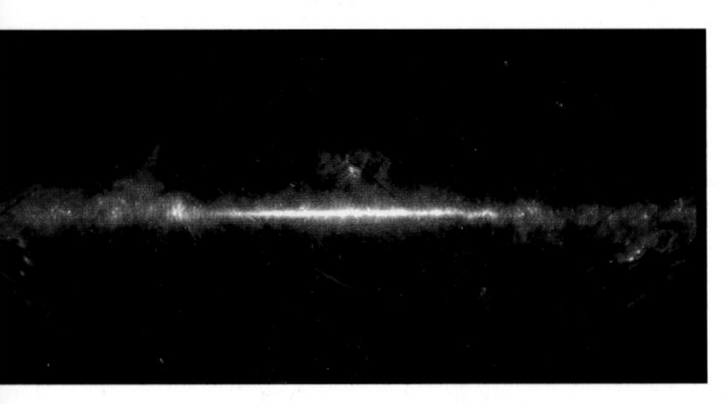

THE MILKY WAY IN INFRARED
This infrared map reveals how the Milky Way wraps around the entire sky. Images such as this one from Japan's Akari Satellite record the emissions of warm dust associated with star formation.

A BAND OF STARS

The name Milky Way and indeed the very word galaxy itself stem from the star system's appearance in Earth's skies. Looking up and down relative to the central disc, a fairly sparse scattering of stars can be seen across one region of space. But looking across the disc, many more stars can be seen. Even though most of these distant stars are not individually visible to the naked eye, together they accumulate into enormous star clouds that create a pale, meandering strip across Earth's skies. Ancient Greek and Roman astronomers likened this path to a river of spilled milk. In Latin, it was called Via Lactea, or "milky road;" in Greek, it was known as Kyklos Galaxias, or "milky circle"—hence it became the Milky Way Galaxy. The Solar System—the Sun and its planets—is embedded within this wide but thin disc of stars.

NIGHT-SKY VIEW
This long-exposure image of the region around the constellation Sagittarius reveals bright star clouds and the pinkish glow of star-forming regions lying toward the center of the Milky Way. Viewed above the mountains of Arizona, the tracery of dark lines across the region is created by silhouetted dust clouds closer to Earth.

A SPIRAL IN SPACE

The Milky Way Galaxy is an enormous spiral, about 100,000 light-years in diameter. The central bulge, or hub, is roughly 27,000 light-years long, 15,000 light-years across, and 6,000 light-years thick. The flattened disc that surrounds it is around 1,000 light-years thick. Above and below the main disc, globular star clusters (see p.220) orbit in an extended halo along with a sparse scattering of lone "runaway" stars that have been kicked out of their normal orbits by close encounters with other stars or by cosmic explosions. Stars in different parts of the galaxy show distinct differences. Those in the central hub are relatively faint and predominantly red and yellow, while those in the outer disc show a mixture of colors and brightnesses.The most brilliant blue-white stars are located in the spiral arms.

Sagittarius Dwarf Galaxy

Far 3 kiloparsec Arm

Central hub
The hub is crossed by a bar of stars, with the spiral arms emerging from either end.

Carina–Sagittarius Arm

younger stars

Galactic center
In the middle of the central hub lies a turbulent region of gas clouds and older, heavyweight stars around a supermassive black hole.

Perseus Arm

dense molecular clouds

THE STRUCTURE OF THE MILKY WAY
According to the latest research, the Milky Way is dominated by two major spiral arms: the Scutum–Crux–Centaurus Arm and the Perseus Arm. There are also several minor arms and spurs. The Solar System lies 26,000 light-years from the center.

THE FIRST MAP

By the late 1600s, astronomers acknowledged that every star was another Sun and that therefore the scale of the Milky Way must be truly enormous. But it was 1785 before anyone dared attempt to map it. English astronomer William Herschel and his sister Caroline based their work on the assumption that all stars are of uniform luminosity (so linking apparent brightness to their distance). They then measured the faintest stars they could see along almost 700 different lines of sight to estimate how far the galaxy stretched.

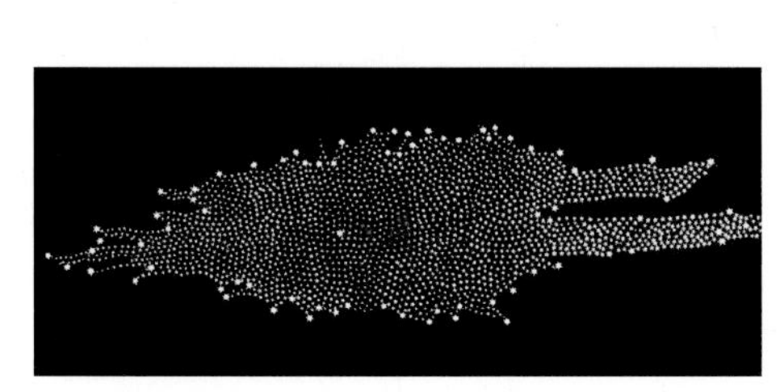

The Herschels' map
Based on flawed assumptions, the Herschels' valiant attempt to map the Milky Way resulted in a flattened, amorphous cloud, with the Sun close to its center.

SECRETS OF THE MILKY WAY

Even though the Milky Way has been studied for centuries, it still yields surprises. Since the 1990s, astronomers have discovered two galaxies (the Sagittarius and Canis Major dwarfs, see p.267) lying close to the disc, but largely obscured by the star clouds of the galactic plane. Both galaxies have been badly distorted by the gravity of the Milky Way and will eventually be absorbed into it. Meanwhile, measurements of the Galaxy's rotation suggest that its halo contains more mass than can be explained from its visible matter. It is therefore probably rich in dark matter (see pp.318–19), which includes some compact dark objects (perhaps rogue planets or black dwarfs). The shape and origin of the spiral arms are still something of a mystery (see below). However, the most intriguing questions of all surround the enormous concentration of mass at the center of the Milky Way (see pp.260–61).

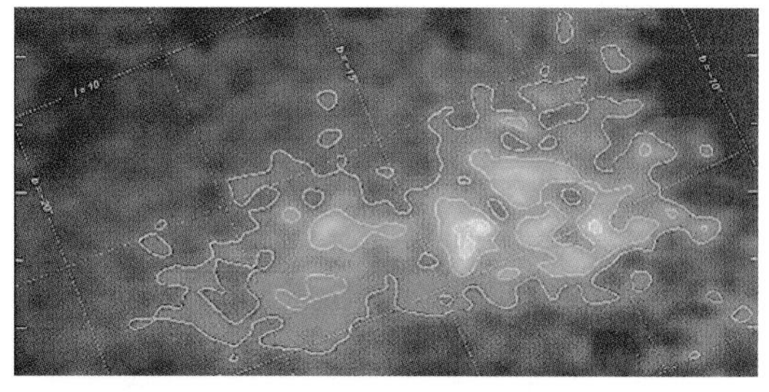

SAGITTARIUS DWARF ELLIPTICAL GALAXY
In 1994 analysis of the density of star clouds in Sagittarius enabled astronomers to identify this galaxy, the light patch in the center, on the other side of the Milky Way.

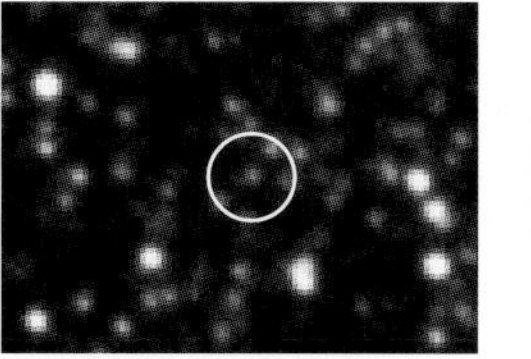

NORMAL VIEW OF STAR

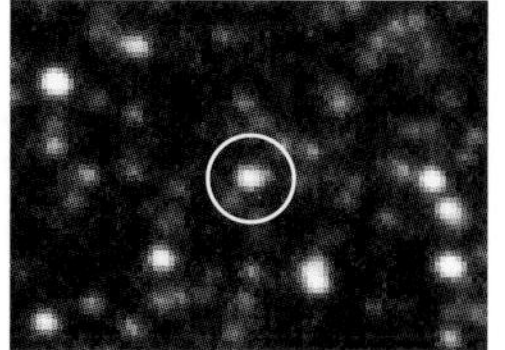

MACHO PASSING STAR

DARK MATTER IN THE MILKY WAY
Some of the Milky Way's mass may be massive compact halo objects (MACHOs). When one passes in front of a star, gravity distorts its light, causing the star to brighten.

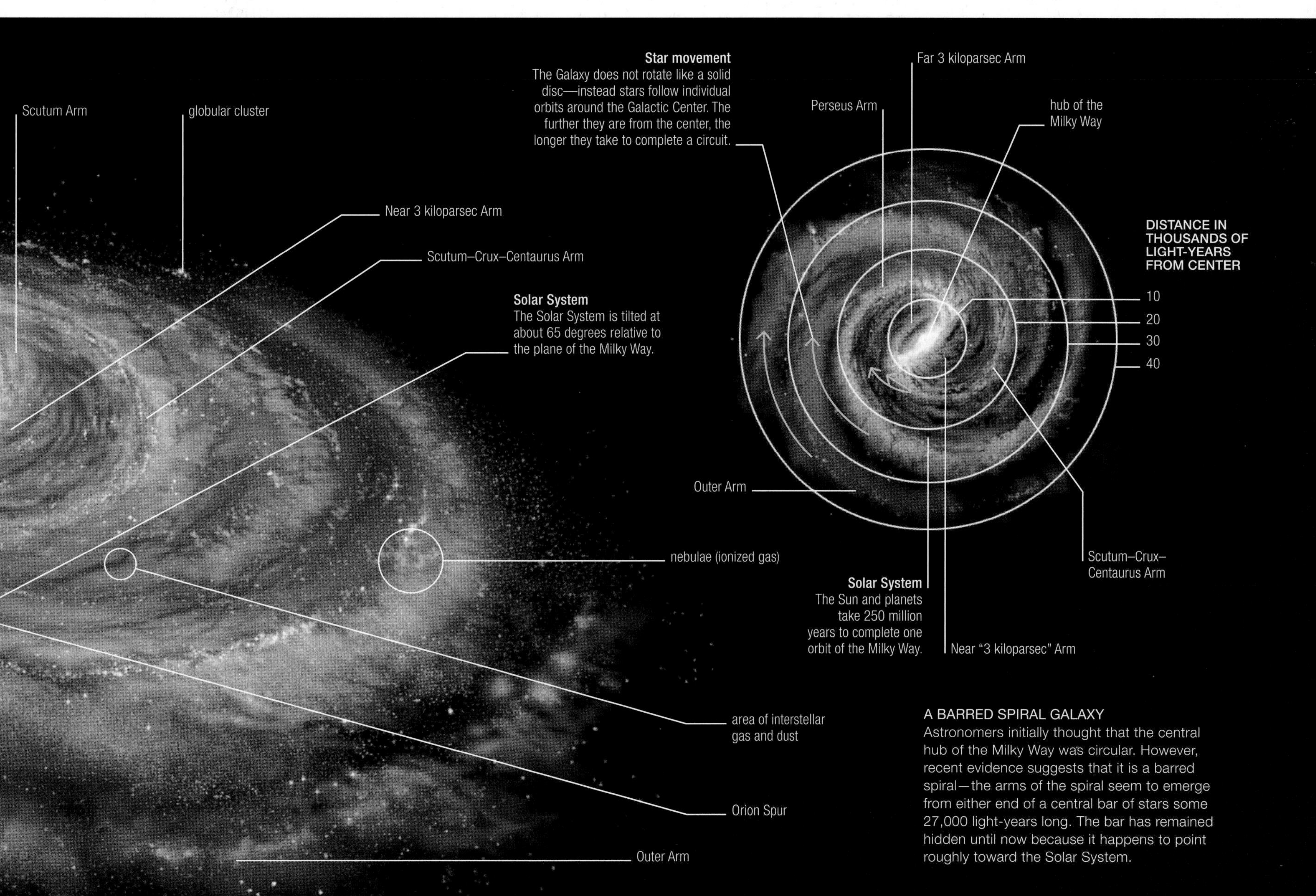
Scutum Arm
globular cluster
Near 3 kiloparsec Arm
Scutum–Crux–Centaurus Arm
Solar System
The Solar System is tilted at about 65 degrees relative to the plane of the Milky Way.
nebulae (ionized gas)
area of interstellar gas and dust
Orion Spur
Outer Arm
Star movement
The Galaxy does not rotate like a solid disc—instead stars follow individual orbits around the Galactic Center. The further they are from the center, the longer they take to complete a circuit.
Far 3 kiloparsec Arm
Perseus Arm
hub of the Milky Way
DISTANCE IN THOUSANDS OF LIGHT-YEARS FROM CENTER
10
20
30
40
Outer Arm
Scutum–Crux–Centaurus Arm
Solar System
The Sun and planets take 250 million years to complete one orbit of the Milky Way.
Near "3 kiloparsec" Arm

A BARRED SPIRAL GALAXY
Astronomers initially thought that the central hub of the Milky Way was circular. However, recent evidence suggests that it is a barred spiral—the arms of the spiral seem to emerge from either end of a central bar of stars some 27,000 light-years long. The bar has remained hidden until now because it happens to point roughly toward the Solar System.

THE MILKY WAY IN EARTH'S SKY
This 360° patchwork of 30 images shows the night sky above Death Valley, CA. The distorting effect of squeezing the entire sky into a rectangle has bent the galactic plane of the Milky Way into an arc; the brighter area of the Galactic Center lies to the right, while dust lanes of the spiral arms form a dark band within the center of the arc.

TOWARD THE STARS

Venturing beyond the Oort Cloud (see pp.178–79) and into the space between the stars, it quickly becomes clear that the void is not as empty as it first appears. In addition to the stars, tenuous clouds of gas and dust, as well as other objects, are all vital parts of the structure and the story of the Milky Way.

INTERSTELLAR DUST CLOUD
Dust clouds, such as NGC 6559 in the constellation Sagittarius, are surprisingly widespread throughout the Milky Way. However, they only become noticeable when they are silhouetted against a background of stars or glowing gas.

INTERSTELLAR SPACE

The space between the stars is filled with a mix of particles collectively known as the interstellar medium. Across the Milky Way on average, this medium has a density of 15 atoms per cubic inch (one atom per cubic centimeter), although it can be a million times thicker than this in places. This medium is dominated by gas, of which roughly 89 percent is hydrogen, 9 percent is helium, and 2 percent is heavier elements. Local conditions in space cause the hydrogen to exist in different forms. This creates a mix of molecular clouds (where pairs of hydrogen atoms are bonded into molecules), neutral atomic regions (where the hydrogen atoms are separated), and ionized media (where the hydrogen atoms are electrically charged).

direction of Galactic Center
Scorpius–Centaurus Association
Gum Nebula
Scorpius–Centaurus Shells
Vela Supernova Remnant
Aquila Rift
the Local Bubble
direction of Sun's movement
Orion Association
Orion Shell
Sun

WANDERER IN THE GALAXY
With a unique chemical composition and an unusual orbit, the short-period Comet

OBJECTS IN A GALAXY

The vast majority of the objects in the Milky Way are dominated by gas. The stars, the most massive individual objects in the galaxy, are essentially just extreme concentrations of gas—their apparently solid surfaces simply marking the edge of a zone within which the gas becomes so dense that it is opaque. Nebulae may appear as glowing clouds but are in fact relatively dense concentrations of interstellar medium containing hundreds of particles in each cubic inch of space. They owe their increased density either to the gravitational collapse associated with star formation (see below) or to the explosive outbursts of star death (see below right). Nebulae shine because they reflect the light of nearby stars or are energized by their own radiation.

Roughly 1 percent of the interstellar medium (by mass) is composed of interstellar dust, which is visible only when silhouetted against a brighter background. When pulled into dense clouds around newborn stars, the dust grains coalesce and stick together to form larger objects such as rocky asteroids and planets. Where temperatures are low enough, chemicals with low melting points (such as water and ammonia) may also form solid ices and mix with the dust to form comets (see pp.180–81), ice-dwarf worlds (see pp.178–79), and the giant outer planets (see pp.124–25).

STARS
The dense concentration of atoms at the heart of each star make the atoms the only place in the Universe where natural nuclear fusion can take place (see p.208). Fusion in the stars generates all of the light and other radiations that illuminate the Milky Way Galaxy.

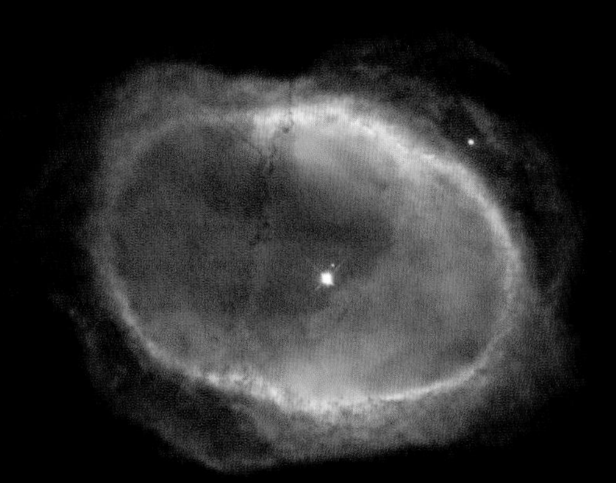

PLANETARY NEBULAE
Dying Sun-like stars, swollen to enormous size, puff off their outer layers to form beautiful and complex planetary nebulae, such as the Eight Burst or Southern Ring Nebula. Their gases glow for a few thousand years before fading away and mixing into the interstellar medium.

STAR-FORMING REGION
When gas is concentrated by gravitational collapse or pushed together by shock waves from exploding stars, it begins to form new stars, as seen here in the Omega Nebula. Ultraviolet radiation from the stars inside nebula breaks hydrogen molecules and atoms apart into ions (electrically charged atoms), creating HII star-forming regions (see p.266).

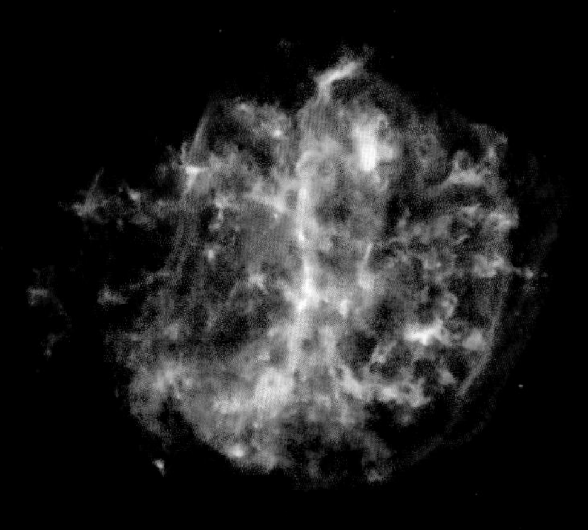

SUPERNOVA REMNANT
Massive stars end their lives in spectacular explosions that create most of the heavier elements in the Universe. Shock waves from supernovae such as G292.0+1.8 (above) can persist for millions of years, sculpting the interstellar medium, but ultimately fade away (see p.248).

PINNING DOWN THE STARS

In order for astronomers to calculate the true properties of stars and other celestial objects, they need to know exactly how far they are from Earth. Interstellar distances are so enormous, this task is much more difficult than it might seem. Fortunately, there are now several different ways to make such measurements.

THE PARALLAX EFFECT

When astronomers first worked out that Earth was orbiting the Sun, they expected to see this motion reflected in the sky. Nearby objects appear to change their position against the more distant background when viewed from two different angles—an effect called parallax—so it seemed that the stars themselves should change their positions when viewed six months apart, and from opposite sides of Earth's orbit. The fact that stars did not change position was evidence that they were immensely far away. Efforts to measure parallax were not successful until, in 1838, German astronomer Friedrich Bessel used the technique to measure the distance of the nearby star 61 Cygni, some 11.4 light-years from Earth.

Earth-based parallax measurements allowed astronomers to identify important patterns among different types of stars, but the difficulty of making such precise observations limited its use to nearby stars. A breakthrough came in 1989 with the launch of the European Space Agency's Hipparcos satellite (opposite), which extended the technique's usefulness to around a thousand light-years from Earth.

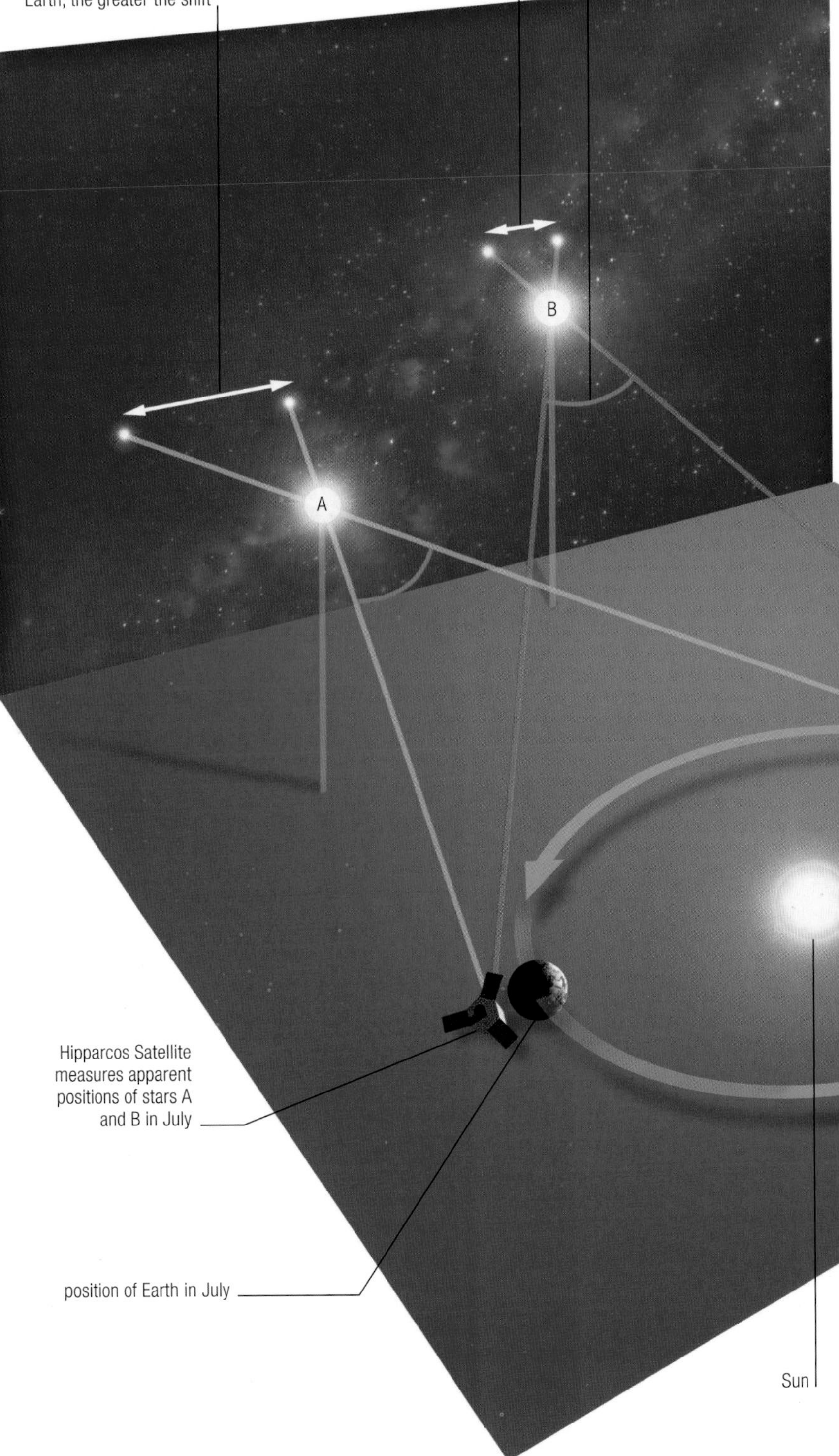
parallax shift of star A; the closer the star is to Earth, the greater the shift
parallax shift of further star B; the further away the star, the smaller the shift
angle between measurements (parallax angle) determines star's distance from Earth
B
A
Hipparcos Satellite measures apparent positions of stars A and B in July
position of Earth in July
Sun

MEASURING DISTANCE WITH PARALLAX
Orbiting parallax satellites, such as Hipparcos, measure the positions of individual stars against the background sky from either side of Earth's orbit. Knowing the diameter of Earth's orbit and the size of the annual parallax, a simple calculation reveals the star's true distance.

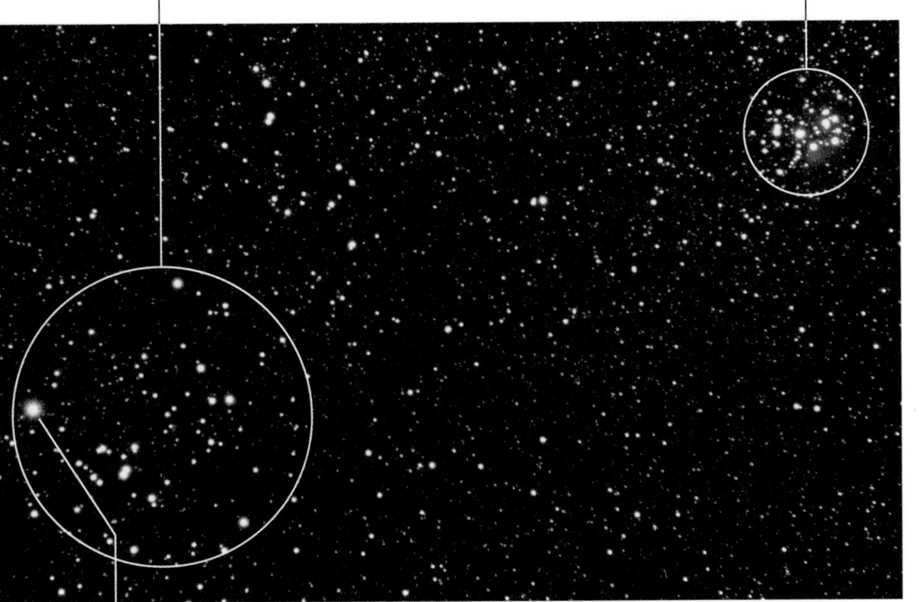
the Hyades Cluster is 150 light-years away
the Pleiades Cluster is 440 light-years away
foreground star Aldebaran is only 65 light-years away

CLUSTER PARALLAX
An ingenious method reveals the distance to star clusters such as the Hyades and Pleiades. Stars in a cluster move in the same direction, but parallax makes them seem to diverge from one point. The amount of apparent divergence depends on the distance from Earth, so measuring divergence reveals distance.

HOW FAR IS A LIGHT-YEAR?

The distances between objects in space are so huge that familiar measurements such as miles, kilometers, and even the astronomical units to measure the Solar System are useless. Astronomers use much larger distance units. The most common is the light-year—the distance light travels in a year—which is equivalent to 5.9 trillion miles (9.5 trillion km). Another widely used unit is the parsec (3.26 light-years)—the distance at which an object has an annual parallax equivalent to one second of arc or 1/3,600th of a degree.

Comparing distance
This diagram illustrates the comparative distances in light-years between Earth and the Sun, the radius of the Solar System, and the Sun to the nearest star. These distances are so great that it impossible to show to scale.

Earth
Earth to Sun: 8.3 light-minutes = 93 million miles (150 million km)
edge of the Oort Cloud
Sun
1 light-year = 5.9 trillion miles (9.5 trillion km)
nearest star Proxima Centauri
4.2 light-years = 25 trillion miles (40 trillion km)

SCALING THE GALAXY

Accurate though it is, the parallax technique only works for a comparatively small region of nearby space. But parallax-based distance measurements can help uncover other useful stellar properties. For example, they reveal that some pulsating variable stars have cycles linked to their average luminosity—a feature that can then be used to establish other stellar distances independently of parallax. The most famous of these stars are the Cepheid variables (see p.276), but other types (named after their "prototype" stars in which the behaviour was discovered) include RR Lyrae and Delta Scuti stars. The variability of these stars allows their true luminosity to be calculated. Astronomers can then compare this with their brightness as seen from Earth in order to work out their distance.

SURVEY SATELLITES

Launched in 1989 and operational for more than three years, the European Space Agency's Hipparcos (High-Precision Parallax-Collecting Satellite) mission measured the parallax of more than 100,000 stars to a precision of about one 4-millionth of a degree, and 2.5 million more at lower precision. Freed from the blurred images caused by Earth's atmosphere, it was able to extend the useful limits of parallax out to a thousand light-years from the Sun. Hipparcos's successor, the Gaia Satellite, will make measurements 100 times more accurate than Hipparcos, cataloging a billion stars, and extending the use of parallax across the Milky Way.

The Gaia Satellite
This satellite from the European Space Agency is set for launch in 2012. It will make measurements by comparing the images from two telescopes separated by a wide, fixed angle. During its five-year mission, it will record the parallax of stars in the Milky Way and the Local Group.

Hipparcos Satellite measures apparent positions of stars A and B in January

position of Earth in January

direction of Earth's orbit

STELLAR MOTIONS

Not all movement of stars in the sky is due to the effect of parallax. Every star is traveling on its own path through space, and some of them move fast enough for their drift to be noticeable from year to year. Any star's path through space can be split into two components: a proper motion that changes its position in the sky; and a radial velocity directly toward or away from Earth. Proper motion can only be detected through precise measurements of a star's position, while radial motion is revealed by Doppler shifts in the wavelength and color of starlight (see p.233). Proper motion complicates attempts to measure parallax, but if it can be combined with radial velocity to reveal a star's true path through space, it can provide astronomers with valuable information. For example, proper motion can help trace back widely separated stars to a common point of origin in an open star cluster, or reveal "runaway" stars sent speeding through space by violent events such as supernova explosions. In addition, proper motion can even reveal the large-scale motions caused by the rotation of the Milky Way.

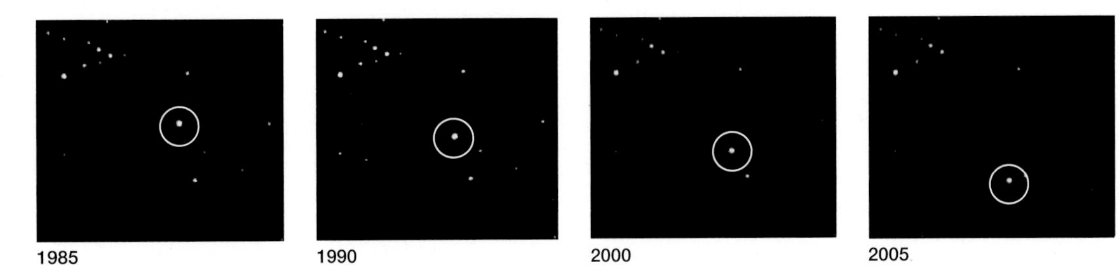

TRACKING BARNARD'S STAR
In 1916, American astronomer E. E. Barnard, discovered that this insignificant red dwarf star (see p.201), currently 5.96 light-years from Earth in the constellation Ophiuchus, has the largest proper motion of any star in the sky, moving by the diameter of the full Moon every 175 years.

NEARBY STARS

Even the closest stars are several light-years from Earth and they are tens of thousands of times further away than the planets of the Solar System. However, they are still much closer than the majority of stars in the sky, and they are among the easiest objects to study in detail.

THE SOLAR NEIGHBORHOOD

Some three dozen star systems lie within 15 light-years of the Sun, and they present a useful cross section of the stars in our region of the galaxy. This random sample contains 14 binary or multiple star systems and stars ranging from feeble brown dwarfs to brilliant white stars far more luminous than the Sun. Despite their proximity, only nine of these stars are bright enough to be seen with the naked eye from Earth; the vast majority of naked-eye stars are much further away but are intrinsically brighter. There are no really luminous giant stars in Earth's immediate neighborhood, the closest being the orange giant Pollux, 33 light-years away in the constellation Gemini, and the red giant Arcturus (the fourth brightest star in the sky), which lies 37 light-years away in Boötes.

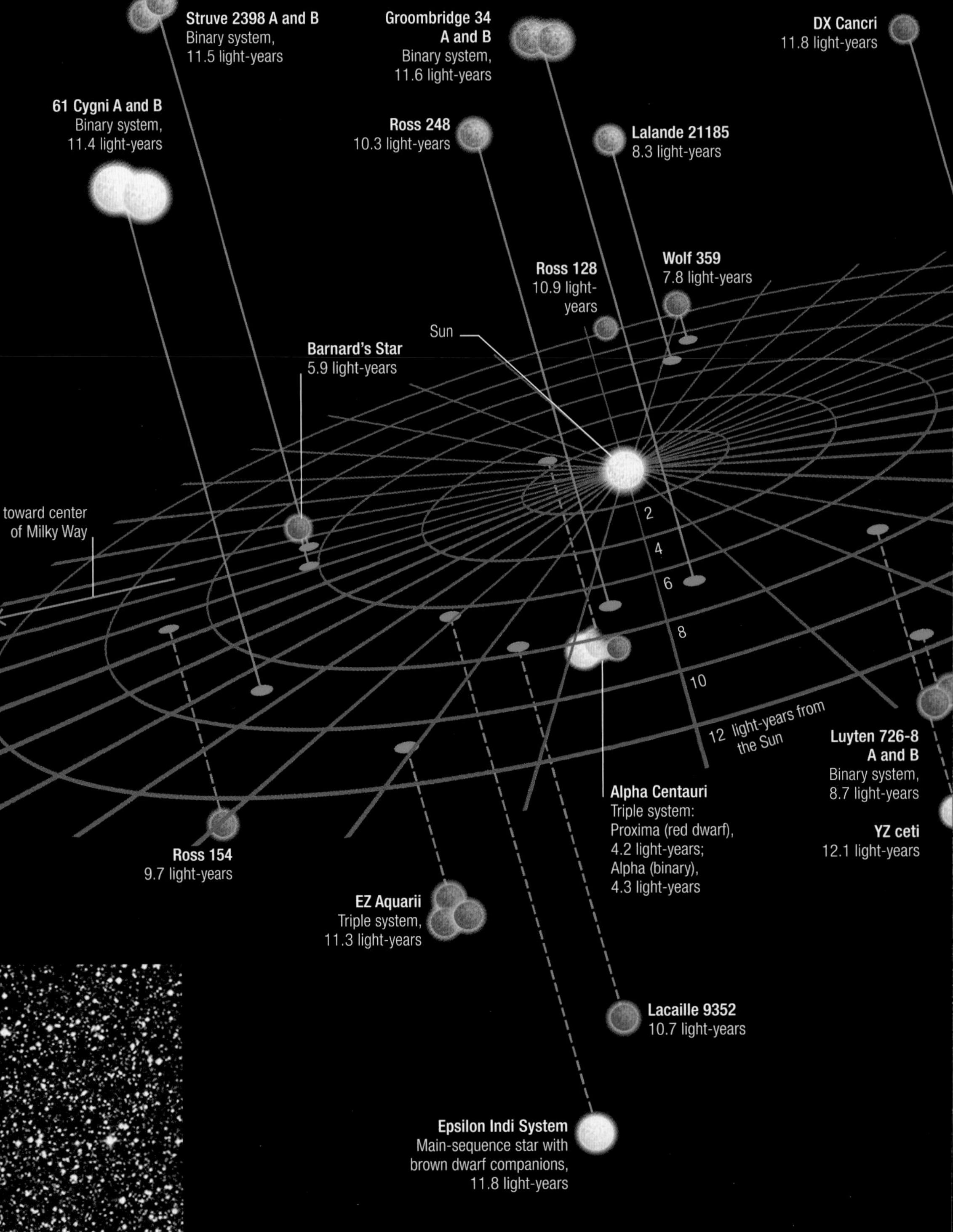

MAPPING THE NEARBY STARS
This chart shows the solar neighborhood out to a distance of 12.5 light-years (the distances to the stars are measured in a straight line from the Sun). The mix of stars ranges from numerous red dwarfs to bright white, and Sun-like stars and tiny white dwarfs.

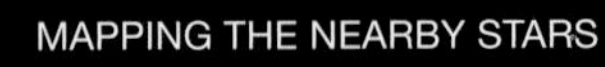

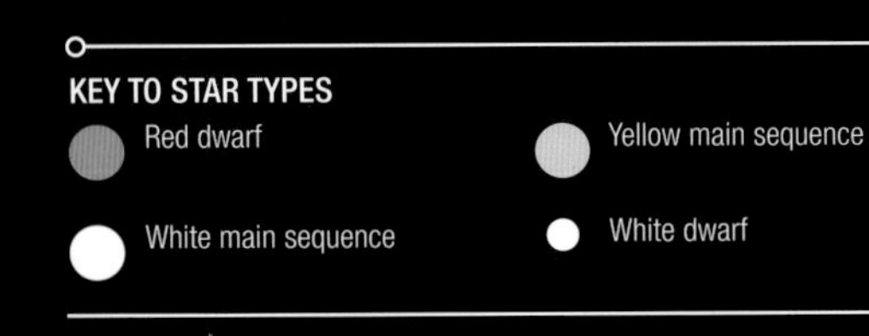

NEAREST NEIGHBORS
The closest star system to the Sun consists of three stars. A central pair of Sun-like stars appears in Earth's sky as a single, brilliant star, called Alpha Centauri. The closest star to Earth (aside from the Sun) is the outlying red dwarf Proxima Centauri, which is only 4.2 light-years away. It is the red star seen at the center of this image from the UK Schmidt Telescope, Australia.

BRILLIANT NEIGHBORS

The most luminous stars in the solar neighborhood are Sirius and Procyon, the brightest stars in the constellations Canis Major and Canis Minor. Each is an average white star that appears brilliant because of its closeness to Earth, at 8.6 and 11.4 light-years respectively. Sirius has twice the mass of the Sun and 25 times its luminosity, while Procyon's mass is 1.5 times greater than the Sun and is almost eight times as bright. Coincidentally, both stars are members of binary systems. The second stars of each system are small, but intensely hot stellar remnants called white dwarfs (see p.239).

Procyon A and B
Binary system,
11.4 light-years

Luyten's Star
12.4 light-years

Teegarden's Star
12.5 light-years

Epsilon Eridani
10.5 light-years

Sirius A and B
Binary system,
8.6 light-years

Tau Ceti
11.9 light-years

GJ 1061
12.0 light-years

SIRIUS A AND B
Despite its apparently tiny size, the white dwarf star Sirius B, far right, still has half the mass of its brighter companion, Sirius A. One hundred and twenty million years ago, Sirius B was a red giant that would have outshone even Sirius A. The pair take a little less than 50 years to orbit one another.

RED DWARFS

The vast majority of stars in the region of space around the Solar System are red dwarfs. These stars typically have less than half the Sun's mass and shine with less than 10 percent of its luminosity—sometimes a great deal less. Red dwarfs generate energy at a much slower rate than stars like the Sun. They also have comparatively low surface temperatures—lower than about 5,800°F (3,200°C)—so they emit mostly red light. Beyond the immediate neighborhood, red dwarfs are extremely hard to detect, but there is good reason to believe that there are more left to find, even on our own doorstep.

Even fainter and smaller than red dwarfs are the brown dwarfs, which have less than 8 percent of the Sun's mass. Astronomers currently know of four nearby brown dwarfs.

VARIABLE STAR CN LEONIS
This star, also known as Wolf 359, is relatively nearby (7.8 light-years away). Because it is so close, its motion in our night sky is easily detected—for example, this can be seen by comparing two images obtained a few years apart with the UK Schmidt Telescope.

NEW NEIGHBOR
Discovered in 2003, Teegarden's Star, shown here in this artist's impression, is a faint red dwarf 12.5 light-years away in the constellation Aries. Astronomers found the star, which is 11,000 times less luminous than the Sun, while searching for faint asteroids.

SOLAR LOOKALIKES

Several of the stars in our locality bear a striking similarity to the Sun. The brightest star of the Alpha Centauri system, known as Alpha Centauri A, is an almost exact match being just 10 percent more massive than the Sun. Alpha Centauri B, meanwhile, has 90 percent of the Sun's mass and is a little smaller and cooler, shining with a distinctly yellow-orange color. Other nearby stars—including Epsilon Eridani, the two members of the 61 Cygni system, Epsilon Indi A, and Tau Ceti—are also just slightly less massive than the Sun.

EPSILON ERIDANI
Slightly fainter and cooler than the Sun, Epsilon Eridani is also much younger, at less than 1 billion years old. It is orbited by two asteroid belts, a huge outer dust ring, and at least one giant planet, which is shown here in this artist's impression.

JEWELS IN THE SKY
This Hubble Space Telescope image shows a dust-free region in the dense star fields in the constellation Sagittarius. It captures stars of an endless variety of color and brightness. These apparently superficial features are nevertheless key to understanding a wider range of stellar properties.

STELLAR CHARACTERISTICS

Stars are the most complex of all astronomical objects as they have a huge range of different properties and a complex evolutionary cycle. They are also immensely distant and live and die on an unimaginable timescale, so piecing their story together has required some ingenious detective work.

WHAT IS A STAR?

Stars are essentially clouds of gas that have condensed under their own gravity until the central regions are incredibly dense and hot and the gases become opaque. At the heart of the star, conditions are so extreme that atomic nuclei are forced together in a process called nuclear fusion, which converts hydrogen to helium, generating huge amounts of energy (see pp.208–09). This energy forces its way out of the star, until it escapes at the surface where the star becomes transparent.

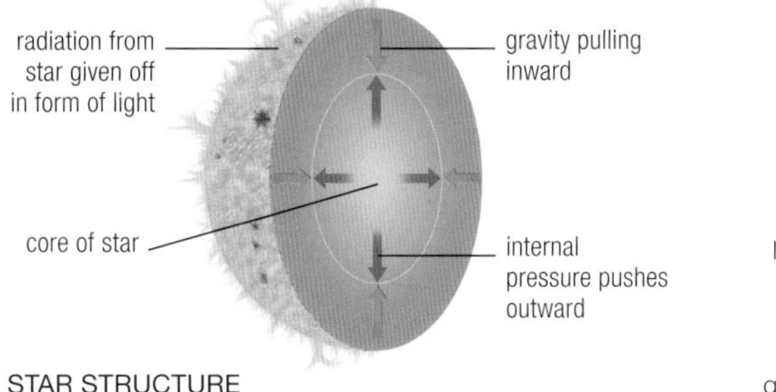

STAR STRUCTURE
Each layer within a star's interior is trapped in a delicate balance between inward and outward forces, known as hydrostatic equilibrium. Outward pressure exerted by hot gas in the star's interior fights against the weight of the outer layers and the inward pull of gravity.

COLOR, TEMPERATURE, AND SIZE

By calculating the distances to stars using parallax and other methods (see pp.198–99), astronomers know that stars vary hugely in their intrinsic brightness, or luminosity. But what do the different colors indicate? The color of any star reflects its surface temperature, which ranges from relatively cool red to hot blue, and the temperature, in turn, is governed by the amount of energy heating each square foot of the star's surface. That amount of energy depends on both the star's overall energy output (its luminosity) and its size—the larger a star is, the greater its surface area and the more spread out its escaping energy. So a combination of color and brightness can reveal a star's size—faint blue-white stars must be smaller than faint red ones to maintain their hot surfaces; conversely, brilliant red stars must be huge compared to equally bright blue ones, to reconcile such luminosity with a cool surface.

STAR SIZES
The stellar balancing act of hydrostatic equilibrium (left), means that more luminous, brighter stars tend to be larger but still have hotter surfaces than fainter and more compact stars. Brilliant but bloated red giants (see p.228) and hot but super-dense white dwarfs (see p.239) are exceptions to this rule.

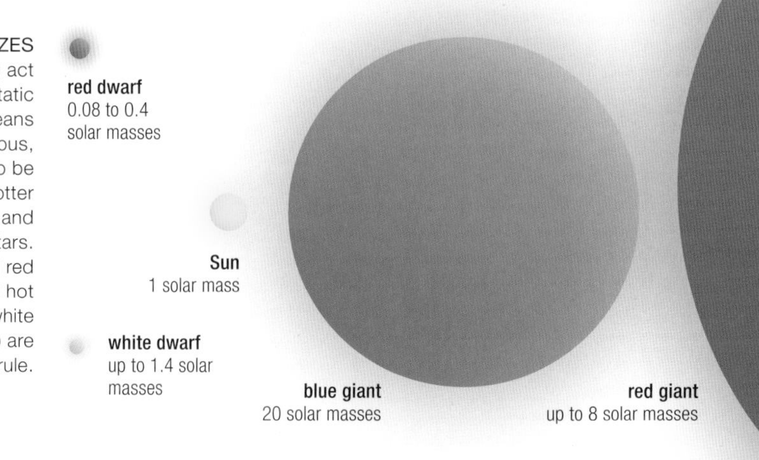

CHEMISTRY AND COLOR

When astronomers began to analyze stellar spectra in the late 19th century, they discovered that a star's color also indicates fundamental differences in its chemical composition (see p.233). Dark absorption lines within a star's spectrum are a result of energy being absorbed by specific atoms in its atmosphere, and different colors tend to match well with different atmospheric make-ups. As a result, stars are now cataloged by spectral class—a letter in the sequence O, B, A, F, G, K, M (see box, below) is followed by a number (0–9) to indicate finer subdivisions.

SPECTRAL CLASSIFICATION OF STARS

TYPE	COLOR	PROMINENT SPECTRAL LINES	AVERAGE TEMPERATURE	EXAMPLE
O	Blue	He^{+}, He, H, O^{2+}, N^{2+}, C^{2+}, Si^{3+}	80,000ºF (45,000ºC)	Regor
B	Blueish white	He, H, C^{+}, O^{+}, N^{+} Fe^{2+}, Mg^{2+}	55,000ºF (30,000ºC)	Rigel
A	White	H, ionized metals	22,000ºF (12,000ºC)	Sirius
F	Yellowish white	H, Ca^{+}, Ti^{+}, Fe^{+}	14,000ºF (8,000ºC)	Procyon
G	Yellow	H, Ca^{+}, Ti^{+}, Mg, H, some molecular bands	12,000ºF (6,500ºC)	The Sun
K	Orange	Ca^{+}, H, molecular bands	9,000ºF (5,000ºC)	Aldebaran
M	Red	Ti0, Ca, molecular bands	6,500ºF (3,500ºC)	Betelgeuse

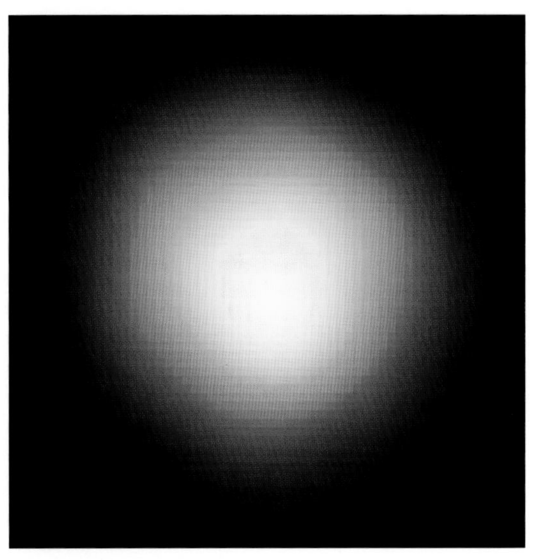

BETELGEUSE
The red supergiant Betelgeuse in the constellation Orion, with a spectral type M2, has a cool extended atmosphere rich in molecules such as water and carbon monoxide. This gives it a complex spectrum, which includes broad, dark absorption bands.

PUTTING IT ALL TOGETHER

While astronomers have deduced much about the properties of individual stars, each star can only be observed at one stage of its incredibly long life. To gain an overview of the full range of stellar properties, it is necessary to compare large numbers of individual stellar snapshots. Around 1910, Swedish scientist Ejnar Hertzsprung and US astronomer Henry Norris Russell independently had the idea of plotting stars on a chart according to their intrinsic luminosities and their spectral classes. The resulting Hertzsprung–Russell, or H–R, diagram, the most important in astronomy, revealed a simple but important pattern: the overwhelming majority of stars lie on a diagonal strip known as the main sequence, which links the countless faint red dwarfs with rarer but brilliant blue giants. Faint blue-white stars (white dwarfs) and brilliant orange-red ones (red giants) are rare exceptions, and super-luminous, multi-colored supergiants are even more rare.

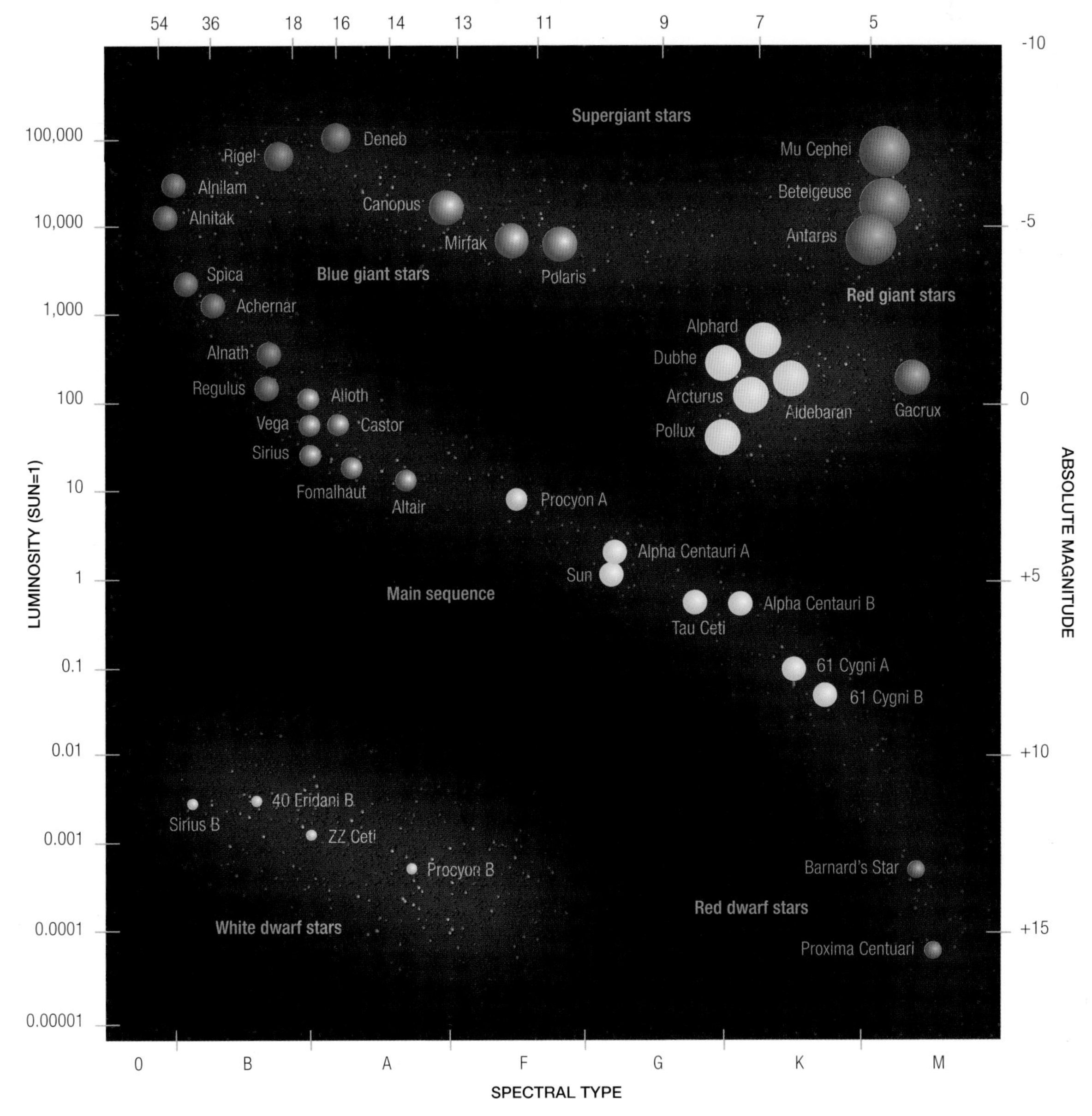

THE HERTZSPRUNG–RUSSELL DIAGRAM
This diagram shows how stars are distributed, and plots the points of some well-known stars, most of which lie along the main sequence. Although bright red giants seem more plentiful than dim red dwarfs, this is due to the fact that red giants are more noticeable in Earth's skies.

MASS AND LUMINOSITY OF STARS

Most stars obey a simple relationship between luminosity and surface temperature as shown on the Hertzsprung–Russell (H–R) diagram (see p.203)—but what actually causes different stars to shine with differing brightness? In 1924, British astronomer Arthur Stanley Eddington looked at the behavior of binary stars (see p.221) in order to estimate the masses of stars within these multiple systems and to compare their other properties. His work revealed that stars on the main sequence have a clear link between luminosity and mass—the heavier a star is, the brighter it burns. Eddington suggested that the higher temperatures and densities achieved at the heart of more massive stars allow them to produce energy more efficiently (see pp.208–09).

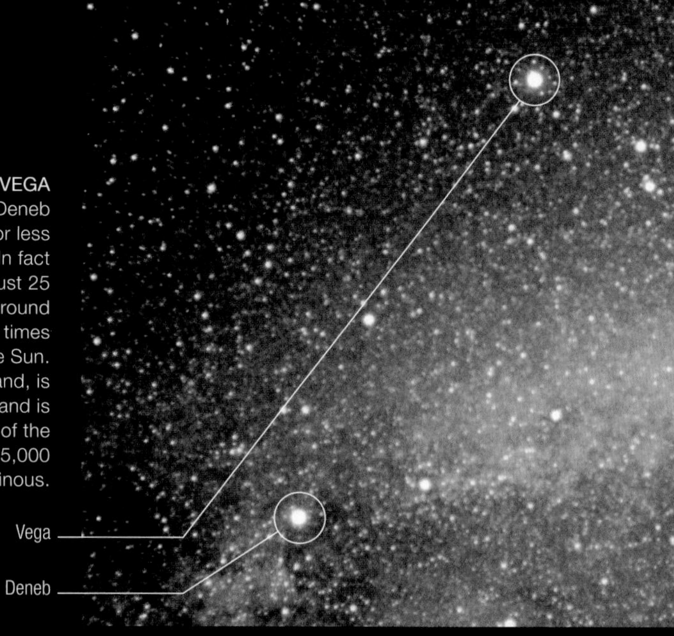
Vega
Deneb

DENEB AND VEGA
The two bright stars Deneb and Vega appear more or less similar in Earth's skies. In fact Vega is a nearby star, just 25 light-years away, with around twice the mass and 37 times the luminosity of the Sun. Deneb, on the other hand, is 1,400 light-years away and is 20 times the mass of the Sun and about 55,000 times more luminous.

STELLAR SNAPSHOTS

A full picture of the patterns of stellar evolution had to wait until the mechanism by which stars are fueled was finally established in the late 1930s (see pp.208–09). This understanding, coupled with Eddington's discoveries and the establishment of the model of hydrostatic equilibrium (see p.202), allowed astronomers to work out the properties of many star types. The slow rate of stellar evolution means that most stars display unchanging properties after centuries of study, but the sheer number of different star systems that can be observed provides a wealth of data. This, along with the spectroscopic observations that track the changing chemistry of stellar atmospheres and with theoretical models of the nuclear-fusion reactions within stars, allows astronomers to work out the progression of one type of star into another.

> “EACH NEW SPECTRUM IS THE GATEWAY TO A WONDERFUL NEW WORLD ... AS IF DISTANT STARS HAD ACQUIRED SPEECH.”
>
> **ANNIE JUMP CANNON**, AMERICAN ASTRONOMER, PIONEER OF STELLAR CLASSIFICATION

PATHS THROUGH LIFE
Most stars start life on the main sequence. The most massive grow into supergiants that ultimately explode. Intermediate-mass stars such as the Sun pass through a red-giant phase and form a planetary nebula. The least massive stars evolve so slowly that even the oldest are still on the main sequence, so astronomers cannot yet observe their final stages.

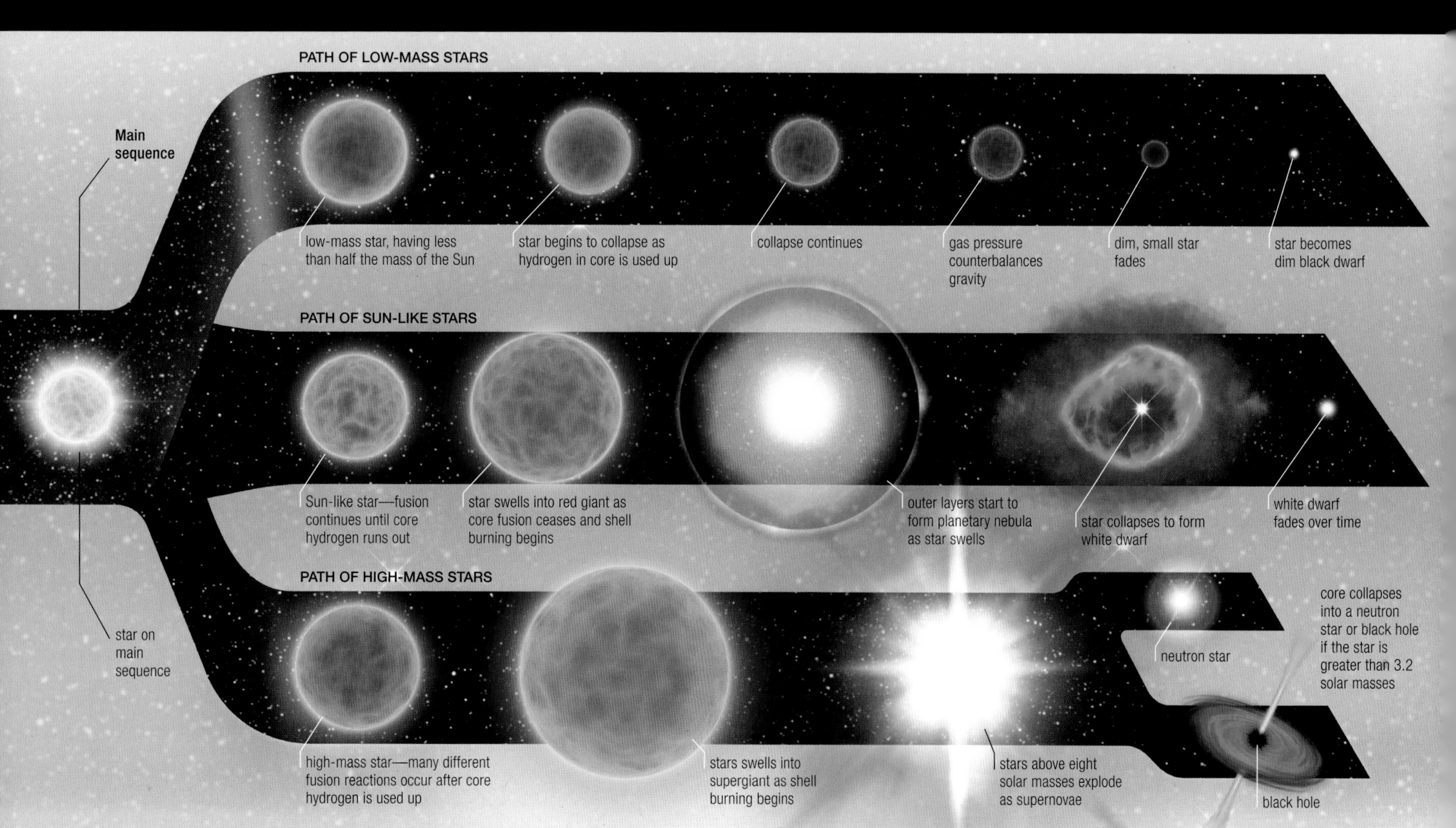
PATH OF LOW-MASS STARS
Main sequence
low-mass star, having less than half the mass of the Sun
star begins to collapse as hydrogen in core is used up
collapse continues
gas pressure counterbalances gravity
dim, small star fades
star becomes dim black dwarf
PATH OF SUN-LIKE STARS
Sun-like star—fusion continues until core hydrogen runs out
star swells into red giant as core fusion ceases and shell burning begins
outer layers start to form planetary nebula as star swells
star collapses to form white dwarf
white dwarf fades over time
PATH OF HIGH-MASS STARS
star on main sequence
high-mass star—many different fusion reactions occur after core hydrogen is used up
stars swells into supergiant as shell burning begins
stars above eight solar masses explode as supernovae
neutron star
core collapses into a neutron star or black hole if the star is greater than 3.2 solar masses
black hole

THE EVOLUTION OF STARS

urrent models of stellar evolution allow astronomers to track the ath of different types of stars across the H–R diagram (see p.203). typical newborn protostar may be bright but cool, due to its lack f internal equilibrium. As it settles down, it grows hotter until it ventually joins the main sequence at a point determined by its mass opposite). A star then remains there through most of its lifetime, ntil it exhausts the supply of hydrogen fuel in its core. Internal hanges will then cause it to evolve off the main sequence. At this oint, stars of different masses follow different paths. For example, a un-like star brightens and cools until it becomes a red giant. It then heds its outer layers, forming a planetary nebulae, after which it is ransformed into a white dwarf. Truly massive stars swell into upergiants and eventually destroy themselves in supernovae.

LIFE OF A STAR
All of the stars in a cluster form at the same time. Open star clusters such as he Jewel Box, NGC 4755, 7,500 light-years away, demonstrate how stars with different masses age. Most of the bright stars in this young cluster are higher-mass, blue-white heavyweights, the largest of which have evolved into red giants. The Sun-like stars will continue to shine on the main sequence for billions of years.

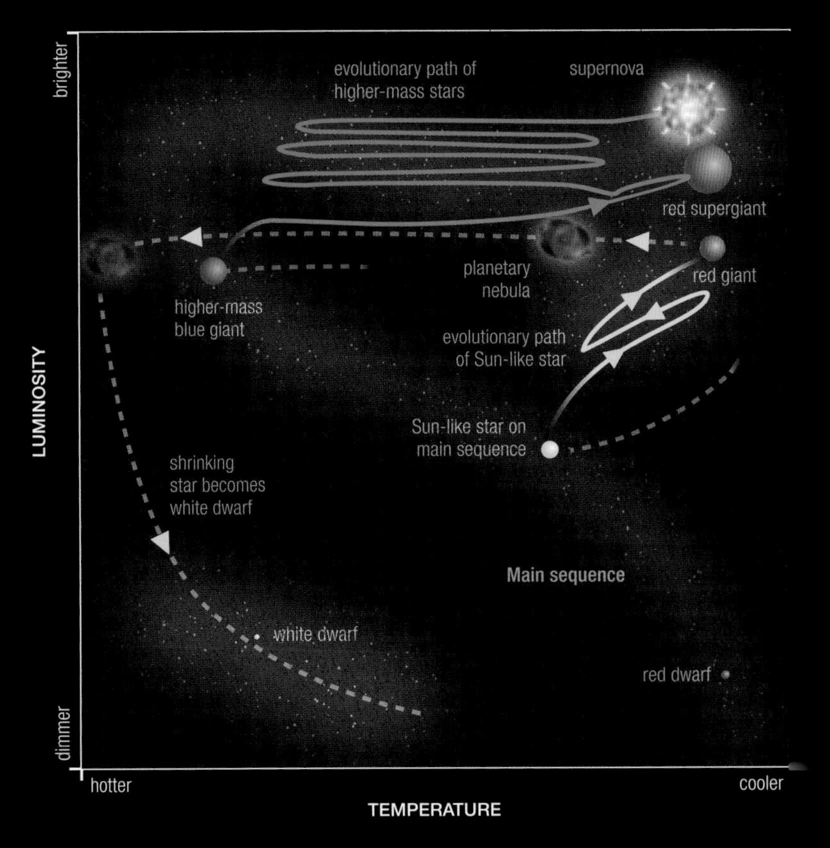

EVOLUTIONARY PATHWAYS
The paths on this H–R diagram show the typical evolutionary tracks taken by two different star types as they age. The smaller, Sun-like star ages relatively slowly (shown in yellow), while the higher-mass, blue-white star with more than eight times the solar mass changes more dramatically (shown in blue).

OUT OF THE ORDINARY

While the H–R diagram and its accompanying model of stellar evolution can explain the properties of the vast majority of stars, there re a few rare exceptions. Most of these are stars that appear older or younger than they should be – for example, the so-called blue-straggler stars found in the heart of dense globular clusters (see pp.176–77). Such oddities are best explained as rare cases of mass transfer, in which large amounts of matter are added to or stripped rom a star during its life, resulting in unexpected properties. Mass transfer is most common in close binary systems but can happen in clusters too (see pp.220–21).

BLUE STRAGGLERS
These are high-mass blue stars that appear to be young. However, they are only found in ancient globular clusters, such as NGC 6397 (shown here) within which star formation ceased billions of years ago. Blue stragglers are thought to have formed from collisions and mergers

THE MILKY WAY'S STAR CLOUDS

This image was produced as part of the near-infrared Two Micron All-Sky Survey. It shows the huge star clouds near the center of the Milky Way, with darker lanes of interstellar dust in between. Every single light source is a star, well over 90 percent of which lie on the main sequence. The glowing pink rosettes scattered throughout are star-forming nebulae.

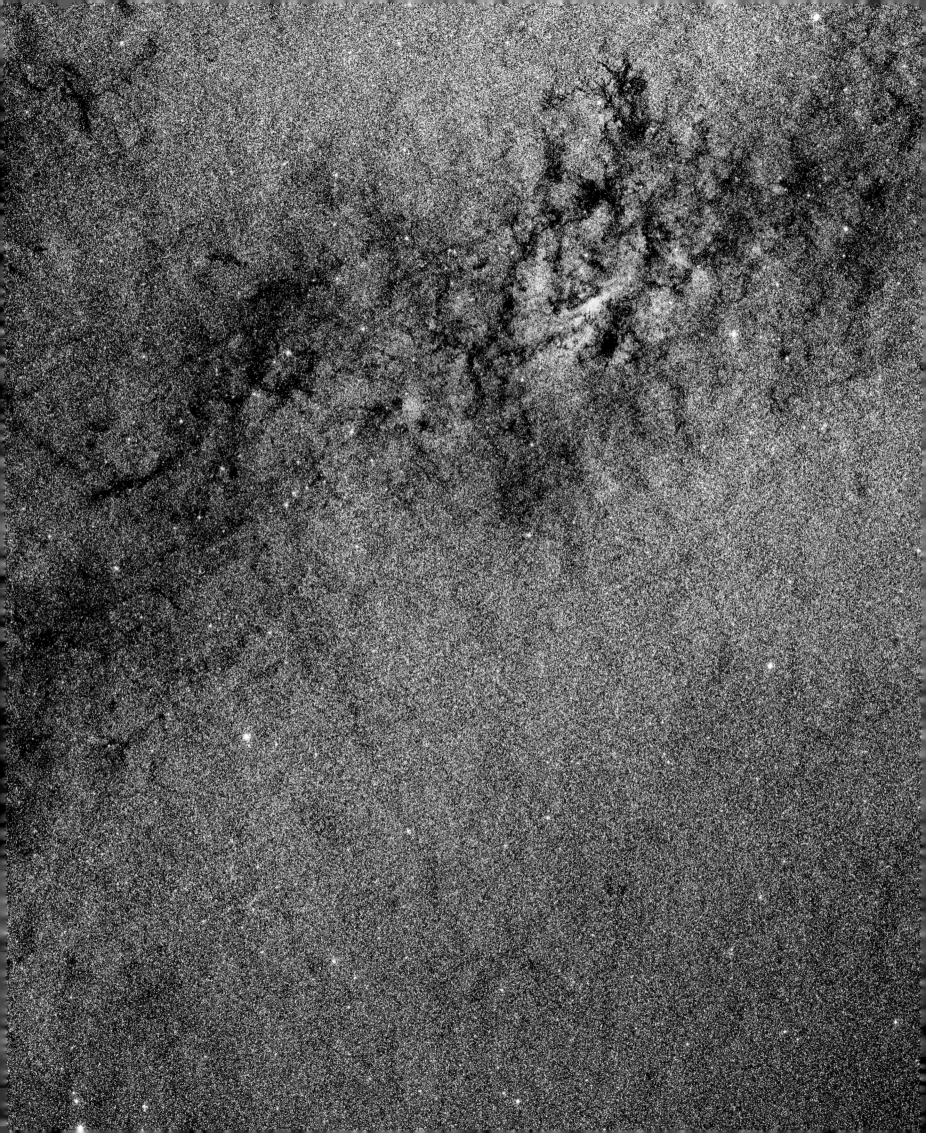

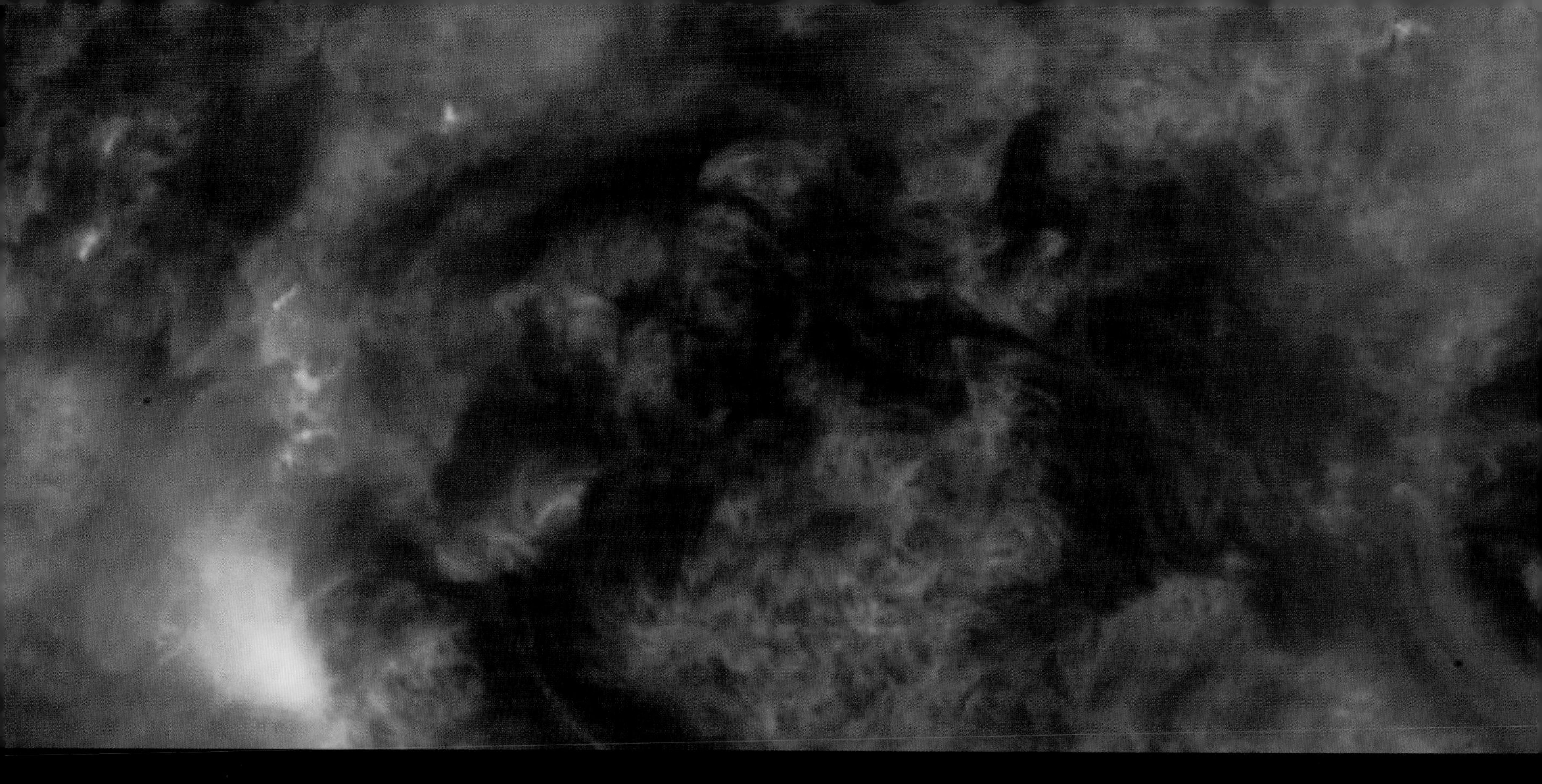

THE FUEL SOURCE OF STARS

At the heart of every star lies a planet-sized nuclear power plant, capable of producing enormous amounts of energy through the interactions of some of the tiniest particles in nature—atomic nuclei. This process not only generates heat and light—it also manufactures the heavier elements in the Universe.

THE SOLAR SURFACE
Radiation bursting from the surface of stars like the Sun is at the end of a long journey that began at the star's core. As it fights its way outward over thousands of years, much of its energy dissipates and ferocious gamma rays and X-rays are largely diminished to lower-energy ultraviolet, visible, and infrared radiation.

HOW STARS SHINE

The mystery of how stars generate so much energy puzzled astronomers for almost a century. Before the mid-19th-century, most assumed that the Sun was burning through a process similar to the chemical combustion seen on Earth. However, this would have rapidly exhausted the Sun's fuel and geological evidence made it clear that the Solar System is billions of years old. In 1920, English astronomer Arthur Stanley Eddington proposed that stars were powered by the nuclear fusion of hydrogen, the lightest element in the Universe.

It was not until 1938 that German and Russian astronomers Hans Bethe and George Gamow worked out the details of the most common nuclear-fusion reaction in stars like the Sun—the proton-proton chain. They calculated that at temperatures of 27 million °F (15 million °C) and under immense pressures, the nuclei of several hydrogen atoms (subatomic particles called protons) combine (fuse together) to create nuclei of the next simplest element, helium. This happens in several steps, over billions of years, and the resulting helium nucleus is slightly lighter than the mass of the particles used to create it. Some of the lost mass is accounted for by the release of subatomic particles, but a large amount is released as high-energy gamma rays.

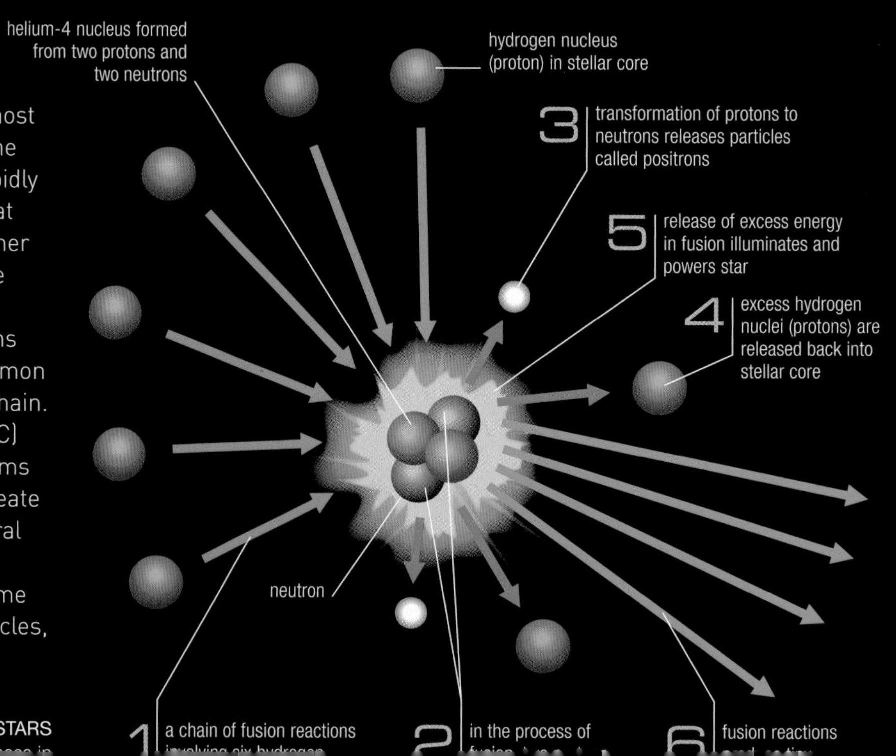

helium-4 nucleus formed from two protons and two neutrons
hydrogen nucleus (proton) in stellar core
3 transformation of protons to neutrons releases particles called positrons
5 release of excess energy in fusion illuminates and powers star
4 excess hydrogen nuclei (protons) are released back into stellar core
neutron
1 a chain of fusion reactions
2 in the process of
6 fusion reactions

FUSION IN STARS
This diagram summarizes the proton-proton fusion process in

MAKING HEAVY ELEMENTS

All stars shine for most of their lives through the fusion of hydrogen into helium. But near the end of their lives when the hydrogen in the core runs out, conditions in their collapsing centers become even more extreme. The helium nuclei themselves begin to fuse together, forming nuclei of elements such as oxygen, carbon, and nitrogen. Stars like the Sun can only push the fusion process this far (see pp.228–29), but higher-mass stars can repeat the trick, fusing the carbon, nitrogen, and oxygen to create even heavier elements, such as iron (see pp.230–31). However, the fusion of these heavier elements generates much less energy than the fusion of lighter ones, and the fusion of iron actually absorbs energy—so no normal star can produce the heaviest elements in the cosmos. Instead, atoms that form precious metals and radioactive elements are created in supernova explosions (see p.248).

HIND'S CRIMSON STAR
This variable red giant (also known as R Leporis) lies 800 light-years from Earth in the constellation Lepus. It gets its particularly intense red coloring from large clouds of carbon in its atmosphere, which absorb the blue component of light emerging from its surface.

385 MILLION BILLION

The energy output of a Sun-like star in gigawatts. A Sun-like main-sequence star pumps out more energy in a single second than all of Earth's power stations could in three-quarters of a million years.

FUSION IN HIGH-MASS STARS

The proton-proton fusion seen in Sun-like stars is a slow process (see opposite). As a result, these stars can shine steadily for billions of years. Higher-mass, hotter stars that are born with small but significant amounts of carbon in their cores follow another, more energy-efficient fusion pathway called the carbon-nitrogen-oxygen (CNO) cycle. In this process, the carbon nuclei act as a catalyst—individual protons bind onto them and build up into nuclei of nitrogen and oxygen, before eventually breaking apart to release the original carbon nuclei and a

BRILLIANT BUT SHORT-LIVED
The second brightest star in the sky, Canopus is a stellar heavyweight, 300 light-years from Earth and 15,000 times more luminous than the Sun. Its brilliant light is powered by the super-efficient reactions of the CNO cycle.

STARBIRTH

Stars are born out of nebulae, the cool clouds of gas and dust that fill much of interstellar space. The process of star formation, in which a single knot of gas collapses and heats up until it is hot and dense enough to shine, may take a few million years, but this is surprisingly rapid on an astronomical timescale.

EMERGENCE OF A PROTOSTAR

Star formation begins when a cold, dark gas cloud in deep space is subjected to a trigger event, such as a nearby supernova explosion, an encounter with a passing star, or passage through a more crowded region of space within a galaxy. The pressure waves or tidal forces that are encountered push and pull at the cloud, compressing parts of it until they become dense enough to exert a significant gravitational pull on their surroundings.

Once the original cloud's density is unbalanced by these forces, gravity does the rest, pulling more and more material onto the developing knot of matter and concentrating most of this material at the very center to form a protostellar nebula. As the material grows denser and more concentrated, random motions within the nebula are transformed into a faster, uniform rotation around a single axis. Collisions between particles jostling in the nebula raise the temperature, most notably in the center, and the protostar begins to glow gently with infrared radiation.

SUPERNOVA SHOCK WAVE
The glowing ring in the upper half of this Spitzer Space Telescope image is the expanding shock wave from a supernova explosion. In the lower half, a burst of star formation is taking place as the shock wave slams into a nearby gas cloud.

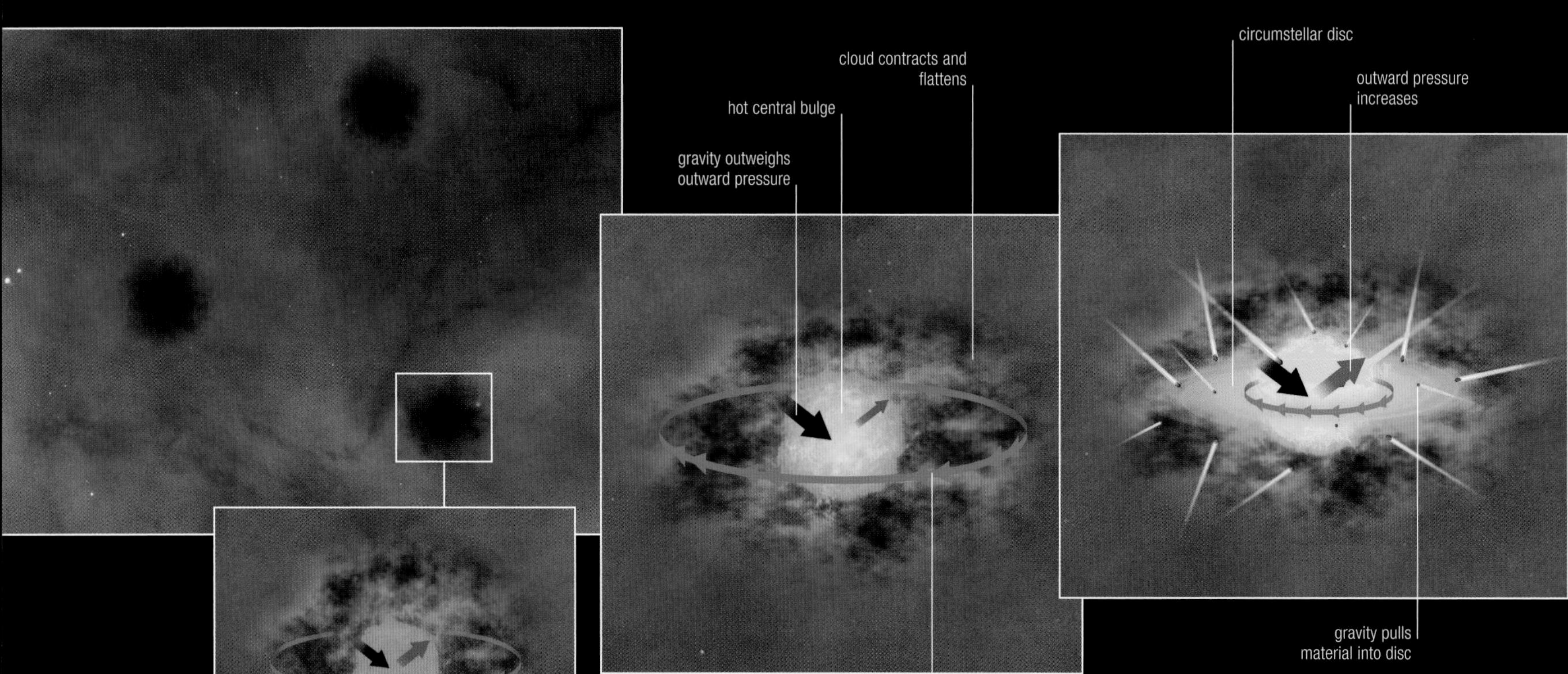

BOK GLOBULES

As individual protostars compete for the available material from the larger nebula, they strip their surroundings bare, emerging from the nebula as discrete, dust-laden dark blobs called Bok globules, perhaps a light-year across. Inside each one, the process of collapse toward the center continues, creating a flattened, spinning disc with a dense bulge, the young star, at the center. Often the star's gravity is not enough to hold onto all the material falling onto it, and large amounts are flung out in jets. Collapse around two or more concentrations of matter to form a binary or multiple star system is also common (see p.221).

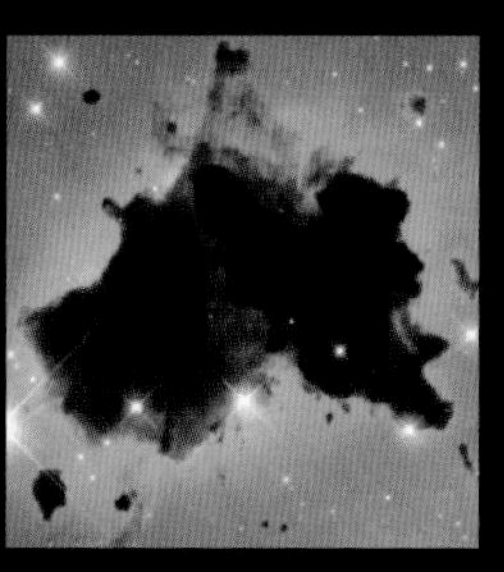

BOK GLOBULE
Dark clouds of gas and dust, such as Thackeray's Globules in the constellation Centaurus (seen here silhouetted against another brighter nebula), conceal the seeds of star formation.

FUSION BEGINS

At first, stars shine through the energy generated by their gravitational contraction. Eventually, temperatures in the core become high enough for some simple, low-energy forms of nuclear fusion (see p.208) to begin. At first these involve the fusion of deuterium (a heavy isotope of hydrogen that requires less extreme conditions for fusion) to make the next lightest element, helium. Eventually, however, the core becomes so hot and dense that true hydrogen fusion can begin. The onset of fusion creates massive changes in the star's interior and generates fierce radiation and stellar winds that light up the surrounding nebula, carving it into fantastical shapes. After a period of fluctuations, the star eventually settles down, joining the main sequence of hydrogen-burning stars (see p.203), where it will spend most of its life.

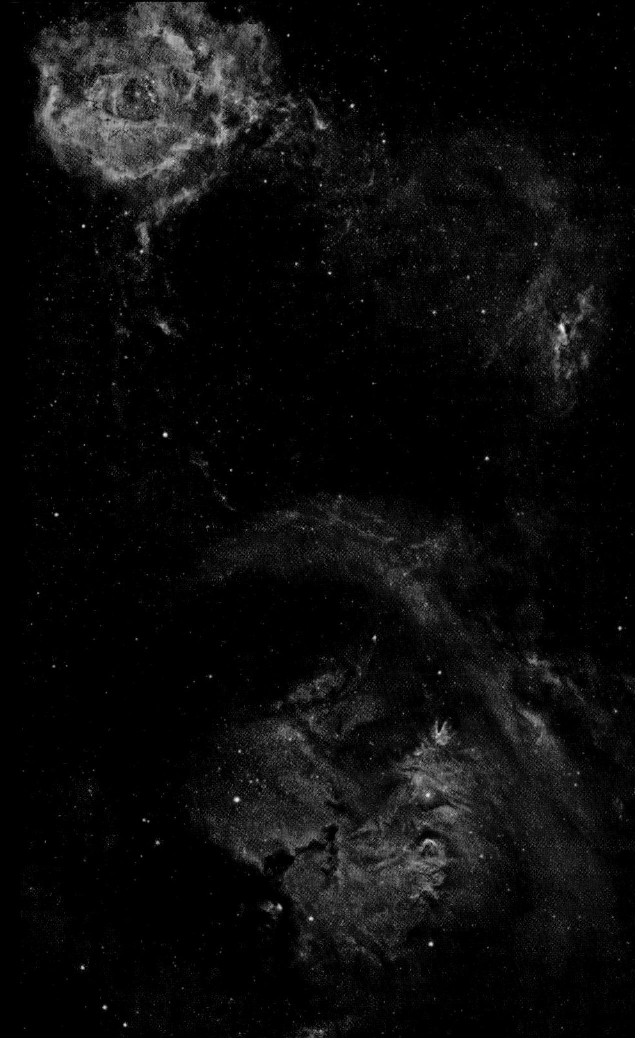

STAR-FORMING REGION
This image reveals two starbirth nebulae in the constellation Monoceros. At the top is the Rosette Nebula, with the newborn stars of cluster NGC 2244 embedded within it. At the bottom is the region around the Snowflake Cluster, NGC 2264.

material continues to fall into disc
outward stellar wind flows along axis of rotation
cloud of gas and dust around star is now smaller and less dense

BIPOLAR OUTFLOW
Eventually the star spins so rapidly that new material falling onto it is almost immediately flung back off. In order to escape from above and below the disc, this excess material forms two tight jets emerging along the axis of rotation. These jets usually only become visible if they collide with more gas in the surrounding region.

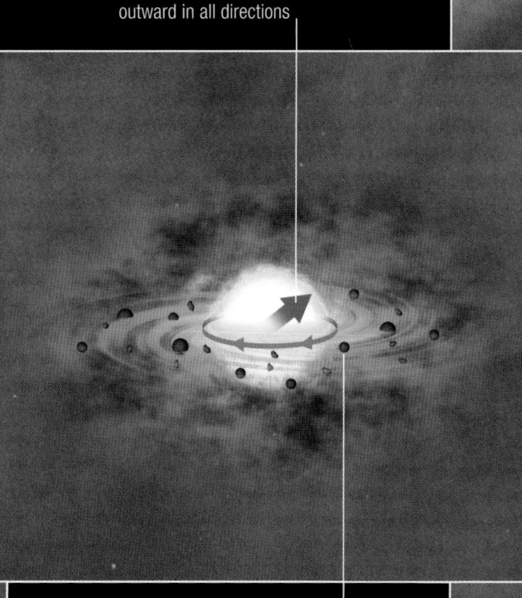

stellar wind now flows outward in all directions
planets begin to form in disc

PRE-MAIN SEQUENCE
As the star begins to shine and generate outward radiation pressure, surrounding gas is blown away, leaving planets to form from the nebula's remnants. Meanwhile, the star's core continues to collapse, until deuterium fusion and finally hydrogen fusion can begin. After a period of instability, the star settles down.

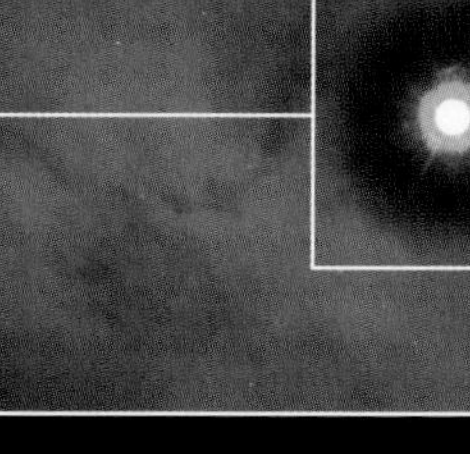

area immediately around star emptied of gas and dust
gas in nebula excited by young star
6

CLUSTER OF YOUNG STARS
Stars form in clusters, and Bok globules often give rise to two or more stars in orbit around each other. Fierce ultraviolet radiation from the hottest, youngest stars can excite gases in remnants of the original nebula, causing them to glow, while the pressure from radiation and stellar wind can erode and compress the nebula's edges.

THE ORION NEBULA

Sometimes known simply as the Great Nebula, the Orion Nebula is the largest and brightest star-forming region visible in the night sky. This vast rosette of glowing gas is about 24 light-years across. The nebula is just the central region of a much larger cloud known as the Orion Molecular Complex, which spans much of the Orion constellation and is hundreds of light-years across. To the naked eye, the Orion Nebula has a distinctive green tinge, but long-exposure images also reveal pinkish-red and blue colors. The red is an emission that occurs when hydrogen in the nebula is excited by high-energy radiation from the central stars, the green is a similar emission from oxygen, while the blue is reflected light.

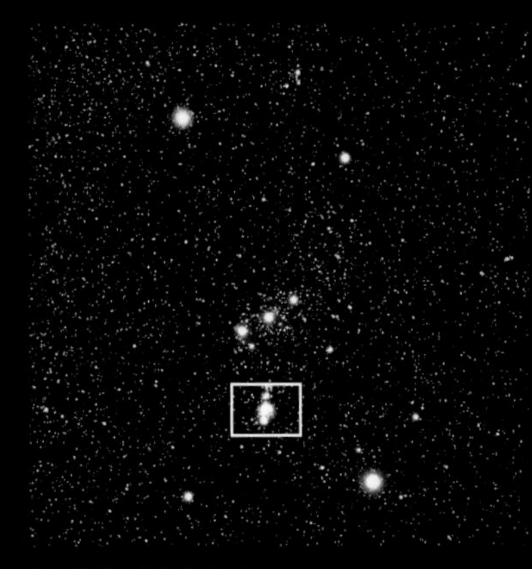

NAKED-EYE VIEW
To the naked eye, the Orion Nebula appears as the central, slightly fuzzy "star" in the sword of Orion. The constellation takes its name from Orion, a hunter in Greek mythology.

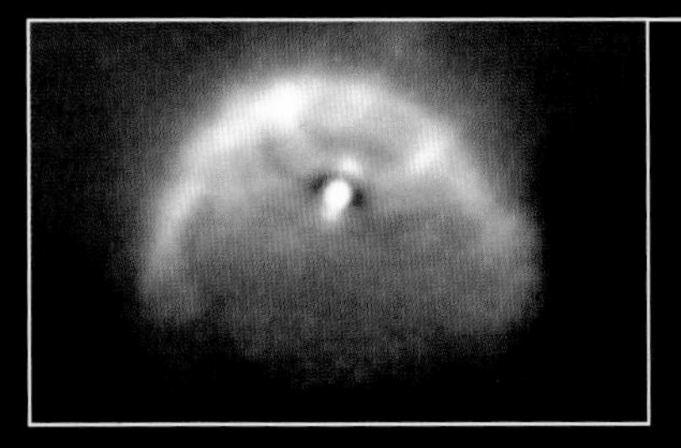

INFANT SOLAR SYSTEMS
Detailed images of the Orion Nebula from the Hubble Space Telescope reveal dozens of structures called proplyds. These are discs of planet-forming material held around young stars. Shock waves form where these infant solar systems plow through the surrounding gas.

INFRARED ORION
This near-infrared image of the Orion Nebula was taken by the European Southern Observatory's VISTA Telescope, Chile. The image reveals a swarm of hitherto unseen stars embedded in the nebula, the youngest of which glow faintly red through their cocoons of gas and dust.

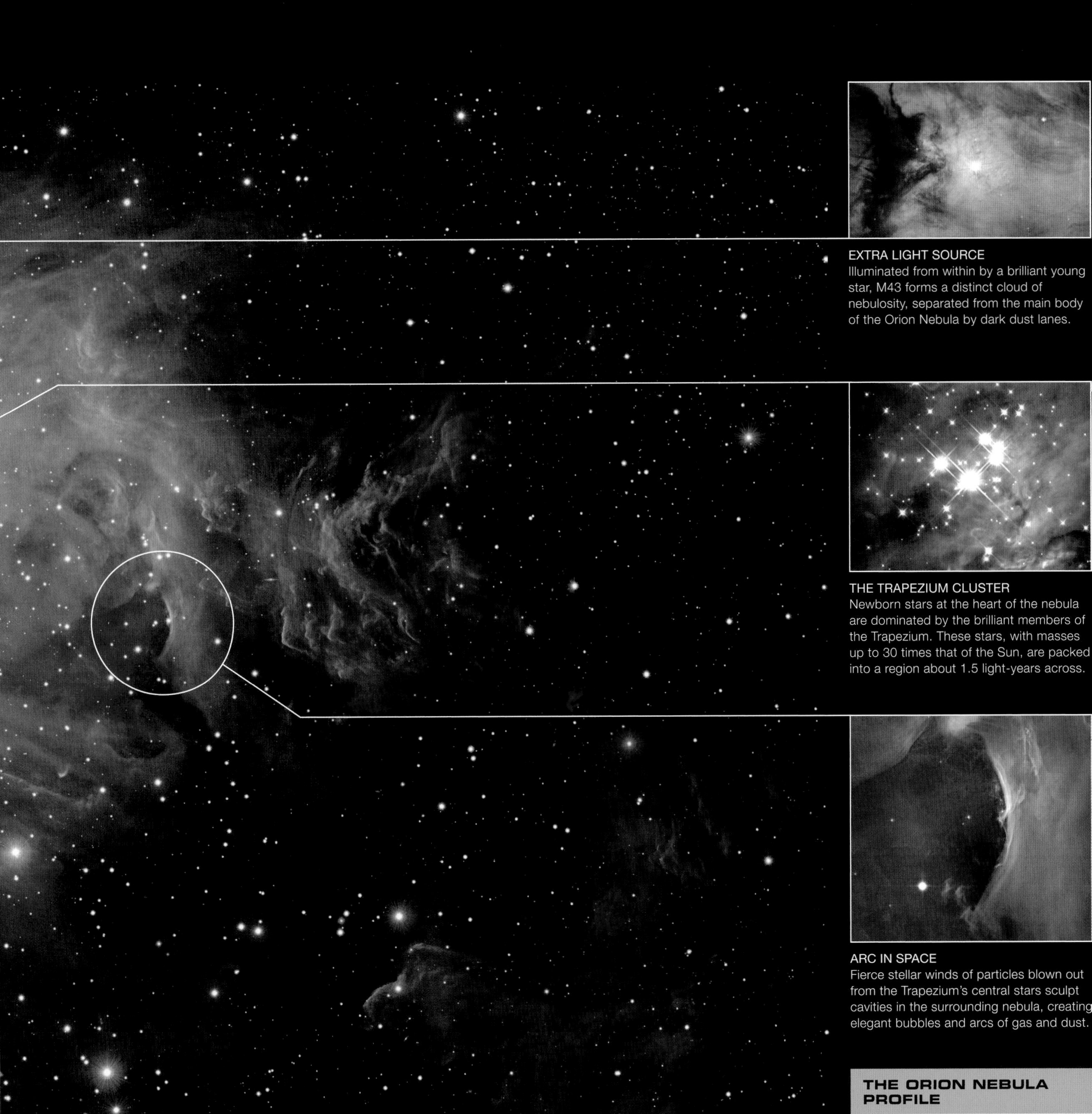

EXTRA LIGHT SOURCE
Illuminated from within by a brilliant young star, M43 forms a distinct cloud of nebulosity, separated from the main body of the Orion Nebula by dark dust lanes.

THE TRAPEZIUM CLUSTER
Newborn stars at the heart of the nebula are dominated by the brilliant members of the Trapezium. These stars, with masses up to 30 times that of the Sun, are packed into a region about 1.5 light-years across.

ARC IN SPACE
Fierce stellar winds of particles blown out from the Trapezium's central stars sculpt cavities in the surrounding nebula, creating elegant bubbles and arcs of gas and dust.

THE ORION NEBULA PROFILE

THE EAGLE NEBULA

One of the sky's best-known star-forming regions, the Eagle Nebula in the constellation Serpens is named after the bird-shaped silhouette formed by the bright nebula of glowing gas against the dark background. The long, fingerlike structures, known as the "pillars of creation," are dense columns in which stars are forming. As fierce radiation from the newborn stars erodes and sculpts their surroundings into bizarre shapes, infant star systems emerge from the edges of the pillars in roughly spherical masses called Bok globules. At the center of the nebula, about 7,000 light-years from Earth, is the young star cluster M16.

HOT CLOUD
Taken by the Spitzer Space Telescope, this infrared image reveals a hot area within the nebula, where dust is heated and excited by a shock wave from the supernova explosion of a massive young star.

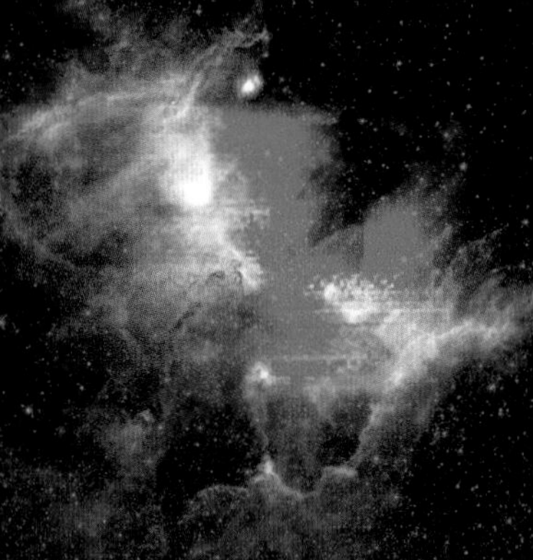

SUPERNOVA SHELL
This false-color infrared view shows the supernova remnant's hot shell (green) against the colder dust of the pillars (blue and purple). After a thousand years, the expanding shock wave will tear through the pillars, exposing the infant stars inside.

THE BIGGER PICTURE
The Eagle Nebula's central region is seen in this wide-field image from the La Silla Observatory in Chile. The bright star cluster M16 can be seen near the top of this picture. The Eagle is in a much wider nebula known as IC 4705.

THE EAGLE NEBULA PROFILE

Catalog number	M16
Type	Star-forming nebula
Constellation	Serpens
Distance from Earth	7,000 light-years
Diameter	70 light-years

X-RAY REVELATION
Here, Hubble's famous "pillars of creation" image is overlaid onto a Chandra X-ray Observatory view of the same region. Piercing the dense veils of dust within the nebula, Chandra uncovers X-rays from young stars in and around the pillars.

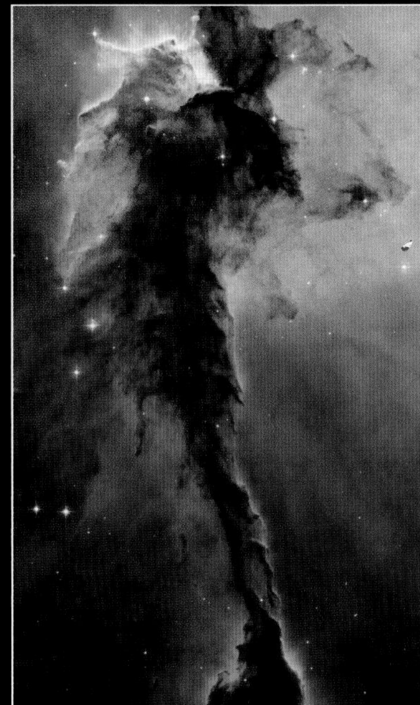

THE SPIRE
Roughly 9.5 light-years long, the Spire is another elegant structure within the Eagle Nebula. It is a lot further along the path of disintegration into individual star-forming knots than the denser pillars nearby.

THE INFRARED UNIVERSE

Telescopes for observing the sky at infrared wavelengths present unique challenges to their builders, but they offer a tantalizing view of a very different Universe to the one seen in visible light. The infrared sky is a wonderland of cool gas and dust, infant stars, and objects too cool to shine in visible light.

WHAT IS THE INFRARED?

Instruments used to observe objects in space use various parts of the electromagnetic spectrum (see p.11). The infrared part is best understood as heat radiation, or rays that have longer wavelengths and lower frequencies than visible red light and therefore require less energy to produce. Most types of matter must be at very high temperatures in order to emit visible light, but every object with a temperature above absolute zero (-459°F/-273°C) emits some form of radiation. Most of this lies in the infrared part of the spectrum. By tuning observations to look at different wavelengths, it is possible to see structures that appear dark in visible light, and to see through materials that are otherwise opaque.

THE TRIFID NEBULA IN VISIBLE LIGHT

THE TRIFID NEBULA IN INFRARED

THE TRIFID NEBULA
A Spitzer Space Telescope infrared image (bottom) exposes hidden detail in this famous nebula, more familiar in visible light (top). Spitzer can see through the dust-laden gas around the nebula, revealing starbirth regions, shown in green, and cooler material, seen in red.

THE HERSCHEL SPACE OBSERVATORY
Launched in 2009, with the Planck satellite, this is the European Space Agency's latest infrared satellite—the largest infrared telescope ever launched. It is carrying a range of ingenious detectors and will observe the infrared sky at a wider range of wavelengths than its predecessors.

THE HERSCHEL SPACE OBSERVATORY PROFILE

MISSION	
Launch date	May 14, 2009
Mission length	Minimum three years; earliest end date 2012
Launch vehicle	Arianne 5 ECA
HERSCHEL	
Agency	European Space Agency (ESA)
Height	24ft 7in (7.5m)
Diameter	13ft (4m)
Weight at launch	7,500lb (3,400kg)
Primary mirror	11ft 4in (3.5m) diameter
Instruments	Heterodyne Instrument for the Far Infrared (HIFI), Photodetector Array Camera and Spectrometer (PACS), Spectral and Photometric Imaging Receiver (SPIRE)
Power source	Solar cells
SCALE	24ft 7in (7.5m); 13ft (4m)

Solar cells
Mounted on the exposed side of the sun shield, the cells generate the satellite's power.

NASA'S INFRARED TELESCOPE FACILITY (IRTF)
Built at an altitude of 13,800ft (4,200m) on Mauna Kea, Hawaii, this is one of the first Earth-based infrared observatories. Unusually for a large telescope, the IRTF mostly observes objects in the Solar System.

COOL OBSERVATORIES

Detecting weak infrared radiation from space creates unique problems. Infrared rays entering Earth's atmosphere are absorbed by water vapor or swamped by the infrared radiation in the air. Even the warmth of the telescope and its detectors can drown out a weak signal from space. So infrared telescopes are generally cooled to extremely low temperatures using low-temperature cryogenic liquids such as liquid helium (boiling point -452°F/-269°C). Earth-based infrared telescopes are also sited on high, dry, mountain tops above most of the atmospheric water vapor. Infrared satellites avoid the water problem but have a limited lifetime because their coolant liquids evaporate and escape over months or years of operation. Nevertheless, they observe a wider range of infrared wavelengths than ground-based telescopes. Since the launch of the InfraRed Astronomical Satellite (IRAS) in 1983, infrared satellites have revolutionized our view of the heavens.

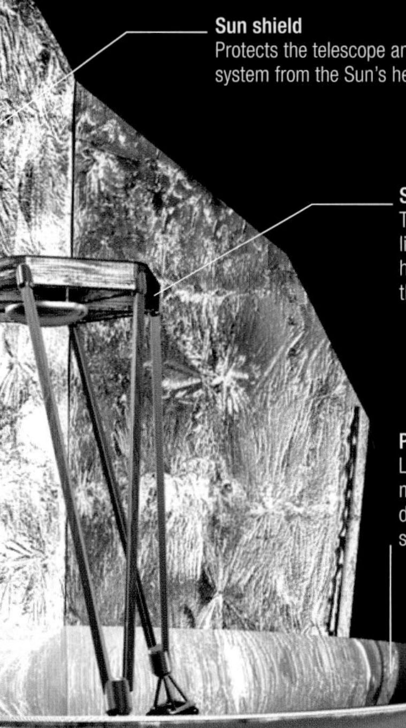
Sun shield
Protects the telescope and cooling system from the Sun's heat and light.
Secondary mirror
This mirror reflects light back through a hole in the center of the primary mirror.
Primary mirror
Large 11ft 6in (3.5m) mirror collects light, directing it toward the secondary mirror.
Instrument housing
All three infrared detectors are housed above the cooling system.
Cryostat tank
Contains over 3,500 pints (2,000 liters) of liquid helium at -458°F (-272°C).
Cooling system
Fueled by liquid helium, this keeps the instruments close to absolute zero (opposite).
Service module
Contains power, attitude control, data handling, and communications systems.

LOOKING INSIDE NEBULAE

Star-forming nebulae are one of the most important sources of infrared radiation in the sky. The use of infrared telescopes has transformed scientists' understanding of these turbulent regions. The dark pillars and globules of dust-laden gas in which stars begin their lives are impenetrable to visible light and produce only a warm glow in the longer infrared wavelengths. But near-infrared telescopes, which can capture the shorter infrared wavelengths emitted by hotter objects, can image the radiation from infant stars embedded within these clouds. These telescopes have revealed the structures and processes of starbirth. For example, they can image protostars that have not yet begun to shine through nuclear fusion but are glowing strongly in the infrared as they contract under their own gravity.

THE RHO OPHIUCHI COMPLEX
One of the closest star-forming regions, the Rho Ophiuchi Nebula lies 430 light-years away from Earth. Much of the region appears dark in visible light, but this Spitzer Space Telescope image reveals the infrared glow of material just 36°F (20°C) above absolute zero. At the heart of the nebula lie around 300 newborn or still-forming stars.

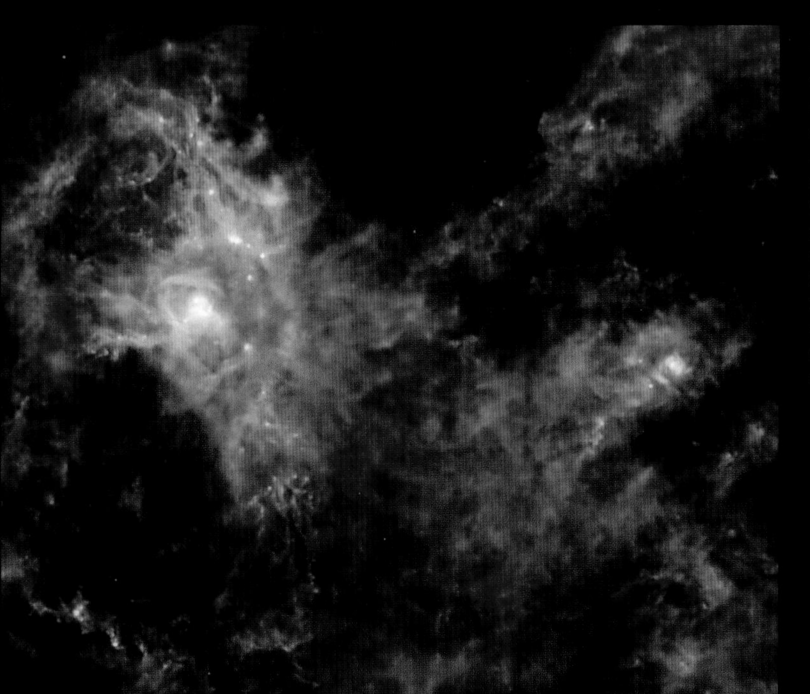

UNSEEN NURSERY
This early infrared image from the Herschel Space Observatory reveals complex structures in the constellation Aquila, around 1,000 light-years from Earth. The two bright regions are newborn stars illuminating their surroundings, while the filaments around them show where new stars are still forming in the same stellar neighborhood.

COOL COSMOS

Infrared radiation also reveals a host of other objects too cool to shine in visible light. Among the most important discoveries made with infrared telescopes have been brown dwarfs, a long-suspected, but hitherto unconfirmed population of "failed stars" that do not have sufficient mass to shine through nuclear fusion but nevertheless are sources of infrared. Equally groundbreaking was the discovery of protoplanetary discs—clouds of dust and gas from which planets may form (see pp.254–55). The discs cannot be seen in visible light, but heat from the central stars warms them enough to glow in infrared.

INFRARED BROWN DWARFS
The Hubble Space Telescope's Near Infrared Camera and Multi-Object Spectrometer revealed a host of hitherto unseen brown dwarf stars embedded within the clouds of the Orion Nebula (see pp.212–13). Such stellar failures are only visible when they are young and relatively hot.

STAR SYSTEMS

Due to stars' enormous mass, gravity ensures that most of them form groups, or systems. These systems range from large groups, or clusters, of hundreds or thousands of stars, to closely bound groups of two or more stars, called doubles (or binaries) and multiples, trapped in orbit around each other.

STAR CLUSTERS

Large associations of stars that share a common origin and are loosely bound together by gravity are known as clusters. Despite their proximity, the stars within clusters do not generally orbit one another. Within the Milky Way, there are two distinct types: open clusters, which can contain anything from dozens to hundreds of relatively young stars; and globular clusters, which can contain fewer than a hundred or up to thousands of much older, red-and-yellow stars.

Open clusters are found along the Milky Way's spiral arms, often near the nebulae from which they formed. Each star or group of stars within a cluster has its own path through space, so the cluster tends to disintegrate within a hundred million years. Globular clusters orbit in the Milky Way's halo, above and below the plane of the galactic disc. The gravity in these clusters is so strong that their close-packed stars are bound together forever. The biggest and brightest of these clusters, Omega Centauri, is some 15,000 light-years away from Earth.

OPEN CLUSTER
The double cluster NGC 884 (left) and NGC 869 (right) is a famous pair of open clusters in the northern constellation Perseus that each contain several hundred bright, young, blue-white stars. The clusters are 6,800 and 7,600 light-years away from Earth respectively and are both about 5 million years old.

GLOBULAR CLUSTER
In the constellation Hercules, M13 is one of the brightest and largest globular clusters visible from Earth. The cluster lies roughly 25,000 light-years away from Earth, within the Milky Way's halo. It contains about 100,000 stars packed into a region of space 160 light-years across.

A NEWBORN CLUSTER
This spectacular image from the Hubble Space Telescope shows the open cluster NGC 3603 in the southern constellation Carina. Just a few million years old, this cluster has recently emerged from a cloud of surrounding nebulosity that is being driven back by the fierce radiation from the hot, bright young stars within it.

MULTIPLE STARS

In contrast to huge star clusters, multiple stars are groups of two or more stars that originate in the same area, or globule, of gas and dust and are locked in orbit around one another. They can occur within larger clusters. The most common grouping, a simple pair of stars, is often called a binary. The stars within multiple systems actually account for most of the stars in the Milky Way—stars that occur singly, such as the Sun, are in the minority.

There is a huge variety of multiple stars: some groups consist of more-or-less identical stars, but in others, the components can be very different from each other. Similarly, the distance between the stars may vary from billions of miles to mere thousands—the latter are often too close to be seen separately through even the most powerful telescope.

PISMIS 24-1
This bright, open cluster has recently emerged from star-forming nebula NGC 6357, roughly 8,000 light-years from Earth in the constellation Scorpius. Measurements of the cluster's brightest star, known as Pismis 24-1, initially suggested that it had a mass of more than 200 Suns.

A TRIPLE STAR
This high-resolution image from the Hubble Space Telescope reveals that Pismis 24-1 is in fact a close pair of binary stars. Spectroscopic measurements of the two stars reveal that one of them is itself a binary, making three stars altogether.

LEARNING FROM STAR SYSTEMS

Stars systems can reveal a great deal about general stellar behavior because the stars within them share certain characteristics. All the stars in a cluster or multiple group are at the same distance from Earth, so any observed differences in apparent brightness are reflections of real differences in their physical luminosity. Astronomers also know that all the stars in these groups formed at the same time, so they can compare the stars' relative stages of evolution. In double and multiple systems, astronomers are able to work out the relative masses of the individual stars from the sizes of their orbits (see right), and in the case of eclipsing binaries (see below), it is possible to calculate the actual mass of each star.

center of mass

EQUAL MASS
If the stars in a binary system have a similar mass, the center of mass around which they both orbit will be midway between the two, and their orbits will be about the same size.

center of mass

UNEQUAL
If one star has a greater mass than the other, then the center of mass will move toward the larger star. The less massive star will have the larger orbit, and the higher-mass star a smaller orbit.

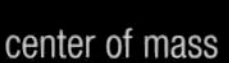

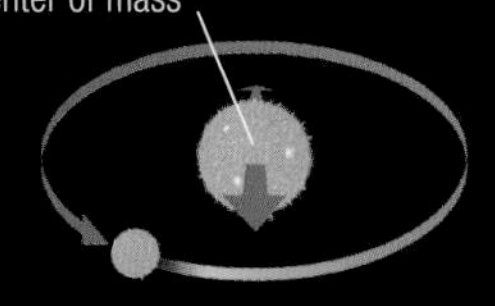

DOMINANT STAR
If one star is very much heavier than the other one, the center of mass is inside the dominant star. The lower-mass star orbits the heavier one like a planet, and the more massive star wobbles back and forth.

DETECTING MULTIPLE STARS

The vast majority of multiple stars are either so tightly grouped or so far from Earth that it is impossible to resolve their individual components with a telescope. However, close study of apparently indivisible stars can reveal their true nature in one of two ways. In rare cases, where two stars pass directly in front of each other as seen from Earth, they can form an eclipsing binary system (see below). More often, studies of stellar spectra reveal multiple sets of spectral lines that shift toward the red or blue end of the spectrum as the stars move toward or away from Earth in their orbits (see p.306).

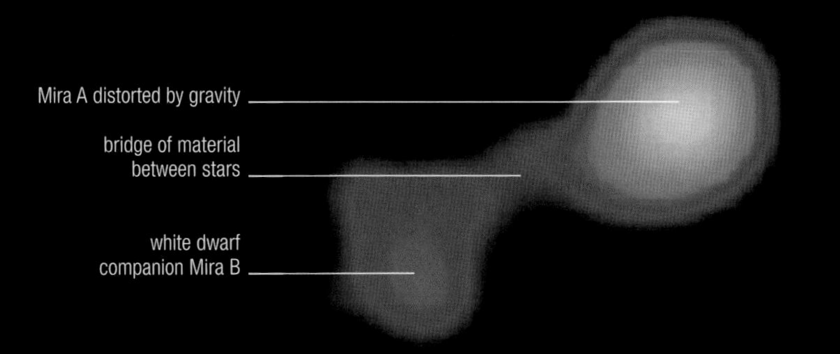

CONTACT BINARIES
Occasionally, stars orbit so closely that they actually touch. One star pulls material away from the other, sometimes in sufficient quantities to alter the evolution of both. These contact-binary systems are often stellar remnants with powerful gravity or bloated giants with a weak grip on their outer layers—as seen in this Chandra X-ray Observatory view of the red giant Mira A and its companion, the white dwarf Mira B.

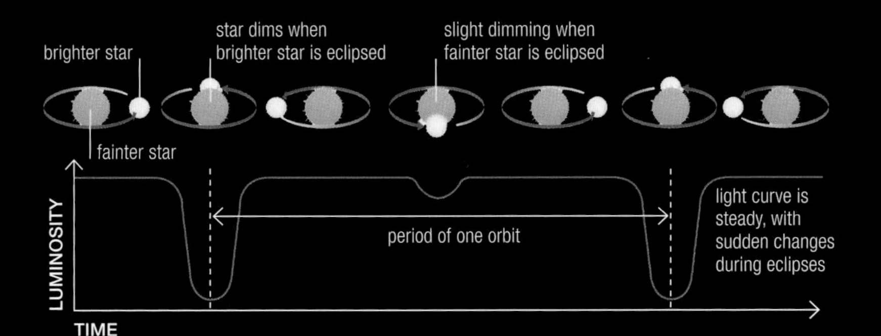

ECLIPSING BINARIES
In an eclipsing binary system, two stars occasionally pass in front of one another as seen from Earth. This results in a drop in the total amount of light from the system, creating one or more distinctive dips in each star's normally steady light. The effect varies according to the luminosity and size of each star.

THE PLEIADES

Probably the most famous star cluster in the sky, the Pleiades is also one of the brightest and closest to Earth. Often called the Seven Sisters, after the nymphs from Greek mythology, it features in folklore around the world. The handful of Pleiades stars that can be seen with the naked eye are just the brightest members of a group that is at least a thousand strong. About 90 light-years in diameter and 440 light-years away, the Pleiades as we see it is thought to be 100 million years old. It will probably exist for another 250 million years before its surviving members are dispersed in different directions. The brightest stars appear to be surrounded by wisps of dust that reflect light from the stars toward Earth, creating a delicate blue reflection nebula.

NAKED-EYE VIEW
At least six of the Pleiades' brilliant blue-white stars are visible to the naked eye on clear nights. They form a small, hook-shaped cloud on the shoulder of the constellation Taurus, the Bull.

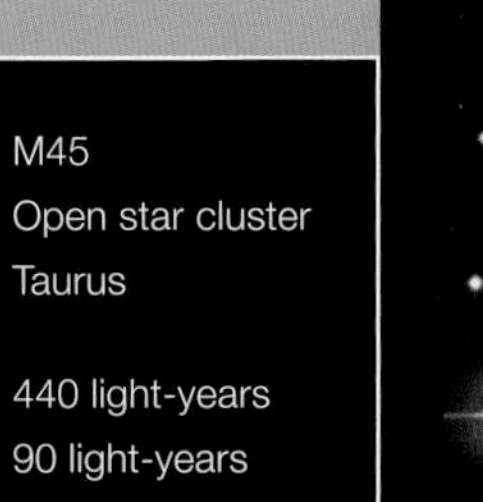

THE PLEIADES PROFILE

Catalog number	M45
Type	Open star cluster
Constellation	Taurus
Distance from Earth	440 light-years
Diameter	90 light-years

PASSING SHOW
Despite appearances, the glowing dust surrounding the Pleiades is not a remnant of the cluster's formation—it is just a cloud in the interstellar medium that the stars happen to be passing through at the moment.

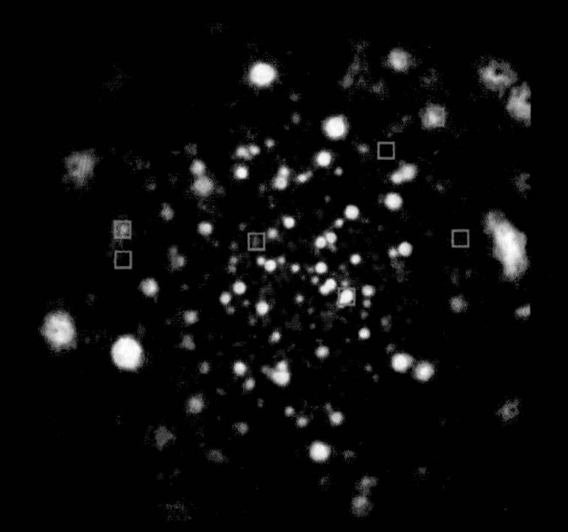

X-RAY VIEW
This ROSAT picture reveals the intensity of X-rays from stars in the Pleiades region. The colors relate to the temperature of each star's outer atmosphere, or corona, ranging from red (comparatively low temperature) through green to blue (high temperature). The green squares indicate the positions of the naked-eye Pleiades.

WEB OF DUST
The weblike structure of the interstellar dust around the central region of the Pleiades can be seen in this infrared view. Infrared also reveals failed brown dwarf stars, too faint to shine in visible light, embedded within the cluster.

MEROPE NEBULA
The fifth-brightest Pleiades star, Merope, is surrounded by the Merope Nebula—the densest and brightest region of reflection nebulosity. The star itself is 600 times more luminous than the Sun.

The second-brightest globular cluster in the sky after Omega Centauri, 47 Tucanae is a huge ball of at least 1 million ancient stars. It has an unusually dense core and a diameter of around 120 light-years. As its name suggests, the cluster lies in the southern constellation of Tucana, about 16,700 light-years away. Studies of 47 Tucanae's crowded core have revealed some previously unknown processes that help to shape the development of globular clusters.

STAR MOTION

In 2006, astronomers used observations of 47 Tucanae's core by the Hubble Space Telescope to measure the motions of 15,000 stars. Their results showed that interactions between crowded stars in a globular cluster cause heavy stars to fall toward the cluster's center.

47 TUCANAE PROFILE

Type	Globular cluster
Constellation	Tucana
Distance from Earth	16,700 light-years
Diameter	120 light-years

SOUTHERN HIGHLIGHT

With an apparent diameter that is roughly equal to that of the full Moon, the globular cluster 47 Tucanae looks to the naked eye like a fuzzy star of middling brightness. In fact, it was originally considered to be a star and given a name like that of other stars.

CLOSE TO THE CENTER

By imaging 47 Tucanae's center over seven years with the Hubble Space Telescope, astronomers tracked individual stars' orbits. This revealed that the heaviest stars in a globular cluster tend to accumulate at its center, where they may fall into orbit around one another or even merge.

EVIDENCE FOR WHITE DWARF STARS

This Hubble image, made shortly after the telescope's initial optics were repaired in 1994, was the first to reveal white dwarfs in a globular cluster. The discovery enabled astronomers to estimate 47 Tucanae's age at an astonishing 13 billion years.

M30

Smaller than 47 Tucanae, but just as beautiful, M30 is a globular cluster comprising several hundred thousand stars packed into a ball roughly 90 light-years in diameter. Along with about 20 percent of the Milky Way's globular clusters, M30 has passed through a stage called core collapse, in which all the massive stars settle at the cluster's center.

HUBBLE WIDE-FIELD VIEW
Located in the constellation Capricornus, M30, the brighter object in the center of the image, can be seen by both northern- and southern-hemisphere observers. However, M30's great distance (around 28,000 light-years away) means that it is invisible to the naked eye.

M30 PROFILE	
Type	Globular cluster
Constellation	Capricornus
Distance from Earth	28,000 light-years
Diameter	90 light-years

REJUVENATED STARS
Some of the bright stars seen at the heart of M30 are blue stragglers. They result from closely packed stars drawing material off each other or merging to form hot, blue-white stars that look out of place compared to their cooler, redder neighbors.

BILLIONS AND BILLIONS
This crowded region of the northern Milky Way reveals countless stars—so many, in fact, that their light blends together into pale clouds. Around 90 percent of these stars lie somewhere along the main sequence of stellar evolution.

MAIN-SEQUENCE STARS

For most stars, a simple relationship links their brightness and surface temperature: the brighter a star is, the hotter its surface will be and the bluer its color. As a result, nearly all stars lie somewhere along a line from faint red dwarfs to brilliant blue-white giants. That line is known as the main sequence.

LIFE ON THE MAIN SEQUENCE

Stars enter the main sequence as their unstable early stages come to an end. Fusion of hydrogen into helium, usually through a mix of the proton-proton chain and the CNO cycle (see pp.208–209), becomes their main source of energy. As the star's radiation output steadies, its color, size, and luminosity stabilize. Most stars find a niche on the main sequence that is determined by their mass and spend most of their life in or around this point. Main-sequence stars are found in a huge variety, ranging from faint red dwarfs with just one ten-thousandth of the Sun's brightness and a surface temperature of perhaps 4,500°F (2,500°C) to luminous giants at least 100,000 times brighter than the Sun and with temperatures up to 36,000°F (20,000°C). A main-sequence star's structure is maintained by hydrostatic equilibrium (see p.202): each layer is held in place because the outward pressure of hot gas is balanced by the inward pull of gravity. Gradual changes to this equilibrium through a star's lifetime lead to a brightening and expansion as the core grows hotter and uses up its hydrogen. Main-sequence stars tend not to pulsate on shorter timescales, and any variability is due to phenomena including magnetic activity, such as the darkening effects of starspots or brightening due to solar flares.

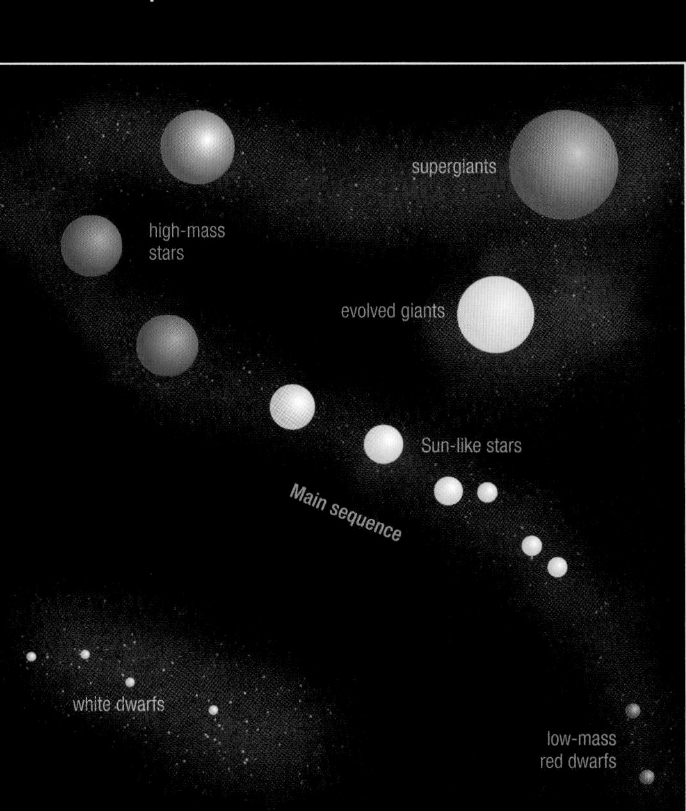

THE MAIN SEQUENCE
When stars are plotted on a graph showing temperature and luminosity, the most massive, hottest, brightest stars are at top left and the faintest, coolest, least massive are at bottom right. Most stars stay on a band called the main sequence (see p.203) for most of their lives.

INTERNAL STRUCTURES

A star is effectively a machine for transferring energy and heat from its core to its outer layers and surface—a journey that may take 100,000 years or more. In order to do this, stars use two different mechanisms. Radiation is the transfer of heat by the emission of electromagnetic waves (similar to light) that are later absorbed elsewhere. Within a typical star, matter is so tightly packed that radiation is almost immediately reabsorbed, only to be re-emitted on a different path. Tiny packages of radiation called photons zigzag back and forth within the star, undergoing countless collisions and only slowly edging their way outward. Convection, in contrast, is the mass movement of hotter, less dense material upward and cooler, denser material downward under gravity. The two mechanisms operate in different layers depending on the temperature and density of the star's material and on its atomic structure. At the star's visible surface, its gases finally become sparse enough for radiation to escape.

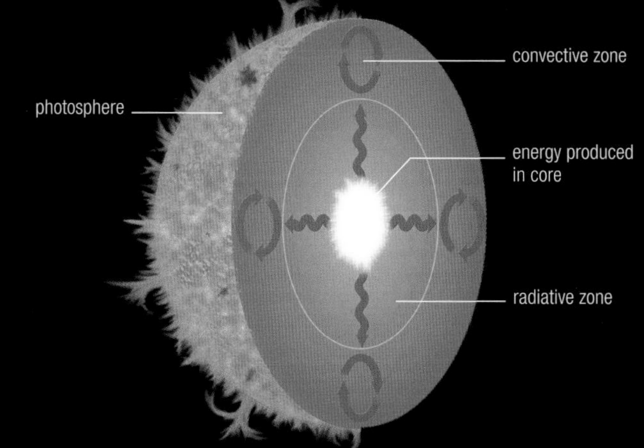

LOW-MASS STAR
In a dwarf star that has less than 40 percent of the Sun's mass, the star's interior is largely opaque. As a result, energy escaping from the star's core through radiative transport (see left) is rapidly absorbed and then carried to the surface by convection currents.

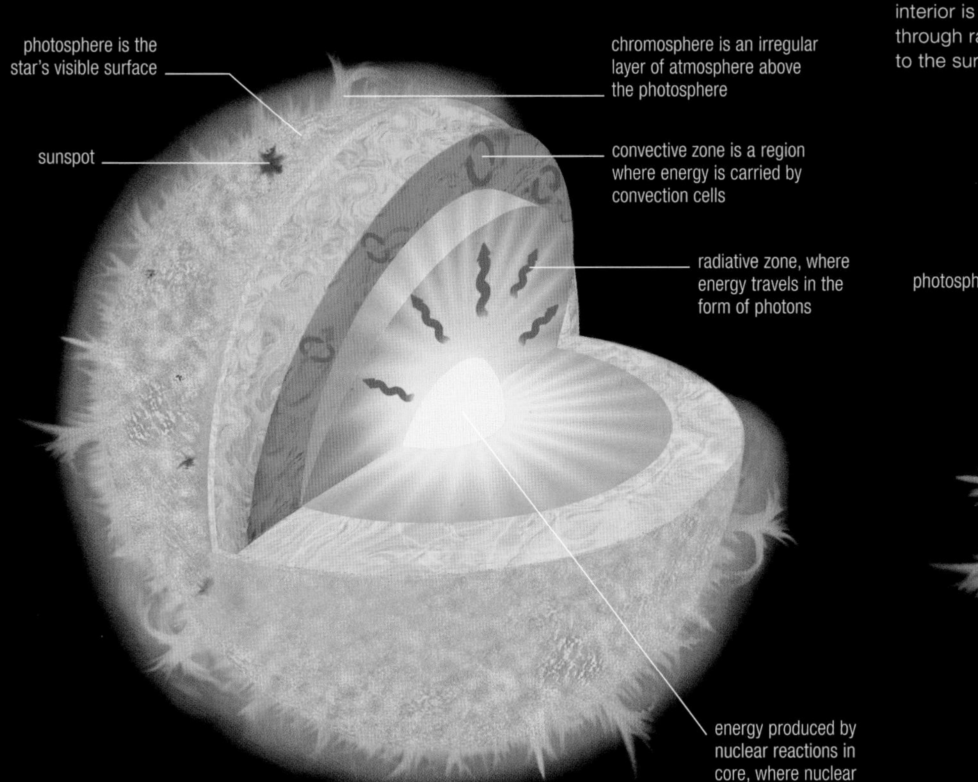

SUN-LIKE STAR
In a star that resembles the Sun, the stellar core is surrounded by a radiative zone that is transparent but "foggy." Radiation bounces around within this zone and migrates outward. At the bottom of the opaque convective zone, radiation is absorbed and carried to the surface by rising convection cells of hot gas.

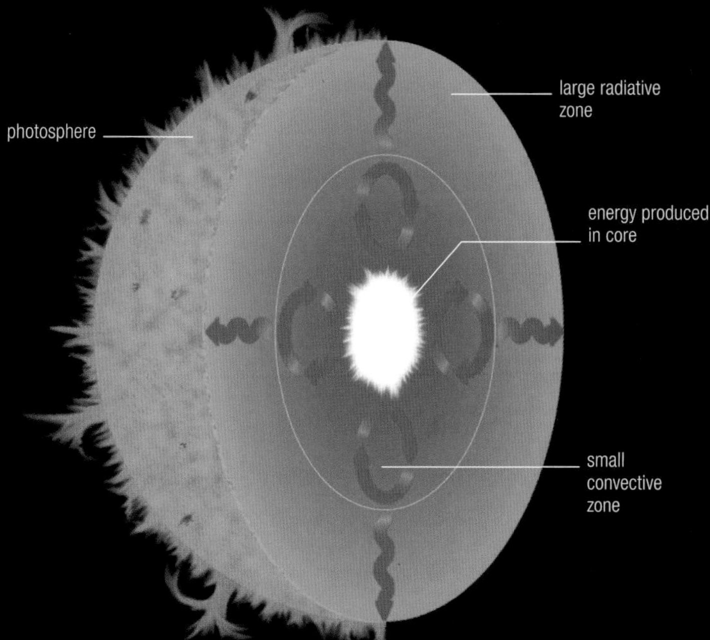

HIGH-MASS STAR
In stars with more than 1.5 times the mass of the Sun, which are dominated by the CNO cycle of nuclear fusion, the region around the core is opaque, so energy is carried outward by convection. As the density of the interior falls, the convection zone gives way to a radiative zone that reaches to the surface.

180 BILLION

The approximate number of main-sequence stars in the Milky Way. This is roughly 90 percent of the galaxy's entire stellar population.

88 PERCENT

The proportion of the Milky Way's entire luminosity generated by B-type stars, which account for just 0.13 percent of all its main-sequence stars.

PLANET-FORMING DISCS

When the first infrared satellites went into orbit in the 1980s, they discovered that some otherwise normal main-sequence stars emitted unusually large amounts of infrared radiation. This infrared excess was eventually traced to extensive discs of warm material surrounding the stars. Such discs are normally too faint to see in visible light. They often occur around young main-sequence stars, such as Vega in the constellation Lyra and Fomalhaut in Piscis Austrinus, and are thought to be the raw material from which solar systems ultimately form.

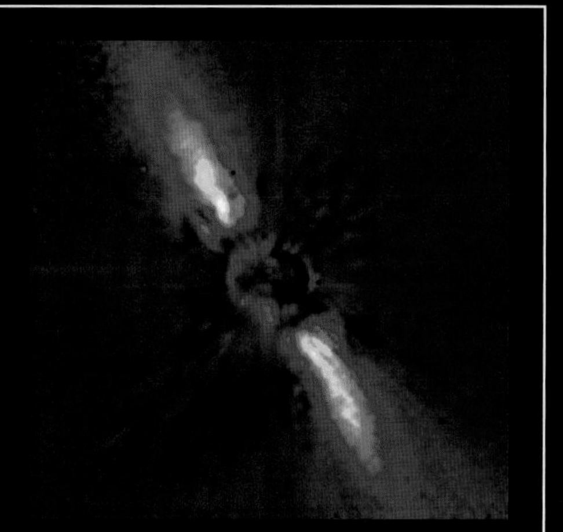

Beta Pictoris
This image shows the disc of dusty material around Beta Pictoris, a young star 64 light-years from Earth. Some astronomers think that a newly formed planet is already orbiting within this disc.

RED GIANTS

When medium-mass, Sun-like stars run out of hydrogen to burn in their cores, their properties begin to change. The stars brighten and swell to enormous size, while at the same time, their surfaces also grow cooler—they are in the process of transforming into red giants.

WHAT IS A RED GIANT?

Throughout its main-sequence phase, a Sun-like star steadily converts the hydrogen in its core into helium (see pp.208–209). However, when all the fuel in its core is used up, the star still has copious amounts of hydrogen in its outer layers. As core fusion falters, the star's interior begins to collapse, becoming denser and hotter until a thin shell of hydrogen around the core is hot enough for fusion to begin there. Boosted by heat from the still-searing core, fusion in this shell accelerates, causing the star to brighten and expand into a red giant.

NGC 2266
This open star cluster, around 10,000 light-years away in the constellation Gemini, has held itself together for an unusually long time, and its stars are thought to be about a billion years old. As a result, many of its heavier stars have evolved into red giants, which stand out among its blue-white stars.

ORBITING A GIANT
An artist's impression shows a planet orbiting the restrained red-orange giant, HD 102272, which is about 1,200 light-years away from Earth in the constellation Leo. The star has a radius about 10 times greater than the Sun, and is about 200 times more luminous.

CHANGES IN THE OUTER LAYERS

Red giants expand to their huge size because the escalating energy output increases the outward push of the gas within them. While the star's energy output may increase a thousandfold, the surface area increases even more, so the average surface temperature drops. The gas in the outer layers becomes so spread out that the star's surface becomes diffuse and blurred at the edges. Red giants often develop large star spots (cool, dark areas formed by the star's magnetic field), and bright hot spots form as hotter material wells up within the star. Astronomers can sometimes calculate the rotation period of a red giant by observing changes in its overall brightness as different parts of the star move in and out of view from Earth.

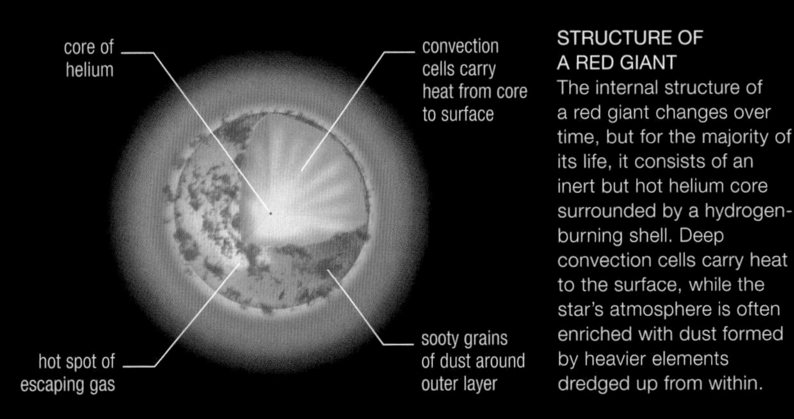

STRUCTURE OF A RED GIANT
The internal structure of a red giant changes over time, but for the majority of its life, it consists of an inert but hot helium core surrounded by a hydrogen-burning shell. Deep convection cells carry heat to the surface, while the star's atmosphere is often enriched with dust formed by heavier elements dredged up from within.

DEVELOPMENT OF A RED GIANT

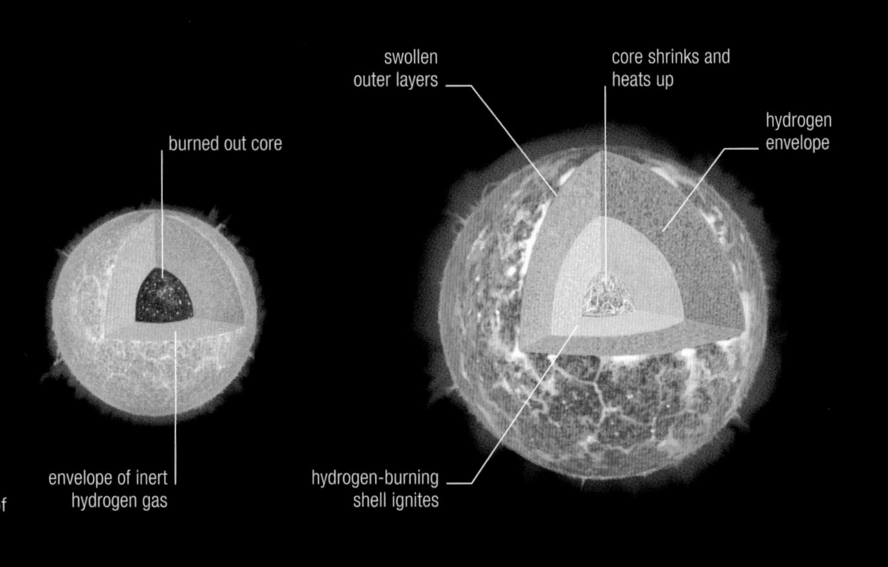

1 EXHAUSTED CORE
At the end of its main-sequence lifetime, a star exhausts the hydrogen supply in its core. The star's inner layers begin to collapse onto the hot core.

2 SHELL IGNITION
A shell of hydrogen around the core becomes hot and dense enough to start nuclear fusion. Meanwhile, the core slowly collapses under its own gravity, growing hotter and more dense.

CHANGES IN THE CORE

While a red giant sustains itself through fusion in a hydrogen-burning shell that steadily moves out through the star, its core slowly collapses, growing denser and hotter. Eventually, the center of the core becomes dense enough to trigger fusion reactions between the helium nuclei, giving the star a new source of power (see below). Helium fusion involves the joining together of two or more helium nuclei to create a nucleus of a heavier element, typically beryllium, carbon, oxygen, or neon. This reaction spreads rapidly through the core in an event called the helium flash, and in the aftermath, the star grows fainter and smaller, but hotter. This behavior is the result of new radiation from the core causing the hydrogen-burning shell to expand and cool, throttling back the fusion reactions in this part of the star and resulting in an overall drop in luminosity. During this phase of its life, the star is known as an asymptotic giant.

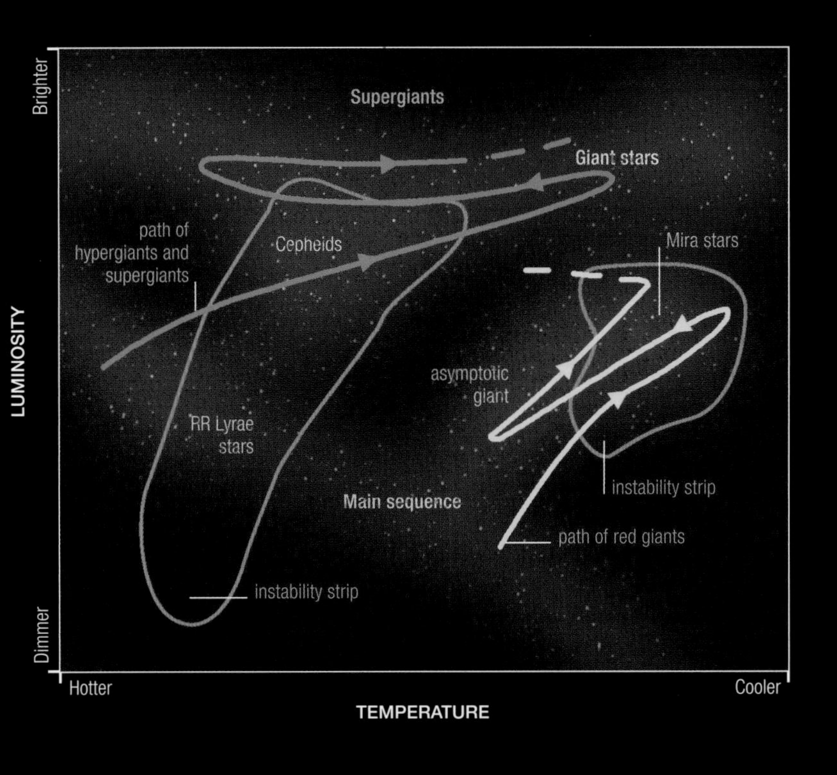
Brighter
Dimmer
LUMINOSITY
Supergiants
Giant stars
path of hypergiants and supergiants
Cepheids
Mira stars
asymptotic giant
RR Lyrae stars
Main sequence
instability strip
path of red giants
instability strip
Hotter
Cooler
TEMPERATURE

RED GIANT EVOLUTION
As a red giant star enters old age, it traces a zigzag path across the Hertzsprung–Russell diagram of stellar properties (see p.203). At points during this evolution, it passes through regions called instability strips, where it develops pulsations and becomes a variable star.

UNSTABLE GIANTS

The final stage in a red giant's evolution sees it become ever more unstable. Supplies of helium in the core are limited, so most stars spend a comparatively short period in this state. When the core helium runs out, the process of core collapse restarts as helium fusion moves out into its own thin shell, fueling itself with the hydrogen shell's by-products. The star increases in luminosity and size again, but as it is dependent on two fusion reactions that are highly temperature sensitive, it is prone to pulsations in size and brightness. Eventually, the outer layers form planetary nebulae as the star swells, before finally collapsing to become a white dwarf (see pp.238–239).

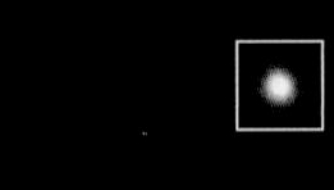

DECEMBER 2004

JANUARY 2008

FAMOUS VARIABLE STAR
Mira is a fairly bright star red giant in the "neck" of the constellation Cetus, the Sea Monster. The star, whose name means wonderful, fluctuates in brightness by several hundred times over a period of about 300 days, from being hardly visible to being clearly visible to the naked eye.

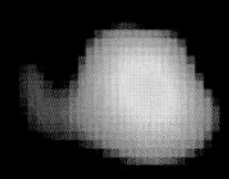

MIRA IN ULTRAVIOLET
This Hubble view reveals a hooklike appendage. This may be material from the star being drawn toward its companion star (not visible) or matter in its atmosphere being heated due to the companion's presence.

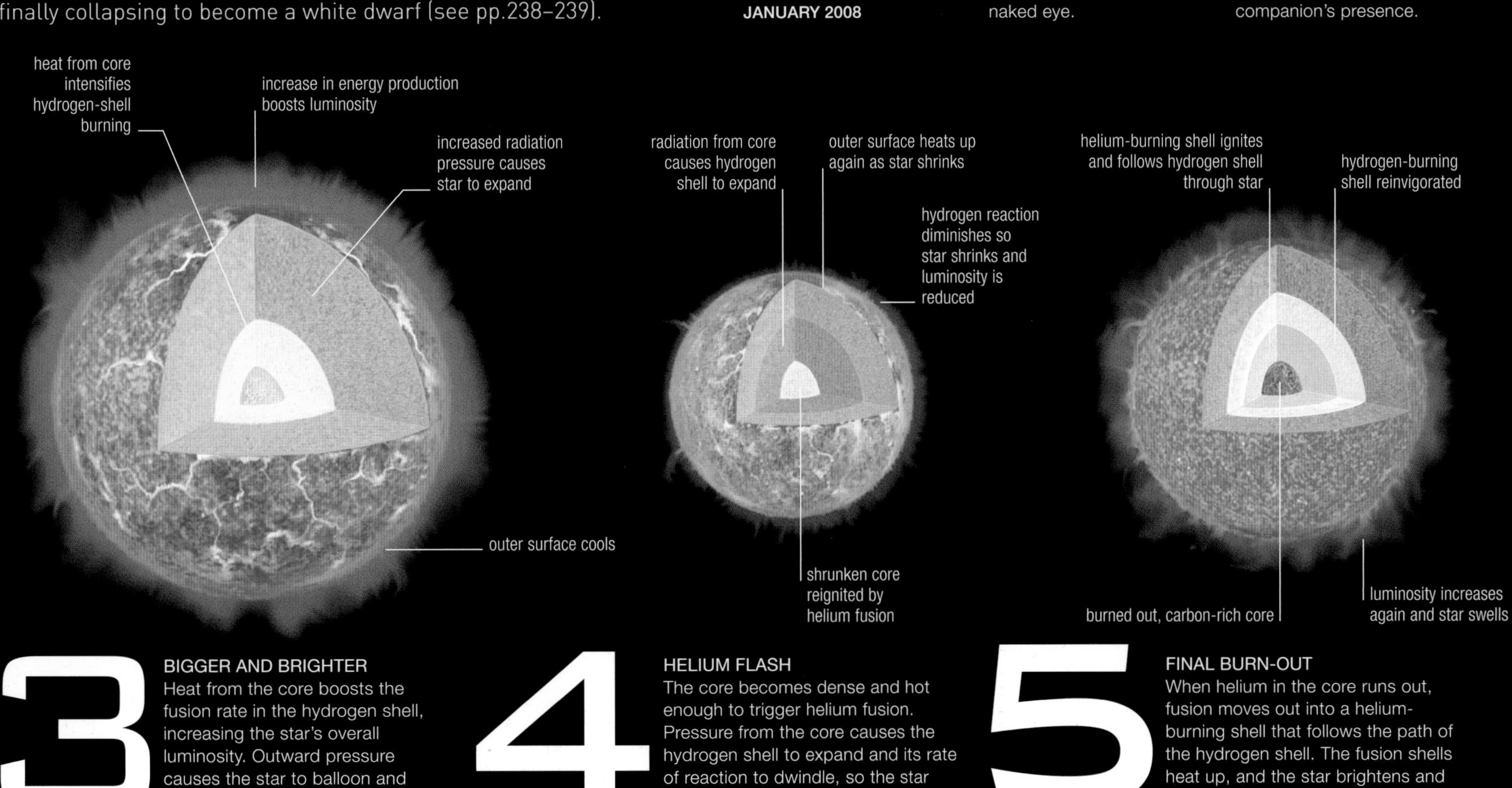
heat from core intensifies hydrogen-shell burning
increase in energy production boosts luminosity
increased radiation pressure causes star to expand
outer surface cools
radiation from core causes hydrogen shell to expand
outer surface heats up again as star shrinks
hydrogen reaction diminishes so star shrinks and luminosity is reduced
shrunken core reignited by helium fusion
helium-burning shell ignites and follows hydrogen shell through star
hydrogen-burning shell reinvigorated
burned out, carbon-rich core
luminosity increases again and star swells

3 BIGGER AND BRIGHTER
Heat from the core boosts the fusion rate in the hydrogen shell, increasing the star's overall luminosity. Outward pressure causes the star to balloon and its surface to cool.

4 HELIUM FLASH
The core becomes dense and hot enough to trigger helium fusion. Pressure from the core causes the hydrogen shell to expand and its rate of reaction to dwindle, so the star grows dimmer and smaller.

5 FINAL BURN-OUT
When helium in the core runs out, fusion moves out into a helium-burning shell that follows the path of the hydrogen shell. The fusion shells heat up, and the star brightens and expands, becoming unstable.

SUPERGIANTS AND HYPERGIANTS

Throughout their lives, the heaviest stars obey a different set of rules from other kinds of stars. Burning brilliantly as short-lived and unstable cosmic beacons, they can be seen across the whole of the Milky Way. Toward the end of their lives, however, these stars can become even more spectacular.

SUPERGIANTS

A star with more than eight times the mass of the Sun is classified as a supergiant. Such stars expand to become even larger than red giants. They begin their lives with huge supplies of hydrogen available for nuclear fusion in their cores. But, because they are usually dominated by the carbon-nitrogen-oxygen cycle (CNO) of fusion reactions rather than the proton-proton chain (see p.208), they burn through this material at a prolific rate. A massive star may drain its core of hydrogen in just a few million years, at the same time shining with at least 100,000 times the luminosity of the Sun. Such blue-white supergiants are so bright they can be seen across thousands of light-years.

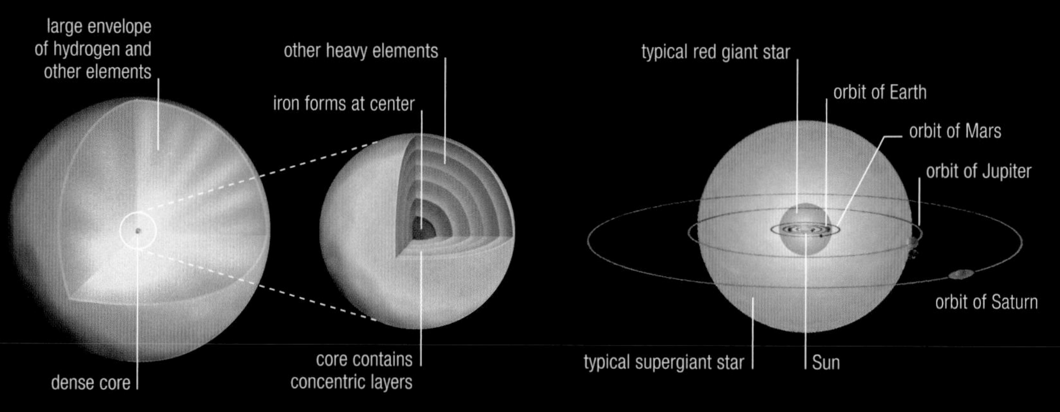

STRUCTURE OF A SUPERGIANT
A supergiant in the later states of its evolution typically has an enormous envelope of hydrogen enriched by heavier elements dredged up from within. The region around the core has an "onion-skin" structure, with a series of thin layers each fusing different elements.

SCALING THE GIANTS
Typical red giant and supergiant stars are shown here. If the Sun was replaced by a red giant with the same mass, it would engulf the planets out to the orbit of Earth, while a supergiant such as Antares (16 solar masses) would encompass almost everything out to Jupiter.

LIVE FAST, DIE YOUNG

Once a supergiant runs out of hydrogen in its core, fusion moves out into a shell, just as in the Sun-like stars (see pp.226–27). However, the resulting unstable, multi-hued supergiants are far larger and more luminous than normal red giants. Helium ignition in the core sees the supergiant fade slightly and heat up, but unlike in the red giant stars, the end of helium fusion does not signal the star's death. Immense temperatures and pressures in these higher-mass stars allow the core to keep shining through the fusion of successively heavier elements left behind by previous waves of fusion. As each element is exhausted, fusion moves into a new shell around the core. It is only when the star begins to create iron in its core that its fate is finally sealed (see pp.258–59).

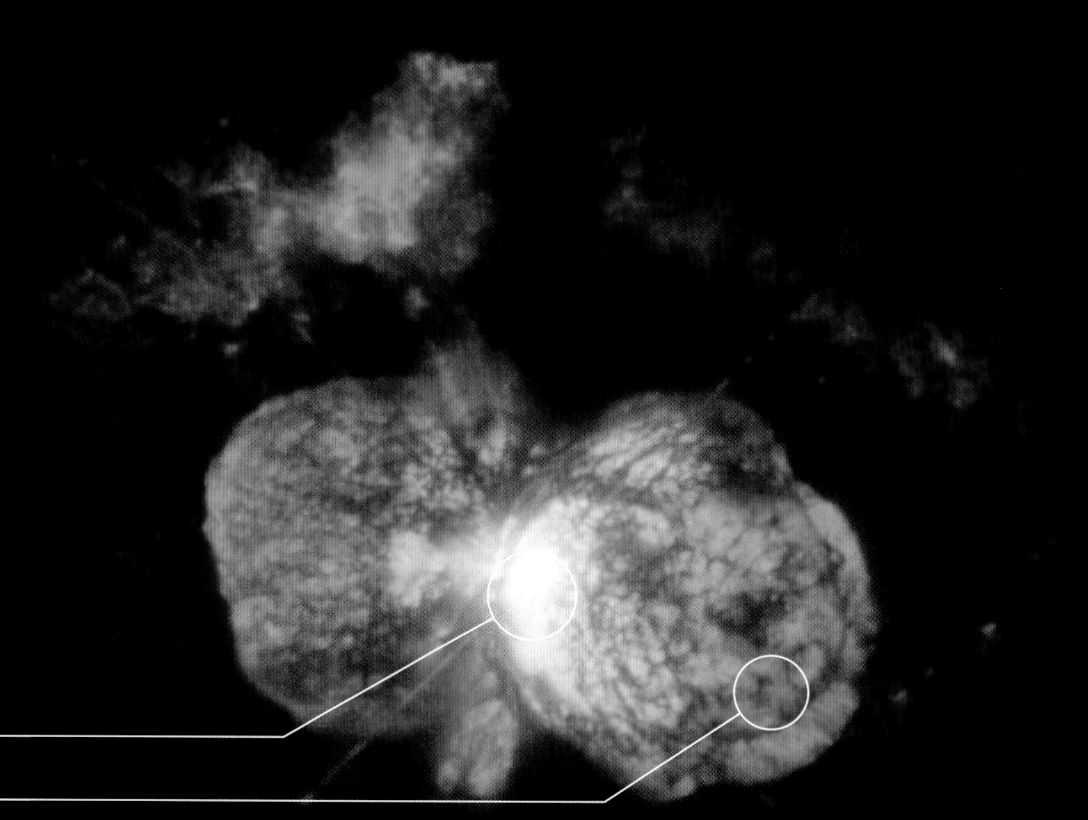

ETA CARINAE
This huge, unstable star is a double supergiant, with one of the two stars rapidly nearing the end of its life. Violent eruptions, seen on Earth in the 1830s, caused Eta Carinae to increase in brightness a hundredfold and to expel a double-lobed cloud of gas and dust that now obscures a direct view of the star from Earth.

SUPERGIANTS AND STELLAR WINDS

The surface temperature of a supergiant star rises as its mass increases, as does the force with which stellar winds blow particles off its surface. In some stars with more than 20 solar masses of material, the rate of loss due to such winds can be a billion times greater than in a star like the Sun, and easily overcomes the powerful gravity of these massive stars. During the few million years of such a star's hydrogen-burning life, it can shed an amount of material with several times the mass of the Sun. In addition, because material blowing away from the outer layers exposes even hotter inner layers, the effect strengthens through the star's lifetime.

These blue-white supergiants are known as Wolf–Rayet stars. They typically have a mass 30 or 40 times greater than the Sun's and are surrounded by the remains of their outer layers. Their surface temperatures can reach up to 90,000°F (50,000°C) when the hotter layers of their interior are exposed to space.

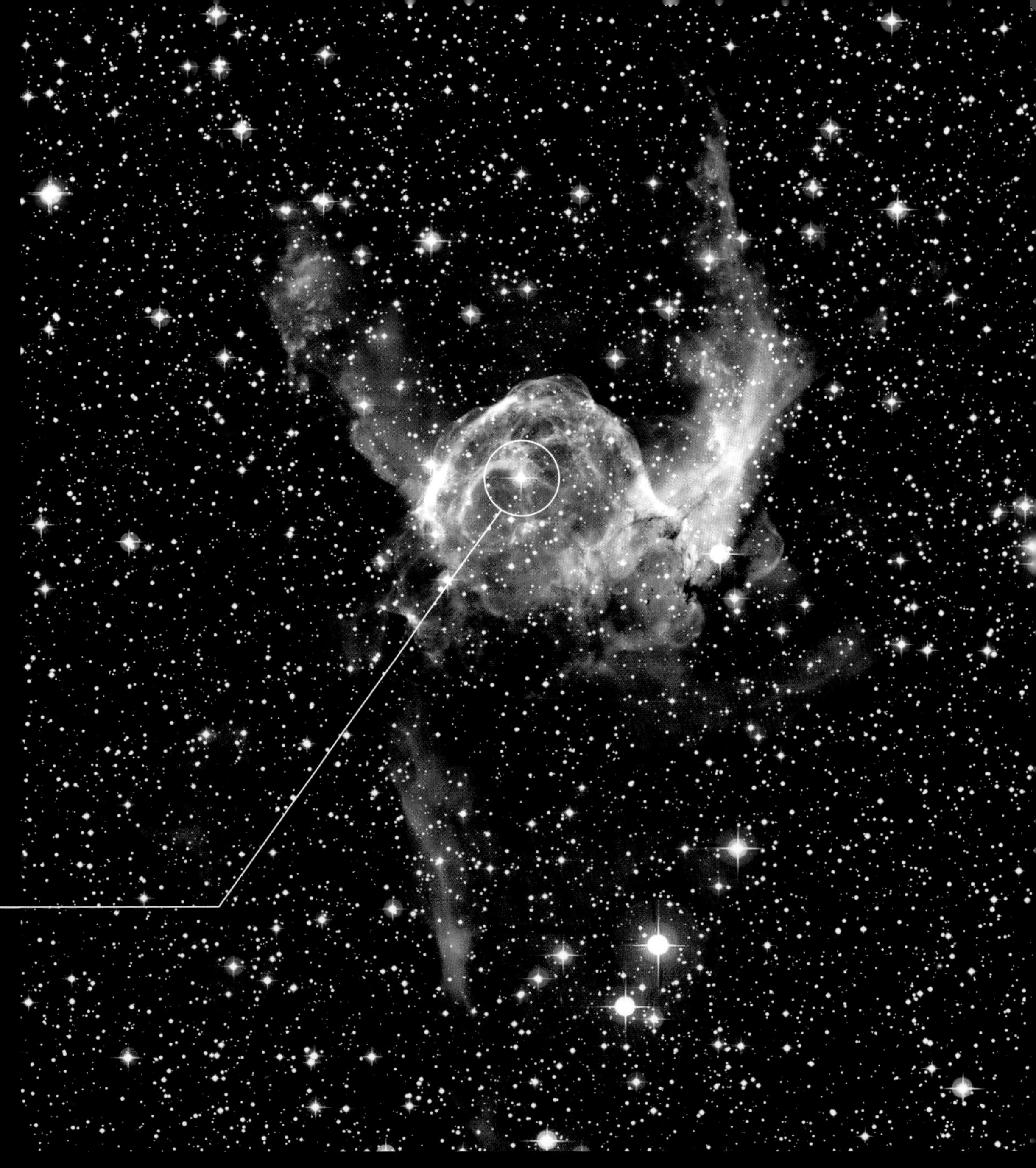

central Wolf–Rayet star—the bluish color results from oxygen in escaping gas

THOR'S HELMET
Resembling the winged helmet of Thor, the Norse god of thunder, the nebula NGC 2359 lies 15,000 light-years from Earth in the constellation Canis Major and is dominated by a huge bubble some 30 light-years across. The bubble and surrounding structures are the result of interactions between interstellar material and the intense stellar winds blowing out from the Wolf–Rayet star near the center.

HYPERGIANTS

The most massive stars in the Universe today are known as hypergiants. They are suspected to have the mass of more than 100 Suns and the luminosity of more than a million Suns. Examples of these hypergiant stars include VY Canis Majoris, a red star in the constellation Canis Major, which has the largest diameter of any known star, and the Pistol Star in Sagittarius (see right).

Is there an upper limit to the size that a star can grow? In the modern cosmos, it seems the answer is yes: collapsing protostars more than 150 times more massive than the Sun generate so much energy that they blow themselves apart before they reach their full potential. Astronomers believe that the deaths of these massive stars may be far more violent events than even traditional supernova explosions (see p.248–49) and that such rare events could be responsible for the occasional short-lived bursts of high-energy gamma rays detected from distant galaxies.

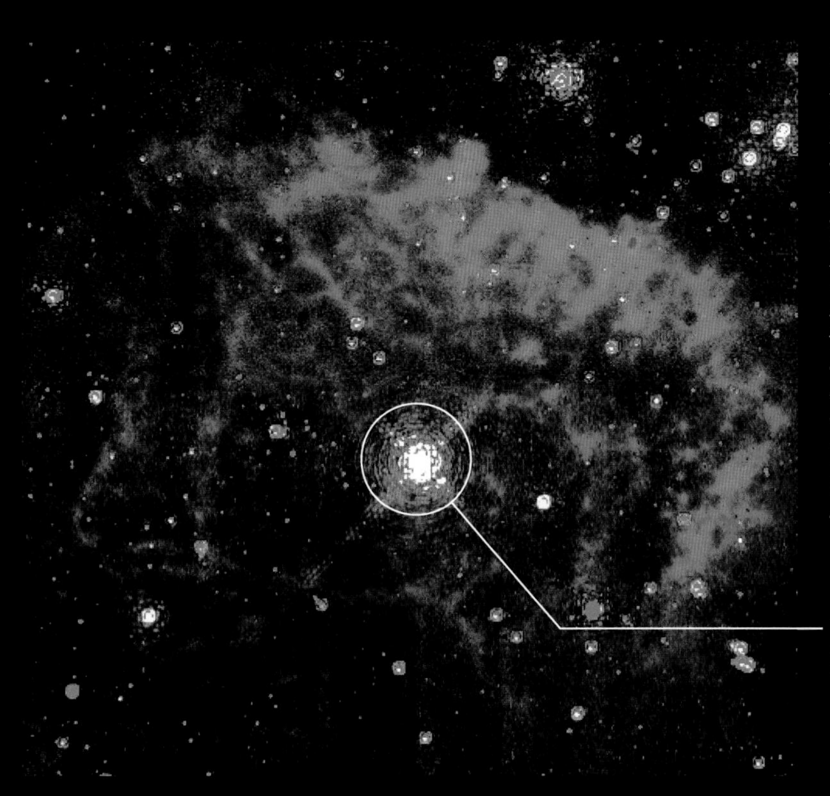

BRIGHT STAR
Lying 25,000 light-years from Earth within the Quintuplet Cluster near the center of the Milky Way, the Pistol Star takes its name from the gun-shaped nebula it illuminates. Dust hides it from view in visible light, but this infrared image taken by the Hubble Space Telescope's Near Infrared Camera and Multi-Object Spectrometer reveals one of the most massive stars known. The Pistol Star radiates as much energy in 15 seconds as the Sun does in a year.

the Pistol Star is large enough to encompass Earth's entire orbit around the Sun

INVESTIGATING STARS

The stars and other celestial bodies beyond the Solar System are so far away that it is unlikely we will ever be able to visit them. However, thanks to ingenious techniques for analyzing light and other signals emanating from stars, astronomers are able to piece together a surprisingly detailed picture of their physical properties.

MESSAGES FROM THE HEAVENS

Earth is constantly bombarded by electromagnetic radiation and particles from the Universe. Electromagnetic radiation consists of waves produced by the interaction of electrical and magnetic fields traveling through space. The properties of this radiation depend on the amount of energy it carries, which is in turn related to its wavelength and frequency. Electromagnetic radiation takes many forms (see p.11). The surfaces of stars release energy as electromagnetic radiation; other objects reflect this radiation or absorb and then re-emit it. Images captured by collecting, intensifying, and recording radiation from space tell us much about the heavens. Particles from space include atoms blown off the surfaces of stars, and pieces of interstellar dust and meteorites.

THE HUBBLE SPACE TELESCOPE
This NASA telescope orbits above Earth's atmosphere. Hubble gives especially bright, sharp visible-light images, as its unique position allows it to avoid the problems of absorption and turbulence that occur when light passes through the air (see p.276).

GAS JETS IN CARINA
Hubble has revealed the Universe in unprecedented clarity. This image shows a star-forming nebula in Carina. in ultraviolet and visible light. Jets of gas evaporate from the top of the nebula as it is heated by the stars within.

MEASURING MOVEMENT AND BRIGHTNESS

Two key techniques for studying stellar properties are astrometry (the measurement of stellar positions) and photometry (the measurement of the brightness of celestial objects). Using astrometry, scientists can create maps of the sky and calculate the sizes of celestial objects and the distances between them. Changes in the position of a star over time can tell us about its motion through space or its orbit within a star system, while parallax measurements (see pp.198–99) indicate the true distance from Earth of relatively nearby stars. Photometry is used to investigate aspects of stellar behavior, including brief dips in brightness that may signal an eclipsing binary system (see p.221) or an orbiting planet (see p.254). Fluctuations in brightness—on a long or short timescale—indicate a true variable star, which pulsates in brightness and size due to its inherent instability.

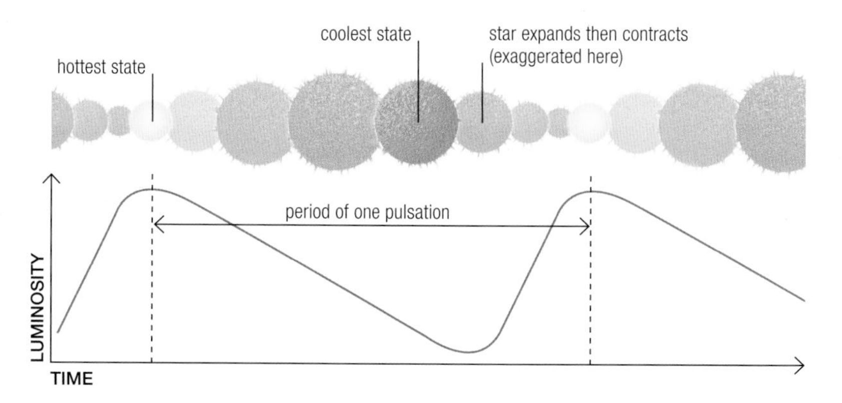

LIGHT CURVE OF A PULSATING VARIABLE
In order to study a star with changing brightness, astronomers construct a light curve by plotting its luminosity (brightness) over a period of time. The graph above shows a typical light curve associated with pulsating variable stars. See p.221 for an example of an eclipsing-binary light curve.

A STAR AND ITS SPECTRUM
This is a spectrum of the double supergiant Eta Carinae (see p.230). The light comes from a thin section of the nebula cast off by an eruption seen in the 1840s. The spectrum's emission lines, below, show that the nebula is rich in nickel and iron. The argon line is probably linked to stellar winds from the two central stars.

SPECTROSCOPY

Perhaps the most revealing astronomical technique is spectroscopy, the analysis of the wavelengths and the amount of energy at which stars and other celestial bodies emit radiation. This usually involves passing light from an object through a diffraction grating—an opaque screen scored with a large number of transparent parallel lines. The grating acts like a traditional glass prism, splitting the light into a spectrum. Every color in a stellar spectrum corresponds to the amount of radiation of a specific wavelength and energy emitted by the star's surface. The spectrum is overlaid with many dark absorption lines that indicate which atoms in the star's atmosphere are absorbing the light. Nebulae have mostly dark spectra with bright emission lines produced by atoms releasing light. Because each element absorbs or emits light at a particular wavelength, producing its own unique pattern of lines, an object's spectrum can be thought of as a kind of fingerprint that reveals its chemical makeup.

helium
argon
iron
nickel
nickel
iron
nickel
iron

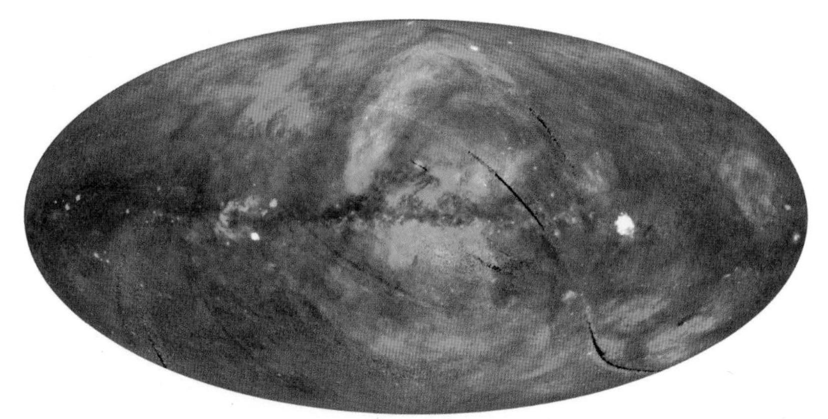

X-RAY IMAGE FROM THE ROENTGEN SATELLITE (ROSAT)

NEAR-INFRARED EMISSION FROM THE COSMIC BACKGROUND EXPLORER (COBE)

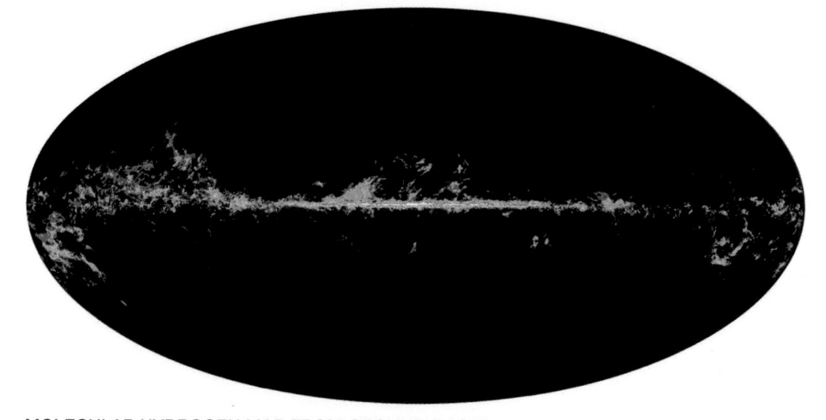

MOLECULAR HYDROGEN MAP FROM GROUND-BASED RADIO OBSERVATORIES

ALL-SKY MULTI-WAVELENGTH VIEWS OF THE MILKY WAY
These three maps show the stars, gas, and dust that fill the heavens at three different wavelengths. In each view, the plane of the Milky Way lies across the center. The X-ray image highlights areas of hot gas around the center of the galaxy (blue and yellow). The near-infrared map shows the distribution of relatively cool stars in the Milky Way's disc and hub. The lower image shows only the wavelength associated with molecular hydrogen, revealing star-forming material.

THE DOPPLER EFFECT

Spectroscopy can be used to study the motion of celestial objects by looking at Doppler shifts in their spectra. The Doppler effect (named after 19th-century Austrian mathematician Christian Doppler) is a shortening or elongation in wavelength when a wave source is moving toward or away from an observer. This is why the pitch of sirens of emergency vehicles changes as they pass by. In astronomy, it explains a shift of the spectral lines in the spectrum of a moving object. If an object is approaching Earth, the lines shift toward the blue end of its spectrum; when the object is moving away, the lines shift toward the red. The amount by which the lines shift show how fast the object is moving.

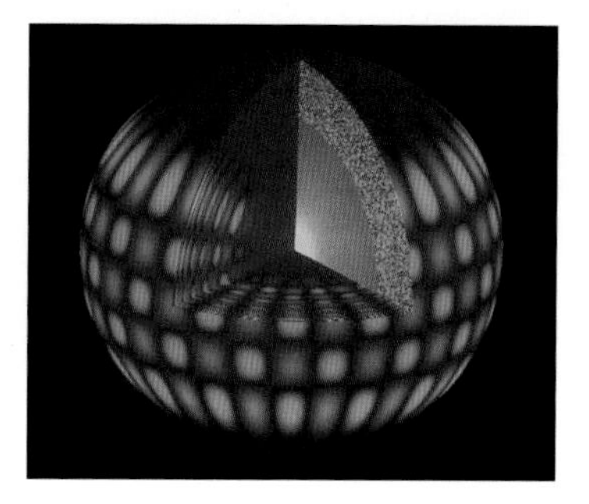

Stellar oscillations
Doppler shifts also occur on the surfaces of stars. This image shows how the Sun's surface oscillates, or vibrates, moving in (red areas) and out (blue areas) by hundreds of miles. These oscillations, are caused by sound waves trapped below the Sun's surface. Studies of oscillations on stars give an insight into the structure of their interiors.

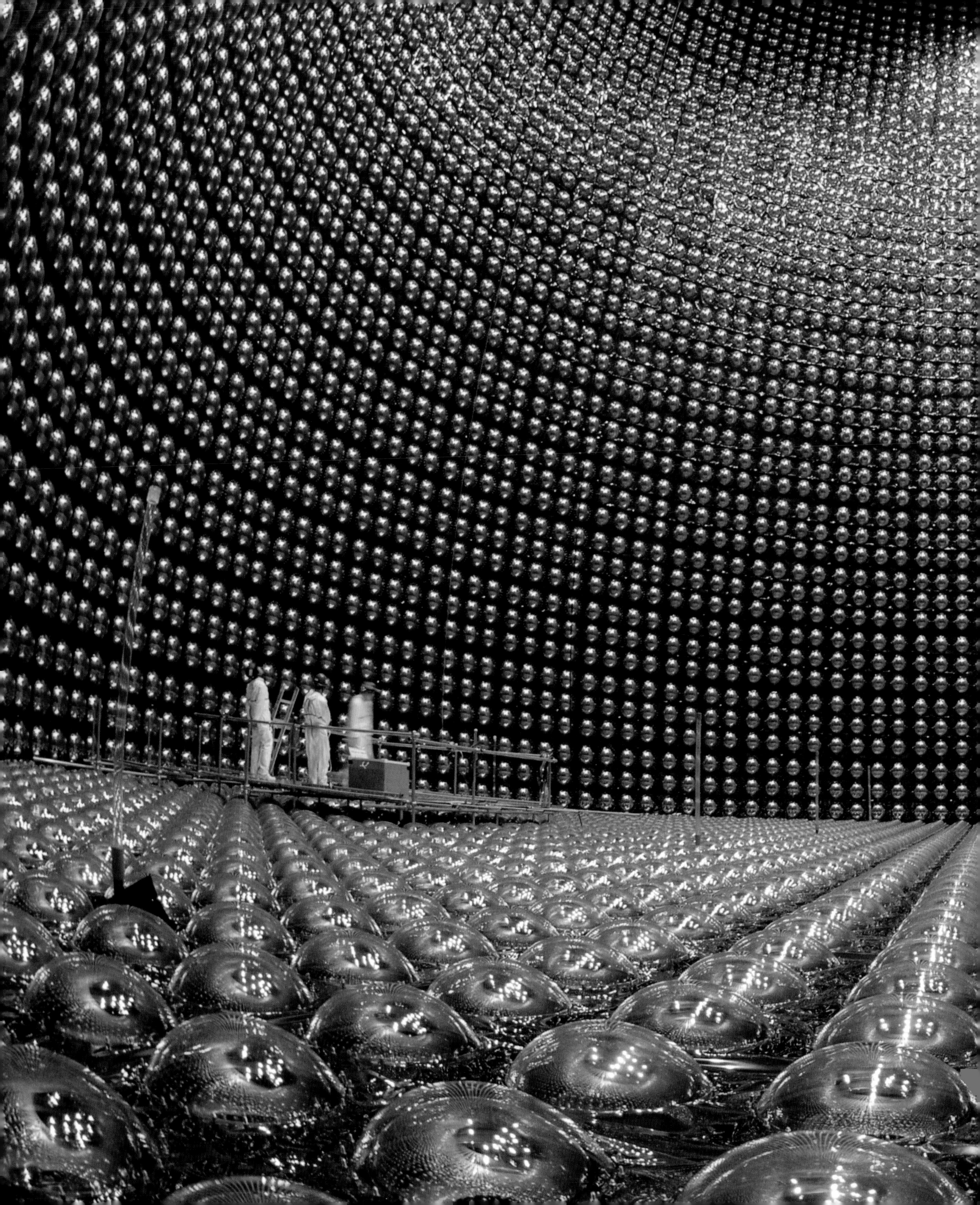

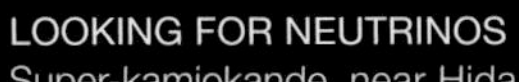

LOOKING FOR NEUTRINOS
Super-kamiokande, near Hida, Japan, is a neutrino observatory, containing thousands of muon-detecting photomultipliers submerged in water. Neutrinos travel through space rarely interacting with matter, but when they reach Earth, they can collide with water molecules to form muon neutrinos, which release information about the stars they came from.

TELESCOPES

The large telescopes in modern observatories are some of the most advanced machines used in any science. Combining complex engineering with sophisticated computing power, they would have been unthinkable just a couple of decades ago. They enable astronomers to study the Universe in more detail than ever before.

the sky above the Gemini North Telescope on Mauna Kea, Hawaii. The laser beam creates a high-altitude guide star that allows the telescope's adaptive optics to compensate for atmospheric turbulence.

EARLY GIANTS

Ever since the first telescopes appeared in the early 1600s, astronomers have recognized that size matters—the larger a telescope's main objective lens or primary mirror, the more light it can gather and the greater the detail it can resolve. Much of modern astronomy has been an "arms race" aimed at building bigger and more powerful telescopes. The 5.1m Hale Telescope on Mount Palomar in California, appeared to mark a natural limit, however. Completed in 1948, it remained the world's largest fully functional telescope for about 30 years. Efforts to build larger instruments faltered due to the sheer weight of the mirrors involved and their tendency to distort under their own weight when tilted to different angles.

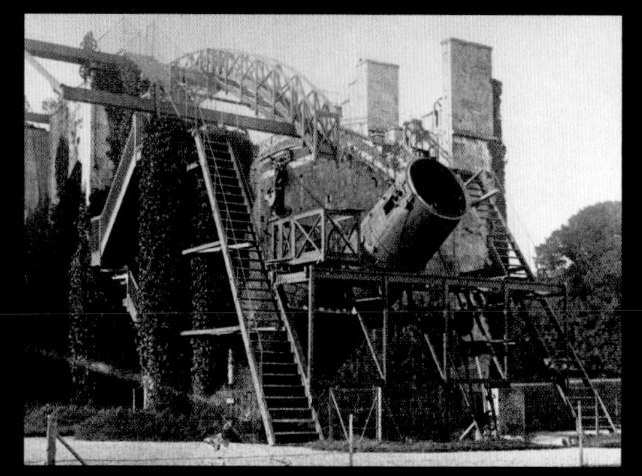

THE LEVIATHAN OF PARSONSTOWN
Enormous for its time, the 1.8m reflecting telescope built by William Parsons, third Earl of Rosse, at Birr Castle, Ireland in 1845 was so large that it had to rely on the rotation of the sky to bring objects into view. Nevertheless, it led to many important discoveries.

NEW-TECHNOLOGY TELESCOPES

In recent decades, a variety of new approaches have been found, and the 5.1m size limit has been shattered. One approach is the multi-mirror telescope—a series of precisely shaped hexagonal mirror "cells" that lock together to form a single, enormous reflecting surface. Such complex mirrors are only possible thanks to computer-aided machines that grind the mirror glass into shape before it is silvered. Pioneered by the Multiple Mirror Telescope (MMT) in Arizona, which became operational in 1979, the technique has since been used by giants up to 10.4m in diameter, and will probably be used for even larger instruments in the future. However, advances in materials science have also opened the way for other solutions, so that larger, stronger, and lighter single mirrors are also now achievable.

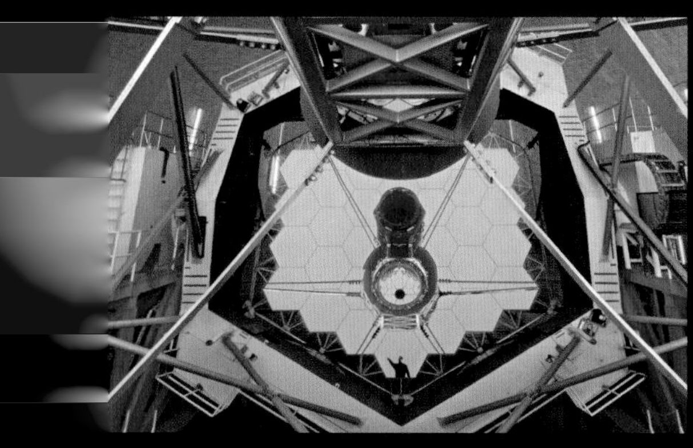

KECK TELESCOPE MIRROR
The 10m mirrors of the twin Keck I and II telescopes on Mauna Kea, Hawaii are made up of 36 individual mirror segments. Each segment is 1.8m in diameter and just 3in (75mm) thick. The segments are ground from a glass-ceramic composite material called Zerodur and coated in

LARGE ZENITH TELESCOPE MIRROR
The remarkable Large Zenith Telescope in Canada collects light with a 6m liquid primary mirror created by a slowly spinning pan of mercury. While limited to looking straight up, liquid mirrors are useful for survey projects and are up to 100 times less expensive to build than conventional

ACTIVE AND ADAPTIVE OPTICS

In order to overcome the inevitable minute distortions that occur as a mirror tilts in different directions, many large telescopes use a technology known as active optics. This involves placing an array of computer-controlled pistons, or actuators, behind a relatively thin mirror surface. As the mirror swivels and distorts, the actuators move in and out by fractions of a millimeter to maintain the mirror's shape. Another technology, adaptive optics, can help telescopes overcome the blurring effects of Earth's atmosphere. A laser beam is aimed at a point very close to the target object, creating a guide star, whose distortions allow a computer to calculate the effects of turbulence in the air. The computer then adjusts the surface of the telescope's secondary mirror to counteract these effects and create a sharper image.

WITHOUT ADAPTIVE OPTICS

WITH ADAPTIVE OPTICS

ADAPTIVE OPTICS IN ACTION
This pair of images from the Keck II Telescope reveals the benefits of adaptive optics. The second image, taken using adaptive optics, provides the sharpest image yet of the region at the center of the Milky Way.

INTERFEROMETRY

A revolutionary technique called interferometry promises to transform future telescopes still further and to deliver the sharpest images yet. Initially developed for use in radio astronomy, interferometry involves combining signals from a source object received by several well-separated telescopes. This reveals tiny differences in the paths traveled by light waves from different parts of the source object to various detectors on Earth. Computers use the information to reconstruct an image with a resolution equivalent to that of a single, much larger telescope. In radio astronomy, interferometry can simulate instruments hundreds of miles across, but in optical interferometry, the telescopes must be linked by complex light tunnels, so its use is limited to purpose-built telescopes on the same site. Nevertheless, interferometry allows Earth-based telescopes to rival the clarity of orbiting instruments.

THE RED SQUARE NEBULA
This unusual, symmetrical planetary nebula was discovered using observations by the Keck telescopes and the Hale Telescope, which is also now equipped with its own set of adaptive optics to bring it into the 21st century. The nebula bears a striking resemblance to the Red Rectangle planetary nebula (see p.244).

DISC AROUND AN AGING STAR
Using the interferometer linking three of the four 8.2m instruments of the Very Large Telescope, Chile, astronomers identified a ring of dust around the old star V390 Velorum. The computer images here show the ring's structure at two different infrared wavelengths.

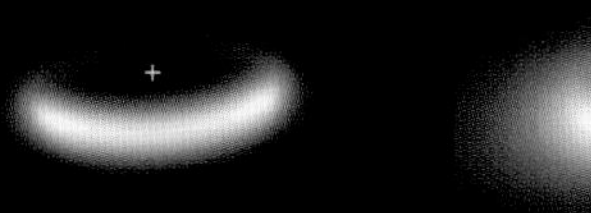

2 MICRONS INFRARED

10 MICRONS INFRARED

PLANETARY NEBULAE

Among the most beautiful of all celestial objects, planetary nebulae are the cosmic equivalent of smoke rings—short-lived, elegant shells and clouds of gas formed in the dying days of stars like the Sun.

UNSTABLE GIANTS

When a red giant star with roughly the mass of the Sun exhausts the supplies of helium fuel in its core (see p.229), its fate is sealed. Now shining only by the fusion in its hydrogen- and helium-burning shells, it becomes increasingly unstable, varying wildly in size and brightness. At the limits of their expansion, the outer layers of such oscillating giants can escape the star's gravity and puff away into space as expanding concentric shells. Stellar winds from the star's surface increase as hotter, deeper layers are exposed to space, and these combine with cast-off material to create fantastical shapes.

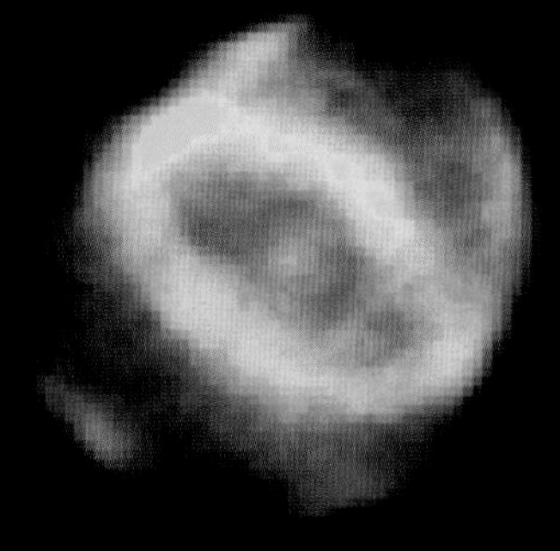

HEN 1357
This planetary nebula is one of the youngest known. Radiation from the central star has only recently grown intense enough to light up the layers of gas already ejected into space. A dense ring of material can clearly be seen around the central star, with bubblelike structures above and below it. About 18,000 light-years away, Hen 1357 is so named because it is the 1,357th object in a list of unusual stars compiled by the American astronomer Karl Henize.

FORMATION OF A PLANETARY NEBULA

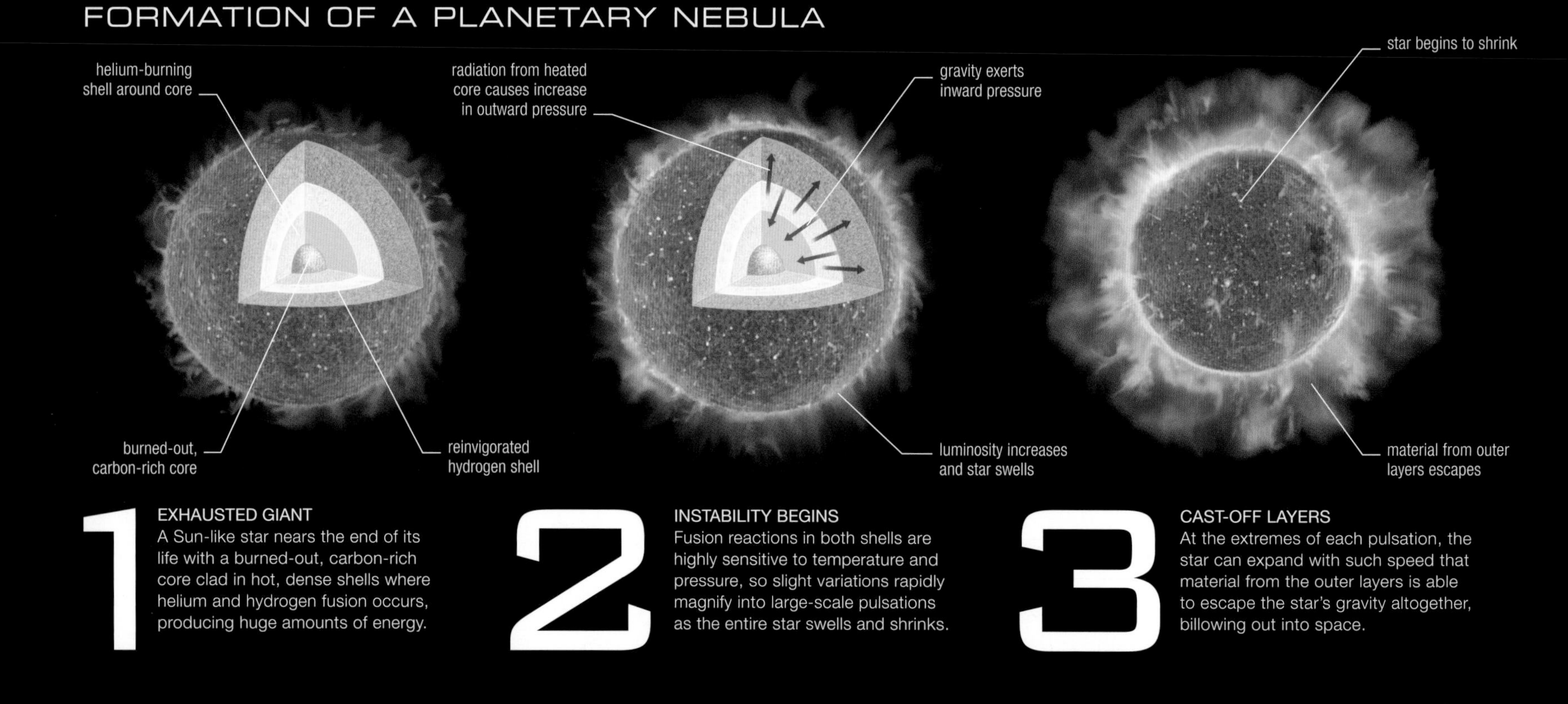
helium-burning shell around core
burned-out, carbon-rich core
reinvigorated hydrogen shell
radiation from heated core causes increase in outward pressure
gravity exerts inward pressure
luminosity increases and star swells
star begins to shrink
material from outer layers escapes

1 EXHAUSTED GIANT
A Sun-like star nears the end of its life with a burned-out, carbon-rich core clad in hot, dense shells where helium and hydrogen fusion occurs, producing huge amounts of energy.

2 INSTABILITY BEGINS
Fusion reactions in both shells are highly sensitive to temperature and pressure, so slight variations rapidly magnify into large-scale pulsations as the entire star swells and shrinks.

3 CAST-OFF LAYERS
At the extremes of each pulsation, the star can expand with such speed that material from the outer layers is able to escape the star's gravity altogether, billowing out into space.

COLOR

The expelled outer layers of a red giant cool rapidly as they escape into space. However, the increasing surface temperature of the exposed star at their center means that it emits increasing amounts of ultraviolet radiation. Absorbed by the surrounding gas and then re-emitted as wavelengths of visible light, the ultraviolet radiation transforms planetary nebulae into multi-colored cosmic beauties. The color of the light emitted by different gases enables astronomers to determine the mix of elements in the star's outer layers.

THE RING NEBULA
Roughly 2,000 light-years from Earth, the Ring Nebula is one of the sky's most prominent planetary nebulae. This Hubble Space Telescope image of the nebula exaggerates its natural colors. The blue central haze is created by helium atoms, the green ring is produced by oxygen emissions, while red hues are generated by nitrogen and hydrogen.

SHAPE

The name planetary nebula comes from the near-spherical, planet-shaped appearance of some of the first such objects discovered. However, modern telescopes have revealed that planetary nebulae occur in a wide range of shapes. Some appear to be genuine rings, or spherical shells that are only opaque around their edges, but "bipolar" nebulae, whose hourglass shape is pinched at the middle by a ring of denser gas and dust around the star, are also common. Depending on the density of material and the speed at which it was emitted, shells of gas released at different times can interact in complex ways. The presence of planets or companion stars can make the shapes even more intricate.

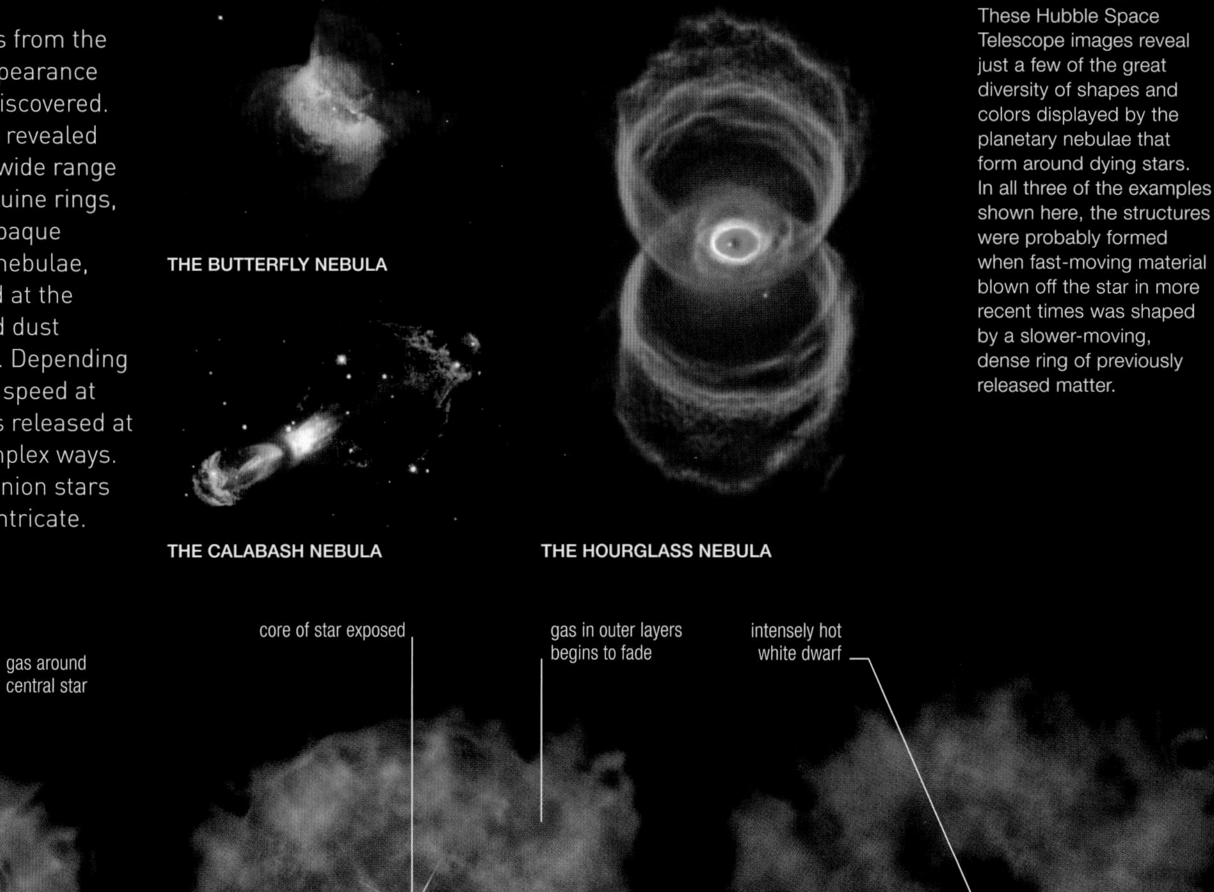

THE BUTTERFLY NEBULA

THE CALABASH NEBULA

THE HOURGLASS NEBULA

INFINITE VARIETY
These Hubble Space Telescope images reveal just a few of the great diversity of shapes and colors displayed by the planetary nebulae that form around dying stars. In all three of the examples shown here, the structures were probably formed when fast-moving material blown off the star in more recent times was shaped by a slower-moving, dense ring of previously released matter.

planetary nebula forms as star sheds outer layers

gas around central star

core of star exposed

gas in outer layers begins to fade

intensely hot white dwarf

gas fades away

LIGHT FROM WITHIN
As the star sheds material into the planetary nebula, its exposed surface grows steadily hotter. Ultraviolet rays from the surface excite the gas shell, causing it to glow in various colors.

SLOW FADE
While the planetary nebula expands further across space, the excitation from the nebula's central star begins to dwindle, and the glow from its gases begins to fade.

SUPERDENSE WHITE DWARF
Finally all that remains is the exhausted core known as a white dwarf. Even though it is intensely hot, it appears faint when seen from a distance because of its small size.

WHITE DWARFS

As the star at the heart of a planetary nebula sheds more material, its fusion shells eventually stop shining and are driven away into space. This leaves just the dense, exhausted remnant of the star's core—a white dwarf—still shining intensely thanks to its tremendous heat. Robbed of the support of radiation pressure generated by fusion, the core collapses until electron pressure—a force generated by the proximity of subatomic particles—eventually stops it. By this point, the star's core will have shrunk to about the size of Earth, with a density of some 36,000lb per cubic inch (1,000kg per cubic centimeter).

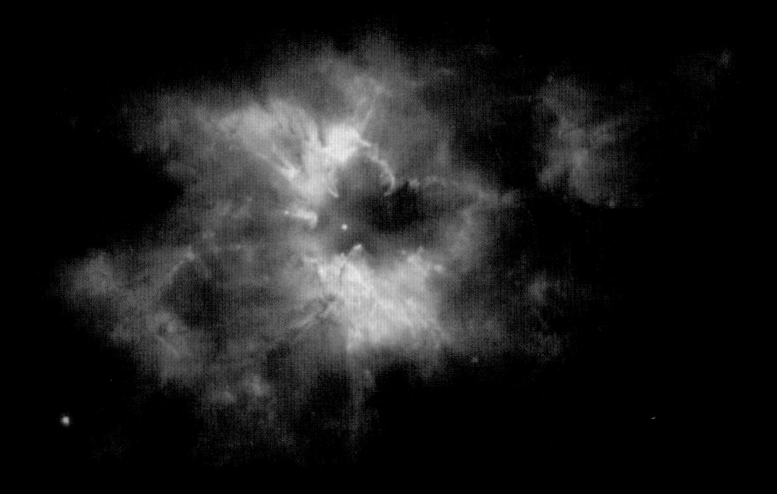

EMERGING DWARF
A newborn white dwarf known as HD 62166 is at the heart of the planetary nebula NGC 2440, some 4,000 light-years away from Earth. With a surface temperature of around 360,000°F (200,000°C) and the luminosity of 1,100 Suns, it is one of the hottest white dwarfs known.

THE HELIX NEBULA

The closest planetary nebula to Earth and the largest in the sky, the intricate Helix Nebula is an expanding shell of gas some 2.5 light-years across. This gas shell surrounds an exposed, hot stellar core. The outer layers of the nebula are expanding at speeds of around 72,000mph (115,000kph), suggesting that the nebula began to form about 12,000 years before it reached the condition in which we now see it.

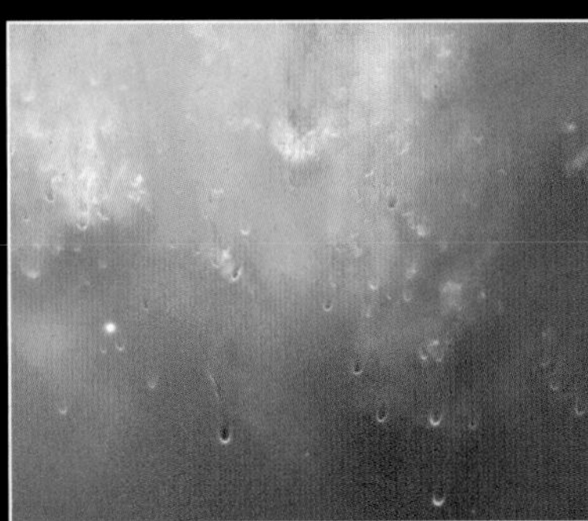

COMETARY KNOTS
As fast-moving, recently released gas interacts with denser gases already surrounding the star, cometlike knots form, each the size of our solar system. The "tails" of these cometary knots stream away from the center of the nebula.

ON THE EDGE
This boundary region highlights the difference between the nebula's outer ring, expelled about 12,000 years before it reached its currently observed state, and the gases of the inner disc, which began to form around 6,000 years later.

COLORED DISCS
In this image, from the La Silla Observatory in Chile, gases can be seen fluorescing colorfully when they receive energy from the hot star at the heart of the Helix Nebula. The blue of the inner disc is emitted by oxygen atoms, while the red is produced by regions that are rich in nitrogen.

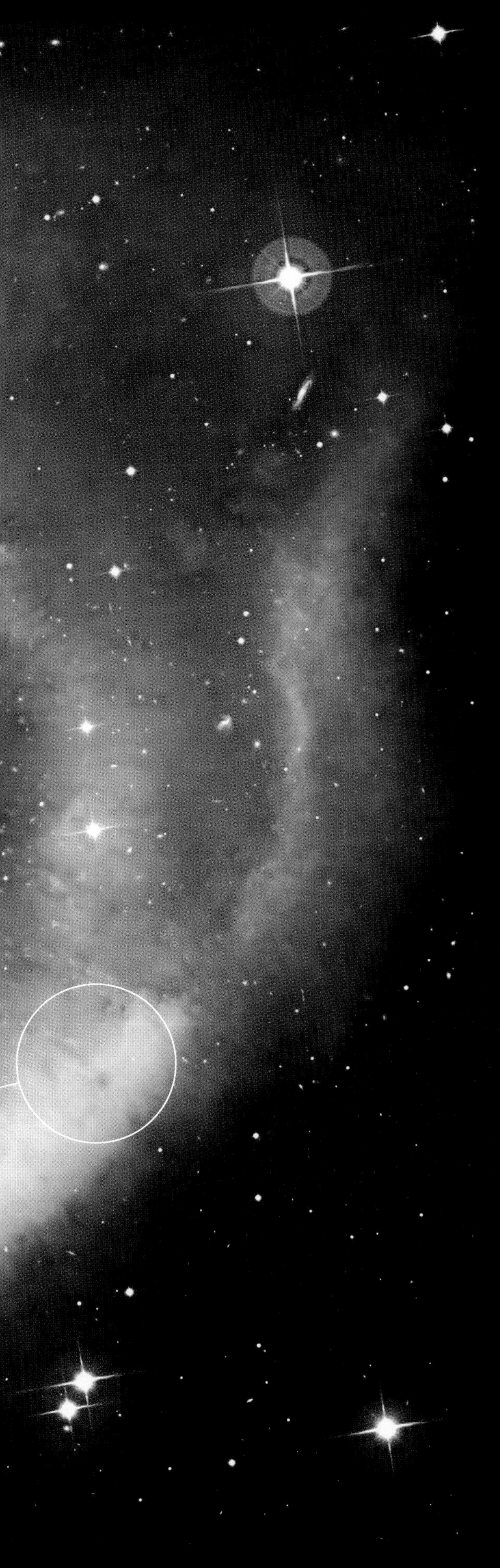

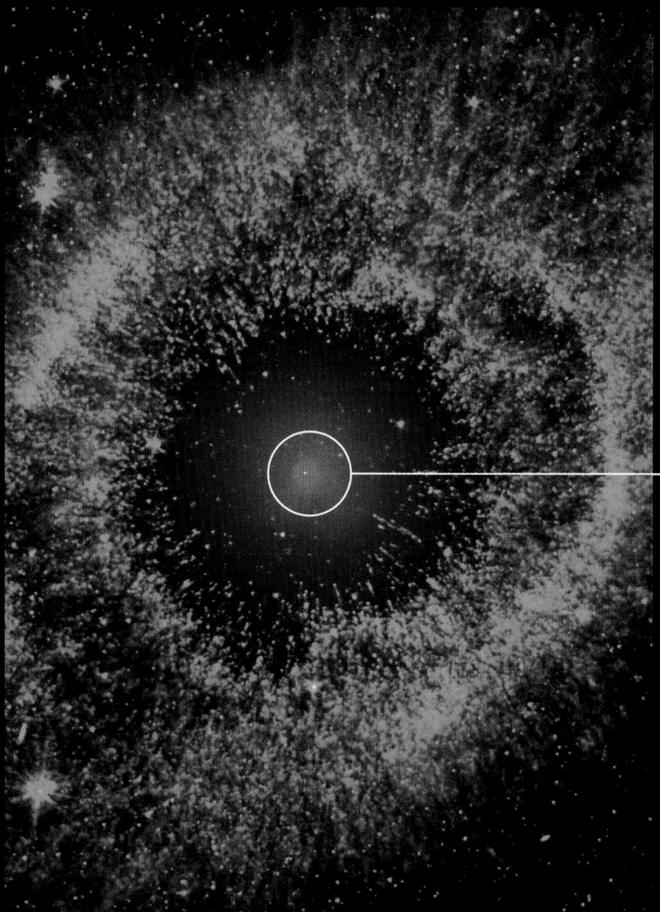

INFRARED VIEW
The colors in this Spitzer Space Telescope view represent infrared emissions of different wavelengths produced by the Helix Nebula. Relatively short-wavelength blue and green colors correspond with temperatures ranging from 2,730°F (1,500°C) around the cometary knots to 1,110°F (600°C) in the outer rings. Longer-wavelength red light comes from cold dust around the central white dwarf star (the tiny white dot at center of the image).

COMET DEBRIS
The cold dust forms a disc of about the same diameter as the Solar System's Kuiper Belt. Astronomers believe this disc may be all that remains of a ring of comets that once orbited the central star.

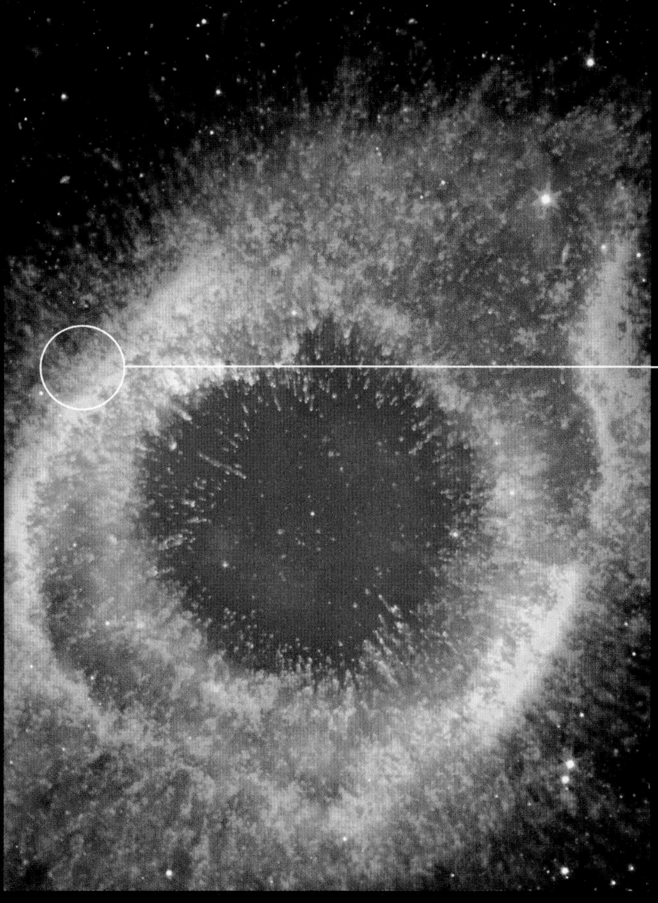

COMPLEX STRUCTURES
One of the difficulties of observing planetary nebulae is that we only see them from one direction, which makes it difficult to discern their true structures. A recent study of the Helix, which combined images from the Hubble and Spitzer space telescopes to produce views such as this one, has revealed that the nebula consists of several interlocking rings and discs, all produced at different times.

SHOCK FRONT
The red and yellow areas in the nebula indicate where the outer rings are slamming into material from the surrounding interstellar medium and heating up.

THE HELIX NEBULA PROFILE

Type	Planetary nebula
Constellation	Aquarius
Distance from Earth	700 light-years
Diameter	2.5 light-years

THE CAT'S EYE NEBULA

Located in the northern constellation Draco, the Cat's Eye Nebula is among the most mysterious and beautiful of all planetary nebulae, with an intricate series of overlapping structures. The Cat's Eye Nebula is still in the process of formation—the central regions formed only 1,000 years before it reached its presently observed state, although the outlying structures are considerably older. The star at the center may be a complex binary system involving an unusually massive star.

SPITZER'S WIDE-ANGLE VIEW
The heart of the Cat's Eye lies at the center of this infrared panorama from the Spitzer Space Telescope. Faint outlying gas clouds produced in previous outbursts are highlighted in green and red.

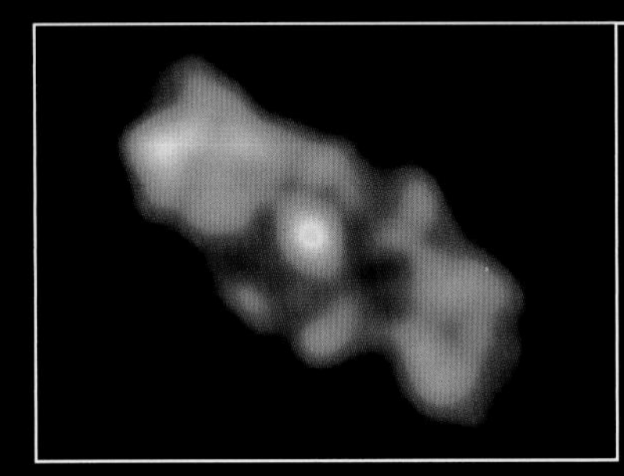

X-RAY HEART
A Chandra view of the heart of the Cat's Eye shows clouds of X-ray-emitting gas with temperatures of several million degrees. The X-rays from the very center may be emissions from a superhot disc of material around one or more stars.

RINGS OF GAS
The Hubble Space Telescope's Advanced Camera for Surveys revealed the presence of these faint concentric rings for the first time. Their regular spacing suggests that they are emitted in pulses at roughly 1,500-year intervals.

COSMIC TWISTER
The "bubbles" in this Hubble view are created by stellar winds blowing outward, inflating the slower-moving gas shells. The winds may be generated by the interaction of binary stars.

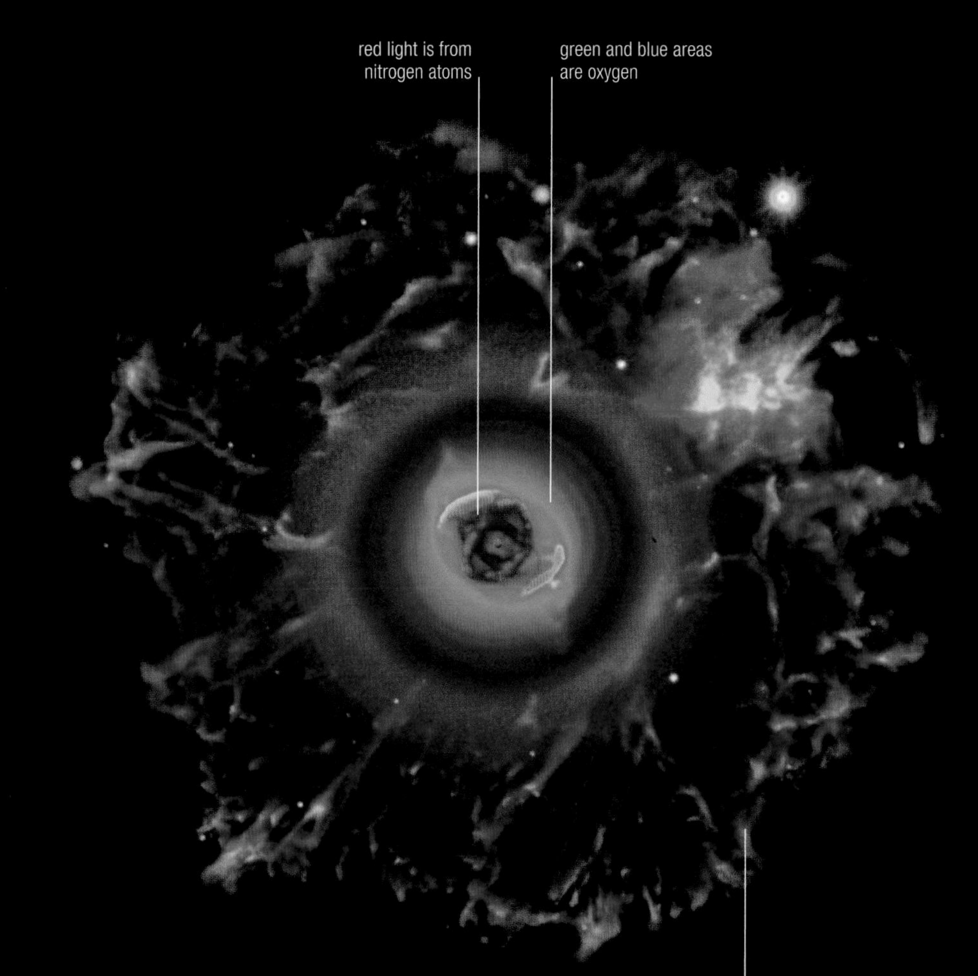
red light is from nitrogen atoms
green and blue areas are oxygen
estimated age of outer portions of halo is 50,000 to 90,000 years

MAGNIFICENT HALO
The huge extent of the nebula's outer structure, or "halo," is evident in this long-exposure image from the Nordic Optical Telescope on the island of La Palma, in the Canaries. The outermost fragments of the halo were probably ejected from their star in excess of 50,000 years ago.

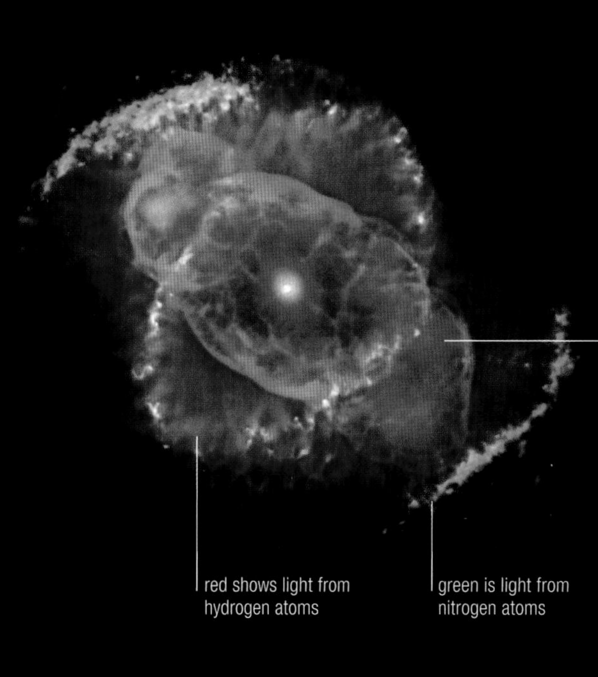
blue area is light from oxygen
red shows light from hydrogen atoms
green is light from nitrogen atoms

COMPOSITE VIEW
A combined Hubble and Chandra view of the Cat's Eye indicates the position of X-ray-emitting gas (purple) in relation to the nebula's visible structures (red and green). The Hubble image has been false-colored to highlight the presence of specific elements.

THE CAT'S EYE NEBULA PROFILE	
Catalog number	NGC 6543
Type	Planetary nebula
Constellation	Draco
Distance from Earth	3,300 light-years
Core diameter	0.2 light-years

THE RED RECTANGLE NEBULA

This unusual planetary nebula has a remarkable geometric appearance, with a series of ladderlike rungs that stretch away from the central star. The binary star system at the nebula's center is surrounded by a thick ring of dust and gas that directs material ejected by one of the stars into two expanding cones. The cones form a distinctive X shape when seen sideways-on.

RECTANGLE N THE SKY
The Red Rectangle is 2,300 light-years away in the constellation Monoceros. A faint object in visible light, it was discovered in 1973 by an infrared telescope carried briefly above Earth's atmosphere by rocket.

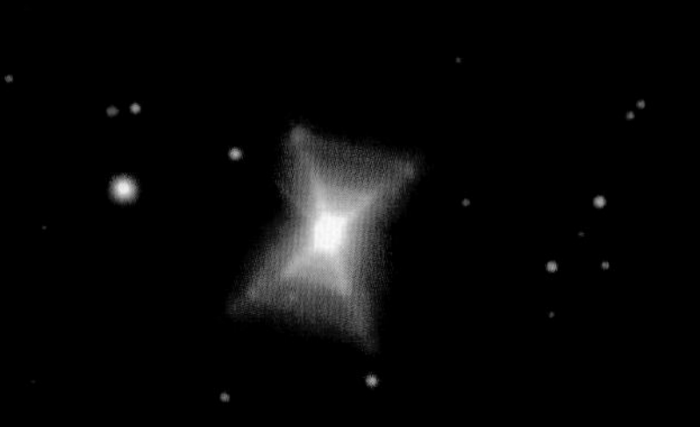

X MARKS THE SPOT
Ground-based images—such as this one from the European Southern Observatory's New Technology Telescope at La Silla, Chile—clearly show the distinctive shape that gives the Red Rectangle its name.

PULSATING CENTER
This Hubble image reveals complex structures within the nebula. The bright rungs are pulses of material that have been ejected from the central star once every few centuries, for roughly 14,000 years.

THE RED RECTANGLE NEBULA PROFILE	
Catalog number	HD 44179
Type	Planetary nebula
Constellation	Monoceros
Distance from Earth	2,300 light-years
Diameter	0.4 light-years

THE ESKIMO NEBULA

Also known as the Clownface, this nebula gets its name from its resemblance to a human face peering out from a furry hood. The "fur" of the outer shell contains many cometlike streamers, each up to one light-year long. Resembling similar structures in the Helix Nebula (see pp.240–41), they probably form as faster-moving hot gas ejected from the central star "catches up" with material released in earlier times.

THE ESKIMO NEBULA PROFILE

Alternative name	The Clownface Nebula
Catalog number	NGC 2392
Type	Planetary nebula
Constellation	Gemini
Distance from Earth	2,900 light-years
Diameter	0.7 light-years

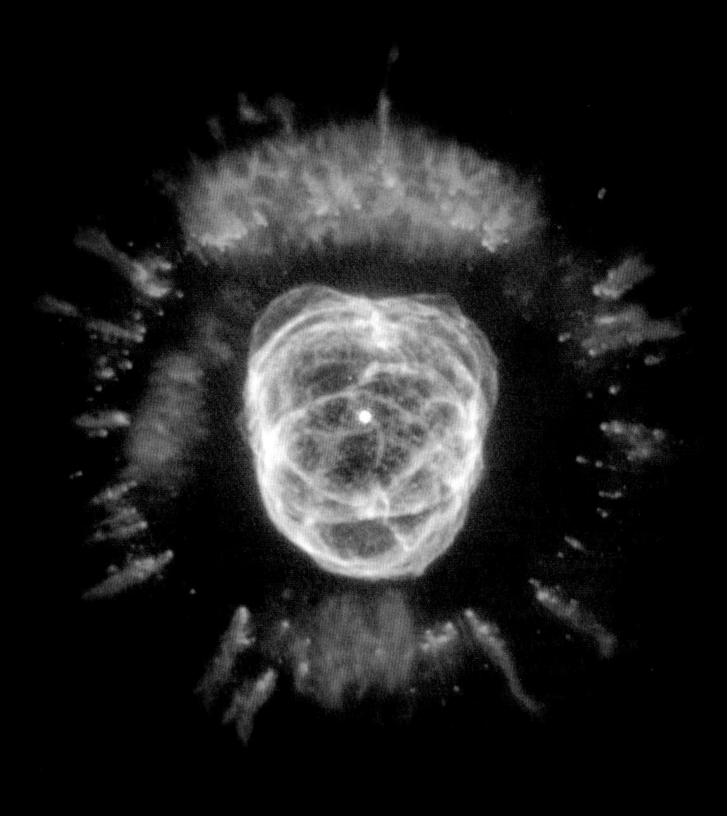

BUBBLES OF GAS
This Hubble image reveals that the features of the Eskimo's "face" are interlocking, expanding bubbles of hot gas. These bubbles are thought to form where fast-moving stellar winds from the central star catch up with a ring of slower-moving gas from the star's equator.

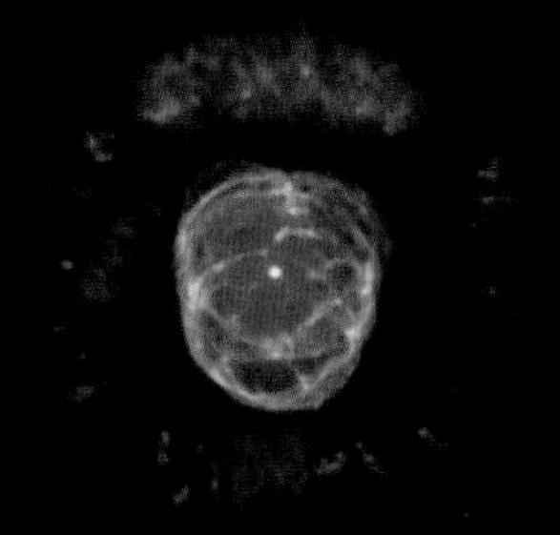

COMPOSITE ESKIMO
This composite image produced by the XMM–Newton Telescope and the Hubble Space Telescope shows hot, X-ray emitting gas (blue) inside the central cavity of the Eskimo Nebula. Around this region is a shell of cooler, ionized gas (green and red).

THE ANT NEBULA

This is one of the most complex planetary nebulae. A series of nested lobes is surrounded by a flat, doughnut-shaped ring called a chakram, most clearly visible in the bright regions beyond the outer lobes. The outflow of material forms an hourglass-shaped structure and a pair of expanding cones. Studies of the nebula's spectrum show distinct chemical differences, suggesting that its gases are from two stars.

THE ANT NEBULA PROFILE

Alternative name	Menzel 3 Mz3
Type	Planetary nebula
Constellation	Norma
Distance from Earth	8,000 light-years
Diameter	2 light-years

HUBBLE'S VIEW
This image of the Ant Nebula reveals a spectacular structure created as material escaping from the central red giant interacts with gas shed by its companion, now a white dwarf.

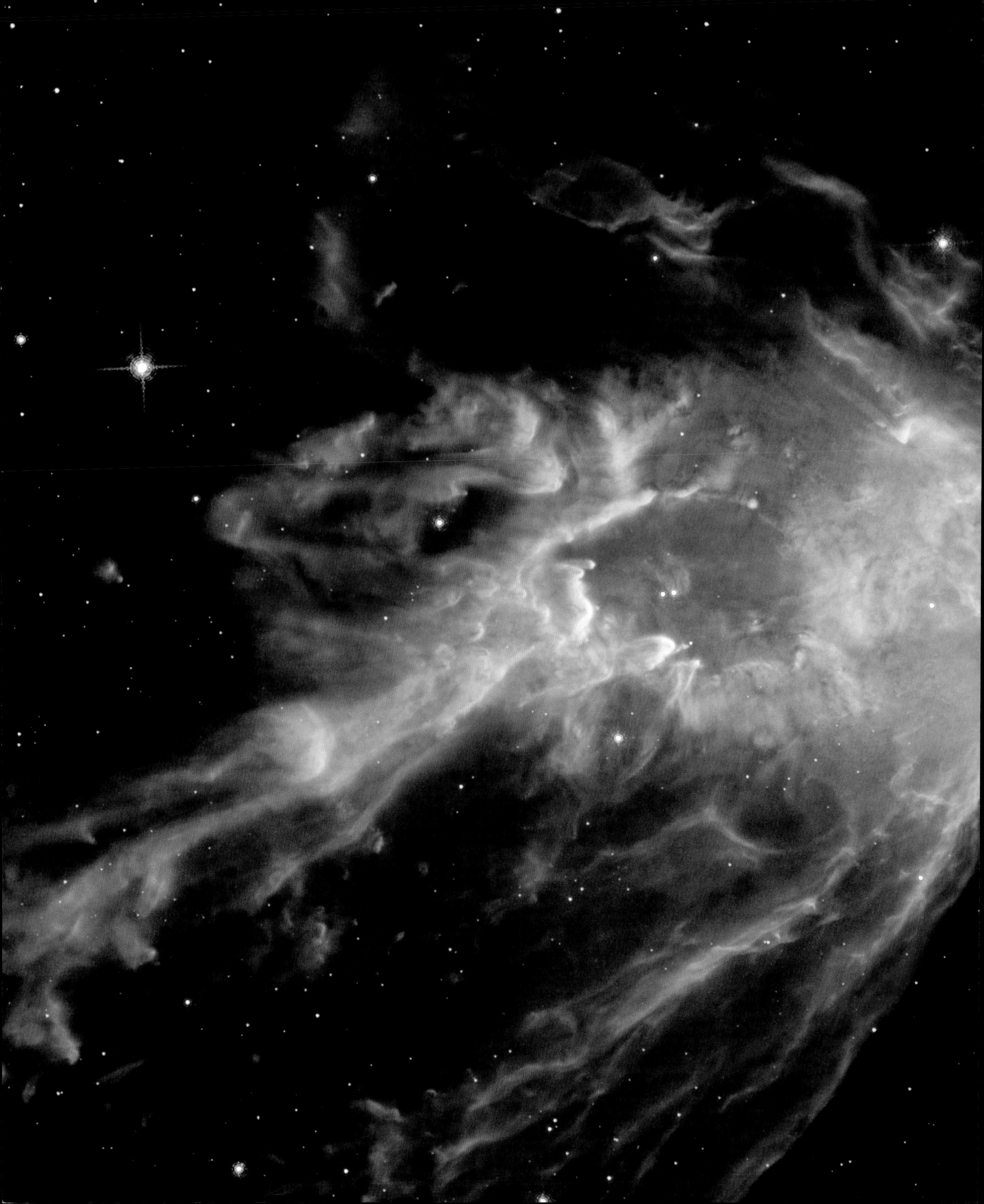

THE BUG NEBULA
NGC 6302, also known as the Bug or Butterfly Nebula, was created when the collapsed star at its core—one of the hottest in the Milky Way—shed its outer layers during its dying red giant stage. Gas ejected at high speed created the wing-shaped structures, which are expanding at a rate of about 1.2 million mph (2 million kph).

THE DEATH OF MASSIVE STARS

Stars with more than eight times the mass of the Sun end their lives in spectacular supernova explosions that can briefly outshine an entire galaxy. They leave behind shredded clouds of scorching gas enriched with heavy elements and superdense stellar remnants that are among the strangest objects in the Universe.

FROM SUPERGIANT TO SUPERNOVA

A supernova (technically a type II supernova) is triggered by the sudden collapse of a supergiant star when the nuclei of iron atoms in its core begin to fuse. Iron-fusion reactions absorb more energy than they release, so the energy source that has until now supported the structure of the massive star is abruptly cut off. The core collapses, and an enormous shock wave rips through the star's outer layers, igniting them in a blaze of fusion. Supernovae are very rare but so brilliant that they are visible at intergalactic distances.

BEFORE EXPLOSION

DURING EXPLOSION

SN 1987A
These two images show an area of the Large Magellanic Cloud (see pp.268–269), before and during the explosion of Supernova 1987A. Named after the year it appeared, it was the brightest supernova since 1604. The eruption resulted from the death of a massive blue supergiant, created by a stellar merger some 20,000 years earlier.

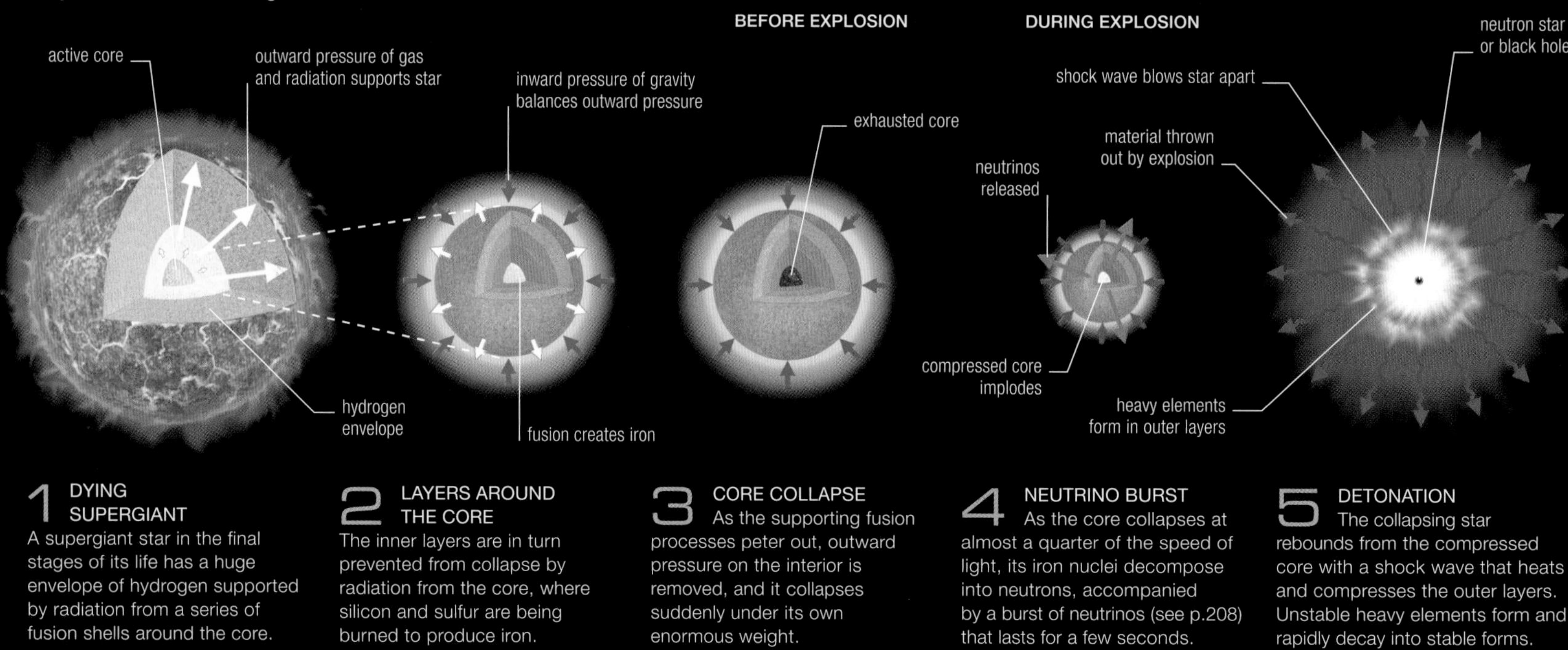

1 DYING SUPERGIANT
A supergiant star in the final stages of its life has a huge envelope of hydrogen supported by radiation from a series of fusion shells around the core.

2 LAYERS AROUND THE CORE
The inner layers are in turn prevented from collapse by radiation from the core, where silicon and sulfur are being burned to produce iron.

3 CORE COLLAPSE
As the supporting fusion processes peter out, outward pressure on the interior is removed, and it collapses suddenly under its own enormous weight.

4 NEUTRINO BURST
As the core collapses at almost a quarter of the speed of light, its iron nuclei decompose into neutrons, accompanied by a burst of neutrinos (see p.208) that lasts for a few seconds.

5 DETONATION
The collapsing star rebounds from the compressed core with a shock wave that heats and compresses the outer layers. Unstable heavy elements form and rapidly decay into stable forms.

NEUTRON STARS AND PULSARS

Most of the material from a supernova ends up scattered across space in a glowing bubble of gas known as a supernova remnant. However, the star's core, with at least 1.4 times the entire mass of the Sun, survives into a strange afterlife. With no internal support from fusion, it collapses with so much force that the atomic nuclei within are shattered into their component subatomic particles and further reduced to a "soup" of uncharged particles called neutrons. Pressure between neutrons eventually halts the core's collapse, but only when it has been reduced to a superdense sphere just a few miles across—a neutron star. Often, such stars have powerful magnetic fields and manifest themselves as celestial beacons of radiation called pulsars.

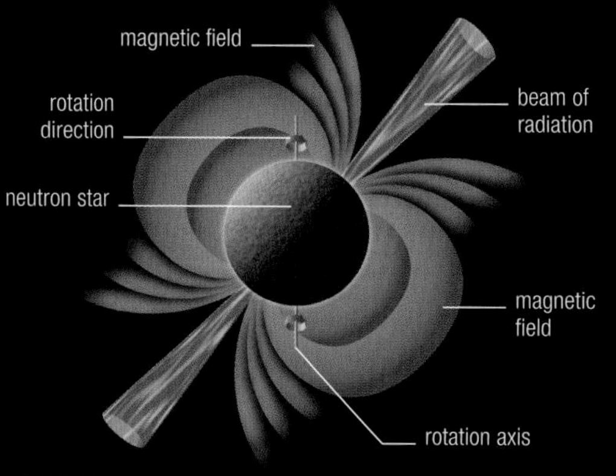

PULSAR
A pulsar retains the magnetic field of the original star, but this is compressed and so is more powerful. The magnetic field channels radiation from the neutron star's surface into two narrow beams that sweep across space as it spins.

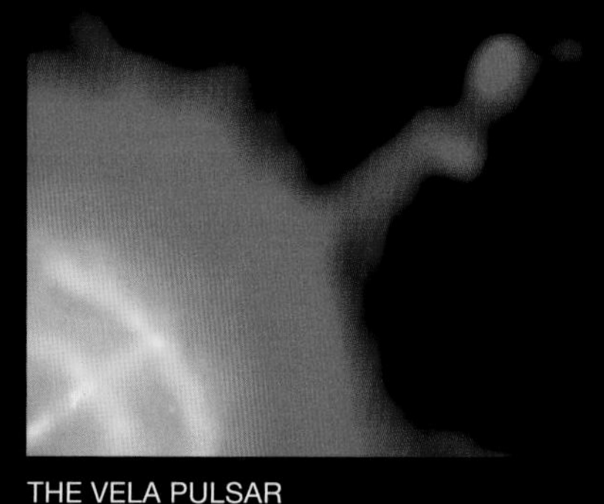

THE VELA PULSAR
This Chandra X-ray Observatory image shows the inner reaches of the Vela Supernova Remnant, where a pulsar is firing a half-light-year-long jet of high-energy particles into space. Kinks and concentrations in the magnetic field cause the particle jet to sputter erratically.

TYPE IA SUPERNOVAE

Not all supernovae are associated with high-mass stars. Type Ia supernovae occur in binary systems where a white dwarf star pulls hot gas off its companion. This gas builds up on the dwarf's surface before burning off in a nova explosion. But if the white dwarf's mass is nearly 1.4 times the Sun's, the extra weight makes it collapse, releasing a huge burst of energy. All type Ia supernovae release the same amount of energy, and astronomers use this fact to measure intergalactic distances (see p.319).

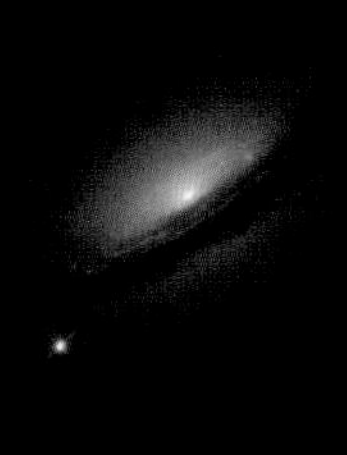

Supernova 1994D
In 1994, the Hubble Space Telescope captured this image of a type Ia supernova in the lenticular galaxy NGC 4526, around 55 million light-years from Earth.

BLACK HOLES

In the heaviest supernovae of all, the stellar core may be so massive—roughly 3.2 times the entire mass of the Sun—that even the pressure between neutrons is not enough to halt the collapse. The subatomic particles within the collapsing star are broken apart into the most basic particles of all, called quarks. The core continues to collapse until it occupies a single, superdense point in space known as a singularity, where the normal laws of physics cease to apply. The singularity's gravity is so strong that within a certain distance, known as the event horizon, nothing—not even light—can escape it. The stellar remnant has become a black hole, a bizarre object that consumes anything that comes too close and that bends the flow of space and time around it.

BLACK-HOLE BINARY
This artist's impression shows Cygnus X-1, one of the best-known black hole candidates. The stellar remnant is one member of a binary system, and it is pulling material from its blue supergiant companion onto a hot, X-ray-emitting accretion disc. The mass of the accreting body suggests that it must be a black hole.

OBSERVING THE EXTREMES

Neutron stars and black holes are some of the hardest objects in the Universe to observe, but fortunately they are not impossible to detect. While neutron stars have intensely glowing surfaces, they are generally so tiny that they cannot be seen unless the effects of a pulsar jet channel their radiation in our direction. Black holes, meanwhile, are invisible by their very nature. However, if either a black hole or a neutron star is in a binary system with a normal star, its intense gravit may be enough to tug matter away from that star and onto itself. Matter spiraling inward forms an accretion disc that can be heated to millions of degrees by tidal forces, causing it to emit X-rays.

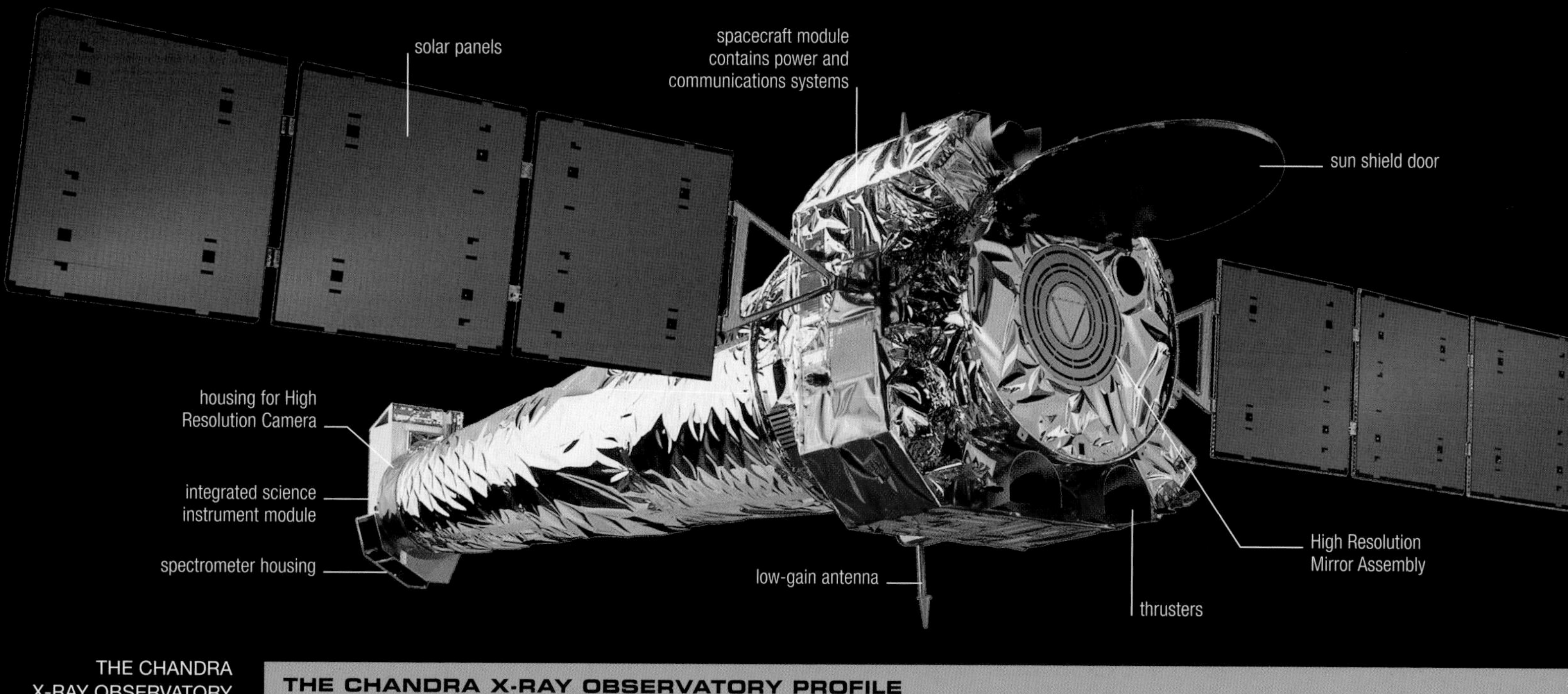

THE CHANDRA X-RAY OBSERVATORY
This is one of the most sophisticated telescopes ever launched for detecting X-rays from violent and energetic objects

THE CHANDRA X-RAY OBSERVATORY PROFILE

MISSION		CHANDRA X-RAY OBSERVATORY		SCALE
Launch date	July 23, 1999	**Agency**	NASA	
Launch vehicle	Space Shuttle *Columbia* on	**Length**	45ft 3in (13.8m)	64ft (19.5m)

THE CRAB NEBULA

1054C.E., Chinese and Arab
stronomers recorded the
udden appearance of a new
ar in the constellation Taurus.
his new celestial body—a
upernova explosion—shone so
rightly that it could be seen in
aylight for 23 days. Today, the
emnant of this supernova is
he Crab Nebula. Some 6,500
ght-years from Earth, this cloud
f stellar wreckage is already
1 light-years across, and is still
xpanding at a rate of 3.4 million
nph (5.4 million kph).

CELESTIAL CRAB
This composite of Hubble, Chandra, and Spitzer images shows the nebula in visible, X-ray, and infrared radiations. The blue X-ray emission is confined to the center. The dot at the center is a pulsar that emits rapid and periodic pulses of radiation.

LEMENTAL COMPOSITION
his false-color Hubble image shows ements within the gas cloud: hydrogen s shown in orange, sulfur in green, and xygen in red and dark blue. The light blue low comes from electrons gyrating within he magnetic field of the central star.

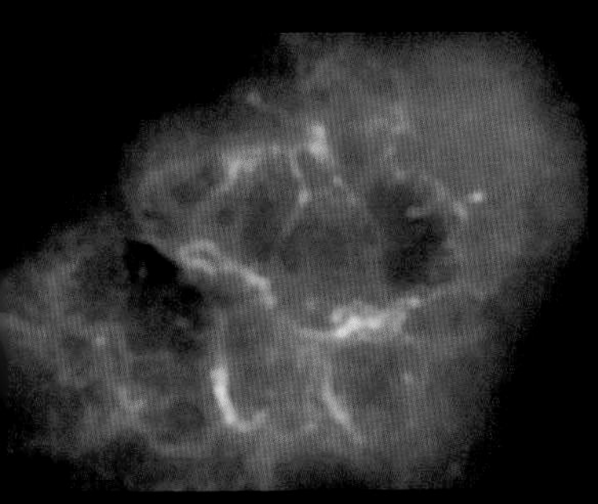

GLOWING GASES
his infrared view of the Crab Nebula from he Spitzer Space Telescope reveals gases hat are still glowing at temperatures of 20,000–32,000°F (11,000–18,000°C) nearly a thousand years after the supernova explosion.

THE CRAB NEBULA PROFILE

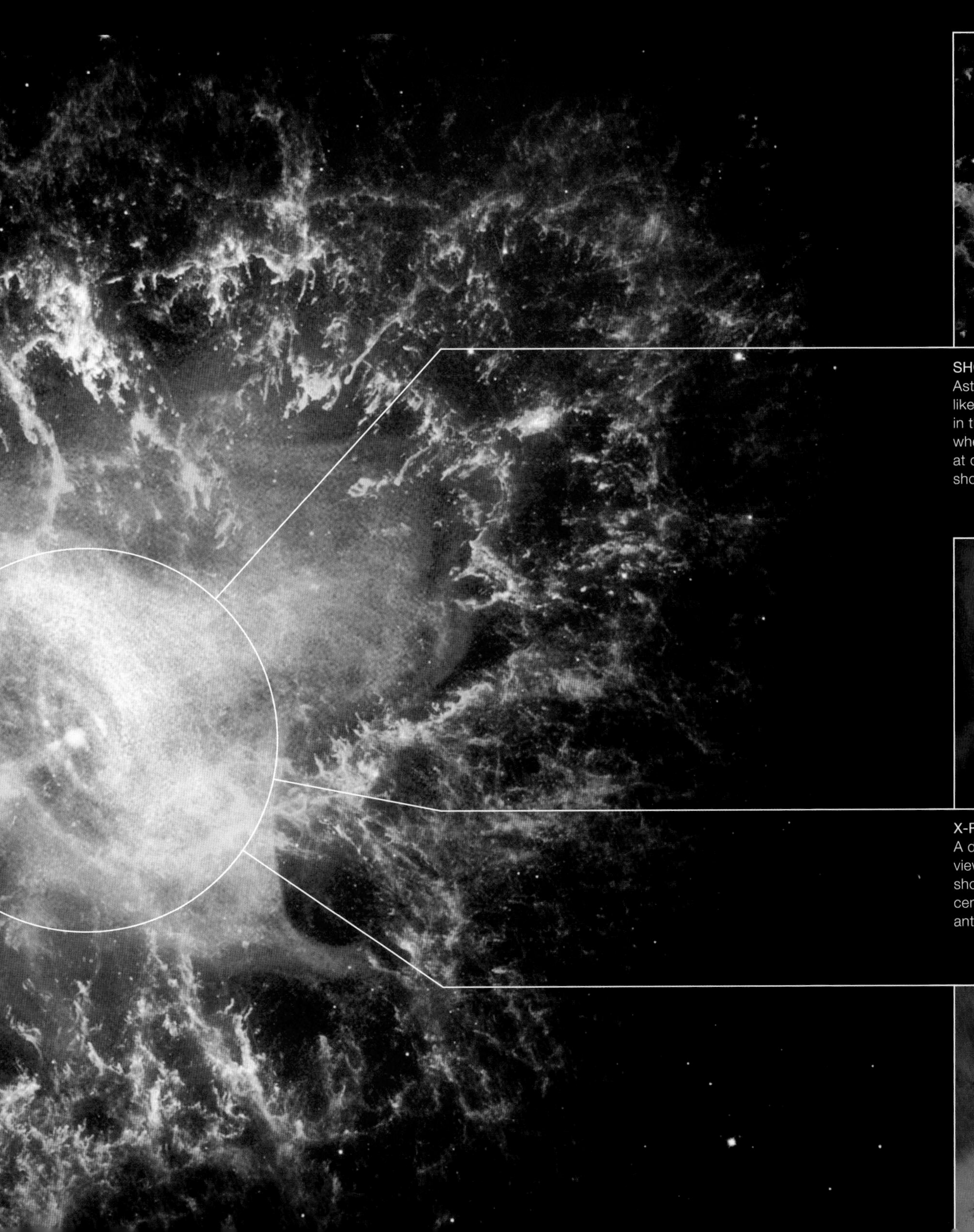

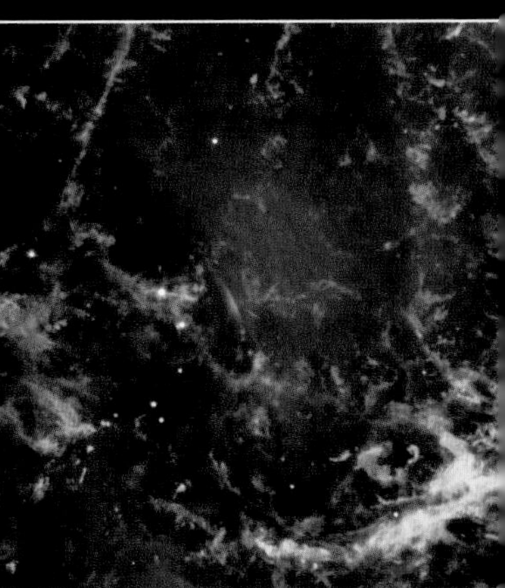

SHOCK-WAVE DETAIL
Astronomers think that the fine, filament-like structure within the Crab Nebula, seen in this false-color Hubble image, is created when gases of different densities traveling at different speeds interact and generate shock waves.

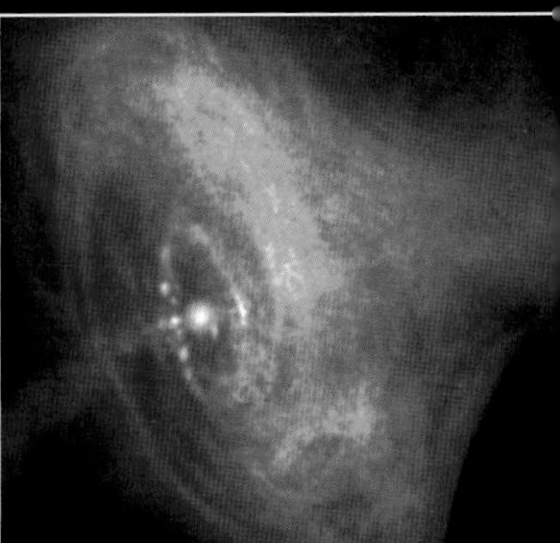

X-RAY EMISSION
A detailed Chandra X-ray Observatory view of the Crab Nebula's X-ray emission shows concentric shock waves around the central pulsar, as well as jets of matter and antimatter gushing from its poles.

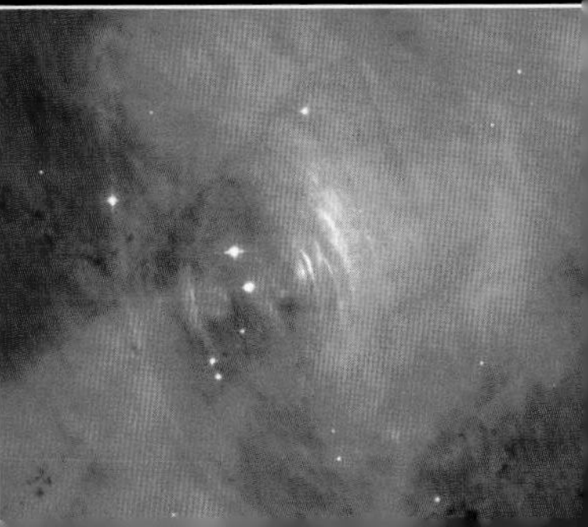

THE VEIL NEBULA

One of the largest supernova remnants in the sky, the Veil Nebula is a cloud of hot gas that is expanding at speeds of up to 370,000mph (600,000kph). It is spreading out from a stellar explosion that occurred about 7,000 years before the nebula reached its presently observed state. As the fast-moving debris collides with neighboring gas clouds, its temperature rises by millions of degrees, creating a bubble of glowing gas approximately 100 light-years in diameter.

52 CYGNI
Although this bright star appears to lie directly in the path of the expanding supernova remnant, it is actually a foreground object—a binary star, just 200 light-years away from Earth.

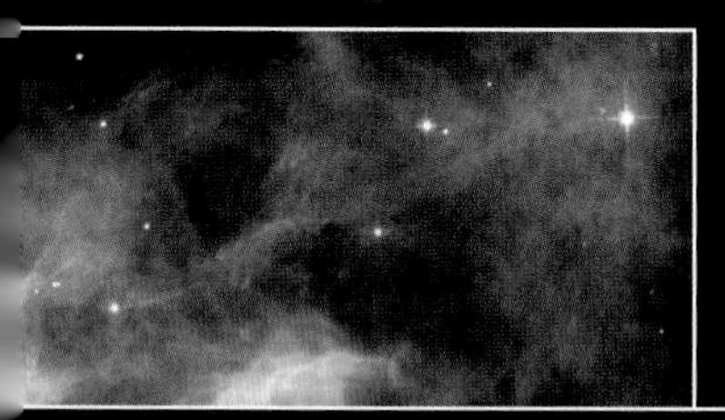

THE EASTERN VEIL
Consisting of NGC 6992 and 6995, the Eastern Veil forms one of the nebula's brightest regions. In this Hubble Space Telescope image, green shows the presence of sulfur, while blue indicates oxygen, and red denotes hydrogen.

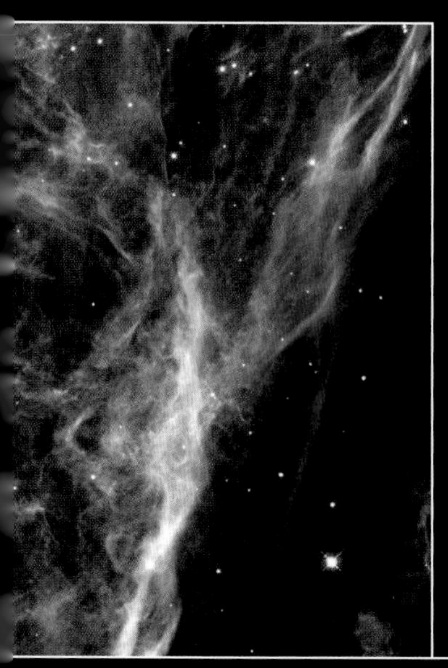

THE WITCH'S BROOM
Named on account of its brushlike structure of glowing filaments, the Witch's Broom (NGC 6960) is an area where we can see the fronts of several expanding shock waves, almost edge-on to Earth.

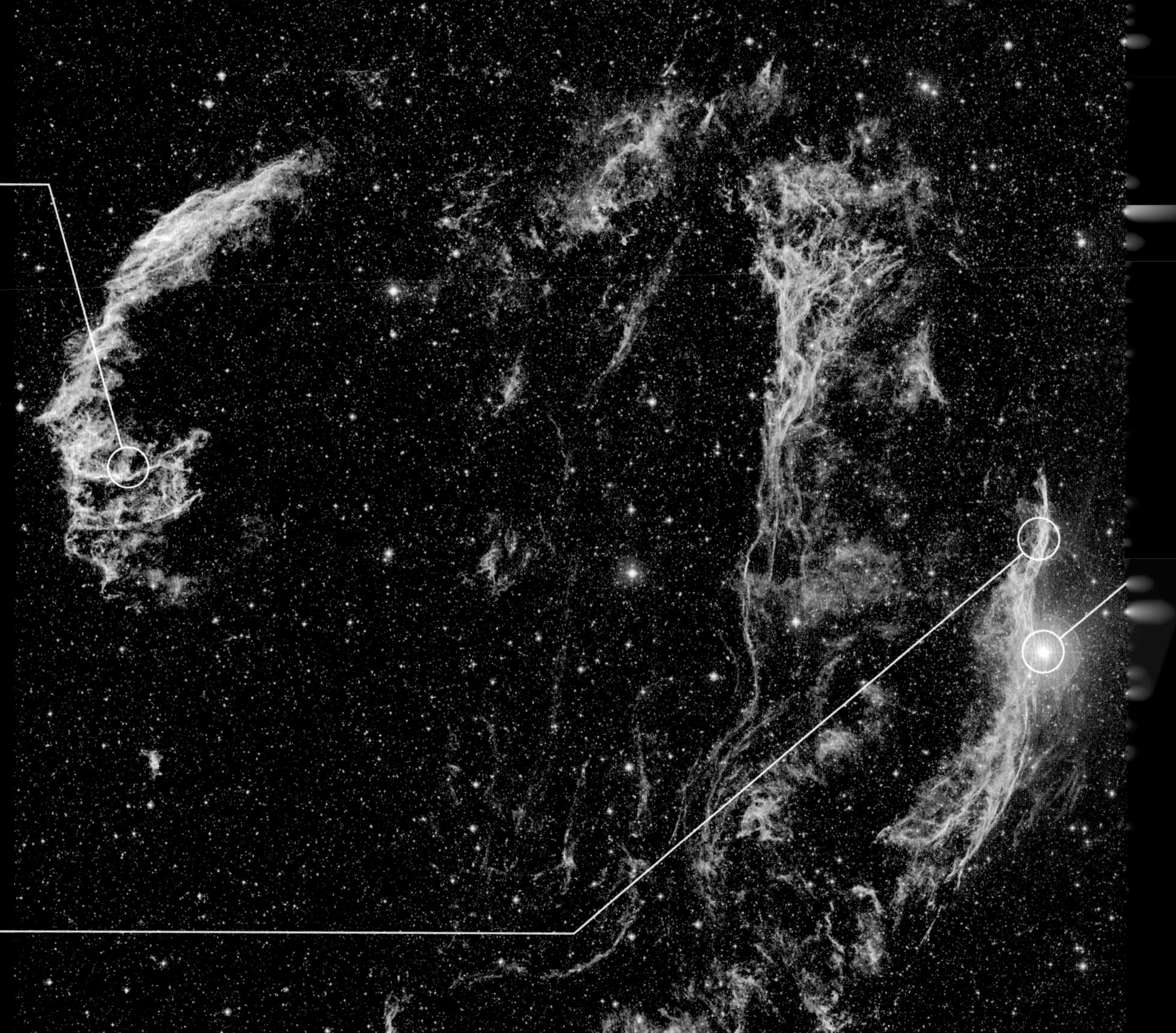

THE VEIL NEBULA
This ground-based image shows the full extent of the Veil Nebula complex in its natural colors in

THE VEIL NEBULA PROFILE

Alternative name	The Cygnus Loop
Catalog	NGC 6960, 6992

CASSIOPEIA A

This supernova remnant is the strongest source of celestial radio waves outside the Solar System. Light from the explosion first reached Earth in 1680, but as the remnant is very faint in visible light, the supernova explosion itself went unseen at the time, perhaps due to its light being blocked by interstellar dust. As a result, this remarkable object was not discovered until 1947, after rockets began carrying radio detectors beyond Earth's atmosphere.

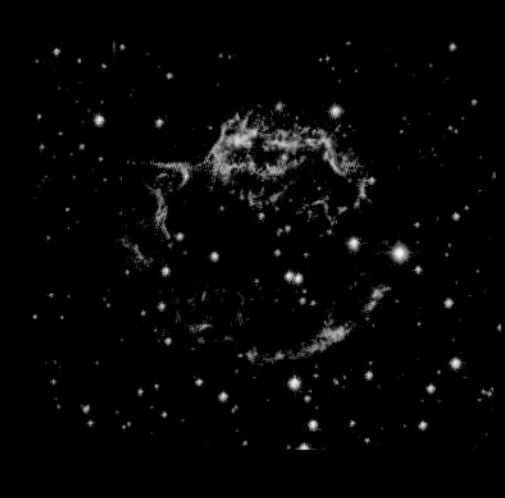

GHOSTLY REMNANT
Cassiopeia A was only detected optically in 1950. Long-exposure images reveal a faint ring of expanding gas that contains features such as jets of material traveling at speeds of up to 30 million mph (50 million kph).

ACROSS THE SPECTRUM
A Hubble Space Telescope view (orange) is combined here with Spitzer and Chandra images of Cassiopeia A's infrared (red) and X-ray emissions (blue and green). The remnant of the original stellar core is compressed into a neutron star or a black hole in the center of the nebula.

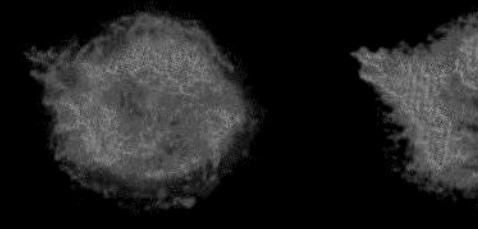

BROADBAND X-RAY

SILICON X-RAY

CALCIUM X-RAY

IRON X-RAY

X-RAY ELEMENTS
These maps generated by the orbiting Chandra X-ray Observatory highlight the presence of ions of different elements emitting high-energy radiation in Cassiopeia A. The colors represent X-ray intensity, with yellow the most intense, then red, purple, and green. The broadband image (top left) includes all the X-rays detected from Cassiopeia A.

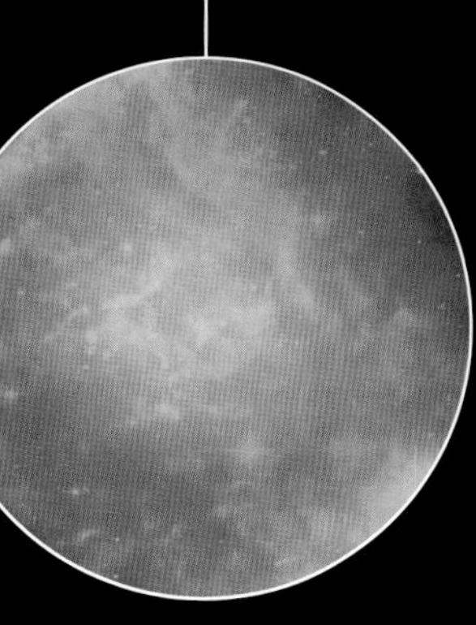

GAMMA-RAY SOURCE
This image shows gamma rays from Cassiopeia A, recorded by NASA's Fermi Space Telescope. The high-energy gamma rays are emitted when charged particles, which have been accelerated by magnetic fields in the hot supernova remnant, interact with cold gas clouds in the space around the remnant.

CASSIOPEIA A PROFILE

Type	Supernova remnant
Constellation	Cassiopeia
Distance from Earth	11,000 light-years
Diameter	10 light-years

PLANETS BEYOND THE SOLAR SYSTEM

In the past two decades, several hundred planets have been discovered orbiting stars beyond the Solar System. These extrasolar planets, or exoplanets, range in size from giant planets many times the mass of Jupiter to rocky worlds a few times heavier than Earth. Astronomers confidently expect to discover Earth-like planets soon.

PLANETS IN WAITING

Throughout the 20th century, astronomers looked for signs of planets. At first, it seemed as if planetary systems like the Solar System were extremely rare in the Milky Way. Then, in the early 1980s, infrared satellites detected evidence of discs of material around some nearby, relatively young stars.

Improved instruments have revealed more about these discs. They cover a volume of space comparable to the Solar System's Kuiper Belt (see p.178) and are made from a mix of dust and ice. They show disturbances that may be due to newborn planets plowing their way through the surrounding material. There is little doubt that these protoplanetary discs, as they are called, are solar systems in the process of formation.

> "IT'S ONLY A MATTER OF TIME BEFORE ... OBSERVATIONS LEAD TO ... THE DISCOVERY OF THE FIRST EARTH ANALOGUE."
>
> **JON MORSE**, US ASTROPHYSICIST, NASA

2006 2004

FOMALHAUT B
When comparing Hubble images taken in 2004 and 2006, astronomers spotted a moving point of light inside Fomalhaut's protoplanetary disc. A Jupiter-sized planet, Formalhaut b is on an 870-year orbit around the central star. It was the first exoplanet to be imaged in visible light.

ALIEN KUIPER BELT
By blocking out light from the central star, the Hubble Space Telescope revealed a disc of debris similar to the Solar System's Kuiper Belt, around a young star, Fomalhaut, lying some 25 light-years from Earth. The inner edge of the disc is 12.4 billion miles (20 billion km) from the star.

LOOKING FOR EXOPLANETS

The 1990s finally brought a breakthrough in the search for extrasolar planets. While any wobbles in a star's track through space caused by orbiting planets are in reality far too small to identify through measurements of its position, they can be detected through the red or blue shifts of its light (see p.233). The so-called radial-velocity method measures changes in the wavelength of a star's spectral lines as it moves toward and away from Earth under the influence of large orbiting planets. In 1995, radial velocity led to the discovery of 51 Pegasi b, the first exoplanet detected around a Sun-like star. Another useful tool used in the search for planets is the transit method (right). Astronomers use this method to watch for dips in a star's light as a planet passes in front of it when seen from Earth. The information available from different techniques varies, but astronomers can typically estimate a planet's orbital parameters and a lower limit for its mass.

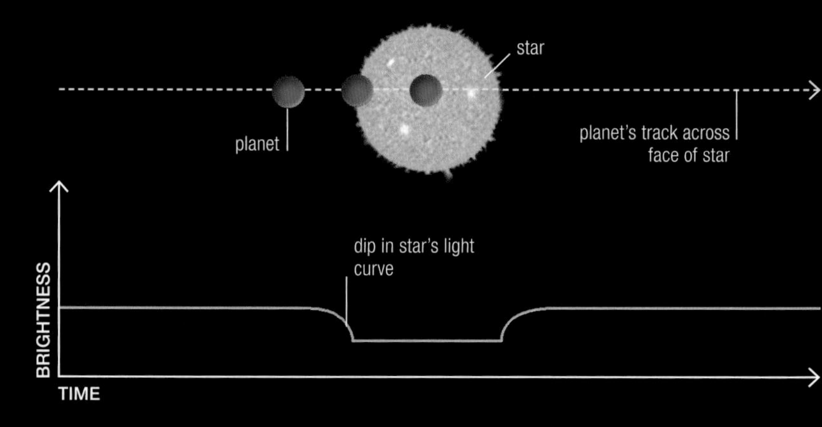

THE TRANSIT METHOD
This method relies on accurately tracking the brightness of distant stars. A planet whose orbit happens to cross the face of its star causes a noticeable dip in the star's brightness as it passes. Such precise alignments, although comparatively rare, are useful as they also allow astronomers to calculate the size of the planet, its mass, and various features of its orbit.

A VARIETY OF WORLDS

The range of extrasolar planets discovered so far reveals that the well-ordered Solar System may be a rarity. The limitations of the radial-velocity and transit methods make it easiest to discover massive planets in shorter orbits, but the number of giant planets orbiting close to their parent stars has come as a surprise. What is more, many of them follow elliptical paths that would disrupt the orbits of smaller, more Earth-like worlds. Other unexpected discoveries include worlds in stable orbits within binary-star systems and even planets that orbit the wrong way compared to their parent stars. Astronomers have also discovered worlds very different from the rocky and giant planets of the Solar System, including so-called hot Jupiters, hot Neptunes, and super-Earths.

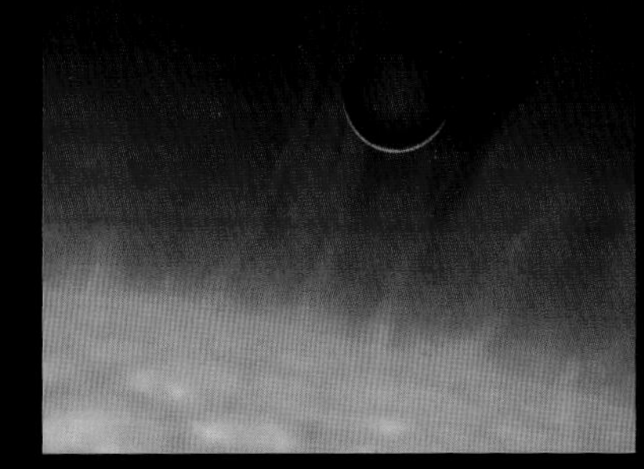

OLD PLANET
A giant planet with a mass three times Jupiter's orbits the 10-billion-year-old star V391 Pegasi. Orbiting 158 million miles (254 million km) from its star, this planet has survived the star's red-giant phase.

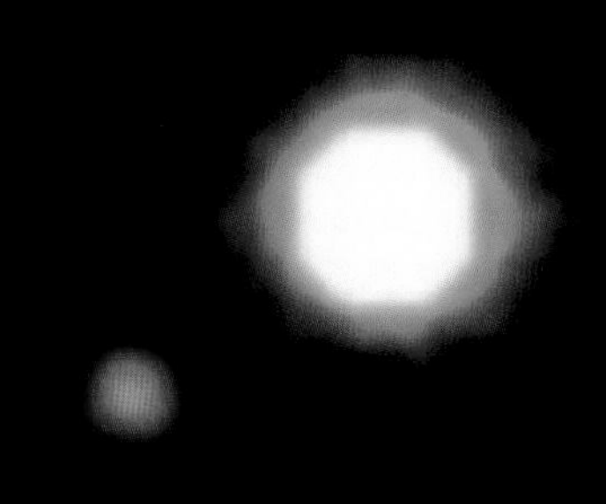

FIRST INFRARED IMAGE
In 2004, the Very Large Telescope in Chile captured this infrared image of the suspected extrasolar planet 2M1207b (red). With a mass of eight Jupiters, it orbits the brown dwarf 2M1207.

THE PLANET HUNTERS

The search for exoplanets continues to accelerate due to improvements in ground-based telescopes, dedicated satellites, and new detection techniques. The radial-velocity method works well with ground-based telescopes, and high-precision spectrographs, such as the High Accuracy Radial-velocity Planet Seacher (HARPS) instrument at the La Silla Observatory, Chile, promises to detect ever smaller stellar wobbles created by lower-mass planets. Meanwhile, orbiting satellites are bringing new precision to the transit method. NASA's Kepler mission will spend at least three years staring at a cloud of stars in the northern Milky Way. The instrument will measure the brightness of 150,000 stars every 30 minutes in search of variations caused by planets passing in front of them.

THE KEPLER MISSION PROFILE

MISSION	
Launch date	March 7, 2009
Launch vehicle	Delta II rocket
Mission duration	3.5 years minimum
THE KEPLER INSTRUMENT	
Agency	NASA
Height	15ft 5in (4.7m)
Diameter	8ft 9in (2.7m)
Weight	2,290lb (1,039kg)
Power source	Solar
Instrument	Photometer featuring 42 charge-coupled devices (CCDs)
Primary mirror	4ft 7in (1.4m)

SCALE
15ft 5in (4.7m)
8ft 9in (2.7m)

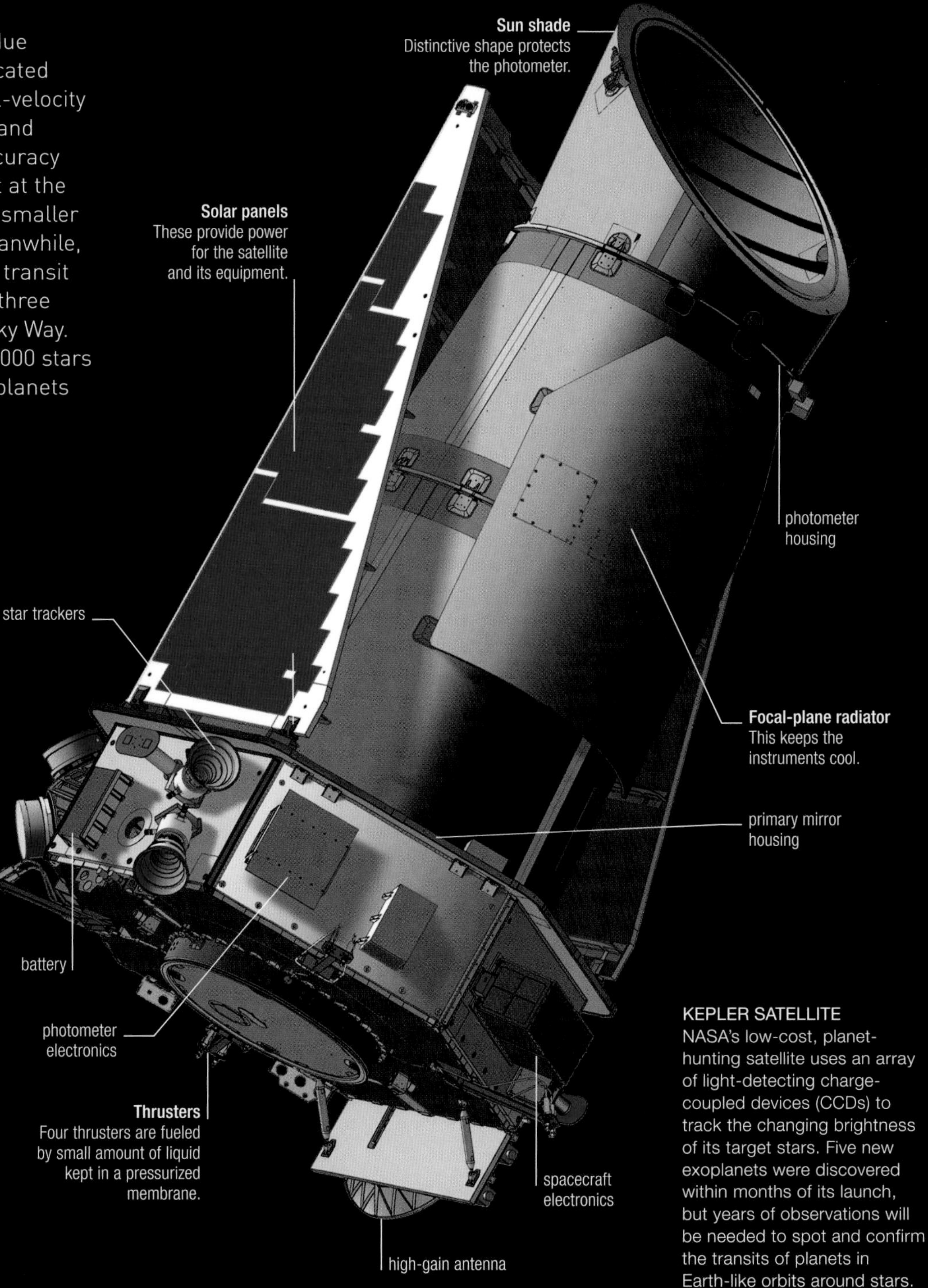
Sun shade
Distinctive shape protects the photometer.
Solar panels
These provide power for the satellite and its equipment.
photometer housing
star trackers
Focal-plane radiator
This keeps the instruments cool.
primary mirror housing
battery
photometer electronics
Thrusters
Four thrusters are fueled by small amount of liquid kept in a pressurized membrane.
spacecraft electronics
high-gain antenna

KEPLER SATELLITE
NASA's low-cost, planet-hunting satellite uses an array of light-detecting charge-coupled devices (CCDs) to track the changing brightness of its target stars. Five new exoplanets were discovered within months of its launch, but years of observations will be needed to spot and confirm the transits of planets in Earth-like orbits around stars.

ALLEN TELESCOPE ARRAY

A revolutionary radio telescope suitable for use both in standard radio-astronomy research and in looking for possible alien signals is being built near San Francisco, CA. Phase one, completed in 2007, consists of 44 relatively small, 20ft (6m) radio dishes. When complete, 350 dishes will be used to create a high-resolution, wide-field image of the sky.

EXTRATERRESTRIAL LIFE

The quest for evidence of life on other planets, whether within the Solar System or beyond, is one of the most exciting areas of modern science and one that pushes astronomy to its limits. Communicating with intelligent extraterrestrial life is potentially the biggest challenge we face in the future.

EXTREMES OF LIFE

There is now overwhelming evidence that the Universe is rich in the carbon-based organic chemicals needed for the development of life as we know it. Most astronomers believe that life also requires a suitable environment in which to grow—in other words, a planet within a star system's habitable zone, where water (necessary for many organic chemical reactions) can exist in liquid form. However, in recent decades, two new trends have widened the possibilities for life in the Universe. One is the discovery of potentially hospitable environments on worlds outside the Solar System's habitable zone, most notably in the subterranean water reservoirs on moons such as Europa (see pp.140–41) and Enceladus (see p.162). The other is the discovery that Earth's life forms are far more versatile than scientists once thought. Living organisms have been found to survive in extreme environments, from undersea volcanic vents to salt lakes—and even in the vacuum of space itself.

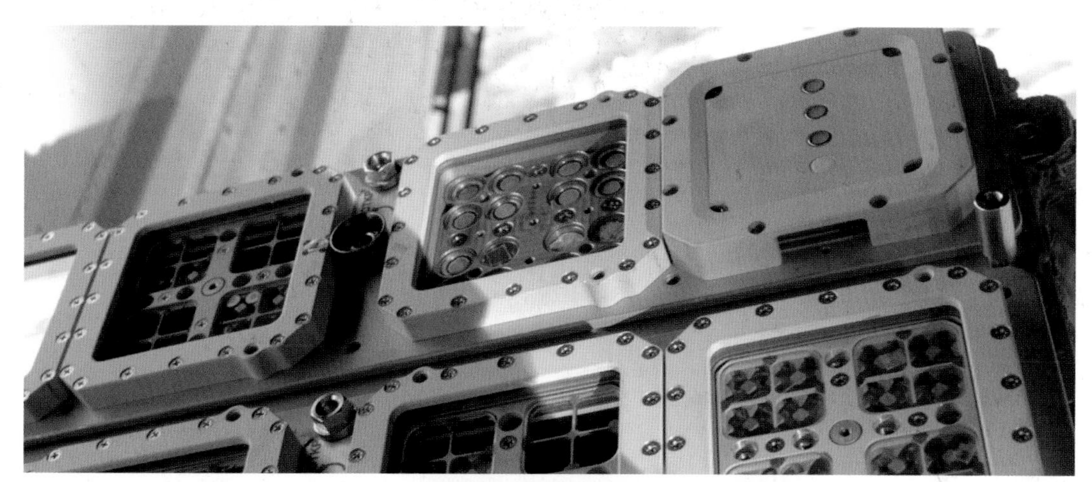

EXPOSE-E EXPERIMENT
Between 2008 and 2010, this European Space Agency experiment on the International Space Station tested how a range of chemical and biological samples, including lichens and insects, fared during 18 months of exposure to the vacuum of space and simulated Mars-like conditions. It showed that a number of organisms can hibernate in a vacuum and revive when they are exposed to water.

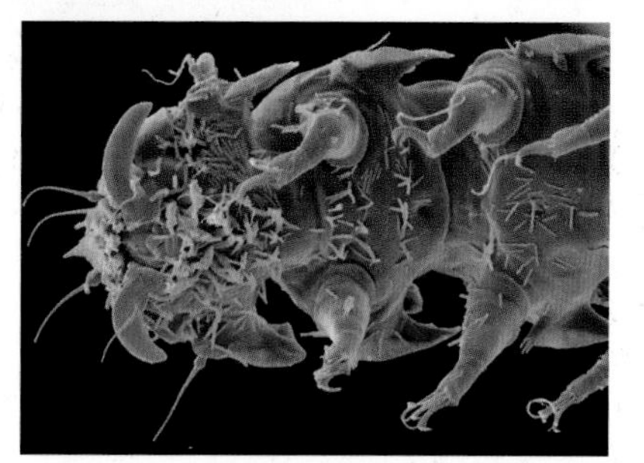

COSMIC TRAVELERS?
The Expose-E experiment revealed that the microscopic Tardigrades, or water bears, are capable of surviving in space. Some astronomers believe it is possible that comets and meteorites may transport microorganisms and bacterial life between habitable planets.

LOOKING AND LISTENING

Careful observations of extrasolar planets may eventually reveal the chemical signatures of life, but finding intelligence is another matter and is the subject of an entire field of astronomy called SETI (the Search for Extraterrestrial Intelligence). SETI scientists use radio telescopes to scan the skies in search of artificial signals from aliens. So far, there have been some intriguing false alarms but no genuine contact. Some astronomers also look for signals in visible light or in the appearance of large-scale interplanetary structures that might have been built by civilizations far more advanced than our own. If an alien signal is detected, it will be a momentous discovery, but it will also present huge challenges to humanity. Who would answer on behalf of our species? Indeed, should we respond at all?

ARECIBO'S MESSAGE
The 1,001ft (305m) Arecibo Telescope in Puerto Rico normally scans the skies for natural radio signals, but it is also used to collect SETI data. In 1974, it sent an experimental coded message toward the globular cluster M13 (see p.220), some 25,000 light-years from Earth.

STELLAR RECYCLING

Each new generation of stars processes the raw materials of the Universe, helping to transform lighter elements into heavier ones, which are then reincorporated into new nebulae, stars, and planets. In this way, every galaxy gradually changes and develops over the billions of years of its evolution.

COSMIC RECYCLING

Stellar recycling involves countless cycles of starbirth and death, but the processes at its heart are surprisingly simple. Matter from the general interstellar medium (see below) accumulates into denser clumps that grow denser and hotter, until they become star-forming nebulae. The nebulae eventually give birth to stars, with their accompanying retinues of planets and cosmic debris. As the stars age, nuclear fusion in their cores transforms large amounts of lighter elements into smaller quantities of heavier ones. At the ends of their lives, Sun-like stars shed much of this processed material as planetary nebulae. High-mass stars continue to produce even heavier elements by fusion, before finally detonating in supernovae. These immense explosions create the heaviest elements, but scatter them across space. The matter released at the end of a star's life provides the raw material for the process to begin again.

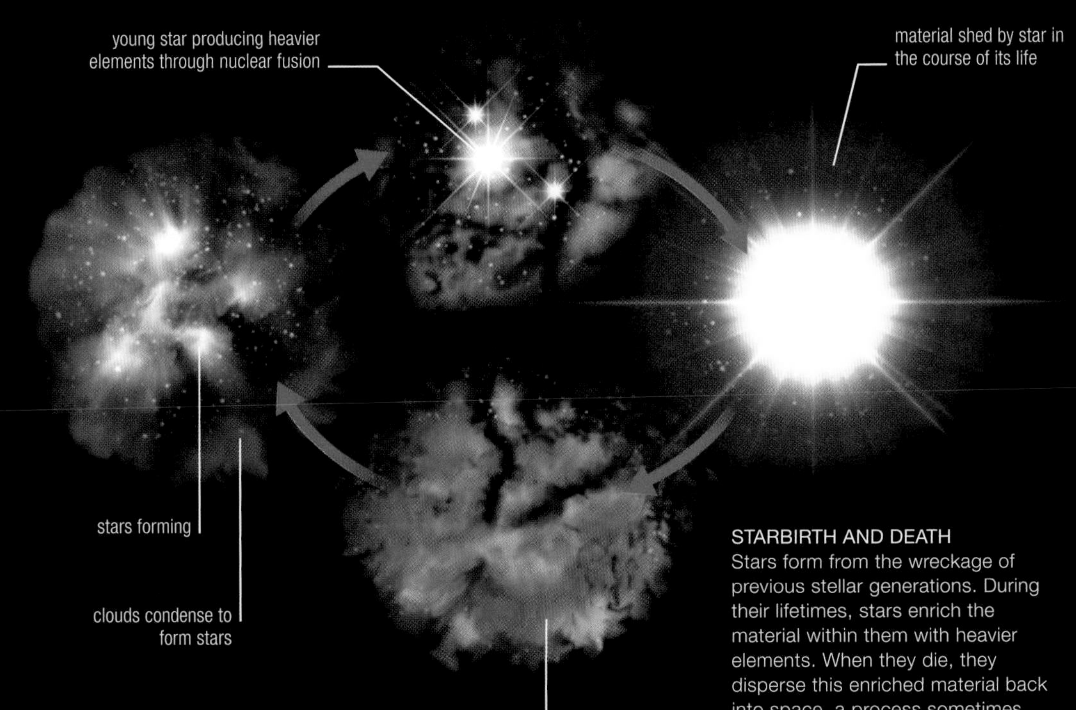

STARBIRTH AND DEATH
Stars form from the wreckage of previous stellar generations. During their lifetimes, stars enrich the material within them with heavier elements. When they die, they disperse this enriched material back into space, a process sometimes accompanied by violent explosions that trigger new waves of starbirth.

THE INTERSTELLAR MEDIUM

The material between the stars is known as the interstellar medium, or ISM. Some of it is the raw material of the Universe, left over from the Big Bang itself. The rest is a mix of materials that have been processed through stars at least once, including heavier elements and complex molecules (groups of atoms), and dust particles thrown off the surfaces of old, cool stars. The dust particles are made largely of carbon or minerals called silicates. In total, the ISM is roughly 89 percent hydrogen, 9 percent helium, and 2 percent heavier elements.

Astronomers divide the ISM into three phases, according to their average temperatures. Cold, dense ISM clouds are dominated by hydrogen atoms and molecules. The second phase—the warmer and more rarefied inter-cloud gas—has temperatures of a few thousand degrees and is a mix of atoms and electrically charged ions (atoms that have been stripped of their electrons). The third phase, called coronal gas, consists of huge clouds of ionized particles with very low densities, heated to millions of degrees by supernova shock waves.

LIGHT ECHOES
In 2002, the unstable star V838 Monocerotis, some 20,000 light-years from Earth at the outer edge of the Milky Way, gave off a huge outburst of brightness. Over the following months and years, light from the initial flare bounced off the surrounding interstellar medium and found its way to Earth as a light echo—a spectacular reminder that even apparently empty space is filled with clouds of material.

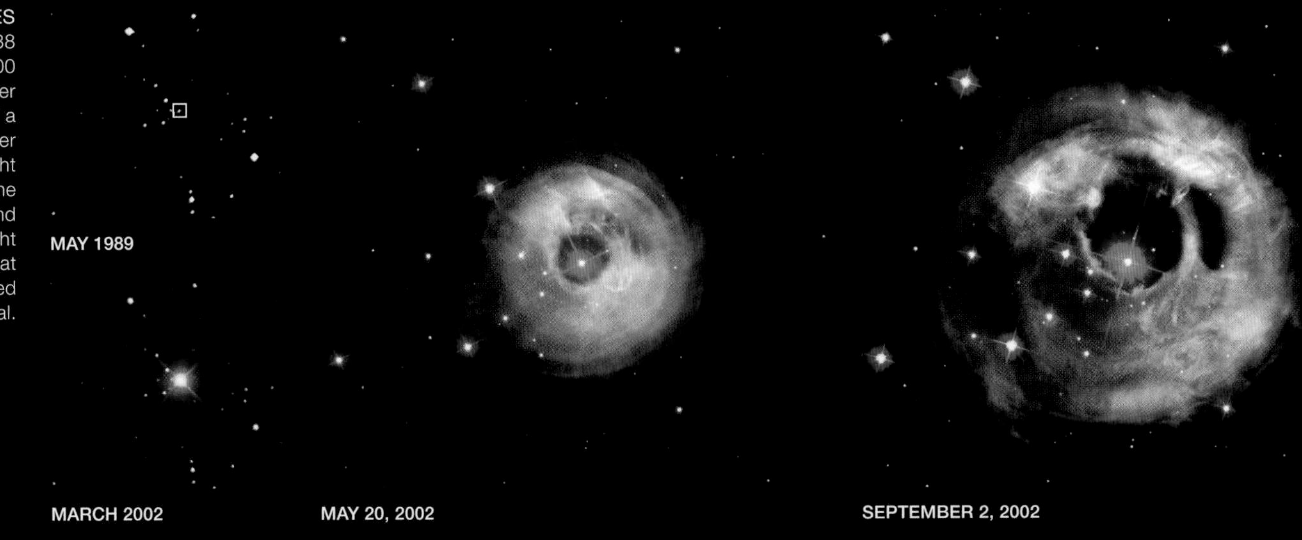

ELEMENTARY PROCESSING

During their main-sequence lifetimes (see pp.226–27), stars transform hydrogen nuclei (protons) into the next lightest element, helium. As the stars age and move off the main sequence, they process helium into heavier elements, such as beryllium, carbon, and oxygen, and smaller amounts of neon and magnesium. Sun-like stars falter and die at this stage, while more massive stars continue fusing carbon, neon, and oxygen, forming elements such as phosphorus, sulfur, and silicon. Finally, these massive stars burn silicon to produce heavier elements such as calcium, titanium, and iron. If the star is large enough, it dies in a supernova explosion, generating conditions suitable for its heavy nuclei to absorb particles called neutrons and grow rapidly into the nuclei of elements heavier than iron, including lead, gold, and uranium.

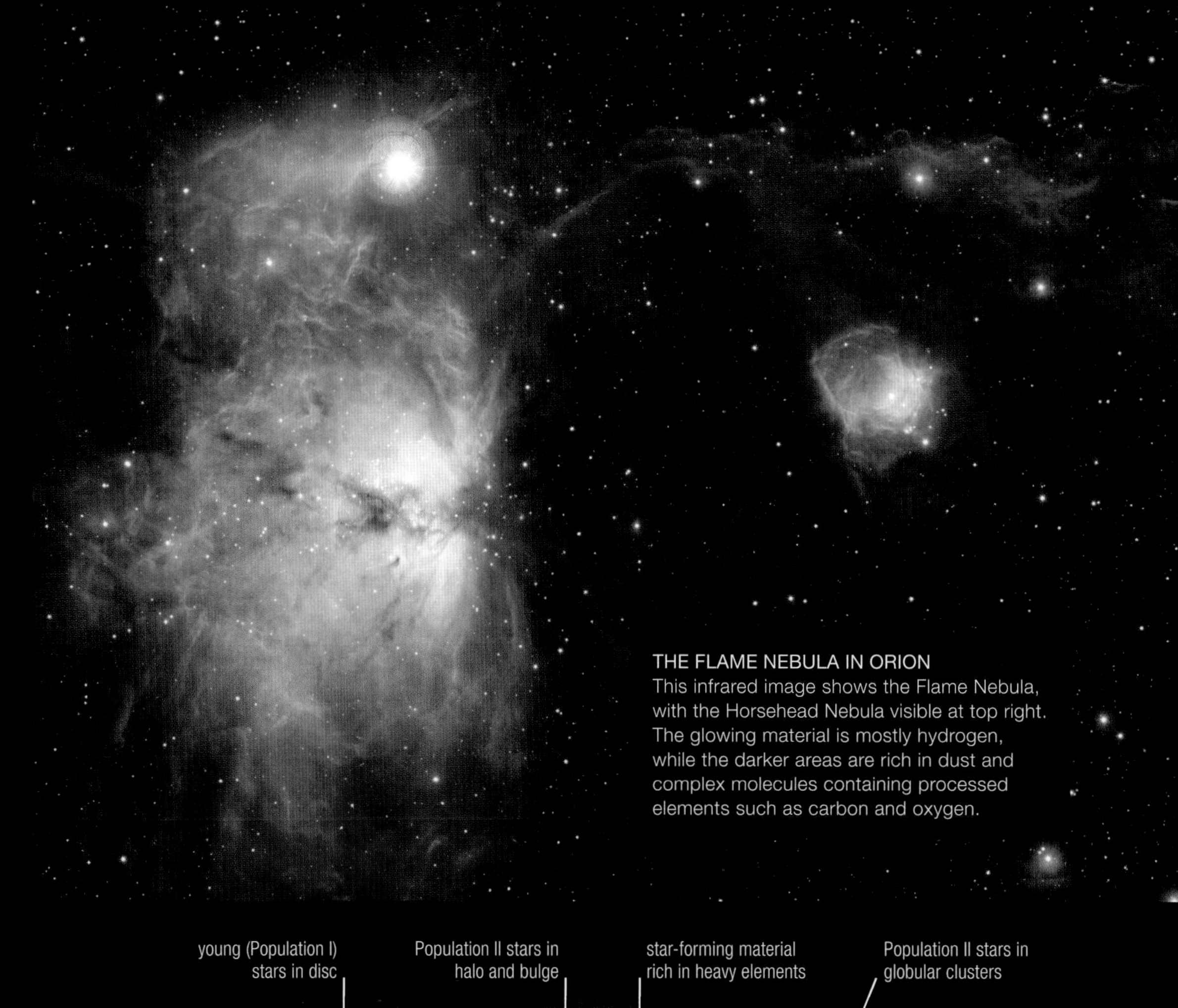

THE FLAME NEBULA IN ORION
This infrared image shows the Flame Nebula, with the Horsehead Nebula visible at top right. The glowing material is mostly hydrogen, while the darker areas are rich in dust and complex molecules containing processed elements such as carbon and oxygen.

STELLAR POPULATIONS

One important result of the gradual enrichment of the interstellar medium with heavier elements is the existence of noticeably different populations of stars. Sun-like stars, which formed in the last few billion years and are relatively rich in heavy elements, are known as Population I stars. Those stars that formed earlier in cosmic history and are poor in heavy elements are known as Population II stars. However, astronomers have yet to detect a star that is completely devoid of heavy elements, so they suspect that the early cosmos was rapidly enriched with material from an initial group of short-lived, truly massive stars called Population III stars.

young (Population I) stars in disc
Population II stars in halo and bulge
star-forming material rich in heavy elements
Population II stars in globular clusters

MILKY WAY MAP
Population II stars in the Milky Way are concentrated in the galaxy's central bulge and after the globular clusters that orbit it. The only survivors from this stellar generation are low-mass, long-lived red-and-yellow stars. Population I stars, meanwhile, orbit in the galaxy's disc and spiral arms and have a range of masses, properties, and ages.

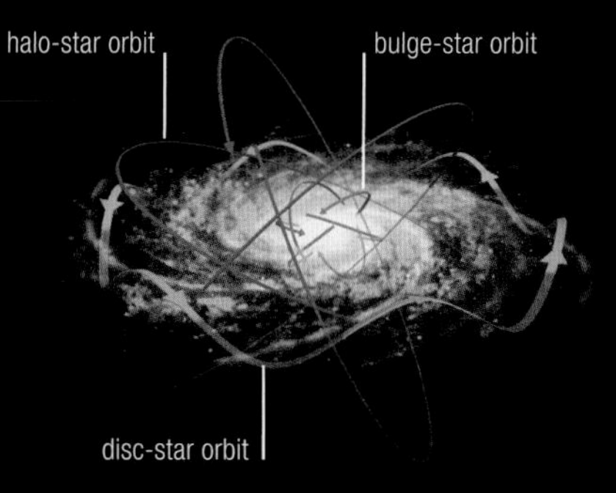

STELLAR MOTIONS
Population I stars in the Milky Way's outer disc follow orbits that weave up and down through the central plane, while Population II stars in the galaxy's bulge have sharply tilted, faster-moving orbits. A small number of stars orbit alone in the galactic halo.

OCTOBER 28, 2002

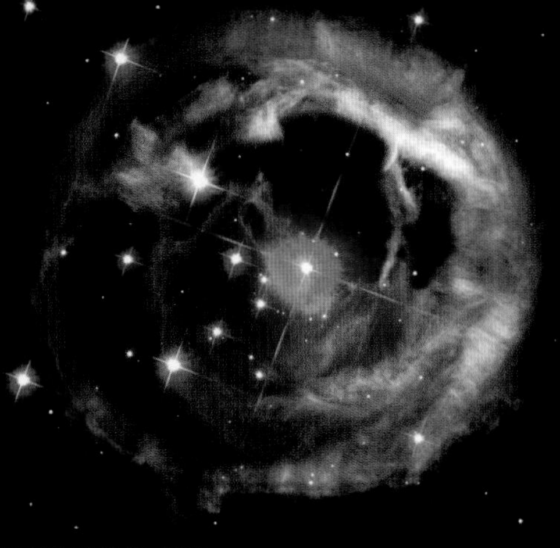

DECEMBER 17, 2002

OCTOBER 23, 2004

THE HEART OF THE MILKY WAY

The central region of the Milky Way is a strange and violent place, home to massive bursts of star formation, enormous star clusters hosting some of the biggest and brightest stars known, plumes of superhot gas, and the dark secret of a supermassive black hole at the center of everything.

SAGITTARIUS A* IN X-RAY
This Chandra X-ray Observatory image reveals dense clusters of young, hot stars (in blue) and clouds of superhot gas from Sagittarius A East (left) and West (right). Stellar winds from the young stars drive gas out of this region, starving the central black hole, Sagittarius A*, of fuel.

TOWARD THE CENTER

The center of the Milky Way is embedded within a bulge of stars, some 27,000 light-years long, 15,000 light-years across, and 6,000 light-years thick. The features inside are hidden behind clouds of stars, gas, and dust, but can be detected at infrared, radio, and X-ray wavelengths. Within the bulge, and surrounding the galactic center at a distance of about five light-years, is a ring of gas, within which lies a dense cluster of 100 or more extremely hot stars.

The center itself, 26,000 light-years from Earth, is known as Sagittarius A (after the constellation in which it lies) and surrounds the brightest radio source in the sky. This source, which occupies the central 10 light-years, has three parts, including a bubblelike supernova remnant called Sagittarius A East and a complex group of gas clouds known as Sagittarius A West. The third part, embedded at the heart of the western feature, is a strong, compact radio source called Sagittarius A*. This is believed to be the location of the Milky Way's central supermassive black hole.

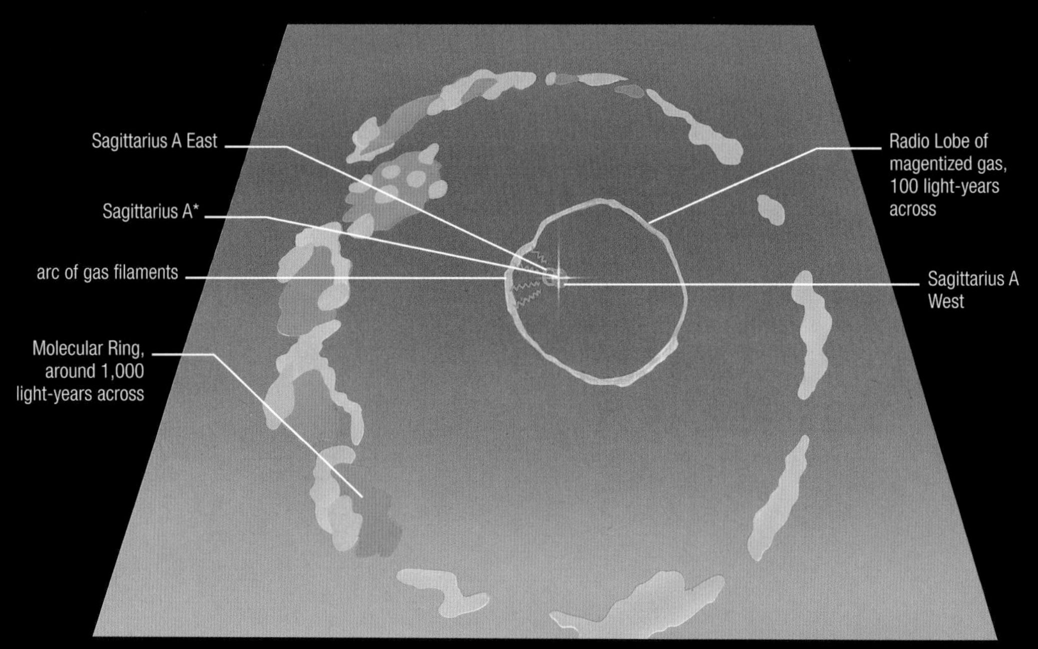

FEATURES OF THE GALACTIC CENTER
The outer features of the galactic center are encircled by a series of cool hydrogen gas clouds, molecular clouds, and nebulae together known as the Molecular Ring. Within the Molecular Ring lies the Radio Lobe—a vast, chimney-shaped region of magnetized gas that is linked by an arc of twisted gas filaments to the Sagittarius A* radio source and the exact galactic center.

EXTREME BEHAVIOR

tars near the galactic center are prone to xtreme behavior. Orbiting at a short distance rom the central supermassive black hole, hey move at high speeds and are closely acked together. As a result, close encounters nd even collisions and mergers between tars are far more common than elsewhere in he galaxy. Conditions in the center also seem o create large star clusters containing some uly enormous stars. The best known of hese stars is the blue hypergiant known s the Pistol Star (see p.231), which has an stimated mass of around 100 Suns. The arge number of short-lived massive stars eads to a relatively high rate of supernova xplosions, and this in turn leaves the galactic enter strewn with plumes and shells of hot as visible at X-ray and radio wavelengths.

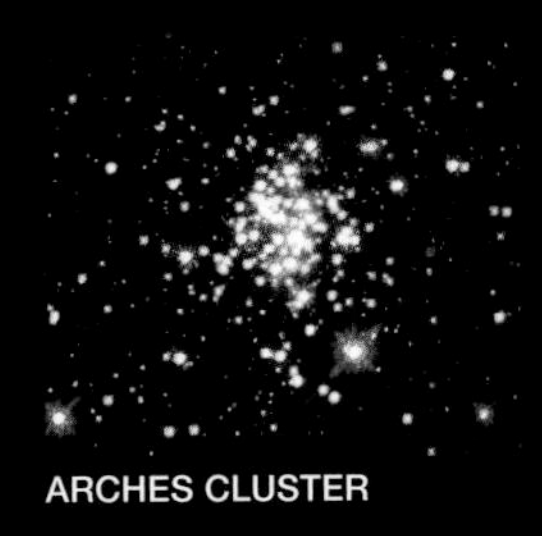

ARCHES CLUSTER

QUINTUPLET CLUSTER

GIANT CLUSTERS
The compact Arches and Quintuplet star clusters both lie within a few hundred light-years of the galactic center. The Arches is the most densely packed open cluster in the Milky Way, while the Quintuplet is home to the galaxy's brightest individual star, known as the Pistol Star.

RADIO VIEW OF THE CENTER
This spectacular image reveals violent conditions in the galactic center. Sagittarius A* is the bright source in the middle, and bubblelike supernova remnants can also be seen. The long arcs and jetlike features are caused by material caught up in the galaxy's magnetic field.

SLUMBERING GIANT

Astronomers had long suspected that our alaxy had a supermassive black hole at its enter, but the proof only came in 2002, when observations of a fast-moving star called S2, close to Sagittarius A*, revealed that it was orbiting one superdense object. The black hole is believed to have the mass of 4.1 million Suns and a diameter of less than 27 million miles (44 million km). It was probably created by the collapse of an enormous gas cloud during our galaxy's formation. At present, the black hole is in a dormant state (see p.295): most of the material within its gravitational grasp has been consumed, and stars and gas clouds in the region orbit at a safe distance. However, material from Sagittarius A West continues to fall onto it, creating its radio glow.

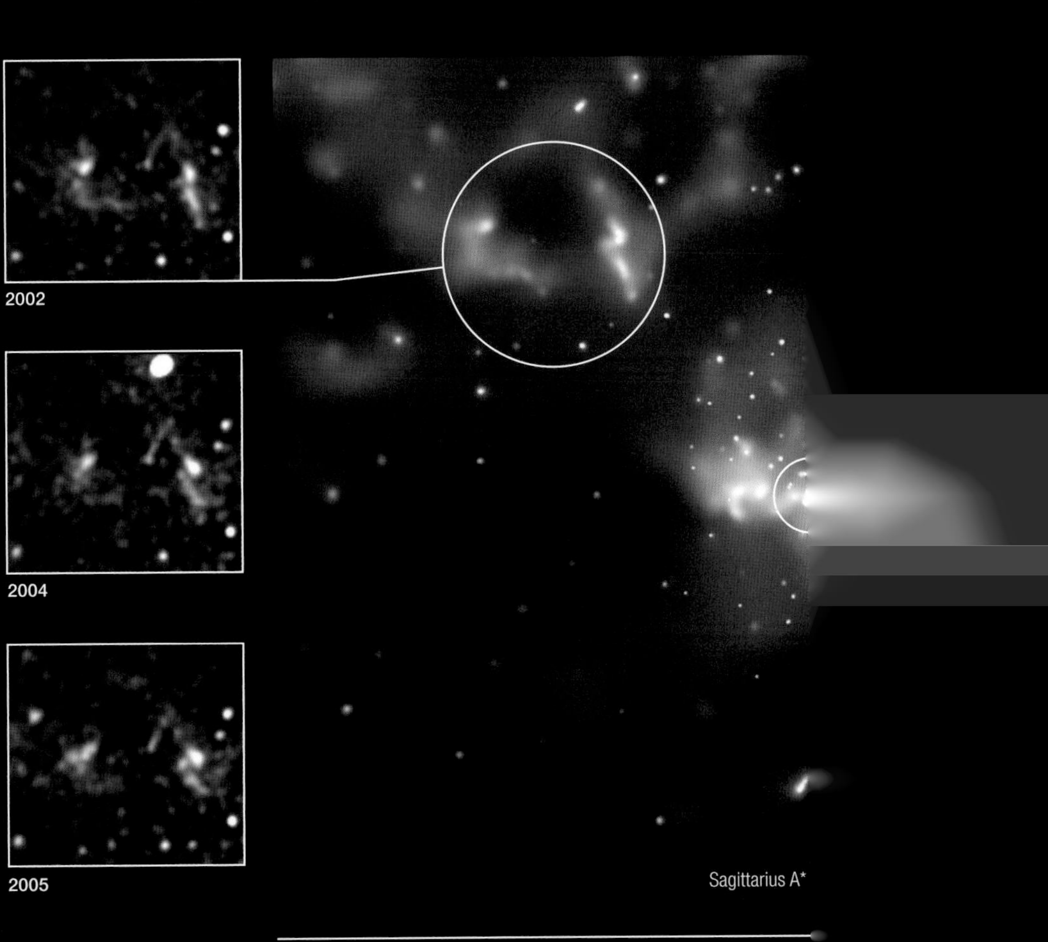

UNEASY SLEEP
A series of X-ray echoes detected by the Chandra X-ray Observatory show that the Milky Way's central black hole was active in the recent past. Coming from a region near Sagittarius A*, they are reflections from an outburst some 50 years ago—equivalent to the light echoes that reach Earth from stellar eruptions (see p.258).

4.1 MILLION SU

The mass of the invisible object at the center of the Milky

1 MILLION SUNS

ANTIMATTER FOUNTAIN
This image from the Compton Gamma Ray Observatory reveals a fountain of antimatter particles rising 3,500 light-years above the center of the Milky Way. This strange material, equivalent to

THE CENTER OF THE MILKY WAY

This infrared image of the center of the Milky Way shows a region 890 light-years across and 26,000 light-years away. The galactic plane appears as a horizontal band of cloud, while the Galactic Center, possibly home to a supermassive black hole, is the bright central spot. The features above and below the galactic plane tend to be closest to Earth.

RSE
AXIES
5

GALAXY ISLANDS

The Milky Way Galaxy is just one among countless vast star systems scattered across the Universe. Galaxies vary greatly in shape and size and even in the type of stars within them. They are often pulled relatively close together by gravity to form galaxy clusters, both large and small.

CITIES OF STARS

A galaxy consists of a huge number of stars mixed with varying amounts of gas and dust. Together, these different elements form an independent system held together by its own gravity and—in most, if not all, cases—centered on a supermassive black hole with the mass of several million Suns. Galaxies range in size and mass from small dwarfs just a few thousand light-years across to giant ellipticals around a quarter of a million light-years in diameter. Smaller galaxies are generally chaotic or at least quite simple in structure, while larger ones are more complex. This complexity is at its most spectacular in the well-developed spiral arms of galaxies such as the Milky Way. The amount of star-forming gas and dust within a galaxy helps determine the nature of the stars it contains. Those galaxies with a plentiful gas supply are rich in young blue-white stars, while galaxies that today lack such interstellar material are composed only of older red-and-yellow stars.

A GRAND DESIGN
This stunning image of the spiral galaxy M94 reveals bright areas of star formation in spiral arms that emerge from the galaxy's core before winding their way through the more sedate stars of its flattened disc region. Galaxies with such well-defined arms are known as grand-design spirals.

GALAXY CLUSTERS

Considering their size, galaxies are crowded relatively close together in space, separated from their neighbors by just tens or hundreds of times their own diameter (compared to millions in the case of most stars). Their immense gravity allows them to reach across the void of intergalactic space and exert influence on one another. As a result, galaxies tend to gather together in clusters, often centered on one or more massive galaxies. The Milky Way Galaxy is a key member of a small cluster called the Local Group, which contains about 50 galaxies. Denser clusters contain many hundreds of galaxies jammed into a roughly similar volume of space, often centered on one or more giant elliptical galaxies. The closest such group to Earth is the Virgo Cluster, around 55 million light-years away. While all the galaxies within an individual cluster form a distinct group drawn together by gravity, clusters tend to blur together at their edges, and the truly massive clusters also gather others around them to form superclusters. The Local Group, for example, is an outlying member of the Virgo Supercluster.

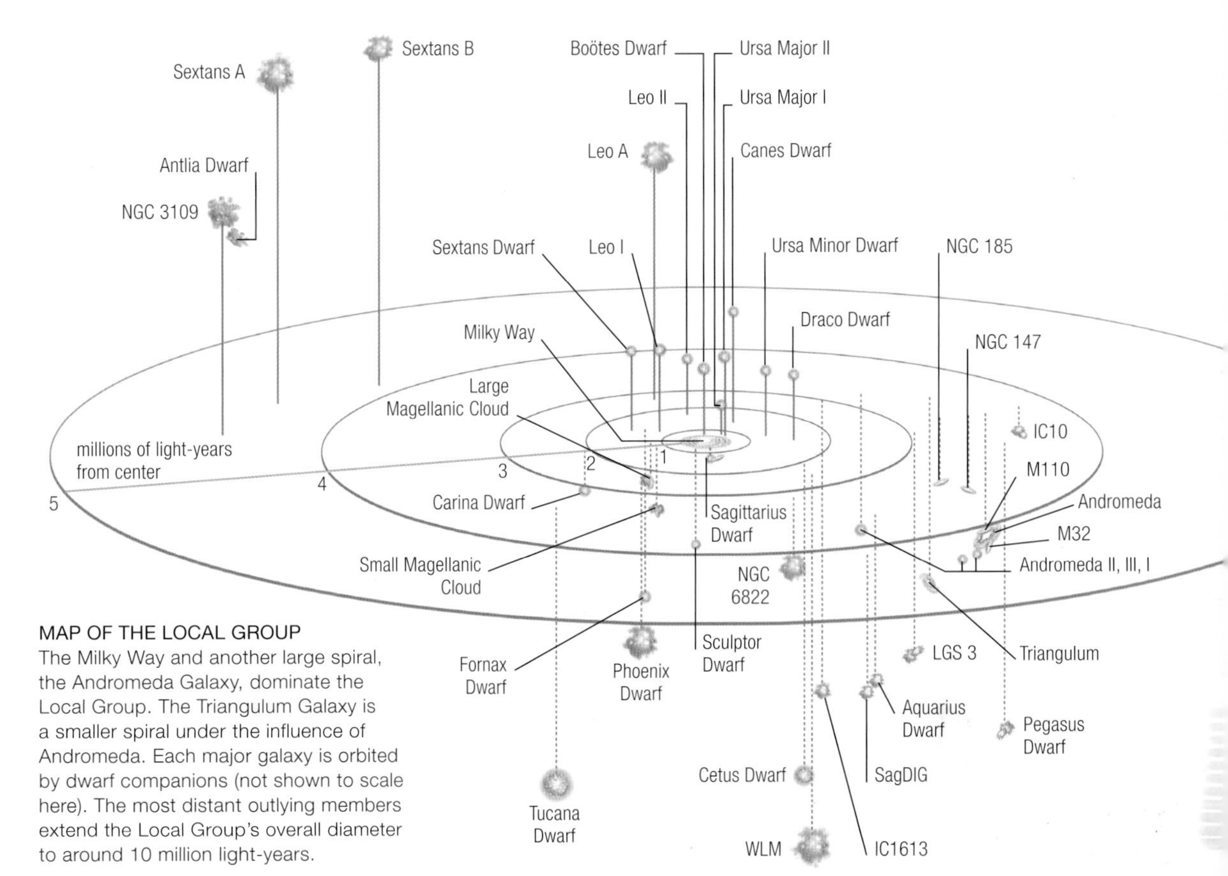

MAP OF THE LOCAL GROUP
The Milky Way and another large spiral, the Andromeda Galaxy, dominate the Local Group. The Triangulum Galaxy is a smaller spiral under the influence of Andromeda. Each major galaxy is orbited by dwarf companions (not shown to scale here). The most distant outlying members extend the Local Group's overall diameter to around 10 million light-years.

ELLIPTICAL GALAXY ESO 325-G004
Elliptical galaxies are classified on a scale ranging from E0 (for perfect spheres) to E6 (for the most elongated cigar shapes). Galaxy ESO 325-G004, shown here, is a type E1.

LENTICULAR GALAXY NGC 2787
Lenticulars (type S0), such as NGC 2787, are like armless spirals or ellipticals with a surrounding gas-and-dust disc. They are a transitional stage in the evolution of galaxies.

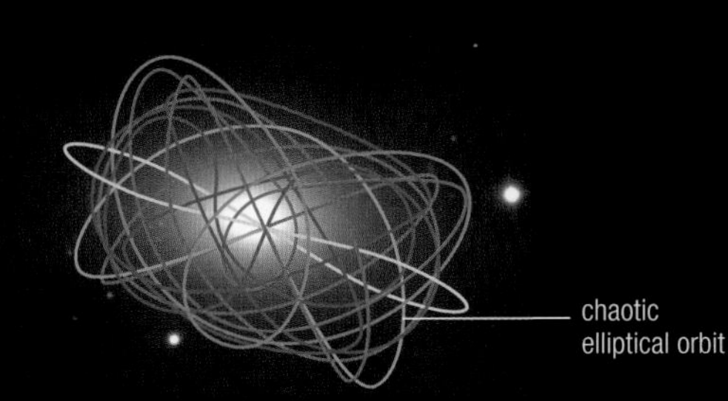

ORBITS IN ELLIPTICAL GALAXIES
Within an elliptical galaxy, individual stars follow chaotic but more-or-less elliptical orbits around the galactic center, tilted at a wide range of angles.

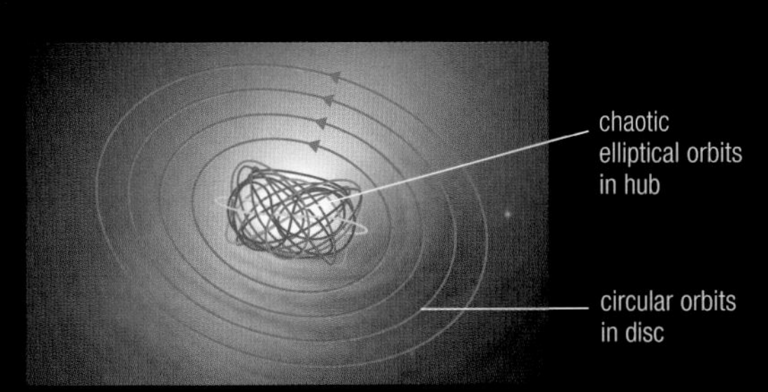

ORBITS IN LENTICULAR GALAXIES
The stars at the hub of a lenticular galaxy follow chaotic elliptical-type orbits. The hub is surrounded by a disc of gas, dust, and faint, Sun-like stars in circular orbits.

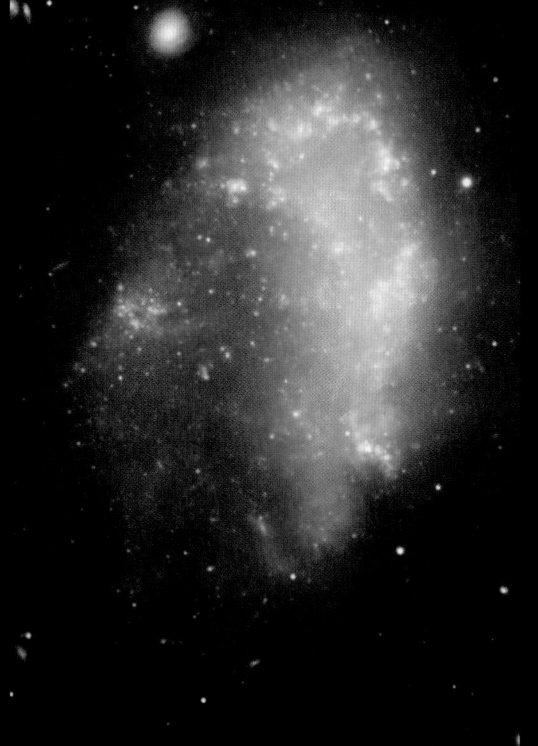

IRREGULAR GALAXY NGC 1427A
Classified as type Irr, irregular galaxies such as NGC 1427A are subdivided into Irr I (with traces of central bars or spiral arms) and completely shapeless Irr II groups.

THE LARGE MAGELLANIC CLOUD

The Large Magellanic Cloud (LMC) is the largest of several galaxies in the immediate vicinity of the Milky Way. Lying 160,000 light-years from Earth, the LMC is an irregular galaxy some 30,000 light-years across. It is packed with gas and dust in hyperactive star-forming nebulae and is dominated by brilliant clusters of young blue-white stars. Seen from Earth, the LMC displays a prominent bar across its center, and there are traces of a single spiral arm, suggesting that the galaxy may be just too small to form into a true spiral. Both the Large and Small Magellanic clouds are named after Portuguese explorer Ferdinand Magellan, one of the first Europeans to observe them. The Magellanic clouds are currently undergoing huge waves of starbirth, triggered by the gravitational forces of the Milky Way as they swing past our own much more massive galaxy.

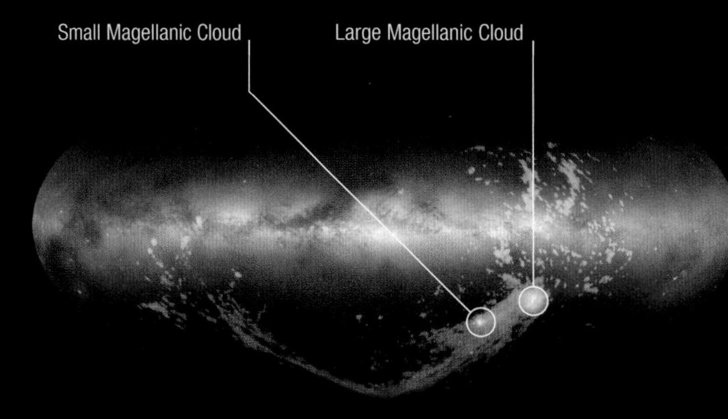

MAGELLANIC STREAM
A long stream of gas trails across the sky in the wake of the two Magellanic clouds. Until recently, the clouds were thought to be trapped in orbit around the Milky Way, but new studies of their motion suggest that they may be just passing by.

THE LMC PROFILE	
Type	Irregular galaxy
Constellation	Dorado
Distance from Earth	160,000 light-years
Diameter	30,000 light-years

TARANTULA'S HEART
The Tarantula Nebula is home to hundreds of newborn, heavyweight stars, formed in at least two distinct waves of activity. While an open cluster called Hodge 301 has already migrated some distance from the center of the nebula, the cluster R136, shown here, still lies at the very heart of the Tarantula.

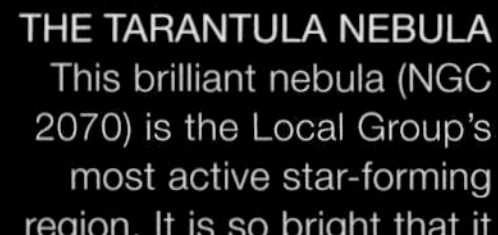

THE TARANTULA NEBULA
This brilliant nebula (NGC 2070) is the Local Group's most active star-forming region. It is so bright that it was originally cataloged as a star, 30 Doradus. The spidery filaments that inspired the Tarantula's name can be seen in this image from the European Southern Observatory's Very Large Telescope at Cerro Paranal, Chile.

NEBULAE IN THE LMC
Viewed from Earth, the Magellanic clouds look like isolated portions of the Milky Way. Here, the Tarantula Nebula (above) and the nebula around the star cluster NGC 2074 (below) are highlighted.

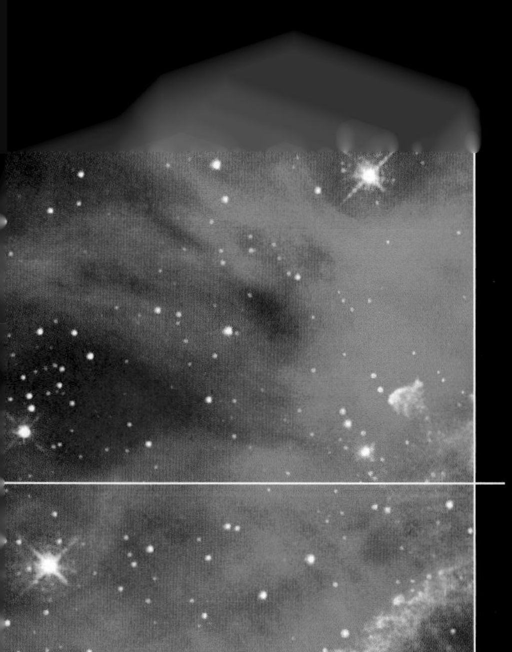

SUPERNOVA 1987A
In 1987, a blue supergiant star in the outskirts of the Tarantula Nebula erupted into a supernova—the brightest and closest exploding star seen from Earth for several centuries. The Hubble Space Telescope and other instruments have kept the developing supernova remnant under observation ever since, as this sequence of images shows.

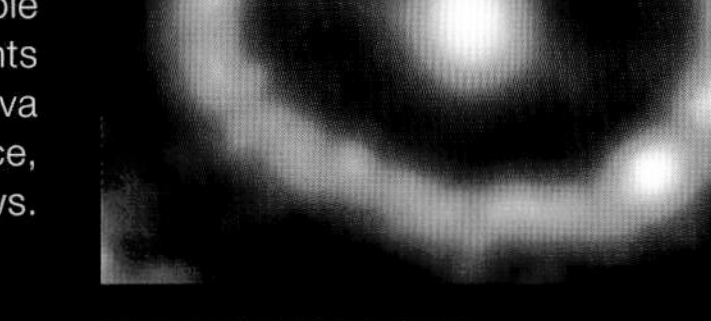

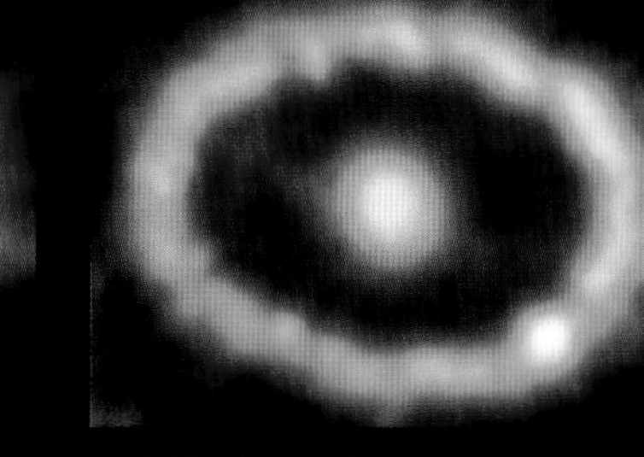

SEPTEMBER 24, 1994

FEBRUARY 6, 1998

NGC 2074
High-energy radiation from hot young stars in NGC 2074 is gradually sculpting and eroding the wall of the surrounding nebula. This same radiation causes the nebula to shine by both reflection and emission.

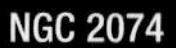

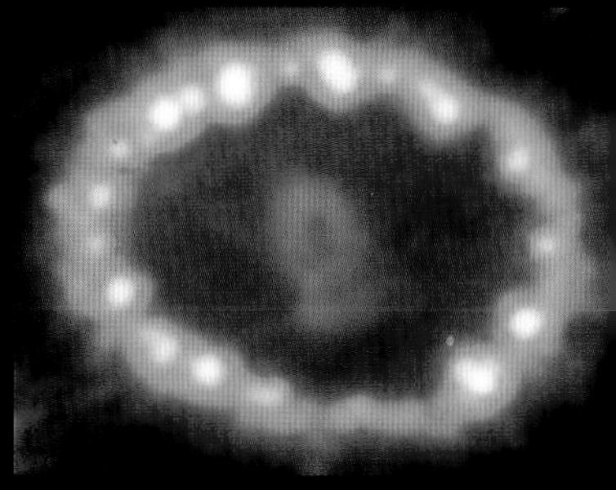

MARCH 23, 2001

JANUARY 5, 2003

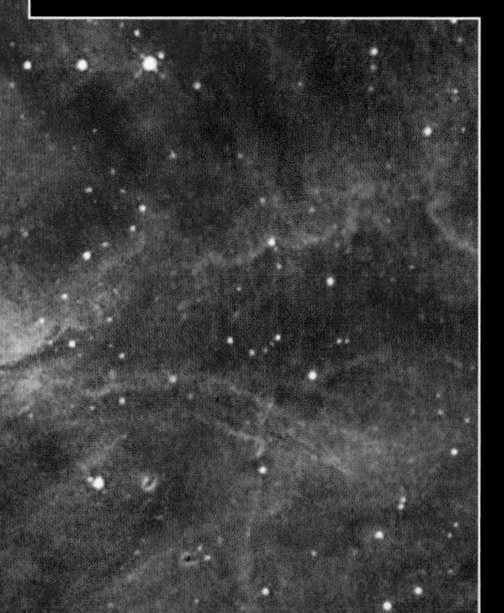

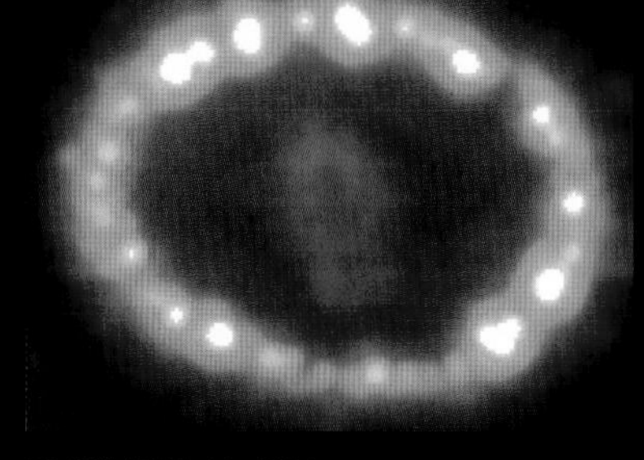

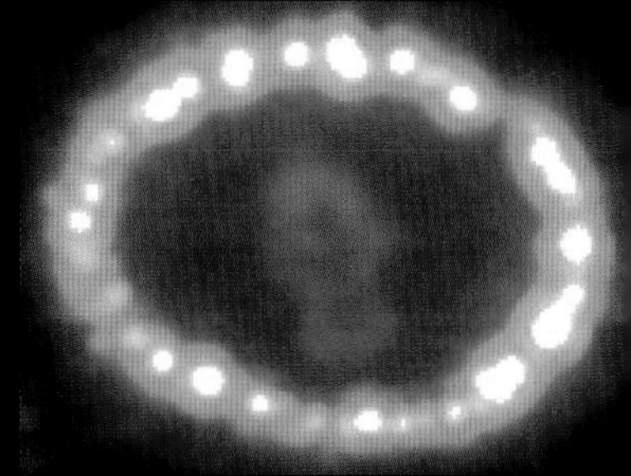

DECEMBER 12, 2004

DECEMBER 6, 2006

NEBULA AROUND NGC 2074
The complex gas clouds of the nebula around NGC 2074 range from delicate wisps and tenuous sheets to dense banks of opaque dust. They are rich in the debris of previous generations of stars, including hydrogen (green), oxygen (blue), and sulfur (red).

THE SMALL MAGELLANIC CLOUD

The Small Magellanic Cloud (SMC) is both physically smaller and further away than the Large Magellanic Cloud, its near neighbor. Lying in the constellation Tucana, this irregular galaxy is 200,000 light-years from Earth, but still easily visible to the naked eye. It contains a total mass of roughly 7 billion Suns, large quantities of gas and dust, and several intense star-forming regions. Much of the cloud's material is concentrated along a distinct bar structure, and some astronomers believe that the SMC is the remnant of a small barred spiral galaxy, battered and distorted by close encounters with the Milky Way.

THE SMC PROFILE	
Type	Irregular galaxy
Constellation	Tucana
Distance from Earth	200,000 light-years
Diameter	7,000 light-years

THE SMALL MAGELLANIC CLOUD IN INFRARED
This Spitzer Space Telescope view highlights two of the SMC's major structures: an extended wing at left, dominated by younger stars; and a long bar at right, containing both mature stars (blue) and young stars embedded in nebulosity (yellow, red, and green). The SMC looks far brighter in infrared than in visible light.

NGC 290
The beautiful open star cluster NGC 290, in the main bar of the SMC, was imaged by Hubble in 2004. Roughly 65 light-years across, it is one of many such clusters whose birth has been triggered by tidal forces from the Milky Way.

NGC 602
This young open cluster lies at the heart of a cavernlike star-forming nebula called N90. Radiation from its hot young stars is beating back and sculpting the interior of the nebula into fantastical shapes, creating waves of compression that trigger new episodes of starbirth.

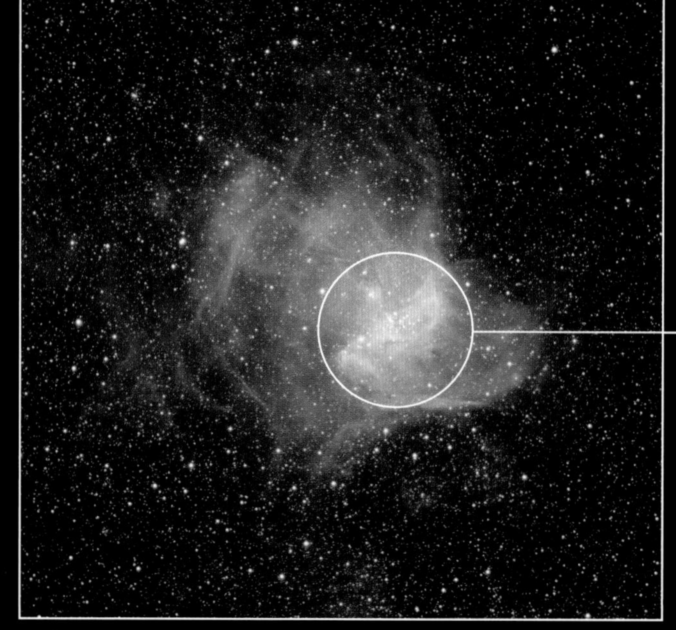

NGC 346
The stars of this open cluster are generating strong radiation and stellar winds that cause the surrounding nebulosity to blow away into space. The result is a delicate, wispy structure, captured in this image from the European Southern Observatory's 87in (2.2m) telescope at La Silla, Chile.

NGC 346 UP CLOSE

This Hubble Space Telescope view of NGC 346 separates the cluster into several distinct knots of brilliant stars surrounded by various complex structures, including a long, dark arc of dust created by the outward pressure of radiation.

THE ANDROMEDA GALAXY

The largest galaxy in the Local Group, the Andromeda Galaxy is also one of the most impressive in the night sky. The most distant object visible to the naked eye, it appears as a pale oval six times the width of the full Moon, with a starlike point of light marking its brilliant hub. Although the shallow angle at which we view the galaxy makes some features hard to distinguish, Andromeda is thought to be a barred spiral with a diameter 40 percent wider than that of the Milky Way, but with a similar mass. A number of smaller satellite galaxies circle Andromeda, drawn to it and held there by gravitational force.

HEAT SOURCES
This false-color mosaic from NASA's Wide-field Infrared Survey (WISE) satellite reveals the distribution of warm dust inside Andromeda. The dust (shown in orange) glows as it is heated by newborn stars close by. This image therefore shows the most intense regions of star formation in Andromeda. The hotter surfaces of mature stars, meanwhile, show up as a bluish haze that defines the galaxy's visible structure.

DOUBLE CORE
An enhanced Hubble image of Andromeda's central region shows a remarkable "double" core. Astronomers once suspected that the fainter secondary core was all that remained of another galaxy consumed by Andromeda, but it now seems likely that the brighter core is just a concentration of bright stars in orbit around Andromeda's true center.

X-RAYS FROM ANDROMEDA'S CORE
This Chandra view shows X-ray sources of different energies (increasing from red through green to blue). The bright central object is a supermassive black hole.

THE ANDROMEDA GALAXY PROFILE

Catalog numbers	M31, NGC 224
Type	Barred spiral galaxy
Constellation	Andromeda
Distance from Earth	2.5 million light-years
Diameter	140,000 light-years

ANDROMEDA IN VISIBLE LIGHT
Dark dust lanes and bright-blue star clusters help to outline the galaxy's spiral structure. Despite appearances, the spiral arms are actually more widely spaced than those in the Milky Way.

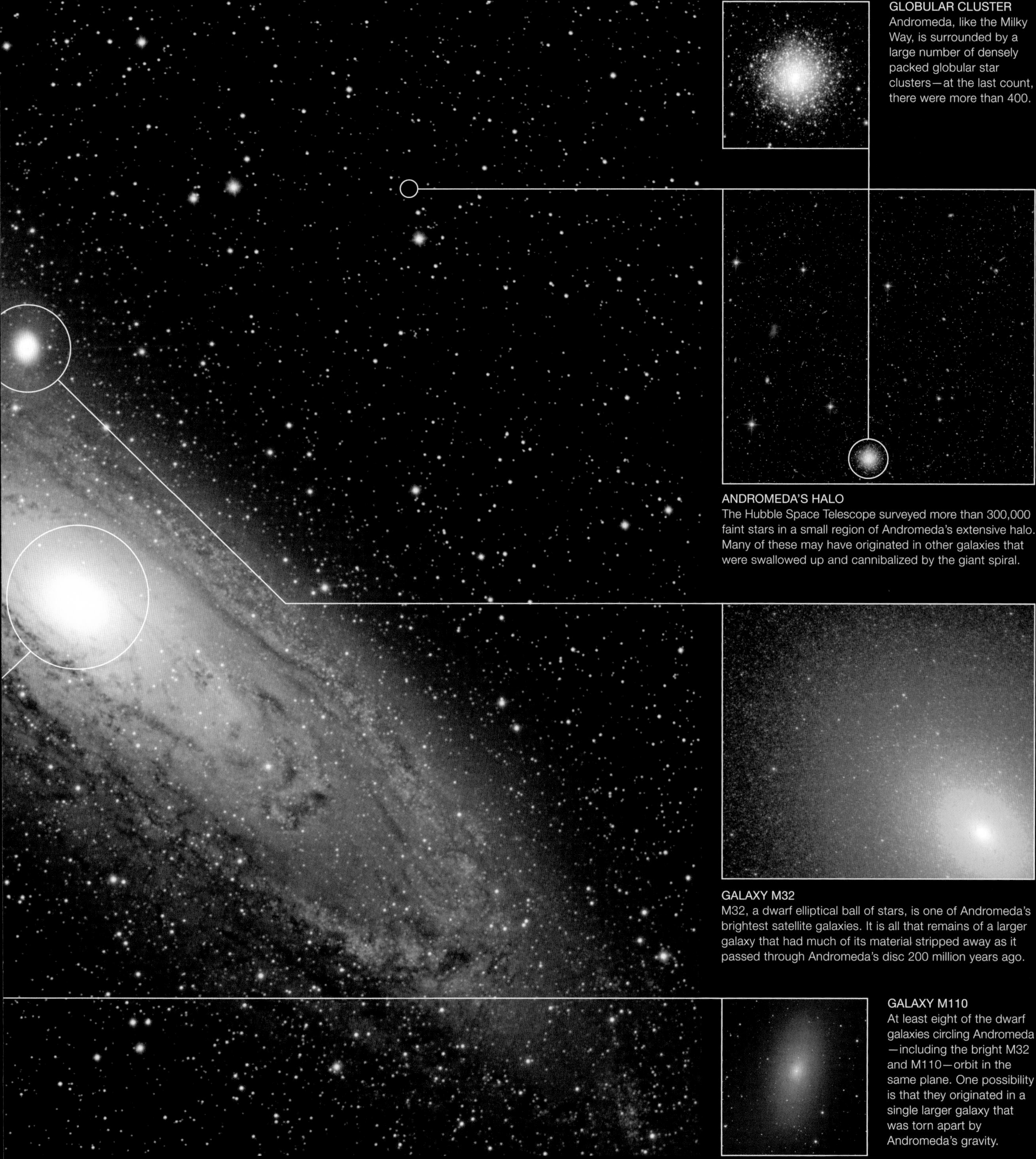
GLOBULAR CLUSTER
Andromeda, like the Milky Way, is surrounded by a large number of densely packed globular star clusters—at the last count, there were more than 400.
ANDROMEDA'S HALO
The Hubble Space Telescope surveyed more than 300,000 faint stars in a small region of Andromeda's extensive halo. Many of these may have originated in other galaxies that were swallowed up and cannibalized by the giant spiral.
GALAXY M32
M32, a dwarf elliptical ball of stars, is one of Andromeda's brightest satellite galaxies. It is all that remains of a larger galaxy that had much of its material stripped away as it passed through Andromeda's disc 200 million years ago.
GALAXY M110
At least eight of the dwarf galaxies circling Andromeda—including the bright M32 and M110—orbit in the same plane. One possibility is that they originated in a single larger galaxy that was torn apart by Andromeda's gravity.

THE TRIANGULUM GALAXY

The third major spiral in the Local Group of galaxies, the Triangulum Galaxy is much smaller than both the Andromeda Galaxy and the Milky Way. As it is less than 3 million light-years from Earth, and quite close to Andromeda in our skies, astronomers suspect that the Triangulum may even be a satellite of Andromeda. Despite its relative closeness and compact size (roughly 50,000 light-years across), from Earth the Triangulum Galaxy appears faint. This is because it contains one-tenth of the number of stars in our own Milky Way, and the light of these stars is dispersed because the galaxy lies almost face-on to us.

NGC 604
This huge cloud of star-forming hydrogen, illuminated by massive stars at its center, is the second-largest emission nebula in the Local Group. Measuring 1,500 light-years in diameter, it is 40 times the size of the Orion Nebula (see p.212) and more than 6,000 times as bright.

STAR-FORMING REGIONS
The Triangulum Galaxy's spiral structure is highlighted by pink knots of star-forming gas scattered across a glowing disc that appears twice the size of the full Moon.

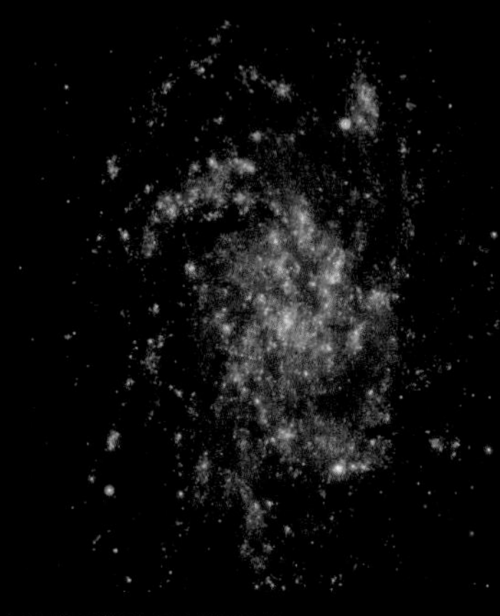

ULTRAVIOLET SPIRAL
This ultraviolet mosaic by NASA's SWIFT satellite highlights hot young stars, revealing a far higher rate of starbirth than in either the Milky Way or Andromeda.

THE TRIANGULUM GALAXY PROFILE

Catalog number	M33
Type	Spiral galaxy
Constellation	Triangulum
Distance from Earth	2.7 million light-years
Diameter	50,000 light-years

UNSEEN EXTREMITIES
In this infrared image, stars are blue, star-forming regions are orange, and green is carbon-rich dust. This proves that the galaxy's material extends well beyond what we can see of it in visible light.

BARNARD'S GALAXY

One of the closest Local Group galaxies beyond the immediate vicinity of the Milky Way, Barnard's Galaxy lies 1.6 million light-years away, in the constellation Sagittarius. This irregular galaxy contains just 10 million stars but has as many as 150 star-orming nebulae in a region 7,000 ight-years across.

STARRY RECTANGLE
NGC 6822 presents a near-perfect rectangle to Earth, framed by wisps of pinkish gas—remnants of starbirth nebulae blown away by infant stars whose stellar winds overcome the galaxy's weak gravity.

BARNARD'S GALAXY PROFILE

Catalog number	NGC 6822
Type	Irregular galaxy
Constellation	Sagittarius
Distance from Earth	1.6 million light-years
Diameter	7,000 light-years

NGC 55

The large, irregular galaxy NGC 55 lies 7.5 million-light years rom Earth in the constellation Sculptor. It is on the boundary between the Local Group and the Sculptor Group of galaxies, traditionally seen as our nearest neighboring galaxy group. But recent measurements have shown that NGC 55 is actually closer than most members of the Sculptor Group, and it may in fact orm an independent pair with the small spiral galaxy NGC 300.

FLAWED SPIRAL

NGC 55 PROFILE

GALAXIES AND THE HUBBLE SPACE TELESCOPE

The Hubble Space Telescope was designed from the outset to help establish Earth's true place in the wider cosmos and is currently the ultimate tool for studying distant galaxies. Launched in 1990, the telescope has also made countless discoveries in the Solar System and among the stars of the Milky Way.

A UNIQUE VIEW OF THE UNIVERSE

The Hubble Space Telescope's location beyond Earth's atmosphere is unique among visible-light telescopes. It gives it a sharper view than any other telescope and enables it to pick up light almost as faint as that detected by the vast telescopes on Earth. This combination enables Hubble to view nearby galaxies in detail as well as to image some of the furthest galaxies ever seen. By looking across billions of light-years, it is detecting light that left its source billions of years ago, and it is imaging galaxies from a younger period of cosmic history. Light from these galaxies is also red shifted by the Doppler effect (see p.233)—the further away the galaxy, the faster it is moving away from us. This is crucial evidence for the expansion of the Universe (see pp.306–307).

ANCIENT LIGHT
This 40-hour exposure from Hubble's Advanced Camera for Surveys shows a small patch of the sky in the constellation Fornax. It reveals a host of galaxies lurking billions of light-years beyond the foreground stars.

Aperture cover
This cover was closed to protect the internal optics during its initial launch.

Heat protection
Insulation protects the telescope from expansion and contraction in changing sunlight.

MEASURING THE HUBBLE CONSTANT

The Hubble Space Telescope Key Project aimed to establish the scale and age of the Universe. It did this using the same technique that American astronomer Edwin Hubble first established (in the 1920s) to determine the distances to other galaxies. The telescope searched a range of galaxies for Cepheid variables—yellow supergiant stars whose cycle of changing brightness reflects their average luminosity. By working out the true luminosity of these stars, astronomers calculated the distances to the host galaxies. These distances could be compared with the rate at which the galaxies were receding (measured from their red shift) to work out the rate of cosmic expansion (the Hubble constant) and therefore the size and age of the Universe. The project concluded that the Universe is 13.7 billion years old.

MEASURING DISTANCE
The spiral galaxy NGC 3021 was one of 18 surveyed by Hubble as part of its Key Project. Cepheid variables within the galaxy were studied to calculate the galaxy's distance from Earth.

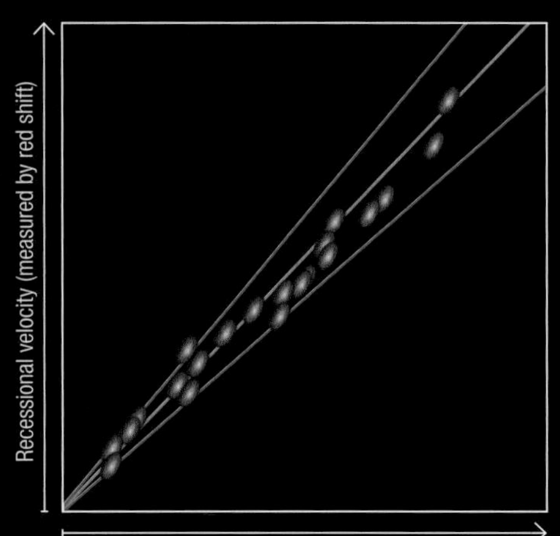

HUBBLE'S LAW
When the distance to galaxies is plotted on a chart against their speed of recession from Earth, they tend to fall within boundaries (blue lines) that show a close relationship. The central slope (orange) yields a value for the Hubble constant—the rate of cosmic expansion. The final Hubble Space Telescope measurements suggest that space is expanding at 13 miles (21.5km) per second for every million light-years of intervening space.

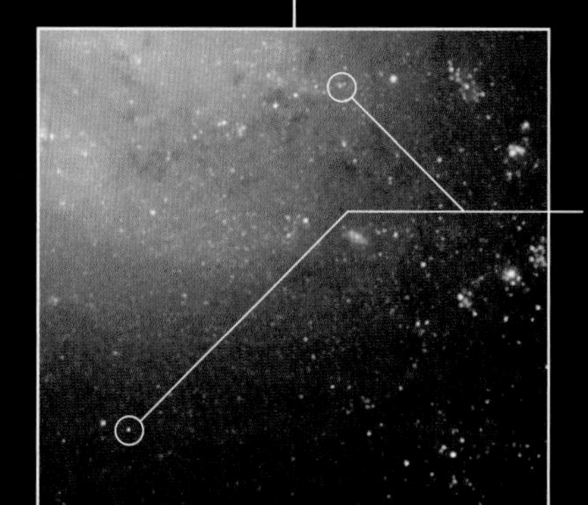

LOOKING AT CEPHEID VARIABLES
During a 60-day observing window each year, Hubble scoured target galaxies for Cepheids. Astronomers observed the distinctive cycle of changes in brightness in 800 stars—the longer the star's period of variability, the more luminous it is.

STARBURST AFTERMATH
This true-color Hubble image reveals dust lanes and clusters of intense blue stars in NGC 2976, a small galaxy 12 million light-years away that has recently undergone an intense burst of star formation that is now slowing down.

UNSEEN DETAILS

As well as measuring intergalactic distances, Hubble's suite of instruments reveals galactic features in unprecedented detail. In its lifetime, the telescope has carried many instruments, including cameras, photometers, and spectroscopes, covering a range of wavelengths, from the near infrared, through visible, to the ultraviolet. Hubble's cameras are monochrome. Color images are constructed by selecting wavelengths of light from objects using filters, then combining the images back on Earth. Some filters only allow very specific wavelengths of light to pass through, while others let through a broad band of light corresponding to a certain color, so that true-color images can be produced.

High-gain antenna
This enables the telescope to communicate with scientists on Earth via NASA's Tracking and Data Relay Satellite.

ACROSS THE SPECTRUM
Captured using different instruments and filters, this gallery shows aspects of the barred spiral galaxy NGC 1512. For example, massive newborn stars are seen with ultraviolet, while infrared reveals cooler dust and star-forming gas.

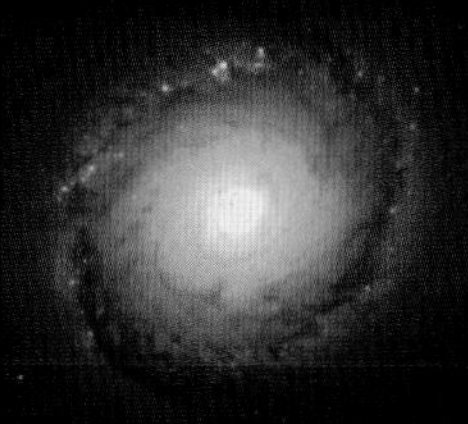

ULTRAVIOLET (FOC)

VISIBLE LIGHT (WFPC2)

NEAR INFRARED (WFPC2)

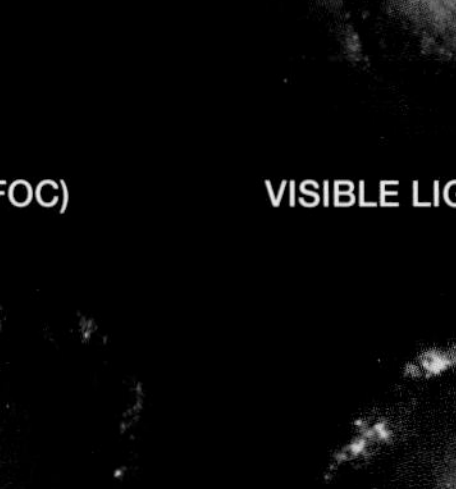

ULTRAVIOLET (WFPC2)

VISIBLE LIGHT (WFPC2)

INFRARED (NICMOS)

Solar panels
Pivoting panels generate electricity for the telescope, which consumes around 3,000 watts of power.

second high-gain antenna

Instrument housing
The rear of the telescope contains the instrument modules and satellite control systems.

THE TELESCOPE IN SPACE
The crew of the Space Shuttle *Atlantis* took this picture of the newly refurbished Hubble Space Telescope shortly after releasing it back into orbit at the end of their most recent and final servicing mission in 2009.

THE HUBBLE SPACE TELESCOPE PROFILE

MISSION

Launch date	April 24, 1990
Launch vehicle	Space Shuttle *Discovery* on mission STS–31
Mission length	20 years and still operational (five servicing missions undertaken to update and repair instruments)

HUBBLE SPACE TELESCOPE

Agency	NASA and European Space Agency
Length	43ft 3in (13.2m)
Width	14ft 1in (4.3m)
Weight	24,500lb (11,110kg)
Power source	Solar panels
Primary mirror	7ft 10in (2.4m) diameter
Instruments	Near Infrared Camera and Multi-Object Spectrometer (NICMOS), Space Telescope Imaging Spectrograph, Advanced Camera for Surveys, Wide Field Camera 3, Cosmic Origins Spectrograph

SCALE

43ft 3in (13.2m)

14ft 1in (4.3m)

REPAIRING HUBBLE

In May 2009, Hubble was brought into the Space Shuttle *Atlantis* for a final overhaul. Astronauts installed a new Wide Field Camera and Cosmic Origins Spectrograph and repaired its Space Imaging Spectrograph and the Advanced Camera for Surveys. In addition, the spacecraft systems, batteries, and outer insulation blankets were replaced.

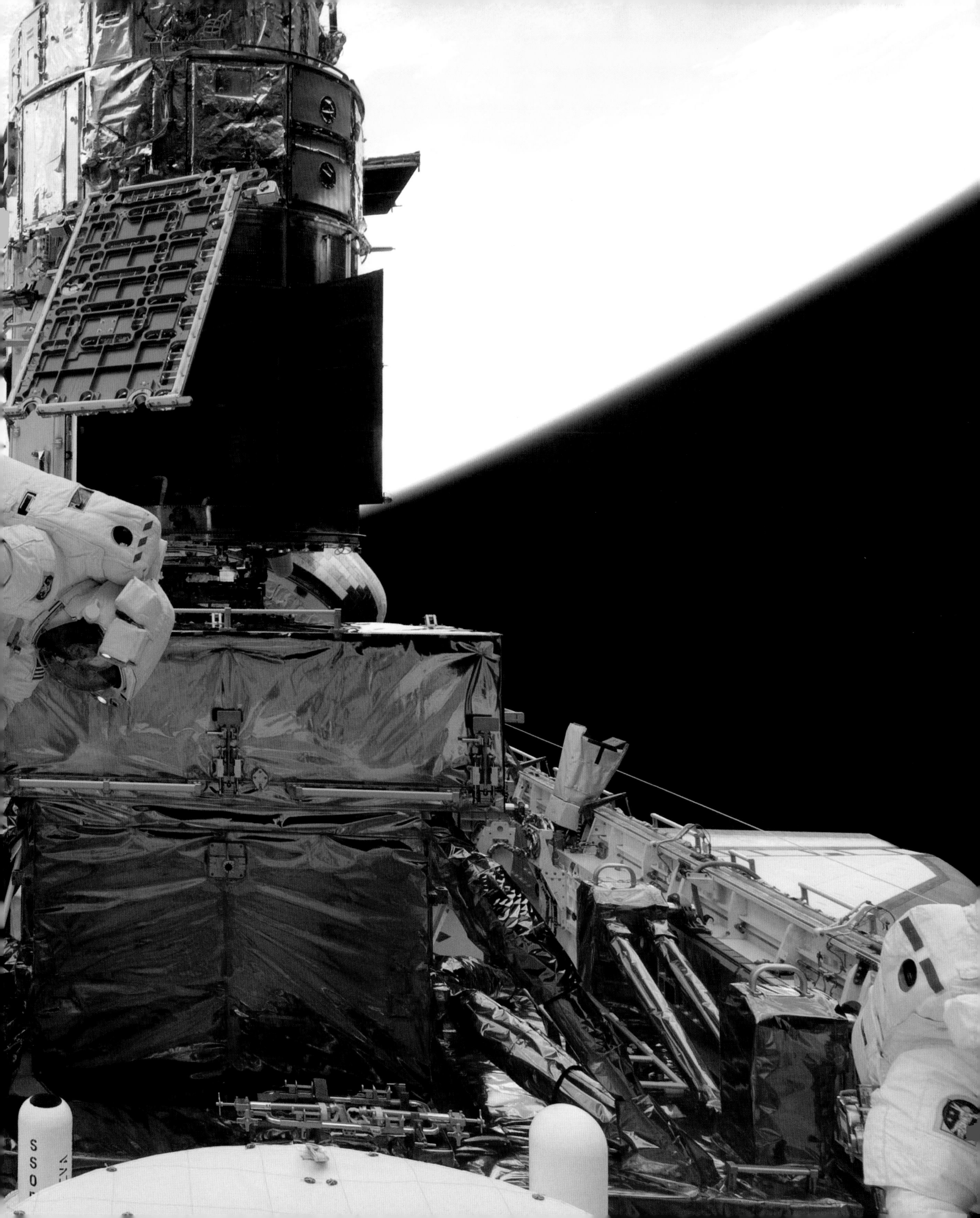

BODE'S GALAXY

Situated almost 12 million light-years from Earth in the constellation Ursa Major, Bode's Galaxy is an elegant spiral with two well-defined arms illuminated by a burst of ongoing star formation. Discovered by the German astronomer Johann Elert Bode in 1775, it has an unusually large, bright core has led some to consider it as a Seyfert galaxy—a subdued form of active galaxy (see pp.294–95). The brightness of the spiral arms and the activity in the galaxy's core are both thought to be due to interactions with the gravity of the nearby Cigar Galaxy (see opposite).

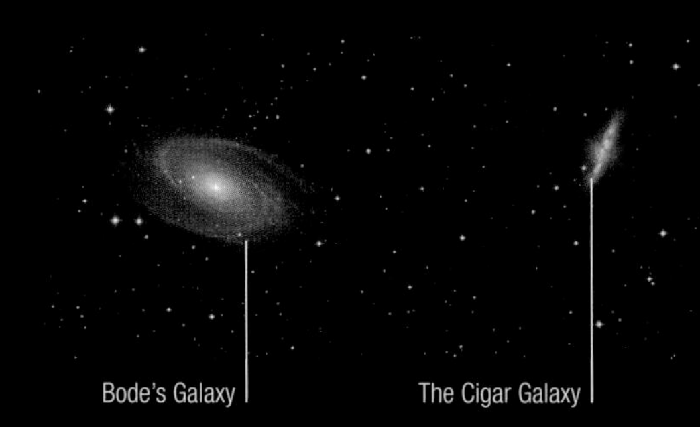

TOGETHER IN SPACE
Viewed from Earth, Bode's Galaxy and the Cigar Galaxy are separated by roughly the width of the apparent full Moon. The two galaxies are located in the northern part of Ursa Major.

OUTSIZED CORE
This Hubble Space Telescope view emphasizes both the fine spiral structure of Bode's Galaxy and its large core. A huge black hole with the mass of 70 million Suns probably lies at its heart.

BODE'S GALAXY PROFILE

Catalog numbers	M81, NGC 3031
Type	Spiral galaxy
Constellation	Ursa Major
Distance from Earth	11.8 million light-years
Diameter	60,000 light-years

BODE'S GALAXY AND ITS NEIGHBORS
Bode's Galaxy is the dominant galaxy in a group of 34. Its strong interactions with two of its nearest neighbors—the Cigar galaxy and NGC 3077—are revealed in this composite image of the region. The visible light of the galaxies is shown in white, and radio-emitting hydrogen gas clouds are in green and red. Repeated encounters between the three galaxies have stripped gas away from them and ejected it into intergalactic space, forming five clouds.

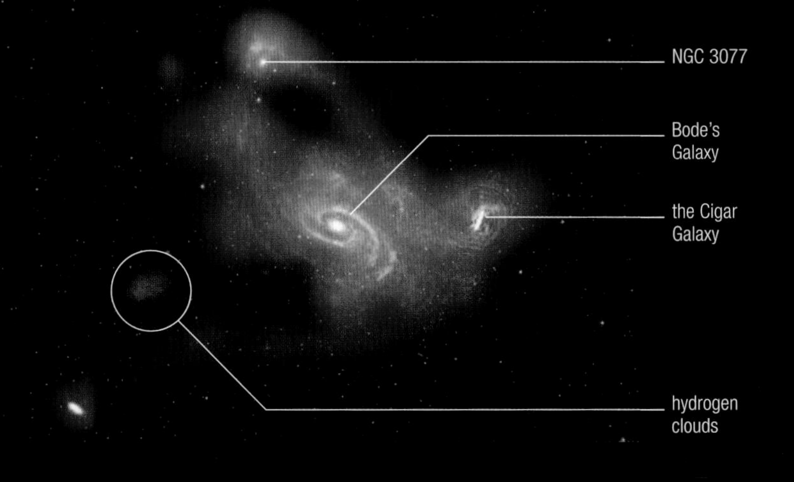

THE CIGAR GALAXY

A close neighbor in space of Bode's Galaxy (see opposite), the Cigar Galaxy (M82) is one of the most famous galaxies and is well known as the prototype starburst galaxy. It is a small and distorted disc- or spiral-shaped galaxy that we happen to see edge-on, with dust lanes silhouetted against its brighter star fields, and plumes of gas emerging from above and below its plane. Situated just 300,000 light-years away from the more massive Bode's Galaxy, the Cigar Galaxy is undergoing massive changes under the influence of its neighbor, which has warped it out of shape and triggered a massive burst of star formation, increasing its rate of starbirth tenfold.

THE CIGAR GALAXY PROFILE

Catalog numbers	M82, NGC 3034
Type	Disc- or spiral-shaped galaxy
Constellation	Ursa Major
Distance from Earth	11.5 million light-years
Diameter	40,000 light-years

MISTAKEN EXPLOSION
At one time, the Cigar Galaxy's intensely bright star-forming core and huge plumes of ejected gas led astronomers to believe that it was ripping itself apart in a galactic explosion.

MULTIWAVELENGTH MONTAGE
A composite of optical, infrared, and X-ray images reveals the Cigar Galaxy's true drama. Hot, X-ray-emitting gas (blue) is escaping the galaxy's weak gravity as massive stars in the central starburst region die in supernova explosions.

STARBURST CORE
Hubble images of the Cigar Galaxy's core reveal a wealth of heavyweight star clusters that formed in a wave during a close encounter with Bode's Galaxy some 600 million years before it reached its presently observed state. The majority of star formation in the Cigar Galaxy as we see it now seems to be triggered as material pulled out of it during this close encounter now falls back into the galaxy.

THE WHIRLPOOL GALAXY

A pair of bright, interacting galaxies, usually cataloged together as M51, can be found in the constellation Canes Venatici. The system is dominated by a compact but bright spiral known as the Whirlpool Galaxy (M51A or NGC 5194). An irregular or elliptical companion (M51B or NGC 5195) lies nearby. The Whirlpool's spiral arms are easily seen from Earth—as a result, this galaxy was the first of the spiral nebulae identified by Irish astronomer Lord Rosse in 1845.

WHIRLPOOL IN SPACE
A stunning Hubble mosaic reveals the true relationship between the two galaxies of M51: the smaller M51B is in fact moving behind the larger galaxy, raising tides of star formation as it passes.

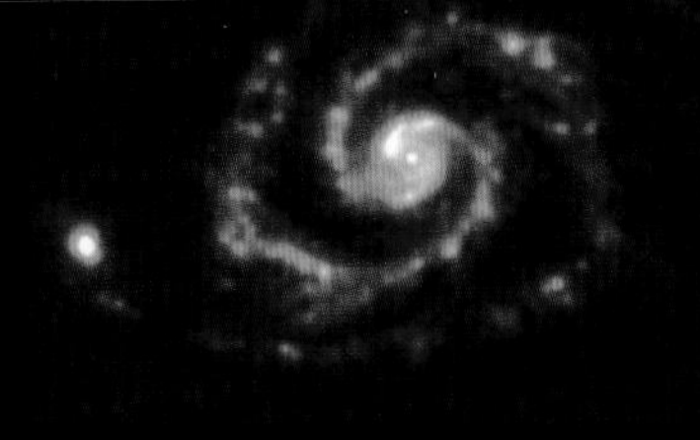

HERSCHEL'S VIEW
An early test image from the infrared Herschel Space Observatory traces concentrations of warm dust around star-forming regions in the two galaxies.

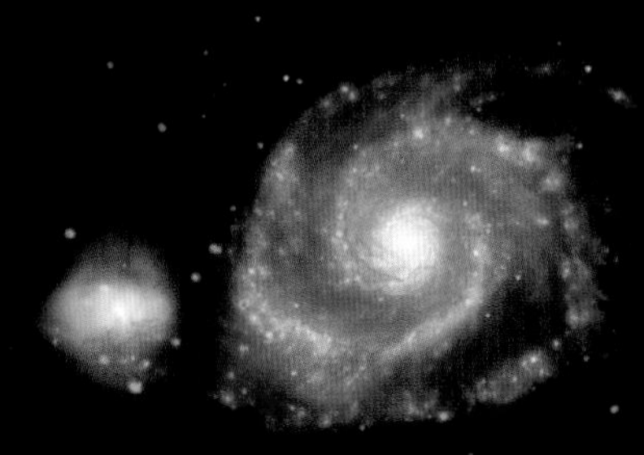

FOUR VIEWS OF THE WHIRLPOOL
This composite combines Chandra X-ray data (purple), a visible-light Hubble image (green), Spitzer's infrared view (red), and a GALEX satellite ultraviolet map (blue).

CONTRASTING COMPANIONS
Viewed in false-color infrared, M51B is clearly dominated by radiation from stars (blue and green) and M51A by radiation from interstellar dust (red and orange).

THE WHIRLPOOL GALAXY PROFILE

Catalog numbers	M51, NGC 5194/5
Type	Spiral galaxy
Constellation	Canes Venatici
Distance from Earth	23 million light-years
Diameter	80,000 light-years

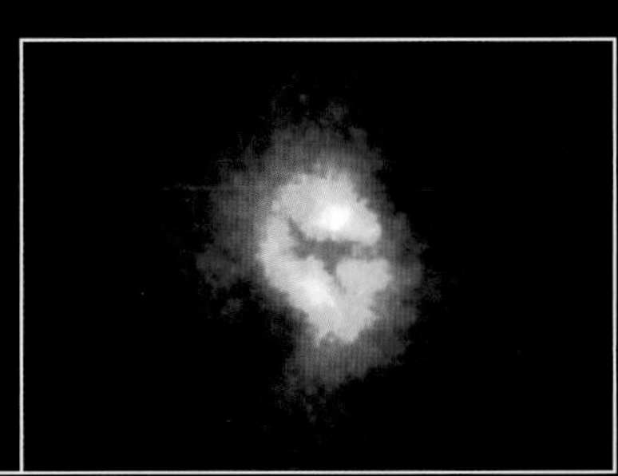

ACTIVE CORE
In this Hubble image of M51A's brilliant core, the "X" is formed by our edge-on view of two intersecting rings of gas and dust, about 100 light-years across. They encircle a supermassive black hole.

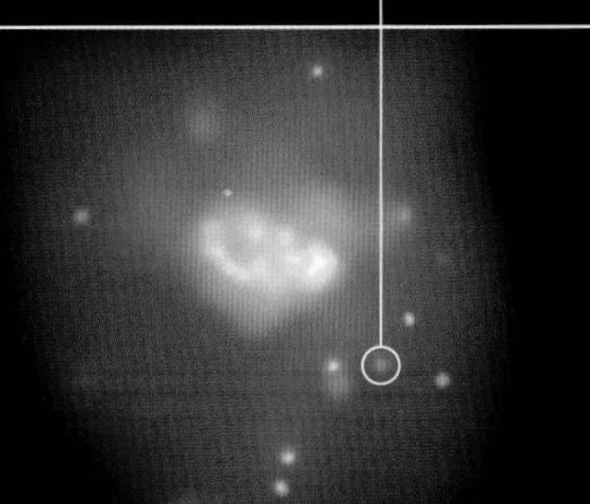

CORE X-RAY
These bright clouds are heated by energetic particles escaping from a central black hole. The smaller points are supernova remnants, such as SN 1994I, in binary star systems.

STAR-FORMING REGION
Across a distance of 23 million light-years, the Hubble Space Telescope is still able to resolve stunning details within M51, such as this star-forming region incorporating a newly formed open cluster still embedded in clouds of glowing nebulosity.

THE PINWHEEL GALAXY

An enormous spiral galaxy that presents itself to us face-on, the Pinwheel Galaxy lies some 27 million light-years from Earth. With a diameter of 170,000 light-years, it is almost twice the size of the Milky Way and covers an area of sky equivalent to the apparent width of the full Moon. The galaxy's spiral shape is noticeably asymmetric, and the Pinwheel is unusually rich in star-forming nebulae. Both of these characteristics may be the result of a relatively recent close encounter with one of the Pinwheel's several companion galaxies.

PINWHEEL IN VISIBLE LIGHT
This overview of the Pinwheel combines data from 51 Hubble images with several Earth-based ones. Released in 2006, it was the largest Hubble image produced up to that time.

HOT YOUNG STAR CLUSTERS
At least 3,000 clusters of brilliant newborn stars have been identified within the Pinwheel, including this massive concentration in the outer reaches of one of its spiral arms.

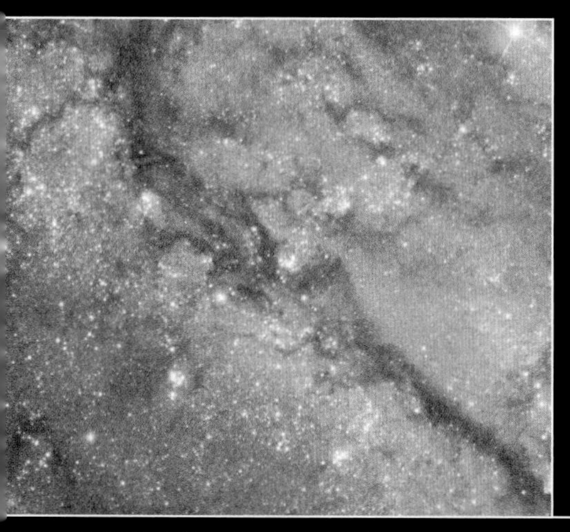

VISIBLE-LIGHT DETAIL OF DUST LANE
Dark dust lanes outline the galaxy's shape, but fine dust is also dispersed across the Pinwheel's disc. Dust particles reflect and scatter short-wavelength starlight, producing a blue sheen and reddening images of the stars themselves.

THE PINWHEEL GALAXY PROFILE

X-RAY MAP
Data for this map of X-ray emissions were collected over a period of 26 hours by the Chandra X-ray Observatory. The pointlike sources include neutron stars, black holes, newborn stars, supernova remnants, and clusters, all wrapped in hot gas clouds.

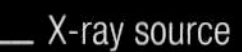

SPITZER SPACE TELESCOPE VIEW
In infrared, the Pinwheel Galaxy's stars are a blue haze, cool dust on the spiral arms looks green, and hot dust in star-forming regions glows orange. Harsh radiation destroys any organic molecules in these star-forming regions.

COMBINED HUBBLE, SPITZER, AND CHANDRA VIEW
The optical view from Hubble is shown in yellow, infrared data from Spitzer are red, and Chandra's X-ray map is blue. Such images help astronomers to understand the relationships between the galaxy's different features.

THE SOMBRERO GALAXY

This unusual spiral galaxy lies nearly edge-on to Earth, in the constellation Virgo. While it appears close to the galaxies of the Virgo Cluster (see pp.290–91) in our skies, the Sombrero is estimated to be considerably closer to Earth, at around 28 million light-years away with a diameter of about 50,000 light-years. The galaxy is believed to be an unbarred spiral, but the very large central bulge and the prominent, dark dust ring around its edge give it a shape like the wide-brimmed Mexican hat from which it takes its popular name. The Sombrero's "brim"—its outer dust ring—seems to be buckled, perhaps due to a close encounter with another galaxy.

THE SOMBRERO GALAXY PROFILE

Catalog numbers	M104, NGC 4594
Type	Spiral galaxy
Constellation	Virgo
Distance from Earth	28 million light-years
Diameter	50,000 light-years

HUBBLE VISIBLE-LIGHT VIEW
This mosaic combines six images taken through three different filters to give a natural-color image that highlights several globular clusters and fine details across the Sombrero's dust ring.

CHANDRA X-RAY IMAGE
This image from the orbiting Chandra X-ray Observatory unveils a host of X-ray sources around the Sombrero. These include black holes and neutron stars within the galaxy's disc and more distant background quasars. At the center is the Sombrero's brilliant core—a supermassive black hole estimated to have the mass of a billion Suns.

SOMBRERO IN INFRARED
In this view from the Spitzer Space Telescope, the Sombrero's outer dust ring is shown in red and pink. Stars appear blue, revealing an inner disc of stars within the dust ring and at the heart of the galaxy's extended bulge.

COLORFUL SOMBRERO
This image combines Hubble, Chandra, and Spitzer data. Chandra reveals a halo of hot, X-ray emitting gas (blue) around the galaxy's bulge; Spitzer's infrared image emphasizes the dust ring (red); and Hubble's optical view shows a sphere of starlight (green) partially blocked by the dust ring. A strong galactic wind is thought to blow out of the galaxy's center, driven by shock waves from supernovae in the densely packed central bulge.

BARRED SPIRAL GALAXY

This Hubble image shows NGC 1300, a barred spiral galaxy some 70 million light-years away, and over 100,000 light-years in diameter. The bar, which appears red in this image, is believed to funnel gas from the outer arms of the galaxy to its core. The bar is also thought to be a temporary phenomenon in the development of a regular spiral galaxy.

THE VIRGO CLUSTER

With an estimated 2,000 galaxies packed into a region of space about 15 million light-years across (only 50 percent larger than the much less crowded Local Group), the Virgo Cluster is the nearest major galaxy cluster to Earth. Its center, which lies roughly 60 million light-years away, is marked by a mix of bright spiral and elliptical galaxies. The Virgo Cluster's huge concentration of mass turns it into the gravitational "anchor" of our Local Supercluster, and the galaxies of the Local Group are being drawn toward it at a relative speed of 870,000 mph (1.4 million kph).

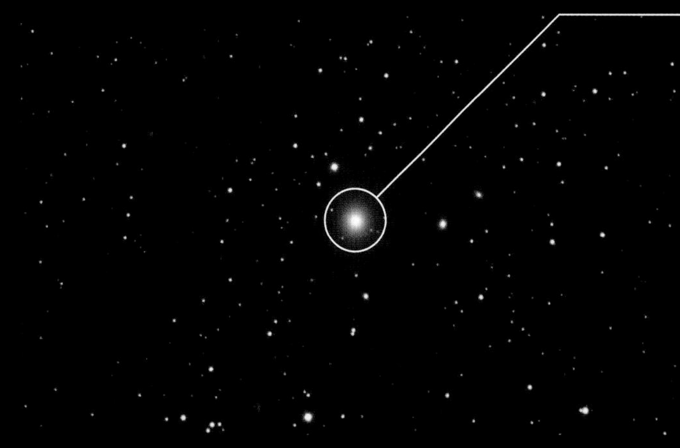

CENTRAL REGION
The central region of the Virgo Cluster covers an area of the sky roughly 16 times the diameter of the full Moon, in the constellation Virgo. Three separate clumps of galaxies are concentrated around the elliptical galaxies M87 and M89 and the lenticular M86.

ELLIPTICAL MONSTER
The giant elliptical galaxy M87 is the biggest galaxy in our region of the Universe, with an estimated mass of 2.4 trillion Suns. It is also violently active, with a jet of particles emerging from its core at close to the speed of light.

GALAXY M88
This spiral galaxy is one of the brightest and closest members of the Virgo Cluster, roughly 47 million light years from Earth. Its spiral structure was deduced in 1850, largely from the dense, light-absorbing dust lanes

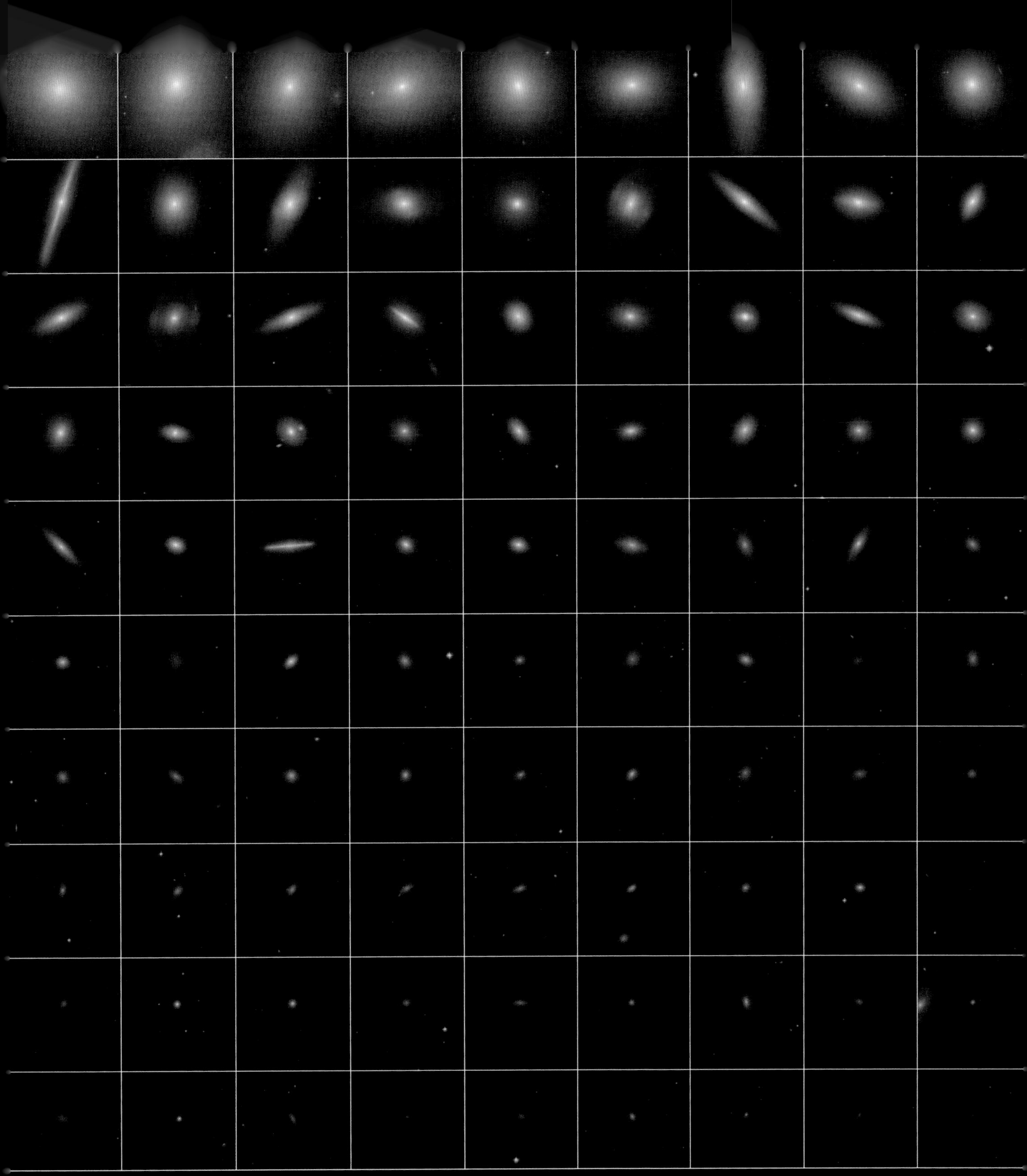

O CLUSTER PROFILE

GALAXIES IN THE VIRGO CLUSTER

Within the Virgo Cluster there are around 160 major

COLLIDING GALAXIES

With galaxies so crowded together on a cosmic scale, it is little wonder that close encounters and collisions between them are quite common. Such galactic interactions create a huge variety of peculiar galaxies and play an important role in the evolution of galaxies from one form to another (see pp.320–21).

CROWDED SPACE

Galaxies are typically separated by anything from tens of thousands to a few million light-years, which means that they are relatively close together compared to their size. What is more, the powerful gravity of the largest galaxies can easily exert itself over these distances. This draws galaxies together in clusters, which makes collisions even more likely. In the 1960s, the American astronomer Halton Arp created a catalog of peculiar galaxies that did not seem to fit into the traditional classification scheme of spirals, lenticulars, ellipticals, and irregulars (see pp.266–67). Thanks to technological advances in telescope construction, it is now clear that many, if not all, of the peculiar objects are in fact close encounters, collisions, and even mergers between two or more galaxies.

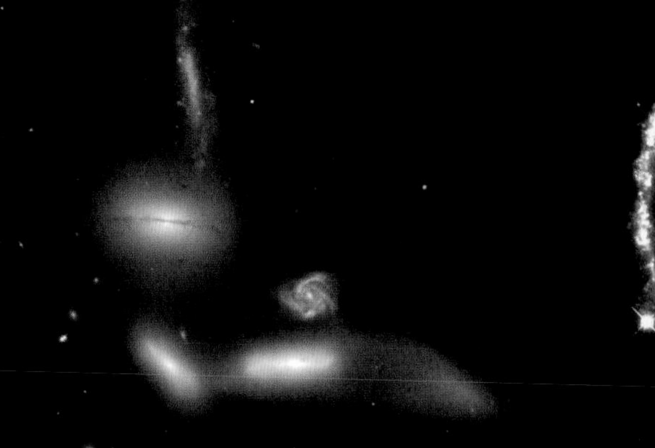

SEYFERT'S SEXTET
Of the six galaxies in this group, only four are interacting. They show clear signs of distortion from the encounters. The other two are an unrelated spiral (center) and a stream of stars (right).

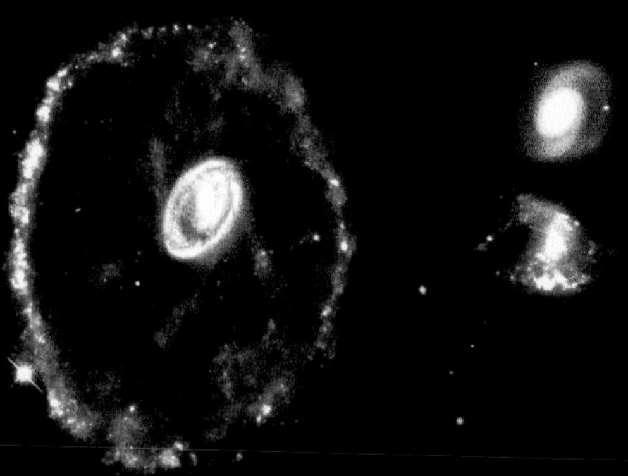

CARTWHEEL GALAXY
This strange galaxy, about 500 million light-years away, suffered a "direct hit" from a smaller galaxy (possibly one of the two on its right) about 200 million years ago, which triggered the formation of the outer ring of young stars.

TIDAL FORCES

As galaxies approach or pass by one another in space, they begin to undergo distortions long before any physical collision takes place, thanks to the gravitational forces exerted by each galaxy. The tidal forces produced cause different regions of the galaxies to experience different gravitational forces because of the varying distances involved. As gravity tugs at parts of a galaxy with varying strength, strange effects can be produced. With spiral galaxies, one or more of the spiral arms may appear to unwind, flinging a streamer of bright open clusters and star-forming nebulae out into intergalactic space. Edge-on spiral galaxies and lenticular galaxies can also show impressive buckling as their discs of gas and dust warp relative to their stars. However, no two galaxy encounters are the same, and each produces its own unique effects.

THE TADPOLE GALAXY
An "unwrapped" spiral arm forms a tail of stars, 280,000 light-years long, behind this galaxy. It is thought to have formed from tidal forces from a smaller passing galaxy (the brighter blue area near the top of the galaxy).

BUCKLED SPIRAL
The spectacular edge-on galaxy ESO 510-G13 lies roughly 150 million light-years away in the constellation Hydra. It displays a badly warped disc of gas and dust after a close encounter with a neighbor.

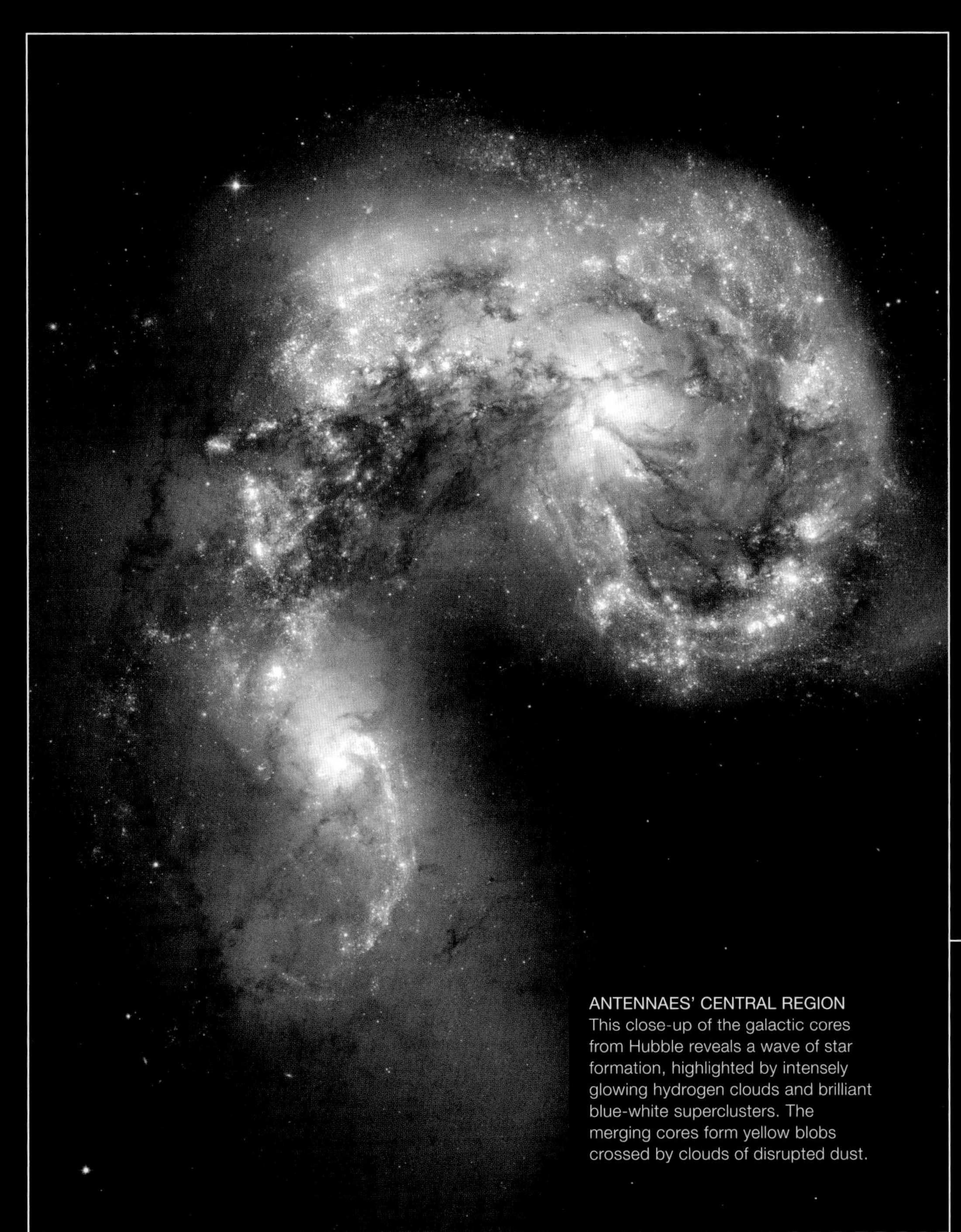

ANTENNAES' CENTRAL REGION
This close-up of the galactic cores from Hubble reveals a wave of star formation, highlighted by intensely glowing hydrogen clouds and brilliant blue-white superclusters. The merging cores form yellow blobs crossed by clouds of disrupted dust.

STARBURSTS

While head-on collisions between individual stars are rare during galaxy mergers, the interstellar clouds of gas and dust within each galaxy inevitably ram straight into each other, with spectacular consequences. The huge shock waves passing through these clouds trigger enormous waves of starbirth, called starbursts, and create brilliant open clusters rich in hot, but short-lived, blue-and-white stars. A closer look at distant colliding galaxies reveals that such events give rise to superclusters—a type of star cluster between the open and globular clusters of the present-day Milky Way (see pp.220–21). Superclusters contain hundreds of thousands, even millions, of young stars held together by gravity. Astronomers believe that as their short-lived blue-and-white stars age and die they will slowly transform into globular clusters. Starbursts may be short-lived—the shock of the collision can heat the gas clouds so much that they rapidly boil away into space, creating the hot, X-ray emitting gas often found within galaxy clusters.

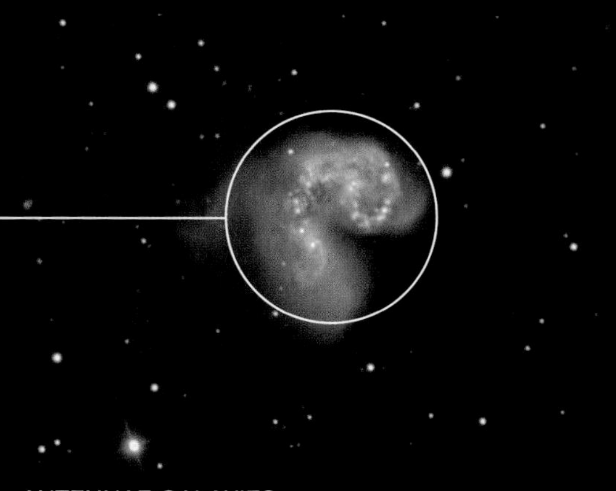

ANTENNAE GALAXIES
This spectacular pair of colliding spiral galaxies, NGC 4038 and 4039, display a long pair of streamers, giving the appearance of an insect's head.

CANNIBAL GALAXIES

While collisions between large galaxies tend to produce the most spectacular results, the vast majority of galaxy interactions involve one large galaxy and one much smaller dwarf galaxy. In these circumstances, the smaller one suffers major disruption at least and complete destruction at worst. The larger partner, meanwhile, can experience regional starbursts and, in the case of spiral galaxies, an overall reinforcement of the spiral pattern. Many galaxies with a well-defined spiral structure have been involved in collisions or close encounters with dwarf galaxies. The Milky Way Galaxy is currently involved in several such collisions, stripping stars away from the Magellanic clouds and shredding the structures of the closer Canis Major and Sagittarius dwarf elliptical galaxies. Eventually, these galaxies will be entirely absorbed into the Milky Way. Images from the Hubble Space Telescope have revealed a host of galaxies in various stages of collision that may take a billion years to complete.

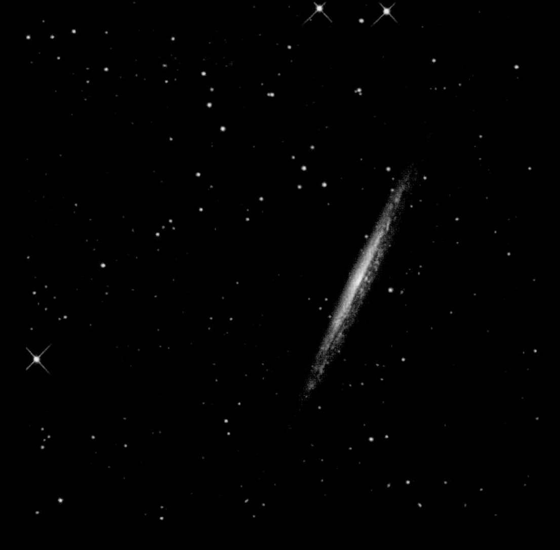

DANCE OF DEATH
This long-exposure image of the so-called "Knife Edge Galaxy" NGC 5907, an edge-on spiral some 40 million light-years from Earth, reveals a ghostly trail of stars left behind by a dwarf galaxy that was torn apart and absorbed into the larger one. The loops extend more than 150,000 light-years from the galaxy. Astronomers estimate the dwarf met its fate around 4 billion years ago.

ACTIVE GALAXIES

Normally, the light from a distant galaxy is the sum of the light from all its stars. But a large minority of galaxies display another source of radiation. Called active galaxies, these all generate huge amounts of energy—perhaps more than the rest of the galaxy—from tiny regions close to their centers.

RADIO GALAXIES

These were the first active galaxies to be identified, following research into prominent radio sources in the sky that could not be linked to any known object in the Milky Way. The long wavelengths of radio waves made it hard to produce detailed images of the objects until the first giant radio telescopes were built in the 1950s. However, once images could be produced, many of these mysterious radio sources turned out to be enormous double-lobed clouds, situated on either side of an apparently normal galaxy. As the resolution of images improved still further, astronomers observed narrow jets emerging from the galaxies, which crossed tens of thousands of light-years before billowing out to form the radio lobes. The so-called radio galaxies, it seems, are formed when a galaxy ejects powerful jets of material that slow down, creating radio waves as they encounter huge clouds of gas in the intergalactic medium (see p.196).

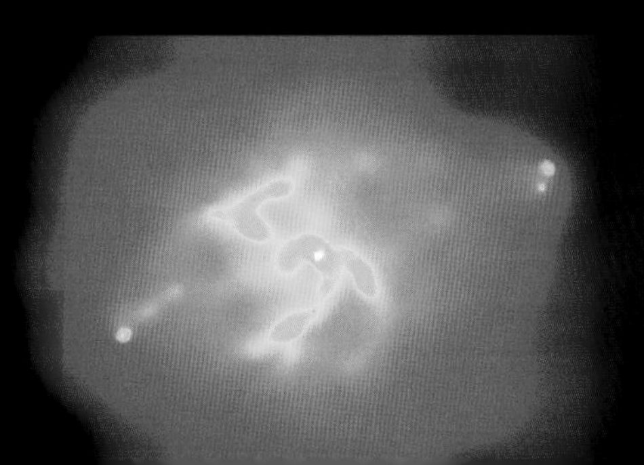

CYGNUS A CAVITY
This image of Cygnus A from the Chandra X-ray Observatory reveals that the galaxy's activity has carved an enormous cavity within a surrounding cloud of hot intergalactic gas. The hot gas piling up around the edge of the cavity creates a bright shell, shown in orange.

QUASARS AND SEYFERT GALAXIES

In the early 1960s, astronomers observed apparently starlike objects, often associated with radio sources, that rapidly changed their brightness. The spectra of these objects—quasi-stellar objects, or quasars—were unlike that of any known object. A breakthrough came when astronomers realized that the spectra were similar to that of hydrogen, only red shifted by an enormous amount due to the Doppler effect (see p.233), suggesting the objects were within remote galaxies. Later, powerful modern telescopes revealed faint galaxies around the central brilliant light sources. Astronomers later realized that Seyfert galaxies (spirals with unusually bright cores) that are found much closer to Earth might be less-active cousins of quasars.

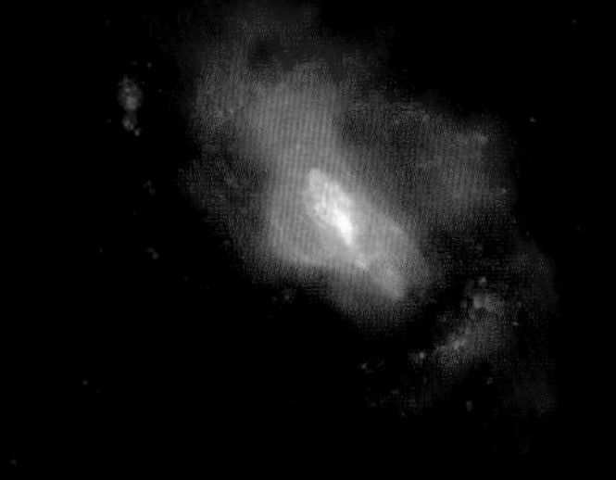

NEARBY SEYFERT GALAXY, M77
At 47 million light-years from Earth, the active nucleus of this barred spiral galaxy emits radio, infrared, and X-ray radiation

QUASAR 3C 273
This X-ray image reveals a jet of superhot gas from the core of the brightest quasar in the sky (also the first to be discovered)

ACTIVE GALACTIC NUCLEI

Astronomers believe that radio galaxies, quasars, and Seyfert galaxies all have an active galactic nucleus. This forms when the supermassive black hole at the center of a galaxy feeds on material from its surroundings. As matter spirals inward, it forms a spinning accretion disc that is heated to extreme temperatures by frictional effects and gives off a wide range of radiations. Some of this material becomes caught up in the black hole's magnetic field and escapes in particle jets traveling at speeds close to that of light. Depending on the level of the galactic nucleus's activity, and the angle at which it and the surrounding galaxy are observed from Earth, the galaxy appears as a radio galaxy, a Seyfert galaxy, a quasar, or a rare form called a blazar (which is aligned so that observers look straight down the jet).

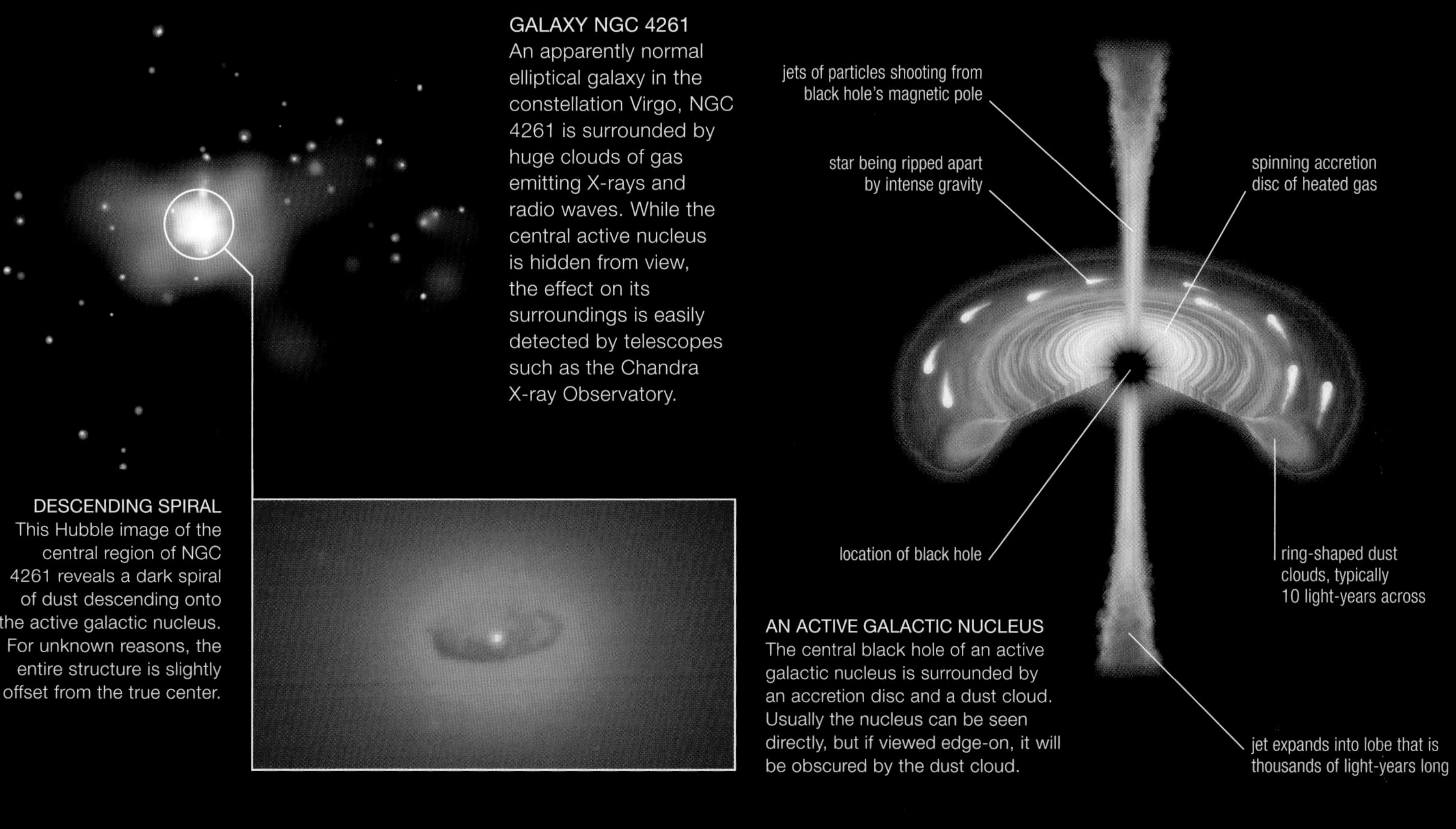

GALAXY NGC 4261
An apparently normal elliptical galaxy in the constellation Virgo, NGC 4261 is surrounded by huge clouds of gas emitting X-rays and radio waves. While the central active nucleus is hidden from view, the effect on its surroundings is easily detected by telescopes such as the Chandra X-ray Observatory.

DESCENDING SPIRAL
This Hubble image of the central region of NGC 4261 reveals a dark spiral of dust descending onto the active galactic nucleus. For unknown reasons, the entire structure is slightly offset from the true center.

AN ACTIVE GALACTIC NUCLEUS
The central black hole of an active galactic nucleus is surrounded by an accretion disc and a dust cloud. Usually the nucleus can be seen directly, but if viewed edge-on, it will be obscured by the dust cloud.

MERGERS AND ACTIVITY

A new generation of telescopes, both on Earth and in orbit, has revealed an important link between active galactic nuclei and intergalactic collisions. Most active galaxies seem to have a black hole at the center, but, in general, material that orbits too close to its gravitational maw, or grasp, has long since been absorbed, leaving gas, dust, and stars to follow circular orbits at a safe distance. This ordered calm is disturbed by close encounters and collisions between galaxies, which push new material toward the center and cause the nucleus to temporarily spark back into life. Such events were more frequent and violent earlier in the history of the Universe, which explains the large numbers of quasars found at greater distances from Earth.

MERGING QUASARS
This composite of a Chandra X-ray Observatory view and a visible-light image from the Magellan Telescopes in Chile captures two quasars, SDSS J1254+0846. The quasars are in the process of merging, some 4.5 billion light-years from Earth.

ONE GALAXY, TWO NUCLEI
This bizarre galaxy, NGC 6240, in the constellation Ophiuchus, is the result of a merger between two smaller galaxies. X-ray images reveal two distinct nuclei, which have both been sparked into activity by the collision.

CENTAURUS A

he object cataloged as Centaurus A is one of
he brightest sources of radio waves in the sky
nd coincides with the bright visible galaxy
IGC 5128. While usually listed as a lenticular
alaxy, NGC 5128 is unusual in that its disc of
tars is silhouetted against a dark foreground
ane of dust. Multiwavelength images reveal
he true nature of Centaurus A. A galactic
ollision between a large elliptical galaxy and
 slightly smaller spiral, just 13.7 million
ght-years from Earth, has triggered violent
ctivity in its core.

ADIO LOBES
radio map of Centaurus A from the Very Large Array in
rizona reveals radio-emitting clouds connected to the
alaxy's core by jets more than 13,000 light-years long.

-RAY EMISSIONS
his Chandra image shows emissions from high-energy
articles in the jets and point sources, color-coded by
ncreasing energy from red through green to blue.

COMPOSITE VIEW
A combination of Chandra X-rays

CENTAURUS A GALAXY PROFILE

Catalog Number	NGC 5128

GROUND-BASED VIEW
The fifth-brightest galaxy in the sky, NGC 5128 was discovered by Scottish astronomer James Dunlop while working in Australia in 1826. The prominent dust lane visible in this image was noted by British astronomer John Herschel some six years later and makes it an easily recognized object for southern-hemisphere observers.

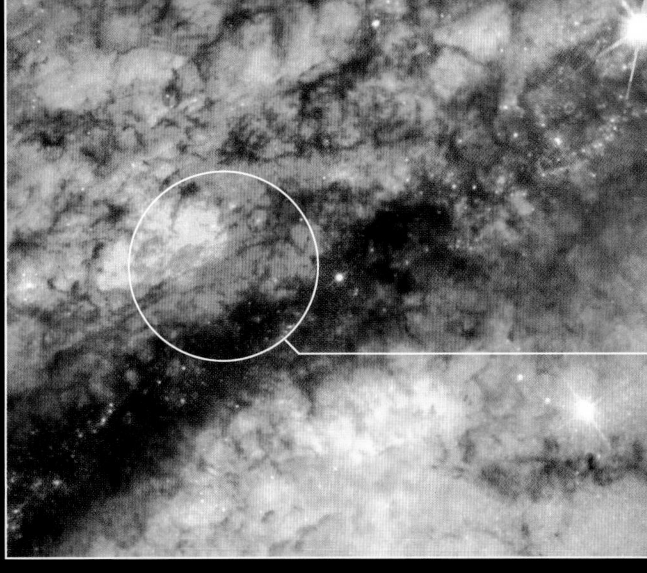

HUBBLE VIEW
This image zooms into the dust lane obscuring the galaxy's active heart. A wave of star formation is in progress, producing blue-white star clusters and creating turbulence that makes the gas and dust look clumpy.

GALAXY CENTER
Using its Near Infrared Camera and Multi-Object Spectrometer, Hubble can see through the obscuring stars and dust to the hot disc of material spiraling down onto the galaxy's central black hole. This black hole is 250 times more massive than the one at the center of the Milky Way.

UNEXPECTED GEOMETRY
This remarkable near-infrared view from the European Southern Observatory's New Technology Telescope reveals that the dusty remains of the spiral galaxy recently cannibalized by Centaurus A have been twisted into a near-perfect parallelogram.

THE CIRCINUS GALAXY

In 1977, astronomers were surprised to discover a previously unknown active galaxy on our own doorstep. The Circinus Galaxy, named after the constellation in which it is found, lies just 13 million light-years away, but it is close to the plane of the Milky Way and so is well hidden behind clouds of stars, gas, and dust. Cataloged as ESO 97–G13, this small spiral combines features of both Seyfert and radio galaxies.

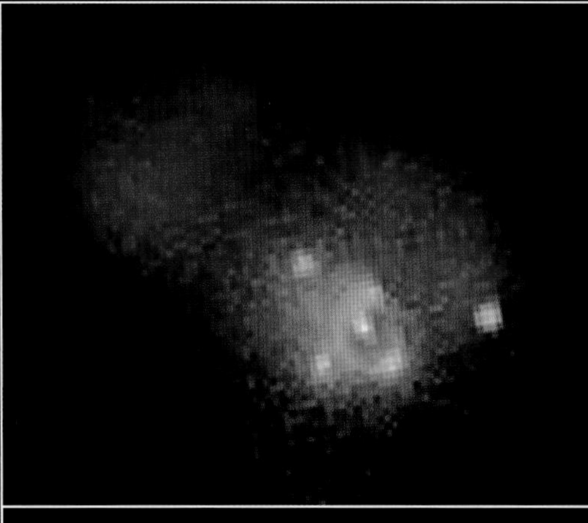

X-RAY VIEW
This Chandra X-ray Observatory image reveals large and small black holes, as well as clouds of X-ray-emitting gas at the center of the Circinus Galaxy. Its central supermassive black hole is marked by the blue region at the center, and the green dots are other black holes or neutron stars. The surrounding clouds of super-hot gas, shown in red, extend for hundreds of light-years.

THE CIRCINUS GALAXY PROFILE

Catalog number	ESO 97-G13
Type	Spiral Seyfert galaxy
Constellation	Circinus
Distance from Earth	13 million light-years
Diameter	40,000 light-years

RINGS OF GAS
This Hubble Space Telescope view reveals a bright inner ring of gas around the center of the galaxy, which is 300 light-years across. The larger, red outer ring is 1,400 light-years in diameter.

GALAXY ESO 0313-192

Nearly all classic double-lobed radio galaxies have either an elliptical galaxy or some kind of galactic merger at their center (see pp.292–93). However, there is one remarkable exception to this rule. ESO 0313–192 is part of a loose cluster called Abell 428. In 2003, astronomers confirmed that the galaxy, flanked by large, bright clouds of radio emissions, is in fact an edge-on spiral. This unexpected discovery has thrown standard models of radio-galaxy formation into question.

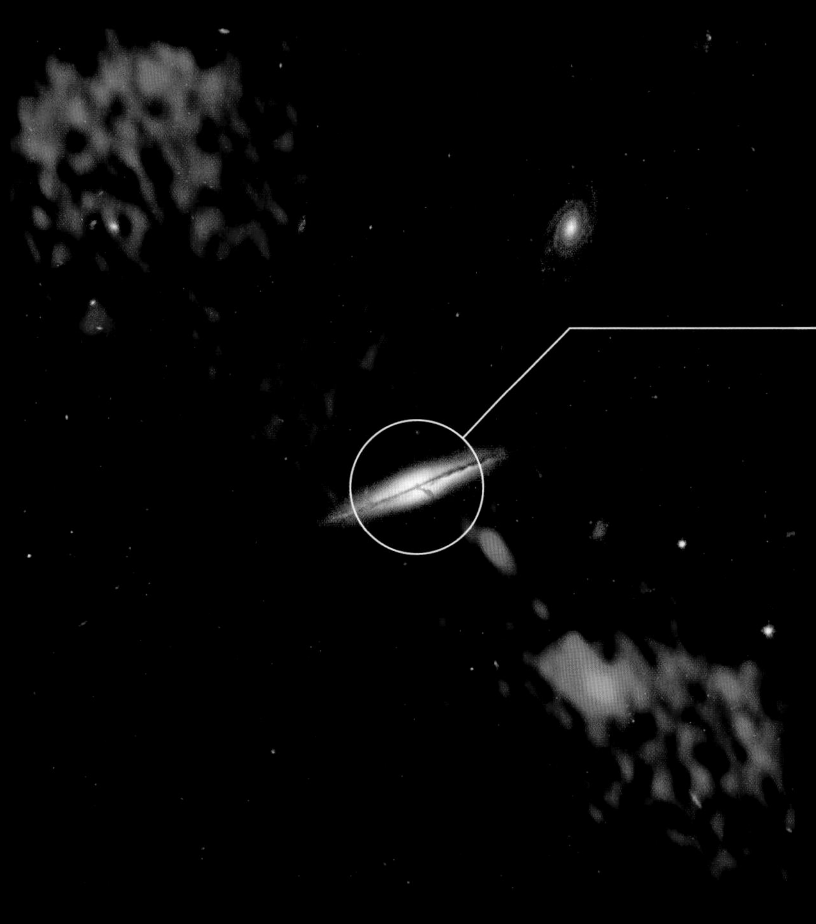

VISIBLE LIGHT AND RADIO WAVES
This image combines a view of the galaxy as seen through Hubble's Advanced Camera for Surveys with a radio map from the Very Large Array. It was the Hubble view that confirmed beyond doubt that it is an edge-on spiral. One of the galaxy's near neighbors can be seen at upper right.

TWISTED DISC
A detailed view from Hubble of the galaxy's center shows that its dark plane of dust is distinctly twisted. This may be due to a collision with a smaller galaxy, which may have sparked the spiral's nucleus into life.

THE ESO 0313-192 GALAXY PROFILE

Type	Spiral radio galaxy
Constellation	Eridanus
Distance from Earth	900 million light-years
Diameter	1.5 million light-years (radio lobe)

THE DEATH STAR GALAXY

In 2007, NASA combined images from the Hubble, Chandra, and Spitzer space observatories with ground-based radio telescopes to study quasar 3C 321. Astronomers discovered a remarkable feature that earned it the nickname the Death Star Galaxy—a jet of material from the quasar is being blasted directly into a neighboring galaxy.

THE DEATH STAR GALAXY PROFILE

FORNAX A

This image shows the giant elliptical galaxy NGC 1316 (center) consuming a smaller galaxy (above center) in the constellation Fornax. The vast orange lobes, each about 600,000 light-years across, are radio emissions picked up by the Very Large Array in New Mexico. Called Fornax A, they are produced by the friction of matter falling into a black hole.

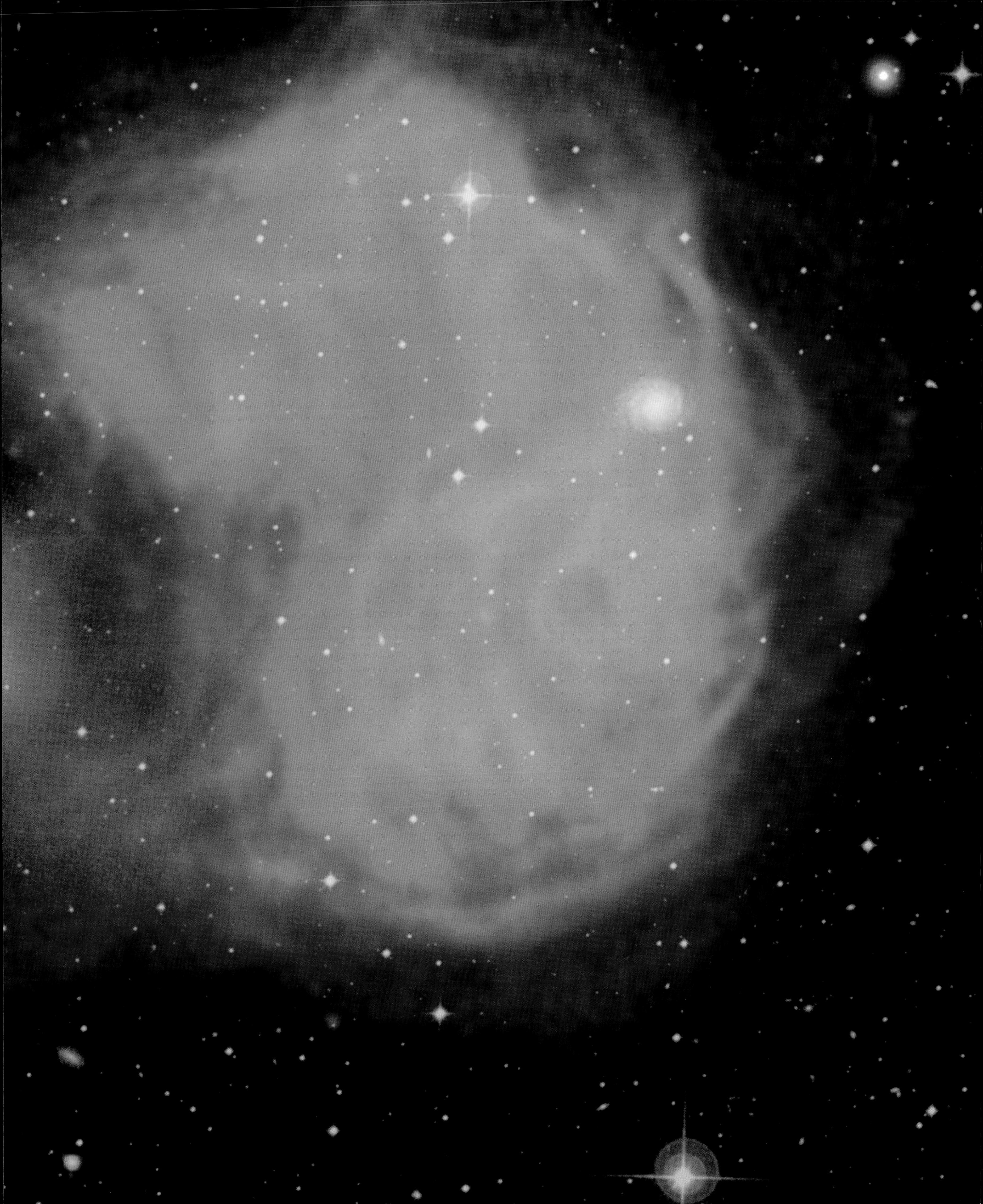

THE VISIBLE EDGE

Modern telescopes can detect faint objects that are billions of light-years from Earth. However, even the largest instruments have limits, and it turns out that the greatest limit of all is determined not by the size of our telescopes, but by the very nature of the Universe itself.

STARING INTO DARKNESS

In 1995, NASA scientists aimed the Hubble Space Telescope at a small, apparently empty patch of sky in the constellation Ursa Major for 11 days. The result was the Hubble Deep Field—an image that revealed thousands of galaxies in this narrow tunnel into the depths of the cosmos. Some of the galaxies were relatively nearby and showed clear signs of spiral or elliptical structures. Others were tiny, misshapen blobs many billions of light-years away. Similar and even more detailed experiments performed on other patches of sky have since proved that the Hubble Deep Field represents a fairly typical region of the sky. They confirm that our Universe is crowded with galaxies—perhaps as many as there are stars in the Milky Way.

THE LIMITS OF THE COSMOS

Astronomers believe that the Universe was created about 13.7 billion years ago in the enormous explosion of the Big Bang (see p.310). As a result, light has not yet had time to reach Earth from any parts of the Universe that are more than 13.7 billion light-years away. In effect, the observable Universe is defined by a bubble of space extending to a distance of 13.7 billion light-years. At the edges of this bubble, in every direction, radio telescopes have detected the microwave afterglow of the Big Bang (see p.316). The most powerful visible-light telescopes cannot see objects from this era because cosmic expansion means that the light from the first generation of stars and galaxies will be red shifted (see p.306) far into the infrared part of the spectrum.

COSMIC TIME MACHINE

When telescopes target objects billions of light-years away, they act as time machines. For objects in the nearby Universe, the so-called look-back time of perhaps tens of millions of years is trivial compared to the lifespan of a galaxy. However, as light from distant galaxies has taken billions of years to reach Earth, we see them as they were much earlier in their history. Consequently, galaxies change the further away we look. Active galaxies such as quasars become more common, as does evidence for galactic mergers and collisions. The evolution of galactic structure unwinds as we go back in time. Nearly all the most distant galaxies detected are irregular and rich in gas, dust, and their first generations of brilliant young stars. It is from these ancient galaxies that all others developed (see p.280).

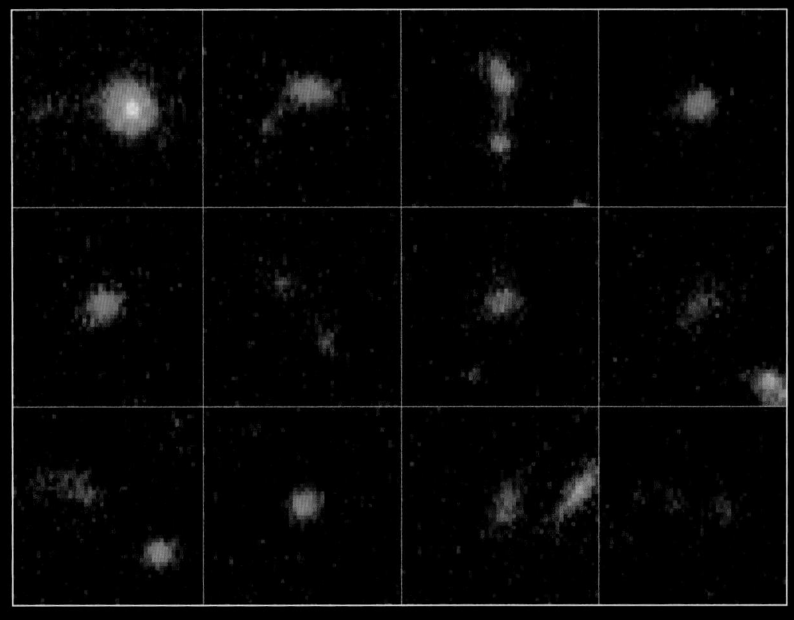

EARLY GALAXIES
Hubble's Deep Field views have revealed more than 500 galaxies from about 1 billion years after the Big Bang. These galaxies are actually blue in color, since they are ablaze with starbirth, but they look red because their light was red shifted on its way to Earth.

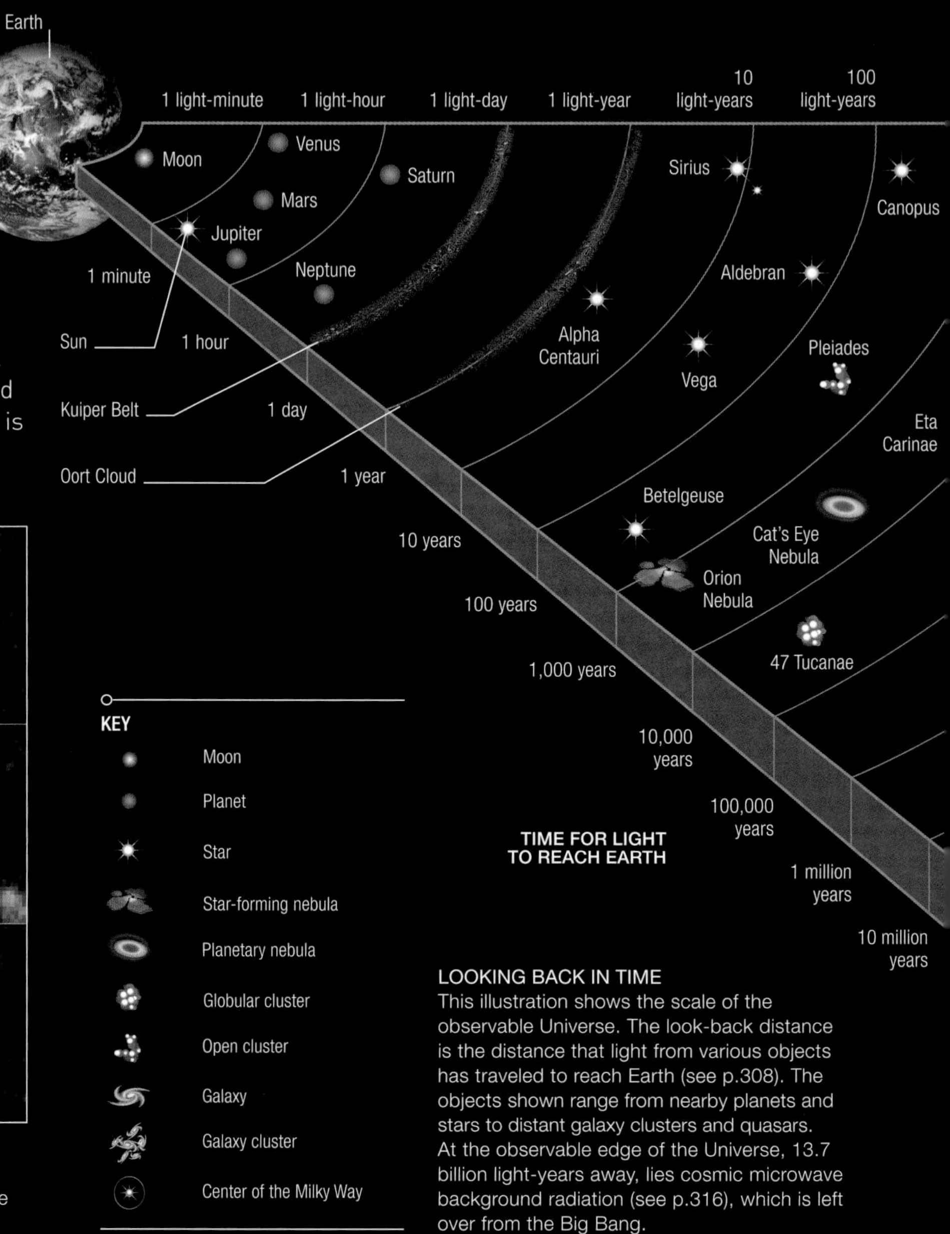

LOOKING BACK IN TIME
This illustration shows the scale of the observable Universe. The look-back distance is the distance that light from various objects has traveled to reach Earth (see p.308). The objects shown range from nearby planets and stars to distant galaxy clusters and quasars. At the observable edge of the Universe, 13.7 billion light-years away, lies cosmic microwave background radiation (see p.316), which is left over from the Big Bang.

NATURE'S MAGNIFIERS

In order to detect objects at the very limits of the Universe, astronomers take advantage of a phenomenon called gravitational lensing. This effect, which was predicted by Einstein's theory of general relativity, comes about when massive objects distort the space and time around them (see p.308). In practice, it means that massive objects such as galaxy clusters can deflect light from more distant objects as the light passes near them—for example, warping the diverging light rays from a distant galaxy so that they come to a focus in our region of the cosmos. In the right configuration, the result can be a distorted but significantly brightened image of a more distant galaxy, appearing as a series of arcs around a foreground cluster. Using computers to process the arc images, astronomers can reconstruct an accurate representation of the original galaxy and even analyze its spectrum.

GRAVITATIONAL LENSING
When a faraway galaxy lies directly beyond a foreground galaxy cluster (as seen from Earth), spreading light rays from the more distant galaxy are deflected at the cluster's edges and bent back toward Earth.

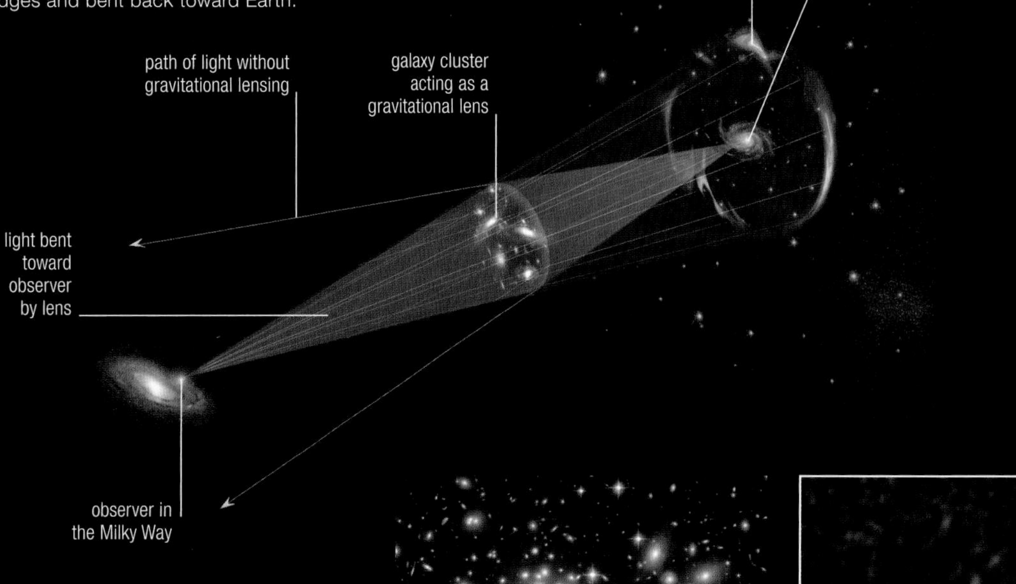

DISTANCE FROM EARTH (LOOK-BACK DISTANCE)

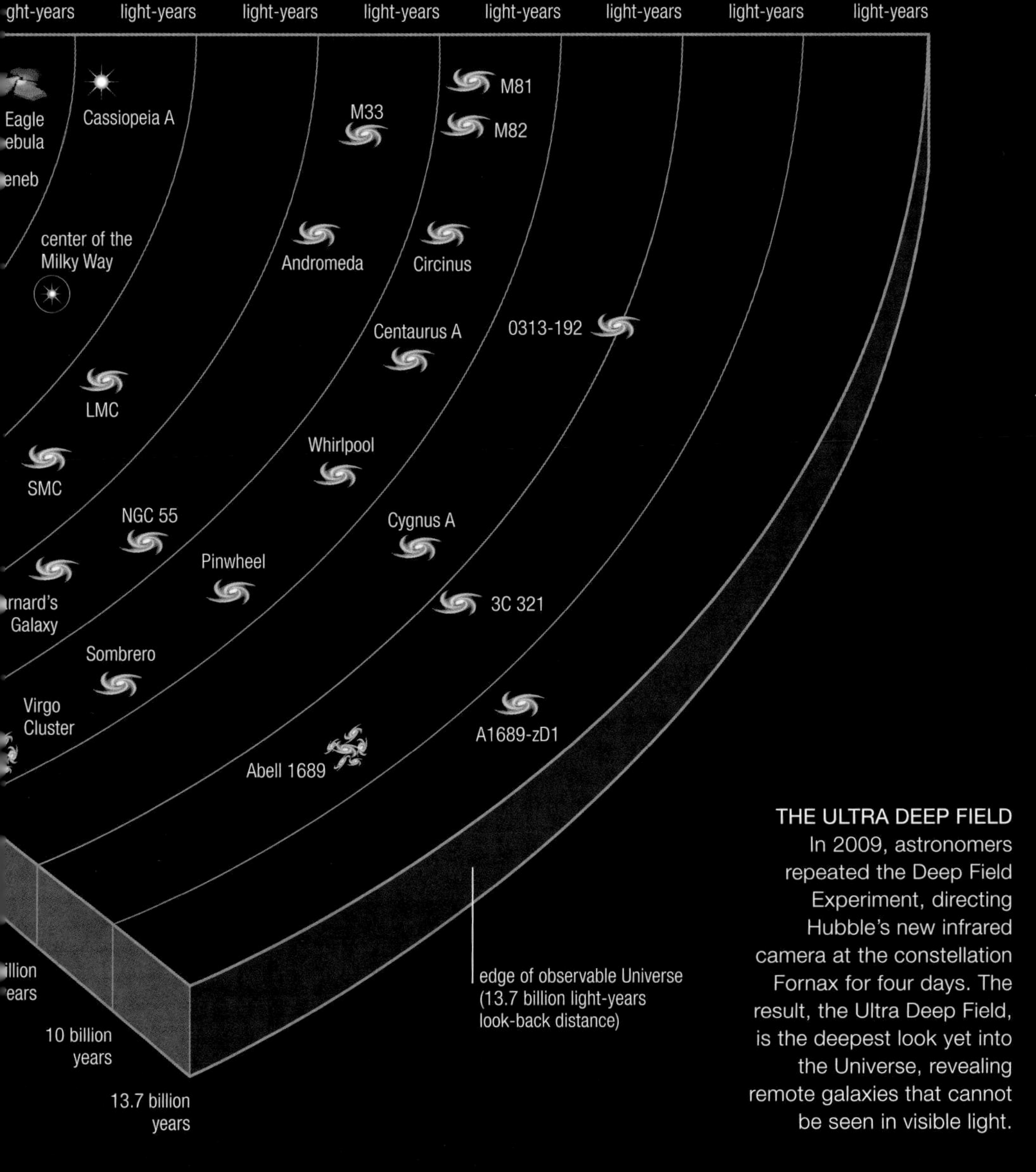

ABELL 1689
This compact galaxy cluster, 2.2 billion light-years away in Virgo, contains a huge amount of matter in a 2-million-light-year-wide region of space, making it an ideal gravitational lens.

COSMIC ZOOM
The faint galaxy A1689-zD1, at the center of this image, is 12.8 billion light-years from Earth and only visible because of the lensing effect of the Abell 1689 galaxy cluster.

FURTHEST OBJECTS
The faintest infrared galaxies captured by the Hubble Ultra Deep Field correspond to "look-back times" of approximately 12.9 billion to 13.1 billion years ago. Hubble sees these galaxies, the most distant ever detected, as they were just 600 to 800 million years after the Big Bang.

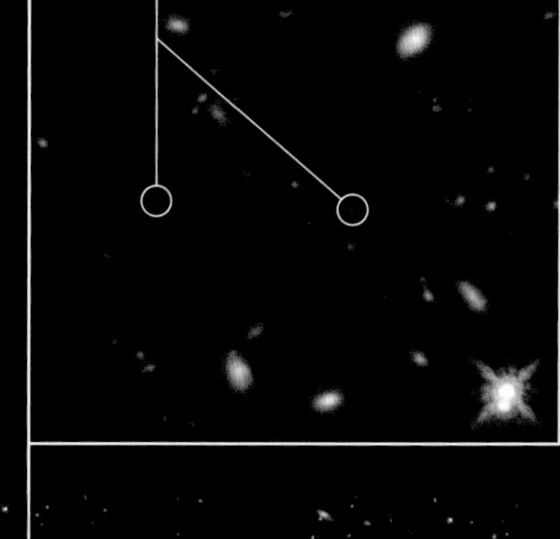

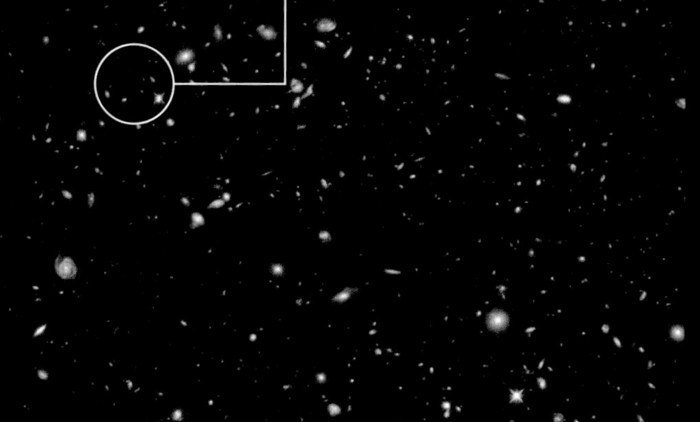

THE ULTRA DEEP FIELD
In 2009, astronomers repeated the Deep Field Experiment, directing Hubble's new infrared camera at the constellation Fornax for four days. The result, the Ultra Deep Field, is the deepest look yet into the Universe, revealing remote galaxies that cannot be seen in visible light.

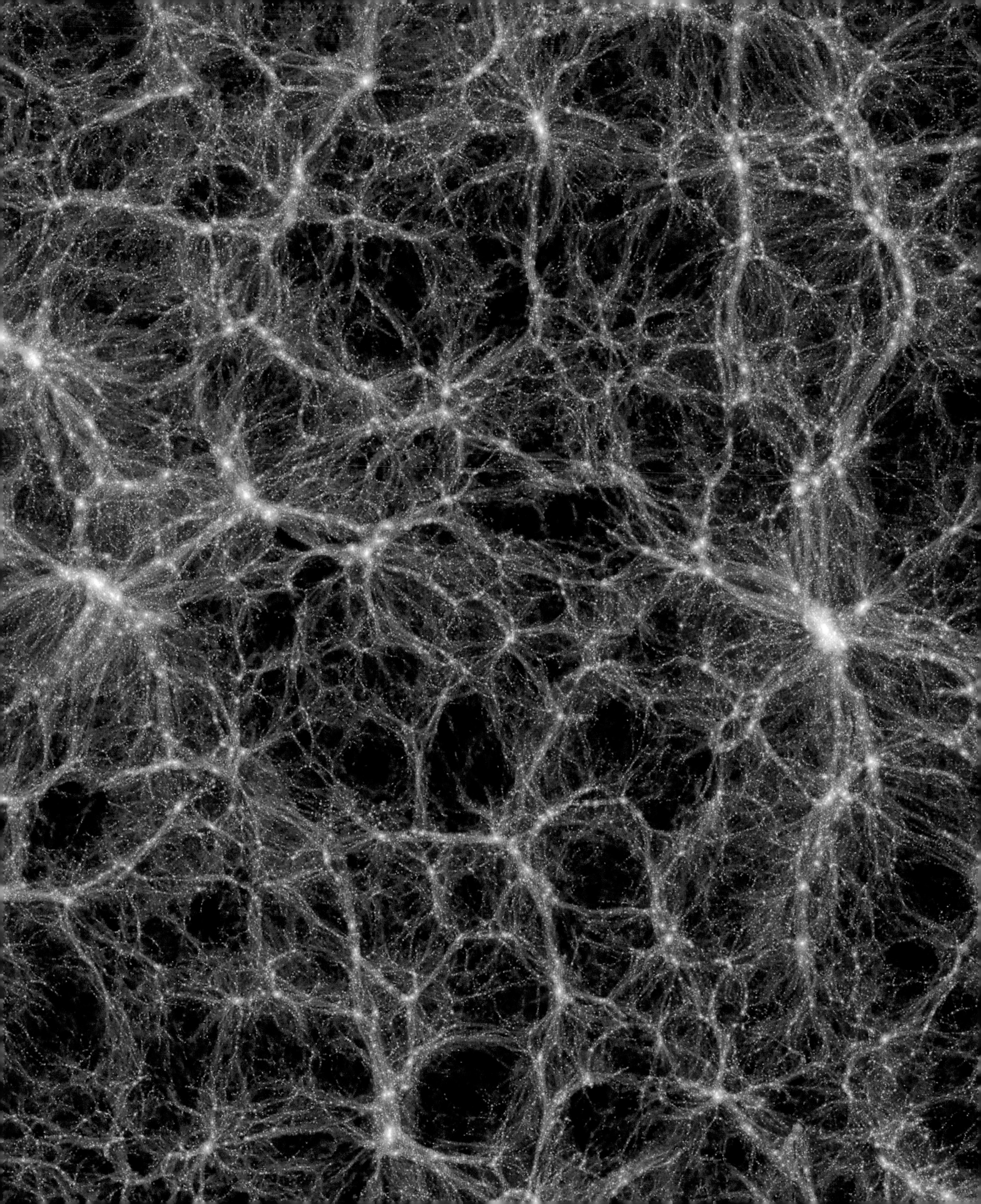

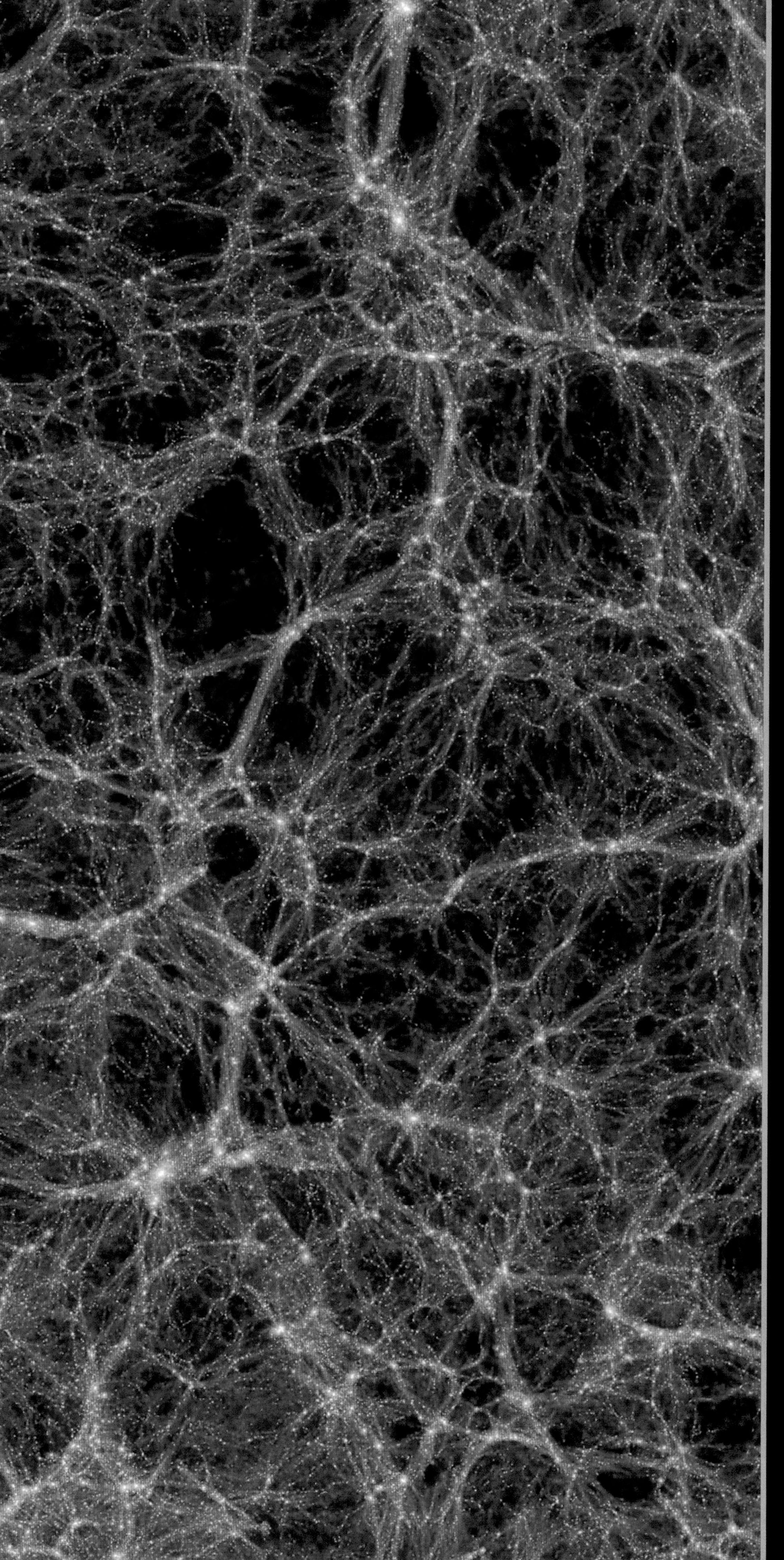

THE OUTER LIMITS

« Dark matter in the local Universe
Dark matter in a region 206 million light-

WHAT IS COSMOLOGY?

Cosmology is the study of the Universe at the largest scale, as seen in features such as galaxy superclusters. As well as being interested in the Universe's overall structure and geometry, cosmologists are concerned with uncovering the details of its origins, evolution, and eventual fate.

MODELING THE UNIVERSE

Much of cosmology is based on the construction of detailed mathematical descriptions of the Universe, known as models, each of which aims to provide an explanation for the current state of the cosmos. The best-accepted model is called the standard model. Some of its main features are that the Universe originated billions of years ago in an event called the Big Bang (see p.310); that it has been expanding ever since; and that it contains vast amounts of mysterious dark matter and ordinary matter (see p.318).

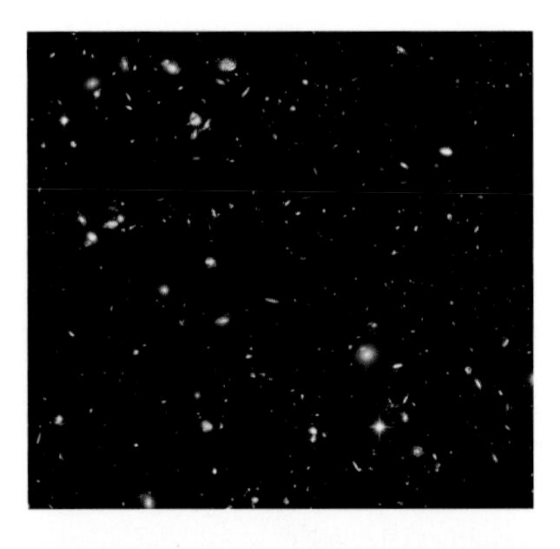

YOUNG AND OLD GALAXIES
A key goal in cosmology is to unravel how galaxies such as these formed and evolved. Observing galaxies of different ages is an important part of this research.

THE COSMOLOGICAL PRINCIPLE

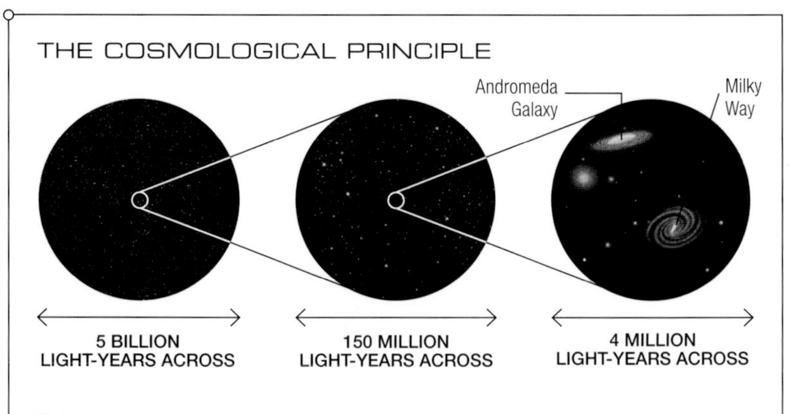

This principle is an assumption that at the largest scales, and looking in all directions, the Universe is uniform in its properties, even though it is clearly not so at smaller scales. This appears to be borne out in practice, when comparing the distribution of galaxies over distances measured in billions of light-years (above left) against smaller distances (above center and right). Two major implications of this principle are that the Universe has no center and no edges.

AN EVER-EXPANDING UNIVERSE

In the early 20th century, astronomers noticed that the spectra of distant galaxies exhibit a feature called a red shift—peaks and troughs in the spectra are shifted from their expected positions toward longer wavelengths (see opposite). Scientists deduced that the galaxies must be rushing away from Earth (see below). In 1929, the American astronomer Edwin Hubble discovered a close relationship between a galaxy's distance from Earth and its red shift—the more remote the galaxy, the faster it is moving away—and concluded that the Universe must be expanding. Since then cosmologists have strived to determine an accurate value for the rate of expansion, which is called the Hubble constant (see p.276). Today, the big question is whether, and how, the rate has varied over the history of the Universe.

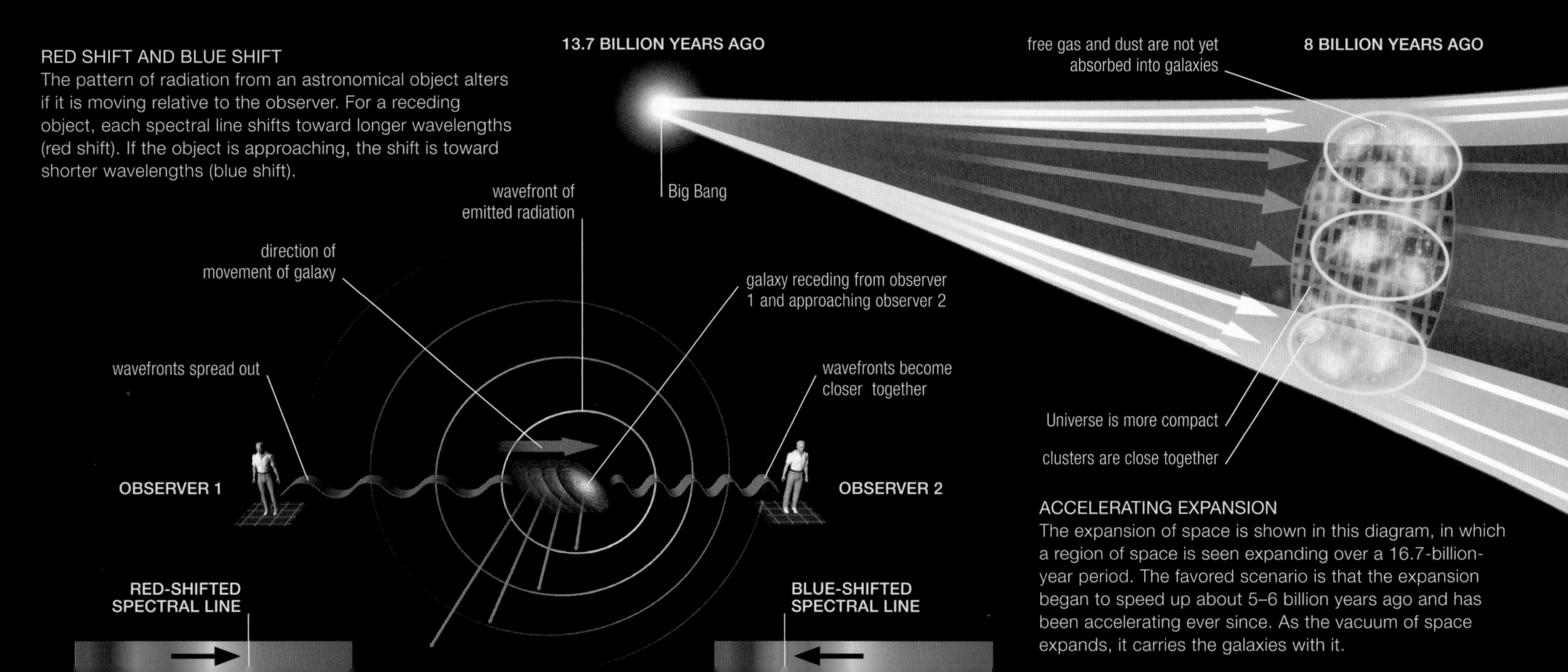

RED SHIFT AND BLUE SHIFT
The pattern of radiation from an astronomical object alters if it is moving relative to the observer. For a receding object, each spectral line shifts toward longer wavelengths (red shift). If the object is approaching, the shift is toward shorter wavelengths (blue shift).

ACCELERATING EXPANSION
The expansion of space is shown in this diagram, in which a region of space is seen expanding over a 16.7-billion-year period. The favored scenario is that the expansion began to speed up about 5–6 billion years ago and has been accelerating ever since. As the vacuum of space expands, it carries the galaxies with it.

> **"THE UNIVERSE FORCES THOSE WHO LIVE IN IT TO UNDERSTAND IT."**
>
> **CARL SAGAN**, ASTROPHYSICIST, BROCA'S BRAIN, 1979

PATTERNS OF RADIATION

Cosmological theories such as the standard model are constantly being refined as new data is gathered about the Universe. Much of this data comes from detecting and analyzing the radiation emitted from distant and often very faint objects. Stars, quasars, galaxies, and gas clouds, for example, each give off radiation in different parts of the electromagnetic spectrum, from radio waves to visible light and X-rays (see p.11). The pattern of emitted radiation depends on factors such as an object's temperature (hotter objects give off more radiation overall, with a bias toward the more energetic, short-wavelength end of the spectrum), its composition, and dynamic processes occurring within it. The pattern of radiation given off by any object—typically a graph of peaks and troughs in the intensity of radiation over a range of wavelengths—is called its spectrum and is a signature for that object. Examining the spectra emitted by stars or galaxies, for example, can provide information about their chemical make-up and temperature. An object's spectrum can indicate the speed at which it is moving away or toward Earth. For a distant galaxy, the speed at which it is moving indicates its distance from Earth (see below).

SPECTRAL SIGNATURES
The graphs below are signatures, captured by the Herschel Space Observatory, from areas in the star-forming cloud DR21 (left). They show peaks of emitted radiation from different substances in objects in the cloud.

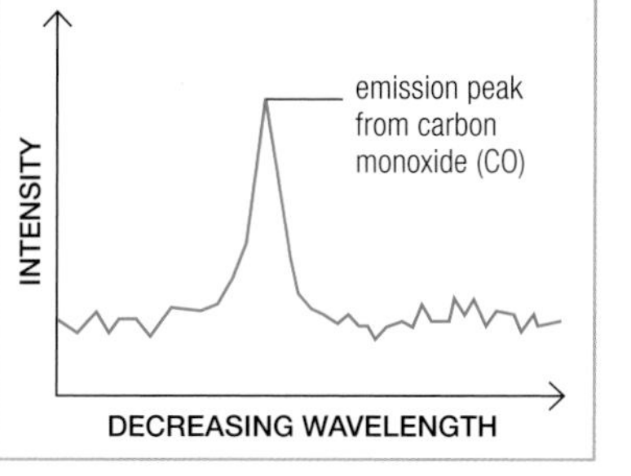
emission peak from carbon monoxide (CO)
INTENSITY
DECREASING WAVELENGTH

AREA OF CLOUD

emission peak from ionized carbon (C)
INTENSITY
DECREASING WAVELENGTH

NEWLY FORMED STAR

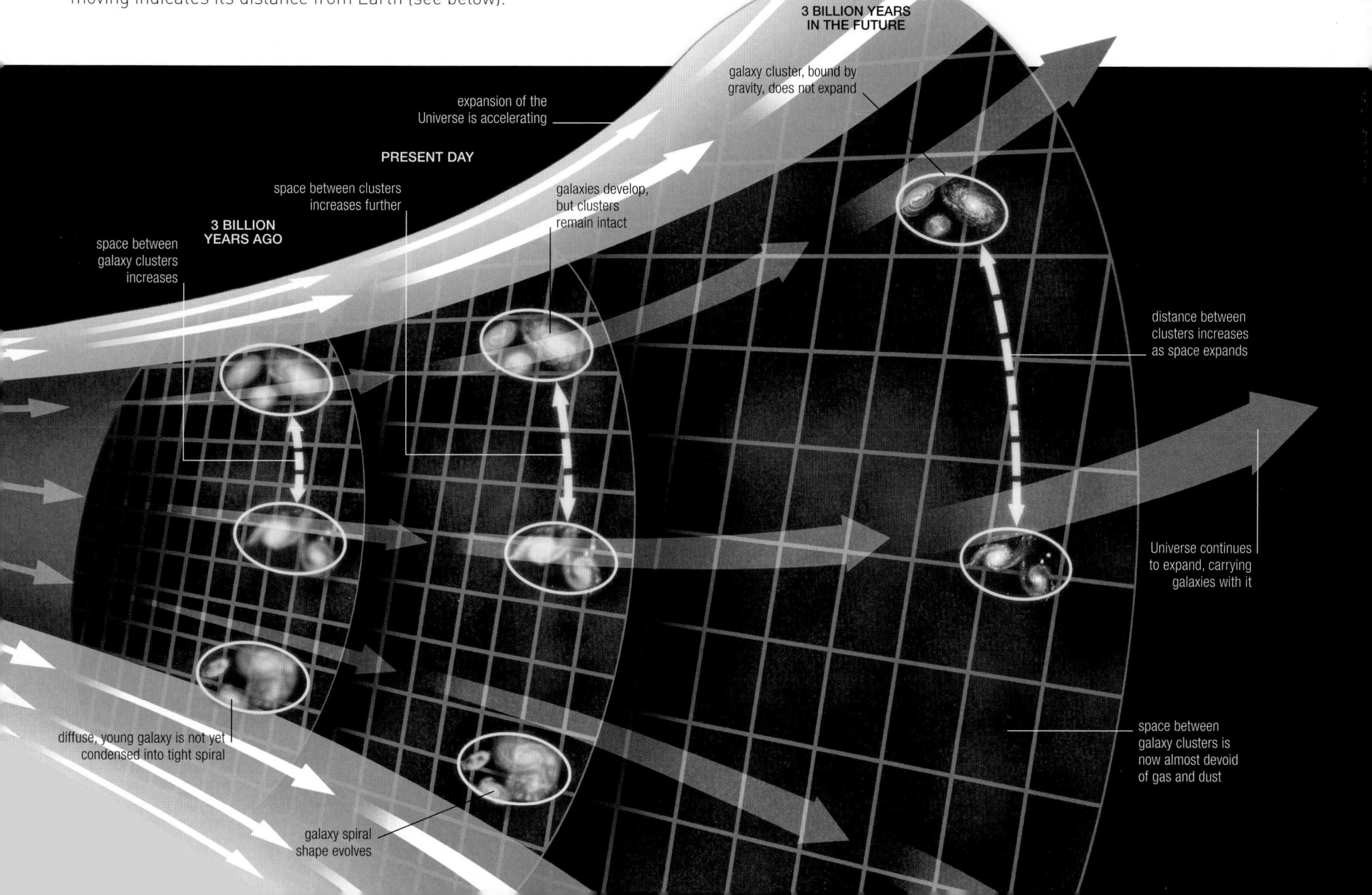
3 BILLION YEARS IN THE FUTURE
galaxy cluster, bound by gravity, does not expand
expansion of the Universe is accelerating
PRESENT DAY
space between clusters increases further
galaxies develop, but clusters remain intact
3 BILLION YEARS AGO
space between galaxy clusters increases
distance between clusters increases as space expands
Universe continues to expand, carrying galaxies with it
diffuse, young galaxy is not yet condensed into tight spiral
space between galaxy clusters is now almost devoid of gas and dust
galaxy spiral shape evolves

DISTANCE AND TIME

In cosmology, the distinction between distance and time is blurred. Light takes a long time to journey from distant objects to Earth, so looking deep into space also means peering back in time, possibly billions of years. The distances to remote objects can be expressed in more than one way. Conventionally, the look-back distance is given—this is the distance that light from a remote object has traveled through space to reach Earth (the distances to objects given in this book are all look-back distances). Because the Universe is expanding, the true current distance—the co-moving distance (see below)—is greater. Some parts of the Universe are so far away that no light from them has ever reached Earth. The part from which light has reached Earth is a spherical region called the observable Universe.

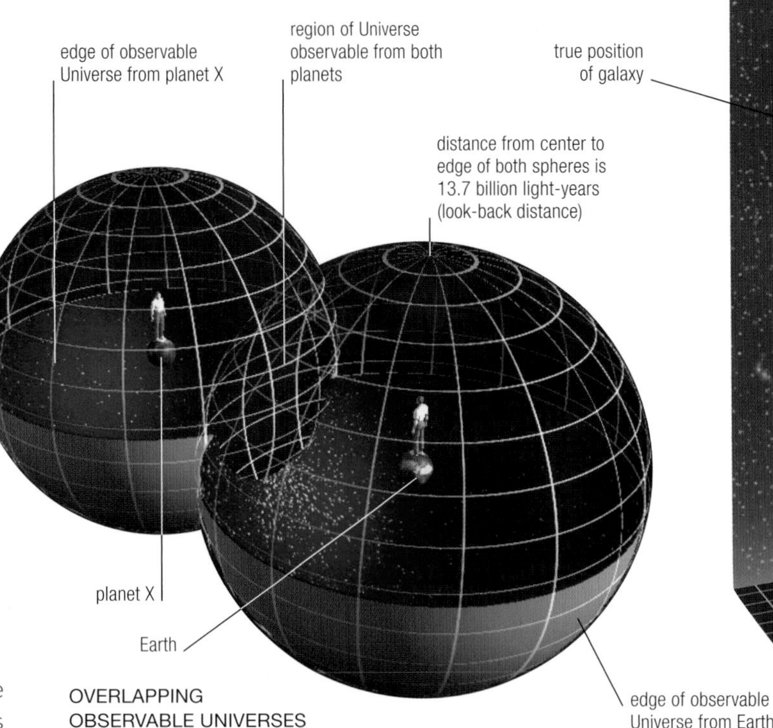

OVERLAPPING OBSERVABLE UNIVERSES
Planet X—an imaginary planet with intelligent life located billions of light-years away from Earth—would have a different observable Universe from Earth's. Its observable Universe might, however, overlap with Earth's, as shown above.

LOOK-BACK AND CO-MOVING DISTANCES

If a galaxy is said to be 11 billion light-years away, this usually refers to its look-back distance—light from it that is just reaching Earth left the galaxy 11 billion years ago and has traveled 11 billion light-years through space. The true current distance to the galaxy, the co-moving distance, is greater.

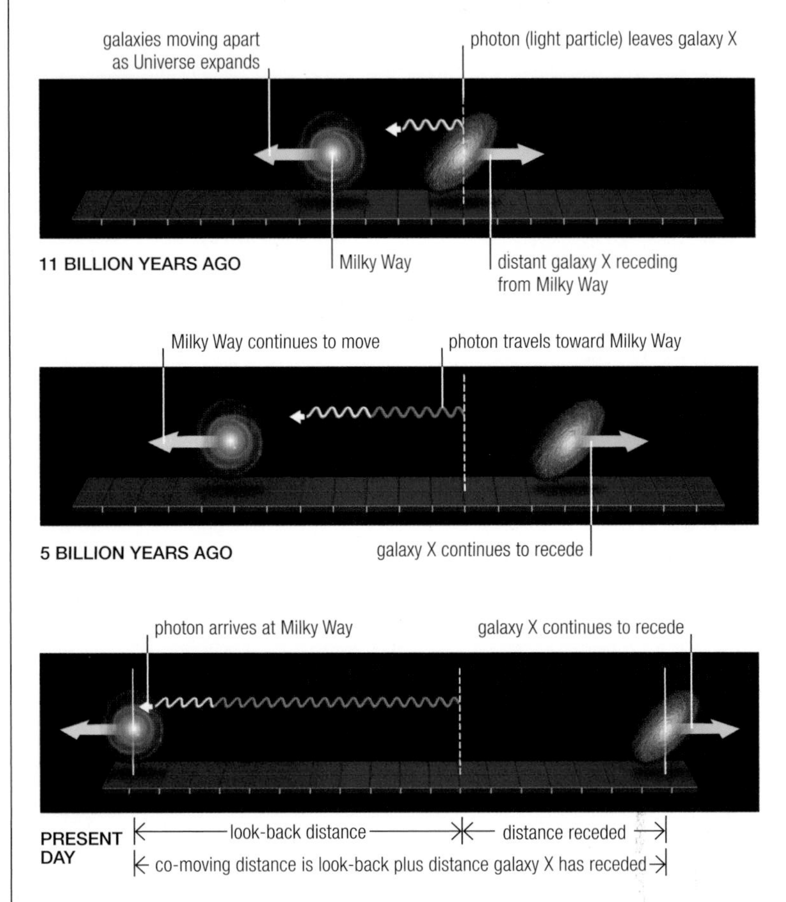

> “THE HISTORY OF ASTRONOMY IS A HISTORY OF RECEDING HORIZONS.”
>
> **EDWIN HUBBLE**, ASTRONOMER, THE REALM OF THE NEBULAE, 1936

MASS AND ENERGY

Two other quantities with a close relationship in cosmology are mass and energy. Although we normally think of these as separate entities, in the 20th century, as part of his theory of special relativity, Albert Einstein showed that they are interchangeable, being two aspects of a single phenomenon called mass-energy. The mass-energy of the Universe is now known to exist in both familiar and less-familiar forms. The well-understood forms include ordinary matter based on atoms, along with light and other electromagnetic radiation. However, these only make up a small part of the total. Other forms are dark matter and a phenomenon that seems to be accelerating the Universe's expansion, called dark energy (see p.319).

Cosmologists are interested in establishing as accurately as possible the overall density of mass-energy in the Universe and the proportions made up by the various forms, since these values can help in calculating the exact age of the Universe and in establishing its eventual fate (see pp.324–25).

DENTED SPACE-TIME

Space-time as described by general relativity can be visualized as a rubber sheet in which concentrations of mass-energy make dents. In this view, planets orbit the Sun because they roll around the dent it produces. Light passing by the Sun has its path deflected by following the local curvature of space-time.

apparent position of galaxy to observers on Earth, who assume light has traveled in a straight line

distortion of space-time caused by the Sun's mass deflects light from galaxy

telescope on Earth

orbiting planet follows elliptical path because space-time is curved in vicinity of the Sun

space-time around the Sun is warped by the Sun's mass, creating a so-called gravitational well

two-dimensional sheet represents four-dimensional space-time; the dents in the sheet represent distortions in space-time

GRAVITY AND SPACE-TIME

Holding everything in the Universe together is gravity. This was first proposed in the 17th century by Isaac Newton, but cosmologists rely on a more complete description of gravity provided by the second of Einstein's relativity theories—general relativity. This theory is of huge importance in cosmology as it helps define possible types of Universe and is a key part of any cosmological model. It provides another way of describing how gravity operates—through concentrations of mass-energy distorting a four-dimensional entity called space-time and having the ability to "bend" light. This way of thinking about gravity has practical implications, as invisible concentrations of matter in space can now be detected because of their light-bending effects.

THE CURVATURE OF THE UNIVERSE

One of cosmology's goals, important to understanding the Universe's possible fate, is to determine its exact "curvature." This has a different sense from the curvature of, say, a Frisbee. It is an intrinsic, hidden property of the Universe that depends on its average mass-energy density. In Universes with different curvatures, represented by the 2-D shapes below, parallel lines may converge or diverge, and the angles in a triangle will add up to more or less than 180°. Current theories indicate the Universe's curvature is close to zero, (or "flat").

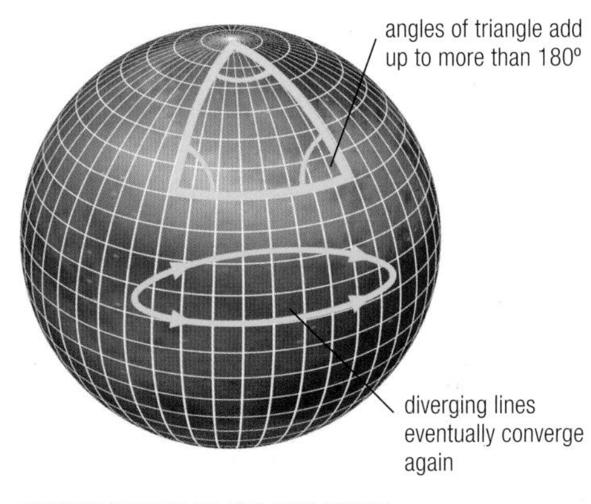

POSITIVELY CURVED UNIVERSE

Also called closed, this type of Universe has rules of geometry similar to those that apply on a spherical surface. It is finite in size and associated with a relatively high mass-energy density.

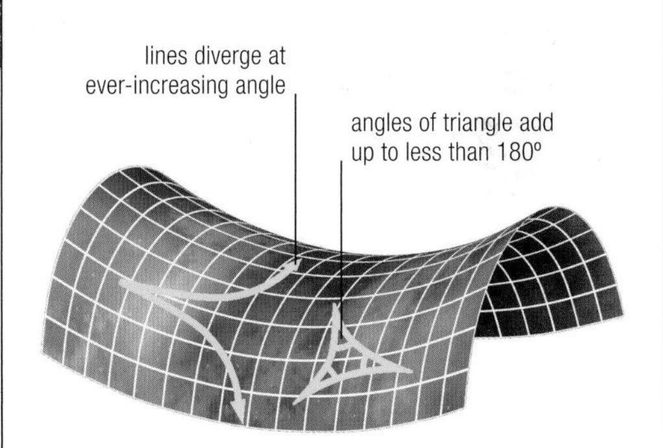

NEGATIVELY CURVED UNIVERSE

Also called open, this type of Universe has geometrical properties similar to those of a saddle-shaped surface. It extends infinitely and has a relatively low mass-energy density.

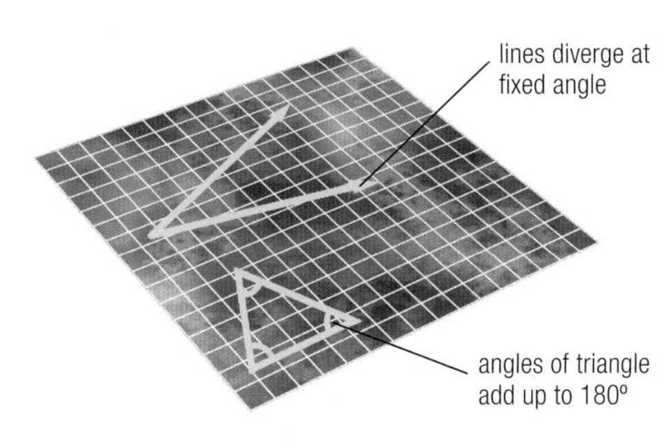

FLAT UNIVERSE

This type of Universe has the same rules of geometry as the surface of a flat plane. In a Universe of this curvature, the average mass-energy density is equally balanced.

THE BIG BANG

According to the standard model of cosmology, the Universe—all of space, time, matter, and energy—had a definite beginning over 13 billion years ago in a stupendously dramatic event known as the Big Bang.

FIRST MOMENTS

At its inception, the Universe was energy concentrated into an exceedingly small but intensely hot point. After a short interlude, a patch of it is thought to have undergone an astoundingly fast expansion known as inflation. This inflated region continued expanding and within a microsecond, was billions of miles in diameter. As the Universe expanded, it cooled, and some of its energy turned into tiny elementary particles. Initially, the mixture of matter particles and their antiparticles (antimatter)—similar to particles, but with opposite electrical charges—was roughy equal. Then, for reasons not fully understood, more particles built up. These supplied the building blocks for the ordinary matter in the Universe today.

INFLATION

The idea that a short but dramatic expansion—inflation—occurred soon after the beginning of the Universe has an important place in modern cosmology. During inflation, a tiny region of the Universe is thought to have expanded by a factor of at least 10^{26}, or to about 100 trillion trillion times its former size. As a consequence, all of what is now the observable Universe (see p.308) could have originated in a region small enough for the properties of each of its parts to have influenced the properties of all of its other parts, which would explain why today the Universe appears to be so uniform on a large scale. The inflation theory helps explain why space-time is so "flat" (see p.309).

SMOOTHING THE UNIVERSE

In a Big Bang without inflation, what are now widely spaced regions of observable space could never have become so similar in density and temperature. The effect of inflation is like expanding a wrinkled sphere—after the expansion, its surface becomes smooth and flat.

THE FIRST SECOND OF THE UNIVERSE

The timeline on these two pages shows some events in the Universe during its first second after the Big Bang. In this first second, the temperature dropped from well over 10^{28} (10,000 trillion trillion)°C to a mere 10^{10} (10 billion)°C. The diameter scale on the timeline refers to the approximate historical diameter of the observable Universe—the part that can currently be observed.

The beginning
Time, space, mass, and energy originated from one tiny, exceedingly hot, dense point

Particle soup
From about 10^{-32} seconds after the Big Bang, the Universe was an exceedingly hot "soup" of elementary particles and antiparticles. These continually formed from energy as particle-antiparticle pairs, then met and annihilated back to energy. Some of the particle types still persist as constituents of matter (for example, quarks and electrons). Others no longer exist at ordinary energy levels or are hard to detect.

quarks

pairs of electrons and positrons forming and then annihilating

gluon

X-boson

The Planck Epoch
What happened in this period, which lasted for only 10^{-43} seconds, is not known. Shortly afterward, the Inflationary Epoch began.

photons (radiant energy)

antiquarks

antineutrino

quark-antiquark pairs forming and then annihilating

Higgs boson (hypothetical)

TIME	A trillionth of a yoctosecond (10^{-36} seconds)	A hundred millionth of a yoctosecond (10^{-32} seconds)
	Inflationary Epoch Part of the Universe expanded from billions of times smaller than a proton to about the size of a badminton court.	**Electroweak Epoch** The Universe was a seething mass of particles and antiparticles in equilibrium with energy. Two basic forces of nature—the electromagnetic and the weak forces—were still unified, hence the name.
DIAMETER	3 x 10^{-51}ft (10^{-51}m)	33ft (10m)
TEMPERATURE	10^{28} (10,000 trillion trillion)°C	10^{27} (1,000 trillion trillion)°C

the Big Bang. The strongest comes from the observation that the Universe is expanding, implying that it must once have been smaller. In addition, the background radiation that pervades all of space (see p.316) indicates that the whole of the observable Universe must once have been uniformly hotter and denser than it is today.

BUILD-UP OF PARTICLES

More evidence comes from the study of galaxies at various distances (and thus of different ages) and from measuring the preponderance of the isotopes of the lightest chemical elements in the Universe (isotopes are different forms of a chemical element). The measurements support the predictions of how the Universe would have developed from a Big Bang origin. The Big Bang also resolves a cosmological puzzle known as Olber's paradox—an argument that in a Universe of infinite age and size, the sky should be uniformly bright in all directions. The Big Bang provides a solution to the paradox because it suggests that the Universe has not always existed.

Evolving galaxies
This Hubble image of deep space reveals remote galaxies with more primitive forms than closer galaxies, reinforcing the idea that they evolved in a Universe of finite age, with a beginning, and that they formed after a period of time and have been evolving ever since.

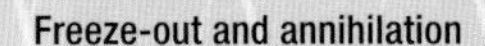

Freeze-out and annihilation
For each particle type, the temperature eventually dropped to a point where particle-antiparticle pairs froze-out so no more could form from the pool of energy. Most existing particles and antiparticles then annihilated, leaving a small excess of particles. But as quarks and antiquarks froze out, instead of annihilating, they grouped to make composite particles, such as protons, neutrons, and their antiparticles.

More matter than antimatter
A particle theorized to have existed is the X-boson, which would have decayed (disintegrated) into other particles and antiparticles. On decaying, it would have produced a tiny excess of particles over antiparticles. When these annihilated, a residue of particles would be left that gave rise to the ordinary matter of the Universe.

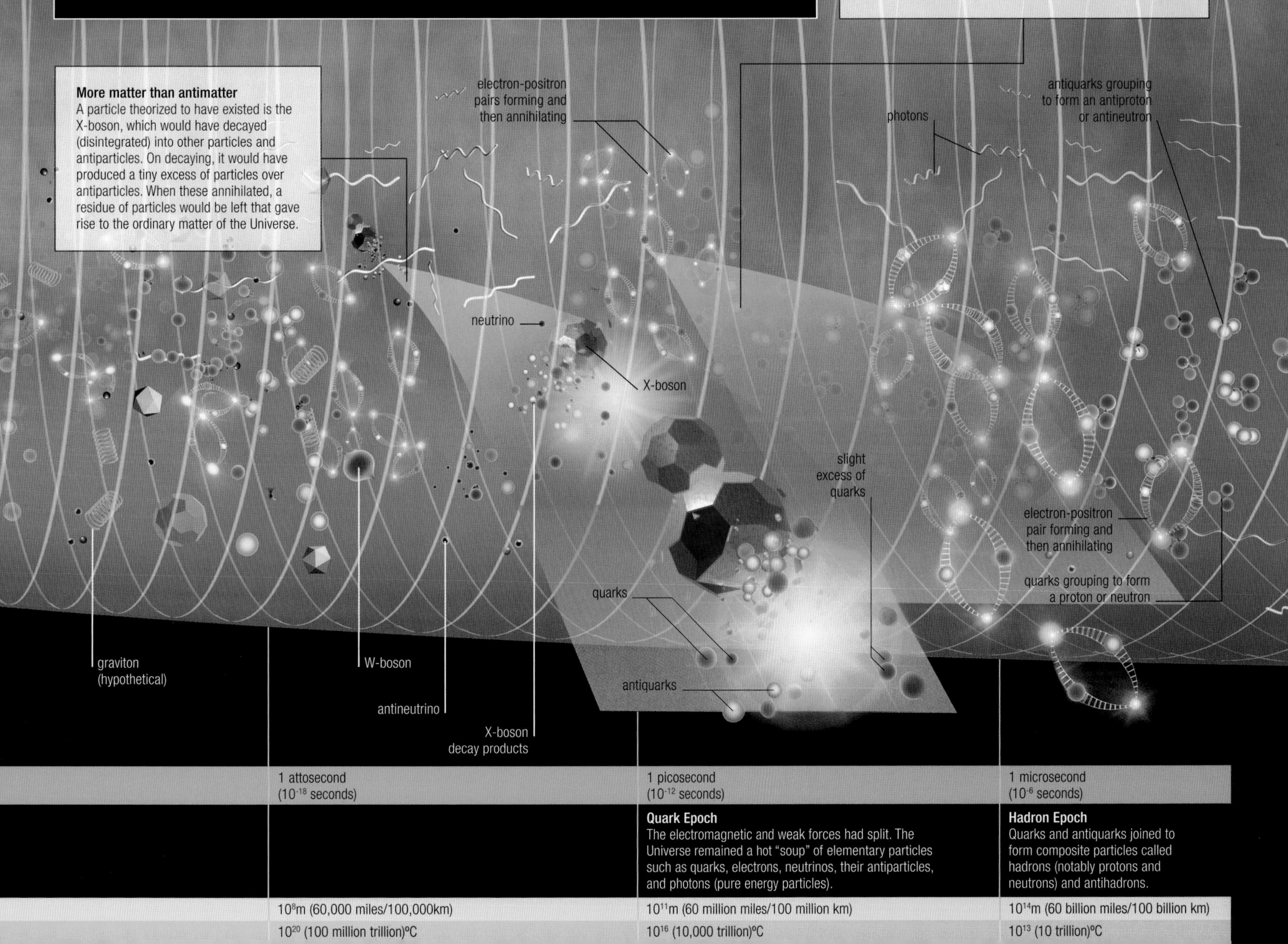

1 attosecond (10^{-18} seconds)	1 picosecond (10^{-12} seconds)	1 microsecond (10^{-6} seconds)
	Quark Epoch The electromagnetic and weak forces had split. The Universe remained a hot "soup" of elementary particles such as quarks, electrons, neutrinos, their antiparticles, and photons (pure energy particles).	**Hadron Epoch** Quarks and antiquarks joined to form composite particles called hadrons (notably protons and neutrons) and antihadrons.
10^{8}m (60,000 miles/100,000km)	10^{11}m (60 million miles/100 million km)	10^{14}m (60 billion miles/100 billion km)
10^{20} (100 million trillion)°C	10^{16} (10,000 trillion)°C	10^{13} (10 trillion)°C

The main matter particles present at 1 second—protons, neutrons, and electrons—came together in stages to form the first atoms.

The first protons and neutrons
By 1 second after the Big Bang, most of the protons and neutrons that had formed from quarks a moment earlier had annihilated with their antiparticles, but a residual amount remained. Some other types of composite particles that had formed, such as pions, decayed too. Because free neutrons are unstable and decay into protons and electrons, a large number of protons soon built up.

Foggy Universe
For over 350,000 years, the Universe continued expanding and cooling, but it was too energetic for atoms to form. Photons (particles of radiant energy) were continually colliding with and bouncing off electrons (a phenomenon called scattering), so they could travel hardly any distance in a straight line. To an outside observer, the Universe at this time would have resembled a dense fog.

The first nuclei
Some 3 minutes after the Big Bang, collisions between protons and neutrons began forming helium-4 nuclei (containing 2 protons and 2 neutrons) as well as tiny amounts of other atomic nuclei, such as helium-3 (2 protons and 1 neutron) and deuterium (1 proton and 1 neutron). Termed Big Bang nucleosynthesis, these reactions finished within about 17 minutes. Many free protons remained.

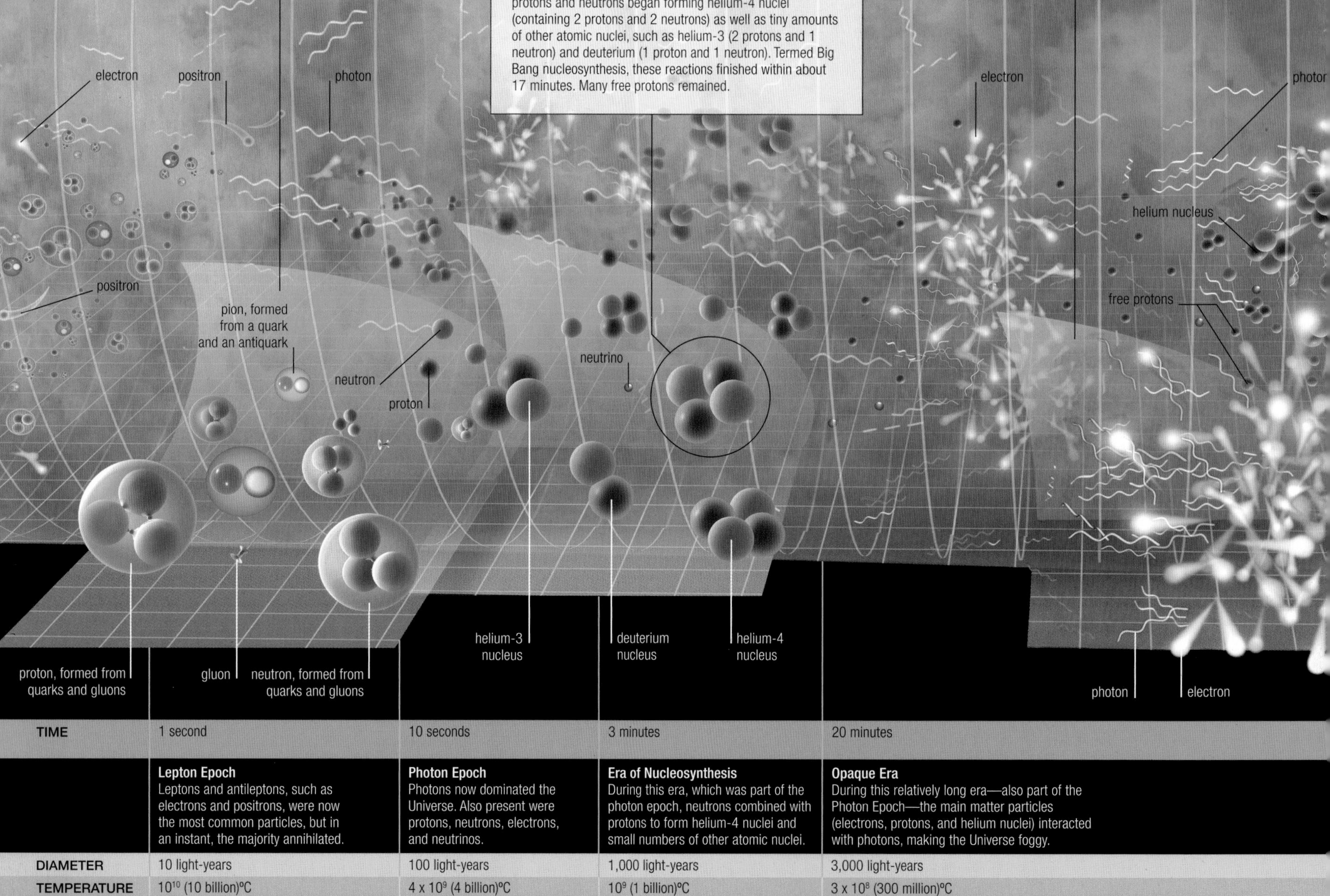

TIME	1 second	10 seconds	3 minutes	20 minutes
	Lepton Epoch Leptons and antileptons, such as electrons and positrons, were now the most common particles, but in an instant, the majority annihilated.	**Photon Epoch** Photons now dominated the Universe. Also present were protons, neutrons, electrons, and neutrinos.	**Era of Nucleosynthesis** During this era, which was part of the photon epoch, neutrons combined with protons to form helium-4 nuclei and small numbers of other atomic nuclei.	**Opaque Era** During this relatively long era—also part of the Photon Epoch—the main matter particles (electrons, protons, and helium nuclei) interacted with photons, making the Universe foggy.
DIAMETER	10 light-years	100 light-years	1,000 light-years	3,000 light-years
TEMPERATURE	10^{10} (10 billion)°C	4×10^9 (4 billion)°C	10^9 (1 billion)°C	3×10^8 (300 million)°C

HOW LONG AGO DID THE BIG BANG HAPPEN?

The age of the Universe is calculated by first establishing how fast it is expanding, then working backwards in time to estimate when all its matter was in one place. To determine the expansion rate, the distances to remote galaxies, measured by the brightness of supernovae within them, are compared with the speeds at which the galaxies are receding, calculated from their red shifts (see p.306).

THE CURRENT BEST ESTIMATE

Once a basic calculation has been made, adjustments are needed to take account of the fact that the Universe's mass-energy must have modified the expansion rate over time. The figure most often quoted for the Universe's age—13.7 billion years—is a best estimate with a margin of error. Analysis of the background radiation (see pp.316–17) has improved the estimate by providing more accurate values for several parameters affecting the calculations.

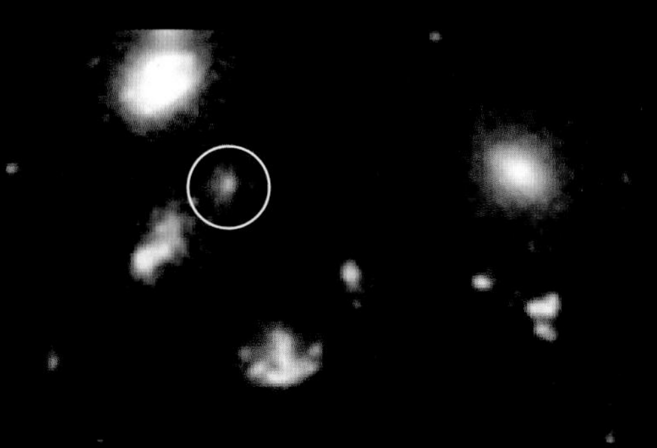

Distant supernova
The "smudge" circled in this Hubble Space Telescope image is some 11 billion light-years away and, as of January 2010, is one of the most distant supernovae ever identified. Studies of supernovae like this are key to determining the Universe's age (see p.319).

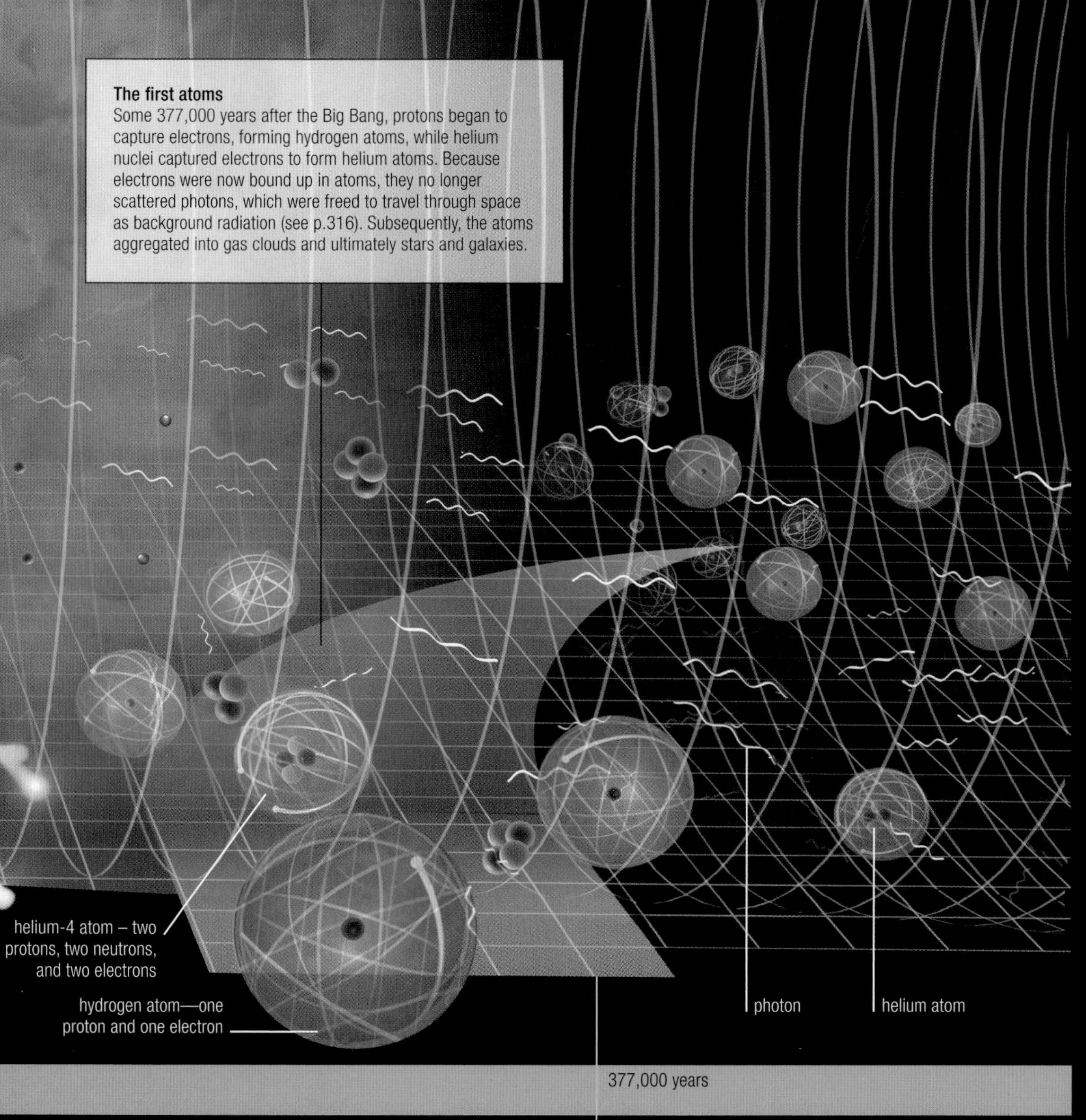

The first atoms
Some 377,000 years after the Big Bang, protons began to capture electrons, forming hydrogen atoms, while helium nuclei captured electrons to form helium atoms. Because electrons were now bound up in atoms, they no longer scattered photons, which were freed to travel through space as background radiation (see p.316). Subsequently, the atoms aggregated into gas clouds and ultimately stars and galaxies.

377,000 years

Balance of elements
At the end of the Opaque Era, many more free protons existed than helium nuclei or other atomic nuclei. The scene was set for the first atoms to form. When they did, about eleven hydrogen atoms were made for each helium atom.

Atomic Epoch
As the first atoms appeared, through atomic nuclei capturing electrons, photons became unattached from matter and were free to travel through space as radiation.

100 million light-years

2,700ºC

STUDYING THE BIG BANG

Much of what is known about the early Universe has come from particle accelerators (or atom smashers). However, what happened in the first 10^{-37} seconds has been largely guesswork, since it has been impossible to generate the levels of energy present at that time. A new accelerator, the Large Hadron Collider (LHC) at CERN (the European Organization for Nuclear Research) is expected to change this. The LHC will collide protons and heavy ions at unprecedented energies in order to look for new particles: one hypothetical particle that may be detected is the Higgs boson. The LHC may also cast more light on the identity of dark matter (see p.318).

Inside the Large Hadron Collider
The LHC occupies a 17-mile- (27km-) long circular tunnel across the Swiss-French border. At its heart is a bundle of pipes, seen here in an opened-up section. Two central pipes carry high-energy particle beams.

UNRESOLVED ISSUES

Although the Big Bang is a reasonably complete theory, a number of flaws in it need addressing. One problem, for example, concerns why so much more matter than antimatter exists in the Universe. The theory that a particle called the X-boson appeared shortly after the Big Bang and decayed to produce an imbalance between matter and antimatter (see p.311) has not convinced all cosmologists. Some speculate that there could be a large collection of antimatter in the Universe, but so far no such region has been detected. Particle-accelerator experiments, if they succeed in producing a particle like the X-boson, may help resolve the issue.

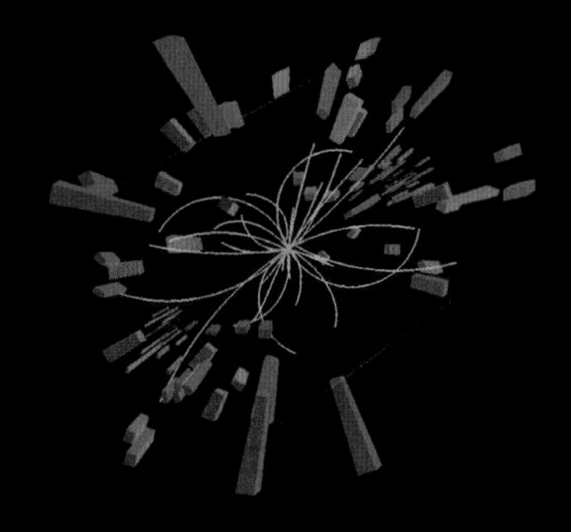

Recreating the Big Bang
These particle tracks (yellow) result from one of the first collision experiments from the LHC in late March 2010. The collision took place at an energy level of 7 trillion electron volts—the type of energy concentration needed to recreate conditions just after the Big Bang.

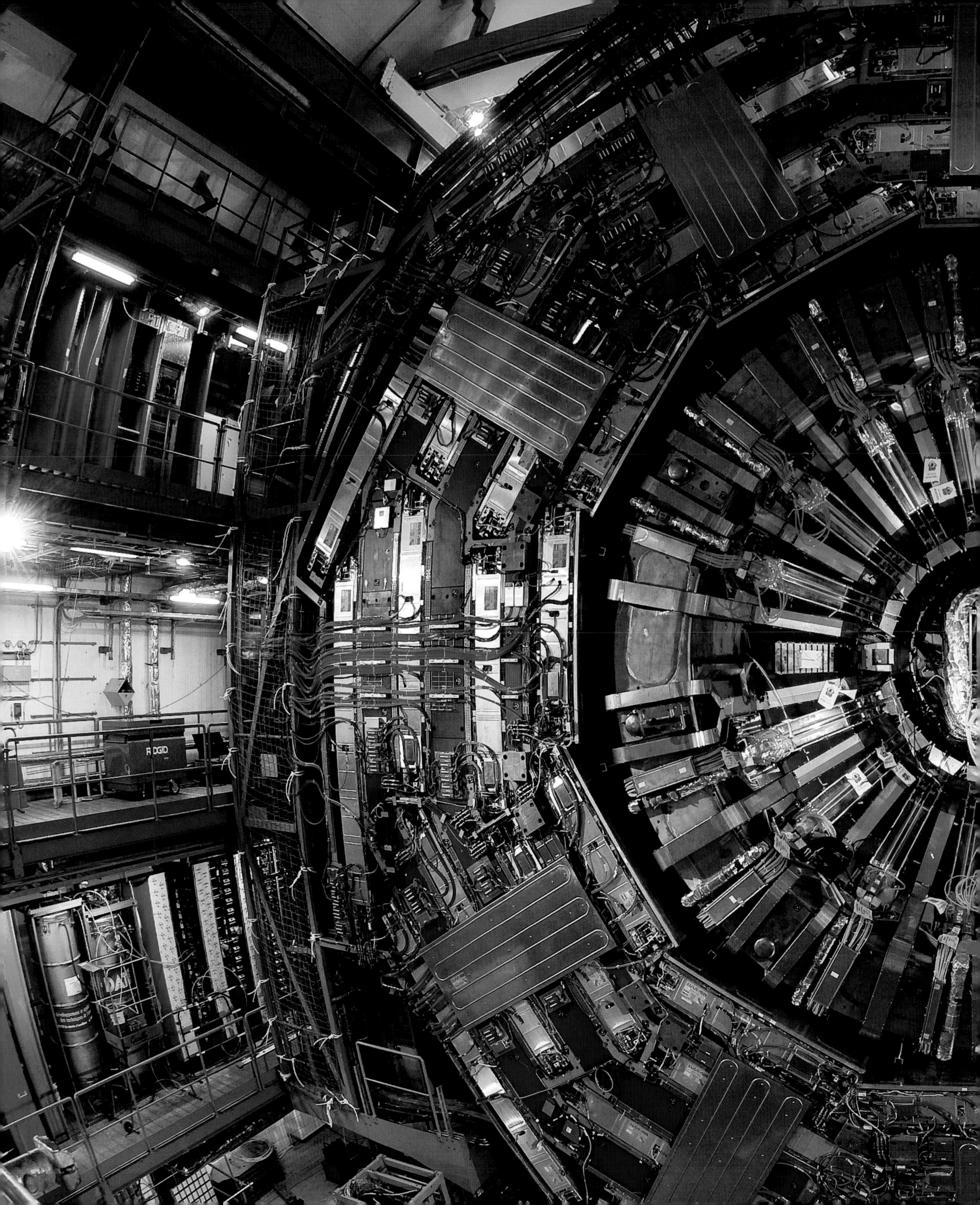
RIDGID

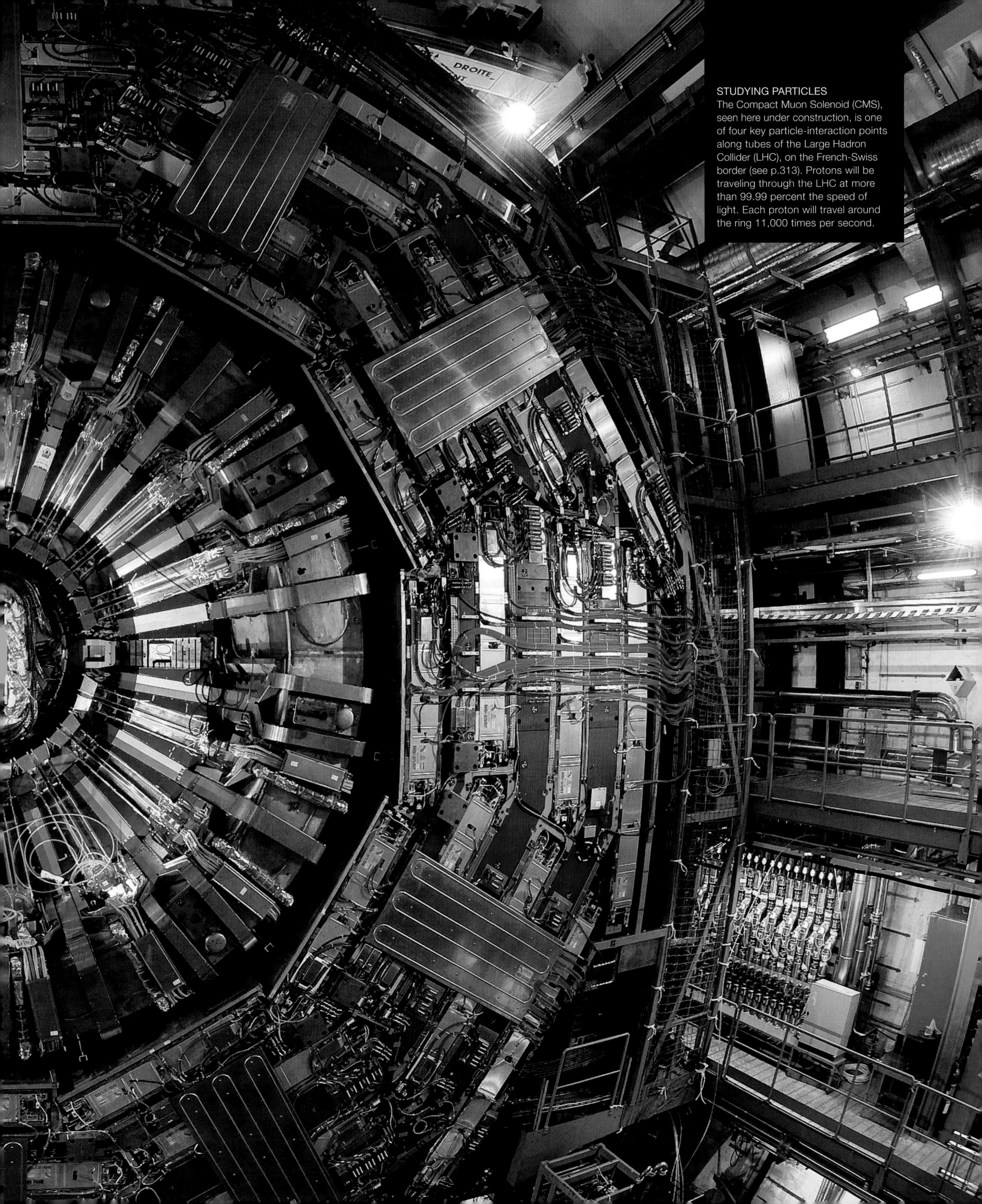

STUDYING PARTICLES

The Compact Muon Solenoid (CMS), seen here under construction, is one of four key particle-interaction points along tubes of the Large Hadron Collider (LHC), on the French-Swiss border (see p.313). Protons will be traveling through the LHC at more than 99.99 percent the speed of light. Each proton will travel around the ring 11,000 times per second.

BACKGROUND RADIATION

Early proponents of the Big Bang theory predicted that a remnant of the Universe's initial hot state, a faint microwave radiation, should permeate space. In 1964 this background radiation was detected, and it has since become a major focus of research.

RELIC OF THE BIG BANG

This elliptical map shows the strength of the background radiation over the whole sky, as measured by the Wilkinson Microwave Anisotropy Probe (WMAP). The tiny variations—anisotropies—have been brought out in the map as colored spots by using an extremely finely graded temperature scale (see opposite).

MAPPING THE RADIATION

The cosmic microwave background radiation (CMBR) is like a picture of the Universe soon after the Big Bang, when photons (particles or packs of radiant energy) were freed from interacting with matter and began to travel unhindered through space. The intensity of the CMBR from every point in the sky has been mapped, most recently by the Wilkinson Microwave Anisotropy Probe (WMAP). Although the CMBR was first emitted by the hot fireball of gas in the early Universe, today its pattern matches that of a cold, dark object with a temperature just above absolute zero. This is because the photon wavelengths have been stretched—and so shifted toward the cooler, long-wavelength end of the spectrum—by the expansion of the Universe.

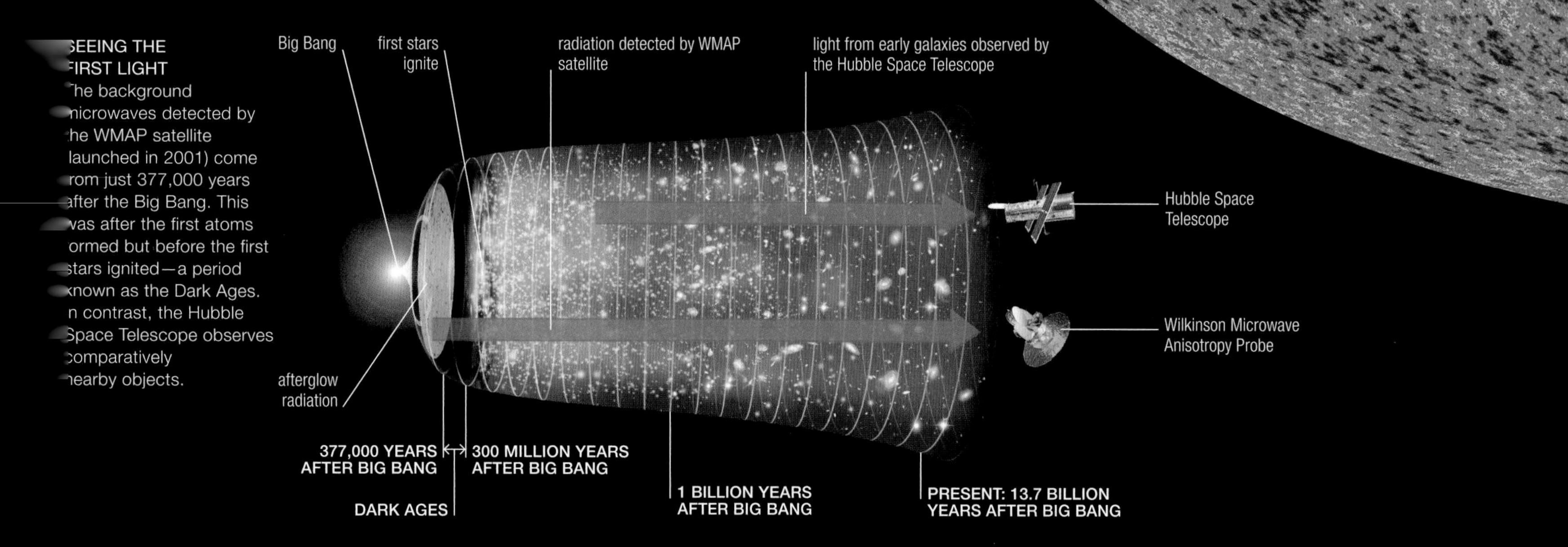

SEEING THE FIRST LIGHT
The background microwaves detected by the WMAP satellite (launched in 2001) come from just 377,000 years after the Big Bang. This was after the first atoms formed but before the first stars ignited—a period known as the Dark Ages. In contrast, the Hubble Space Telescope observes comparatively nearby objects.

RESEARCH DEVELOPMENTS

The Planck Spacecraft, a 4,200lb (1,900kg) space observatory, was launched in May 2009 specifically to study the CMBR and improve upon observations made by the WMAP satellite. Its primary goal is to map the variations in the CMBR more precisely, in order to more accurately determine the Hubble constant (see p.276), the curvature of space, and the relative proportions of matter, dark matter, and dark energy (see pp.318–19). It will also test the theory that the Universe underwent a colossal expansion (inflation) soon after the Big Bang and will explore theories of how galaxies are formed.

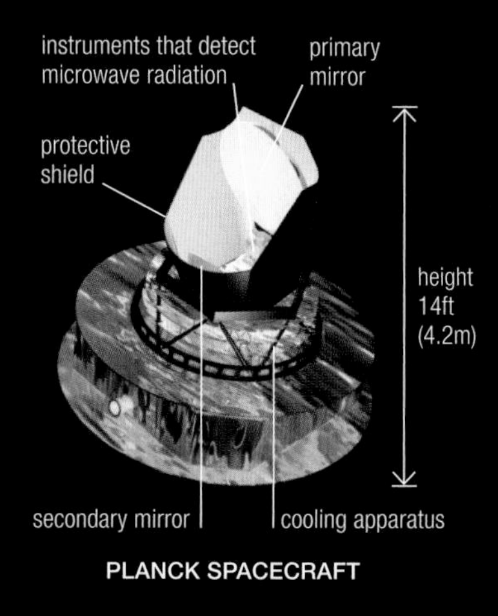

PLANCK SPACECRAFT

RIPPLES IN THE RADIATION

The CMBR is extremely uniform across the sky, but there are tiny variations in its intensity, called anisotropies. Most maps of the radiation, such as that from the WMAP satellite (above), display these as colored spots. The patterning of the spots is as expected from an exceedingly hot gas containing extremely small temperature variations that has been expanded billions of times over, as occurred after the Big Bang. The radiation in the CMBR is also slightly polarized in that the orientations of the little waveforms that make up the radiation are not completely random. The pattern of polarization carries hidden information about a process called reionization that occurred a few hundred million years after the Big Bang. During reionization, some of the Universe's hydrogen atoms were ripped apart into their constituent protons and electrons, possibly by ferocious ultraviolet radiation from the first stars. The electrons freed up by this process interfered with the movement of individual CMBR photons. It was this interference that produced the pattern of polarization detectable today.

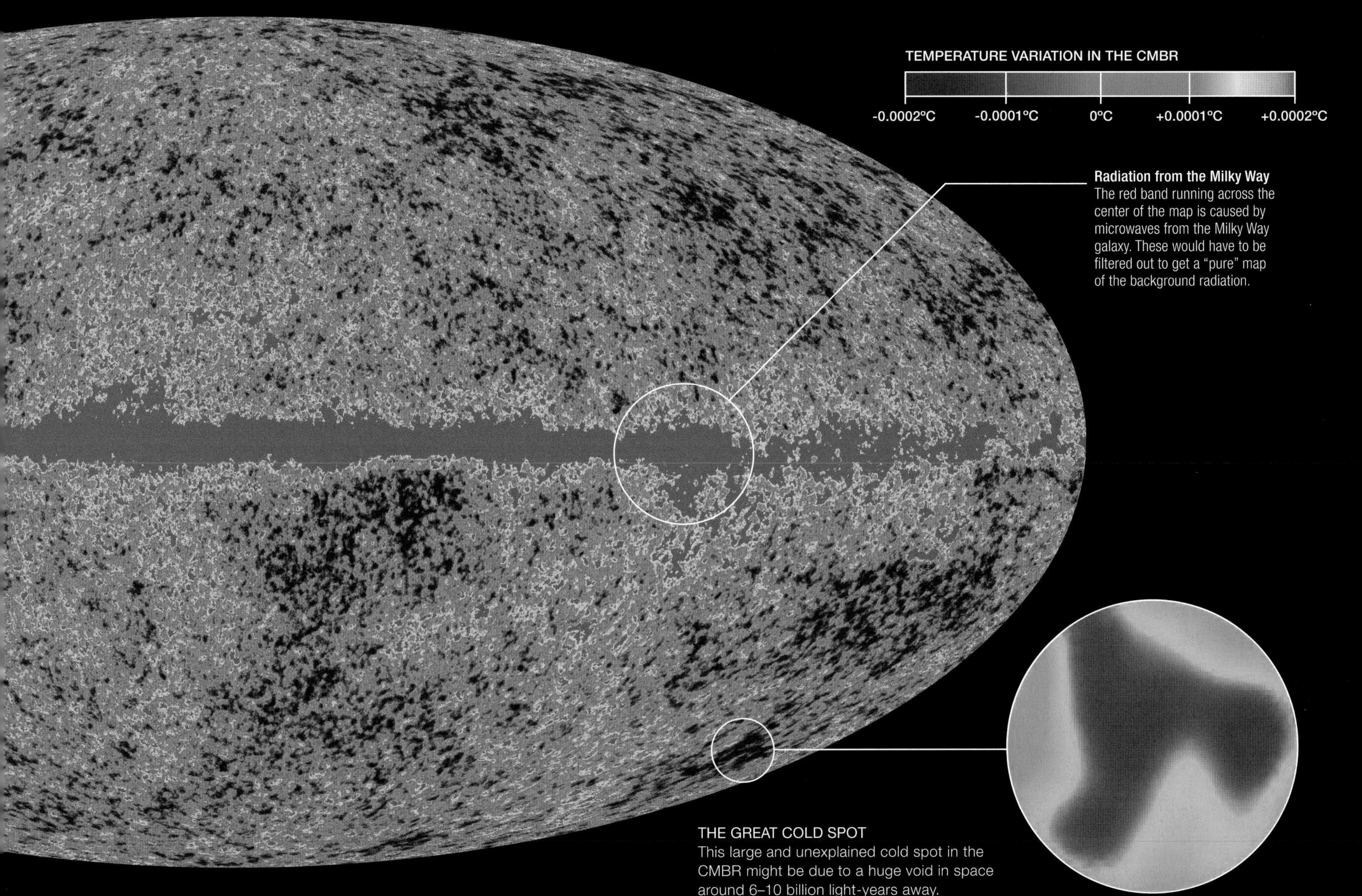

TEMPERATURE VARIATION IN THE CMBR
-0.0002°C
-0.0001°C
0°C
+0.0001°C
+0.0002°C
Radiation from the Milky Way
The red band running across the center of the map is caused by microwaves from the Milky Way galaxy. These would have to be filtered out to get a "pure" map of the background radiation.
THE GREAT COLD SPOT
This large and unexplained cold spot in the CMBR might be due to a huge void in space around 6–10 billion light-years away.

WHAT THE RADIATION REVEALS

Detailed mathematical analysis of CMBR data obtained by the WMAP satellite has yielded an astonishing amount of information about the Universe. For example, examination of the hot and cold spots in the CMBR has confirmed that the curvature of the Universe is very close to zero, or flat (see below). The data has helped to determine a more accurate value for the Hubble constant (the rate at which the Universe is expanding), as well as the relative proportions of ordinary matter, dark matter, and dark energy in the Universe (see pp.318–19). This in turn has helped establish a more accurate age for the Universe and narrowed down the theories about its possible fate. Finally, analysis of CMBR has helped supply a more accurate time frame for when reionization of the Universe occurred.

RADIATION AND GEOMETRY

The observed size of the hot and cold spots in the CMBR accords closely with the actual size of the spots as predicted by the Big Bang theory. This provides strong evidence that the Universe is flat, or has a near-zero curvature (see p.309). If it had a significant positive or negative curvature, the spots would appear bigger or smaller than their actual size.

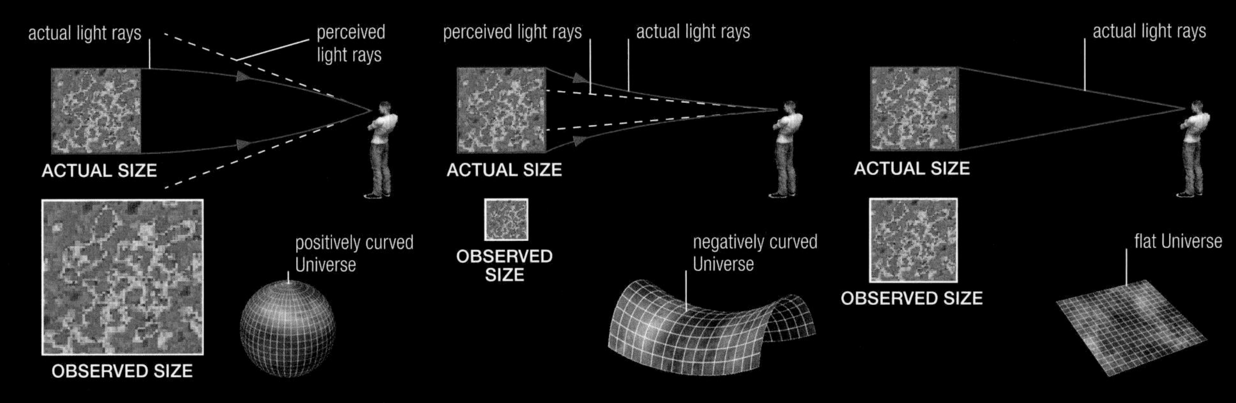

IF POSITIVELY CURVED
Light rays coming from opposite sides of a spot would bend toward each other, and the area would look bigger than it is.

IF NEGATIVELY CURVED
Light rays coming from opposite sides of a spot would bend away from each other, so the area would look smaller than it is.

FLAT UNIVERSE
In a flat Universe, light rays coming from opposite sides of a spot do not bend, so the area is seen at its actual size.

DARK COSMOLOGY

Mysterious cosmic phenomena whose existence is not in doubt, but whose underlying nature is not understood, are labeled "dark" by cosmologists, pending further explanation. Three examples are dark matter, dark energy, and the more recently discovered dark flow.

DARK MATTER

This is a form of matter that is hard to detect as it gives off no electromagnetic radiation. The main evidence for it comes from the study of galaxies and galaxy clusters. The amount of radiation-emitting matter in these structures does not account for the gravitational attraction that holds them together. In addition, they bend light more than can be accounted for by the gravity of their visible matter. Some dark matter may be contained in black holes or collections of ordinary matter (atoms) that are giving off an insignificant amount of radiation. But most, called exotic dark matter, is thought to be something else. Candidates are classified into hot varieties (subatomic particles moving at close to light speed) or cold varieties (unidentified, slower-moving, heavier particles). The standard model of cosmology proposes that most dark matter is cold and that clumps of it played a role in the formation and evolution of galaxies.

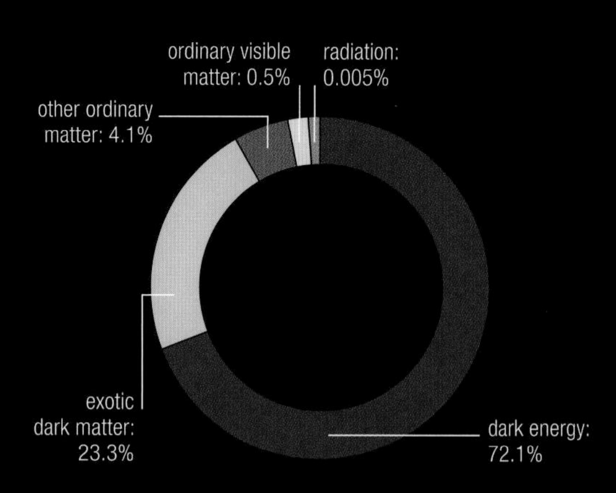

TYPES OF MASS-ENERGY IN THE UNIVERSE
Data from the WMAP satellite (see p.316) indicates that most of the Universe's mass-energy is dark energy, and much of the rest is exotic dark matter (something other than atoms). Of the ordinary matter (atoms), some is visible, for example, stars; the rest emits no visible light, but some produces X-rays or other non-visible radiation.

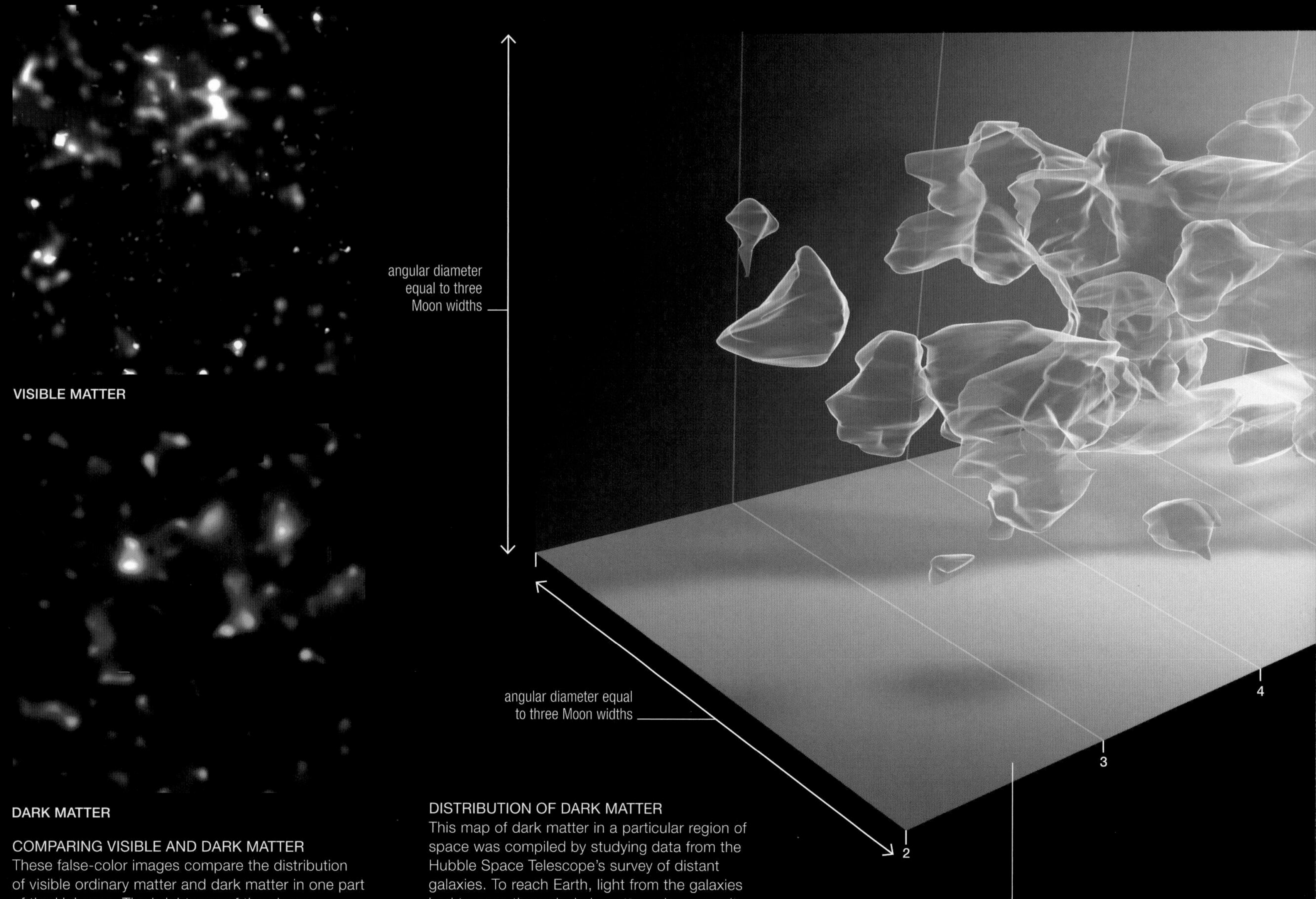

VISIBLE MATTER

DARK MATTER

COMPARING VISIBLE AND DARK MATTER
These false-color images compare the distribution of visible ordinary matter and dark matter in one part of the Universe. The brightness of the clumps corresponds to the density of matter in each case. The presence of dark matter was deduced from its light-bending characteristics.

DISTRIBUTION OF DARK MATTER
This map of dark matter in a particular region of space was compiled by studying data from the Hubble Space Telescope's survey of distant galaxies. To reach Earth, light from the galaxies had to pass through dark matter, whose gravity deflected the light rays. By analyzing distortions in the galaxy shapes, the location of dark matter clumps could be calculated.

DARK ENERGY

This is a form of mass-energy that appears to be causing the expansion of the Universe to accelerate. Although dark energy is poorly understood, evidence for its presence comes from the study of supernovae (explosions that occur at the end of a star's life, see p.248) in distant galaxies. It is thought to account for most of the mass-energy of the Universe (see left). Nothing is known about its fundamental nature except that it is evenly spread through space. Some scientists think it may be a vacuum energy—a type of outward pressure created in empty space from fleeting appearances and disappearances of virtual particles. NASA is planning a project (the Joint Dark Energy Mission) that is expected to launch a telescope designed to improve the understanding of dark energy.

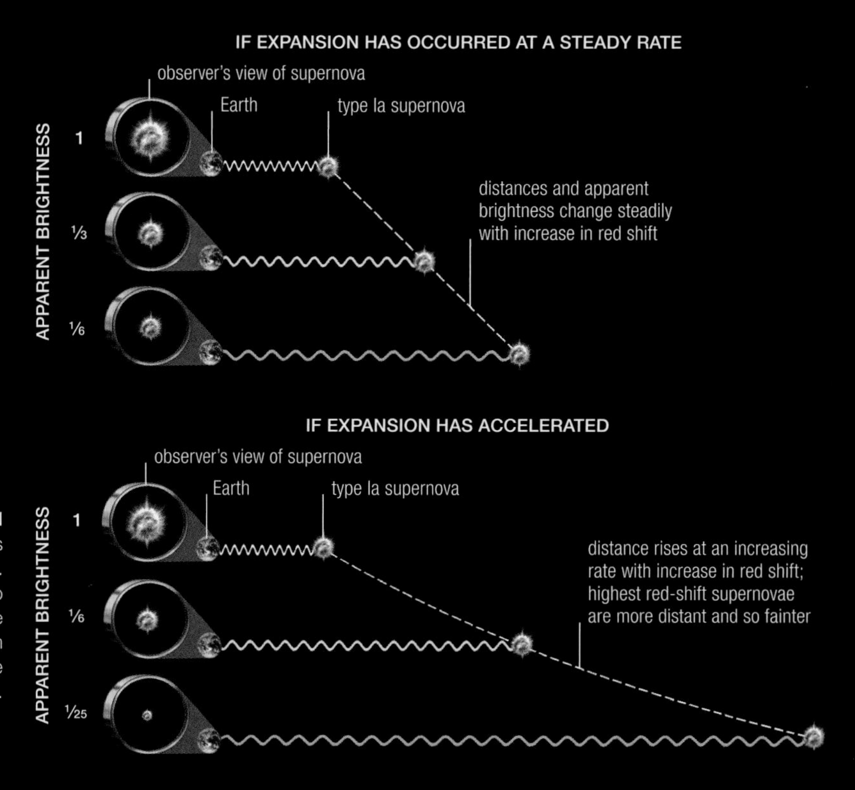

EVIDENCE FOR THE ACCELERATING EXPANSION
The study of type Ia supernovae in remote galaxies provides the evidence for the accelerating expansion of the Universe. The brightness of the supernovae indicates the distances to the galaxies, while their red shifts reveal how fast they are receding from Earth. Galaxies with the highest red shifts seem to be more distant than would be expected if the Universe had been expanding at a steady rate.

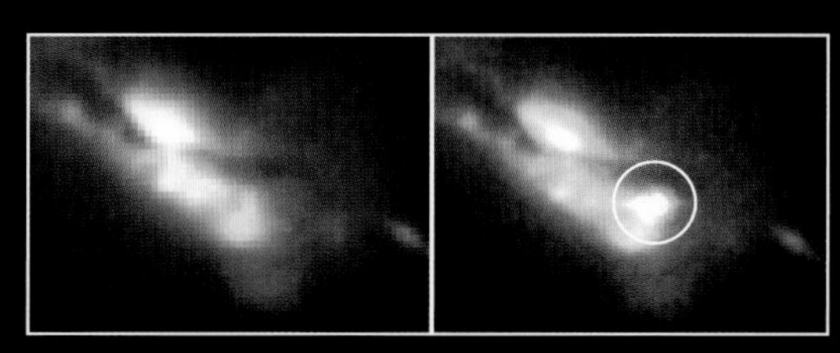

COSMIC CANDLE
These photographs show a galaxy, some 5 billion light-years away from Earth, before and after a type 1a supernova (circled) occurred in it. Because all such supernovae have a similar luminosity, their observed brightness can be used to determine their distances from Earth; they are often known as standard candles.

DARK FLOW

First detected in 2008, dark flow refers to an apparent and unexplained movement of some galaxy clusters billions of light-years from Earth. These clusters seem to be moving up to 620 miles (1,000km) per second in one direction relative to the overall background of movement caused by the expansion of space. The movement is puzzling, as it conflicts with an underlying assumption of cosmology: at large scales the Universe is uniform and there is no "preferred" direction for behaviors such as movement. Some cosmologists think dark flow provides evidence for a relatively dense part of the Universe beyond the particle horizon (the boundary that separates the observable Universe from the rest of the cosmos).

remote region of space

7

6

5

LOOK-BACK DISTANCE FROM EARTH (BILLIONS OF LIGHT-YEARS)

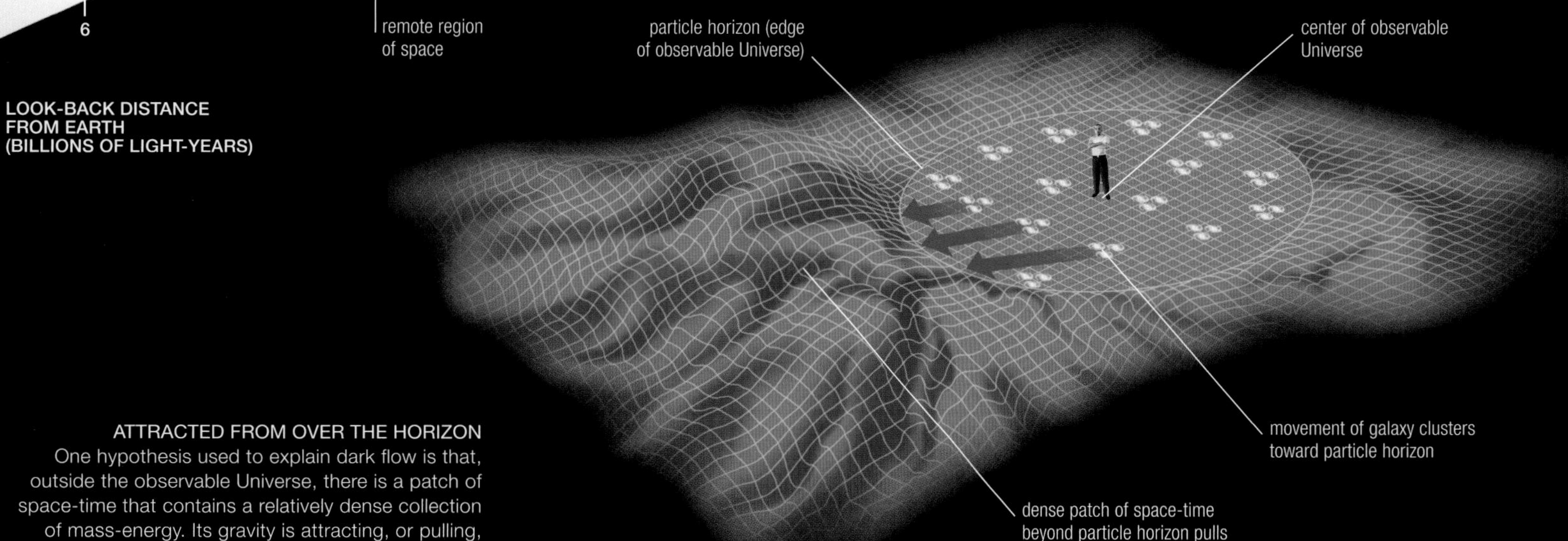

ATTRACTED FROM OVER THE HORIZON
One hypothesis used to explain dark flow is that, outside the observable Universe, there is a patch of space-time that contains a relatively dense collection of mass-energy. Its gravity is attracting, or pulling, galaxy clusters from inside the observable Universe.

GALAXY FORMATION

Images of deep space show that the first galaxies had developed by 500 million years after the Big Bang. Astronomers are now trying to solve the riddle of what caused these early galaxies to form and how they subsequently evolved and interacted.

THE COSMIC WEB

The three-stage computer simulation shows how large-scale structure evolves in space. It starts with a cube of near-uniformly distributed matter with tiny irregularities then allows gravity to produce its effects.

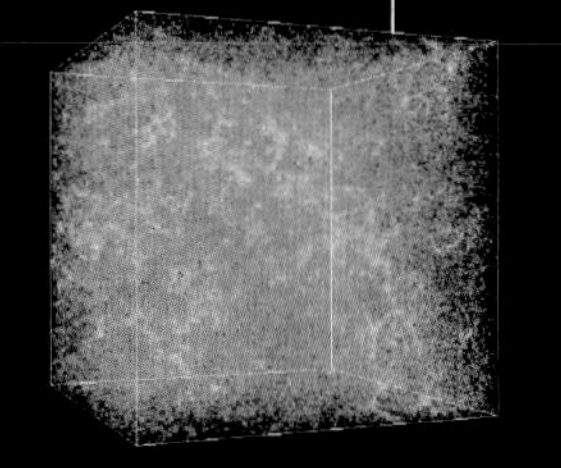

500 MILLION YEARS OLD
Even at this early stage, some faint clumping is visible within the region modeled by the simulation.

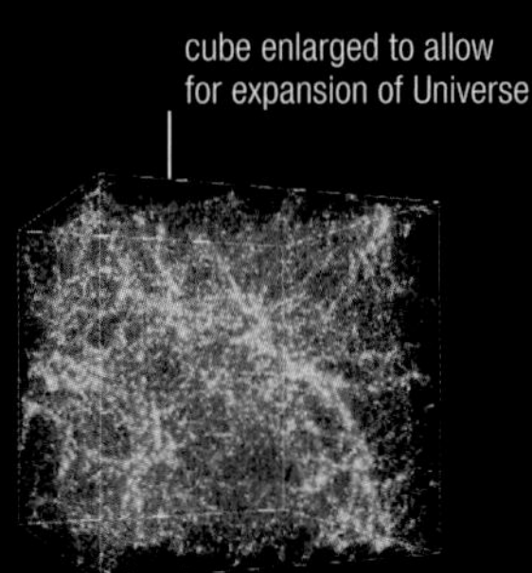

2.3 BILLION YEARS OLD
Some two billion years later, much more obvious clumping and filament formation can be seen.

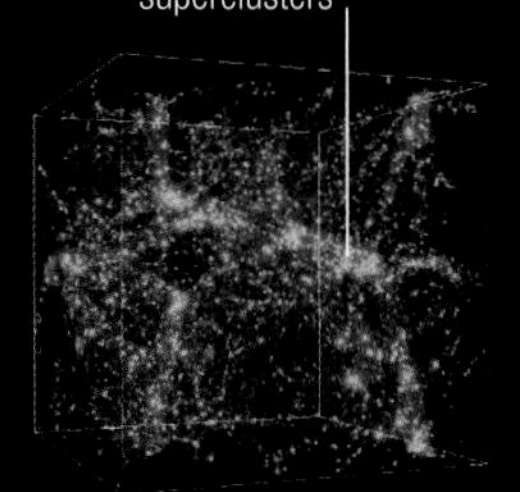

13.7 BILLION YEARS OLD
The matter distribution in the simulation eventually resembles that in our area of the Universe.

SEEDS OF STRUCTURE

Galaxies are not just solitary objects in space, but are dispersed in a gigantic network of filaments—strings of galaxy clusters and superclusters—that pervade the cosmos. The cosmic web of matter is thought to have originated as tiny spatial variations in the temperature and mass-energy density of the Universe that developed during the first fractions of a second after the Big Bang. The colossally fast expansion, or inflation, that happened immediately afterward blew these small variations up to cosmic size. Subsequently, gravitational attraction worked on the irregularities, causing matter gradually to coalesce into clumps and strands, which eventually led to galaxy formation. Some of the main evidence for this hypothesis comes from supercomputer simulations (see left). Cosmologists run these to study how well mathematical models of the evolution of the Universe agree with current surveys of galaxy distribution (see p.322) and to help them understand any discrepancies.

> “NO MATTER WHAT SIZE ... ALL GALAXIES SEEMED TO BE SITTING IN THE SAME AMOUNT OF DARK MATTER.”
>
> **MARK WILKINSON**, BRITISH ASTRONOMER, CA.2007

THEORIES OF GALAXY FORMATION

Although the formation of a cosmic web of matter helps explain the current distribution of galaxies, there remains the still unanswered question: what processes caused galaxies to develop from this web? For many years there were two main competing models of how galaxies first formed. The top-down theory proposed that they developed directly from the collapse and break-up of large clumps of matter. In contrast, the bottom-up theory maintained that smaller objects (stars and small clumps of cold gas) formed first, then drew together to form small galaxies, some of which later merged to make larger galaxies. As telescopes became more powerful and looked deeper into space, they revealed a preponderance of small galaxies in the early Universe, so the bottom-up theory tended to gain the upper hand. In recent years, however, the focus has switched onto different ways in which dark matter (see p.318) may have influenced galaxy formation.

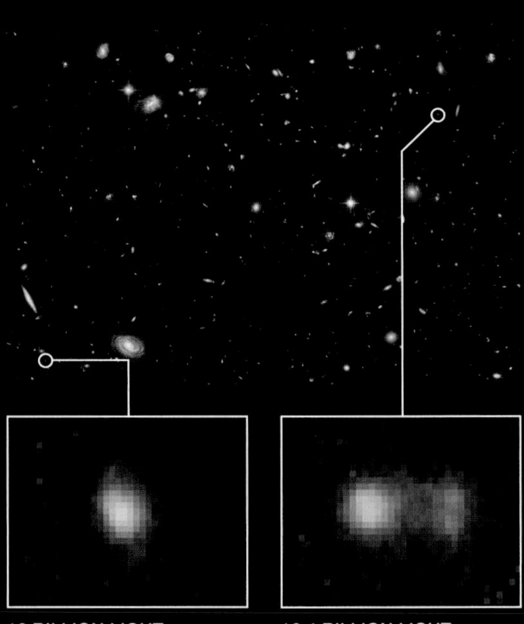

BUILDING-BLOCK GALAXIES
Small galaxies, like those seen in this Spitzer Space Telescope image, formed about a billion years after the Big Bang and may have merged to create larger galaxies.

THE ROLE OF DARK MATTER

Many cosmologists now think that in the early Universe, it was dark matter, not ordinary matter, that first began to clump together. The dark matter attracted ordinary matter (then a cold gas consisting of hydrogen and helium atoms) into it, producing structures (haloes) connected by filaments of dark and ordinary matter. Stars and galaxies may have formed from the network of haloes and filaments. Computer simulations support this, but there are discrepancies: for example, they predict fewer spiral galaxies than are observed in the Universe.

GALAXY EVOLUTION

Collisions and mergers between galaxies have undoubtedly had an important role in galaxy evolution, particularly in the formation of elliptical galaxies. The result of a galaxy collision depends on factors such as the relative size of the galaxies (see pp.292–93). A goal of future missions looking at deep space is to observe the processes involved in galaxy evolution in the early Universe. Also still to be resolved is whether the supermassive black holes at the centers of galaxies had a role in galaxy formation (see pp.294–95).

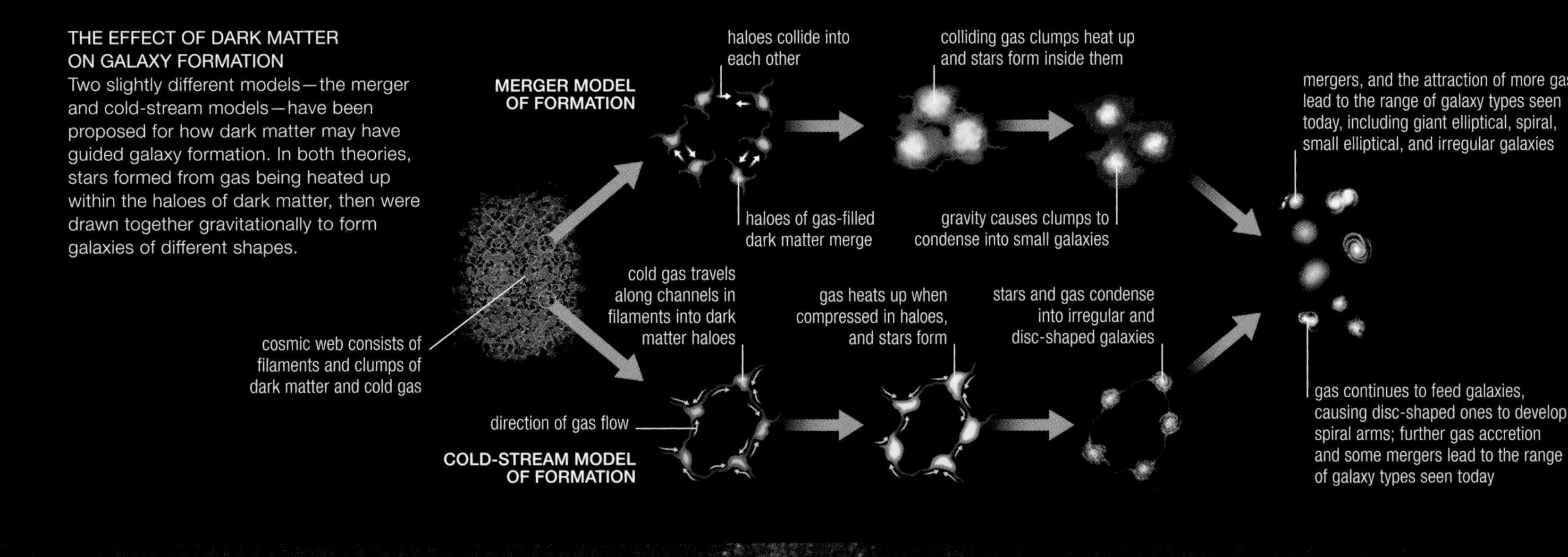
MERGER MODEL OF FORMATION
haloes collide into each other
colliding gas clumps heat up and stars form inside them
mergers, and the attraction of more gas lead to the range of galaxy types seen today, including giant elliptical, spiral, small elliptical, and irregular galaxies
haloes of gas-filled dark matter merge
gravity causes clumps to condense into small galaxies
cold gas travels along channels in filaments into dark matter haloes
gas heats up when compressed in haloes, and stars form
stars and gas condense into irregular and disc-shaped galaxies
cosmic web consists of filaments and clumps of dark matter and cold gas
direction of gas flow
COLD-STREAM MODEL OF FORMATION
gas continues to feed galaxies, causing disc-shaped ones to develop spiral arms; further gas accretion and some mergers lead to the range of galaxy types seen today

THE EFFECT OF DARK MATTER ON GALAXY FORMATION
Two slightly different models—the merger and cold-stream models—have been proposed for how dark matter may have guided galaxy formation. In both theories, stars formed from gas being heated up within the haloes of dark matter, then were drawn together gravitationally to form galaxies of different shapes.

ADVANCED GALAXY MERGER
This is a Hubble Space Telescope view of NGC 520, which is about 100 million light-years away from Earth. Although classified as a single object, it is thought to be the result of a collision 300 million years ago between two spiral galaxies, now in an advanced stage of merger, to form an elliptical galaxy.

BIRD'S-EYE VIEW

f it were possible to move outside the observable Universe, galaxies could be seen to be grouped into clusters and superclusters that form ong filaments separated by huge voids. This has been confirmed by a series of detailed surveys of deep space.

A TRIO OF GALAXY SURVEYS

One of the first surveys of galaxy distribution was carried out in the ate 1980s and early 1990s using photographic plates produced by a elescope in Australia. The plates were digitized by the Automated Plate Measurement (APM) facility in Cambridge, England. This APM survey resulted in a sky map showing the positions and magnitudes (or brightnesses) of about 3 million galaxies in a large volume of space, mainly concentrated on the southern sky. Revealed as small, bright patches on the map were numerous galaxy clusters, each containing hundreds of closely packed galaxies. Many of the patches were joined ogether to form filaments, or galaxy superclusters. Also visible on the map were many darker voids containing few or no galaxies.

The Two-degree Field (2dF) Red-shift Survey, performed between 997 and 2002, plotted the distribution of 200,000 galaxies in two regions of the local Universe. The distance to each galaxy was calculated from its red shift (see p.306). The result was a map of galaxy superclusters out to nearly 2 billion light-years from Earth.

The Two Micron All Sky Survey (2MASS), completed in 2003, only surveyed galaxies up to about 1.3 billion light-years away, but it covered the whole sky. The survey concentrated on detecting galaxies rom their infrared radiation (based around wavelengths of two microns) rather than their visible light (see p.11). As a result, various elliptical whole-sky maps were produced revealing some 1.5 million galaxies and numerous galaxy clusters and superclusters.

GALAXY MAP FROM THE SLOAN SURVEY

Maps from the Sloan Digital Sky Survey, such as the one below, look like radar images. The point from which the survey has been conducted (in this case Earth) is at the center, but dust in the Milky Way obscures the telescope's view of part of the sky. Each dot on the map represents a galaxy. The observed galaxies are plotted at distances from the center proportional to their remoteness from Earth.

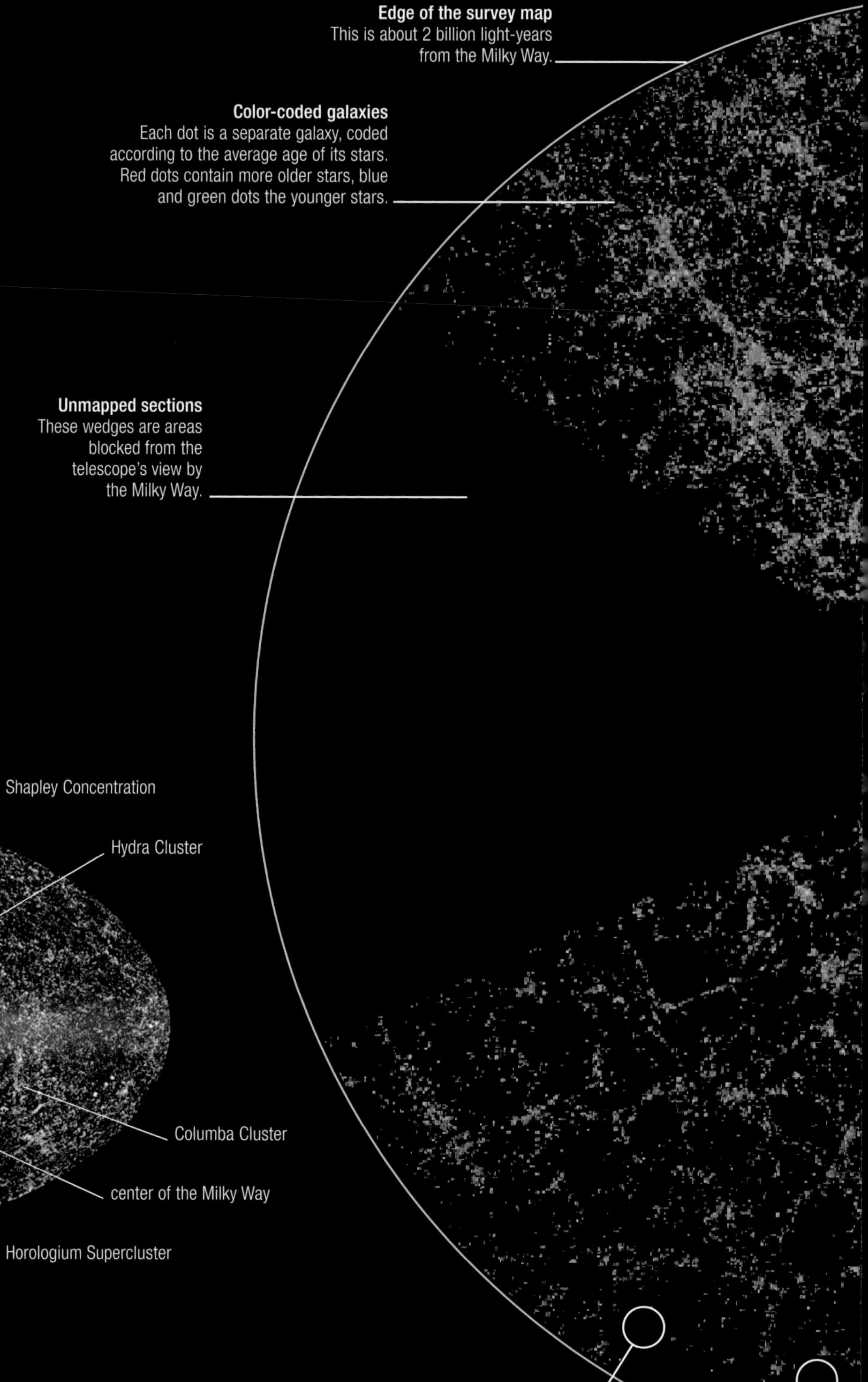

Edge of the survey map
This is about 2 billion light-years from the Milky Way.
Color-coded galaxies
Each dot is a separate galaxy, coded according to the average age of its stars. Red dots contain more older stars, blue and green dots the younger stars.
Unmapped sections
These wedges are areas blocked from the telescope's view by the Milky Way.
Voids in space

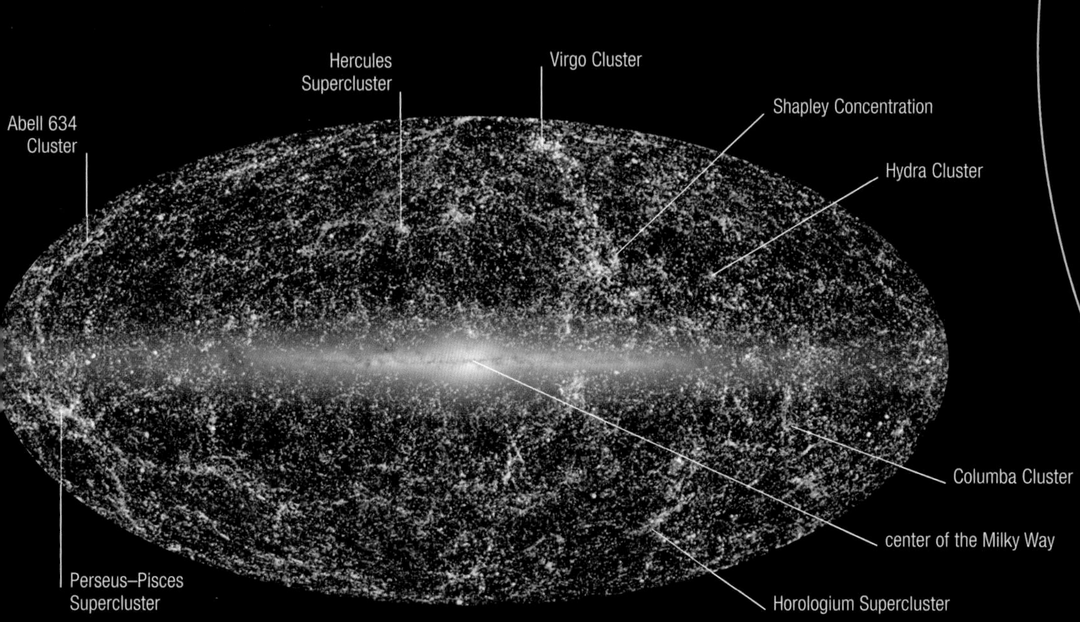

LOCAL GALAXY CLUSTERS AND SUPERCLUSTERS

n this all-sky infrared image from the Two Micron All Sky Survey, different galaxies are color coded by their red shift (see key, right). The tiny

RED SHIFT

REGION DETAILED BY GALAXY MAP

The galaxy map below covers two thin, wedge-shaped regions of space, still only representing a small fraction of the observable Universe.

Earth

edge of observable Universe—13.7 billion light-years away (look-back distance)

area depicted in survey galaxy map—diameter is a seventh of the observable Universe

THE SLOAN DIGITAL SKY SURVEY

The largest galaxy survey so far, begun in 2000, is the Sloan Digital Sky Survey (SDSS). This aims to map about a quarter of the sky. The average distance to the galaxies it has imaged is about 1.3 billion light-years, but some quasars (energetic galactic nuclei) have been surveyed out as far as 12 billion light-years. Maps have been created using data from about 800,000 galaxies and 100,000 quasars. Some maps are three-dimensional, since data is gathered from wedge-shaped volumes, rather than flat slices, of space. A major discovery has been the Sloan Great Wall, about 1 billion light-years from Earth.

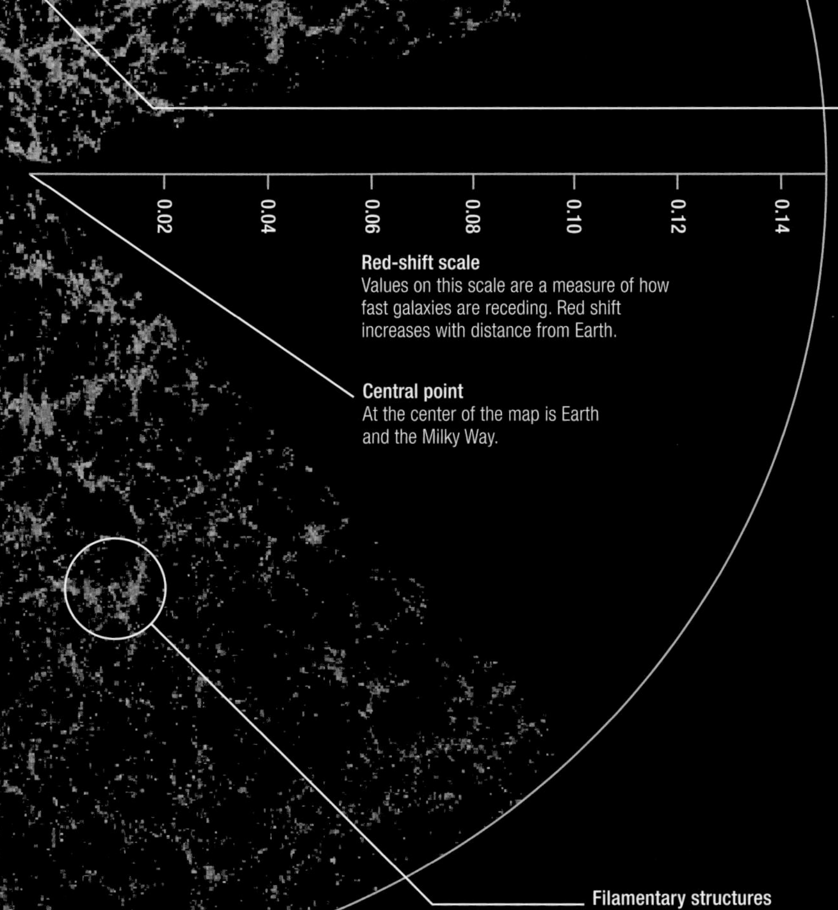

Sloan Great Wall
A giant hypercluster, this is the largest known structure in the Universe. It is 1 billion light-years across.

Part of the Shapley Concentration
Also known as the Shapley Supercluster, this is a huge group of about 25 clusters of galaxies.

Red-shift scale
Values on this scale are a measure of how fast galaxies are receding. Red shift increases with distance from Earth.

Central point
At the center of the map is Earth and the Milky Way.

Filamentary structures
These consist of strings of galaxy clusters that are only partially mapped.

SURVEY TELESCOPE
The Sloan Digital Sky Survey relies on this visible-light telescope at the Apache Point Observatory, New Mexico, which has a 8ft 2in (2.5m) mirror, linked to a 120-megapixel camera. The telescope images 1.5 square degrees of the sky at a time (about eight times the area of the Moon).

CENTER OF ABELL 3558
This bright galaxy cluster is at the center of the Shapley Concentration, or Supercluster, of galaxies. It is also known as Shapley 8. Near its center is a supergiant elliptical galaxy.

> “EVENTUALLY WE REACH THE DIM BOUNDARY ... THERE, WE MEASURE SHADOWS, AND WE SEARCH AMONG GHOSTLY ERRORS OF MEASUREMENT FOR LANDMARKS.”
>
> **EDWIN HUBBLE**, AMERICAN ASTRONOMER, THE REALM OF THE NEBULAE, 1936

THE FUTURE

Most of the observable Universe has yet to be mapped, and surveys are being developed that will concentrate, for example, on the distribution of quasars (see p.294). But the surveys to date have reached far enough into space to begin to see what cosmologists call "the End of Greatness." This means that at scales larger than about a billion light-years, the lumpiness of structures seen locally—superclusters, filaments, walls, and voids—is replaced by a more uniform distribution of matter, as predicted by the cosmological principle (see p.306). So, although much of the observable Universe is still to be surveyed, structures larger than those already known are unlikely to be found.

HOW MIGHT THE UNIVERSE END?

The Universe began with the Big Bang, but how will it end? Four competing possibilities, of varying likelihood, are a Big Crunch (recollapse), a Big Rip (the Universe is torn apart), or either of two types of Big Chill (a long, slow fade-out).

THE BIG CRUNCH

The fate of the Universe depends on the outcome of a struggle between two tendencies. One of these is its current expansion and a factor that seems to be encouraging that expansion—dark energy (see p.308). The other is the force of gravity, which opposes expansion and whose strength depends on the Universe's average mass-energy density. If gravity were eventually to win out in this struggle, the Universe would start contracting, eventually collapsing to an infinitely dense point (a black hole) in the outcome known as a Big Crunch. This scenario could happen only if space is positively curved—which means that its mass-energy density is relatively high (see p.310)—and if dark energy is weaker than currently believed, or if it reverses its effects in the future. Since none of these conditions is thought to apply, a Big Crunch currently looks a highly unlikely ending for the Universe. But some cosmologists have suggested that should this ever happen, it might trigger another Big Bang (a scenario called a Big Bounce), with a new Universe starting up and a new phase of expansion (see below).

BIG BANG AGAIN AND AGAIN
A highly speculative theory is that the Big Bang was preceded by a Big Crunch as part of a series of bangs and crunches, in what is called an oscillatory Universe. The massive inflation after the Big Bang is followed by a deceleration as gravity becomes dominant, causing a collapse to a Big Crunch, triggering a new Big Bang.

A Big Chill
Space might continue to expand at a rate that slowly decreases but never quite stops. This could occur if the Universe has a geometry close to flat (its mass-energy density is near a critical value) and if dark energy is really rather weak or its effects tail off. The Universe would suffer a cold fade-out lasting through eternity.

A modified Big Chill
If dark energy continues to behave as at present, the Universe would expand at a rate that speeds up, whatever its mass-energy density and curvature. Structures not held together by gravity would eventually fly apart, resulting in a cold fade-out lasting through eternity.

A Big Rip
If dark energy is sufficiently strong or becomes stronger, it might overwhelm the fundamental forces of nature, causing the Universe to disintegrate cataclysmically. Galaxies, then stars, planets, and finally atoms would be torn apart.

A Big Crunch
In this scenario, the Universe continues expanding for a while, then contracts back to a hot, infinitely dense point in tens of billions of years' time. This outcome is considered highly unlikely at present because of the space-expanding effect of dark energy.

FOUR ENDINGS
All four theoretical fates of the Universe—a Big Crunch, a Big Rip, a Big Chill, and a modified Big Chill—are shown in this conceptual interpretation, starting from the Big Bang. Currently the consensus is that a modified Big Chill is the most likely ending. But cosmologists continuously monitor new research on phenomena such as dark energy for evidence that this view might need to be changed.

THE BIG RIP

If dark energy is of a certain hypothetical form (called phantom energy), it will cause the Universe to expand at an ever-accelerating rate, whatever its mass-energy density and curvature might be. In this scenario, the Universe has a finite age, and a truly spectacular ending would occur in a few billion years' time. Matter will be ripped apart, with incredible rapidity in the later stages. The distance to the edge of the observable Universe will shrink as nearby regions of space begin moving away at the speed of light, the sky will darken to complete blackness, and atoms will be torn apart into elementary particles that will rush away from one another. This sounds dire, but it is not considered to be one of the more likely endings, given what is known about dark energy.

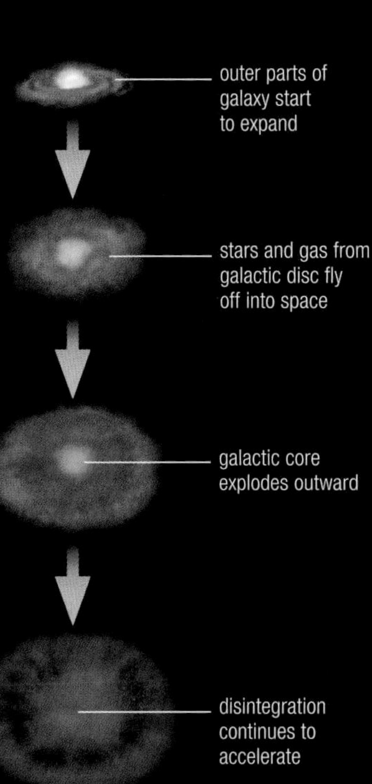
outer parts of galaxy start to expand
stars and gas from galactic disc fly off into space
galactic core explodes outward
disintegration continues to accelerate

GALAXIES TORN APART BY DARK ENERGY
The first objects to disintegrate in a Big Rip will be galaxy clusters, then galaxies themselves, in stages as shown here. Then stars and planets will fall apart, and finally atoms.

TIME

present day

BIG BANG

THE BIG CHILL

In this scenario, gravity is unable to reverse the Universe's expansion, but the expansion is much less extreme than in a Big Rip and goes on forever. A Big Chill is a more likely outcome if the Universe has a flat or negatively curved geometry (see p.309) and dark energy is anything from absent to moderately strong in its effects (as it seems to be). There are two varieties: a classic Big Chill in which expansion gradually slows, and a modified one whereby it speeds up (thought to be the most likely). Either way, the Universe undergoes a slow fade-out. Eventually, galaxies will exhaust their supply of gas to make new stars, and existing stars will degenerate to black dwarfs and black holes. Atoms disintegrate to elementary particles, and the temperature of the Universe will approach absolute zero.

LINGERING DEATH OF GALAXIES
In a Big Chill, a trillion years from now, the Universe would contain just old galaxies such as these. Most stars in the galaxies would have died or be fading away.

THE FUTURE OF DEEP-SPACE EXPLORATION

Many mysterious aspects of the Universe, from how galaxies form to the nature of dark energy, await resolution. New space observatories will help scientists try to answer such questions.

ADVANCES IN SPACE TELESCOPES

From the year 2014, the most advanced observatory in space will be the James Webb Space Telescope (JWST). One of the main purposes of this space telescope will be to search for infrared radiation and visible light from the very first stars and galaxies, including some that may have formed as little as 500 million years after the Big Bang. These objects are currently beyond the reach of either ground-based instruments or the Hubble Space Telescope. By obtaining images of these very faint and distant objects, along with more detailed images of somewhat closer objects, this new observatory is expected to provide fresh information about galaxy formation and the characteristics of the first stars.

TECHNICAL INNOVATIONS

Several innovative technologies have been developed for the JWST. For example, one of the infrared-detecting instruments in the telescope will be equipped with 62,000 microshutters—tiny cells that can be individually opened or closed to permit or block off radiation coming from small parts of the sky. This will help the instrument image extremely faint objects even if they are close to a much brighter object in the sky. Other innovations, such as the thermal management system and the sun shield, will keep the JWST's mirrors at extremely low temperatures, close to absolute zero. Without these, the faint infrared radiation (heat) from distant galaxies would be swamped by heat from the mirrors.

PRIMARY MIRROR
This is made up of 18 segments, each coated with gold to enhance infrared light reflection. The segments are tested in a cryogenic facility to ensure that they can withstand the cold.

Instrument module
Houses the science instruments: camera, spectrograph, mid-infrared instrument, and fine-guidance sensor

Momentum flap
Part of the system for controlling the telescope's orientation in space

thermal management system

solar panel

Spacecraft bus
Provides electrical power, attitude control, communication, command and data handling, propulsion, and thermal control

antenna for communications with Earth

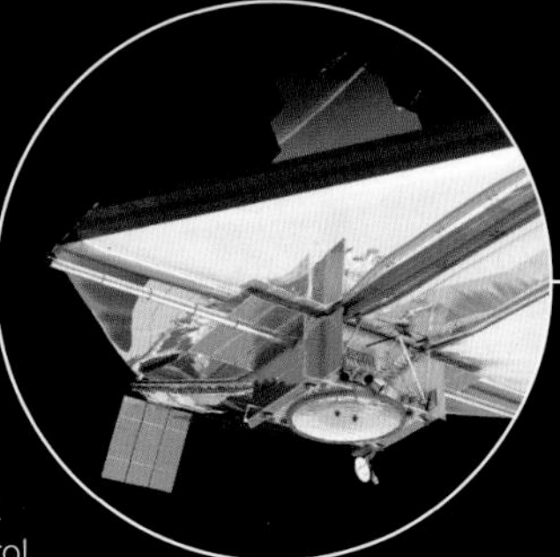

UNDERNEATH THE SUN SHIELD
The sun shield protects the telescope's mirrors from the Sun's heat. Beneath it, on the "warm" side of the observatory, are solar panels and instruments involved in functions such as data handling and communication with mission control.

COMPARING SPACE TELESCOPES

The JWST primary mirror will have a surface area almost six times larger than the Hubble Space Telescope's mirror, and its instruments will collect light about nine times faster. Ultra-lightweight materials such as beryllium have been incorporated into the design, so the mirror will weigh only one-tenth of Hubble's per unit area.

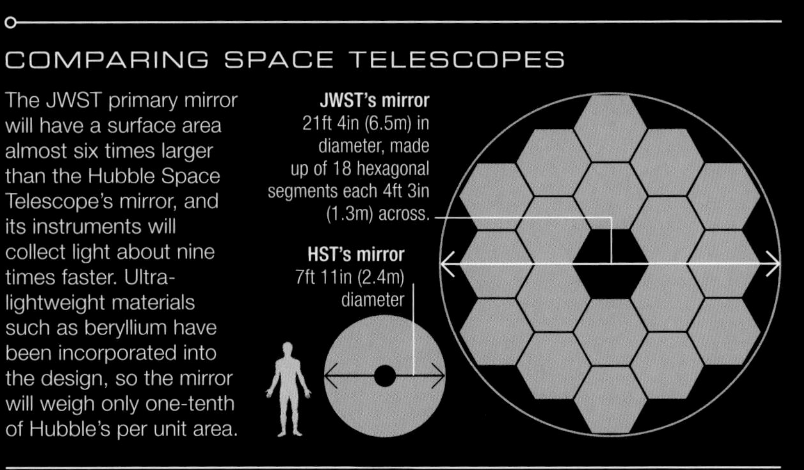

THE JWST ORBIT

The telescope will be much further out than the Hubble Space Telescope. It will be placed on the far side of the Earth from the Sun, 930,000 miles (1.5 million km) away, near Lagrangian point two (L2). From there, it will orbit the Sun at the same speed as Earth and will be shielded from both the Earth's and the Sun's radiation.

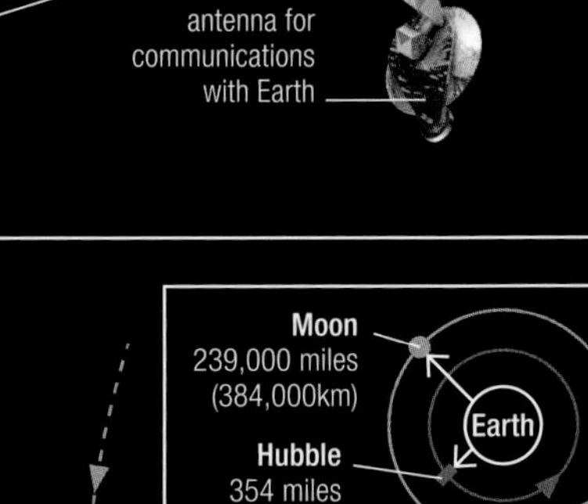

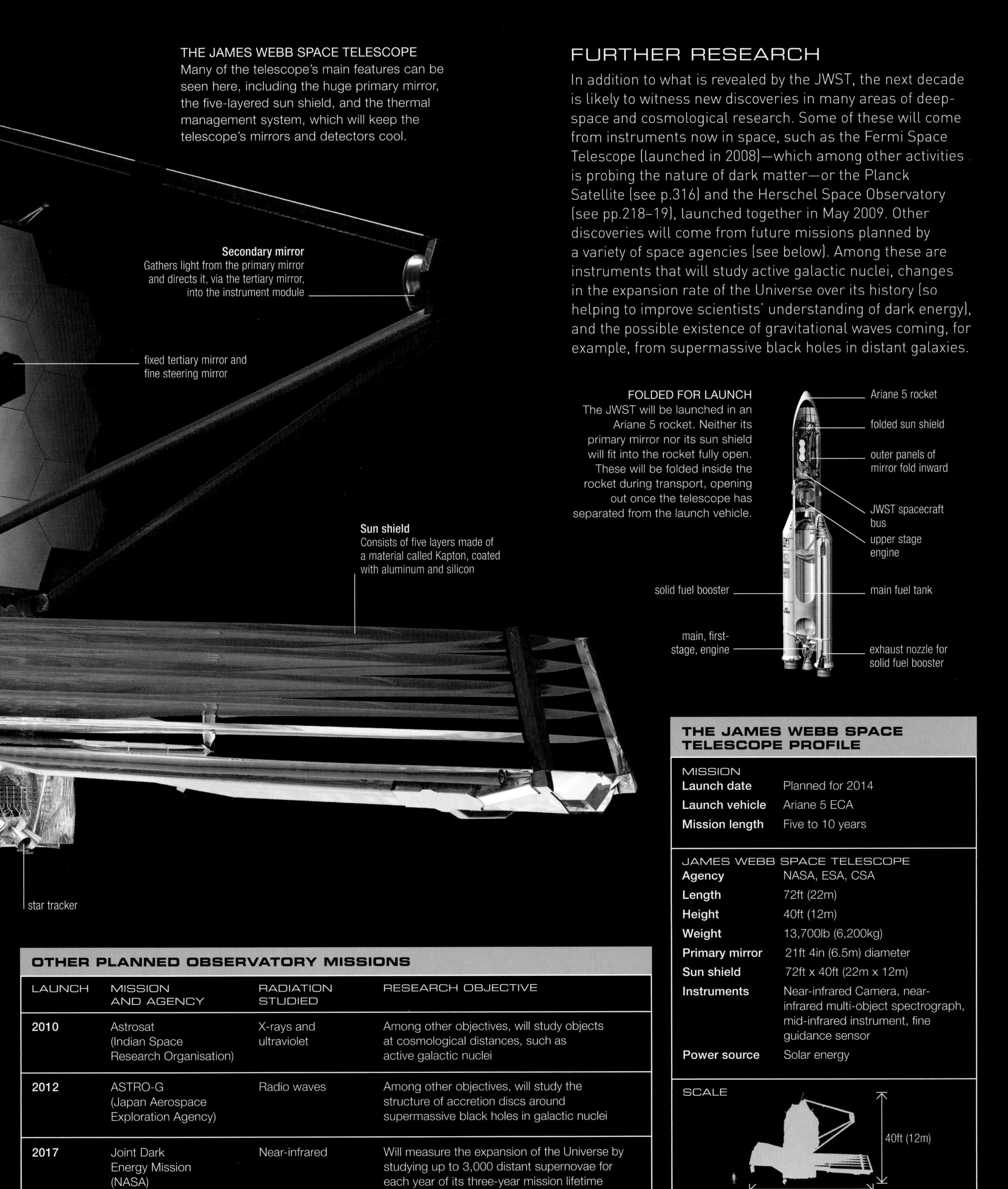

THE JAMES WEBB SPACE TELESCOPE
Many of the telescope's main features can be seen here, including the huge primary mirror, the five-layered sun shield, and the thermal management system, which will keep the telescope's mirrors and detectors cool.

FURTHER RESEARCH

In addition to what is revealed by the JWST, the next decade is likely to witness new discoveries in many areas of deep-space and cosmological research. Some of these will come from instruments now in space, such as the Fermi Space Telescope (launched in 2008)—which among other activities is probing the nature of dark matter—or the Planck Satellite (see p.316) and the Herschel Space Observatory (see pp.218–19), launched together in May 2009. Other discoveries will come from future missions planned by a variety of space agencies (see below). Among these are instruments that will study active galactic nuclei, changes in the expansion rate of the Universe over its history (so helping to improve scientists' understanding of dark energy), and the possible existence of gravitational waves coming, for example, from supermassive black holes in distant galaxies.

FOLDED FOR LAUNCH
The JWST will be launched in an Ariane 5 rocket. Neither its primary mirror nor its sun shield will fit into the rocket fully open. These will be folded inside the rocket during transport, opening out once the telescope has separated from the launch vehicle.

THE JAMES WEBB SPACE TELESCOPE PROFILE

MISSION	
Launch date	Planned for 2014
Launch vehicle	Ariane 5 ECA
Mission length	Five to 10 years
JAMES WEBB SPACE TELESCOPE	
Agency	NASA, ESA, CSA
Length	72ft (22m)
Height	40ft (12m)
Weight	13,700lb (6,200kg)
Primary mirror	21ft 4in (6.5m) diameter
Sun shield	72ft x 40ft (22m x 12m)
Instruments	Near-infrared Camera, near-infrared multi-object spectrograph, mid-infrared instrument, fine guidance sensor
Power source	Solar energy

SCALE

OTHER PLANNED OBSERVATORY MISSIONS

LAUNCH	MISSION AND AGENCY	RADIATION STUDIED	RESEARCH OBJECTIVE
2010	Astrosat (Indian Space Research Organisation)	X-rays and ultraviolet	Among other objectives, will study objects at cosmological distances, such as active galactic nuclei
2012	ASTRO-G (Japan Aerospace Exploration Agency)	Radio waves	Among other objectives, will study the structure of accretion discs around supermassive black holes in galactic nuclei
2017	Joint Dark Energy Mission (NASA)	Near-infrared	Will measure the expansion of the Universe by studying up to 3,000 distant supernovae for each year of its three-year mission lifetime

ORBIUM PLA
TERRAM COM
TIUM SCENO
SPHÆRA ZODIACI
SPHÆRA SATVRNI
SPHÆRA IOVIS
SPHÆRA MARTIS
SPHÆRA SOLIS
SPHÆRA VENERIS
LE O
SCOR PIVS
CIRCVLVS
CAPRI CORNVS
AQVA RIVS

7

REFERENCE

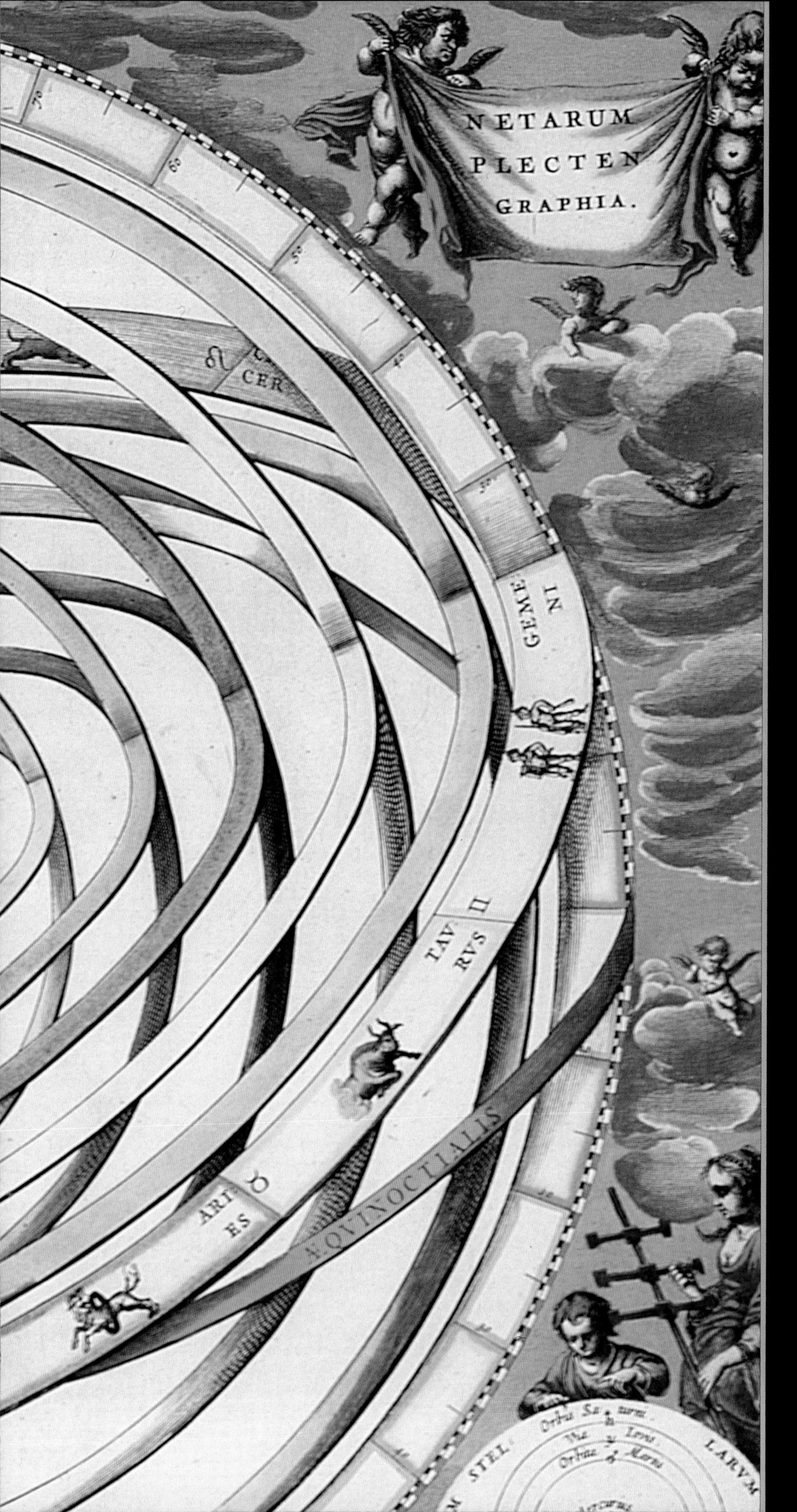

« Ptolemy's system of orbits
Taken from *The Celestial Atlas*, published in 1660–61

THE SOLAR SYSTEM

PLANETARY PHYSICAL CHARACTERISTICS

Eight major planets orbit the Sun. They range in size from tiny Mercury, roughly one-third the diameter of the Earth, to giant Jupiter, over 11 times wider than our home planet. The four planets closest to the Sun —Mercury, Venus, Earth, and Mars—are relatively small and rocky. They are termed the terrestrial planets. The outer four planets—Jupiter, Saturn, Uranus, and Neptune—are much larger and consist mostly of gas. The temperatures in this table refer to surface temperatures of the inner planets, and cloud-top temperatures of the outer planets.

Planet	Radius miles (km)		Mass Earth equals 1	Surface gravity Earth equals 1	Average temperature °F (°C)		Escape velocity mph (kph)		Number of moons
MERCURY	1,516	(2,440)	0.06	0.38	333	(167)	9,619	(15,480)	0
VENUS	3,761	(6,052)	0.82	0.91	867	(464)	23,175	(37,296)	0
EARTH	3,963	(6,378)	1.00	1.00	59	(15)	25,024	(40,270)	1
MARS	2,110	(3,396)	0.11	0.38	-81	(-63)	11,252	(18,108)	2
JUPITER	44,423	(71,492)	317.83	2.36	-162	(-108)	133,098	(214,200)	63
SATURN	37,449	(60,268)	95.16	0.92	-218	(-139)	79,411	(127,800)	62
URANUS	15,882	(25,559)	14.54	0.89	-323	(-197)	47,647	(76,680)	27
NEPTUNE	15,388	(24,764)	17.15	1.12	-328	(-201)	52,568	(84,600)	13

PLANETARY ORBITAL CHARACTERISTICS

These diagrams present orbital data for the eight major planets. Their orbital periods increase with distance from the Sun, from innermost Mercury, which speeds around the Sun every 88 days, to distant Neptune, which takes almost 165 years. All orbits are slightly eccentric (elliptical) in shape, so the planet's distance to the Sun varies—the closest point is known as perihelion, and the most distant is aphelion.

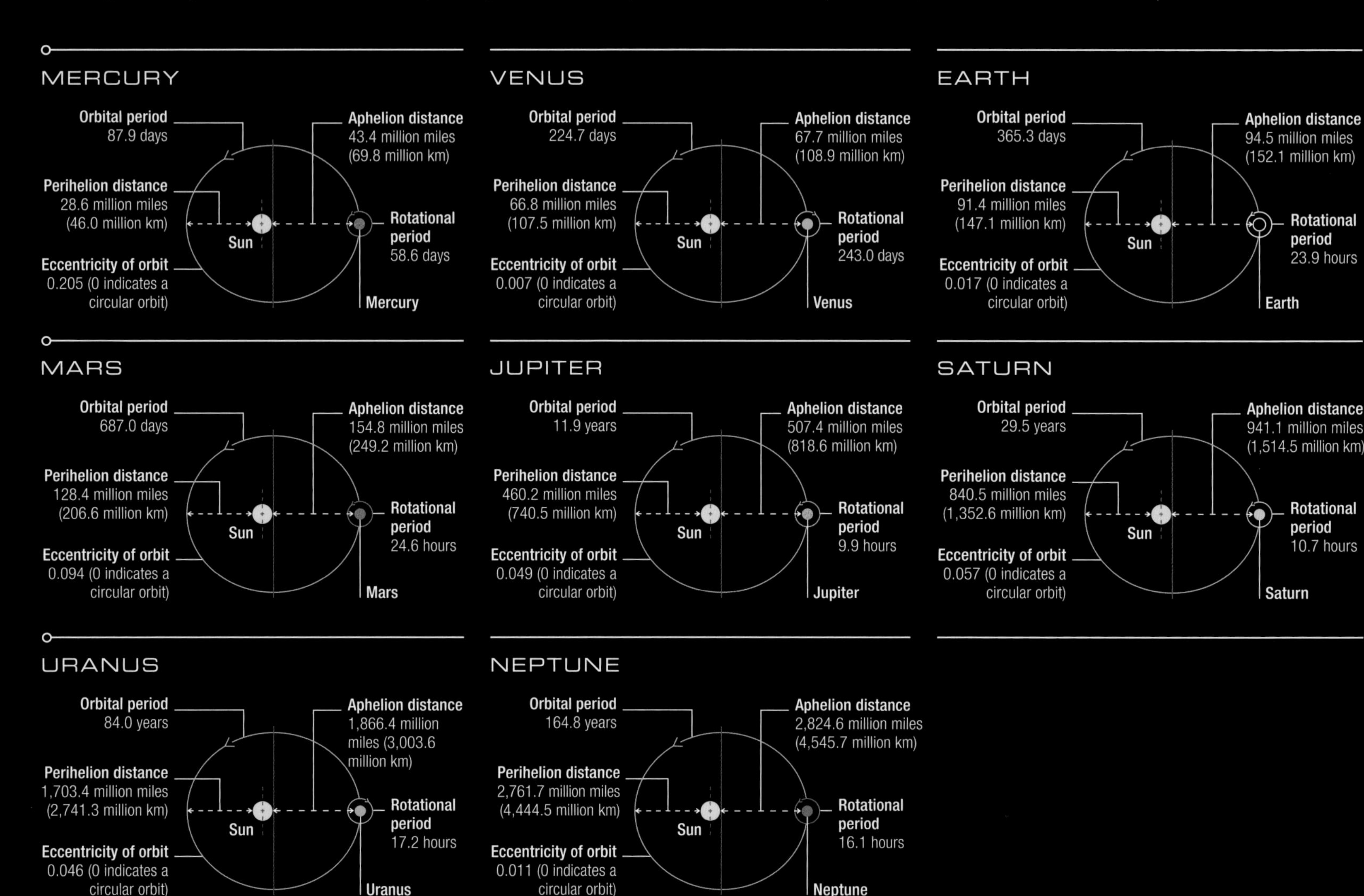

ATMOSPHERES OF THE PLANETS

All of the planets have gaseous atmospheres, although Mercury's is so thin as to be almost nonexistent. The atmospheres of Venus and Mars consist primarily of carbon dioxide. On Venus, the atmosphere is so dense that heat from the Sun has difficulty escaping, boosting the surface temperature to higher than in a kitchen oven. The composition of Earth's atmosphere is strongly affected by the existence of plant life, which breaks down carbon dioxide to release oxygen. The four outer planets are virtually all atmosphere. Their composition is predominantly the same as the Sun's—mostly hydrogen and helium gas, although in the planets' interiors, the gas is compressed into a liquid. Both Uranus and Neptune have a couple of percent of methane gas in their atmospheres, which gives them their bluish-green color.

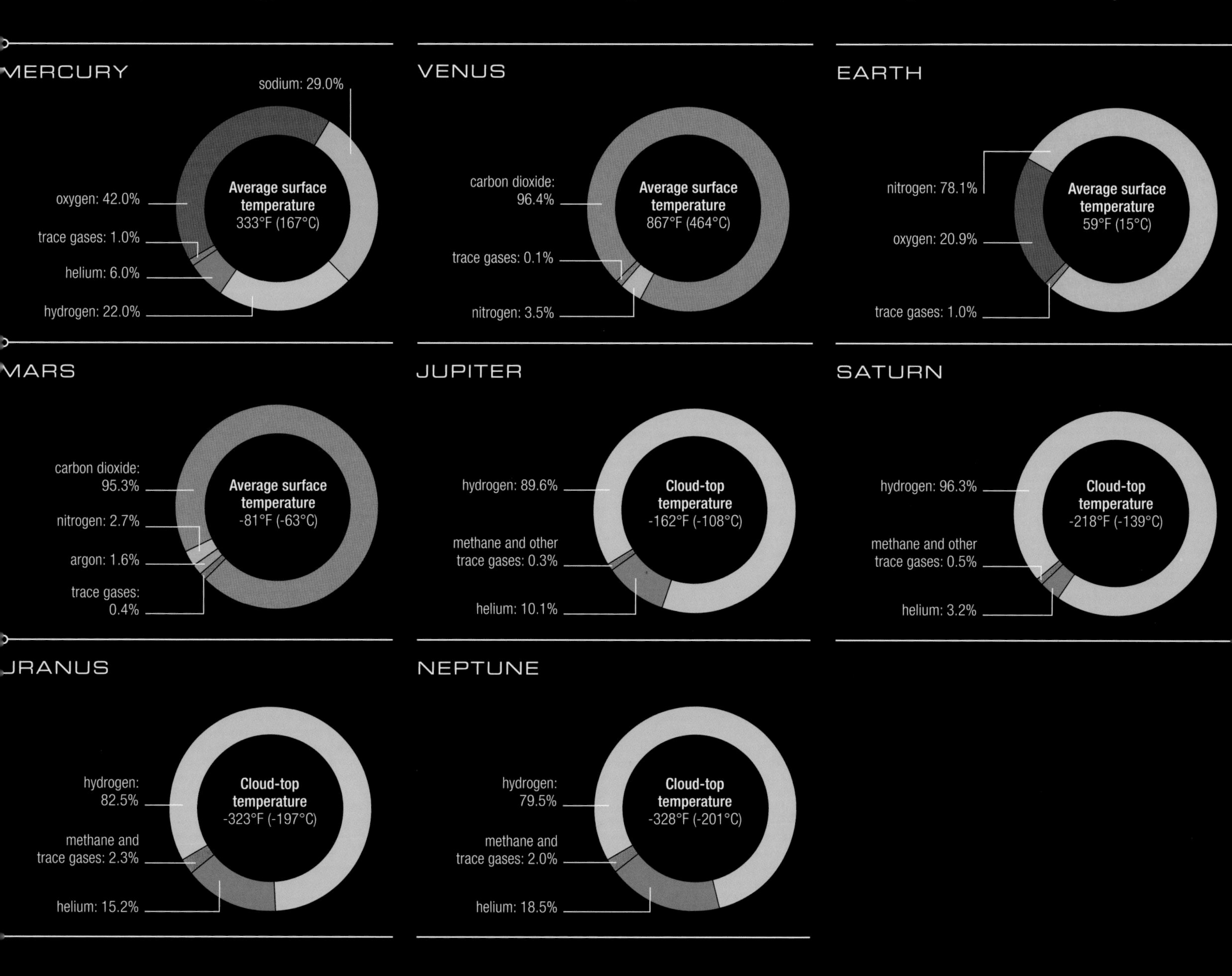

RINGS OF THE OUTER PLANETS

Rings of orbiting debris encircle the four outer planets. Only those around Saturn are bright enough to be visible through small telescopes. The rings of Jupiter, Uranus, and Neptune are very faint and can be seen only by spacecraft or at infrared wavelengths through large telescopes. Each system of rings is divided into several bands with intervening gaps. In this table, the width of the main structure refers to: Jupiter's main ring; the inner edge of Saturn's B ring to the outer edge of the A ring; the width of Uranus's Epsilon ring; and Neptune's Adams ring. The radius is the distance from the planet's center to the outermost edge of the main rings. The thickness is the depth from top to bottom.

Planet	Width of main structure miles (km)		Radius miles (km)		Thickness miles (km)		Particle size range ft (m)		Total mass Mt (1 million tonnes)
JUPITER	4,350	(7,000)	80,700	(129,900)	20–190	(30-300)	0.003	(0.001)	10,000
SATURN	28,000	(45,000)	85,100	(137,000)	0.02	(0.03)	0.03–30	(0.01–10)	100 billion
URANUS	60	(100)	31,800	(51,100)	0.006–0.6	(0.01–0.1)	0.7–70	(0.2–20)	1 million
NEPTUNE	30	(50)	39,100	(62,900)	30	(50)	Unknown		Unknown

Six of the eight major planets are accompanied by satellites; Mercury and Venus are the two without. The Earth has just one, known as the Moon (with a capital M). Jupiter and Saturn have more than 60 satellites each, although most of these are small and irregularly shaped. Jupiter's largest moon, Ganymede, is the biggest in the Solar System, measuring 3,270 miles (5,262km) in diameter, although Saturn's moon Titan is not far behind. Titan is unique in being the only planetary moon with a dense atmosphere. Jupiter's four largest moons are known as the Galileans, because they were discovered by Galileo in 1610. They are bright enough to be seen with binoculars.

Planet	Satellite name	Radius or dimensions miles (km)		Mass Mt (1 million tonnes)	Distance from planet miles (km)		Orbital period days	Year of discovery
EARTH	Moon	1,080	(1,738)	73.5 trillion	238,855	(384,400)	27.32	n/a
MARS	Phobos	8 x 7 x 8	(13 x 11 x 9)	10.6 million	5,827	(9,378)	0.32	1877
	Deimos	6 x 4 x 3	(8 x 6 x 5)	2.4 million	14,577	(23,459)	1.26	1877
JUPITER	Io	1,132	(1,822)	89 trillion	261,970	(421,600)	1.77	1610
	Europa	970	(1,561)	48 trillion	416,878	(670,900)	3.55	1610
	Ganymede	1,635	(2,631)	148 trillion	665,116	(1,070,400)	7.15	1610
	Callisto	1,498	(2,410)	108 trillion	1,169,855	(1,882,700)	16.69	1610
SATURN	Mimas	130 x 122 x 119	(209 x 196 x 191)	37.9 billion	115,277	(185,520)	0.94	1789
	Enceladus	159 x 154 x 152	(256 x 247 x 245)	108 billion	147,899	(238,020)	1.37	1789
	Tethys	333 x 328 x 327	(536 x 528 x 526)	618 billion	183,093	(294,660)	1.89	1684
	Dione	348	(560)	1.1 trillion	234,505	(377,400)	2.74	1684
	Rhea	475	(764)	2.3 trillion	327,487	(527,040)	4.52	1672
	Titan	1,598	(2,575)	135 trillion	759,210	(1,221,830)	15.95	1655
	Hyperion	115 x 87 x 70	(185 x 140 x 113)	5.5 billion	920,313	(1,481,100)	21.28	1848
	Iapetus	446	(718)	1.8 trillion	2,212,889	(3,561,300)	79.33	1671
URANUS	Miranda	149 x 145 x 145	(240 x 234 x 233)	66 billion	80,399	(129,390)	1.41	1948
	Ariel	361 x 359 x 359	(581 x 578 x 578)	1.4 trillion	118,694	(191,020)	2.52	1851
	Umbriel	363	(585)	1.2 trillion	165,471	(266,300)	4.14	1851
	Titania	490	(789)	3.5 trillion	270,862	(435,910)	8.71	1787
	Oberon	473	(761)	3.0 trillion	362,582	(583,520)	13.46	1787
NEPTUNE	Proteus	137 x 129 x 125	(220 x 208 x 202)	50 billion	73,102	(117,647)	1.12	1989
	Triton	841	(1,353)	21 trillion	220,438	(354,760)	5.87	1846
	Nereid	106	(170)	30 billion	3,425,868	(5,513,400)	360.14	1959

KUIPER BELT OBJECTS

Beyond Neptune orbits a swarm of bodies known as the Kuiper Belt (see p.178), an icy version of the Main Belt. The four largest of these bodies—Eris, Pluto, Makemake, and Haumea—are now classified as dwarf planets. as they are large enough for their own gravity to have shaped them into spheres. Other dwarf planets may await discovery in this region. The Kuiper Belt was not known about until the 1990s, and its existence was one reason why Pluto was demoted from planetary status. Absolute magnitude refers to the actual light-output of the bodies.

Name	Year of discovery	Diameter miles (km)		Orbital period years	Absolute magnitude	Discoverer(s)
PLUTO	1930	1,413	(2,274)	247.74	-0.7	Clyde Tombaugh, Flagstaff, AZ
CHARON	1978	730	(1,172)	Moon of Pluto	1.0	J. Christy, Washington
LOGOS	1997	50	(80)	302.26	6.55	Mauna Kea
CHAOS	1998	465	(745)	309.41	4.9	Deep Ecliptic Survey, Kitt Peak
BORASISI	1999	105	(170)	292.73	5.9	C. Trujillo, J.X. Luu, and D.C. Jewitt, Mauna Kea
DEUCALION	1999	130	(210)	292.41	6.6	Deep Ecliptic Survey, Kitt Peak
RHADAMANTHUS	1999	125	(200)	242.53	6.7	Deep Ecliptic Survey, Kitt Peak
VARUNA	2000	500	(800)	280.55	3.6	Spacewatch, Kitt Peak
ALTJIRA	2001	210	(340)	294.80	5.77	Deep Ecliptic Survey, Kitt Peak
IXION	2001	455	(730)	249.46	3.2	Deep Ecliptic Survey, Kitt Peak
TEHARONHIAWAKO	2001	110	(176)	294.77	5.60	Deep Ecliptic Survey, Kitt Peak
QUAOAR	2002	710	(1,140)	286.61	2.66	M. Brown and C. Trujillo, Palomar
ERIS	2003	1,660	(2,670)	550.90	-1.17	M. Brown, C. Trujillo, and D. Rabinowitz, Palomar
HAUMEA	2003	785	(1,265)	282.29	0.18	Sierra Nevada Observatory
SEDNA	2003	995	(1,600)	11,518.40	1.56	M. Brown, C. Trujillo, and D. Rabinowitz, Palomar
ORCUS	2005	565	(910)	245.11	2.3	T.-A. Suer and M. Brown, Palomar
MAKEMAKE	2005	930	(1,500)	305.45	-0.45	M. Brown, C. Trujillo, and D. Rabinowitz, Palomar

SIGNIFICANT ASTEROIDS

Between Mars and Jupiter orbits a band of rubble known as the Main Belt (see pp.126–27), consisting of debris from the formation of the inner planets. The largest member of the Main Belt, Ceres, is nearly 600 miles (1,000km) across and is now classified as a dwarf planet. In all, over 1 million asteroids larger than 0.6 miles (1km) are thought to exist in the Main Belt. Some asteroids stray from the belt and cross Earth's orbit, posing a potential hazard to our planet. The first such Earth-crossing asteroid to be discovered was Apollo, in 1932. Some asteroids are highly elongated, such as Eros, or even two closely aligned bodies named a double asteroid (Toutatis), and some have small moons (Ida).

Number and name	Dimensions miles (km)		Absolute magnitude	Rotation period days	Orbital period years	Average distance from the Sun million miles (million km)	
1 CERES	606 x 565 x 565	(975 x 909 x 909)	3.36	0.38	4.60	257	(414)
2 PALLAS	362 x 345 x 311	(582 x 556 x 500)	4.13	0.33	4.62	258	(415)
3 JUNO	199 x 166 x 124	(320 x 267 x 200)	5.33	0.30	4.37	248	(400)
4 VESTA	354 x 348 x 285	(570 x 560 x 458)	3.20	0.22	3.63	219	(353)
6 HEBE	127 x 115 x 106	(205 x 185 x 170)	5.71	0.30	3.78	226	(363)
10 HYGIEA	329 x 253 x 230	(530 x 407 x 370)	5.43	1.15	5.56	292	(470)
16 PSYCHE	149 x 115 x 90	(240 x 185 x 145)	5.90	0.18	4.99	271	(437)
31 EUPHROSYNE	159	(256)	6.74	0.23	5.59	293	(471)
48 DORIS	173 x 88	(278 x 142)	6.90	0.50	5.49	289	(465)
52 EUROPA	224 x 118 x 157	(360 x 302 x 252)	6.31	0.24	5.46	288	(464)
65 CYBELE	188 x 180 x 144	(302 x 290 x 232)	6.62	0.17	6.36	319	(514)
243 IDA	33.3 x 14.9 x 9.4	(53.6 x 24.0 x 15.2)	9.94	0.19	4.84	266	(428)
433 EROS	21.4 x 7.0 x 7.0	(34.4 x 11.2 x 11.2)	11.16	0.22	1.76	135	(218)
951 GASPRA	11.3 x 6.5 x 5.5	(18.2 x 10.5 x 8.9)	11.46	0.29	3.28	206	(331)
1862 APOLLO	1.1	(1.7)	16.25	0.13	1.78	135	(218)
4179 TOUTATIS	2.8 x 1.5 x 1.2	(4.5 x 2.4 x 1.9)	15.30	5.4–7.3	4.03	235	(379)
25143 ITOKAWA	0.3 x 0.2 x 0.1	(0.5 x 0.3 x 0.2)	19.2	0.51	1.52	123	(198)

GREAT COMETS 1900 TO PRESENT

Occasionally a comet becomes so bright that it is easily visible to the naked eye, and can remain so for weeks or months. Such great comets are mostly unpredictable, although Halley's Comet can become prominent if it passes close to Earth. Three spectacular comets have appeared in recent years: Hyakutake, Hale–Bopp, and McNaught, although McNaught was well seen only from the southern hemisphere.

Comet name	Perihelion passage	Perihelion distance million miles (million km)	
VISCARA	April 24, 1901	22.7	(36.6)
1P/HALLEY	April 20, 1910	54.6	(87.8)
GREAT JANUARY	January 17, 1910	12.0	(19.3)
SKJELLERUP–MARISTANY	December 18, 1927	16.4	(26.4)
AREND–ROLAND	April 8, 1957	29.4	(47.3)
SEKI–LINES	April 1, 1962	2.9	(4.7)
IKEYA–SEKI	October 21, 1965	0.7	(1.2)
BENNETT	March 20, 1970	50.0	(80.4)
WEST	February 25, 1976	18.3	(29.4)
HYAKUTAKE	May 1, 1996	21.4	(34.4)
HALE–BOPP	April 1, 1997	85.0	(136.8)
MCNAUGHT	January 12, 2007	15.8	(25.5)

SELECTED PERIODIC COMETS

Comets that have been seen on more than one orbit around the Sun are known as periodic comets (see p.181). The comet with the shortest orbital period is Comet Encke, which goes around the Sun every 3.3 years; however, it is now very faint. Halley's Comet is the brightest of the periodic comets. One periodic comet, Biela, broke up in the 19th century, and a shower of meteors appeared in its place.

Comet name	Orbital period years	Perihelion distance million miles (million km)	
2P/ENCKE	3.30	31.3	(50.4)
1P/HALLEY	75.32	54.5	(87.7)
3D/BIELA	6.65	81.7	(131.5)
6P/D'ARREST	6.54	125.8	(202.5)
9P/TEMPEL 1	5.52	140.3	(225.8)
19P/BORRELLY	6.85	125.9	(202.6)
81P/WILD 2	6.42	148.6	(239.1)
46P/WIRTANEN	5.44	98.3	(158.2)
7P/PONS–WINNECKE	6.36	116.5	(187.5)
26P/GRIGG–SKJELLERUP	5.31	103.8	(167.1)
29P/SCHWASSMANN–WACHMANN 1	14.65	531.5	(855.4)
21P/GIACOBINI–ZINNER	6.62	96.5	(155.3)
28P/NEUJMIN 1	18.17	144.3	(232.2)
50P/AREND	8.27	178.8	(287.8)
39P/OTERMA	19.53	506.7	(815.5)
32P/COMAS–SOLAR	8.80	175.1	(281.8)

THE MILKY WAY AND OTHER GALAXIES

THE SUN'S VITAL STATISTICS

Our Sun is an average star, important to us because it is so close. Light from the Sun takes 8.3 minutes to reach us, but over four years from the nearest nighttime star, Alpha Centauri (also known as Rigil Kentaurus). Stars are of different temperatures, which determines their color. The coolest stars appear reddish and the hottest ones bluish. The Sun, of medium temperature, appears yellow-white.

432,164mi
The radius of the Sun, which is approximately 109 times the radius of Earth.

332,900
If Earth is equal to one, this is the mass of the Sun. The Sun contains 99.7 percent of the Solar System's mass.

90lbs per cubic foot
The density of the Sun. It is 1.4 times the density of water and about one-quarter of Earth's average density. Unlike Earth, however, the Sun mainly consists of hydrogen and helium.

225,000,000 years
The time taken for the Sun, and with it the whole Solar System, to complete one orbit of the Milky Way.

4,600,000,000 years
The age of the Sun. It is roughly halfway through its life, and in 5 billion years, it will start the end of its life cycle.

92,957,130 miles
The average distance from Earth to the Sun. It changes slightly throughout the year. It is closest in January and furthest in July.

385 million billion gigawatts
The amount of light, or luminosity, emitted by the Sun. This figure can vary slightly with the occurence of sunspots.

9,941°F
The Sun's mean surface temperature. The corona is 3.6 million°F hotter, probably due to the magnetic field.

28,260,032°F
The average temperature at the very center of the Sun. It is here, in the core, where nuclear fusion-reactions occur.

25.05 days
The Sun's equatorial rotation period in Earth days. The polar rotation period is slower, taking 34.3 Earth days.

THE BRIGHTEST 20 STARS

The brightness that a star appears in our sky is known as its apparent magnitude. This depends on two factors: the star's actual light-output (termed its absolute magnitude) and its distance from us. This table lists the 20 brightest stars in our night sky, with their apparent and absolute magnitudes, distances, and spectral types. For a guide to their surface temperatures and colors, see pp.202–03.

Name	Constellation	Apparent magnitude	Absolute magnitude	Distance from Earth light-years	Spectral class
SUN		-26.78	4.82	0.000016	G2 V
SIRIUS	Canis Major	-1.44	1.45	8.6	A0 V
CANOPUS	Carina	-0.72	-5.53	313.0	FO Ia
RIGIL KENTAURUS	Centaurus	-0.28	4.07	4.4	G2 V
ARCTURUS	Boötes	-0.05	-0.31	36.7	K1 III
VEGA	Lyra	0.03 (variable)	0.58	25.3	A0 V
CAPELLA	Auriga	0.08	-0.48	42.2	G6 + G2 III
RIGEL	Orion	0.18 (variable)	-6.69	773.0	B8 Ia
PROCYON	Canis Minor	0.40	2.68	11.4	F5 IV
ACHERNAR	Eridanus	0.45	-2.77	144.0	B3 V
BETELGEUSE	Orion	0.45 (variable)	-5.14	427.0	M1 Ia
HADAR	Centaurus	0.61 (variable)	-5.42	525.0	B1 III
ALTAIR	Aquila	0.76 (variable)	2.20	16.8	A7 V
ACRUX	Crux	0.77	-4.19	321.0	B1 V
ALDEBARAN	Taurus	0.87	-0.63	65.1	K5 III
SPICA	Virgo	0.98 (variable)	-3.55	262.0	B1 III
ANTARES	Scorpius	1.05 (variable)	[illegible]	[illegible]	[illegible]

THE CLOSEST STARS AND GROUPS

Stars are widely separated by many light-years, but some come in families of two, three, or more. The closest star system to the Sun consists of two bright stars, Alpha Centauri A and B, and a fainter red dwarf, Proxima Centauri. Sirius, the brightest star, and Procyon both have white dwarf companions. Most of the other nearby stars are red dwarfs that cannot be seen without a telescope.

Name	Group	Component stars	Apparent magnitude	Absolute magnitude	Distance from Earth light-years	Spectral class
SUN	Single		-26.78	4.82	0.000016	G2 V
ALPHA CENTAURI	Triple	Proxima	11.09	15.53	4.2421	M5 V
		Alpha Cen A	0.01	4.38	4.3650	G2 V
		Alpha Cen B	1.34	5.71	4.3650	K0 V
BARNARD'S STAR	Single		9.53	13.22	5.9630	M4 V
WOLF 359	Single		13.44	16.55	7.7825	M6 V
LALANDE 21185	Single		7.47	10.44	8.2904	M2 V
SIRIUS	Double	Alpha CMa A	-1.43	1.47	8.5828	A1 V
		Alpha CMa B	8.44	11.34	8.5828	D A2
LUYTEN 726-8	Double	BL Ceti	12.54	15.40	8.7279	M5.5 V
		UV Ceti	12.99	15.85	8.7279	M6 V
ROSS 154	Single		10.43	13.07	9.6813	M3.5 V
ROSS 248	Single		12.29	14.79	10.3216	M5.5 V
EPSILON ERIDANI	Single		3.73	6.19	10.5217	K2 V
LACAILLE 9352	Single		7.34	9.75	10.7418	M1.5 V
ROSS 128	Single		11.13	13.51	10.9187	M4 V
EZ AQUARII	Triple	EZ Aqu A	13.33	15.64	11.2664	M5 V
		EZ Aqu B	13.27	15.58	11.2664	M5 V
		EZ Aqu C	14.03	16.34	11.2664	M5 V
PROCYON	Double	Alpha CMi A	0.38	2.66	11.4023	F5 V
		Alpha CMi B	10.70	12.98	11.4023	DA
61 CYGNI	Double	61 Cyg A	5.21	7.49	11.4027	K5 V
		61 Cyg B	6.03	8.31	11.4027	K7 V

EXTRASOLAR PLANETARY SYSTEMS

One of the most exciting areas of research in astronomy is the search for planetary systems around other stars. The first planets to be found were mostly large, like the giant planets in our own Solar System, but as techniques improve, it is expected that smaller planets like Earth will be discovered as well. One day, it will be possible to analyze their atmospheres for signs of life.

Name	Lower limit of planet mass Jupiter is equal to one	Orbital period days	Orbital eccentricity 0 indicates a circular orbit	Year of discovery
51 PEGASI b	0.47	4.2	0.01	1995
16 CYGNI B b	1.68	798.5	0.68	1996
55 CANCRI e	0.03	2.8	0.07	2004
b	0.82	14.7	0.01	1996
c	0.17	44.3	0.09	2002
f	0.11	260.0	0.20	2007
d	3.84	5,218.0	0.03	2002
UPSILON ANDROMEDAE b	0.69	4.6	0.01	1996
c	1.92	241.3	0.22	1999
d	10.29	1,302.1	0.32	1999
47 URSAE MAJORIS b	2.53	1,078.0	0.03	1996
c	0.54	2,391.0	0.10	2001
d	1.64	14,002.0	0.16	2010
70 VIRGINIS b	7.49	116.7	0.40	1996
23 LIBRAE b	1.59	258.2	0.23	1999
c	0.82	5,000.0	0.12	2009
PI MENSAE	10.31	2,151.0	0.64	2001
61 VIRGINIS b	0.02	4.2	0.12	2009
c	0.06	38.0	0.14	2009
d	0.07	123.0	0.35	2009

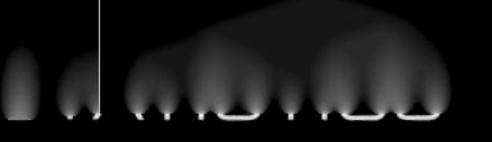

Our Sun is but one star in a huge stellar spiral popularly known as the Milky Way, or simply the Galaxy (with a capital G). The name Milky Way comes from the dim band of light seen crossing the sky on dark nights. Through a telescope, this band breaks up into countless faint stars. All of these stars are members of the Galaxy and the Milky Way band is in fact the disc of the Galaxy seen from within.

200,000,000,000
The number of stars in the Milky Way. The exact number is uncertain because we do not know what portion of the Milky Way's mass is in the form of stars.

100,000
The diameter of the Milky Way in light-years. The Sun is 26,000 light-years from its center.

2,000
The thickness of the galactic disc of stars in light-years. The surrounding gas makes it at least 6,000 light-years thick.

4,100,000 solar masses
The mass of the black hole, known as Sagittarius A*, at the center of the Milky Way. It is about 15 million miles (24 million km) across.

13,200,000,000 years
The age of the Milky Way, in Earth years. Its age is determined by measuring the age of the oldest stars in the Galaxy.

580,000,000,000
The mass of the Milky Way in solar masses. Most of the mass is dark matter that can only be detected through its gravitational force.

180
The number of globular clusters in the Milky Way. Some galaxies have thousands of such clusters.

25,000
The distance in light-years to the Canis Major Dwarf Galaxy, a galactic remnant being absorbed by the Milky Way.

GALAXY CLUSTERS

Galaxies huddle together in clusters, from small groups of a few dozen members, such as the Local Group in which we live (opposite), to those containing thousands of galaxies. The largest nearby cluster is the Virgo Cluster, nearly 60 million light-years away, which spreads across the sky in the constellations Virgo and Coma Berenices. In large clusters, the brightest galaxies are supergiant ellipticals. These are thought to have been formed by the merger of several smaller galaxies. The brightest galaxy in the Virgo Cluster is the supergiant elliptical known as M87, which is also a powerful radio source and is emitting a bright jet of gas.

Name	Distance millions of light-years	Additional information
LOCAL GROUP	0	Five galaxies are visible to the naked eye: M31, M33, the Milky Way, the Large Magellanic Cloud, and the Small Magellanic Cloud.
M81 GROUP	11	Bode's Galaxy (M81) may be glimpsed with the naked eye under ideal conditions. The group also contains M82.
CENTAURUS A GROUP	12	Contains M83, NGC 5128, and NGC 5253.
SCULPTOR GROUP	12.7	Its brightest member is the Sulptor Galaxy (NGC 253).
CANES VENATICI I CLOUD	13	Contains M64 and M94.
CANES VENATICI II GROUP	26	Contains M106.
M51 GROUP	31	Contains M51 and M63.
LEO TRIPLET	35	Contains M65, M66, and NGC 3628. May be physically related to Leo I Group.
LEO I GROUP	38	Contains M95, M96, and M105.
DRACO GROUP	40	Contains NGC 5866, NGC 5907, and NGC 5879.
URSA MAJOR CLOUD	55	Contains M108 and M109, and over 80 other galaxies.
VIRGO CLUSTER	59	The core of our galaxy supercluster, containing 16 Messier objects (see pp.338–39), and 2,000 other galaxies.

THE LOCAL GROUP OF GALAXIES

Our home galaxy, the Milky Way, is part of a small cluster of about 40 galaxies known as the Local Group (see p.267). The two largest members of the Local Group are the Milky Way and the Andromeda Galaxy (M31), a great spiral galaxy that can be seen with the naked eye on clear nights. The third-largest member of the Local Group, a spiral galaxy in Triangulum (M33), can be seen with binoculars. Two satellite galaxies of the Milky Way, the Magellanic clouds, appear like detached portions of the Galaxy in southern skies. The other member galaxies of the Local Group are small and faint. Most are satellites of either the Milky Way or the Andromeda Galaxy.

Name	Type	Visual magnitude	Distance from Solar System light-years	Diameter light-years	Year of discovery
MILKY WAY	Barred spiral		0	100,000	Prehistory
SAGITTARIUS DWARF	Dwarf spheroidal/elliptical	15.5	78,000	20,000	1994
URSA MAJOR II	Dwarf spheroidal	14.3	100,000	1,000	2006
LARGE MAGELLANIC CLOUD	Irregular	0.1	165,000	25,000	Prehistory
SMALL MAGELLANIC CLOUD	Irregular	2.3	195,000	15,000	Prehistory
BOÖTES DWARF	Dwarf spheroidal	13.6	197,000	2,000	2006
URSA MINOR DWARF	Dwarf spheroidal	10.9	215,000	2,000	1954
SCULPTOR DWARF	Dwarf spheroidal	10.5	258,000	3,000	1937
DRACO DWARF	Dwarf spheroidal	9.8	267,000	2,000	1954
SEXTANS DWARF	Dwarf spheroidal	12.0	280,000	3,000	1990
URSA MAJOR I	Dwarf spheroidal	Unknown	325,000	3,000	2005
CARINA DWARF	Dwarf spheroidal	20.9	329,000	2,000	1977
FORNAX DWARF	Dwarf spheroidal	8.1	450,000	5,000	1938
LEO II	Dwarf spheroidal	12.6	669,000	3,000	1950
LEO I	Dwarf spheroidal	9.8	815,000	3,000	1950
PHOENIX DWARF	Irregular	13.1	1,450,000	2,000	1976
NGC 6822	Irregular	9.3	1,520,000	8,000	1884
NGC 185	Dwarf elliptical	9.2	2,010,000	8,000	1787
ANDROMEDA II	Dwarf spheroidal	13.5	2,165,000	3,000	1970
LEO A	Irregular	12.9	2,250,000	4,000	*c*1940
IC 1613	Irregular	9.2	2,365,000	10,000	*c*1890
NGC 147	Dwarf elliptical	9.5	2,370,000	10,000	*c*1830
ANDROMEDA III	Dwarf spheroidal	13.5	2,450,000	3,000	1970
CETUS DWARF	Dwarf spheroidal	14.4	2,485,000	3,000	1999
ANDROMEDA I	Dwarf spheroidal	13.2	2,520,000	2,000	1970
LGS 3	Irregular	15.4	2,520,000	2,000	1978
ANDROMEDA GALAXY (M31)	Barred spiral	3.4	2,560,000	140,000	*c*964
M32	Dwarf elliptical	8.1	2,625,000	8,000	1749
M110	Dwarf elliptical	8.5	2,690,000	15,000	1773
IC 10	Irregular	10.3	2,690,000	8,000	1889
TRIANGULUM GALAXY (M33)	Spiral	5.7	2,735,000	55,000	1654
TUCANA DWARF	Dwarf spheroidal	15.7	2,870,000	2,000	1990
PEGASUS DWARF	Irregular	13.2	3,000,000	6,000	*c*1950
WLM	Irregular	10.9	3,020,000	10,000	1909
AQUARIUS DWARF	Irregular	13.9	3,345,000	3,000	1959
SAGDIG	Irregular	15.5	3,460,000	3,000	1977
ANTLIA DWARF	Dwarf elliptical	14.8	4,030,000	3,000	1997
NGC 3109	Irregular	10.4	4,075,000	25,000	*c*1836
SEXTANS A	Irregular	11.9	4,350,000	10,000	*c*1942
SEXTANS B	Irregular	11.8	4,385,000	8,000	*c*1942

MESSIER OBJECTS

In 1758, Charles Messier, a French astronomer, began to compile a list of objects that could be mistaken for comets when seen through a small telescope. When his final list was published in 1781, Messier and his colleagues had logged over 100 objects. Later astronomers have added to this tally, creating the final Messier catalog of 109 objects shown below. M102 is missing, as it is regarded as a duplicate of M101.

Messier number	New General Catalog (NGC) number	Constellation	Type (popular name)
1	1952	Taurus	Supernova remnant (Crab Nebula)
2	7089	Aquarius	Globular cluster
3	5272	Canes Venatici	Globular cluster
4	6121	Scorpius	Globular cluster
5	5904	Serpens	Globular cluster
6	6405	Scorpius	Open cluster (Butterfly Cluster)
7	6475	Scorpius	Open cluster (Ptolemy Cluster)
8	6523	Sagittarius	Emission nebula (Lagoon Nebula)
9	6333	Ophiuchus	Globular cluster
10	6254	Ophiuchus	Globular cluster
11	6705	Scutum	Open cluster
12	6218	Ophiuchus	Globular cluster
13	6205	Hercules	Globular cluster
14	6204	Ophiuchus	Globular cluster
15	7078	Pegasus	Globular cluster
16	6611	Serpens	Emission nebula (Eagle Nebula)
17	6618	Sagittarius	Emission nebula
18	6613	Sagittarius	Open cluster
19	6273	Ophiuchus	Globular cluster
20	6514	Sagittarius	Emission nebula (Trifid Nebula)
21	6531	Sagittarius	Open cluster
22	6656	Sagittarius	Globular cluster
23	6494	Sagittarius	Open cluster
24	6604	Sagittarius	Star cloud
25	IC 4725	Sagittarius	Open cluster
26	6694	Scutum	Open cluster
27	6853	Vulpecula	Planetary nebula (Dumbbell Nebula)
28	6626	Sagittarius	Globular cluster
29	6913	Cygnus	Open cluster
30	7099	Capricornus	Globular cluster
31	224	Andromeda	Spiral galaxy
32	221	Andromeda	Dwarf elliptical galaxy
33	598	Triangulum	Spiral galaxy
34	1039	Perseus	Open cluster
35	2168	Gemini	Open cluster
36	1960	Auriga	Open cluster
37	2099	Auriga	Open cluster
38	1912	Auriga	Open cluster
39	7092	Cygnus	Open cluster
40		Ursa Major	Double star
41	2287	Canis Major	Open cluster
42	1976	Orion	Emission nebula (Orion Nebula)

Messier number	New General Catalog (NGC) number	Constellation	Type (popular name)
43	1982	Orion	Emission nebula (De Mairan's Nebula)
44	2632	Cancer	Open cluster (Beehive Cluster)
45		Taurus	Open cluster (Pleiades)
46	2437	Puppis	Open cluster
47	4222	Puppis	Open cluster
48	2548	Hydra	Open cluster
49	4472	Virgo	Elliptical galaxy
50	2323	Monoceros	Open cluster
51	5194	Canes Venatici	Spiral galaxy (Whirlpool Galaxy)
52	7654	Cassiopeia	Open cluster
53	5024	Coma Berenices	Globular cluster
54	6715	Sagittarius	Globular cluster
55	6809	Sagittarius	Globular cluster
56	6779	Lyra	Globular cluster
57	6720	Lyra	Planetary nebula (Ring Nebula)
58	4579	Virgo	Barred spiral galaxy
59	4621	Virgo	Elliptical galaxy
60	4649	Virgo	Elliptical galaxy
61	4303	Virgo	Spiral galaxy
62	6266	Ophiuchus	Globular cluster
63	5055	Canes Venatici	Spiral galaxy (Sunflower Galaxy)
64	4826	Coma Berenices	Spiral galaxy (Black Eye Galaxy)
65	3623	Leo	Barred spiral galaxy
66	3627	Leo	Barred spiral galaxy
67	2682	Cancer	Open cluster
68	4590	Hydra	Globular cluster
69	6637	Sagittarius	Globular cluster
70	6681	Sagittarius	Globular cluster
71	6838	Sagitta	Globular cluster
72	6981	Aquarius	Globular cluster
73	6994	Aquarius	Star group
74	628	Pisces	Spiral galaxy
75	6864	Sagittarius	Globular cluster
76	650/651	Perseus	Planetary nebula (Little Dumbbell Nebula)
77	1068	Cetus	Spiral galaxy
78	2068	Orion	Diffuse nebula
79	1904	Lepus	Globular cluster
80	6093	Scorpius	Globular cluster
81	3031	Ursa Major	Spiral galaxy (Bode's Galaxy)
82	3034	Ursa Major	Barred spiral galaxy (Cigar Galaxy)

essier mber	New General Catalog (NGC) number	Constellation	Type (popular name)
	5236	Hydra	Barred spiral galaxy (Southern Pinwheel Galaxy)
	4374	Virgo	Lenticular galaxy
	4382	Coma Berenices	Lenticular galaxy
	4406	Virgo	Lenticular galaxy
	4486	Virgo	Elliptical galaxy
	4501	Coma Berenices	Spiral galaxy
	4552	Virgo	Elliptical galaxy
	4569	Virgo	Spiral galaxy
	4548	Coma Berenices	Barred spiral galaxy
	6341	Hercules	Spiral galaxy
	2447	Puppis	Open cluster
	4736	Canes Venatici	Spiral galaxy
	3351	Leo	Barred spiral galaxy

Messier number	New General Catalog (NGC) number	Constellation	Type (popular name)
96	3368	Leo	Spiral galaxy
97	3587	Ursa Major	Planetary nebula (Owl Nebula)
98	4192	Coma Berenices	Spiral galaxy
99	4254	Coma Berenices	Spiral galaxy
100	4321	Coma Berenices	Spiral galaxy
101	5457	Ursa Major	Spiral galaxy
103	581	Cassiopeia	Open cluster
104	4594	Virgo	Spiral galaxy
105	3379	Leo	Elliptical galaxy
106	4258	Canes Venatici	Spiral galaxy
107	6171	Ophiuchus	Globular cluster
108	3556	Ursa Major	Spiral galaxy
109	3992	Ursa Major	Barred spiral galaxy
110	205	Andromeda	Dwarf elliptical galaxy

HE CONSTELLATIONS

tronomers divide the night sky into 88 regions called constellations, hich fit together like the pieces of a celestial jigsaw. Of these, 48 have me down to us from the ancient Greeks. Gaps between the ancient eek constellations, and the area around the south celestial pole that was invisible from ancient Greece, were filled in by European astronomers from the late 16th century to the mid-18th century. The list of 88 constellations shown here was officially adopted by the International Astronomical Union, astronomy's governing body, in 1922

ame	Abbreviation	Rank in size order
NDROMEDA	And	19
NTLIA	Ant	62
PUS	Aps	67
QUARIUS	Aqr	10
QUILA	Aql	22
RA	Ara	63
RIES	Ari	39
URIGA	Aur	21
OÖTES	Boö	13
AELUM	Cae	81
AMELOPARDALIS	Cam	18
ANCER	Can	31
ANES VENATICI	CVn	38
ANIS MAJOR	CMa	43
ANIS MINOR	CMi	71
APRICORNUS	Cap	40
ARINA	Car	34
ASSIOPEIA	Cas	25
ENTAURUS	Cen	9
EPHEUS	Cep	27
ETUS	Cet	4
HAMAELEON	Cha	79
RCINUS	Cir	85
OLUMBA	Col	54
OMA BERENICES	Com	42
ORONA AUSTRALIS	CrA	80
ORONA BOREALIS	CrB	73
ORVUS	Crv	70
RATER	Crt	53
RUX	Cru	88

Name	Abbreviation	Rank in size order
CYGNUS	Cyg	16
DELPHINUS	Del	69
DORADO	Dor	72
DRACO	Dra	8
EQUULEUS	Equ	87
ERIDANUS	Eri	6
FORNAX	For	41
GEMINI	Gem	30
GRUS	Gru	45
HERCULES	Her	5
HOROLOGIUM	Hor	58
HYDRA	Hya	1
HYDRUS	Hyi	61
INDUS	Ind	49
LACERTA	Lac	68
LEO	Leo	12
LEO MINOR	LMi	64
LEPUS	Lep	51
LIBRA	Lib	29
LUPUS	Lup	46
LYNX	Lyn	28
LYRA	Lyr	52
MENSA	Men	75
MICROSCOPIUM	Mic	66
MONOCEROS	Mon	35
MUSCA	Mus	77
NORMA	Nor	74
OCTANS	Oct	50
OPHIUCHUS	Oph	11
ORION	Ori	26

Name	Abbreviation	Rank in size order
PAVO	Pav	44
PEGASUS	Peg	7
PERSEUS	Per	24
PHOENIX	Phe	37
PICTOR	Pic	59
PISCES	Psc	14
PISCIS AUSTRINUS	PsA	60
PUPPIS	Pup	20
PYXIS	Pyx	65
RETICULUM	Ret	82
SAGITTA	Sge	86
SAGITTARIUS	Sgr	15
SCORPIUS	Sco	33
SCULPTOR	Scl	36
SCUTUM	Sct	84
SERPENS	Ser	23
SEXTANS	Sex	47
TAURUS	Tau	17
TELESCOPIUM	Tel	57
TRIANGULUM	Tri	78
TRIANGULUM AUSTRALE	TrA	83
TUCANA	Tuc	48
URSA MAJOR	UMa	3
URSA MINOR	UMi	56
VELA	Vel	32
VIRGO	Vir	2
VOLANS	Vol	76
VULPECULA	Vul	55

ince Yuri Gagarin of the Soviet Union became the first man in space in 961, over 500 humans have orbited the Earth. Some of the American pollo astronauts went even further, to the Moon. More recently, representatives from Europe, China, and Japan have flown into space. In years to come, it will be possible for private citizens to take short trips into space as fare-paying passengers on commercial flights.

Name	Date	Craft	Nation	Additional information
YURI GAGARIN	April 12, 1961	Vostok 1	Soviet Union	The first man in space, and the first astronaut to orbit Earth.
ALAN SHEPARD	May 5, 1961	Freedom 7	USA	The first American in space, the second man in space, and the fifth person to walk on the Moon.
GHERMAN TITOV	August 6, 1961	Vostok 2	Soviet Union	The youngest person in space, aged 25. He was also the second man to orbit Earth, and the first man to sleep in space.
VALENTINA TERESHKOVA	June 16, 1963	Vostok 6	Soviet Union	The first woman in space. She stayed in space for two days and 23 hours.
ALEXEI LEONOV	March 18, 1965	Voskhod 2	Soviet Union	The first tethered spacewalk; also flew on Soyuz 19 and was selected to be the first Soviet astronaut to land on the Moon.
NEIL ARMSTRONG	July 20, 1969	Apollo 11	USA	The first man on the Moon; he also flew on Gemini 8.
HARRISON SCHMITT	December 11, 1972	Apollo 17	USA	The only scientist to walk on the Moon.
ULF MERBOLD	November 28, 1983	STS 9	Germany	The first ESA astronaut in space; he also flew on STS 42 and Soyuz TM-20 and was the first ESA astronaut to fly with Soviets.
MAMORU MORI	September 12, 1992	STS 47	Japan	The first Japanese astronaut in space; he also flew on STS 99.
VALERI POLYAKOV	January 1994–March 1995	Mir	Soviet Union	The longest single stay in space—437 days and 18 hours. He also flew on Soyuz 6, 18, and 20.
JOHN GLENN	October 29, 1998	STS 95	USA	The oldest person in space, aged 77. He also flew on Friendship 7 in 1962.
DENNIS TITO	April 28, 2001	Soyuz TM 32	USA	As the first space tourist, he spent one week on the International Space Station.
YANG LIWEI	October 15, 2003	Shenzhou 5	China	The first Chinese astronaut in space.
MIKE MELVILL	June 21, 2004	SpaceShipOne	South Africa/USA	The first commercial astronaut.
SERGEI KRIKALEV	October 11, 2005	Soyuz TMA-6	Soviet Union	He has completed 6 space missions and holds the record for the most overall time in space (803 days, 9 hours, 39 minutes).

NOTABLE MISSIONS TO THE MOON

As our closest neighbor in space, the Moon was a natural target for the first spacecraft to leave Earth. Early craft gave us the first views of the Moon's far side and close-ups of the surface. These were followed by orbiters and soft landers (craft designed to land intact) in preparati for the first manned missions. Interest in the Moon is now growing again, with recent spacecraft from the US, China, Japan, and India.

Name	Lauch date	Additional information
LUNA 2	September 12, 1959	The first crash landing onto the lunar surface.
LUNA 3	October 4, 1959	Took the first images of the lunar far side.
RANGER 7	July 28, 1964	Took 4,316 images of Mare Nubium (the Sea of Clouds) before impact.
LUNA 9	January 31, 1966	The first craft to land undamaged; it also took images of the surface.
SURVEYOR 1	May 30, 1966	Soft landed on Oceanus Procellarum (the Ocean of Storms).

Name	Lauch date	Additional information
LUNAR ORBITER	August 10, 1966	Orbited the Moon, taking 413 images to map potential manned landing sites.
SURVEYOR 3	April 17, 1967	The first use of a shovel to dig into the lunar surface.
APOLLO 8	December 21, 1968	The first humans to orbit the Moon and take images of its unseen face.
APOLLO 11	July 16, 1969	The first manned mission to the Moon. Forty eight lb (22kg) of soil and rock were returned, and a seismometer was set up on the surface.
LUNA 16	September 12, 1970	A 3.6oz (101g) sample was returned robotically to Earth from Mare Fecunditatis (the Sea of Fertility).
LUNA 17	November 10, 1970	Landed on the Moon and released the Lunokhod 1 rover, which traveled 6.5 miles (10.5km) around Sinus Iridum (the Bay of Rainbows).
APOLLO 15	July 26, 1971	The fourth manned landing on the Moon. A lunar vehicle returned 170lb (77kg) of samples; a sub-satellite investigated the lunar gravity, magnetism, and plasma environment.
APOLLO 17	December 7, 1972	The sixth (and last) manned mission carried Harrison Schmitt, the first scientist to walk on the Moon, and collected 243lb (110kg) of material.
LUNA 24	August 14, 1976	Drilled into the Mare Crisium (The Sea of Crises) and returned a sample from a depth of 6.6ft (2m).
CLEMENTINE	January 25, 1994	Orbited the Moon, mapping topography and mineralogy.
LUNA PROSPECTOR	January 7, 1998	Orbited the Moon in search of polar water.
SELENE (KAGUYA)	September 14, 2007	Orbited the Moon, producing detailed gravity maps of the far side.
LUNAR RECONNAISSANCE ORBITER	June 18, 2009	Continues to orbit the Moon, mapping its surface and collecting data for future manned missions.

EARTH-ORBITING SPACE STATIONS

Salyut 1, launched by the former Soviet Union in 1971, was the first Earth-orbiting space station. A series of Salyuts followed, culminating in the improved and enlarged Mir space station. During this time, the US had one space station of its own, Skylab. Since 1998, the US, the Soviet Union, Europe, Canada, and Japan have been jointly building and operating the International Space Station.

Name	Nation	Launch	Re-entry	Additional information
SALYUT 1	Soviet Union	April 19, 1971	October 11, 1971	The three-man crew of Soyuz 11 stayed for 23 days, but died on return to Earth.
SALYUT 2	Soviet Union	April 3, 1973	May 28, 1973	This military space station lost pressure, its flight control failed, and its solar panels fell off after 11 days. It was never occupied.
SKYLAB	USA	May 14, 1973	July 11, 1979	NASA's only space station; visited three times by three-man crews and occupied for 171 days.
SALYUT 3	Soviet Union	June 25, 1974	January 24, 1975	This second military space station maintained constant orientation to Earth's surface. It was occupied by a two-man crew for 15 days.
SALYUT 4	Soviet Union	December 26, 1974	February 2, 1977	Occupied by two two-man crews, first for 30 days, then for 63 days. They observed the Sun, X-ray stars, and Earth.
SALYUT 5	Soviet Union	June 22, 1976	August 8, 1977	Occupied for 67 of its 412 days. Visited by crews in Soyuz 21, 23, and 24; Soyuz 23 failed to dock.
SALYUT 6	Soviet Union	September 29, 1977	July 29, 1982	Second-generation craft with two docking ports; it was visited by 16 crews. Re-supplied by the unmanned Progress craft.
SALYUT 7	Soviet Union	April 19, 1982	February 7, 1991	In orbit for 3,216 days and occupied for 816 of these. Its damaged fuel pipe was repaired during a spacewalk.
MIR	Soviet Union	February 19, 1986 (first part)	March 23, 2001	Modular design, built in orbit. Continuously occupied for nearly 10 years; visited by NASA's Space Shuttle.
THE INTERNATIONAL SPACE STATION (ISS)		November 20, 1998 (first part)	Operational untill approximately 2020	Built in space; The ISS is the most expensive object ever constructed and the largest satellite to orbit Earth.

All eight major planets in our Solar System have now been visited by spacecraft. The first missions were simply flybys, starting with our two closest neighbors, Venus and Mars. These brief encounters were followed by orbiting spacecraft and, in the case of Venus and Mars, landers. Some craft have visited more than one planet, such as the two Voyagers launched by the US in 1977, which flew to Jupiter and Saturn. Voyager 2 then went on to the other giant planets, Uranus and Neptune. A craft called New Horizons is scheduled to arrive at Pluto in 2015. As well as these planetary explorers, there have also been missions to comets (see pp.184–85) and asteroids (see pp.128–29).

Planet	Spacecraft	Space agency	Encounter date	Additional information
MERCURY	Mariner 10	NASA	March 1974	Flew by on March 29, 1974, September 21, 1974, and March 16, 1975. It took 10,000 images of 57 percent of the planet's surface and detected a magnetic field.
	Messenger	NASA	January 2008	Flew by on January 14, 2008, October 6, 2008, and September 29, 2009. Moves into orbit in March 2011. Topography and composition are measured and imaged.
VENUS	Mariner 5	NASA	October 1967	Flew by at 2,480 miles (3,990km), measuring surface temperature, atmospheric pressure, and density.
	Venera 9	USSR	October 1975	Orbiter and lander. Lander lasted 53 minutes and took images of the surface in the Beta Regio region, the first images to be returned from the surface of another planet.
	Pioneer Venus	NASA	December 1978	Consisted of two spacecraft: an orbiter with a surface mapper and cloud detector; and a multiprobe that released four probes into the atmosphere in November 1978.
	Magellan	NASA	August 1990	This orbiter used synthetic-aperture radar to produce a map of 98.3 percent of the surface and collected extensive gravity data.
	Venus Express	ESA	April 2006	This polar orbiter measures atmospheric dynamics and clouds, surrounding plasma, and magnetic fields. It also uses radio waves to probe the surface.
MARS	Mariner 4	NASA	July 1965	Flyby produced the first close-up images of another planet and revealed a dead, red, dusty, cratered world.
	Mariner 9	NASA	November 1971	Orbited Mars for 11 months. Initially the planet was covered by dust storms. The craft mapped 90 percent of its surface.
	Viking 1 and 2	NASA	June/August 1976	Two orbiters and two landers. The orbiters monitored temperature and mapped water vapor. The landers analyzed the soil and looked for evidence of life.
	Mars Pathfinder	NASA	July 1997	Landed in Ares Valles, an ancient flood plain. It concentrated on meteorology. A rover, Sojourner, examined nearby rocks using an X-ray spectrometer.
	Mars Express	ESA	December 2003	This orbiter contains a high–resolution camera and a radar system to search for subsurface water. The Beagle-2 geology and exobiology lander failed to land.
	Mars Exploration Rover Mission	NASA	January 2004	Two rovers, *Spirit* and *Opportunity*, landed on Mars. They roamed around and investigated rocks, searching for evidence of processes associated with water.
JUPITER	Pioneer 10	NASA	December 1973	First flyby of Jupiter, at a distance of 81,000 miles (130,000km). It imaged the cloud layers and investigated the radiation.
	Pioneer 11	NASA	December 1974	Flew by Jupiter at 21,000 miles (34,000km). It imaged the Great Red Spot and the polar regions. Its path was changed by Jovian gravity so that it flew by Saturn.
	Voyager 1	NASA	March 1979	Flew by Jupiter at 217,000 miles (349,000km), before continuing on to Saturn, as did Voyager 2. It is now the most distant manmade object, heading to interstellar space.
	Voyager 2	NASA	July 1979	Flew by at 350,000 miles (570,000km). The two Voyagers imaged the visible surface and ring system, discovered two satellites, and revealed details of Io and Europa.
	Galileo	NASA	December 1995	Orbited Jupiter until September 2003; flew by the Galileans and Amalthea; saw impact of Comet Shoemaker Levy 9; released probe into Jupiter's atmosphere.
SATURN	Pioneer 11	NASA	September 1979	Flew by at 12,500 miles (20,000km). Discovered the F ring and solar backscatter from material in the ring gaps. Measured Titan's surface temperature.
	Voyager 1	NASA	November 1980	Found that 7 percent of the atmosphere is helium and that Saturn's subdued color contrast is due to cloud mixing. It also observed polar aurorae.
	Voyager 2	NASA	August 1981	Used radio to measure atmospheric temperature and pressure variation with depth and latitude. Cameras detected atmospheric features and winds.
	Cassini-Huygens	NASA/ESA	July 2004	Cassini orbited Saturn and made close flybys of its moons, particlularly Titan and Europa. Huygens probe landed on Titan's surface, discovering hydrocarbon lakes.
URANUS	Voyager 2	NASA	January 1986	Imaged the five largest moons and discovered 10 more. Investigated Uranus's atmosphere, magnetic field, radiation belts, rings, and radio emission.
NEPTUNE	Voyager 2	NASA	August 1989	Flew by Neptune and Triton. Discovered Neptune's Great Dark Spot (subsequently disappeared) and measured planet's mass to be 5 percent less than thought.
PLUTO	New Horizons	NASA	July 2015	New Horizons will make the first flyby of this dwarf planet before moving on to fly by a Kuiper Belt object.

ASTRONOMICAL SATELLITES

Telescopes in space have vastly improved our knowledge of the Universe, from the Sun to distant galaxies. Above the blurring and filtering effect of Earth's atmosphere, they can see the sky far more clearly than from the ground. Additionally, space observatories can detect a wide range of wavelengths that do not pass through the Earth's atmosphere, from the infrared to X-rays and gamma rays.

Satellite name	Space agency	Launch date	Date of termination	Additional information
COS-B	ESA	August 1975	April 1982	Observed the gamma-ray sky for more than six years, detecting 25 sources and mapping the Milky Way.
IUE (INTERNATIONAL ULTRAVIOLET EXPLORER)	NASA/ESA/SERC	January 1978	September 1996	The ultraviolet spectrometer observed galaxies, stars, planets, and comets.
HUBBLE SPACE TELESCOPE	NASA/ESA	April 1990		An optical, ultraviolet, and near-infrared telescope fitted with cameras and spectrometers.
ROSAT	NASA/DLR	June 1990	February 1999	Completed a six-month survey of the X-ray sky followed by observations of specific targets.
COMPTON GAMMA RAY OBSERVATORY	NASA	April 1991	June 2000	Its telescopes and spectrometers made a systematic survey of the cosmic sources of gamma rays.
SOHO (SOLAR AND HELIOSPHERIC OBSERVATORY)	NASA	December 1995		Has 12 instruments that carry out continuous observations of the Sun, corona, and heliosphere.
CHANDRA X-RAY OBSERVATORY	NASA	July 1999		An advanced X-ray telescope, which is sensitive to X-rays 100 times fainter than previous X-ray telescopes.
WMAP (WILKINSON MICROWAVE ANISOTROPY PROBE)	NASA	June 2001		Made accurate measurements of the temperature and detailed properties of the Big Bang's remnant radiation.
SPITZER SPACE TELESCOPE	NASA	August 2003		An infrared, cryogenically cooled telescope used for imaging, photometry, and spectrometry.
HERSCHEL SPACE OBSERVATORY	ESA/NASA	May 2009		A far infrared and submillimeter telescope with cameras and spectrometers. The largest-aperture telescope yet launched.

EARTH-OBSERVATION SATELLITES

Space is the ideal location for observing Earth's surface, atmosphere, and oceans. Weather satellites keep watch on clouds, atmospheric temperatures, and humidity, while other satellites follow ocean currents, waves, and ice. Earth-resources satellites monitor agriculture, forests, rivers, and land use. Satellites are important contributors to managing our use of Earth's finite resources.

Satellite name	Space agency	Date operational	Description
NIMBUS 1–7	NASA	1964–1994	The Nimbus satellites were the primary Earth-observation satellites for the US for 30 years. They collected atmospheric data, and Nimbus 7 monitored sea ice over a nine-year period.
LANDSAT	NASA/NOAA	1972–present	A series of satellites; each lasts about five years, and seven have been launched so far. The images have moderate resolution that allows scientists to assess changes in Earth's landscape.
GOES 1-14	NASA	1975–present	The Geostationary Operational Environmental Satellites. They are the mainstay of weather forecasting, storm tracking, and meteorological research in the US.
GMS	Japan	1977–1995	A series of Geostationary Meteorological Satellites that observed the surface and atmosphere of Earth in order to provide information on the weather.
METEOSAT 1-9	ESA/EUMETSAT	1977–present	Geostationary satellites stationed over different Earth regions, producing images in visual, infrared, and water-vapor channels. They also transmit data from remote observation stations.
SPOT 1-15	France	1986–present	A low Earth orbit (LEO) imaging satellite that takes high-resolution pictures for climatology, oceanography, human activity, and natural phenomena studies.
RADARSAT 1 AND 2	Canada/NASA	1995–present	A Sun-synchronous LEO satellite with a range of image modes. Its images monitor agriculture, hydrology, oceanography, and ice on Earth.
EOS	NASA plus	1999–present	A team of satellites, such as Aqua (EOS PM-1), Terra (EOS AM-1), and Aura (EOS CH-1), all in LEO, that monitor water, land, and atmospheric characteristics.
ENVISAT	ESA	2002–present	An Earth-observation satellite in LEO; nine instruments measure such things as surface height, reflectance, temperature, water vapor, and ozone.
ICESAT	NASA	2003–2010	Measured ice thickness, cloud height, and land elevation, especially over Greenland and the Antarctic, using a visual and infrared laser altimeter.
CRYOSAT	ESA	2010–present	Precisely monitors the changing thickness, and variations in thickness, of Earth's polar ice sheets and of floating sea ice.

GLOSSARY

A

absolute magnitude A measure of the true or intrinsic brightness of an object, defined as the apparent magnitude it would have at a distance from Earth of 10 parsecs (32.6 light-years). See also *apparent magnitude*, *luminosity*, *parsec*.

accretion The colliding and sticking together of small, solid objects and particles to make larger ones.

accretion disc A disc of gas that revolves around a star or black hole, after it has been drawn in from a companion star or from neighboring gas clouds.

active galaxy A galaxy that emits an exceptional amount of radiant energy over a wide range of wavelengths, from radio waves to X-rays. The compact, highly luminous core of an active galaxy is thought to be powered by the accretion of gas onto a supermassive black hole and, in many cases, varies markedly in brightness. See also *galaxy*, *supermassive black hole*.

albedo A measure of how reflective the surface of an object is. Albedo values range from 0, for a perfectly dark object that reflects nothing, to 1, for a perfect reflector.

antimatter Material composed of antiparticles. See also *antiparticle*.

antiparticle A fundamental particle that has the same mass as a particle of ordinary matter, but opposite values of other quantities, such as electrical charge. For example, the antiparticle of the negatively charged electron is the positively charged positron.

aphelion The point on its elliptical orbit at which a body such as a planet, asteroid, or comet is at its greatest distance from the Sun. See also *perihelion*.

apogee The point on its elliptical orbit around Earth at which a body such as the Moon or a spacecraft is at its greatest distance from Earth. See also *perigee*.

apparent magnitude A measure of the brightness of a celestial object as seen from Earth. This depends on the object's luminosity and its distance from Earth. The brighter the object, the smaller the numerical value of its apparent magnitude. Very bright objects have negative apparent magnitudes. See also *absolute magnitude*, *luminosity*.

arachnoid A type of structure found on the surface of Venus. It consists of a series of concentric circular or oval fractures or ridges, resembling a spider's web.

asterism A conspicuous pattern of stars that is not itself a constellation. An example is the Big Dipper—part of the constellation Ursa Major (the Great Bear). See also *constellation*.

asteroid A small, irregular Solar System object, with a diameter of less than 600 miles (1,000km). Asteroids are made of rock and/or metal and are thought to be detritus left over from the formation of the planets. See also *Main Belt*, *near-Earth asteroid*.

astronomical unit (AU) A unit of distance, defined as the average distance between Earth and the Sun. 1 AU = 92,956,000 miles (149,598,000km).

atom A building block of ordinary matter. It consists of a central nucleus surrounded by a cloud of electrons.

aurora A glowing display of light in Earth's upper atmosphere (and the atmospheres of some other planets), caused by particles in the solar wind being trapped by Earth's (or another planet's) magnetic field and funneled down close to the polar regions. The particles collide with gas atoms, stimulating them to emit light. See also *solar wind*.

B

background radiation See *cosmic microwave background radiation*.

barred spiral galaxy A galaxy that has spiral arms emanating from the ends of an elongated, bar-shaped nucleus. See also *galaxy*, *spiral galaxy*.

Big Bang The event in which the Universe was born. According to Big Bang theory, the Universe originated a finite time ago in an extremely hot, dense initial stage and has been expanding ever since.

binary star A pair of stars, bound together gravitationally, that orbit a common center of mass. See also *center of mass*.

black dwarf star A white dwarf star that has cooled so much that it emits no detectable light. So far, the Universe has not existed long enough for any black dwarf stars to have formed. See also *brown dwarf star*, *white dwarf star*.

black hole A compact region of space, surrounding a collapsed mass, within which gravity is so powerful that no object or radiation can escape. See also *active galaxy*, *singularity*, *stellar-mass black hole*, *supermassive black hole*.

blazar The most variable type of active galaxy, characterized by a compact and highly variable energy source (a supermassive black hole) at its center. Blazars include the most violently variable quasars. See also *active galaxy*, *black hole*, *quasar*, *supermassive black hole*.

blue shift The displacement of spectral lines to shorter wavelengths when a light source approaches an observer. See also *red shift*, *spectral line*.

Bok globule A compact dark nebula that contains up to 1,000 solar masses of gas and dust. Globules of this kind are thought sometimes to collapse to form stars. See also *dark nebula*.

brown dwarf star A body that forms out of a contracting cloud of gas in the same way as a star, but which, because it contains too little mass, never becomes hot enough to ignite the nuclear-fusion reactions that power a normal star.

C

caldera A bowl-shaped depression caused by the collapse of a volcanic structure into an empty magma chamber. See also *crater*.

celestial poles The celestial equivalent of Earth's poles. The night sky appears to rotate on an axis through the celestial poles.

celestial sphere An imaginary sphere, surrounding Earth, on which all celestial objects appear to lie.

center of mass The point within a system of bodies around which those bodies revolve. Where the system consists of two bodies (for example, a binary star), it is located on a line joining their centers.

Cepheid variable A type of variable star with a regular pattern of brightness changes linked to the star's luminosity. Cepheids vary in brightness as they expand and contract. The more luminous the Cepheid, the longer its period of variation. See also *variable star*.

chondrite A stony meteorite that contains many small, spherical objects called chondrules. One group of chondrites, carbonaceous chondrites, are thought to be some of the least-altered remnants of the protoplanetary disc from which the Solar System formed. See also *meteorite*, *protoplanetary disc*.

chromosphere A thin layer in the Sun's atmosphere that lies between the photosphere and the corona. See also *corona*, *photosphere*.

circumstellar disc A flattened, disc-shaped cloud of gas and dust that surrounds a star. A disc of this kind is usually associated with a young or newly forming star. See also *protoplanetary* disc.

closed universe A universe that is curved in such a way that space is finite but has no boundary (analogous to the surface of a sphere). In the absence of a repulsive force, a closed universe will eventually cease to expand and then will collapse. See also *flat universe*, *open universe*.

cloud band A belt of clouds surrounding a giant planet parallel to its equator.

coma See *comet*.

comet A small body composed mainly of dust-laden ice that revolves around the Sun, usually in a highly elongated, elliptical orbit. When a comet enters the inner Solar System, heating causes gas and dust to evaporate from its solid nucleus, forming an extensive glowing cloud called the coma and one or more tails. See also *tail*.

co-moving distance The estimated true current distance to a remote celestial object, taking into account the expansion of space that has occurred over the time it has taken light to travel from the object to Earth. See also *look-back distance*.

conjunction A close alignment in the sky of two or more celestial bodies, which occurs when they lie in the same direction as viewed from Earth. When a planet lies directly on the opposite side of the Sun from Earth, it is said to be at superior conjunction. When either Mercury or Venus passes between Earth and the Sun, the planet is said to be at inferior conjunction. See also *opposition*.

constellation A named pattern of stars, or an area of the night sky with boundaries that are determined by the International Astronomical Union. See also *asterism*.

convection The transportation of heat by rising bubbles or plumes of hot liquid or gas.

core The central region of a star or planet.

corona The outermost region of the atmosphere of a star. The Sun's corona can only be seen directly during a total solar eclipse. See also *chromosphere*, *photosphere*.

coronal mass ejection A huge, rapidly expanding bubble of plasma that is ejected from the Sun's corona. A typical coronal mass ejection propagates outward through interplanetary space at a speed of several hundred miles per second. See also *corona*, *ion*, *plasma*.

cosmic microwave background radiation (CMBR) The radiation left over from the Big Bang, appearing from all directions in the sky.

cosmic rays Highly energetic subatomic particles, such as electrons, protons, and atomic nuclei, that hurtle through space at velocities close to the speed of light.

cosmology The study of the nature, structure, origin, and evolution of the Universe.

crater A bowl- or saucer-shaped depression in the surface of a planet or satellite. An impact crater is one excavated by a meteorite, asteroid, or comet impact, whereas a volcanic crater develops at the summit of a volcano.

crust The thin, rocky, outermost layer of a planet or large moon that, like Earth, has separated into layers.

D

dark energy A little-understood phenomenon that appears to account for about 70 percent of the total mass-energy in the Universe. It is thought to be causing the expansion of the Universe to accelerate. See also *mass-energy*.

dark matter Matter that exerts a gravitational pull, but does not emit detectable amounts of radiation. Dark matter appears to make up a large proportion of the total amount of matter in the Universe.

dark nebula A dust-laden cloud that blocks out the light from background stars and appears as a dark patch in the sky. See also *nebula*.

deep-sky object Any celestial object external to the Solar System but excluding stars.

diffuse nebula A luminous cloud of gas and dust. The term "diffuse" refers to the cloud's fuzzy appearance. See also *nebula*.

double star Two stars that are close together in the sky. If they orbit each other, the system is called a binary. An optical double star consists of two stars that appear to be close together only because they happen to lie in the same direction when viewed from Earth. See also *binary star*.

dwarf planet A body orbiting a star that is large enough to be nearly round, but which has not cleared its neighborhood of other objects.

dwarf star A star with a mass similar to, or less than, the Sun's. Most dwarf stars are main-sequence stars. See also *main sequence*.

E

eccentricity The extent to which a body's orbit differs from a circle. An eccentricity of 0 means a circular orbit, with larger values indicating more elongated ellipses up to a maximum of 1. See also *ellipse*.

eclipse An alignment of a planet or moon with the Sun that casts a shadow on another body. In a lunar eclipse, Earth's shadow is cast on the Moon. In a solar eclipse, the Moon's shadow is cast on Earth.

eclipsing binary A binary star system in which each star alternately passes in front of the other, cutting off all or part of its light and causing a periodic variation in the combined light of the two stars as seen from Earth.

ecliptic 1) The plane on which Earth's orbit around the Sun is situated. **2)** The track along which the Sun travels around the celestial sphere, relative to the background stars, in the course of a year. See also *celestial sphere*.

ejecta Material thrown outward by the blast of an impact. Sometimes the ejected material—which may be markedly brighter than the adjacent surface—forms extensive streaks, or rays, radiating outward from the point of impact.

electromagnetic (EM) radiation Oscillating electric and magnetic disturbances that propagate energy through space in the form of waves (electromagnetic waves). Examples include light and radio waves.

electromagnetic (EM) spectrum The entire range of energy emitted by different objects in the Universe, from the shortest wavelengths (gamma rays) to the longest wavelengths (radio waves). Our eyes can see a specific range within the spectrum called visible light.

electron A lightweight fundamental particle with a negative electrical charge. A cloud of electrons surrounds the nucleus of an atom. See also *antiparticle*, *atom*.

electron volt The additional amount of energy acquired by an electron when it is accelerated through a potential difference of 1 volt.

ellipse A shape like a flattened circle. The maximum diameter of an ellipse is the major axis, and half of this diameter is the semimajor axis. See also *eccentricity*, *orbit*.

elliptical galaxy A galaxy that is round or elliptical in shape. Elliptical galaxies lack large amounts of gas and dust; therefore, unlike spiral galaxies, they are normally devoid of star formation. See also *galaxy*.

elongation The angular separation between the Sun and a planet, or other Solar System body, as viewed from Earth. Greatest elongation is the maximum possible elongation of a body, such as Mercury or Venus, that lies inside the orbit of Earth. See also *conjunction*, *opposition*.

emission nebula A cloud of gas and dust that contains one or more extremely hot, young, high-luminosity stars; ultraviolet radiation emitted by these stars causes the surrounding gas to glow. See also *nebula*.

eruptive variable See *variable star*.

escape velocity The minimum speed at which a projectile must be launched in order to recede forever from a massive body and not fall back. Earth's escape velocity is 7 miles (11.2km) per second.

extrasolar planet (exoplanet) A planet that orbits a star other than the Sun. Since the first confirmed detection of one in 1992, more than 450 exoplanets have been detected.

F

facula A patch of enhanced brightness on the solar photosphere. See also *photosphere*, *sunspot*.

flare star A faint red dwarf star that displays a sudden, short-lived increase in luminosity caused by extremely powerful flares that occur above its surface. See also *red dwarf star*, *solar flare*.

flat universe A universe in which the overall curvature of space is zero. In such a universe, space is flat in the sense that, aside from near to localized distortions caused by massive bodies, light rays travel in straight lines. See also *closed universe*, *open universe*, *relativity*.

frequency The number of crests of a wave that pass a given point in one second. See also *electromagnetic radiation*, *wavelength*.

fusion (nuclear fusion) A process whereby atomic nuclei join to form heavier atomic nuclei. Stars are powered by fusion reactions that take place in their cores and release large amounts of energy. See also *main sequence*.

G

galaxy A large aggregation of stars and clouds of gas and dust, held together by gravity. Galaxies may be elliptical, spiral, or irregular in shape. They may contain from a few million to several trillion stars. See also *Milky Way*.

galaxy cluster An aggregation of around 50 to 1,000 galaxies held together by gravity. See also *galaxy supercluster*.

galaxy supercluster A cluster of galaxy clusters. A supercluster may contain up to about 10,000 galaxies, spread through a volume of space with a diameter of up to about 200 million light-years. See also *galaxy cluster*.

Galilean moon Any of the four largest of Jupiter's moons—Io, Europa, Ganymede, and Callisto—which were discovered by Galileo Galilei.

gamma radiation Electromagnetic radiation with extremely short wavelengths (shorter than X-rays) and very high frequencies. See also *electromagnetic radiation*, *electromagnetic spectrum*.

gamma-ray burst (GRB) A sudden burst of gamma radiation from a source in a distant galaxy. Gamma-ray bursts are the most powerful explosive events in the Universe. They may be triggered by collisions between neutron stars or black holes, or by an extreme supernova called a hypernova. See also *black hole*, *neutron star*, *supernova*.

geocentric Having Earth at the center.

giant planet A large planet like Jupiter or Saturn that consists mainly of hydrogen and helium. See also *rocky planet*.

giant star A star that is larger and much more luminous than a main-sequence star of the same surface temperature. See also *main sequence*, *supergiant star*.

globular cluster A near-spherical cluster of between 10,000 and more than 1 million stars, bound together gravitationally. See also *open cluster*.

gravitational lens A massive body, or a distribution of mass (such as a galaxy cluster), whose gravity deflects light rays from a more distant background object, thereby acting as a lens to produce a magnified or distorted image, or images, of that object.

gravitational wave A wavelike distortion of space that propagates at the speed of light. Although waves of this kind have not yet been detected directly, there is indirect evidence that they exist.

gravity The attractive force that acts between objects, particles, and photons. See also *relativity*.

greenhouse effect The process by which atmospheric gases make the surface of a planet hotter than it would otherwise be. Incoming sunlight is absorbed at the surface of a planet and re-radiated as infrared radiation, which is then absorbed by greenhouse gases, such as carbon dioxide. Part of this trapped radiation is re-radiated back down toward the ground, so raising its temperature.

H

hadron A subatomic particle made up of smaller particles called quarks and/or the antiparticles of quarks (antiquarks). Protons and neutrons belong to a class of hadron called baryons. See also *antiparticle*, *quark*.

halo A spherical region surrounding a galaxy that contains globular clusters, thinly scattered stars, and some gas. A dark-matter halo is an accumulation of dark matter within which a galaxy is embedded.

heliocentric Having the Sun at the center.

heliosphere The region of space around the Sun within which the solar wind and interplanetary magnetic field are confined by the pressure of the interstellar medium. See also *interstellar medium*, *magnetic field*, *solar wind*.

helium burning The generation of energy in the cores of red giant stars by means of fusion reactions that convert helium into other elements. See also *fusion*.

Hertzsprung–Russell (HR) diagram A diagram on which stars are plotted as points according to their luminosity (or absolute magnitude) and surface temperature (or spectral class or color). Astrophysicists use the diagram to classify stars. See also *luminosity*, *main sequence*, *spectral class*.

Higgs boson A hypothetical particle that may have given other particles their masses.

Hubble constant See *Hubble's law*.

Hubble's law The observed relationship between the red shifts in the spectra of remote galaxies and their distances, which implies that the speeds at which galaxies are receding are directly proportional to their distances. The Hubble constant is a number that relates speed of recession to distance. See also *red shift*.

hydrogen burning The generation of energy by means of fusion reactions that convert hydrogen into helium. Hydrogen burning takes place in the core of a main-sequence star. See also *fusion*, *main sequence*.

hypernova See *gamma-ray burst*.

I

inferior conjunction See *conjunction*.

inflation A sudden, short-lived episode of accelerating expansion thought to have occurred at a very early stage in the history of the Universe. See also *Big Bang*.

infrared radiation Electromagnetic radiation with wavelengths longer than visible light but shorter than microwaves or radio waves. Infrared is the dominant form of radiation emitted from many cool astronomical objects. See also *electromagnetic radiation*.

intergalactic medium (IGM) The matter in the vast regions of space between galaxies, consisting mainly of a thin hydrogen plasma—an equal mixture of protons and electrons. See also *plasma*.

interstellar medium (ISM) The gas and dust that permeates the space between the stars in a galaxy.

ion A particle or group of particles with a net electrical charge. The process by which ions form from atoms is called ionization. See also *electron*, *plasma*.

irregular galaxy A galaxy that has no well-defined structure or symmetry. See also *galaxy*.

isotope Any one of two or more forms of a particular chemical element, the atoms of which contain different numbers of neutrons. See also *atom*, *nucleus*.

K

Kepler's laws of planetary motion Three laws that describe the orbits of planets around the Sun. The first states that each planet's orbit is an ellipse, the second shows how a planet's speed varies as it travels around its orbit, and the third links its orbital period to its average distance from the Sun.

Kuiper Belt A region of the Solar System, outside the orbit of the planet Neptune, containing icy planetesimals. See also *Oort Cloud*, *planetesimal*.

L

lenticular galaxy A galaxy shaped like a convex lens. It has a central bulge that merges into a disc, but no spiral arms. See also *galaxy*.

lepton A fundamental particle with a low mass, such as an electron or a neutrino.

light-year The distance that light travels through a vacuum in one year —that is, 5,878 billion miles (9,460 billion km).

limb The outer edge of the observed disc of the Sun, a moon, or a planet.

Local Group A small cluster of over 40 galaxies that includes our own galaxy, the Milky Way. The other major members are the spiral galaxies M31 (the Andromeda Galaxy) and M33. Most of the members are small elliptical or irregular galaxies. See also *galaxy cluster*.

look-back distance The distance that the light currently arriving at Earth from a remote object has traveled through space since it left the object. See also *co-moving distance*.

luminosity A measure of the amount of light produced by a celestial object. See also *magnitude*.

lunar eclipse See *eclipse*.

M

magnetic field The region of space surrounding a magnetized body within which its magnetic influence affects the motion of electrically charged particles.

magnetosphere The region of space around a planet within which the motion of charged particles is controlled by the planetary magnetic field. See also *magnetic field*, *solar wind*.

magnitude A measure of the brightness of a celestial object. See also *absolute magnitude*, *apparent magnitude*.

Main Belt A region of the Solar System, lying between the orbits of Mars and Jupiter, that contains a high concentration of asteroids.

main sequence A band that slopes diagonally across the Hertzsprung–Russell diagram and contains about 90 percent of stars. Main-sequence stars, such as the Sun, convert hydrogen to helium in their cores. See also *Hertzsprung–Russell diagram*, *main sequence*.

mantle The rocky layer that lies between the core and the crust of a rocky planet or large moon. See also *core*, *crust*.

mare A dark, low-lying area of the Moon, filled with lava.

mass-energy A measure of the energy possessed by anything from a subatomic particle to the entire observable Universe, taking into account that mass is convertible into energy and so has an energy equivalence.

Messier catalogue A catalog of nebulous objects (most of them nebulae, star clusters, and galaxies) that was first published in 1781. Objects contained in this catalog are designated by the letter M followed by a number. See also *New General Catalog*.

meteor The short-lived streak of light, also known as a shooting star, seen when a meteoroid hits Earth's atmosphere and is heated by friction. See also *meteorite*, *meteoroid*.

meteorite A meteoroid that reaches the ground and survives impact. Meteorites are usually classified according to their composition as stony, iron, or stony-iron. See also *meteor*, *meteoroid*.

meteoroid A lump or small particle of rock, metal, or ice orbiting the Sun in interplanetary space. See also *asteroid*, *comet*, *meteor*, *meteorite*.

microwave Electromagnetic radiation with wavelengths longer than infrared and visible light but shorter than radio waves.

Milky Way 1) The spiral galaxy that contains the Sun. **2)** A band of light across the night sky that consists of the combined light of vast numbers of stars and nebulae in our galaxy. See also *galaxy*.

Mira variable A class of variable stars named after the star Mira. A Mira variable is a cool, giant pulsating star that varies in brightness over a period ranging from 100 days to more than 500 days. See also *variable star*.

molecular cloud A cool, dense cloud of dust and gas, within which the temperature is low enough for atoms to join together to form molecules, such as molecular hydrogen or carbon monoxide, and within which conditions are suitable for stars to form.

moon A natural satellite orbiting a planet. The Moon is Earth's natural satellite.

multiple star A system consisting of two or more stars bound together by gravity and orbiting around each other. See also *binary star*.

N

near-Earth asteroid (NEA) An asteroid whose orbit comes close to, or intersects, the orbit of Earth. See also *asteroid*.

nebula A cloud of gas and dust in interstellar space, visible either because it is illuminated by embedded or nearby stars, or because it is obscuring more distant stars. See also *dark nebula*, *diffuse nebula*, *emission nebula*, *planetary nebula*, *reflection nebula*, *solar nebula*.

neutrino A fundamental particle, of exceedingly low mass and with zero electrical charge, that travels at close to the speed of light.

neutron A particle, composed of three quarks and with zero electrical charge, found in the nuclei of all atoms except those of hydrogen. See also *atom*, *nucleus*, *quark*.

neutron star An exceedingly dense, compact star that is composed almost entirely of tightly packed neutrons. A neutron star forms when the core of a high-mass star collapses, triggering a supernova explosion. See also *neutron*, *pulsar*, *supernova*.

New General Catalog (NGC) A catalog of nebulae, clusters, and galaxies that was first published in 1888. Objects in this catalog are denoted by "NGC" followed by a number. See also *Messier catalog*.

nova A star that suddenly brightens, then fades back to its original brightness over a period of weeks or months. The flare-up occurs when a fusion reaction is triggered on the surface of a white dwarf star by gas flowing from a companion star. See also *fusion*, *supernova*, *white dwarf star*.

nucleus 1) The compact central core of an atom. **2)** The solid, ice-rich body of a comet. **3)** The central core of a galaxy, within which stars are relatively densely packed together.

O

observable universe The part of the Universe from which light has had time, since the Big Bang, to reach Earth.

occultation The passage of one body in front of another, which causes the more distant one to be wholly or partially hidden.

Oort Cloud A vast spherical region in the outer reaches of the Solar System, thought to contain a huge number of icy planetesimals and comets. See also *comet*, *planetesimal*.

open cluster A loose group of stars that formed at the same time.

Open clusters are found in the arms of spiral galaxies. See also *globular cluster*.

open universe A universe in which the average density is less than that needed to halt expansion and which therefore will expand forever. See also *closed universe*, *flat universe*.

opposition The time when Mars or one of the giant planets lies on the exact opposite side of Earth from the Sun, and is highest in the sky at midnight. The planet is closest to Earth, and appears at its brightest, at this time. See also *conjunction*.

optical double star See *double star*.

orbit The path a celestial body takes in space under the influence of the gravity of other, relatively nearby, objects. Closed orbits, such as those of planets going around the Sun, are elliptical in shape, although some are almost circular.

orbital period The time an orbiting body takes to travel once around the object it is orbiting.

P

parallax The apparent shift in the position of an object when it is observed from different locations. Annual parallax is the maximum angular shift of a star from its average position due to parallax.

parsec (pc) The distance at which a star would have an annual parallax of one second of angular measurement. A parsec is 3.26 light-years, or 19.2 trillion miles (30.9 trillion km). See also *parallax*.

penumbra 1) The lighter, outer part of the shadow cast by an opaque body. An observer within the penumbra can see part of the illuminating source. **2)** The less-dark and less-cool outer region of a sunspot. See also *eclipse*, *sunspot*, *umbra*.

perigee The point on its orbit where a body orbiting Earth is at its closest to Earth. See also *apogee*.

perihelion The point on its orbit where a planet, or other Solar System body, is at its closest to the Sun. See also *aphelion*.

phase The proportion of the visible hemisphere of the Moon or a planet that is illuminated by the Sun at any particular instant.

photon A tiny package of electromagnetic radiation. See also *electromagnetic radiation*.

photosphere The thin, gaseous layer at the base of the Sun's atmosphere, from which visible light is emitted and which corresponds to the visible surface of the Sun. See also *chromosphere*, *corona*.

planet A celestial body that orbits around a star, is massive enough to have cleared away any debris from its orbit, and is roughly spherical. See also *dwarf planet*.

planetary nebula A glowing shell of gas ejected by a star of similar mass to the Sun toward the end of its evolutionary development. In a small telescope, it resembles a planet's disc. See also *nebula*.

planetesimal One of the large number of small bodies, composed of rock or ice, that formed within the solar nebula and from which the planets were eventually assembled through the process of accretion. See also *solar nebula*.

plasma A mixture of equal numbers of positively charged ions and negatively charged electrons that behaves like a gas, but which conducts electricity and is affected by magnetic fields. Examples include the solar corona and solar wind. See also *corona*, *solar wind*.

positron See *antiparticle*.

precession A slow change in the orientation of a body's rotational axis, caused by the gravitational influence of neighboring bodies.

prominence A vast, flamelike plume of plasma reaching out from the Sun's photosphere. See also *photosphere*, *plasma*.

proton A positively charged particle, composed of three quarks, that is a constituent of every atomic nucleus. See also *atom*, *nucleus*, *quark*.

protoplanet A precursor of a planet, which forms through the aggregation of planetesimals. Protoplanets collide to form planets. See also *planetesimal*, *protoplanetary disc*.

protoplanetary disc A flattened disc of dust and gas surrounding a newly formed star, within which matter may be aggregating to form the precursors of planets. Many protoplanetary discs have been observed around stars. See also *planetesimal*, *protoplanet*.

protostar A star in the early stages of formation, consisting of the center of a collapsed cloud that is heating up and growing through the addition of surrounding matter, but inside which hydrogen fusion has not yet begun.

pulsar A rapidly rotating neutron star with a powerful magnetic field. If the poles of the magnetic field are not aligned with the rotation axis of the star, jets of radiation sweep around space at high speed. See also *magnetic field*, *neutron star*.

pulsating variable See *variable star*.

Q

quark A fundamental particle. Quarks join in groups of three to make composite particles called baryons (such as protons or neutrons) or in quark-antiquark pairs to form particles called mesons. Baryons and mesons together are called hadrons. See also *antiparticle*, *hadron*.

quasar A compact but extremely powerful source of radiation that is almost starlike in appearance, but which is believed to be the most luminous kind of active galactic nucleus. See also *active galaxy*.

R

radio galaxy A galaxy that is exceptionally luminous at radio wavelengths. A typical radio galaxy contains an active nucleus from which jets of energetic charged particles are propelled toward huge clouds of radio-emitting material. See also *active galaxy*.

radio telescope An instrument designed to detect radio waves from astronomical sources. The most familiar type is a concave dish that collects radio waves and focuses them onto a detector.

red dwarf star A cool, red, low-luminosity star.

red giant star A large, highly luminous star with a low surface temperature and a reddish color. A red giant has evolved away from the main sequence, is "burning" helium in its core rather than hydrogen, and is approaching the final stages of its life. See also *helium burning*, *main sequence*.

red shift The displacement of spectral lines to longer wavelengths that is observed when a light source is moving away from an observer. The shift in wavelength is proportional to the speed at which the source is receding. See also *blue shift*, *Hubble's law*, *spectral line*.

red supergiant star An extremely large star of very high luminosity and low surface temperature.

reflection nebula A nebula containing tiny dust particles that are lit up by light from a neighboring bright star. See also *nebula*.

reflecting telescope (reflector) A telescope that collects and focuses light by using a concave mirror.

refracting telescope (refractor) A telescope that collects and focuses light by using a lens.

regolith A layer of dust and loose rock fragments that covers the surface of a planet or planetary satellite.

relativity Theories developed in the early 20th century by Albert Einstein. The special theory of relativity describes how the relative motion of observers affects their measurements of mass, length, and time. One consequence is that mass and energy are equivalent. The general theory of relativity treats gravity as a distortion of space-time associated with the presence of matter or energy. See also *gravitational lens*, *space-time*.

retrograde motion 1) An apparent temporary reversal in the direction of motion of a planet, such as Mars, when it is being overtaken in its orbital motion by Earth. **2)** Orbital motion in the opposite direction to that of Earth and the other planets of the Solar System. **3)** The motion of a satellite along its orbit in the opposite direction to the rotation of its parent planet.

retrograde rotation The rotation of a planet or moon in the opposite direction to its orbit. All the planets orbit the Sun in the same direction that the Sun rotates. Most planets also rotate (spin) in the same direction, but Venus and Uranus have retrograde rotation.

ring A flat belt of small particles and lumps of material that orbits a planet, usually in the plane of the planet's equator. Jupiter, Saturn, Uranus, and Neptune each have many rings.

rocky planet A planet composed mainly of rock, with similar basic characteristics to Earth. The four rocky planets in the Solar System are Mercury, Venus, Earth, and Mars. See also *giant planet*.

S

satellite A body that orbits a planet, otherwise known as a moon. An artificial satellite is an object deliberately placed in orbit around Earth or another Solar System body.

semimajor axis See *ellipse*.

Seyfert galaxy A spiral galaxy with an unusually bright, compact nucleus that in many cases exhibits brightness fluctuations. Seyfert galaxies comprise one of the several categories of active galaxy. See also *active galaxy*.

shepherd moon A small moon that, through its gravitational influence, confines orbiting particles into a well-defined ring around a planet. See also *moon*, *ring*.

singularity A point of infinite density into which matter has been compressed by gravity. A singularity exists at the center of a black hole. See also *black hole*.

solar cycle A cyclic variation in solar activity (for example, the production of sunspots and flares), which reaches a maximum at intervals of about 11 years. The sunspot cycle is the 11-year variation in the number (and overall area) of sunspots. See also *solar flare*, *sunspot*.

solar eclipse See *eclipse*.

solar flare A violent release of huge amounts of energy—in the form of electromagnetic radiation, subatomic particles, and shock waves—from a site located just above the Sun's surface. See also *electromagnetic radiation*.

solar mass A unit of mass equal to the mass of the Sun.

solar nebula The cloud of gas and dust from which the Sun and planets formed. See also *accretion, nebula, protoplanetary disc.*

Solar System The Sun, along with the family of four rocky planets, four giant planets, and smaller bodies (dwarf planets, moons, asteroids, comets, meteoroids, dust, and gas) that orbit the Sun.

solar wind A continuous stream of fast-moving charged particles, mainly electrons and protons, that escapes from the Sun and flows outward through the Solar System.

space-time The combination of the three dimensions of space (length, breadth, and height) and the single dimension of time. See also *relativity.*

spectral class A class into which a star is placed according to the lines that appear in its spectrum. See also *spectral line, spectrum.*

spectral line A bright or dark feature that appears at a particular wavelength in an object's spectrum, due to emission or absorption of radiation by that object at a distinct wavelength. Spectral lines can be thought of as the fingerprints of different chemicals in an object.

spectroscopic binary A system of binary stars in which the two stars are too close to be resolved into separate points of light, but whose binary nature is revealed by its spectrum. See also *binary star.*

spectroscopy The science of obtaining and studying the spectra of objects. Because the appearance of a spectrum is influenced by factors such as chemical composition, temperature, velocity, and magnetic fields, spectroscopy can reveal a wealth of information about the properties of various celestial bodies. See also *spectrum.*

spectrum The range of wavelengths of light emitted by a celestial object. The spectrum, and the presence of any spectral lines, gives clues about the chemical and physical properties of the object. See also *spectral line.*

spiral arm A spiral-shaped structure extending outward from the central bulge of a spiral or barred spiral galaxy. It consists of gas, dust, emission nebulae, and hot young stars.

spiral galaxy A galaxy that consists of a central concentration of stars surrounded by a flattened disc of stars, gas, and dust, within which the major visible features are clumped together into spiral arms. See also *galaxy, spiral arm.*

star A huge sphere of glowing plasma that generates energy by means of nuclear fusion reactions at its center. Our Sun is a star of medium size. See also *fusion, plasma.*

starburst galaxy A galaxy within which star formation is taking place at an exceptionally rapid rate.

star cluster A gravitationally bound group of between a few tens and around 1 million stars, all of which are thought to have formed from the same original massive cloud of gas and dust. See also *globular cluster, open cluster.*

stellar-mass black hole A type of black hole that forms when the core of a high-mass star collapses. See also *black hole.*

stellar wind An outflow of charged particles from the atmosphere of a star. See also *solar wind.*

sunspot A region of intense magnetic activity in the Sun's photosphere that appears dark because it is cooler than its surroundings. See also *photosphere, solar cycle.*

supergiant star An exceptionally large, luminous star. Supergiant stars can be many hundreds of times larger than the Sun and thousands of times brighter. See also *giant star.*

superior conjunction See *conjunction.*

supermassive black hole A type of black hole that forms in the core of a galaxy when a few million to a few billion solar masses of material collapses, or a number of black holes merge. See also *black hole.*

supernova An exceptionally violent explosion of a star during which it expels the great bulk of its material, and its brightness increases enormously. See also *neutron star, type Ia supernova, type II supernova, white dwarf star.*

supernova remnant The expanding cloud of debris created by a supernova explosion.

synchronous rotation The rotation of a body around its axis in the same period of time that it takes to orbit another body. The orbiting body always keeps the same face turned toward the object around which it is orbiting. Earth's moon displays synchronous rotation. See also *orbital period, satellite.*

synchrotron radiation Electromagnetic radiation emitted when charged particles gyrate at very high speed in a magnetic field. Astronomical sources of synchrotron radiation include supernova remnants and radio galaxies. See also *electromagnetic radiation, magnetic field.*

T

tail (of a comet) A stream, or streams, of ionized gas and dust swept out of a comet when it approaches, and begins to recede from, the Sun. See also *comet.*

tectonic plate One of the large rigid sections into which Earth's crust and upper mantle are divided. Their relative motions and interactions give rise to phenomena such as continental drift, earthquakes, volcanic activity, and mountain building. See also *crust, mantle.*

tektite A small, rounded, glassy object formed when a large meteorite or asteroid strikes a rocky planet, melting the rocks at the surface and throwing molten drops of rock into the atmosphere. See also *asteroid, meteorite.*

terminator The edge of the sunlit area of a moon or planet's surface, where the surface falls into shadow.

terrestrial planet See *rocky planet.*

transit The passage of a smaller body in front of a larger one (for example, the passage of Venus across the face of the Sun).

T Tauri star A young star, surrounded by gas and dust, that varies in brightness and usually shows evidence of a strong stellar wind. T Tauri stars are named after the first star of this kind to be identified. See also *stellar wind.*

type Ia supernova A type of supernova that involves the complete destruction of a white dwarf star. See also *white dwarf star.*

type II supernova A type of supernova that occurs when the core of a massive star collapses and the rest of the star's material is blasted away; the collapsed core usually becomes a neutron star. See also *neutron star.*

U

ultraviolet radiation Electromagnetic radiation with wavelengths shorter than visible light but longer than X-rays.

umbra 1) The dark central part of the shadow cast by an opaque body. The illuminating source will be completely hidden from view at any point within the umbra. **2)** The darker, cooler, central region of a sunspot. See also *eclipse, penumbra, sunspot.*

V

variable star A star that varies in brightness. A pulsating variable star expands and contracts in a periodic way, varying in brightness as it does so. An eruptive variable star brightens and fades abruptly. See also *Cepheid variable, nova.*

W

wavelength The distance between two succesive crests in a wave motion. See also *electromagnetic radiation, frequency.*

white dwarf star The dense, intensely hot glowing star left when a star of similar mass to our Sun dies. As the star dies, it sheds its outermost layers and eventually it becomes a white dwarf. See also *black dwarf star, planetary nebula.*

Wolf–Rayet star A hot, massive star from which gas is escaping at an exceptionally rapid rate, producing an expanding gaseous envelope.

X

X-ray Electromagnetic radiation with wavelengths shorter than ultraviolet radiation, but longer than gamma rays.

X-ray burster An object that emits strong bursts of X-rays, lasting from a few seconds to a few minutes. The bursts are believed to occur when gas drawn from an orbiting companion star accumulates on the surface of a neutron star and triggers a nuclear-fusion chain reaction. See also *fusion, neutron star.*

Z

zenith The point on the sky directly above an observer.

zodiac The area of the celestial sphere around the ecliptic through which the Sun, the Moon, and the planets move. Over the course of each year, the Sun passes through 13 constellations in this region, 12 of which correspond to the signs of the zodiac. See also *celestial sphere, ecliptic.*

INDEX

Page numbers in **bold** indicate feature profiles or extended treatments of a topic. Page numbers in *italics* indicate pages on which the topic is illustrated. Celestial objects and spacecraft whose names begin with a number can be found at the end of the index.

T

ACKNOWLEDGMENTS

Dorling Kindersley would like to thank the following people for their work on this book: Nigel Wright (of XAB Design) and Sarah Larter for initial design and editorial work; and Jenny Baskaya for compiling the picture credits.

Picture Credits
The publisher would like to thank the following for their kind permission to reproduce their photographs:

(Key: a-above; b-below/bottom; c-center; f-far; l-left; r-right; t-top)

1 James N. Brown. 2–3 Corbis: NASA/ Science Faction. **4–5 NASA:** JPL-Caltech/ESA/CXC/STScI. **6–7 ESA:** 2009 MPS for OSIRIS Team. **8 European Southern Observatory (ESO):** Y. Beletsky (ca). **NASA:** N. Benitez (JHU) et al. and ESA (bc); ESA (tr); JPL/USGS (bl). **9 NASA:** JPL (tr). **10 Courtesy of TMT Observatory Corporation:** rendering Todd Mason (b). **11 Corbis:** Lester Lefkowitz (tc). **Ryan Keisler:** (ca). **NASA:** MSFC (bc). **Laurie Hatch Photography:** (cb). **Science Photo Library:** Royal Observatory, Edinburgh (c). **12 Corbis:** Bettmann (bl). **NASA:** Amanda Diller (br). **13 ESA:** S. Corvaja (br). **Getty Images:** AFP (l). **NASA:** Bill Ingalls (cra). **14–15 Getty Images:** 2006 NASA. **15 Corbis:** Yuri Kochetkov/EPA (cr); Jim Sugar (br). **NASA:** (cra). **16 Corbis:** NASA/Roger Ressmeyer (tr). **NASA:** (bl) (br). **17 NASA:** (bc) (br); Jim Grossmann (l). **Science Photo Library:** European Space Agency (cr). **18 Getty Images:** AFP (br). **18–19 Corbis. 19 NASA:** JSC (br) (crb). **SOHO (ESA & NASA):** Alex Lutkus (bl/satellite). **20 Corbis:** Roger Ressmeyer (bc). **NASA:** JSC (ca). **Science Photo Library:** RIA Novosti (bl). **20–21 NASA. 21 NASA:** (tc); JSC (br). **22–23 NASA:** (all). **24 NASA:** (tl) (bl) (cl). **24–25 NASA. 26 Corbis:** Bettmann (ca). **NASA:** (cra) (bc). **26–27 NASA. 27 Corbis:** NASA/Roger Ressmeyer (tr). **28 NASA:** (bl). **28–29 NASA. 29 NASA:** (cr) (br); STS-116 Shuttle Crew (fcr). **30–31 NASA:** STS-114 Crew, ISS Expedition 11 Crew. **32 NASA:** (bl). **32–33 Corbis. 33 NASA:** (cra); JPL/UCSD/JSC (fbr); JSC-ES&IA (br). **34 NASA:** Hal Pierce, SSAI/GSFC (t). **34–35 NASA:** Marit Jentoft-Nilsen. **35 ESA:** (b). **NASA:** (tl) (tc); Reto Stockli, Earth Observatory team (crb); Reto Stockli, GSFC (cra). **36 NASA:** VAL/GSFC (bl). **36–37 ESA. 37 NASA:** (tr); Jacques Descloitres, MODIS Land Science Team (crb); JPL Ocean Surface Topography Team (br); Richard Ray, GSFC (cra). **38–39 Corbis:** NASA. **40 NASA:** GSFC/METI/ERSDAC/JAROS and U.S./Japan ASTER Science Team (b); GSFC U.S. Geological Survey (cl) (c). **USGS:** (tc). **40–41 NASA. 41 Corbis:** NASA (cr). **NASA:** Jesse Allen (tl); Liam Gumley, Space Science and Engineering Center, University of Wisconsin-Madison and the MODIS science team (br). **42 ESA:** (br). **Courtesy of JAXA:** EORC (bl). **NASA:** Reto Stockli, GSFC (tl). **42–43 NASA:** Jeffrey Kargel, USGS/JPL/AGU. **43 NASA:** GSFC Scientific Visualization Studio (bl); Jeff Schmaltz MODIS Land Rapid Response Team, NASA GSFC (br). **44 NASA:** (bl); GISS (ca). **44–45 Corbis:** NASA. **45 Corbis:** NASA (br). **NASA:** Jesse Allen, based on data provided by Shannon Brown, JPL (cr); JPL (tr). **46 NASA:** JHUAPL (bl). **46–47 Corbis:** moodboard. **47 ESA:** (br). **48–49 NASA:** GSFC/SDO AIA Team. **52 NASA:** GSFC (tr); JPL/MSSS (tl). **53 NASA:** JHUAPL/ Arizona State University/Carnegie Institution of Washington (tr); JHUAPL/ Smithsonian (br); JPL (tl). **Science Photo Library:** Manfred Kage (bl). **54 NASA:** GSFC (tl); ISS Expedition 20 (cb). **54–55 NASA:** GSFC. **55 Corbis:** Ralph White (br). **Dorling Kindersley:** Courtesy of the Natural History Museum, London (crb). **Getty Images:** Brian Bailey (bc). **56–57 NASA:** GSFC/ORBIMAGE. **58 NASA:** JHUAPL (tr). **59 Thomas Jäger:** (cla). **NASA:** (clb) (tr). **60 Science Photo Library:** Eckhard Slawik (b). **61 Getty Images:** ChinaFotoPress (br). **Johannes Schedler (panther-observatory.com) :** (crb). **Science Photo Library:** Laurent Laveder (t). **62 Corbis:** (bl). **Science Photo Library:** Detlev Van Ravenswaay (cl). **62–63 NASA:** JPL/USGS. **63 Lunar and Planetary Institute:** ACT Corporation and University of Hawai'i (b). **NASA:** (c). **USGS:** Courtesy USGS Astrogeology Science Center (t). **64 Getty Images:** NASA/Newsmakers (r). **65 NASA:** (all). **66 Getty Images:** Rex Stucky (bl). **NASA:** GSFC/Arizona State University (ca). **66–67 NASA. 67 NASA:** (br); JPL/USGS (tr). **68 NASA:** JSC (bl). **68–69 NASA:** JSC. **69 NASA:** JPL (all). **70 NASA:** JPL (t). **70–71 Moonpans. com:** Mike Constantine (b). **71 NASA:** JPL (t); Scans organized by Ken Glover from a chart provided by David Portree, USGS Flagstaff (c). **72–73 Moonpans. com:** Mike Constantine. **74 NASA:** NSSDC (cl). **Wikipedia, The Free Encyclopedia:** (bl). **74–75 NASA:** JPL/ USGS. **75 Lunar and Planetary Institute:** Clementine Science Group (b). **NASA:** (t) (c). **76 NASA:** JHUAPL/ Carnegie Institution of Washington (l); JPL (tc). **76–77 ESA:** VIRTIS/INAF-IASF/ Obs. de Paris-LESIA. **77 ESA:** (cr); VIRTIS/INAF-IASF/Obs. de Paris-LESIA/ Univ. of Oxford (t). **78 NASA:** (bl). **Soviet Planetary Exploration Program, NSSDC:** (cla). **The Art Agency:** Terry Pastor (br). **79 NASA:** JPL (cra); JPL/ USGS (l) (fcra). **Science Photo Library:** David P. Anderson, SMU/NASA (br). **80 NASA:** (bl); JPL (cla) (clb). **80–81 NASA:** JPL. **81 NASA:** JPL (ca) (b) (cra); JPL/ USGS (c). **82 NASA:** JHUAPL/Arizona State University/Carnegie Institution of Washington (tl). **Science Photo Library:** Fred Espenak (cla). **82–83 NASA:** JHUAPL/Arizona State University/ Carnegie Institution of Washington. **83 NASA:** JHUAPL/Carnegie Institution of Washington (ca) (cra) (crb); JHUAPL/ Arizona State University/Carnegie Institution of Washington. Image reproduced courtesy of Science/AAAS (c). **84 Tim Loughhead:** (all). **85 NASA:** JHUAPL/Arizona State University/ Carnegie Institution of Washington (cla/ true & false colour) (clb); JHUAPL/ Carnegie Institution of Washington (r); JHUAPL/Smithsonian Institution/ Carnegie Institution of Washington (b). **86 NASA:** JAXA/ISAS (cl). **SOHO (ESA & NASA):** (c). **86–87 SOHO (ESA & NASA). 87 Corbis:** Jay Pasachoff/Science Faction (b). **88 Corbis:** Michael Benson/ Kinetikon Pictures (b). **The Art Agency:** Stuart Jackson-Carter (c). **88–89 Science Photo Library:** Scharmer et al, Royal Swedish Academy Of Sciences. **89 Corbis:** Fred Hirschmann/Science Faction (cr). **ESA:** (b). **90–91 SST, Royal Swedish Academy of Sciences, LMSAL:** L. Rouppe van der Voort (University of Oslo), picture recorded with the Swedish 1-m Solar Telescope. **92 NASA:** JPL (cb). **USGS:** (tl). **92–93 USGS. 93 ESA:** DLR/FU Berlin (G. Neukum) (tr) (cb). **NASA:** JPL/University of Arizona (crb). **94 ESA:** DLR/FU Berlin (G. Neukum) (br). **NASA:** JPL (cla) (bl); JPL/GSFC (cra). **95 NASA:** JPL (t); JPL/JHUAPL/ Brown University (br); JPL/University of Arizona (bl). **96 NASA:** JPL/Arizona State University (b); Viking Project, USGS (tl). **96–97 NASA:** JPL/USGS. **97 ESA:** DLR/FU Berlin (G. Neukum) (tl) (br); **NASA:** JPL/MSSS (bl) (bc); JPL/ University of Arizona (tr). **98 ESA:** DLR/ FU Berlin (G. Neukum) (bl). **NASA:** JPL (cl); JPL/University of Arizona (c) (cr); JPL/MSSS (tl) (br) (tr). **99 ESA:** DLR/FU Berlin (G. Neukum) /astroarts.org (l). **NASA:** JPL/MSSS (r). **100 ESA:** DLR/FU Berlin (G. Neukum) (bl). **NASA:** JPL-Caltech/University of Rome/Southwest Research Institute/University of Arizona (t); MGS/JPL/MSSS (b); JPL-Caltech/ University of Arizona (crb); JPL/MSSS (c). **101 NASA:** GSFC (b); JPL-Caltech/ University of Arizona (t); JPL/MSSS (c). **102–103 NASA:** JPL/University of Arizona. **104 NASA:** HiRISE, MRO, LPL (U. Arizona) (br); JPL/University of Arizona (bc) (crb). **104–105 NASA:** JPL-Caltech/Cornell. **105 NASA:** JPL-Caltech/Cornell/UNM (b); JPL/ University of Arizona (cr) (cb). **106 NASA:** Arizona State University TES Team (c); JPL/MSSS (ca) (b). **106–107 NASA:** JPL/MSSS. **107 NASA:** (br); JPL-Caltech/Cornell (tr); JPL-Caltech/ University of Arizona (bc) (cr). **108–109 NASA:** JPL/University of Arizona. **111 NASA:** (br); JPL (t); JPL-Caltech (bc); JPL/GSFC (c). **112–113 NASA:** JPL/ Cornell (all). **114 NASA:** JPL-Caltech (cl) (c); JPL-Caltech/Cornell/Panoramic camera (tr). **114–115 NASA:** JPL; **115 NASA:** JPL/University of Arizona (t); JPL-Caltech/University of Arizona/ Cornell/Ohio State University (cl); JPL/ Cornell (cra). **116 NASA:** JPL/Cornell (b). **116–117 NASA:** JPL/Cornell. **117 NASA:** JPL-Caltech/Cornell (br); JPL/ Cornell (cr) (bl). **118–119 NASA:** JPL/ Texas A&M/Cornell. **120 Corbis:** Michael Benson/Kinetikon Pictures (t). **ESA:** DLR/FU Berlin (G. Neukum) (bl). **Tim Loughhead:** (c). **121 NASA:** JPL/ University of Arizona (cla) (tr); JPL-Caltech/University of Arizona (cb); JPL/ Cornell (b). **122–123 NASA:** JPL/Space Science Institute. **124 W.M. Keck Observatory:** Lawrence Sromovsky, University of Wisconsin-Madison (tc). **NASA:** Erich Karkoschka, University of Arizona (tl). **124–125 NASA:** JPL/Space Science Institute. **125 NASA:** JPL (br); JPL/Space Science Institute (tr). **126 The Art Agency:** Terry Pastor (tr). **126–127 The Art Agency:** Stuart Jackson-Carter. **127 Corbis:** Sanford/Agliolo (br). **The Art Agency:** Terry Pastor (bl) (bc). **128 Tim Loughhead:** (b). **NASA:** JPL/ USGS (ca) (cra). **129 Courtesy of JAXA:** (cla). **NASA:** courtesy of Steve Ostro, JPL (cra); JPL/JHUAPL (bl) (br) (crb). **130 NASA:** JHUAPL/Southwest Research Institute (cra); JPL/Space Science Institute (tl). **131 NASA:** (clb); JPL/Space Science Institute (tr). **132 NASA:** ESA/I. de Pater & M. Wong (University of California, Berkeley) (cl); Amy Simon (Cornell University)/Reta Beebe (NMSU)/Heidi Hammel (Space Science Institute, MIT) (bl) (bc) (br). **132–133 NASA:** JPL. **133 NASA:** JPL (br/infrared images); JPL/Cornell (cr); JPL-Caltech (bc/lightning); Amy Simon (Cornell University)/Reta Beebe (NMSU) /Heidi Hammel (Space Science Institute, MIT) (bl) (bc). **134–135 Corbis:** NASA JPL-Caltech/Science Faction. **136 Tim Loughhead:** (bl). **136–137 NASA:** JPL/ DLR (galilean moons). **137 NASA:** JPL/ Cornell University (tr/inner moons). **138 Henrik Hargitai (Eötvös Loránd University, Cosmic Materials research Group, Budapest, Hungary); Paul Schenk (Lunar and Planetary Institute, Houston, Texas, USA) :** (clb); **NASA:** JPL (bl); JPL/University of Arizona (cla). **138–139 Corbis:** NASA JPL/Science Faction. **139 NASA:** JHUAPL/Southwest Research Institute (tc); JPL (br); JPL/ University of Arizona (cb) (tr). **140 NASA:** JPL (bl); JPL/University of Arizona (cla) (clb). **140–141 NASA:** ESA/ DLR. **141 NASA:** (br); JPL/University of Arizona (cra); JPL/University of Arizona/ University of Colorado (ca). **142–143 Corbis:** Michael Benson/Kinetikon Pictures. **144 NASA:** JPL/Brown University (cl); JPL/LPI (bl); JPL/ University of Arizona (cla). **144–145 NASA:** JPL. **145 NASA:** JPL (tr) (bc); JPL/Brown University (cr) (br); JPL/DLR (c). **146 NASA:** JPL (b); JPL/Arizona State University (cla) (clb). **146–147 Corbis:** NASA JPL-Caltech/Science Faction. **147 NASA:** JPL (tr) (bl); JPL/ Arizona State University, Academic Research Lab (br); JPL/ASU (tc) (bc); JPL/DLR (c). **148 NASA:** Hubble Heritage Team (STScI/AURA) (clb); JPL/ Space Science Institute (tl). **148–149 NASA:** JPL/Space Science Institute. **149 NASA:** JPL/University of Arizona (tr); X-ray: NASA/MSFC/CXC/A.Bhardwaj et al.; Optical: NASA/ESA/STScI/AURA (crb). **150–151 NASA:** Cassini Imaging Team, SSI, JPL, ESA; JPL/University of Arizona (b). **152 NASA:** JPL/Space Science Institute (t) (cr); JPL/University of Arizona (b). **153 NASA:** ESA, J. Clarke (Boston University), and Z. Levay (STScI) (br); JPL/Space Science Institute (cra); JPL/University of Arizona (t) (bl). **154 NASA:** JPL/University of Colorado (t). **154–155 NASA:** JPL/Space Science Institute. **155 NASA:** JPL/Space Science Institute (tc) (tr); JPL-Caltech/R. Hurt (SSC) (cr); JPL-Caltech/Univ. of Virginia (cl). **156 NASA:** JPL/Space Science Institute (fbr) (bl); JPL/University of

Arizona (fbl); JPL/University of Colorado (br). **Science Museum/Science & Society Picture Library:** Science Museum (t). **157 Dorling Kindersley:** (r). **Tim Loughhead:** (tl) (bl). **158–159 NASA:** JPL/Space Science Institute. **160 NASA:** JPL/Space Science Institute (all). **161 NASA:** JPL/Space Science Institute (main image: saturn & 4 moons) (clb/ tethys moves behind Titan) (clb); W. Purcell (NWU) et al., OSSE, Compton Observatory (b). **162 NASA:** JPL/GSFC/ SwRI/SSI (cl); JPL/Space Science Institute (r) (bc) (bl). **163 NASA:** (all). **164 Max Planck Institute for Solar System Research:** (clb). **NASA:** JPL/ Space Science Institute (rest). **165 NASA:** JPL/Space Science Institute (all). **166 NASA:** JPL (tr); JPL/Space Science Institute (cla); JPL/University of Arizona (br); JPL/USGS (bl). **167 ESA:** NASA/ JPL/University of Arizona (br). **NASA:** JPL/University of Arizona (tc); JPL (bc); JPL/Space Science Institute (l) (tr). **168 W.M. Keck Observatory:** Lawrence Sromovsky, University of Wisconsin-Madison (r). **NASA:** JPL (tl); JPL/STScI (clb). **169 Tim Loughhead:** (tr). **NASA:** JPL (cr). **170 NASA:** ESA, M. Showalter (SETI Institute) (r); JPL (l); Keck Observatory (b). **171 European Southern Observatory (ESO) :** (c/ Uranus & moons). **Calvin J. Hamilton:** (clb). **NASA:** JPL (tr) (br) (cra) (crb). **P. Rousselot and O. Moussis (Observatoire de Besancon, France) and B. Gladman (University of British Columbia, Canada) :** (bc). **172 NASA:** Erich Karkoschka, University of Arizona (tl); JPL (cra) (c) (cr). **172–173 NASA:** Erich Karkoschka, University of Arizona. **173 NASA:** JPL (r). **174 NASA:** ESA, E. Karkoschka (University of Arizona), and H.B. Hammel (Space Science Institute, Boulder, Colorado) (cb); JPL (cra) (bl) (br) (cl) (cr). **175 NASA:** JPL/USGS (tr) (clb); JPL (crb); JPL/Universities Space Research Association/Lunar & Planetary Institute (br). **176–177 Ted Stryk**. **179 NASA:** ESA, M. Brown (Caltech) (tr); ESA, M. Buie (SwRI) (c); ESA, H. Weaver (JHU/APL), A. Stern (SwRI), and the HST Pluto Companion Search Team (tc); JHUAPL/SwRI (b). **180–181 Science Photo Library:** Dan Schechter. **181 © Stéphane Guisard:** (cla). **Tim Loughhead:** (t). **NASA:** Dr. Hal Weaver and T. Ed Smith (STScI) (cb); ESA, H. Hammel (Space Science Institute, Boulder, Colo.) , and the Jupiter Impact Team (b); JPL-Caltech/M. Kelley (Univ. of Minnesota) (cra). **182–183 2006, Ray Gralak**. **184 NASA:** JPL-Caltech/UMD (br); JPL/UMD (bc); JPL (cb) (bl). **185 ESA:** European Southern Observatory (cr); J. Huart (t). **Tim Loughhead:** (b). **186 Corbis:** Walter Geiersperger (clb/ gibeon). **Getty Images:** Time & Life Pictures (c). **NASA:** (cl). **The Natural History Museum, London:** (clb) (cb). **186–187 Getty Images:** National Geographic. **187 NASA:** (ca); JPL (bl) (br); JSC (tr). **188 NASA:** JPL/JHUAPL (bl). **188–189 NASA:** IBEX/Adler Planetarium. **189 NASA:** JPL (r/all). **190–191 © Stéphane Guisard**. **192 Richard Payne (Arizona Astrophotography):** (tr). **Courtesy of JAXA:** (cla). **192–193 The Art Agency:** Stuart Jackson-Carter. **193 NASA:** ESA, D. Bennett (University of Notre Dame) (cra) (fcra). **R. Ibata, M. Irwin, and G. Gilmore:** (tr). **The Art Agency:** Stuart Jackson-Carter (crb). **194–195 courtesy of the National Park Service:** Dan Duriscoe. **196 Paolo Candy of the Cimini Astronomical Observatory and Planetarium, Italy:** (bl). **Gemini Observatory:** (t). **The Art Agency:** Mark Garlick (br). **197 European Southern Observatory (ESO) :** (bl). **NASA:** CXC/ UCSC/L. Lopez et al. (br); JPL-Caltech/ University of Virginia/R. Schiavon (Univ. of Virginia) (t); The Hubble Heritage Team (STScI/AURA/NASA) (cr). **198 Science Photo Library:** Jerry Schad (l). **198–199 The Art Agency:** Mark Garlick. **199 ESA:** (cra). **Wikipedia, The Free Encyclopedia:** Steve Quirk (b/all). **200 Anglo-Australian Observatory:** Royal Observatory, Edinburgh (bl). **200–201 The Art Agency:** Terry Pastor. **201 European Southern Observatory (ESO):** Digitized Sky Survey (c/CN Leonis images). **NASA:** ESA, H. Bond (STScI) , and M. Barstow (University of Leicester (t); Walt Feimer (cr); JPL-Caltech (b). **202 NASA:** Hubble Heritage Team (AURA/STScI/NASA/ESA) (t). **203 NASA:** Andrea Dupree (Harvard-Smithsonian CfA), Ronald Gilliland (STScI), NASA and ESA (tr). **204 Andy Steere:** (t). **205 European Southern Observatory (ESO):** Y. Beletsky (cla). **NASA:** ESA, H. Richer (University of British Columbia) (b). **206–207 2MASS:** G. Kopan, R. Hurt. **208–209 SOHO (ESA & NASA):** (t). **209 Joseph Brimacombe:** (c). **NASA:** (b). **210 NASA:** JPL-Caltech/V. Gorjian (JPL) (t). **211 Rogelio Bernal Andreo (Deep Sky Colors) :** (tr). **NASA:** The Hubble Heritage Team (STScI/AURA) (cla). **212 NASA:** ESA and L. Ricci (ESO) (bl); A. Fujii (cl). **212–213 European Southern Observatory (ESO) :** J. Emerson/VISTA. Acknowledgment: Cambridge Astronomical Survey Unit. **213 NASA:** ESA/M. Robberto (STScI/ESA) et al. (tr) (crb); K. Luhman (Harvard-Smithsonian Center for Astrophysics) (cra). **214 NASA:** JPL-Caltech/N. Flagey (IAS/SSC) & A. Noriega-Crespo (SSC/Caltech) (l/2 images). **214–215 European Southern Observatory (ESO)**. **215 NASA:** ESA, The Hubble Heritage Team (STScI/AURA) (br); X-ray: NASA/CXC/U.Colorado/ Linsky et al.; Optical: NASA/ESA/STScI/ ASU/J.Hester & P.Scowen. (cra). **216–217 NASA:** ESA, N. Smith (University of California, Berkeley), and The Hubble Heritage Team (STScI/AURA). **218 Corbis:** Roger Ressmeyer (bl). **NASA:** JPL-Caltech/J. Rho (SSC/Caltech) (cl). **NOAO/AURA/NSF:** Todd Boroson (cla). **218–219 ESA:** AOES medialab. **219 ESA:** SPIRE & PACS consortia, Ph. André (CEA Saclay) for the Gould's Belt Key Programme Consortia (c). **NASA:** JPL-Caltech/L. Allen (Harvard-Smithsonian Center for Astrophysics) (t); K. Luhman (Harvard-Smithsonian Center for Astrophysics) et al. (b). **220 Thomas V. Davis:** (t); ESA, J. Maíz Apellániz (Instituto de Astrofísica de Andalucía, Spain) (tr). **NASA:** ESA, The Hubble Heritage Team (STScI/AURA) (b). **NOAO/ AURA/NSF:** T. Bash, J. Fox, and A. Block (c). **221 NASA:** CXC/SAO/M. Karovska et al. (crb); ESA, J. Maíz Apellániz (Instituto de Astrofísica de Andalucía, Spain) (tc). **222–223 NASA:** ESA and AURA/Caltech. **223 NASA:** JPL-Caltech/J. Stauffer (SSC/Caltech) (cr); T. Preibisch (MPIfR), ROSAT Project, MPE (tr); The Hubble Heritage Team (STScI/AURA) (br). **224 Anglo-Australian Observatory:** David Malin Images (bl). **European Southern Observatory (ESO) :** Very Large Telescope/R. Kotak and H. Boffin (ESO) (c) (bc). **NASA:** Dr. R. Jedrzejewski (STScI) NASA, ESA (br). **225 European Southern Observatory (ESO) :** Digitized Sky Survey 2 (t). **NASA:** ESA (b). **226 Science Photo Library:** Jerry Lodriguss (t). **227 European Southern Observatory (ESO) :** A.-M. Lagrange et al. (br). **228 T. Credner & S. Kohle, Allthesky.com:** (cl). **William Lile:** (cra). **228–229 The Art Agency:** Stuart Jackson-Carter (b/ sequence). **229 Galaxy Picture Library:** Robin Scagell (c/mira images). **NASA:** Margarita Karovska (Harvard-Smithsonian Center for Astrophysics) (cr). **230 NASA:** ESA and the Hubble SM4 ERO Team (b). **231 Canada-France-Hawaii Telescope:** J.-C. Cuillandre & G. Anselmi (t). **NASA:** Don F. Figer (UCLA) (b). **232 NASA:** ESA and the Hubble SM4 ERO Team (cr); STScI (l). **233 Global Oscillation Network Group (GONG):** NSO/AURA/NSF/MLSO/ HAO (br). **NASA:** ESA and the Hubble SM4 ERO Team (tl); GSFC (c/3 all-sky maps). **234 Corbis:** Hulton-Deutsch Collection (ca). **NASA:** Paul Hickson (UBC) (bc). **Laurie Hatch Photography:** 2007 Laurie Hatch.com/image and text (bl). **234–235 Gemini Observatory**. **235 European Southern Observatory (ESO) :** Stéphane Guisard (br/2 images). **W.M. Keck Observatory:** Peter Tuthill\ Palomar (crb); UCLA Galactic Center Group (tr) (cra). **236–237 Kamioka Observatory, ICRR (Institute for Cosmic Ray Research), The University of Tokyo**. **238 NASA:** Matt Bobrowsky (CTA INCORPORATED) (t); The Hubble Heritage Team (STScI/AURA/NASA) (b). **238–239 The Art Agency:** Stuart Jackson-Carter (c/sequence). **239 ESA:** Valentin Bujarrabal (OAN, Spain) (ca). **NASA:** R. Ciardullo (PSU) /H. Bond (STScI) (br); Raghvendra Sahai and John Trauger (JPL) and the WFPC2 science team (tr); The Hubble Heritage Team (AURA/STScI) (tc). **240 ESA:** NASA, NOAO, ESA, the Hubble Helix Nebula Team, M. Meixner (STScI), and T.A. Rector (NRAO) (cla) (clb). **240–241 European Southern Observatory (ESO)**. **241 NASA:** ESA/JPL-Caltech/J. Hora (CfA) and C.R. O'Dell (Vanderbilt) (b); JPL-Caltech/K. Su (Univ. of Ariz.) (t). **242 NASA:** JPL-Caltech/J. Hora (Harvard-Smithsonian CfA) (cla); UIUC/Y.Chu et al. (clb). **242–243 NASA:** ESA, HEIC, and The Hubble Heritage Team (STScI/ AURA). **243 NASA:** X-ray: UIUC/Y.Chu et al., Optical: HST (b). **Nordic Optical Telescope, Spain:** Romano Corradi (t). **244 ESA:** Digitized Sky Survey 2 (cla); ESO and Hans van Winckel (Catholic University of Leuven, Belgium) (clb). **NASA:** ESA, Hans Van Winckel (Catholic University of Leuven, Belgium) and Martin Cohen (University of California, USA) (r). **245 ESA:** M. A. Guerrero (IAA-CSIC) (tr). **NASA:** ESA and The Hubble Heritage Team (STScI/AURA) (b); Andrew Fruchter and the ERO Team (tc). **246–247 NASA:** ESA and the Hubble SM4 ERO Team. **248 Anglo-Australian Observatory:** David Malin Images (t/ before & after). **NASA:** CXC/PSU/G. Pavlov et al. (br). **The Art Agency:** Stuart Jackson-Carter (c/sequence). **249 NASA:** ESA, Martin Kornmesser (ESA/ Hubble) (tr); ESA, The Hubble Key Project Team, and The High-Z Supernova Search Team (cla); NGST (cb). **250 NASA:** ESA, J. Hester and A. Loll (Arizona State University) (cla); JPL-Caltech/Univ. Minn./R.Gehrz (clb). **250–251 NASA:** X-ray: NASA/CXC/SAO/F. Seward; Optical: NASA/ESA/ASU/J. Hester & A.Loll; Infrared: NASA/JPL-Caltech/Univ. Minn./R.Gehrz. **251 NASA:** CXC/ASU/J. Hester et al. (cr); Jeff Hester and Paul Scowen (Arizona State University) (br); The Hubble Heritage Team (STScI/AURA) (tr). **252 NASA:** ESA and The Hubble Heritage Team (STScI/ AURA) (tl) (cl); ESA, The Hubble Heritage Team (STScI/AURA) and the Digitized Sky Survey 2 (r). **253 NASA:** DOE/Fermi LAT Collaboration (bc); ESA and The Hubble Heritage Team (STScI/ AURA) (tr); GSFC/U.Hwang et al. (bl/4 x-ray maps); JPL-Caltech/STScI/CXC/ SAO/O. Krause (Steward Observatory) (c). **254 NASA:** ESA, P. Kalas et al. (University of California, Berkeley), M. Clampin (GSFC), M. Fitzgerald (Lawrence Livermore National Laboratory), and K. Stapelfeldt and J. Krist (JPL) (cra) (c). **255 European Southern Observatory (ESO):** (tr). **Mark A. Garlick/space-art.co.uk:** (tc). **NASA:** (br). **256 Science Photo Library:** Dr Seth Shostak. **257 Corbis:** David Scharf/ Science Faction (c). **ESA:** (t). **Science Photo Library:** Dr Seth Shostak (b). **258 NASA:** Anglo-Australian Observatory, U.S. Naval Observatory and Z. Levay (STScI) (bl/1989 & March 2002). **258–259 NASA:** ESA and H.E. Bond (STScI) (sequence: 20 May - 17 December). **259 European Southern Observatory (ESO):** J. Emerson/VISTA (tr). **NASA:** The Hubble Heritage Team (AURA/STScI) (br). **Science Photo Library:** Chris Butler (cr). **260 NASA:** CXC/MIT/F.K. Baganoff et al. (t). **261 NASA:** CXC/Caltech/M. Muno et al. (c/4 X-ray echo images); Don Figer (STScI) (tc/giant clusters). **Naval Research Lab.:** N. E. Kassim, D. S. Briggs, T. J. W. Lazio, T. N. LaRosa, J. Imamura (NRL/RSD) (tr). **262–263 NASA:** JPL-Caltech/S. Stolovy (SSC/ Caltech). **264–265 NASA:** ESA and The Hubble Heritage Team (STScI/AURA). **266 R Jay GaBany, Cosmotography. com:** (t). **NASA:** The Hubble Heritage Team (AURA/STScI/NASA) (cb). **267 European Southern Observatory (ESO):** (crb). **NASA:** ESA and The Hubble Heritage Team (STScI/AURA) (clb) (cb). **The Art Agency:** Terry Pastor (t). **268 European Southern Observatory (ESO):** (bl). **NASA:** ESA, F. Paresce (INAF-IASF), R. O'Connell (U. Virginia), & the HST WFC3 Science Oversight Committee (c). **NRAO/AUI/NSF:** David L. Nidever et al. & A. Mellinger, LAB Survey, Parkes Obs., Westerbork Obs., Arecibo Obs. (t). **Wei-Hao Wang (IfA, U. Hawaii) :** (br). **268–269 NASA:** ESA and M. Livio (STScI). **269 NASA:** P. Challis, R. Kirshner (Harvard-Smithsonian Center for Astrophysics) and B. Sugerman (STScI) (tr/ whole sequence). **270 European Southern Observatory (ESO):** (br). **NASA:** ESA, E. Olszewski (University of Arizona) (cl);

ESA and The Hubble Heritage Team (STScI/AURA) (bl); JPL-Caltech/K Gordon (STScI) (cr). **270–271 NASA:** ESA and A. Nota (STScI/ESA). **272 NASA:** ESA and T. Lauer (NOAO/AURA/NSF) (cb); JPL-Caltech/UCLA (cla); UMass/Z. Li & Q.D. Wang (bc). **272–273 Science Photo Library:** Adam Block. **273 NASA:** ESA and T.M. Brown (STScI) (tr) (cra); Thomas M. Brown et al. (GSFC) and Henry C. Ferguson (STScI) (crb). **NOAO/AURA/NSF:** (br). **274 NASA:** JPL-Caltech/J. Hinz (Univ. of Arizona) (br); Swift Science Team/Stefan Immler (clb); The Hubble Heritage Team (AURA/STScI) (t). **NOAO/AURA/NSF:** T.A. Rector and M. Hanna (cla). **275 NOAO/AURA/NSF:** Local Group Galaxies Survey Team (t); T.A. Rector (b). **276 NASA:** ESA and A. Riess (STScI/JHU) (crb); ESA and The Hubble Heritage Team (STScI/AURA) (ca) (br). **276–277 NASA**. **277 NASA:** (cra/6 images); ESA and J. Dalcanton and B. Williams (University of Washington, Seattle) (tl). **278–279 NASA**. **280 Leonardo Orazi:** (t). **NASA:** ESA and The Hubble Heritage Team (STScI/AURA) (c). **NRAO/AUI/NSF:** Chynoweth et al., Digital Sky Survey (b); **281 NASA:** ESA and the Hubble Heritage Team (c) (bc); ESA, CXC, and JPL-Caltech (bl). **282 ESA:** PACS Consortium (cla). **NASA:** JPL-Caltech/R. Kennicutt (Univ. of Arizona) (bl); X-ray: NASA/CXC/Wesleyan Univ./R. Kilgard et al; UV: NASA/JPL-Caltech; Optical: NASA/ESA/S. Beckwith & Hubble Heritage Team (STScI/AURA); IR: NASA/JPL-Caltech/ Univ. of AZ/R. Kennicutt (clb). **282–283 NASA:** ESA, S. Beckwith (STScI) , and The Hubble Heritage Team STScI/AURA). **283 NASA:** CXC/UMd./A.Wilson et al. (crb); ESA and The Hubble Heritage Team (STScI/AURA) (cra); ESA, S. Beckwith (STScI), and The Hubble Heritage Team (STScI/AURA) (br). **284 ESA:** NASA (cla) (clb). **284–285 ESA:** NASA. **285 NASA:** CXC/SAO/R.DiStefano et al. (tr); ESA, CXC, SSC, and STScI (br); JPL-Caltech/Potsdam Univ. (cr). **286 NASA:** JPL-Caltech/R. Kennicutt (University of Arizona), and the SINGS Team (bl); UMass/Q.D.Wang et al (br). **286–287 NASA:** The Hubble Heritage Team (STScI/AURA). **287 NASA:** X-ray: NASA/UMass/Q.D.Wang et al.; Optical: NASA/STScI/AURA/Hubble Heritage; Infrared: NASA/JPL-Caltech/Univ. AZ/R. Kennicutt/SINGS Team (br). **288–289 NASA:** ESA and The Hubble Heritage Team (STScI/AURA). **290 Adam Block/Mount Lemmon SkyCenter/University of Arizona (Board of Regents):** (b). **Corbis:** Stocktrek Images (tl). **NASA:** The Hubble Heritage Team (STScI/AURA) (tr). **291 NASA:** ESA and E. Peng (Peking University, Beijing). **292 NASA:** Kirk Borne (STScI) (cra); H. Ford (JHU) et al. (bl); J. English (U. Manitoba) et al. (ca); The Hubble Heritage Team (STScI/AURA) (br). **293 R Jay GaBany, Cosmotography.com:** (b). **NASA:** ESA and The Hubble Heritage Team (STScI/AURA) (tl) (cr). **294 NASA:** CXC/SAO/H. Marshall et al. (br); A. Wilson & A. Young (UMD), P. Shopbell (Caltech), CXC (bl); X-ray (NASA/CXC/ MIT/C.Canizares, D.Evans et al), Optical (NASA/STScI), Radio (NSF/ NRAO/VLA) (bc). **NRAO/AUI/NSF:** (t). **295 NASA:** CXC/A. Zezas et al (cla); JPL-Caltech/STScI-ESA (br); STScI (cl); X-ray: NASA/CXC/SAO/P. Green et al., Optical: Carnegie Obs./Magellan/W.Baade Telescope/J.S.Mulchaey et al. (bl). **296 NASA:** CXC/CfA/R.Kraft et al. (clb). **National Science Foundation, USA:** VLA/Univ.Hertfordshire/M.Hardcastle (cla). **296–297 NASA:** X-ray: NASA/CXC/CfA/R.Kraft et al.; Submillimeter: MPIfR/ESO/APEX/A.Weiss et al.; Optical: ESO/WFI. **297 Tim Carruthers:** (t). **European Southern Observatory (ESO):** Y. Beletsky (b). **NASA:** E.J. Schreier (STScI) (c) (cr). **298 NASA:** Andrew S. Wilson (University of Maryland) et al. (b); Penn State/F.Bauer et al. (t). **299 NASA:** ESA, D. Evans (Harvard-Smithsonian Center for Astrophysics), X-ray: NASA/CXC/CfA/D. Evans et al.; Optical/UV: NASA/STScI; Radio: NSF/VLA/CfA/D.Evans et al., STFC/JBO/MERLIN (b); NRAO/AUI/NSF and W. Keel (University of Alabama, Tuscaloosa) (tc) (cra). **300–301 NRAO/AUI/NSF:** J. M. Uson. **302 NASA:** ESA, R. Bouwens and G. Illingworth (University of California, Santa Cruz) (bl); GSFC (c). **303 NASA:** ESA, G. Illingworth and R. Bouwens (University of California, Santa Cruz), and the HUDF09 Team (crb) (br); ESA, L. Bradley (JHU) et al. (cra) (fcra). **304–305 Science Museum/Science & Society Picture Library:** Volker Springel/Max Planck Institute For Astrophysics. **306 NASA:** ESA/Hubble & S. Beckwith (STScI) & HUDF Team (cla). **The Art Agency:** Terry Pastor (cra/cosmological principle). **306–307 The Art Agency:** Barry Croucher. **307 NASA:** JPL-Caltech/A. Marston (ESTEC/ESA) (tr). **309 The Art Agency:** Barry Croucher (r/3 curvature images). **311 NASA:** ESA, R. Windhorst (Arizona State University) and H. Yan (Spitzer Science Center, Caltech) (tc). **313 © CERN :** Maximilien Brice (cra); Marzena Lapka (br). **NASA:** ESA/Hubble & Adam Riess (STScI) (tc). **314–315 © CERN :** Maximilien Brice. **316 ESA:** Alcatel (b). **The Art Agency:** Barry Croucher (c/seeing the first light). **316–317 NASA:** LAMBDA. **317 NASA:** LAMBDA (b/w-band map details). **NRAO/AUI/NSF:** Rudnick et al./NASA (cr). **The Art Agency:** Barry Croucher (b/universe curvature diagrams). **318 NASA:** ESA/Hubble & R. Massey (California Institute of Technology) (l/visible & dark matter). **318–319 NASA:** ESA and R. Massey (Caltech). **319 NASA:** ESA and A. Riess (STScI) (c/before & after). **The Art Agency:** Terry Pastor (b) (t/steady & accelerated expansion). **320 NASA:** ESA/Hubble, N. Pirzkal (ESA/STScI) & HUDF Team (STScI) (tr) (cra) (fcra). **320–321 NASA:** ESA/Hubble, Hubble Heritage Team (STScI/AURA) & B. Whitmore (STScI). **321 The Art Agency:** Terry Pastor (t). **322 2MASS:** Thomas Jarrett (bl). **322–323 Sloan Digital Sky Survey**. **323 2MASS:** Atlas Image mosaic obtained as part of the Two Micron All Sky Survey (2MASS), a joint project of UMASS and IPAC/Caltech, funded by NASA and NSF (cr). **Sloan Digital Sky Survey:** (cra). **324 The Art Agency:** Terry Pastor (bl). **325 Science Photo Library:** Royal Observatory, Edinburgh/AATB (br). **The Art Agency:** Terry Pastor (tr). **326 NASA:** (clb); MSFC/Emmett Givens (tr). **326–327 NASA**. **327 NASA:** ESA/Arianespace (cr). **328–329 The Bridgeman Art Library:** Private Collection

Endpapers: **NASA**

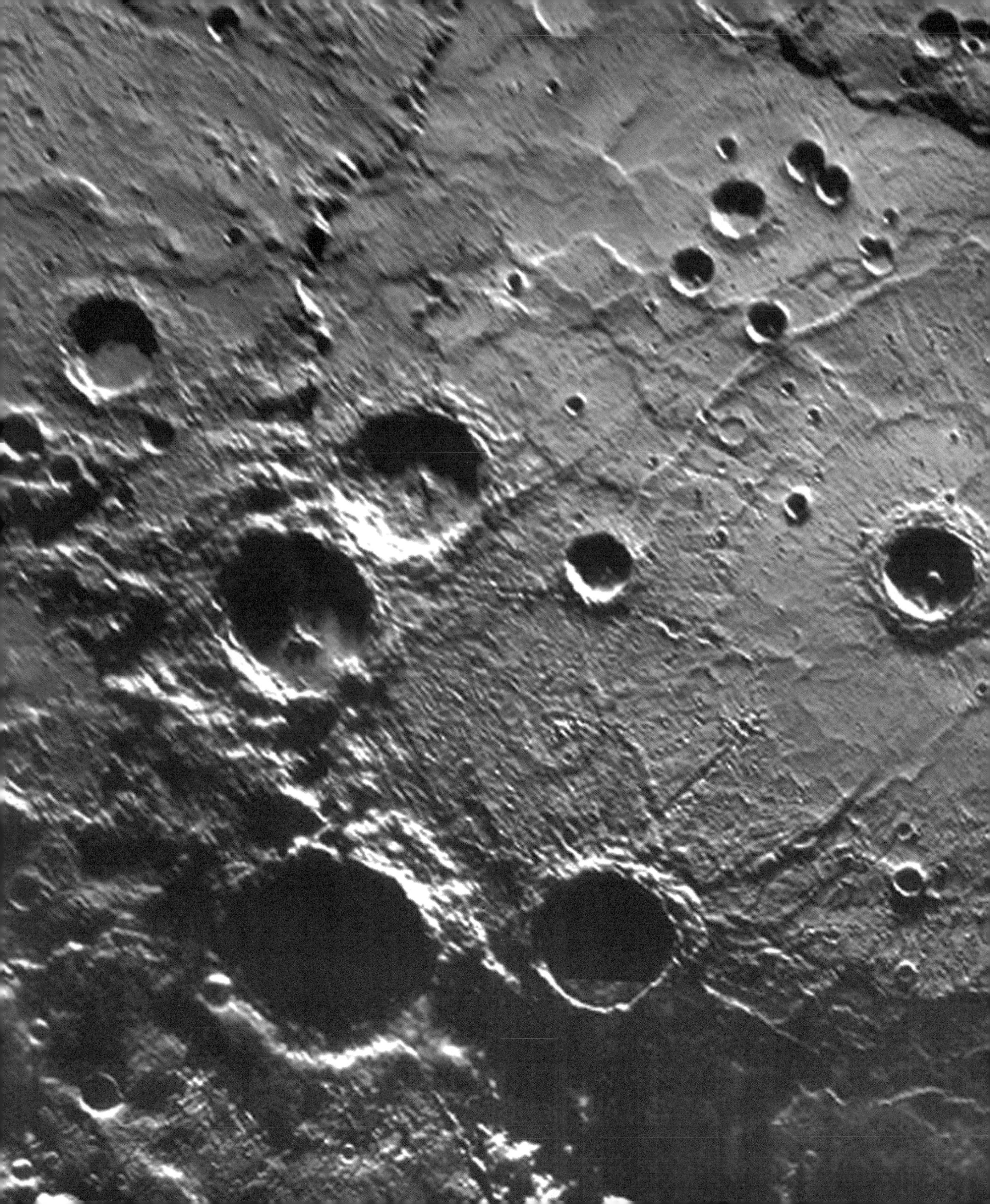